Sergey T. Surzhikov

Theoretical and Computational Physics of Gas Discharge Phenomena

Texts and Monographs in Theoretical Physics

Sergey T. Surzhikov

Theoretical and Computational Physics of Gas Discharge Phenomena

A Mathematical Introduction

2nd Edition

DE GRUYTER

Physics and Astronomy Classification 2010
52.30.-q, 52.30.Cv, 52.80.-s, 02.70.-c.

Author
Prof. Dr. Sergey T. Surzhikov
Russian Academy of Sciences
Institute for Problems in Mechanics
Prospect Vernadskogo 101
MOSCOW
119526
Russia
surg@ipmnet.ru

ISBN 978-3-11-064635-1
e-ISBN (PDF) 978-3-11-064883-6
e-ISBN (EPUB) 978-3-11-064771-6
ISSN 2627-3934

Library of Congress Control Number: 2019955971

Bibliographic information published by the Deutsche Nationalbibliothek
The Deutsche Nationalbibliothek lists this publication in the Deutsche Nationalbibliografie; detailed bibliographic data are available on the Internet at http://dnb.dnb.de.

Cover image: kertlis / E+ / gettyimages.de
Typesetting: Integra Software Services Pvt. Ltd.
Printing and binding: CPI books GmbH, Leck

www.degruyter.com

Preface

This book is dedicated basically to numerical modeling of gas discharge phenomena in gas flows. To define the specificity of computing models of discharge phenomena presented in the book, it is necessary to give some classification of computational models of discharges of different kinds.

Unfortunately, the problems considered in the framework of the book arising at numerical modeling of gas discharge physical mechanics processes are far from complete. A wide class of problems of physical and chemical kinetics of gas discharge plasma is out of this review. Hybrid, kinetic, and stochastic models of gas discharge processes are also not featured. For acquaintance to those models, studies such as Capitelli M., et al. (2000), Birdsall C.K., et al. (1985), Hockney R.W., et al. (1981), and Boeuf J.P., et al. (1982) can be recommended. The reason for such restriction of considered problems was that problems of electrodynamics and physical mechanics of discharges were covered as a rule separately from problems of the physicochemical kinetics so far. Undoubtedly, it is a principal question of the modern state of the computational physics and physical mechanics of gas discharges. It is obvious that the problem of combining the specified two concepts is extremely real. In this book, the author hopes that the explained models of the combination of discharge and gas dynamic processes will promote their further combination with models of physical and chemical kinetics.

Discussing the problems of combining computational models of a different class, it is necessary to mention the studies devoted to the numerical modeling of electrodischarge and gas dynamic lasers (Losev S.A., 1980; Anderson G., 1989). However, modern lines in the development of aerophysics of electrodischarge processes demand to review many similar problems about some other positions. Interest in studying electrodischarge effects in subsonic, supersonic, and especially, hypersonic gas flows has arisen in the last decades in view of attempts at developing hypersonic flying vehicles (HFV). The following lines of research are discussed for the uses of electric discharges in hypersonic aerodynamics:

- problems of global modification of the aerodynamic characteristics of HFV;
- control of the flow under conditions of external flow and in internal flows, for example, in the circuits of hypersonic scramjet engines; and
- raising the efficiency of energetic devices utilizing plasma chemical methods of combustion at high velocities of flow.

A detailed analysis of the processes of interaction of gas flows with discharges of various types has been performed in the book, as well as analysis of the prospects for application of magnetoplasmadynamic methods in aerophysics. Of the problems of discharge effects in hypersonic aerophysics identified earlier, a separate class of problems is associated with the use of electromagnetic actuators for the purpose of local modification and control of rarefied flow, which is of fundamental importance

https://doi.org/10.1515/9783110648836-202

from the standpoint of developing HFV capable of performing flights at altitudes of 30–50 km.

The book as a whole is intended for those who wish to develop without assistance computing models of physical mechanics of glow discharges or other types of discharge. For this reason, the book contains typical statements of problems, modes of buildup and testing finite-difference schemes, some methods of solution of the finite-difference equations, algorithms of numerical modeling, and a series of test variants. The statement of the given material does not claim for completeness at all, but it is quite enough to create one's own theoretical model.

The author thanks professors Yu. P. Raizer and J. S. Shang for their support of studies in making physical models and computer models for solving of problems of physics and mechanics of gas discharges, and also for long-term cooperation in this field.

The author thanks the staffs of the Department of Physicochemical Mechanics of the Moscow Institute of Physics and Technology and to scientists of the scientific schools of the A. Ishlinsky Institute for Problems in Mechanics of the Russian Academy of Sciences and to staffs of the All-Russian Scientific Research Institute of Automatics called after N. Dukhov.

A significant part of scientific tasks presented in the book was initiated and supported by the Russian Scientific Foundation under grants numbers 16-11-10275 and 16-11-10275-П.

Contents

Part I: **Elements of the theory of numerical modeling of gas discharge phenomena**

In the first part of this book, computational models of electric discharges and gas dynamic processes used in gas discharge physics and physical gas dynamics are considered. In the first chapter, the classification of hydrodynamic models of plasma physics accompanying electric discharges is described. The set of equations of the ideal (one-liquid and one-temperature) radiative magnetic gas dynamics is given.

Examples for some computational models are considered:
- the inductive radiofrequency plasma generator;
- the arc discharge plasma generator;
- the microwave plasma generator.

Several models of the inhomogeneous chemically equilibrium and nonequilibrium plasma of gas discharges are discussed.

The significant attention is given to a problem of laser combustion waves as one of the most intricate problems of radiation–gas dynamics of optical discharges. The problem of the unsteady subsonic movements of laser plasma is investigated.

It is essential that in the first chapter only a few computational models developed for the study of gas discharge plasma dynamics are given. Principal assigning of the review is to give a common representation of the typical numerical simulation models and important tasks accompanying with the numerical investigation of gas discharge and physical gas dynamic problems.

The second chapter gives some examples of the application of numerical simulation models for investigation of laser-supported waves.

The third chapter reviews some important singularities of the magnetic hydrodynamic equations, which should be taken into account at numerical simulation gas discharge plasma dynamics.

https://doi.org/10.1515/9783110648836-001

1 Models of gas discharge physical mechanics

In a large variety of so-called hydrodynamic models, the whole plasma or its components are considered as a continuous medium, which is used in numerical modeling of plasma and gas discharge physics. The plasma dynamics and its thermodynamic state are characterized by velocity, pressure, density, temperature, thermal capacities, and also by a set of transport properties (viscosity, thermal conduction, and electric conduction) and optical properties (spectral coefficients of absorption and emission).

Plasma is characterized essentially by a greater number of defining parameters than liquid or gas; therefore, the choice of a hydrodynamic model for an adequate description of plasma behavior is the first problem, which should be solved before practical numerical realization of a chosen model.

Creation or selection of the plasma hydrodynamic model is based on comparison of characteristic scales of space L and time t, dictated by a problem under consideration, with the spatial and timescales describing separate elementary processes. The basic elementary physical and chemical processes that are necessary to consider the formulation of such a model are as follows:

- collisional processes of plasma particles, characterized by the mean free path l_c and the collision frequency ν_c;
- shielding of elementary charge located in plasma, which is characterized by the Debye screening radius r_D and Maxwellian time of a volume charge relaxation t_M;
- electronic and ionic plasma oscillations, which are characterized by frequencies ω_e and ω_i accordingly;
- electronic and ionic cyclotron oscillation in an exterior magnetic field, which are characterized by cyclotron frequencies $\omega_{B,e}$ and $\omega_{B,i}$, accordingly. If velocities of electronic and ionic motions are known, then in a magnetic field the major spatial scales of plasma are values of radiuses of the particles' cyclotron rotation trajectories. These are the Larmor radiuses for electrons $R_{B,e}$ and ions $R_{B,i}$;
- collisional and radiative processes resulting in ionization (recombination), dissociation (association), excitation (depletion) of excited energy levels of particles, and other chemical reactions of various types;
- physical and chemical processes resulting in collision energy transfer (electronic heating of atoms; translational, rotational, vibrational, and electronic heating of molecules);
- processes of energy and momentum transfer characterizing propagation of sound, viscous dissipation, oscillating and electronic thermal conduction, diffusion, and ambipolar diffusion;
- magnetohydrodynamic waves (the magnetosonic waves and the Alfven waves).

https://doi.org/10.1515/9783110648836-002

A strong definition of all the introduced parameters can be found in textbooks on plasma and discharge physics (Bittencourt J.A., 2004; Krall N.A. and Trivelpiece A.W., 1973; Chen F.F., 1984; Raizer Yu.P., 1991; Shang J.S. and Surzhikov S.T., 2018).

Unfortunately, the choice of plasma hydrodynamic model becomes more complicated because a lot of above-enumerated parameters can be authentically defined only after a solution to the problem.

Classification of plasma hydrodynamic models can be presented as follows:

I.	One-liquid models
I.1.	Models of collisionless plasma
I.2.	Models of collision plasma (perfectly conducting plasma and plasma with finite conductivity)
I.2.1.	Single-temperature models: models of equilibrium (Boltzmann) population of particles at excited states
I.2.1.1.	Models of homogeneous chemically equilibrium plasma
I.2.1.2.	Models of inhomogeneous chemically equilibrium plasma
I.2.1.3.	Models of inhomogeneous chemically nonequilibrium plasma
I.2.2.	Two-temperature models of homogeneous plasma
I.2.3.	Two-temperature models of inhomogeneous plasma
I.2.3.1.	Models of chemically equilibrium plasma
I.2.3.2.	Models of chemically nonequilibrium plasma
I.2.4.	Multitemperature models of chemically nonequilibrium plasma with equilibrium (Boltzmann) population of particles at excited states
I.2.5.	Multitemperature models of chemically nonequilibrium plasma with nonequilibrium population of particles at excited states

II.	Multiliquid models of low-temperature plasma
II.1	Models with equilibrium (Boltzmann) population of particles at excited states
II.1.1.	Two-liquid model of partially ionized media [electrons + (atoms + ions + molecules)]
II.1.2.	Two-liquid model of fully ionized media [ions + electrons]
II.1.3.	Three-liquid model [(atoms + molecules) + ions + electrons]
II.1.4.	Multiliquid models [atoms ++ molecules ++ ions + electrons]

(continued)

II.2.	Models with nonequilibrium population of particles at excited states
II.2.1.	Two-liquid model of partially ionized media [electrons + (atoms + ions)]
II.2.2.	Two-liquid model of fully ionized media [ions + electrons]
II.2.3.	Three-liquid model [atoms + ions + electrons]
II.2.4.	Multiliquid models [atoms + . . . + molecules + . . . + ions + electrons]

This classification will be considered later in more detail, which allows bringing the hydrodynamic models in line with specific peculiarities of plasma dynamics and physics of gas discharges.

1.1 Models of the homogeneous chemically equilibrium plasma

Such models represent the most widespread class of the models applied to research of plasma dynamics and its stationary states when considered spatial and time-scales noticeably surpass all other scales. It is assumed that the plasma particles` collision frequency is so high that the locally equilibrium thermodynamic state has formed significantly faster than any considered gas dynamic function is changed. In these conditions, the external electric field strength is not so great that electrons, intensively heated up by it, can be characterized by the temperature which strongly differs from the temperature of heavy particles (atoms, molecules, and ions).

Used here a term of the homogeneous chemically equilibrium plasma should be understood so that to solve the problem stated, there is no necessity to define concretely a chemical composition of the plasma in each point of the investigated area. That is the whole plasma is considered as a single liquid characterized by any two thermodynamic parameters (temperature and pressure, density and pressure, etc.), and for the definition of all its properties, there is no necessity to specify its chemical composition. Certainly, it is well to bear in mind that thermodynamic, thermal-physical, and optical properties of the plasma can vary in the investigated volume, and it is impossible to define the specified properties without calculation of plasma chemical composition. However, the basic singularity of such a model is the possibility of separation of the problem–solution procedure on two independent constituents. At the preparatory stage, calculation of equilibrium chemical composition and definition of the thermodynamic, transport, and optical properties is performed. At the second stage, the solution of the stated plasma dynamics problem itself is fulfilled.

Models of the homogeneous chemically equilibrium plasma are used, for example, at a solution to the following problems:

- super- and hypersonic flow around returning into atmosphere space vehicles; and
- calculation of arc, inductive radiofrequency, and optical plasma generator (PG) performance.

The set of equations for the homogeneous chemically equilibrium plasma model has the following form:
- continuity equation

$$\frac{\partial \rho}{\partial t} + \mathrm{div}\rho\mathbf{V} = 0, \tag{1.1}$$

- equation of motion

$$\frac{\partial \rho\mathbf{V}}{\partial t} + \mathrm{div}[(\rho\mathbf{V})\cdot\mathbf{V}] = \\ = -\ \mathrm{grad}\left(p + p^{\mathrm{R}}\right) - \mu_0^{-1}[\mathbf{J}\times\mathbf{B}] + \mathbf{F}_\tau + \mathbf{F}_\tau^{\mathrm{R}} + \mathbf{g}\rho + \rho_{\mathrm{e}}\mathbf{E}, \tag{1.2}$$

- energy conservation equation

$$\frac{\partial}{\partial t}\left(\rho\varepsilon + \frac{\rho\mathbf{V}^2}{2} + \frac{\mathbf{B}^2}{2\mu_0}\right) + \mathrm{div}\left[\rho\mathbf{V}\left(\varepsilon + \frac{\mathbf{V}^2}{2} + \frac{p}{\rho} + \frac{p^{\mathrm{R}}}{\rho}\right)\right] = \\ = \mathrm{div}(\lambda\,\mathrm{grad}\,T) + \mathrm{div}\mathbf{W} + A_\tau + A_\tau^{\mathrm{R}} + \rho(\mathbf{g}\cdot\mathbf{V}) + (\mathbf{J}\cdot\mathbf{E}), \tag{1.3}$$

- set of the Maxwell equations

$$\mathrm{rot}\,\mathbf{E} = -\frac{\partial \mathbf{B}}{\partial t}, \tag{1.4}$$

$$\mathrm{div}\,\mathbf{D} = \rho_{\mathrm{e}}, \tag{1.5}$$

$$\mathrm{div}\,\mathbf{B} = 0, \tag{1.6}$$

$$\left(\mu_0^{-1}\right)\mathrm{rot}\,\mathbf{B} = \mathbf{J}, \tag{1.7}$$

- generalized Ohm's law

$$\mathbf{J} + P_{\mathrm{H}}[\mathbf{J}\times\mathbf{b}] = \sigma\left(\mathbf{E} + [\mathbf{V}\times\mathbf{B}] + \frac{1}{en_{\mathrm{e}}}\mathrm{grad}p_{\mathrm{e}}\right), \tag{1.8}$$

- thermal equation of state

$$p = \frac{R_0}{M}\rho T, \tag{1.9}$$

where t is the time; **B**, **D**are the magnetic and electric inductions related to corresponding strengths by the following expressions: $\mathbf{B} = \mu_0\mathbf{H}$, $\mathbf{D} = \varepsilon_0\mathbf{E}$; μ_0, ε_0

are the plasma magnetic inductivity and permittivity: $\mu_0 = 4\pi \times 10^{-7}$ Gs/m, $\varepsilon_0 = 8.854 \times 10^{-12}$ F/m; **J** is the current density; **g** is the acceleration of gravity; $\mathbf{F}_\tau$, $\mathbf{F}_\tau^{\mathrm{R}}$ are the forces of viscous friction caused by gas dynamic and radiation processes; A_τ, A_τ^{R} are the work of viscous friction caused by gas dynamic and radiation processes; **W** is the vector of integral radiation flux; $P_{\mathrm{H}} = \sigma|\mathbf{B}|/en_{\mathrm{e}}$ is the Hall parameter (for the case $n_{\mathrm{i}} = n_{\mathrm{e}}$); $\mathbf{b} = \mathbf{B}/|\mathbf{B}| = \mathbf{B}/B$; **V** is the vector of plasma velocity; ρ, p are the density and pressure; p^{R} is the radiation pressure;λ is the thermal conductivity coefficient; μ is the coefficient of viscosity; M is the molecular weight; R_0 is the universal gas constant; $\varepsilon = \int_{T_0}^{T} c_v \mathrm{d}T + \varepsilon_0$ is the specific internal energy of plasma; c_v is the specific thermal capacity at constant volume; T_0, ε_0 are the temperature and energy taken for a reference point; $\rho_{\mathrm{e}} = e(n_{\mathrm{i}} - n_{\mathrm{e}})$ is the charge density; n_{i}, n_{e} are the volume ion and electron densities; e is the electron charge; σ is the electrical conduction.

In the general case, formulas for calculation of viscous friction force and its work are represented as follows:

$$\mathbf{F}_\tau = (2\mu + \eta)\nabla \cdot (\nabla \cdot \mathbf{V}) - \mu[\nabla \times (\nabla \times \mathbf{V})] +$$
$$+ \{2(\nabla\mu \cdot \nabla)\mathbf{V} + (\nabla \cdot \mathbf{V}) \cdot (\nabla \cdot \eta) + [(\nabla \cdot \mu) \times [\nabla \cdot \mathbf{V}]]\},$$
$$A_\tau = \nabla \cdot \left\{\mu\left[\left(\nabla \cdot 0.5V^2\right) + (\mathbf{V} \cdot \nabla)\mathbf{V}\right] + \eta\mathbf{V}(\nabla \cdot \mathbf{V})\right\},$$

where besides coefficient of dynamic viscosity μ the second viscosity coefficient η is also considered, which however is practically very rarely taken into account at a solution of plasma dynamic problems at moderate pressure.

The set of eqs. (1.1)–(1.11) gives the common representation on the basic model of radiative hydrodynamics in the nonrelativistic formulation. At the practical realization of such a model in the form of computing code, it is more preferable to rewrite the equations in the conservative flux form. In the case of Cartesian rectangular coordinates, such set of equations has the following form:

$$\frac{\partial \mathbf{U}}{\partial t} + \frac{\partial \mathbf{F}_x}{\partial x} + \frac{\partial \mathbf{F}_y}{\partial y} + \frac{\partial \mathbf{F}_z}{\partial z} = G, \tag{1.10}$$

$$\mathbf{U} = \left| \begin{array}{c} \rho \\ \rho u \\ \rho v \\ \rho w \\ \rho E + p_{\mathrm{m}} \\ B_x \\ B_y \\ B_z \end{array} \right|,$$

$$\mathbf{F}_x = \mathbf{F}_x^{\mathrm{Eu}} + \mathbf{F}_x^{\mathrm{NS}} + \mathbf{F}_x^{\mathrm{R}} + \mathbf{F}_x^{\mathrm{MGD}} + \mathbf{F}_x^{\mathrm{VMGD}},$$

$$\mathbf{F}_y = \mathbf{F}_y^{\mathrm{Eu}} + \mathbf{F}_y^{\mathrm{NS}} + \mathbf{F}_y^{\mathrm{R}} + \mathbf{F}_y^{\mathrm{MGD}} + \mathbf{F}_y^{\mathrm{VMGD}},$$

$$\mathbf{F}_z = \mathbf{F}_z^{\mathrm{Eu}} + \mathbf{F}_z^{\mathrm{NS}} + \mathbf{F}_z^{\mathrm{R}} + \mathbf{F}_z^{\mathrm{MGD}} + \mathbf{F}_z^{\mathrm{VMGD}},$$

$$\mathbf{F}_x^{\mathrm{Eu}} = \begin{vmatrix} \rho u \\ \rho u^2 + p \\ \rho uv \\ \rho uw \\ \rho u\left(E + \frac{p}{\rho} + \frac{p^{\mathrm{R}}}{\rho}\right) + W_x \\ 0 \\ 0 \\ 0 \end{vmatrix}, \quad \mathbf{F}_x^{\mathrm{NS}} = \begin{vmatrix} 0 \\ -\tau_{xx} \\ -\tau_{xy} \\ -\tau_{xz} \\ -(u\tau_{xx} + v\tau_{xy} + w\tau_{xz})q_x \\ 0 \\ 0 \\ 0 \end{vmatrix},$$

$$\mathbf{F}_x^{\mathrm{R}} = \begin{vmatrix} 0 \\ -\tau_{xx}^{\mathrm{R}} + p^{\mathrm{R}} \\ -\tau_{xy}^{\mathrm{R}} \\ -\tau_{xz}^{\mathrm{R}} \\ -\left(u\tau_{xx}^{\mathrm{R}} + v\tau_{xy}^{\mathrm{R}} + w\tau_{xz}^{\mathrm{R}}\right) \\ 0 \\ 0 \\ 0 \end{vmatrix}, \quad \mathbf{F}_x^{\mathrm{MGD}} = \begin{vmatrix} 0 \\ -\left(\frac{1}{2\mu_0}\right)\left(B_x^2 - B_y^2 - B_z^2\right) \\ -(1/\mu_0)B_xB_y \\ -(1/\mu_0)B_xB_z \\ 2up_{\mathrm{m}} - (1/\mu_0)B_x(uB_x + vB_y + wB_z) \\ 0 \\ -vB_x + uB_y \\ -wB_x + uB_z \end{vmatrix},$$

$$\mathbf{F}_y^{\mathrm{Eu}} = \begin{vmatrix} \rho v \\ \rho vu \\ \rho v^2 + p \\ \rho vw \\ \rho v\left(E + \frac{p}{\rho} + \frac{p^{\mathrm{R}}}{\rho}\right) + W_y \\ 0 \\ 0 \\ 0 \end{vmatrix}, \quad \mathbf{F}_y^{\mathrm{NS}} = \begin{vmatrix} 0 \\ -\tau_{yx} \\ -\tau_{yy} \\ -\tau_{yz} \\ -(u\tau_{yx} + v\tau_{yy} + w\tau_{yz}) + q_y \\ 0 \\ 0 \\ 0 \end{vmatrix},$$

$$\mathbf{F}_y^{\mathrm{R}} = \begin{vmatrix} 0 \\ -\tau_{yx}^{\mathrm{R}} \\ -\tau_{yy}^{\mathrm{R}} + p^{\mathrm{R}} \\ -\tau_{yz}^{\mathrm{R}} \\ -\left(u\tau_{yx}^{\mathrm{R}} + v\tau_{yy}^{\mathrm{R}} + w\tau_{yz}^{\mathrm{R}}\right) \\ 0 \\ 0 \\ 0 \end{vmatrix}, \quad \mathbf{F}_y^{\mathrm{MGD}} = \begin{vmatrix} 0 \\ -(1/\mu_0)B_yB_z \\ -(1/2\mu_0)(B_y^2 - B_x^2 - B_z^2) \\ -(1/\mu_0)B_yB_z \\ 2vp_{\mathrm{m}} - (1/\mu_0)B_y(uB_x + vB_y + wB_z) \\ -uB_y + vB_x \\ 0 \\ -wB_y + vB_z \end{vmatrix},$$

$$\mathbf{F}_y^{\mathrm{VMGD}} = \begin{vmatrix} 0 \\ 0 \\ 0 \\ 0 \\ \dfrac{1}{\mu_0^2\sigma}\left[B_z\left(\dfrac{\partial B_y}{\partial z} - \dfrac{\partial B_z}{\partial y}\right) + B_x\left(\dfrac{\partial B_y}{\partial x} - \dfrac{\partial B_x}{\partial y}\right)\right] \\ -\dfrac{1}{\mu_0\sigma}\left(\dfrac{\partial B_y}{\partial x} - \dfrac{\partial B_x}{\partial y}\right) \\ 0 \\ \dfrac{1}{\mu_0\sigma}\left(\dfrac{\partial B_z}{\partial y} - \dfrac{\partial B_y}{\partial z}\right) \end{vmatrix},$$

$$\mathbf{F}_z^{\mathrm{Eu}} = \begin{vmatrix} \rho w \\ \rho wu \\ \rho wv \\ \rho w^2 + p \\ \rho w\left(E + \frac{p}{\rho} + \frac{p^{\mathrm{R}}}{\rho}\right) + W_z \\ 0 \\ 0 \\ 0 \end{vmatrix}, \quad \mathbf{F}_z^{\mathrm{NS}} = \begin{vmatrix} 0 \\ -\tau_{zx} \\ -\tau_{zy} \\ -\tau_{zz} \\ -\left(u\tau_{zx} + v\tau_{zy} + w\tau_{zz}\right) + q_z \\ 0 \\ 0 \\ 0 \end{vmatrix},$$

$$\mathbf{F}_z^{\mathrm{NS}} = \begin{vmatrix} 0 \\ -\tau_{zx} \\ -\tau_{zy} \\ -\tau_{zz} \\ -\left(u\tau_{zx} + v\tau_{zy} + w\tau_{zz}\right) + q_z \\ 0 \\ 0 \\ 0 \end{vmatrix},$$

$$\mathbf{F}_z^{\mathrm{R}} = \begin{vmatrix} 0 \\ -\tau_{zx}^{\mathrm{R}} \\ -\tau_{zy}^{\mathrm{R}} \\ -\tau_{zz}^{\mathrm{R}} + p^{\mathrm{R}} \\ -\left(u\tau_{zx}^{\mathrm{R}} + v\tau_{zy}^{\mathrm{R}} + w\tau_{zz}^{\mathrm{R}}\right) \\ 0 \\ 0 \\ 0 \end{vmatrix}, \quad \mathbf{F}_z^{\mathrm{MGD}} = \begin{vmatrix} 0 \\ -(1/\mu_0)B_zB_x \\ -(1/\mu_0)B_zB_y \\ -\left(\frac{1}{2\mu_0}\right)\left(B_z^2 - B_x^2 - B_y^2\right) \\ 2wp_{\mathrm{m}} - (1/\mu_0)B_z\left(uB_x + vB_y + wB_z\right) \\ -uB_z + wB_x \\ -vB_z + wB_y \\ 0 \end{vmatrix},$$

$$\mathbf{F}_z^{\mathrm{VMGD}} = \begin{vmatrix} 0 \\ 0 \\ 0 \\ 0 \\ \dfrac{1}{\mu_0^2\sigma}\left[B_x\left(\dfrac{\partial B_z}{\partial x} - \dfrac{\partial B_x}{\partial z}\right) + B_y\left(\dfrac{\partial B_z}{\partial y} - \dfrac{\partial B_y}{\partial z}\right)\right] \\ \dfrac{1}{\mu_0\sigma}\left(\dfrac{\partial B_x}{\partial z} - \dfrac{\partial B_z}{\partial x}\right) \\ -\dfrac{1}{\mu_0\sigma}\left(\dfrac{\partial B_z}{\partial y} - \dfrac{\partial B_y}{\partial z}\right) \\ 0 \end{vmatrix},$$

$$\mathbf{G} = \begin{vmatrix} 0 \\ \rho f_x + \rho_{\mathrm{e}}E_x \\ \rho f_y + \rho_{\mathrm{e}}E_y \\ \rho f_z + \rho_{\mathrm{e}}E_z \\ Q_v + \rho\left(ug_x + vg_y + wg_z\right) \\ 0 \\ 0 \\ 0 \end{vmatrix},$$

where $$E = \varepsilon + \frac{u^2 + v^2 + w^2}{2},$$

$$\mathbf{q} = -\lambda \nabla T = -\mathbf{i}\lambda \frac{\partial T}{\partial x} - \mathbf{j}\lambda \frac{\partial T}{\partial y} - \mathbf{k}\lambda \frac{\partial T}{\partial z},$$

$$P_{\alpha\beta} = p\delta_{\alpha\beta} - \tau_{\beta\alpha},$$

$$P^{\mathrm{R}}_{\alpha\beta} = p^{\mathrm{R}}\delta_{\alpha\beta} - \tau^{\mathrm{R}}_{\beta\alpha},$$

$$\tau_{\beta\alpha} = \mu\left[\left(\frac{\partial u_\beta}{\partial x_\alpha} + \frac{\partial u_\alpha}{\partial x_\beta}\right) - \frac{2}{3}\delta_{\alpha\beta}\mathrm{div}\mathbf{V}\right],$$

$$\mathbf{V} = \mathbf{i}u + \mathbf{j}v + \mathbf{k}w,$$

x, y, z are the rectangular Cartesian coordinates; $P_{\alpha\beta}$ are the components of gas dynamic pressure tensor; $P^{\mathrm{R}}_{\alpha\beta}$ are the components of radiation pressure tensor; $\tau_{\beta\alpha}$ are the components of viscous stress tensor; E is the total specific energy; ε is the specific internal energy; $\mathbf{q}$ is the heat flux vector caused by thermal conduction; $\mathbf{f}$is the mass force acting on the unit of volume (e.g., in case of only gravity force acting $\mathbf{f} = \mathbf{g}$); W_x, W_y, W_z are the components of integral radiation heat flux vector $\mathbf{W}$; Q_v is the power source caused by external heat sources; $T$is the temperature; B_x, B_y, B_z are the components of magnetic induction vector $\mathbf{B}$; $p_{\mathrm{m}} = \left(B_x^2 + B_y^2 + B_z^2\right)/2\mu_0$ is the magnetic pressure.

The set of equations of gas dynamics that is close to the caloric equation of state can be formulated in general form as follows:

$$\varepsilon = \varepsilon(p, T) = \varepsilon(\rho, T).$$

Using the notation of the equations of radiative magnetohydrodynamics in the form of eq. (1.10), it is possible to receive any necessary form of the equations for one-, two-, and three-dimensional cases in any coordinate system, which supposes uniqueness of transformation of rectangular Cartesian coordinates.

To determine the radiative heat flux vector $\mathbf{W}$, the equation for transfer of spectral thermal radiation is used. This equation is formulated for the spectral intensity of heat radiation $J_v(s, \mathbf{\Omega}, t)$ along an arbitrary ray s with a unit vector $\mathbf{\Omega}$, which was eradiated from a spatial point $\mathbf{r}$ (Figure 1.1):

$$\frac{1}{c}\frac{\partial J_v(s, \mathbf{\Omega}, t)}{\partial t} + \frac{\partial J_v(s, \mathbf{\Omega}, t)}{\partial s} + [\kappa_v(s) + \sigma_v(s)]J_v(s, \mathbf{\Omega}, t) = J_v^{\mathrm{em}}(s, t) +$$
$$+ \frac{1}{4\pi}\sigma_v(s)\int_{v'=0}^{\infty}\int_{\Omega'=4\pi}\gamma(s;\mathbf{\Omega}', \mathbf{\Omega}; v', v)J_{v'}(s, \mathbf{\Omega}', t)\mathrm{d}\mathbf{\Omega}'\mathrm{d}v', \tag{1.11}$$

where $J_v(s, \mathbf{\Omega}, t)$ is the spectral intensity of radiation; v is the frequency of radiation; $s$is the physical coordinate along a ray; $\mathbf{\Omega}$ is the unit vector of radiation transfer

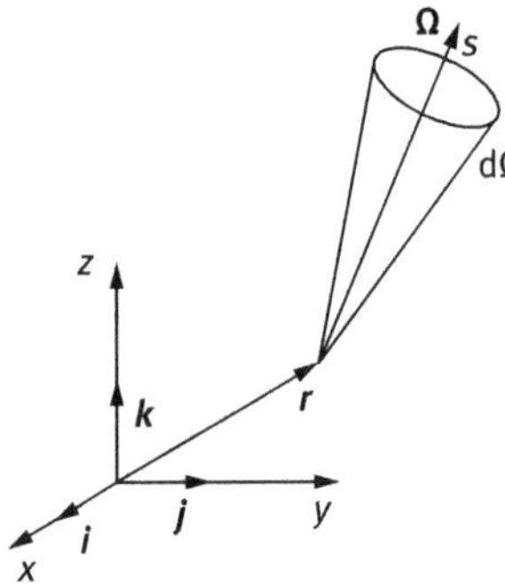

Figure 1.1: System of spatial and angular coordinates for definition of spectral intensity of radiation.

direction; $\kappa_\nu(s)$ is the volume absorptivity (the spectral volume absorption coefficient); $\sigma_\nu(s)$ is the spectral volume scattering coefficient; $J_\nu^{\mathrm{em}}(s,t)$ is the spectral emissivity of unit volume; $\gamma(s;\mathbf{\Omega}',\mathbf{\Omega};\nu',\nu)$ is the spectral scattering indicatrix by direction and frequency of radiation; $\mathbf{\Omega}'$ is the unit vector for rays specifying a direction of thermal radiation propagation, which falls on elementary physical volume with coordinate s and afterward dispersed by it with probability $\gamma(s;\mathbf{\Omega}',\mathbf{\Omega};\nu',\nu)$; c is the speed of light. It is recommended to use definitions and description of the functions and coefficients presented in books of Siegel R. and Howell J.R. (1972), Mihalas D. and Mihalas B.W. (1984), and Ozisik M.N. (1973).

In the case of a local thermodynamic equilibrium (LTE):

$$J_\nu^{\mathrm{em}}(s,t) = \kappa_\nu(s) J_{b,\nu}[T(s,t)],$$

where $J_{b,\nu}[T(s,t)]$ is the spectral intensity radiation of the absolute black body (the Planck function).

The radiation heat flux vector used in the set of eq. (1.10) is calculated under the formula:

$$\mathbf{W}(\mathbf{r},t) = \int_0^\infty \mathrm{d}\nu \int_{4\pi} J_\nu(\mathbf{r},\mathbf{\Omega},t)\mathbf{\Omega}\,\mathrm{d}\mathbf{\Omega}. \tag{1.12}$$

It follows from eqs. (1.2) and (1.3) that energy transfer by heat radiation is presented in the equations of radiation–gas dynamics only in the form of radiation stress tensor and in the form of divergence of integral radiation heat flux vector. After application of the operator divto eq. (1.31), one can determine this term under the formula

$$\operatorname{div}\mathbf{W}(\mathbf{r},t) = 4\pi \int_0^\infty J_\nu^{\mathrm{em}}\mathrm{d}\nu - \int_0^\infty \kappa_\nu(\mathbf{r}) \int_{\mathbf{\Omega}=4\pi} J_\nu(\mathbf{r},\mathbf{\Omega},t)\mathbf{\Omega}\,\mathrm{d}\nu. \tag{1.13}$$

As to components of radiation pressure tensor, they are negligible in comparison with components of gas dynamic pressure tensor in the overwhelming majority of cases, which are of interest for physics and dynamics of the low-temperature plasma. Note that these terms can be significant for radiative gas dynamic problems issued from astrophysical applications (Mihalas D. and Mihalas B.W., 1984).

The comparative estimation for values of gas dynamics and radiation pressure tensor components is connected with the radiation pressure parameter:

$$R_r = \frac{\tilde{\sigma} T^4}{3p},$$

where $\tilde{\sigma}$ is the Stefan–Boltzmann constant. At $R_r \ll 1$ the radiation pressure can be neglected.

Several physically justified assumptions are used at a solution to problems of gas discharge physics and radiative gas dynamics. They allow simplifying the radiative transfer equation. Some of them are as follows:

a) as the velocity of thermal radiation propagation is many times more than velocities of all other processes, it is possible to calculate the radiative transfer in a stationary approximation;
b) light scattering over directions is essentially a more probable process than scattering over frequencies (this is the so-called approximation of the coherent scattering);
c) in the case of lack of condensed scattering centers in plasma it is possible to neglect processes of scattering.

An approximation of LTE is also in common use.

The radiation heat transfer equation gets the following form in view of the mentioned assumptions:

$$\frac{\partial J_\nu(s, \mathbf{\Omega})}{\partial s} + \kappa_\nu(s) J_\nu(s, \mathbf{\Omega}) = \kappa_\nu(s) J_{b,\nu}[T(s,t)]. \tag{1.14}$$

Despite the relative simplicity of eq. (1.14), actually, it is not used in radiative gas dynamic models in such form, according to the definition of a vector of radiative flux density as well as its divergence (see eqs. (1.12) and (1.13)). Note that besides the integration in eq. (1.14) along the physical coordinate, it is necessary to perform integration on angular variables and on the frequency of radiation. Laboriousness of this procedure is extremely great; therefore, usually the approximate methods of definition of functions are used. In detail, the computational methods of radiation heat transfer with reference to problems of radiative gas dynamics are viewed in Siegel R. and Howell J.R. (1972) and Mihalas D. and Mihalas B.W. (1984).

Systems of equations of radiative magnetic gas dynamics (1.1)–(1.9) should be added by the corresponding boundary conditions. Examples of such boundary conditions will be considered below.

Three model problems of gas discharge physical mechanics are considered below as an example of models of the homogeneous chemically equilibrium plasma.

1.1.1 Mathematical model of inductive radiofrequency plasma generator

Inductive radiofrequency (IRF) PGs are applied in plasma chemical and aerophysical investigations, as well as in different plasma technologies. Characteristic parameters of the radiofrequency electromagnetic radiation are as follows: $\lambda_{RF} \sim 3 - 3,000$m and frequency of $f \sim 10^5 - 10^8$ Hz. A distinctive feature of the inductive PGs is the purity of generated plasma jets, which has made them one of the most widespread PGs used in thermophysical and aerophysical applications.

Computing model of the IRF PG operating in steady regime is formulated in a two-dimensional cylindrical coordinate frame $r - x$ (Figure 1.2) (Dresvin S.V., 1977):

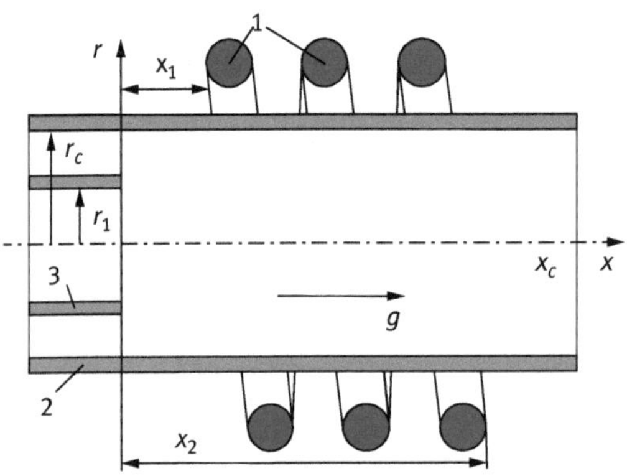

Figure 1.2: Schematic of IPG: 1, inductor; 2, the gas discharge channel; 3, the coaxial channel for making of a near-wall gas stream (possibly with vortex); r_1, r_c, radiuses of the interior and exterior channel of the plasmatron; x_c, the axial size of the discharge chamber; x_1, x_2, axial coordinates of the inductor.

– continuity equation

$$\frac{\partial \rho u}{\partial x} + \frac{1}{r}\frac{\partial r \rho v}{\partial r} = 0, \tag{1.15}$$

– equation of motion

$$\rho u \frac{\partial u}{\partial x} + \rho v \frac{\partial v}{\partial r} = -\frac{\partial p}{\partial x} + S_u + F_x - \rho g, \tag{1.16}$$

$$\rho u\frac{\partial v}{\partial x}+\rho v\frac{\partial v}{\partial r}=-\frac{\partial p}{\partial r}+S_v+F_r, \tag{1.17}$$

$$S_u=-\frac{\partial}{\partial x}\left(\frac{2}{3}\mu\operatorname{div}\mathbf{V}\right)+\frac{1}{r}\frac{\partial}{\partial r}\left[r\mu\left(\frac{\partial u}{\partial r}+\frac{\partial v}{\partial x}\right)\right]+\frac{\partial}{\partial x}\left(2\mu\frac{\partial u}{\partial x}\right),$$

$$S_v=-\frac{\partial}{\partial r}\left(\frac{2}{3}\mu\operatorname{div}\mathbf{V}\right)+\frac{\partial}{\partial x}\left[\mu\left(\frac{\partial u}{\partial r}+\frac{\partial v}{\partial x}\right)\right]+\frac{1}{r}\frac{\partial}{\partial r}\left(r2\mu\frac{\partial v}{\partial r}\right)-2\mu\frac{v}{r^2},$$

- energy conservation equation

$$\rho c_p u\frac{\partial T}{\partial x}+\rho c_p v\frac{\partial T}{\partial r}=\frac{\partial}{\partial x}\left(\lambda\frac{\partial T}{\partial x}\right)+\frac{1}{r}\frac{\partial}{\partial r}\left(r\lambda\frac{\partial T}{\partial r}\right)+Q_{\mathrm{E}}-Q_{\mathrm{R}}, \tag{1.18}$$

- thermal equation of state

$$p=\frac{R_0}{M}\rho T,$$

where x, $r$are the axial and radial coordinates; $\mathbf{V}=(u,v)$ is the velocity of a plasma flow and its x and r components, accordingly; ρ, p are the density and pressure; c_p is the specific heat capacity at constant pressure; T is the temperature; Q_{E} is the Joule thermal emission; Q_{R} is the power of volumetric radiative losses; $\mathbf{F}=c^{-1}[\mathbf{j}\times\mathbf{H}]=(F_x,F_r)$ is the average electromagnetic force for a field oscillations period.

The definition of energy release power Q_{E} and components of a magnetic force system of eqs. (1.15)–(1.19) are supplemented with the Maxwell equations. Two formulations of the Maxwell equations for calculations of these parameters are in common use, and they are presented below.

The local one-dimensional formulation

The main guess of the given model (Boulos M.I., 1976) is that in each section perpendicular to the symmetry axis, the electromagnetic part of the problem is solved as for the infinite cylinder. In this case, only the azimuth component of the electric field E_φ and the axial component of the magnetic field H_x are subjected to the definition. The following system of equations is considered in the case:

$$\frac{\mathrm{d}H_x}{\mathrm{d}r}=-\sigma E_\varphi\cos\chi, \tag{1.20}$$

$$\frac{\mathrm{d}E_\varphi}{\mathrm{d}r}=-\frac{E_\varphi}{r}-\mu_a\omega H_x\sin\chi \tag{1.21}$$

together with the equation for a phase shift between the specified field components:

$$\frac{d\chi}{dr} = \frac{\sigma E_\varphi}{H_x}\sin\chi - \frac{\mu_a \omega H_x}{E_\varphi}\cos\chi, \tag{1.22}$$

where $\omega = 2\pi f$ is the circular frequency of the electromagnetic field; $\chi = \chi_H - \chi_E$ is the phase angle between H_x and E_φ; E_φ and H_x are the components of the peak values of the field.

Power of a thermal emission in a unit volume and radial component of electromagnetic force are defined under formulas:

$$Q_E = \sigma E_\varphi^2,$$

$$F_r = -\mu_a j_\varphi H_x \cos\chi,$$

where $j_\varphi = \sigma E_\varphi$ is the azimuth component of current density.

For the solution of eqs. (1.36)–(1.41), the following boundary conditions are used:

$$x = 0: \ u = V_0(r), \quad v = 0, \quad T = T_0(r);$$

$$r = 0: \ v = 0, \quad \frac{\partial u}{\partial r} = \frac{\partial T}{\partial r} = 0;$$

$$r = r_e: \ u = v = 0, \quad \lambda\frac{\partial T}{\partial r} = \frac{\lambda_c}{\delta_c}(T_w - T);$$

$$x = x_c: \ \frac{\partial u}{\partial x} = \frac{\partial v}{\partial x} = \frac{\partial T}{\partial x},$$

where $V_0(r)$, $T_0(r)$ are the velocity and temperature in the inlet cross section of the discharge chamber; λ_c, δ_c are the thermal conductivity and thickness of the discharge chamber (as a rule, a quartz tube); T_w is the temperature of an exterior surface of a tube.

For the solution of eqs. (1.21) and (1.22) for components of electromagnetic field, the following boundary conditions are used:

$$r = 0: \ H = H_0(x), \quad E = E_\varphi = 0, \quad \chi = \pi/2.$$

The magnetic intensity $H_0(x)$ is calculated by the approximate formulas

$$H_0(x) = \frac{H_\infty}{2}\left[\frac{x_2 - x}{\sqrt{r_i^2 + (x_2 - x)^2}} - \frac{x_1 - x}{\sqrt{r_i^2 + (x_1 - x)^2}}\right],$$

$$H_\infty = \frac{I\omega}{x_2 - x_1},$$

where I is the current in the inductor; r_i is the radius of the inductor. The axial coordinate dependence of the magnetic intensity is considered here as it is created by the inductor without plasma, while the inductor is considered as the homogeneous finite cylinder.

In the given model it is necessary to correct H_∞ value, considering the influence of plasma. For this purpose, the full thermal emission in induction plasma is calculated as:

$$P^* = \int_{x_1}^{x_2}\int_{0}^{r_c} 2\pi Q_E r\,\mathrm{d}r\mathrm{d}x.$$

Then the specified correction could be made:

$$H_\infty = H_\infty\sqrt{\frac{P_0}{P^*}},$$

where P_0 is the given value of heat input into the plasma (this value can be measured experimentally).

Being returned to the discussion of a class of the homogeneous chemically equilibrium plasma models, we shall pay attention that in viewed IRF PG problem definition all thermodynamic and thermophysical properties are defined by local values of temperature and pressure (or density).

Two-dimensional axisymmetric formulations

System of the two-dimensional electromagnetic equations can be formulated relating the real E_1 and imaginary E_2 components of the complex amplitude of an electric field intensity ($E = E_1 + iE_2$):

$$\frac{\partial}{\partial r}\left(\frac{1}{r}\frac{\partial rE_1}{\partial r}\right) + \frac{\partial^2 E_1}{\partial x^2} = \frac{4\pi\sigma\omega}{c^2}E_2, \tag{1.23}$$

$$\frac{\partial}{\partial r}\left(\frac{1}{r}\frac{\partial rE_2}{\partial r}\right) + \frac{\partial^2 E_2}{\partial x^2} = \frac{4\pi\sigma\omega}{c^2}E_1. \tag{1.24}$$

With the purpose of the formulation of boundary conditions, a series of simplifying assumptions is being done. It is assumed that the inductor consists of N_k ring currents I_i with radius r_i. Then the axially oriented intensity of the magnetic field created by such inductor with a discharge chamber's radius r_c is equal:

$$H_x = \frac{I_i}{c}\sum_{k}^{N_k}\varphi(l_k),$$

$$\varphi(l_k) = \frac{2}{\sqrt{(r_i + r_c)^2 + (x - x_k)^2}} \times \left[K^*(l_k) + \frac{r_i^2 - r_c^2 - (x - x_k)^2}{(r_i - r_c)^2 + (x - x_k)^2} E^*(l_k) \right],$$

$$l_k = \frac{4 r_c r_i}{(r_i + r_c)^2 + (x - x_k)^2},$$

where k is the number of coils with a current; x_k is the axial coordinate of the kth coil; $K^*(l_k)$, $E^*(l_k)$ are the full elliptic integrals of the first and second kind.

One more physical assumption is done, namely, the inductor is located far enough from the input ($x = 0$) and output ($x = x_c$) sections of the discharge channel; therefore, axial gradients of an electric field are small.

Considering the above-mentioned assumptions, boundary conditions for eqs. (1.23) and (1.24) are written as follows:

$$x = 0: \ \frac{\partial E_1}{\partial x} = \frac{\partial E_2}{\partial x} = 0;$$

$$r = 0: \ E_1 = E_2 = 0;$$

$$r = r_c: \frac{1}{r}\frac{\partial}{\partial r}(rE_2) = \frac{\omega}{c^2} I_i \sum_k^{N_k} \varphi(l_k),$$

$$\frac{1}{r}\frac{\partial}{\partial r}(rE_1) = 0;$$

$$x = x_c: \ \frac{\partial E_1}{\partial x} = \frac{\partial E_2}{\partial x} = 0.$$

Another formulation of an electromagnetic part of this problem was offered in Dresvin S.V. (1977):

$$\frac{1}{r}\frac{\partial r E_\varphi}{\partial r} = -\mu_a \frac{\partial H_x}{\partial t}, \tag{1.25}$$

$$\frac{\partial H_r}{\partial x} - \frac{\partial H_x}{\partial r} = \sigma E_\varphi, \tag{1.26}$$

$$\frac{\partial E_\varphi}{\partial x} = \mu_a \frac{\partial H_r}{\partial t}, \tag{1.27}$$

where H_x, H_r are the axial and radial components of the electric field intensity; E_φ is the azimuth component of the electric field intensity.

In the same study, the method of the integration of the equations with the use of the vector potential was stated. Recommendations were given based on the investigation on the application of approximating dependences for the magnetic field axial component on the boundary of the discharge channel:

$$H_x(x,r=r_c)=\frac{I\omega}{x_2-x_1}\frac{1}{\pi}\sqrt{\frac{r_i}{r_c}}\left\{\operatorname{arctg}\left[\frac{0.5(x_2-x_1)-x}{r_i-r_c}\right]+\right.$$
$$+\operatorname{arctg}\left[\frac{0.5(x_2-x_1)+x}{r_i+r_c}\right]-\left(1-\sqrt{\frac{r_c}{r_i}}\right)\operatorname{arctg}\left[\frac{0.5(x_2-x_1)-x}{r_i}\right]-$$
$$\left.-\left(1-\sqrt{\frac{r_c}{r_i}}\right)\operatorname{arctg}\left[\frac{0.5(x_2-x_1)+x}{r_i}\right]\right\},$$

where r_i is the radius of the PG inductor.

1.1.2 Mathematical model of electric arc plasma generator

The electroarc (EA) PGs are widely applied in aerophysical investigations along with inductive PGs, as well as in plasma technologies.

The physical mechanics of the low-temperature plasma of EA PG are considered in detail in some studies (see, e.g., Raizer Yu.P., 1987). The EA plasma is usually considered thermally equilibrium. However, at small pressures and currents, and near walls of the PGs the use of LTE approach becomes unjustified. In these cases, two-temperature and more complex models of plasma should be used.

EAs are characterized by currents of $I\sim10^0-10^5$ A, densities of a current on the cathode of $j\sim10^2-10^7$ A/cm^2, and voltage drop between electrodes of $V\sim20-30$ V. Typical diagram of an EA generator is shown in Figure 1.3.

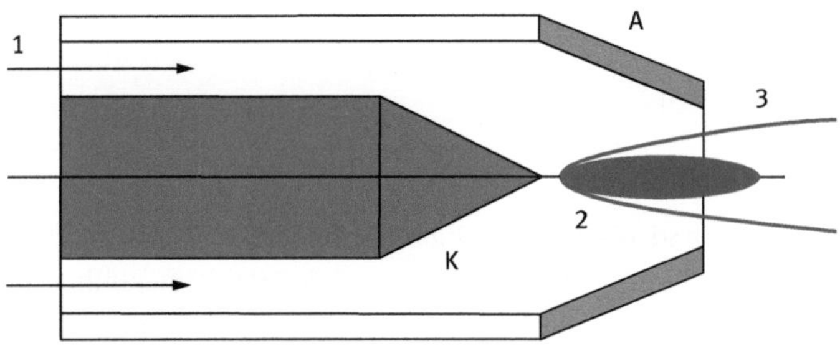

Figure 1.3: Schematic of electroarc generator: K is the cathode, A is the anode, 1 the gas stream, 2 the electroarc plasma, and 3 the plasma jet.

The equations of the homogeneous chemically equilibrium EA plasma are formulated in the following form:

– continuity equation

$$\frac{\partial\rho}{\partial t}+\operatorname{div}\rho\mathbf{V}=0; \tag{1.28}$$

- equation of motion

$$\frac{\partial \rho \mathbf{V}}{\partial t} + \operatorname{div}[(\rho \mathbf{V}) \cdot \mathbf{V}] =$$
$$= -\operatorname{grad}(p + p^{\mathrm{R}}) - \frac{1}{\mu_0}[\mathbf{J} \times \mathbf{B}] + \mathbf{F}_\tau + \mathbf{F}_\tau^{\mathrm{R}} + \mathbf{g}\rho + \rho_{\mathrm{e}}\mathbf{E}; \qquad (1.29)$$

- energy conservation equation

$$\frac{\partial}{\partial t}\left(\rho\varepsilon + \frac{\rho \mathbf{V}^2}{2} + \frac{\mathbf{B}^2}{2\mu_0}\right) + \operatorname{div}\left[\rho \mathbf{V}\left(\varepsilon + \frac{\mathbf{V}^2}{2} + \frac{p}{\rho} + \frac{p^{\mathrm{R}}}{\rho}\right)\right] =$$
$$= \operatorname{div}(\lambda \operatorname{grad} T) + \operatorname{div} \mathbf{W} + A_\tau + A_\tau^{\mathrm{R}} + \rho(\mathbf{g} \cdot \mathbf{V}) + (\mathbf{J} \cdot \mathbf{E}); \qquad (1.30)$$

- set of the Maxwell equations

$$\operatorname{rot} \mathbf{E} = -\frac{\partial \mathbf{B}}{\partial t}, \qquad (1.31)$$

$$\operatorname{div} \mathbf{D} = \rho_{\mathrm{e}}, \qquad (1.32)$$

$$\operatorname{div} \mathbf{B} = 0, \qquad (1.33)$$

$$\frac{1}{\mu_0} \operatorname{rot} \mathbf{B} = \mathbf{J}; \qquad (1.34)$$

- the Ohm law

$$\mathbf{J} + P_{\mathrm{H}}[\mathbf{J} \times \mathbf{b}] = \sigma\left(\mathbf{E} + [\mathbf{V} \times \mathbf{B}] + \frac{1}{en_{\mathrm{e}}}\operatorname{grad} p_{\mathrm{e}}\right); \qquad (1.35)$$

- thermal equation of state

$$p = \frac{R_0}{M}\rho T,$$

where **B**, **D** are the magnetic and electrical inductions related to corresponding strengths as $\mathbf{B} = \mu_0 \mathbf{H}, \mathbf{D} = \varepsilon_0 \mathbf{E}; \mu_0$, ε_0 are the magnetic and dielectric permeabilities of the plasma; $\mathbf{F}_\tau$, $\mathbf{F}_\tau^{\mathrm{R}}$ are the forces of viscous friction caused by gas dynamic and radiative processes; A_τ, A_τ^{R} are the work of the viscous friction caused by gas dynamic and radiative processes; **W** is the vector of integrated radiation heat transfer flux; $P_{\mathrm{H}} = \sigma|\mathbf{B}|/en_{\mathrm{e}}$ is the Hall parameter (at $n_{\mathrm{i}} = n_{\mathrm{e}}$); $\mathbf{b} = \mathbf{B}/|\mathbf{B}| = \mathbf{B}/B$; **V** is the velocity of plasma; ρ, p are the density and pressure; M is the molecular weight; R_0 is the universal gas constant; c_v is the specific heat capacity at constant volume; T_0, $\tilde{\varepsilon}_0$ are the temperature and energy taken for a reference thermodynamic point; $\rho_{\mathrm{e}} = e(n_{\mathrm{i}} - n_{\mathrm{e}})$ is the density of electric charge; $n_{\mathrm{i}}, n_{\mathrm{e}}$ are the volume concentrations of ions and electrons; e is the electron charge; p_{e} is the pressure of electronic gas.

It is assumed that the homogeneous chemically equilibrium plasma model of arc discharges at atmospheric pressure is well founded if the following inequality is satisfied:

$$\frac{T_e - T_i}{T_e} \approx \frac{3\pi}{32}\frac{m_i}{m_e}\left[\frac{\Lambda_e eE}{(3/2)kT_e}\right]^2 \approx \frac{3\pi}{32}\frac{m_i}{m_e}\left[\frac{\Lambda_e e\lambda}{(3/2)kT_e}\right]^2 \ll 1,$$

where T_e, T_i are the temperatures of electrons and ions; m_e, m_i are the masses of electrons and ions; Λ_e is the Coulomb logarithm.

Typical parameters entering into this inequality for air plasma at atmospheric pressure are $T{\sim}10,000\,\text{K}$, $\Lambda_e \approx 10^{-6}\,\text{m}$, and $\lambda{\sim}10^{-1}\text{W}/(\text{m}\cdot\text{K})$. Therefore, at a typical current in arc discharge of $I \approx 10\text{A}$, one can get

$$\frac{T_e - T_i}{T_e} \approx 10^{-4}.$$

The following assumptions simplifying eqs. (1.28)–(1.35) are also used:

1. Neglect of a gravity $\mathbf{g}\rho$. However, it is necessary to mean that for open-flame arcs of great volumes, located horizontally, gravity leads to their curvature.
2. Neglect of electrostatic force $\rho_e\mathbf{E}$. The estimate of the ratio of electrostatic force to inertial force shows that

$$\frac{\rho_e E}{\rho V^2} \sim \frac{\varepsilon_0 E^2}{L\rho V^2} \sim 10^{-8},$$

where typical values for electric force $E = 10^3\text{V/m}$, velocity $V = 10^2\text{m/s}$, density $\rho = 10^{-1}\,\text{kg/m}^3$, and characteristic scale distance $L = 0.1\,\text{m}$ are used.

3. Neglect of the Hall effect (the second term in the left-hand part of eq. (1.35)). For a substantiation of this approach it is necessary to consider two possible cases: the self- and the exterior magnetic fields. In the case of the self-magnetic fields, the Hall parameter is estimated as follows:

$$P_H = \frac{\sigma B}{en_e} \sim \frac{\sigma E}{Ven_e} \sim 10^{-6},$$

where $n_e = 10^{22}\,\text{m}^{-3}$, $E = 10^3\,\text{V/m}$, $V = 10^2\text{m/s}$, $\sigma = 10^4(\text{Ohm m})^{-1}$.

It is obvious from the definition of the Hall parameter that the induction of the exterior magnetic field also should be rather high to affect appreciably on dynamics of the plasma.

4. Neglect of an electronic pressure gradient and the induced electric field in comparison with electric intensity E:

$$\frac{1}{en_e E}\nabla p_e \sim \frac{kT_e}{LEe} \ll 1,$$

$$\frac{[\mathbf{V}\times\mathbf{B}]}{E} \sim \frac{VB}{E} \ll 1.$$

Assumptions 3 and 4 give grounds to use the Ohm law in the following simplified form:

$$\mathbf{J} = \sigma\mathbf{E}$$

Note that the using the full system of eqs. (1.70)–(1.72) is extremely rare, while the gas dynamic models based on the approach of a boundary layer are of considerable use at present.

1.1.3 Models of microwave plasma generators

The typical schematic diagram of the microwave (MW) PG is shown in Figure 1.4.

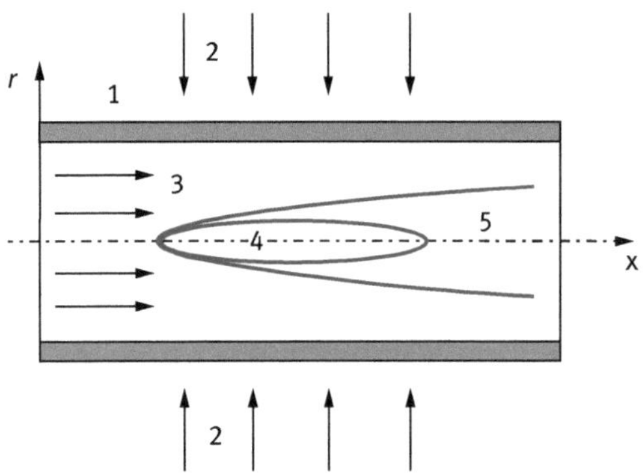

Figure 1.4: Schematic diagram of the RF axial type plasma generator: 1, the gas discharge channel; 2, axially symmetric flow of RF radiation; 3, gas stream (possibly with vortex); 4, discharge plasma; 5, plasma jet.

Calculation models of the MW discharge are similar to models of equilibrium EA and RF PGs. The difference is only in that part of the equations which features an energy release in the low-temperature plasma. In case of an MW PG, this is the Maxwell equation formulated in the most convenient view for geometry explored, subject to features of interaction between electromagnetic MWs and plasma. For MW discharges, the frequency band of $f_{MW} \sim 10^9 - 10^{11}$ Hz and wavelengths $\lambda_{MW} \sim 0.3 - 30$ cm are typical. An overwhelming majority of cases under consideration, the physical conditions, in which the electron concentration appreciably surpasses the critical electron concentration

$$n_{c,e} = \frac{\varepsilon_0 m_e \omega^2}{e^2},$$

are realized in the MW discharges.

Therefore, as a rule the valid requirement is

$$\omega < \omega_e,$$

where ω is the circular frequency of the electromagnetic field; $\omega_e = \sqrt{n_e e^2/\varepsilon_0 m_e}$ is the plasma frequency. Note that the frequency of collisions of electrons with heavy particles ν can surpass circular frequency *omega* or be less than it as well.

A distinctive feature of MW discharges is that the wavelength of electromagnetic radiation is usually close to the characteristic size of plasma volume. Therefore, the geometrical optics approximations, which allow simplifying an electromagnetic formulation of the problem, are in most cases inapplicable. In this sense, the formulation of MW discharge mathematical model represents greater difficulty than the formulation of such a model for RF discharge. Nevertheless, practically explored models of the MW PGs allow apply some simplifications. Some of such models are presented below.

The elementary one-dimensional computing model of MW discharge

The MW discharge is considered in one-dimensional cylindrical geometry. It means that change of its parameters in the radial direction is many times more than the changes in the axial direction. If gas dynamic processes are negligible for such discharge, the following governing equations can be used:

- energy conservation equation

$$\frac{1}{r}\frac{\mathrm{d}}{\mathrm{d}r}\left(r\lambda\frac{\mathrm{d}T}{\mathrm{d}r}\right) = -\frac{1}{2}\sigma|E|^2 - \mathrm{div}\,\mathbf{W}, \tag{1.36}$$

- equation for electric field intensity

$$\frac{1}{r}\frac{\mathrm{d}}{\mathrm{d}r}\left(\lambda\frac{\mathrm{d}E}{\mathrm{d}r}\right) + \omega^2\varepsilon_0\mu_0\varepsilon_k E = 0, \tag{1.37}$$

where $\varepsilon_k = \varepsilon + i(\sigma/\varepsilon_0\omega)$ is the complex permeability; ε, σ are the dielectric permeability and electrical conductivity of the medium; μ_0 is the magnetic permeability of free space; E is the complex electric field intensity.

Boundary conditions for eqs. (1.36) and (1.37) are formulated along a symmetry axis and on the external radial boundary of the discharge channel:

$$r = 0\colon\ \frac{\partial T}{\partial r} = 0, \qquad \frac{\partial E}{\partial r} = 0,$$

$$r = R\colon\ T = T_w, \quad Q_{\sum} = Q_R + Q_A,$$

where $Q_A = \pi \int_0^R \sigma |E|^2 \, r dr$ is the absorbed power of electromagnetic radiation; Q_Σ, Q_R are the radiation power falling on the plasma and the reflected by it. Value of Q_Σ is the prescribed parameter of the problem.

Two-dimensional axisymmetric model of MW discharge

Several assumptions used in this model are as follows:
- gas flow in the MW PG is stable and axially symmetric,
- gas discharge plasma is in chemical equilibrium,
- external electromagnetic energy falls to plasma in the radial direction (see Figure 1.4), and
- change of electromagnetic functions in the radial direction appreciably surpasses their change in the axial direction.

Governing equations of the model are formulated in the following form:
- continuity equations

$$\frac{\partial \rho}{\partial t} + \operatorname{div} \rho \mathbf{V} = 0; \tag{1.38}$$

- equations of motion

$$\rho u \frac{\partial u}{\partial x} + \rho v \frac{\partial v}{\partial r} = -\frac{\partial p}{\partial x} + S_u + F_x - \rho g, \tag{1.39}$$

$$\rho u \frac{\partial v}{\partial x} + \rho v \frac{\partial v}{\partial r} = -\frac{\partial p}{\partial r} + S_v + F_r, \tag{1.40}$$

$$S_u = -\frac{\partial}{\partial x}\left(\frac{2}{3}\mu \operatorname{div} \mathbf{V}\right) + \frac{1}{r}\frac{\partial}{\partial r}\left[r\mu\left(\frac{\partial u}{\partial r} + \frac{\partial v}{\partial x}\right)\right] + \frac{\partial}{\partial x}\left(2\mu \frac{\partial u}{\partial x}\right),$$

$$S_v = -\frac{\partial}{\partial r}\left(\frac{2}{3}\mu \operatorname{div} \mathbf{V}\right) + \frac{\partial}{\partial x}\left[\mu\left(\frac{\partial u}{\partial r} + \frac{\partial v}{\partial x}\right)\right] + \frac{1}{r}\frac{\partial}{\partial r}\left(2r\mu \frac{\partial v}{\partial r}\right) - 2\mu \frac{v}{r^2};$$

- energy conservation equations

$$\rho c_p u \frac{\partial T}{\partial x} + \rho c_p v \frac{\partial T}{\partial r} = \frac{\partial}{\partial x}\left(\lambda \frac{\partial T}{\partial x}\right) + \frac{1}{r}\frac{\partial}{\partial r}\left(r\lambda \frac{\partial T}{\partial r}\right) + Q_E - Q_R; \tag{1.41}$$

- thermal equations of state

$$p = \frac{R_0}{M}\rho T;$$

- the equations of an electromagnetic field

$$\frac{1}{r}\frac{\mathrm{d}}{\mathrm{d}r}\left(r\frac{\partial E_x}{\partial r}\right)+\frac{\omega^2}{\varepsilon_0\mu_0}\varepsilon_k E_x=0, \tag{1.42}$$

$$H_\varphi=-\frac{i}{\omega\mu_0}\frac{\partial E_x}{\partial r}, \tag{1.43}$$

$$E_r=\frac{i}{\omega\varepsilon_0\varepsilon_k}\frac{\partial H_\varphi}{\partial x}, \tag{1.44}$$

where x, r are the axial and radial coordinates; $\mathbf{V}=(u, v)$ is the velocity of the plasma flow and its projections to axes$x$and$r$accordingly; ρ, p are the density and pressure; c_p is the specific heat capacity at constant pressure; T is the temperature; $Q_\mathrm{E}=0.5\sigma|E|^2$ is the Joule thermal emission; Q_R is the power of volumetric radiative losses; $\mathbf{E}=(E_r, E_\varphi=0, E_x)$, $\mathbf{H}=(H_r=0, H_\varphi, H_x=0)$ are the intensities of electrical and magnetic fields; ε_k is the complex inductivity.

Features of the solution of the gas dynamic and thermal parts of the problem (eqs. (1.38)–(1.41)) will be discussed in Section 1.1.6.

Feature of the solution of the electrodynamic part of the problem (eqs. (1.42)–(1.44)) is that components of electrical and magnetic fields are complex functions. Therefore, these equations should be rewritten in the form of the system of equations concerning the real and imaginary parts, but for all that it is necessary to take into account that the boundary conditions are set for incident radiation energy only on the exterior radial boundary of the channel.

1.1.4 Models of laser-supported plasma generators

In many respects, the schematic diagram of the laser-supported PG (LSPG) considered in the present part (Figure 1.5) is analogous to the classical schemes of arc and high-frequency PGs. The plasma, which is the source of the plasma jet used for various technological and research purposes, is created in a cylindrical channel through which a gas is pumped. In the case of LSPG, the plasma is formed by

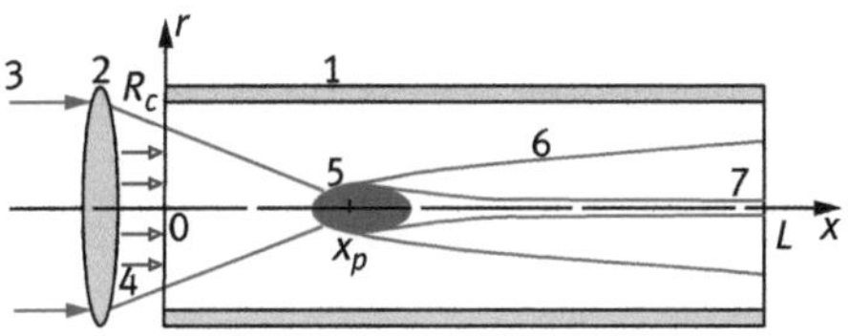

Figure 1.5: Schematic diagram of a laser plasma generator: cylindrical chamber of the generator (1), focusing lens (2), unfocused laser beam (3), undisturbed gas flow in the inlet cross section of the chamber (4), steady-state laser plasma (5), gas and generated plasma jet mixing layer (6), and plasma jet (7).

absorption of the radiation of a continuous laser, most frequently, a CO_2 laser with the radiation wavelength $\lambda = 10.6$ μm. The plasma thus obtained is called a continuous (steady-state) optical discharge or simply a steady-state laser plasma, and the process of its propagation through a stationary or moving gas is called a laser combustion wave or laser-supported wave (Raizer Yu.P., 1991).

LSPG has unique properties, which make it possible to regard it as a promising source for solving important practical problems of radiation–gas dynamics: at the atmospheric air pressure, the temperature reaches 1.6×10^4 K in the laser radiation absorption zone, and the high-temperature plasma zone has no contact with the surfaces. This ensures a high-purity plasma flow.

In several experimental and theoretical investigations (Klosterman E.L., et al., 1974; Jackson J.P., et al., 1974; Fowler M.S., et al., 1975; Keefer D.R., et al., 1975; Su F.Y., et al., 1976; Conrad R., et al., 1996; Surzhikov S.T., 1997, 2005), the conditions of steady-state existence of the laser plasma for a stationary or moving gas and focused or unfocused laser beams were considered. In Jones L.W., et al., (1982), Mirabo L., et al., (1995), and Surzhikov S.T. (1995), continuous or pulse-periodic laser PGs and laser ramjets were analyzed.

In continuous laser engines, gas is heated inside the combustion chamber by means of continuous laser radiation. The heated gas flows out through a nozzle as in an ordinary jet engine. The operating principle of the pulse-periodic engine is different. In this case, pulse-periodic lasers and specially profiled channels are used. They ensure that the optical gas breakdown shock wave pulse is transferred with maximum efficiency to the vehicle which, as it were, is pushed along by the shock waves.

Clearly, the power of the laser radiation used and its wavelength are the most important parameters determining the regimes of the existence of the laser plasma and the working characteristics of the engine.

Accurate focusing of the several kilowatt CO_2 laser radiation at a wavelength of 10.6 μm makes it possible to generate laser plasma with a characteristic spatial scale of the order of 0.01–0.03 cm. At such small dimensions, the laser plasma exists in the so-called heat conduction regime. This means that the laser radiation energy absorbed in the plasma is mainly transferred to the surrounding gas by heat conduction. The plasma radiation zone is so small that the thermal radiation power is insufficient to heat the surrounding medium up to temperatures of the single ionization at which intensive absorption of the laser radiation begins. The transition from the heat conduction to the radiative regime of the existence of the laser plasma is possible only if the laser radiation power increases significantly.

In order for the radiative regime of the existence of the laser plasma (or its propagation along an unfocused or weakly focused laser beam) to be realized in atmospheric air, it is necessary for the plasma dimensions to be of the order of 1 cm. It is precisely these dimensions of the laser plasma that were observed in 1974 in experiments (Klosterman E.L., et al., 1974) on plasma propagation along an unfocused

CO_2 laser beam and investigated numerically with reference to problems of laser physics (Surzhikov S.T., 2000b).

In the radiative regime of the existence of the laser plasma, the energy of the absorbed laser radiation is transferred to the surrounding gas by thermal radiation whose power is sufficient to heat the adjacent gas layers to the single-ionization temperature. Hence, there follows the principal importance of correctly taking into account the effect of the thermal radiation transfer processes on the formation of the laser plasma and the plasma jet. Clearly, in this case, the problem under consideration should be assigned to the class of radiation–gas dynamic problems with strong radiation–gas dynamic interaction.

Another basic feature of the radiative regime considered is the selectivity of the thermal radiation. This significantly complicates the theoretical analysis of the phenomena and the numerical realization of the radiation–gas dynamic models since the spectral absorption coefficient varies by orders of magnitude on different spectral ranges and at different points of the laser plasma and the gas flow.

When investigating the characteristics of the laser PG one more significant problem should also be considered. This problem is the radiation heating of the walls of the gas dynamic channel in which the laser plasma is ignited and from which the plasma jet flows out. Solving this problem is important for two reasons. First, it is enabled to determine the energy losses, integral over the spectrum, to the generator channel walls. Second, there is a risk of thermal damage to the channel walls by the thermal radiation at such high plasma temperatures. This is directly associated with the problem of generator design.

The radiation–gas dynamic model should be developed and used in the case of LSPG in order to upgrade the calculation of the radiative thermal fluxes on the internal surface of the generator.

Self-oscillations of the laser plasma and bifurcations of the gas-plasma structures accompanying the processes of gas flow past a laser plasma can develop in both the heat conduction and radiative regimes (see Chapter 2). Some working regimes of laser PGs and of subsistence of the atmospheric pressure laser combustion waves will be considered below in connection with the transition from the heat conduction mode to the radiation regime of the existence of the laser plasma. The effects of the initial free stream gas flow turbulence and reabsorption of the thermal radiation of the laser plasma on flow stabilization are also considered.

Summing up this brief analysis of the problem of creating a radiation–gas dynamic model of the laser PG, note that this model (or its elements) can also be used for solving problems of strong radiation–gas dynamic interaction in the shock layer near the surface of space vehicles re-entering at super-orbital velocities in the denser layers of the atmosphere. In fact, the unique properties of the laser plasma, namely, its temperature $T{\sim}(1.5 \div 2) \times 10^4$ K at atmospheric pressure, are making it possible to simulate the conditions in the shock layer in the neighborhood of space vehicles entering the Earth's atmosphere at velocities higher than 16 km/s. However, this model must be supplemented by taking the physical–chemical processes that are important for hypersonic flow problems into account.

In the given section, the task of the radiative–gas dynamics of the laser-supported plasma in gas flow is formulated in the two-dimensional cylindrical geometry corresponding to the conditions of symmetry of the process considered. As the gas medium, let us consider an air at atmospheric pressure. A continuous CO_2 laser beam with a radiation wavelength of 10.6 μm falls on the plasma from the left (Figure 1.5).

For describing the thermodynamic state of the low-temperature plasma, the LTE approximation is used. As before, it is assumed that particle collision processes predominate in the atmospheric pressure plasma considered and that its thermodynamic state can be completely characterized by the pressure and the temperature. Of course, this is not actually entirely true at such high temperatures and densities of the laser and thermal self-radiation and there are zones in the neighborhood of the laser combustion wave in which the LTE approximation is certainly not satisfied. For example, the powerful ultraviolet radiation from the high-temperature zone leads to photodissociation and photo-ionization of the relatively cold gas layers, as this takes place ahead of the front of a strongly radiating shock wave (action of "precursor" radiation). Nevertheless, preliminary investigations showed that the nonequilibrium processes have no appreciable effect on the laser combustion wave dynamics or on the parameters of the steady-state laser plasma.

In order to solve the task of the laser plasma dynamics, the following system of equations of continuity, Navier–Stokes, and energy conservation together with radiation heat transfer and laser radiation transfer equations will be used:

$$\frac{\partial \rho}{\partial t} + \operatorname{div}(\rho \mathbf{V}) = 0, \tag{1.95}$$

$$\frac{\partial \rho u}{\partial t} + \operatorname{div}(\rho u \mathbf{V}) = -\frac{\partial p}{\partial x} + S_u, \tag{1.96}$$

$$\frac{\partial \rho \mathrm{v}}{\partial \mathrm{t}} + \operatorname{div}(\rho \mathrm{v} \mathbf{V}) = -\frac{\partial p}{\partial \mathrm{r}} + \mathrm{S_v}, \tag{1.97}$$

$$\rho c_p \frac{\partial T}{\partial t} + \rho c_p \mathbf{V} \operatorname{grad} T = \operatorname{div}(\lambda_\Sigma \operatorname{grad} T) - Q_\mathrm{R} + Q_\mathrm{L}, \tag{1.98}$$

$$Q_\mathrm{L} = \chi_\omega(x, r=0) P_\mathrm{L} \exp\left(-\frac{r^n}{R_\mathrm{L}^n}\right) \exp\left[-\int_0^x \chi_\omega(x', r=0)\mathrm{d}x'\right] \frac{1}{\pi R_\mathrm{L}^2}, \tag{1.99}$$

$$Q_\mathrm{R} = \sum_{k=1}^{N_k} \kappa_k (U_{b,k} - U_k) \Delta\omega_k, \tag{1.100}$$

$$\mathrm{div}\left(\frac{1}{3\kappa_k}\mathrm{grad}U_k\right) = -\kappa_k(U_{b,k} - U_k), \quad k = 1, 2, \ldots, N_k, \tag{1.101}$$

$$S_u = -\frac{2}{3}\frac{\partial}{\partial r}(\mu_\Sigma \,\mathrm{div}\,\mathbf{V}) + \frac{1}{r}\frac{\partial}{\partial r}\left[r\mu_\Sigma\left(\frac{\partial u}{\partial r} + \frac{\partial u}{\partial x}\right)\right] + 2,\frac{\partial u}{\partial x}\left(\mu_\Sigma\frac{\partial u}{\partial x}\right), \tag{1.102}$$

$$S_v = -\frac{2}{3}\frac{\partial}{\partial r}(\mu_\Sigma \,\mathrm{div}\,\mathbf{V}) + \frac{\partial u}{\partial x}\left[\mu_\Sigma\left(\frac{\partial u}{\partial r} + \frac{\partial u}{\partial x}\right)\right] + 2\frac{\partial}{\partial r}\left(\mu_\Sigma\frac{\partial u}{\partial r}\right) + 2\mu_\Sigma\frac{\partial u}{\partial r}\left(\frac{u}{r}\right). \tag{1.103}$$

where x and r are the axial and radial coordinates; p, c_p, and T are the density, specific heat at constant pressure, and temperature, respectively; u and v are the axial and radial components of the velocity $\mathbf{V}$; p is the pressure; μ_Σ and λ_Σ are the effective dynamic viscosity and thermal conductivity coefficients; Q_R and Q_L are the volume energy release powers associated with the transport of selective thermal and laser radiation, respectively; κ, U, and U_b are the volume thermal absorption coefficient and the volume radiation densities for the medium and an absolutely black body; χ_Ω is the laser radiation absorption coefficient; P_L is the laser power; R_L is the radial boundary of the laser beam, and n is the index of the intensity distribution over the laser beam cross section. The subscripts ω and k denote the spectral and group characteristics determined by averaging the corresponding spectral characteristics over each of N_k spectral wavenumber ranges $\Delta\omega_k$.

In eqs. (1.98), (1.102), and (1.103), the turbulent viscosity and thermal conductivity coefficients are calculated using the Boussinesq hypothesis, according to which the effective gas flow viscosity can be determined from the formula

$$\mu_\Sigma = \mu_\mathrm{m} + \mu_\mathrm{t}, \tag{1.104}$$

where μ_m is the dynamic viscosity coefficient, which takes into account the molecular collision processes, μ_t is the turbulent viscosity coefficient, for determining which one can use a turbulent mixing model. In the case under consideration, the $k-\varepsilon$ model (Wilcox D.C., 1993; Cebeci T., 2004) is used. Assume that the turbulent Prandtl number, $\mathrm{Pr}_\mathrm{t} = \mu_\mathrm{t}c_p/\lambda_\mathrm{t}$, is equal to unity. This makes it possible to find the corresponding thermal conductivity coefficient $\lambda_\mathrm{t} = c_p\mu_\mathrm{t}$. The equations of the $k-\varepsilon$ model have the form

$$\frac{\partial\rho k}{\partial} + \frac{1}{r}\frac{\partial}{\partial r}\left[r\left(\rho kv - \frac{\mu_t}{\sigma_k}\frac{\partial k}{\partial r}\right)\right] + \frac{\partial}{\partial x}\left(\rho ku - \frac{\mu_t}{\sigma_k}\frac{\partial k}{\partial x}\right) = p - \rho\varepsilon, \tag{1.105}$$

$$\frac{\partial\rho\varepsilon}{\partial t} + \frac{1}{r}\frac{\partial}{\partial r}\left[r\left(\rho\varepsilon v - \frac{\mu_t}{\sigma_\varepsilon}\frac{\partial\varepsilon}{\partial r}\right)\right] + \frac{\partial}{\partial x}\left(\rho\varepsilon u - \frac{\mu_t}{\sigma_\varepsilon}\frac{\partial\varepsilon}{\partial x}\right) = (C_1P - C_2\rho\varepsilon)\frac{\varepsilon}{k}, \tag{1.106}$$

$$\mu_\mathrm{t} = \rho C_\mu\frac{k^2}{\varepsilon}, \tag{1.107}$$

$$P=\mu_t\left\{2\left[\left(\frac{\partial v}{\partial r}\right)^2+\left(\frac{\partial u}{\partial x}\right)^2+\left(\frac{v}{r}\right)^2\right]+\left(\frac{\partial v}{\partial x}+\frac{\partial u}{\partial r}\right)^2\right\}-$$

$$-\frac{2}{3}\rho k\left(\frac{\partial u}{\partial x}+\frac{1}{r}\frac{\partial rv}{\partial r}\right)-\frac{\lambda_t}{\rho^2}\left(\frac{\partial\rho}{\partial r}\frac{\partial p}{\partial r}+\frac{\partial\rho}{\partial x}\frac{\partial p}{\partial x}\right), \quad (1.108)$$

$$C_\mu=0.09;\ C_1=1.44;\ C_2=1.92;\ \sigma_k=1.0;\ \sigma_\varepsilon=1.3.$$

In the immediate neighborhood of the surface, different modifications of the turbulent mixing model can be used (Cebeci T., 2004).

The laser radiation propagation is described in the geometric optics approximation. The continuum absorption mechanism, which is the inverse of the electron bremsstrahlung mechanism under LTE conditions, is assumed to be determining for the laser radiation absorption coefficient in eq. (1.99):

$$\chi_\omega=2.82\times10^{-29}n_e(n_++4n_{++})T^{-3/2}\lg\left(2.17\times10^3Tn_e^{-1/3}\right),\ 1/\text{cm} \quad (1.109)$$

where n_e, n_+, and n_{++} are the number densities of electrons and singly and doubly charged ions, and T is the temperature in K.

The power of the energy release associated with the thermal radiation transport process can be found by integrating the radiation heat transfer equation for selective heat radiation in the form of the multigroup Pl approximation of the spherical harmonics method, that is, as a result of solving the system of N_k eq. (1.101), each of which is assigned its own $\kappa_k(T,p)$ function.

The following boundary conditions can be used for solution of the task: for the cooled surface (at a temperature T_0), no-slip conditions on the surface, axial symmetry of the functions T, u, v, and U_k, and absence of flow perturbations at the channel inlet: $u=u_0$, $v=0$, and $T=T_0$. In the exit cross section of the cylindrical channel, there is possibility to use boundary condition $f_x(x\to L)=0$ (a) or $f_{xx}(x\to L)=0$ (b), where $f=\{T,\ u,\ v,\ U_k\}$. The distance to the exit boundary of the computation domain should be taken so as to reduce as for as possible its effect on the laser plasma gas dynamics. At the practical realization, the Neumann boundary conditions (a) turned out to be acceptable in computation variants with a high-temperature laminar plasma jet.

As initial conditions, a spherical plasma cloud of 0.5 cm radius at a temperature of 20,000 K against the background of unperturbed gas flow can be used.

The thermophysical properties of equilibrium air should be calculated over the entire temperature range from normal conditions to double ionization using, for example, the approximations (Giordano D., et al., 1994). These approximations were obtained over wide pressure and temperature range using compact formulas, which are based on data on the transport properties in the third approximation of the Chapman–Enskog method.

To solve the radiation heat transfer equation the multigroup spectral optical model can be recommended. The spectral optical model of hot air (thermal radiation absorption coefficient in eqs. (1.100) and (1.101)) was calculated using ASTEROID software (Surzhikov S.T., 2000a). Initially, it was found the absorption coefficients of high-temperature air over the temperature range from 300 to 20,000 K at million points along the spectrum. In this case, all the most significant elementary radiation processes that contribute to the total absorption coefficient of air under LTE conditions were taken into account. Nonuniform computation grid in the wavelength was used. This made it possible to describe the contours of about 4,000 atomic lines of nitrogen and oxygen atoms and ions (with a resolution of the multiplet structure) at not less than 10 spectral points. This method is usually called as the line-by-line method. By averaging the results of the line-by-line calculations, it is possible to create any group spectral models. Three group spectral models with 37, 74, and 148 spectral groups are demonstrated here. Figure 1.6 reproduces the line-by-line and group absorption coefficients at temperatures of 4,000, 10,000, and 16,000 K. These temperature points give a representation of the features of the spectral absorption coefficients of hot air for dissociation temperatures (4,000 K), for conditions of the almost complete breakdown of the molecular components (10,000 K), and developed ionization (16,000 K).

Let us note the main structural features of the hot air absorption spectrum over the mentioned temperature range investigated. Strong radiation absorption increasing with a decrease in the temperature is observed over the ultraviolet spectrum range, in particular, in its “vacuum” part (80,000 cm^{-1}). In this range, the absorption coefficient is attributable to atomic photoionization from the ground states and from the low-excitation energy states and molecular photodissociation and photoionization. In this part of the spectrum, the atomic and ionic spectral lines formed in quantum transitions from the ground and low-excited levels radiate and absorb at high temperatures.

As the temperature increases, the effect of the atomic lines in the integral energy balance also increases. At the centers of the atomic lines, the absorption coefficients are appreciably greater than 1 cm^{-1} so that under the conditions considered, their radiation is absorbed in the plasma itself without leaving it. From Figure 1.6a, one can see that the radiation with the wavenumber $\omega > 90,000\ cm^{-1}$ is almost completely absorbed by the plasma layers at a temperature $T = 4,000$ K since the coefficient $\kappa_\omega > 10\ cm^{-1}$ and, consequently, the characteristic free path of the thermal photons $l_r < 0.1$ cm. Colder gas layers absorb the ultraviolet radiation even more strongly.

The molecular plasma components, which dissociate as the temperature increases (as a result, the plasma becomes more “transparent”), absorb appreciable amounts of radiation over the near-ultraviolet spectrum range ($\omega = 25,000-60,000\ cm^{-1}$).

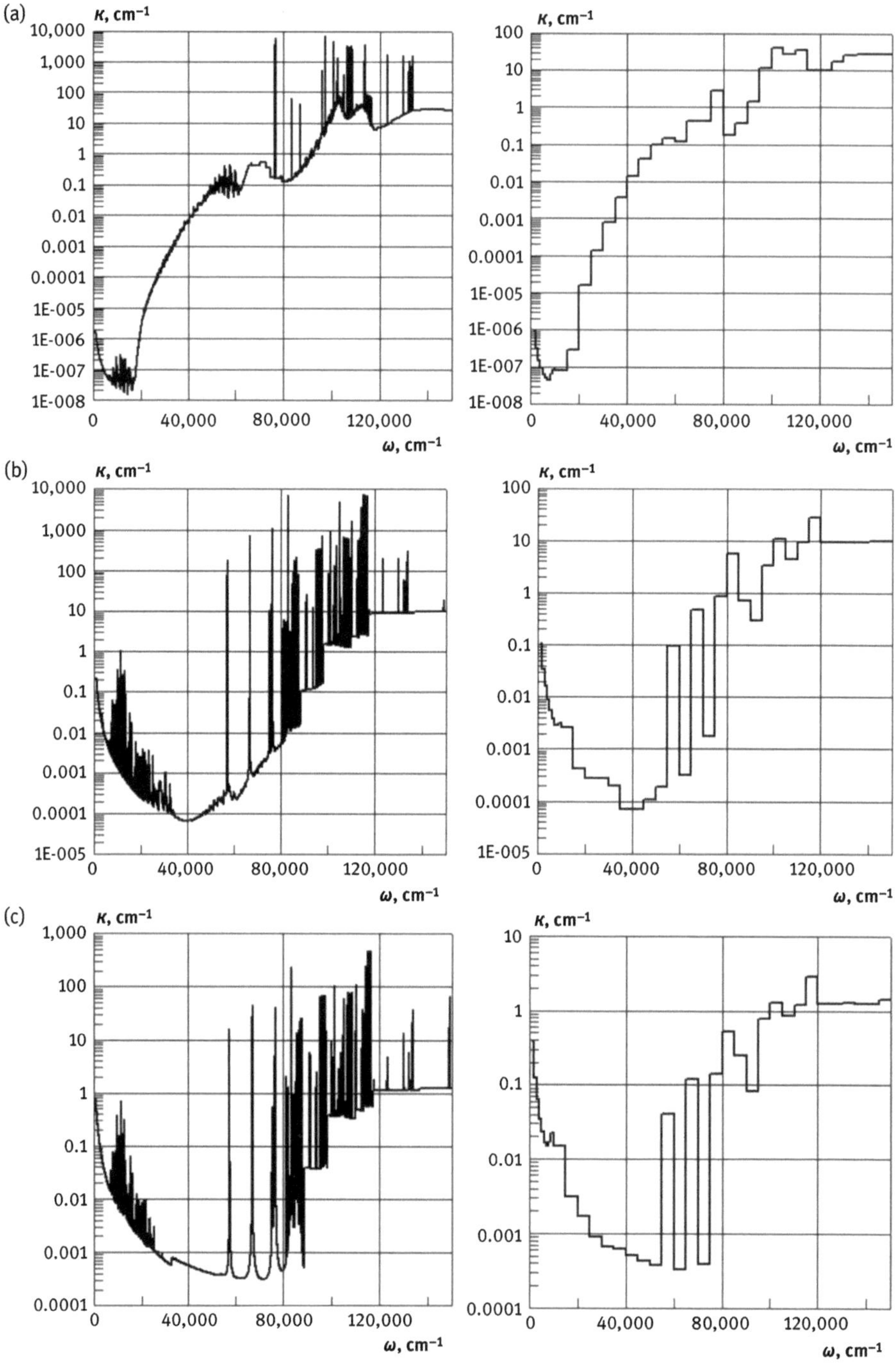

Figure 1.6: Line-by-line (left) and group (right) absorption coefficients of high-temperature air for $T = 4,000$ K (a), 10,000 K (b), and 16,000 K (c).

Over the visible ($\Omega = 13,160 - 25,000 \ \text{cm}^{-1}$) and near-infrared spectrum ranges, the molecular components absorb radiation mainly in electron-vibrational quantum transitions and also in atomic lines formed in quantum transitions from excited energy states. On this spectral range, the continuous absorption spectrum is associated with the interaction between free electrons and plasma ions. Therefore, at low temperatures, that is, at low degrees of ionization, the heated gas is almost transparent for radiation.

The chosen temperature points are important for investigating the laser plasma dynamics. In a laser plasma located in a gas flow, the temperature distribution has the following behavior: the temperature reaches a maximum of $(1.5 - 2.0) \times 10^4$ K in the zone of the strong absorption of laser radiation. Then sharply decreases to 10^4 K, while the resulting plasma jet has a temperature of the order of 5×10^3 K. Note that if the zone with the maximum temperature is localized in space of focused laser beam, then the relatively low-temperature zones occupy a significant volume and are mainly determined by the gas dynamics.

The need to analyze various group models of heat radiation is due to the following. Ideally, in order to solve radiation–gas dynamic problems, one should use the line-by-line models. The tendency to use the most detailed spectral models is expressed in world practice for solving various radiation–gas dynamic problems as computer power increases and the parallel algorithms are developed. However, in the overwhelming majority of cases, the use of these models is not justified, not to mention their extreme computational inefficiency. Therefore, the problem of choosing an adequate multigroup optical model, in particular, for analyzing processes admitting the use of the LTE model remains fairly topical for the majority of radiation–gas dynamic problems. The laser PG calculations carried out in the present part using various group models demonstrate the possibility of using optical models with approximately 37 groups, within the framework of which it is possible to describe the most important structural features of the absorption spectrum, including the groups of strongest atomic lines. Nevertheless, note that the multigroup models presented here still cannot solve the problems of adequately calculating the radiation transport in atomic and ion lines.

1.1.5 Numerical simulation models of steady-state radiative gas dynamics of RF, EA, MW, and LSW plasma generators

Computing two-dimensional models of radiative-gas dynamic processes in different kinds of PGs is considered in the present section. Governing equations of gas dynamic processes in PGs were formulated above in two forms. The first one was the formulation of eqs. (1.36)–(1.48) in the explicitly stationary form. The second one was the formulation of gas dynamic equations in the nonstationary form (see eqs. (1.70)–(1.72) and (1.95)–(1.98)). Here we will consider numerical simulation models for the solution of the equations for steady-state regimes. If the equations do not contain time derivatives, they will be modified by adding artificial time derivatives of corresponding functions. These artificial time derivatives will be used, as in the case of explicit nonstationary formulation, as an auxiliary summand for the time-marching numerical procedure.

The above-considered gas dynamic models of PGs are based on the process of the subsonic gas motion through the area of intensive absorption of electromagnetic energy where abrupt changes (by a factor of a ten or hundred!) of a flow rate and temperature, and also of corresponding thermodynamic, thermophysical, and optical properties, take place.

The energy conservation equation in all models contains a volume rate of energy release caused by electromagnetic radiation of different wavelengths. This equation is classified as the elliptic equation in stationary form or as the parabolic equation in nonstationary form. The basic difficulty of its solution is connected with the creation of optimum iterative process between the functions defining a plasma heating degree (temperature, enthalpy, internal energy), and nonlinear rates of energy exchange. Remind that radiant loss is proportional approximately to T^4, and laser radiation absorption is defined by an exponential function of temperature.

The significant methodical complexity of the task consists in the realization of a numerical integration method of the gas dynamic equations. In the practice of the computational fluid dynamics, two basic approaches for numerical simulation of subsonic gas flows with any arbitrary changes of density, caused by changing of temperature at practically constant pressure in subsonic flows, are well known. These are the approaches based on the natural form of the gas dynamic equations, that is, with the use of velocity components and pressure as unknown functions, and the approach based on the so-called dynamic variables. First of these methods allows to calculate the unsteady and stationary gas flows; however, it demands special arrangements for accurate definition of small pressure field disturbances. Use of the second method allows to manage without a definition of a pressure field but corresponds to stationary flows only. Both of these approaches are considered in this chapter.

Computing model of the natural form of the equations for gas flow

Governing equations of gas dynamic processes in the channel of a PG in a two-dimensional axisymmetric geometry are formulated in the model as follows:

$$\frac{\partial \rho}{\partial t} + \operatorname{div}(\rho \mathbf{V}) = 0, \tag{1.110}$$

$$\frac{\partial \rho u}{\partial t} + \operatorname{div}(\rho u \mathbf{V}) = -\frac{\partial p}{\partial x} + \frac{S_u}{\mathrm{Re}}, \tag{1.111}$$

$$\frac{\partial \rho v}{\partial t} + \operatorname{div}(\rho v \mathbf{V}) = -\frac{\partial p}{\partial r} + \frac{S_v}{\mathrm{Re}}, \tag{1.112}$$

$$\frac{\partial \rho H}{\partial t} + \operatorname{div}(\rho \mathbf{V} H) = \frac{1}{\mathrm{Pr}\cdot\mathrm{Re}}\left(\frac{\operatorname{div}\lambda}{c_p \operatorname{grad} H} + \frac{Q^* L^2}{\lambda_0 T_0}\right), \tag{1.113}$$

$$Q^* = W - Q_{\mathrm{R}}, \tag{1.114}$$

$$Q_{\mathrm{R}} = \int_0^{\omega} c\kappa_\lambda (U_{b\lambda} - U_\lambda)\mathrm{d}\lambda = \sum_1^{N_k} c\kappa_k (U_{bk} - U_k)\Delta\lambda_k, \qquad c = 3\times 10^{10}\mathrm{cm/s}, \tag{1.115}$$

$$\operatorname{div}\left(\frac{1}{3\kappa_k}\right)\operatorname{grad} U_k = \kappa_k (U_k - U_{b,k}), k = 1, 2, \ldots, N_k, \tag{1.116}$$

where

$$S_u = -\frac{2}{3}\frac{\partial}{\partial x}(\mu \operatorname{div} \mathbf{V}) + \frac{1}{r}\frac{\partial}{\partial r}\left[r\mu\left(\frac{\partial u}{\partial r} + \frac{\partial v}{\partial x}\right)\right] + 2\frac{\partial}{\partial x}\left(\mu\frac{\partial u}{\partial x}\right),$$

$$S_v = -\frac{2}{3}\frac{\partial}{\partial r}(\mu \operatorname{div} \mathbf{V}) + \frac{\partial}{\partial x}\left[\mu\left(\frac{\partial u}{\partial r} + \frac{\partial v}{\partial x}\right)\right] + 2\frac{\partial}{\partial r}\left(\mu\frac{\partial v}{\partial r}\right) + 2\mu\frac{\partial v}{r\partial r} - \frac{2mv}{r^2},$$

where x, r are the axial and radial variables; ρ, c_p, T are the density, specific heat capacity at constant pressure, and temperature; u, v are axial and radial velocity components $\mathbf{V}$; p is the pressure; μ, λ are the coefficients of dynamic viscosity and thermal conductivity; Q_{R} is the volume power flux related to thermal radiation transfer; κ, U, U_b are the volume absorption coefficient and volume radiation density of a medium and of a blackbody; ω, k are the indexes of spectral and group characteristics; W is the energy release due to the absorption of external electromagnetic energy in plasma. The group absorption coefficient κ_k is the average in each of N_k spectral ranges $\Delta\omega_k$ covering spectral region $2,000 - 250,000\ \mathrm{cm}^{-1}$.

All variables in eqs. (1.110)–(1.113), namely velocity $\mathbf{V} = \{u, v\}$, density ρ, pressure p, enthalpy H, viscosity μ, and thermal conduction λ_t are referred to corresponding values in undisturbed input gas stream with temperature T_0, density ρ_0, and velocity V_0. The similarity parameters $\mathrm{Pr} = \mu_0 c_{p0}/\lambda_0$, $\mathrm{Re} = \rho_0 V_0 L/\mu_0$ are defined by the thermodynamic and thermophysical properties $(c_{p,0}, \mu_0, \lambda_0)$ at the same conditions. Inclusion of the spatial scale L allows defining also the timescale $t_0 = L/V_0$.

Attempts to get a self-consistent solution of eqs. (1.110)–(1.116) encounter with the complexity proper for all systems of stiff equations, where the strong distinction of the characteristic timescales of running processes is observed. Let's make explaining estimates. The characteristic processes for the first two equations are the convective transfer with the velocity V_0 and propagation of sound perturbations with the acoustic speed a_0. For the characteristic spatial scale of $L = 1$ cm, we shall estimate a computational mesh step with the value $h = 0.01$ cm. Therefore, supposing $V_0 = 1,000$ cm/s, $T_0 = 500$ K, $\rho_0 = 0.00124$ g/cm^3, $p_0 = 1$ atm $= 10^6$ g/(cm $\times$ s^2), one can estimate the characteristic times of convective and sound processes as $t_c = L/V_0 \sim 10^{-3}$s, $t_s = L/a_0 \sim 3\times 10^{-6}$ s ($a_0 = 3.36\times 10^4$ cm/s, $\gamma = c_p/c_V = 1.4$).

Typical time of thermoconductive wave propagation can be estimated under the formula:

$$t_T = \frac{L^2}{\chi} = \frac{L^2}{\lambda/c_p\rho},$$

where, however, values of functions should be taken at the representative temperature of discharge $T_* = 15,000$ K, $\rho_* = 7.8\times 10^{-6}$g/sm^3, $c_p = 21.4$ J/(g × K), $\lambda_* = 3.62\times 10^{-2}$W/(cm × K), when $t_t \cong 4.5\times 10^{-5}$ s. However, at high temperatures it is necessary to consider heat transfer by a radiative heat conductivity, then $\chi_R \cong l_R c/3$. For ultraviolet photons, a free length is $l_R \cong 10^{-2}$ cm and $\chi_R \cong 10^8$ cm^2/s. Therefore, the characteristic time of the radiative heat conductivity is about $t_R \cong 10^{-10}$ s, that is much less than other characteristic times. Thus, the limiting process for phenomena under consideration is the radiative heat exchange.

Two stages of numerical simulation procedure are used for the solution of eqs. (1.110)–(1.116). At the first step, called the "energy stage," an iterative procedure for the self-consistent solution of the energy conservation eq. (1.113) and radiation transfer eq. (1.116) is used. The implicit finite-difference scheme is preferable here.

It is easy to determine by means of numerical experiments that a maximum computational time step for the execution of the first stage is significantly less than necessary for the satisfaction of a stability condition of explicit schemes with reference to gas dynamic equations. Therefore, the Navier–Stokes eqs. (1.111, 1.112) together with continuity eq. (1.110) can be solved with the use of explicit schemes. The solution of these equations makes the second stage (the so-called gas dynamic stage).

Let us assume that at the point of time t^{p+1} the enthalpy field H^{p+1}, as well as fields of other thermodynamic and thermophysical functions $(\rho^{p+1}, \mu^{p+1}, \kappa_k{}^{p+1}, \lambda_t^{p+1})$ have been calculated. It is necessary to calculate $G^{p+1} = \{\rho u, \rho v\}^{p+1}$, p^{p+1} with the use of these functions at points of time t^p, and thermodynamic and thermophysical functions at t^{p+1}. The splitting method (Chorin A.J., 1967) can be used for these purposes. In accordance with this method, the calculation procedure is represented in the form of three sequential steps.

The first calculation step. It is supposed in eqs. (1.111) and (1.111) that

$$\frac{\partial p}{\partial x} = \frac{\partial p}{\partial r} = 0,$$

then

$$\frac{\partial G_u}{\partial t} + \mathrm{div}\ G_u \mathbf{V} = \frac{S_u}{\mathrm{Re}}, \tag{1.117}$$

$$\frac{\partial G_v}{\partial t} + \text{div } G_v\mathbf{V} = \frac{S_v}{\text{Re}}, \tag{1.118}$$

where $G_u = \rho u$, $G_v = \rho v$.

Required finite-difference scheme is created with the use of the finite-volume method, which is applied for elementary grid volume shown in Figure 1.7. The integral operator

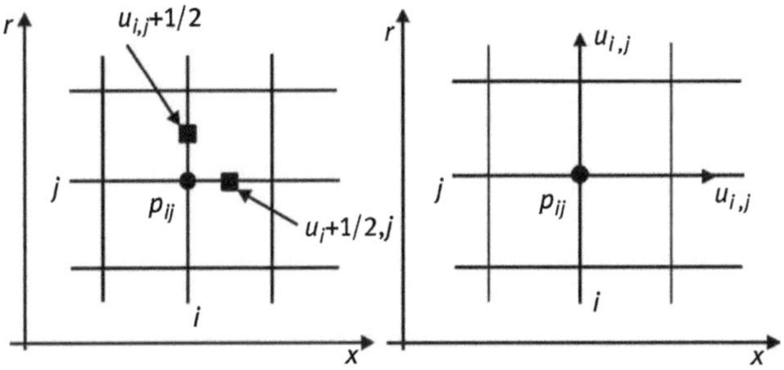

Figure 1.7: Finite-difference grids.

$$O_{\tau,ij}\{*\} = \int_p^{p+1} \mathrm{d}t \int_{j-1}^{j+1} \mathrm{d}x \int_{i-1/2}^{i+1/2} r\{*\}\mathrm{d}r, \tag{1.119}$$

being applied to eqs. (1.117) and (1.118) results in the following relations for preliminary values of mass fluxes:

$$\tilde{G}_{u,i,j} = G^p_{u,i,j} + \tau\left[-\frac{b_r^- G_{u,i,j+1} + (b_r^+ - b_l^-)G_{u,i,j} - b_l^+ G_{u,i,j-1}}{h_j} - \right.$$
$$\left. - \frac{a_r^-(rG_u)_{i+1,j} + (a_r^+ - a_l^-)(rG_u)_{i,j} - a_l^+(rG_u)_{i-1,j}}{\Delta_i R_i}\right] + \frac{O_{\tau,i,j}\{S_u\}}{\text{Re}h_j\Delta_i R_i}, \tag{1.120}$$

$$\tilde{G}_{v,i,j} = G^p_{v,i,j} + \tau\left[-\frac{b_r^- G_{v,i,j+1} + (b_r^+ - b_l^-)G_{v,i,j} - b_l^+ G_{v,i,j-1}}{h_j} - \right.$$
$$\left. - \frac{a_r^-(rG_v)_{i+1,j} + (a_r^+ - a_l^-)(rG_v)_{i,j} - a_l^+(rG_v)_{i-1,j}}{\Delta_i R_i}\right] + \frac{O_{\tau,i,j}\{S_v\}}{\text{Re}h_j\Delta_i R_i}, \tag{1.121}$$

where

$$h_j = 0.5(x_{j+1} - x_{j-1}), \quad \Delta_i = 0.5(r_{i+1} - r_{i-1}); \quad R_i = 0.25(r_{i-1} + 2r_i + r_{i+1}),$$

$$b_r^\pm = 0.5(u_r \pm |u_r|), \quad b_l^\pm = 0.5(u_l \pm |u_l|);$$

$$a_r^{\pm} = 0.5(v_r \pm |v_r|), \quad a_l^{\pm} = 0.5(v_l \pm |v_l|);$$
$$u_r = 0.5(u_{ij} + u_{i,j+1}), \quad u_l = 0.5(u_{ij} + u_{i,j-1});$$
$$v_r = 0.5(v_{ij} + v_{i+1,j}), \quad v_l = 0.5(v_{ij} + v_{i-1,j}).$$

The finite-difference approximations of dissipative summands in the Navier–Stokes equations are expressed as follows:

$$O_{\tau,i,j}\{S_u\} = \tau h_j \mu_{i,j}\left\{ r_{i+1/2}\left[\frac{u_{i+1,j} - u_{ij}}{r_{i+1} - r_i} + \frac{0.25(v_{i,j+1} - v_{i,j-1} + v_{i+1,j+1} - v_{i+1,j-1})}{h_j}\right] - \right.$$
$$\left. - r_{i-1/2}\left[\frac{u_{i,j} - u_{i-1,j}}{r_i - r_{i-1}} + \frac{0.25(v_{i,j+1} - v_{i,j-1} + v_{i-1,j+1} - v_{i-1,j-1})}{h_j}\right]\right\} +$$
$$+ \tau R_i \Delta_i \mu_{i,j}\left[2\left(\frac{u_{i,j+1} - u_{i,j}}{x_{j+1} - x_j} - \frac{u_{i,j} - u_{i,j-1}}{x_j - x_{j-1}}\right) + \frac{1}{3}(\text{div}\,\mathbf{V})_{i,j-1} - \frac{1}{3}(\text{div}\,\mathbf{V})_{i,j+1}\right], \tag{1.122}$$

$$O_{\tau,i,j}\{S_v\} = \tau \Delta_i R_i \mu_{i,j} \times$$
$$\times \left[\frac{v_{i,j+1} - v_{i,j}}{x_{j+1} - x_j} - \frac{v_{i,j} - v_{i,j-1}}{x_j - x_{j-1}} + \frac{0.25(u_{i+1,j+1} + u_{i-1,j-1} - u_{i-1,j+1} - u_{i+1,j-1})}{\Delta_i}\right] +$$
$$+ 2\tau h_j \mu_{i,j}\left[\frac{r_{i+1/2}(v_{i+1,j} - v_{i,j})}{r_{i+1} - r_i} - \frac{r_{i+1/2}(v_{ij} - v_{i-1,j})}{r_i - r_{i-1}}\right] -$$
$$- 2\tau h_j \mu_{i,j} v_{i,j} \ln\left(\frac{r_{i+1/2}}{r r_{i-1/2}}\right) - \frac{2}{3}\tau h_j \mu_{i,j}\left\{0.5 r r_{i+1/2}\left[(\text{div}\,\mathbf{V})_{i+1,j} + (\text{div}\,\mathbf{V})_{i,j}\right] - \right.$$
$$\left. - 0.5 r_{i-1/2}\left[(\text{div}\,\mathbf{V})_{i,j} + (\text{div}\,\mathbf{V})_{i-1,j}\right] - (\text{div}\,\mathbf{V})_{i,j}\Delta_i\right\}. \tag{1.123}$$

At deriving approximations (1.122) and (1.123), it was supposed that $\mu_{i\pm1/2,j\pm1/2} = \mu_{i,j}$.

The second calculation step. Intermediate values of the mass fluxes $\tilde{G}_u$ and $\tilde{G}_v$, obtained at the first calculation step, are used here to find the pressure disturbances. Following the basic ideas of the splitting method, the initial system of equations of the second step is formulated in the following form:

$$\frac{\partial G_u}{\partial t} = -\frac{\partial p}{\partial x}, \tag{1.124}$$

$$\frac{\partial G_v}{\partial t} = -\frac{\partial p}{\partial r}. \tag{1.125}$$

Let us act on eq. (1.124) with the operator $\partial\{\ \}/\partial x$, and also on eq. (1.125) with the operator $r^{-1}\partial\{r\}/\partial r$, then:

$$\frac{\partial(\text{div}\,\mathbf{G})}{\partial t} = -\,\text{div}(\text{grad}\,p). \tag{1.126}$$

Following the splitting scheme and transforming to finite-time discretization, one can receive the Poisson equation relative to the pressure

$$\text{div}(\text{grad}\,p) = -\,\frac{\text{div}\,\mathbf{G}^{p+1} - \text{div}\,\tilde{\mathbf{G}}}{\tau}. \tag{1.127}$$

This equation gives birth to two groups of possible calculation models. The first group includes stationary models, which correspond to identical conditions at each time step, $\text{div}\,\mathbf{G}^{p+1} = 0$. The second group includes the nonstationary models $\text{div}\,\mathbf{G}^{p+1} \neq 0$. In the latter case for determination of function $\text{div}\,\mathbf{G}^{p+1}$ it is necessary to attract an equation of continuity.

Let us consider the equation for pressure in the general form (1.127), but take into account specific properties of slow subsonic motion of heated gas in PGs. For these purposes, the thermal equation of state $p' = \rho' R_0 T'/M'$ (all values are dimensional; R_0 is the universal gas constant) can be presented in the following form:

$$\rho' = \frac{(p_0' + p_g')M'}{R_0 T'} = \frac{(1 + p_g'/p_0')M' p_0'}{R_0 T'} = \frac{(1 + p_g'/p_0')M' T_0' \rho_0'}{M_0' T'}, \tag{1.128}$$

where p_0' is the average in space pressure and p_g' is the local perturbation of pressure p_0'.

Considering $a'^2 = \gamma R_0 T_0'/M_0'$, where $\gamma = c_{p,0}/c_{v,0}$, $c_{p,0}$, $c_{v,0}$ are the specific heat capacities at constant pressure and volume at temperature T_0, one can get the dimensionless thermal equation of state:

$$\rho = \frac{(1 + \gamma M_\infty^2 p)M}{T}, \qquad M_\infty^2 = \frac{\gamma p_0}{\rho_0}, \tag{1.129}$$

and at requirements $M_\infty^2 \ll 1$ considered here, it assumes even more simple form $\rho \approx M/T$.

In our case, at practically constant pressure, the density can be considered as a function of an enthalpy $\rho = \rho(H)$ only, therefore:

$$\text{d}\rho = \frac{\partial \rho}{\partial H}\text{d}H = c_p^{-1}\frac{\partial \rho}{\partial T}\text{d}H = -\,\frac{\rho}{c_p T}\text{d}H. \tag{1.130}$$

Now, combining the continuity equation $\text{d}\rho/\text{d}t = -\rho\,\text{div}\,\mathbf{V}$ and the energy conservation equation $\rho\,\text{d}H/\text{d}t = Q_\Sigma$, where Q_Σ is the right-hand side of eq. (1.113), one can rewrite eq. (1.130) in the following form:

$$\rho\,\text{div}\,\mathbf{V} = \frac{Q_\Sigma}{c_p T} \quad \text{or} \quad \text{div}(\rho\mathbf{V}) = \mathbf{V}\,\text{grad}\,\rho + \frac{Q_\Sigma}{c_p T}. \tag{1.131}$$

Taking into account that

$$\mathbf{V}\ \mathrm{grad}\,\rho = -\rho\frac{\left(u\frac{\partial H}{\partial x}\right)+\left(v\frac{\partial H}{\partial r}\right)}{c_pT},$$

the analog of the continuity equation becomes:

$$\mathrm{div}(\rho\mathbf{V}) = \frac{Q_\Sigma - \mathbf{G}\ \mathrm{grad}\,H}{c_pT}, \qquad \mathbf{G} = \{\rho u, \rho v\}. \tag{1.132}$$

Now, the five-point finite-difference scheme for the calculation of pressure by the iterative method can be presented in the form:

$$A_{i,j}p_{i-1,j}^{p+1} + B_{i,j}p_{i+1,j}^{p+1} + \bar{A}_{i,j}p_{i,j-1}^{p+1} + \bar{B}_{i,j}p_{i,j+1}^{p+1} - C_{i,j}p_{i,j}^{p+1} + F_{i,j} = 0, \tag{1.133}$$

with the following coefficients:

$$A_{i,j} = \frac{r_{i-1/2}}{\Delta_i R_i(r_i - r_{i-1})}, \quad B_{i,j} = \frac{r_{i+1/2}}{\Delta_i R_i(r_{i+1} - r_i)},$$

$$\bar{A}_{i,j} = \frac{1}{h_j h_j^-}, \quad \bar{B}_{i,j} = \frac{1}{h_j h_j^+}, \quad C_{i,j} = A_{i,j} + B_{i,j} + \bar{A}_{i,j} + \bar{B}_{i,j},$$

$$F_{i,j} = \frac{\left(\mathrm{div}\,\mathbf{G}^{p+1}\right)_{i,j}}{\tau} - \frac{1}{2\tau}\left[\frac{\tilde{G}_{u,i,j+1} - \tilde{G}_{u,i,j-1}}{h_j} + \frac{r_{i+1}\tilde{G}_{v,i+1,j} - r_{i-1}\tilde{G}_{v,i-1,j}}{\Delta_i R_i}\right].$$

Note that at the finite-difference approximation of eq. (1.127) we used the grid shown in Figure 1.8b. It is assumed that the numerical errors generated by the two kinds of grids will be neutralized at the third calculation step.

Boundary conditions for pressure are defined in the form of $\partial p/\partial x = \partial p/\partial r = 0$.

For solution of the difference eq. (1.133), any of the used known methods (Anderson D.A., et al., 1997; Hoffmann K.A., et al., 2000) can be recommended, namely the successive over-relaxation method and alternating direction implicit method. Some of them will be considered below.

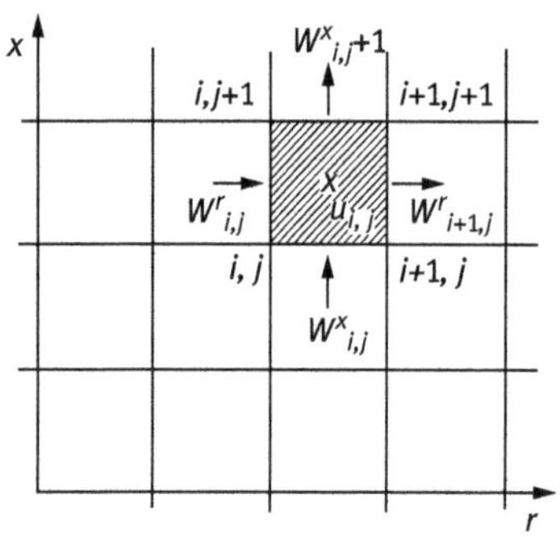

Figure 1.8: Radiative fluxes in a computational mesh.

The third calculation step. Final correction of the components of the mass flux is performed in this step. Calculation relations that are in use follow from eqs. (1.124) and (1.125):

$$G_u^{p+1} = \tilde{G}_u - \tau \frac{\partial p}{\partial x} \quad \text{and} \quad G_v^{p+1} = \tilde{G}_v - \tau \frac{\partial p}{\partial r}, \tag{1.139}$$

with the approximation for the pressure gradient:

$$G_{u,i,j}^{p+1} = \tilde{G}_{u,i,j} - \frac{\tau}{2}\left(\frac{p_{i,j+1}^{p+1} - p_{i,j}^{p+1}}{x_{j+1} - x_j} + \frac{p_{i,j}^{p+1} - p_{i,j-1}^{p+1}}{x_j - x_{j-1}}\right),$$

$$G_{v,i,j}^{p+1} = \tilde{G}_{v,i,j} - \frac{\tau}{2}\left(\frac{p_{i+1,j}^{p+1} - p_{i,j}^{p+1}}{r_{i+1} - r_i} + \frac{p_{i,j}^{p+1} - p_{i-1,j}^{p+1}}{r_i - r_{i-1}}\right).$$

Note that the use of mass flux vector components as required functions is appeared to be more preferable for considered problems of radiative gas dynamics of PGs than using density and velocity components separately.

Computing model in dynamic variables

Flow equations in the dynamic variables can be obtained by the act of the Curl operator on the original Navier–Stokes equations in the vector form. In the two-dimensional axisymmetric formulation, this mathematical procedure is simplified significantly. We need only to act with the operator $\partial\{\ \}/\partial r$ on the x-projection of the impulse conservation eq. (1.111), and with the operator $\partial\{\ \}/\partial x$ on the r-projection of the impulse conservation eq. (1.112). System of governing equations in the dynamic variables has the following form:

$$\operatorname{div}\frac{1}{\rho r^2}\operatorname{grad}\psi = -\frac{\omega}{r}, \qquad \frac{\partial \psi}{\partial r} = r\rho u, \qquad \frac{\partial \psi}{\partial x} = -r\rho v, \tag{1.140}$$

$$\begin{aligned}
\frac{\partial}{\partial t}\left(\frac{\rho\omega}{r}\right) + \operatorname{div}\left(\frac{\rho\nabla\omega}{r}\right) &= \frac{1}{r^2}\operatorname{div}\left[\frac{1}{r^2}\operatorname{grad}\left(\frac{\mu\omega}{r}\right)\right] + \\
&+ \frac{2}{r}\frac{\partial}{\partial x}\left[\frac{\partial}{\partial x}\left(\mu\frac{\partial u}{\partial r}\right) - \frac{\partial}{\partial r}\left(\mu\frac{\partial u}{\partial x}\right)\right] + \frac{2}{r^2}\frac{\partial}{\partial r}r\left[\frac{\partial}{\partial x}\left(\mu\frac{\partial v}{\partial r}\right) - \frac{\partial}{\partial r}\left(\mu\frac{\partial v}{\partial x}\right)\right] - \\
&- \frac{2v}{r^3}\frac{\partial \mu}{\partial x} + \frac{1}{r}\left[\frac{\partial \rho}{\partial r}\frac{\partial}{\partial x}\left(\frac{u^2+v^2}{2}\right) - \frac{\partial \rho}{\partial x}\frac{\partial}{\partial r}\left(\frac{u^2+v^2}{2}\right)\right],
\end{aligned} \tag{1.141}$$

$$\omega = \frac{\partial v}{\partial x} - \frac{\partial u}{\partial r}, \tag{1.142}$$

$$\rho\frac{\partial H}{\partial t}+\operatorname{div}(\rho\mathbf{V}H)=\operatorname{div}\left(\frac{\lambda_t}{c_p}\operatorname{grad}H\right)+W-Q_{\mathrm{R}},\tag{1.143}$$

$$Q_{\mathrm{R}}=\int_0^{\infty}c\kappa_\lambda(U_{b\lambda}-U_\lambda)\mathrm{d}\lambda=\sum_{k=1}^{N_k}c\kappa_\lambda(U_{bk}-U_k)\mathrm{d}\lambda_k,\tag{1.144}$$

$$\operatorname{div}\frac{1}{3\kappa_k}\operatorname{grad}U_k=\kappa_k(U_k-U_{bk}),\tag{1.145}$$

where u, v are the projections of velocity $\mathbf{V}=\{u,v\}$ on axial x and radial r coordinates; ρ, c_p, μ, λ_t are the density, heat capacity at constant pressure, viscosity, and thermal conduction; $H=\int_{T_\infty}^{T}c_p\mathrm{d}T$ is the enthalpy; T, T_∞ are the temperature in the discharge and in undisturbed input stream; κ, U, U_b are the volumetric absorption coefficient and volumetric density of heat radiation of heated gas and of the blackbody.

Equation (1.144) corresponds to a diffusion approach to the solution of the radiation heat transfer equation. The gas stream and gas discharge plasma, as before, are in the LTE. A smallness of pressure changes in gas stream allows to use functional relationships of thermodynamic, thermophysical, and optical properties in the form of $f=f(T)$, where $f=\{\rho,\mu,\lambda_t,H,\kappa_k\}$.

Time derivatives in the eqs. (1.141) and (1.143) are used for the solution of the equations by a time-marching method and in the case under consideration do not have the physical meaning.

Remind that usually all numerical methods suppose the following generalized formulation of boundary conditions:

$$a+bf+c\frac{\partial f}{\partial n}=0,\tag{1.146}$$

where n is the normal coordinate to a boundary and f is the function to be defined. The concrete definition of boundary conditions (definition of a,b,c coefficients) and the assignment of initial conditions are carried out at the solution of specific problems.

The task can be solved with nonuniform regular meshes in the lines of x and r. The finite-difference implicit scheme can be applied not only for the energy conservation equation (as before) but also for gas flow eqs. (1.140) and (1.141). Numerical analogs of all solved eqs. (1.140), (1.141), (1.143), (1.145) are reduced to the canonical form of the five-point finite-difference equation:

$$A_{i,j}\Phi_{i-1,j}+B_{i,j}\Phi_{i+1,j}+\bar{A}_{i,j}\Phi_{i,j-1}+\bar{B}_{i,j}\Phi_{i,j+1}-C_{i,j}\Phi_{i,j}+F_{i,j}=0,\tag{1.147}$$

where indexes $i=1,2,\ldots,NI$ set nodes of the computational mesh in a radial direction, and indexes $j=1,2,\ldots,NJ$ are used for axial direction; $\Phi=\{\psi,\mathbf{\Omega}/r,H,U_k\}$. Formulas for calculation of the coefficients in eq. (1.147) for each of the equations (in the dimensionless form for (1.140) and (1.141)) are given below.

1. The finite-difference equation for a stream function ψ:

$$A_{i,j} = \frac{r_{i-1/2}a_{i-1/2,j}}{p\hat{p}p^-},\ B_{i,j} = \frac{r_{i+1/2}a_{i+1/2,j}}{p\hat{p}p^+}$$

$$\bar{A}_{i,j} = \frac{a_{i,j-1/2}}{qq^-},\quad \bar{B}_{i,j} = \frac{a_{i,j+1/2}}{qq^+},\quad F_{i,j} = \frac{\omega_{i,j}}{\hat{p}},$$

$$C_{i,j} = A_{i,j} + B_{i,j} + \bar{A}_{i,j} + \bar{B}_{i,j},$$

where

$$p = 0.5(r_{i+1} - r_{i-1});\ q = 0.5(x_{j+1} - x_{j-1});\ q^{\pm} = \pm(x_{j\pm 1} - x_j);$$

$$p^{\pm} = \pm(r_{i\pm 1} - r_i);\ \hat{p} = 0.25(r_{i-1} + 2r_i + r_{i+1});$$

$$a_{i+m/2,j+n/2} = \frac{1}{2}\left[\left(\frac{1}{\rho r^2}\right)_{i,j} + \left(\frac{1}{\rho r^2}\right)_{i+m,j+n}\right]$$

or

$$a_{i+m/2,j+n/2} = \frac{1}{2r^2_{i+m/2}}\left(\frac{1}{\rho_{ij}} + \frac{1}{\rho_{i+m,j+n}}\right),\quad m,n = 0, \pm 1.$$

2. The finite-difference equation for vortex function ω/r:

$$A_{i,j} = \frac{a^+_{l,i,j} + r^2_{i-1}\mu_{i-1,j}}{\mathrm{Re} r_{i-1/2}p^- p\hat{p}},\qquad B_{i,j} = \frac{-a^-_{r,i,j} + r^2_{i+1}\mu_{i+1,j}}{\mathrm{Re} r_{i+1/2}p^+ p\hat{p},}$$

$$\bar{A}_{i,j} = \frac{b^+_{l,i,j} + (\mu_{i,j-1}/\mathrm{Re}q^-)}{q},\qquad \bar{B}_{i,j} = \frac{-b^-_{r,i,j} + (\mu_{i,j+1}/\mathrm{Re}q^+)}{q},$$

$$C_{i,j} = \frac{\rho_{i,j}}{\tau} + \frac{a^+_{r,i,j} - a^-_{l,i,j}}{p\hat{p}} + \frac{b^+_{r,i,j} - b^-_{l,i,j}}{q} +$$

$$+ \frac{\mu_{i,j}}{\mathrm{Re}}\left[r_i^2 \frac{1/(p^+ r_{i+1/2}) + 1/(p^- r_{i-1/2})}{p\hat{p}} + \frac{1/q^+ + 1/q^-}{q}\right],$$

$$F_{i,j} = \frac{\rho_{i,j}\omega^s_{i,j}}{r_i\tau} + \frac{1}{\mathrm{Re}\hat{p}}\left[\frac{M_{i,j+1} - M_{i,j-1}}{q} + \frac{N_{i+1,j} - N_{i-1,j}}{p} +\right.$$

$$\left. + \frac{2N_{i,j}\ln\left(r_{i+1/2}/r_{i+1/2}\right)}{p} + \frac{v_{i,j}(\mu_{i,j+1} - \mu_{i,j-1})\left(r^{-1}_{i+1/2} - r^{-1}_{i-1/2}\right)}{pq}\right] +$$

$$+ \left(\frac{\rho_{i+1,j} - \rho_{i-1,j}}{8qp\hat{p}}\right)\left[(u_{i,j+1} - u_{i,j-1})(u_{i,j-1} + 2u_{i,j} + u_{i,j+1}) +\right.$$

$$\left. + (v_{i,j+1} - v_{i,j-1})(v_{i,j-1} + 2v_{i,j} + v_{i,j+1})\right] -$$

$$-\left(\frac{\rho_{i,j+1}-\rho_{i,j-1}}{8p\hat{p}q}\right)\left[\left(u_{i+1,j}-u_{i-1,j}\right)\left(u_{i+1,j}+2u_{i,j}+u_{i+1,j}\right)+\right.$$

$$\left.+\left(v_{i+1,j}-v_{i-1,j}\right)\left(v_{i-1,j}+2v_{i,j}+v_{i+1,j}\right)\right],$$

where

$$\mathrm{Re}=\frac{u_\infty \rho_* L}{\mu_*};\quad r_{i\pm 1/2}=\frac{r_i+r_{i\pm 1}}{2};\quad \tau=\frac{tL}{u_\infty},$$

and functions

$$M=\frac{\partial \mu}{\partial x}\frac{\partial u}{\partial r}-\frac{\partial \mu}{\partial r}\frac{\partial u}{\partial x}\quad \text{and}\quad N=\frac{\partial \mu}{\partial x}\frac{\partial v}{\partial r}-\frac{\partial \mu}{\partial r}\frac{\partial v}{\partial x}$$

are approximated by the central differences; s is the index of the previous time layer.

Here density ρ_* and viscosity μ_* at the characteristic temperature of the gas discharge (e.g., for LSW PG, $T_* = 18,500$ K) together with u_∞ and L were used as the reference parameters. Functions $a = \rho v$ and $b = \rho u$ are approximated with the use of the simplest approach:

$$a_r = 0.5\left(r_i a_{i,j} + r_{i+1} a_{i+1,j}\right),\quad a_l = 0.5\left(r_i a_{i,j} + r_{i-1} a_{i-1,j}\right);$$

$$b_r = 0.5\left(b_{i,j} + b_{i,j+1}\right),\quad b_l = 0.5\left(b_{i,j} + b_{i,j-1}\right);$$

$$a_{r,l}^{\pm} = 0.5(a_{r,l} \pm |a_{r,l}|),\quad b_{r,l}^{\pm} = 0.5(b_{r,l} \pm |b_{r,l}|).$$

3. The finite-difference equation for enthalpy H

$$A_{i,j}=\frac{1}{p\hat{p}}\left(\hat{a}_{l,i,j}^{+}+\frac{r_{i-1/2}\tilde{a}_{i-1/2,j}}{p^-}\right),$$

$$B_{i,j}=\frac{1}{p\hat{p}}\left(-\hat{a}_{r,i,j}^{-}+\frac{r_{i+1/2}\tilde{a}_{i+1/2,j}}{p^+}\right),$$

$$\bar{A}_{i,j}=\frac{1}{q}\left(b_{l,i,j}^{+}+\frac{\tilde{a}_{i,j-1/2}}{q^-}\right),\quad \bar{B}_{i,j}=\frac{1}{q}\left(-b_{r,i,j}^{+}+\frac{\tilde{a}_{i,j+1/2}}{q^+}\right),$$

$$C_{i,j}=\frac{\rho_{i,j}}{\tau}+\frac{\hat{a}_{r,i,j}^{+}-\hat{a}_{l,i,j}^{-}}{p\hat{p}}+\frac{b_{r,i,j}^{+}-b_{l,i,j}^{-}}{q}+\frac{a_{i,j-1/2}}{qq^-}+\frac{\tilde{a}_{i,j+1/2}}{qq^+}+$$

$$+\frac{r_{i-1/2}\tilde{a}_{i-1/2,j}}{p\hat{p}p^-}+\frac{r_{i+1/2}\tilde{a}_{i+1/2,j}}{p\hat{p}p^+},$$

$$F_{i,j}=\frac{\rho_{i,j}H_{i,j}^{s}}{\tau}+W_{i,j}-Q_{R,i,j},$$

where $\tilde{a} = \lambda_t/\lambda_t c_p$; $\hat{a} = r\rho v$; $b = \rho u$;

$$\tilde{a}_{i+m/2,j+n/2}=\frac{\tilde{a}_{i,j}+\tilde{a}_{i+m,j+n}}{2r^2},\qquad m,n=0,\pm 1.$$

4. The finite-difference equation for group radiative energy density U_k.

As mentioned earlier, eq. (1.116) corresponds to the P_1 approximation of the spherical harmonics method, or to the so-called diffusion approximation to the radiation heat transfer equation (Pomraning G.C., 1973; Ozisik M.N., 1973). Let us rewrite this equation in the form acceptable for the further derivation of a finite-difference scheme (spectral index k is omitted for simplicity):

$$-\frac{1}{r^p}\cdot\frac{\partial}{\partial r}\left(\frac{r^p}{3\kappa}\cdot\frac{\partial U}{\partial r}\right)-\frac{\partial}{\partial z}\left(\frac{1}{3\kappa}\cdot\frac{\partial U}{\partial x}\right)+\kappa U=\kappa U_b. \tag{1.148}$$

with the boundary conditions on the symmetry axis:

$$\frac{\partial U}{\partial r}=0,$$

and on the exterior boundaries:

$$\frac{2}{3}\cdot\frac{1}{\kappa}\cdot\frac{\partial U}{\partial \mathbf{n}}=-U,$$

or with asymptotic boundary conditions of the form

$$U(\mathbf{r})|_s=U_b(\mathbf{r})|_s.$$

Let the density of heat radiation energy U is determined in a center of elemental calculation volume shown in Figure 1.9, and the projections of the radiation heat flux vector (W_x, W_r) is determined in its sides. With the use of the finite-volume approach, one can get the following canonical five-point scheme:

$$A_{i,j}U_{i-1,j}+B_{i,j}U_{i+1,j}+\bar{A}_{i,j}U_{i,j-1}+\bar{B}_{i,j}U_{i,j+1}-C_{i,j}U_{i,j}+F_{i,j}=0, \tag{1.149}$$

$$A_{i,j}=\frac{r^p_{i-1/2}}{3\kappa_{i-1/2,j}p\bar{p}^p(r_i-r_{i-1})},\qquad \bar{A}_{i,j}=\frac{r^p_{i+1/2}}{3\kappa_{i+1/2,j}p\bar{p}^p(r_{i+1}-r_i)},$$

$$B_{i,j}=\frac{1}{3\kappa_{i,j-1/2}q\left(x_j-x_{j-1}\right)},\qquad \bar{B}_{i,j}=\frac{1}{3\kappa_{i,j+1/2}q\left(x_{j+1}-x_j\right)},$$

$$C_{i,j}=\kappa_{i,j}+\frac{r^p_{i+1/2}}{3\kappa_{i+1/2,j}p\bar{p}^p(r_{i+1}-r_i)}+\frac{r^p_{i-1/2}}{3\kappa_{i-1/2,j}p\bar{p}^p(r_i-r_{i-1})}+$$
$$+\frac{1}{3\kappa_{i,j+1/2}q\left(x_{j+1}-x_j\right)}+\frac{1}{3\kappa_{i,j-1/2}q\left(x_j-x_{j-1}\right)},$$

$$F_{i,j}=\kappa_{i,j}U_{b,i,j},$$

where

$$p=\frac{r_{i+1}-r_{i-1}}{2},\quad \bar{p}=\frac{r_{i-1}+2r_i+r_{i+1}}{4},\quad q=\frac{x_{j+1}-x_{j-1}}{2}$$

$$\kappa_{i\pm1/2,j} = \frac{\kappa_{i\pm1,j} + \kappa_{i,j}}{2}, \quad \kappa_{i,j\pm1/2} = \frac{\kappa_{i,j\pm1} + \kappa_{i,j}}{2}, \quad r^p_{i\pm1/2} = \left(\frac{r_i + r_{i\pm1}}{2}\right)^p.$$

Boundary conditions for computing models in dynamic variables

Boundary conditions of the first kind (Farlow S.J., 1982) are stated for mesh functions $\psi_{i,j}$, $H_{i,j}$ at $x=0$ and $r=R_c$. Function of velocity vortex $\mathbf{\Omega}_{i,j}/r_i$ is defined from undisturbed flow requirements at $x=0$ and $r=R_c$ with the first-order accuracy (for calculation cases in free space). At $r=0$ for all functions, conditions of axial symmetry are used.

Formulation of the boundary conditions for the exit cross section is traditionally the most difficult problem for subsonic flows with localized heat release regions. Numerical experiments performed in a series of papers (Surzhikov S.T., 1997) show that the following conditions are acceptable:

$$\frac{\partial f}{\partial x} = 0 \quad \text{or} \quad \frac{\partial^2 f}{\partial x^2} = 0,$$

if the exit cross section is located far from the region of heat release. More accurate boundary conditions can be formulated if we use special prohibitive conditions to admit any disturbances transferred from outside of the calculation domain.

1.1.6 Method of numerical simulation of nonstationary radiative gas dynamic processes in subsonic plasma flows: the method of unsteady dynamic variables

This method will be described in the example of computing model of unsteady laser-supported wave considered in Section 1.1.4. As before, at some fixed stage of the nonstationary process ($t=t^p$), all desired functions ($u^p, v^p, \rho^p, T^p, U^p_k$) and other constitutive functions $c^p_p, \mu^p, \lambda^p, \chi^p_\Omega, k^p_k$ are known.

Here we suppose the explicit method of consecutive integration of system of equations for gas dynamics and energy conservation. In cases of the subsonic flow, the plasma density is defined by its temperature; therefore, it is possible to suppose that ρ^{p+1} has been defined together with T^{p+1} (hence, $\partial\rho/\partial t \approx (\rho^{p+1} - \rho^p)/\tau$ has been defined also, where τ is the time step).

Let us exclude pressure from the system of gas dynamic equations as it is usually done at the transition from the native to dynamic variables, by cross-differentiation on r and x of eqs. (1.96) and (1.97) accordingly, with the subsequent subtraction of one from another. As a result one can get the equation:

$$\frac{\partial \omega}{\partial t} + \frac{\partial}{\partial x}\text{div}(\rho v\mathbf{V}) - \frac{\partial}{\partial r}\text{div}(\rho u\mathbf{V}) = \left(\frac{\partial S_v}{\partial x} - \frac{\partial S_u}{\partial r}\right) + g\frac{\partial \rho}{\partial r}, \tag{1.150}$$

where

$$\omega = \frac{\partial \rho v}{\partial x} - \frac{\partial \rho u}{\partial r}. \tag{1.151}$$

It is obvious that for incompressible flows, this equation is equivalent to the usual equation for "vortex" function

$$\omega = \frac{\partial v}{\partial x} - \frac{\partial u}{\partial r},$$

that is for the azimuth component of the vector function Curl **V**.

The second equation of the desired system is usually derived with the use of the continuity eq. (1.95). There is a possibility to introduce function ψ from the requirement of identical satisfaction to this equation. However, it is well known that in such a way the stream function ψ can be entered only in two special cases when the density is not changed in time:

a) for the incompressible flow, when div $\mathbf{V} = 0$ and
b) for the steady-state motion of compressible gas, when div $\rho\mathbf{V} = 0$.

In the plasma model under consideration, these conditions are not satisfied, and it is impossible to enter the stream function in the usual way.

Let us enter new scalar function E by means of equation

$$\text{div}[\rho E \mathbf{V}] = \frac{\partial \rho}{\partial t}, \tag{1.152}$$

Then the continuity equation can be rewritten in the form of

$$\text{div}[(1+E)\rho\mathbf{V}] = 0,$$

Now it is possible to enter analog of a stream function with following relations:

$$\frac{\partial \Psi}{\partial r} = (1+E)r\rho u, \quad \frac{\partial \Psi}{\partial x} = -(1+E)r\rho v. \tag{1.153}$$

Now eq. (1.153) will serve for calculation of projections of a mass flux vector on coordinate axes:

$$\rho u = \frac{1}{r(1+\mathrm{E})}\frac{\partial \Psi}{\partial r}, \quad \rho v = -\frac{1}{r(1+\mathrm{E})}\frac{\partial \Psi}{\partial x}. \tag{1.154}$$

Substituting these relations in eq. (1.151), we shall find the relation between functions ω and Ψ:

$$\frac{\partial}{\partial x}\left[\frac{1}{r^2(1+\mathrm{E})}\frac{\partial \Psi}{\partial x}\right] + \frac{1}{r}\frac{\partial}{\partial r}\left[\frac{1}{r^2(1+\mathrm{E})}\frac{\partial \Psi}{\partial r}\right] = -\frac{\mathbf{\Omega}}{r}. \tag{1.155}$$

Before the formulation of the final system of equations, we shall use one formal replacement of the newly entered function E which, however, will make computing procedure more effective. Let $E = \exp(\varepsilon) - 1$, where ε is the unknown scalar function, then instead of eq. (1.152) it is possible to derive a new equation (using the continuity equation):

$$\rho u \frac{\partial \varepsilon}{\partial x} + \rho v \frac{\partial \varepsilon}{\partial r} = \frac{\partial \rho}{\partial t}. \tag{1.156}$$

Having made corresponding replacement in eqs. (1.154) and (1.155) we shall formulate equations for gas dynamic stage of the calculation procedure in the final form

$$\frac{\partial \omega}{\partial t} + \mathrm{div}(\omega \mathbf{V}) + \frac{\partial}{\partial x}\left(\rho v \frac{\partial u}{\partial x} - \rho u \frac{\partial u}{\partial r}\right) + \frac{\partial}{\partial r}\left(\rho v \frac{\partial v}{\partial x} - \rho u \frac{\partial v}{\partial r}\right) +$$

$$+ \rho v \frac{\partial (v/r)}{\partial x} - \rho u \frac{\partial (v/r)}{\partial r} = \left(\frac{\partial S_v}{\partial x} - \frac{\partial S_u}{\partial r}\right) + g \frac{\partial \rho}{\partial r}, \tag{1.157}$$

$$\mathrm{div}\left[\frac{\exp(-\varepsilon)}{r^2} \mathrm{grad}\Psi\right] = -\frac{\omega}{r}, \tag{1.158}$$

$$\rho \mathbf{V} \mathrm{grad}\varepsilon = \frac{\partial \rho}{\partial t}, \tag{1.159}$$

$$\frac{\partial \Psi}{\partial r} = \exp(\varepsilon) r \rho u, \qquad \frac{\partial \Psi}{\partial x} = -\exp(\varepsilon) r \rho v. \tag{1.160}$$

Thus, by the introduction of a new scalar function ε (or E) it is possible to formulate a system of *nonstationary equations for viscous compressible fluid in dynamic variables* "$\varepsilon - \Psi - \omega$," which have been called as nonstationary dynamic variables (Surzhikov S.T., et al, 2001).

Boundary conditions for functions ω and Ψ are formulated as for usual dynamic variables. The formulation of boundary conditions for ε does not cause difficulties also:

$$x = 0: \ \varepsilon(x = 0, r) = 0, \tag{1.228}$$

$$r = R: \ \varepsilon(x, r = R) = 0, \tag{1.229}$$

$$r = 0: \ \frac{\partial \varepsilon}{\partial r} = 0, \tag{1.230}$$

$$x = X: \ \frac{\partial \varepsilon}{\partial x} = 0. \tag{1.231}$$

Concerning the derived system of equations, it is possible to make some remarks of the general character.

1. For simplified cases of incompressible fluid or stationary flow of a compressible fluid $\varepsilon \equiv 0$, eqs. (1.157)–(1.160) automatically transfer to equations for usual dynamic variables.

2. The structure of the new system of equations is similar to those for usual dynamic variables, which allows using practically without changing numerical methods developed before for solving the Navier–Stokes equations in dynamic variables.
3. Numerical integration of the new eq. (1.159) does not present the fundamental difficulties as it is the equation well investigated in the theory of numerical methods.

1.2 Models of nonuniform chemically equilibrium and nonequilibrium plasma

The necessity for the development of the nonuniform chemically equilibrium plasma models arises from the analysis of the complex chemical composition plasma dynamics when preliminary computation of thermodynamic, thermophysical, and optical properties in a wide range of pressures and temperatures is obviously not possible. Such a case is often observed when the LTE approach is valid, but the equilibrium plasma composition is very complex. As an example, the problems of entering into atmosphere space vehicles with an ablative thermoprotection system can be considered.

A distinctive feature of the nonuniform plasma models is the necessity of definition of its chemical composition for all grid nodes in the whole calculation domain. It also allows to calculate thermodynamic, thermophysical, and optical plasma characteristics in the same points, which appreciably raises universality of the computing models. A negative feature of such models is the essential complication of computing procedure so it is necessary to develop special high-performance algorithms.

The system of governing equations of the nonuniform chemically reacting plasma has the following form:

- continuity equation

$$\frac{\partial \rho}{\partial t} + \operatorname{div} \rho \mathbf{V} = 0; \tag{1.161}$$

- continuity equation for individual components (or, so-called the diffusion equations)

$$\frac{\partial n_i}{\partial t} + \operatorname{div}(n_i \mathbf{V}) + \operatorname{div}(n_i \mathbf{V}_{d,i}) = \dot{\omega}_i, \quad i = 1, 2, \ldots, N_c; \tag{1.162}$$

- momentum equation (the impulse conservation equation)

$$\frac{\partial \rho \mathbf{V}}{\partial t} + \operatorname{div}[(\rho \mathbf{V}) \cdot \mathbf{V}] = -\operatorname{grad}\left(p + p^{\mathrm{R}}\right) + \mathbf{F}_g + \mathbf{F}_e + \mathbf{F}_\tau + \mathbf{F}_\tau^{\mathrm{R}}; \tag{1.163}$$

- energy conservation equations

$$\frac{\partial}{\partial t}\left(\rho\varepsilon+\frac{\rho\mathbf{V}^2}{2}+\frac{\mathbf{B}^2}{2\mu_0}\right)+\operatorname{div}\left[\rho\mathbf{V}\left(\varepsilon+\frac{\mathbf{V}^2}{2}+\frac{p}{\rho}+\frac{p^{\mathrm{R}}}{\rho}\right)\right]=$$
$$=\operatorname{div}\mathbf{q}+\operatorname{div}\mathbf{W}+A_g+A_e+A_\tau+A_\tau^{\mathrm{R}} \tag{1.164}$$

or

$$\frac{\partial}{\partial t}\left(\rho h+\frac{\rho\mathbf{V}^2}{2}+\frac{\mathbf{B}^2}{2\mu}\right)+\operatorname{div}\left[\rho\mathbf{V}\left(h+\frac{\mathbf{V}^2}{2}\right)\right]=\frac{\partial(p+p^{\mathrm{R}})}{\partial t}+$$
$$=\operatorname{div}\mathbf{q}+\operatorname{div}\mathbf{W}+A_g+A_e+A_\tau+A_\tau^{\mathrm{R}};$$

– system of the Maxwell equations

$$\operatorname{rot}\mathbf{E}=-\frac{\partial\mathbf{B}}{\partial t}, \tag{1.165}$$

$$\operatorname{div}\mathbf{D}=\rho_e, \tag{1.166}$$

$$\operatorname{div}\mathbf{B}=0, \tag{1.167}$$

$$\frac{1}{\mu_0}\operatorname{rot}\mathbf{B}=\mathbf{J}; \tag{1.168}$$

– generalized Ohm's law

$$\mathbf{J}+P_{\mathrm{H}}[\mathbf{J}\times\mathbf{b}]=\sigma\left(\mathbf{E}+[\mathbf{V}\times\mathbf{B}]+\frac{1}{en_e}\operatorname{grad}p_e\right); \tag{1.169}$$

– thermal equation of state

$$p=\frac{R_0}{M}\rho T; \tag{1.170}$$

where

$$\rho=\sum_{i=1}^{N_S}m_in_i,\quad p=\sum_{i=1}^{N_S}p_i=kT\sum_{i=1}^{N_S}n_i;$$
$$\rho\varepsilon=\sum_{i=1}^{N_S}\rho_i\varepsilon_i,\quad \varepsilon_i=\int_{T_0}^{T}c_{v,i}\mathrm{d}T+\varepsilon_{i,0};\quad \rho h=\sum_{i=1}^{N_S}\rho_ih_i,\quad h_i=\int_{T_0}^{T}c_{p,i}\mathrm{d}T+h_{i,0};$$
$$M=\frac{R_0}{k}\sum_{i=1}^{N_S}m_iX_i=\sum_{i=1}^{N_S}M_iX_i,\quad X_i=n_i\Big/\sum_{i=1}^{N_S}n_i;$$
$$\mathbf{F}_g=\sum_{i=1}^{N_S}F_{g,i}=\sum_{i=1}^{N_S}\rho_i\mathbf{g}=\rho\mathbf{g},$$

$$\mathbf{F}_e = \sum_{i=1}^{N_S} F_{e,i} = \sum_{i=1}^{N_S} \rho_i q_i (\mathbf{E} + [\mathbf{V}_i \times \mathbf{B}]) = \rho_e \mathbf{E} + [\mathbf{J} \times \mathbf{B}];$$

$$A_g = \sum (\mathbf{F}_{g,i} \cdot \mathbf{V}_i) = \rho(\mathbf{g} \cdot \mathbf{V}), \quad A_e = \sum (\mathbf{F}_{e,i} \cdot \mathbf{V}_i) = (\mathbf{J} \cdot \mathbf{E}) + \rho_e (\mathbf{V} \cdot \mathbf{E}).$$

where $\mathbf{F}_g$ is the gravity force acting on a unit mass; $\mathbf{F}_e$ is the electromagnetic force (the Lorentz force) acting on a unit mass; A_g is the work of gravities; A_e is the work of the Lorentz force; $M_i = (R_0/k)m_i = M_0 \tilde{m}_i$ is the molecular weight of the ith species; $M_0 = 1.66 \times 10^{-24}$ g is the atomic mass unit; $\tilde{m}_i$ is the relative atomic weight of the ith species; m_i is the particle mass of the ith species; n_i is the volume concentration of the ith species; p_i is the partial pressure of the ith species; $\mathbf{V}_i$ is the velocity of the ith species; $\dot{\omega}_i$ is the volume formation rate of species i, caused by chemical reactions; ε_i, h_i are the specific internal energy and enthalpy of the ith species; $c_{v,i}, c_{p,i}$ are the specific heat capacities of the ith species at constant volume and pressure; $\varepsilon_{i,0}, h_{i,0}$ are the specific internal energy and enthalpy of the ith species at temperature $T = T_0$, used for reference; N_s is the number of chemical species. Refer Section 1.1 for other designations.

Using a symbolic formulation of the nth chemical reaction

$$\sum_{j=1}^{N_S} a_{j,n} [X_j] = \sum_{j=1}^{N_S} b_{j,n} [X_j], \quad n = 1, 2, \ldots, N_r, \tag{1.171}$$

the mole rate of formation of species i can be written for nth chemical reaction as follows:

$$\left(\frac{\mathrm{d}X_i}{\mathrm{d}t}\right)_n = k_{f,n}\left(b_{i,n} - a_{i,n}\right) \prod_{j=1}^{N_S} X_j^{a_{j,n}} - k_{r,n}\left(b_{i,n} - a_{i,n}\right) \prod_{j=1}^{N_S} X_j^{b_{j,n}} = \left(b_{i,n} - a_{i,n}\right)\left(S_{f,i}^n - S_{r,i}^n\right), \tag{1.172}$$

where $a_{i,n}, b_{i,n}$ are the stoichiometric coefficients of the nth reaction; $[X_j]$ is the chemical symbol of reagents and products of reactions; N_r is the number of chemical reactions; $k_{f,n}$, $k_{r,n}$ are the rate constants of the forward and reverse reactions; $S_{f,i}^n$, $S_{r,i}^n$ are the rates of the forward and reverse reactions.

The mole formation rate for species j is written in the following form:

$$W_i = \sum_{n=1}^{N_r} \left(b_{i,n} - a_{i,n}\right)\left(S_{f,i}^n - S_{r,i}^n\right) \mathrm{mol}/(\mathrm{cm}^3 \cdot \mathrm{s}), \tag{1.173}$$

and the volume formation rate of species i can be calculated as follows:

$$\dot{\omega}_i = N_\mathrm{A} W_i, \quad 1/(\mathrm{cm}^3 \cdot \mathrm{s}) \tag{1.174}$$

where N_A is the Avogadro number.

To calculate the rates of formation it is necessary to know the forward and reverse reaction constants for each nth reaction

$$k_{f(r),n} = A_{f(r),n} T^{n_{f(r),n}} \exp\left(-\frac{E_{f(r),n}}{kT}\right), \tag{1.175}$$

where $A_{f(r),n}, n_{f(r),n}, E_{f(r),n}$ are the approximation coefficients for the forward (f) and reverse (r) reaction rates. Note that the equilibrium constant for each nth chemical reaction is determined as follows:

$$K_{\text{eq},n} = k_{f,n}/k_{r,n}, \tag{1.176}$$

which are determined in ones turn with thermodynamic data.

Vector of a heat flux density q in the case of the nonuniform chemically active plasma includes not only a component caused by the molecular thermal conduction (in the form of the Fourier law) but also the components related to a concentration diffusion and a thermodiffusion:

$$\mathbf{q} = -\lambda \text{grad} T + \sum_{i=1}^{N_C} \rho_i h_i \mathbf{V}_{d,i} + p \sum_{i=1}^{N_C} k_i^T \mathbf{V}_{d,i}, \tag{1.177}$$

where λ is the molecular thermal conduction; $\mathbf{V}_{d,i} = \mathbf{V}_i - \mathbf{V}$ is the vector of diffusion velocity of the ith species. The diffusion velocities satisfy the requirement

$$\sum_{i=1}^{N_C} \rho_i V_{d,i} = 0,$$

where k_i^T is the thermal diffusion coefficient of the ith species.

To define the diffusion velocities of plasma components, the matrix equation of the following kind is necessary to solve (Hirschfelder J.O., et al., 1966; Ferziger J.H., et al., 1972)

$$\mathbf{G}_i = \sum_{j=1}^{N_S} \frac{X_i X_j}{D_{i,j}} \left(\mathbf{V}_{d,j} - \mathbf{V}_{d,i}\right), \quad i = 1, 2, \ldots, N_S,$$

where vectors of diffusion forces $\mathbf{G}_i$ are defined as

$$\mathbf{G}_i = \text{grad} X_i - \left(\frac{\rho_i}{\rho} - X_i\right) \frac{\text{grad} p}{p} - k_i^T \frac{\text{grad} T}{T} - \frac{X_i q_i \mathbf{E}}{kT}, \quad i = 1, 2, \ldots, N_S,$$

where coefficients of binary diffusion $D_{i,j}$ are defined with the use of the molecular kinetic theory.

Formulated equations suppose essential simplification in two asymptotic cases: "frozen" chemical reactions and "equilibrium" chemical reactions.

In the case of "frozen" chemical reactions, gas dynamic process rate is many times more than chemical reaction rates so no chemical transformations have time to occur, that is, $\dot{\omega}_i = 0$.

In the case of "equilibrium" chemical reactions, their rates essentially surpass gas dynamic process rate so in any spatial point of the gas flow, the local thermochemical equilibrium has time to become steady. It allows formulating an algebraic system of equations for calculating species partial pressures:

$$K_{\mathrm{eq}}^{j}(T) = \prod_{i=1}^{N_c} p_i^{\nu_{i,j}}, \quad j = 1, 2, \ldots, N_p,$$

where $K_{\mathrm{eq}}^{j}(T)$ is the equilibrium constant of jth chemical reaction. At the solution of this system of equations, it is possible to take into account the Dalton law and a quasi-neutrality requirement (in the case of the charged intermixture of gases).

In both cases of the relationship between rates of chemical reactions and gas dynamic processes, it is necessary to calculate thermodynamic and thermophysical properties (viscosity, thermal conduction, and electrical conductivity) (Giordano D., et al., 1994).

Examples of the nonuniform nonequilibrium gas discharge plasma models are given below.

1.2.1 Model of the five-component RF plasma generator

Let us consider five-species plasma mixture, consisting of electrons (e), atoms (a), diatomic homonuclear molecule ($m = 2a$), atomic ions (ai), and molecular ions (mi). It can be, for example, the molecular nitrogen plasma, which is widely spread in plasma investigations: N_2, N, N_2^+, N^+, e^-. Kinetic processes that will be taken into account are presented in Table 1.1.

The system of defining equations includes the Navier–Stokes equations and continuity equation for determination of the mass-averaged gas dynamic functions, the system of energy conservation equations for individual species, and the system of kinetic equations. To reduce total number of used equations it is advisable to suppose that temperatures of atoms and atomic ions $T_a = T_{\mathrm{ai}}$ as well as temperatures of molecules and the molecular ions $T_a = T_{\mathrm{ai}}$ are equal.

Computational model of the five-component RF PG has the following form:

- continuity equation

$$\operatorname{div} \rho \mathbf{V} = 0; \tag{1.178}$$

- equations of impulse conservation (with reference to two-dimensional axisymmetric geometry)

$$\operatorname{div}(\rho u \mathbf{V}) = -\frac{\partial p}{\partial x} + F_x + S_x + \rho g_x, \tag{1.179}$$

$$\operatorname{div}(\rho v \mathbf{V}) = -\frac{\partial p}{\partial r} + F_r + S_r, \tag{1.180}$$

– energy conservation equation of electronic gas

$$\mathrm{div}(\rho \mathbf{V}_e H_e) = \mathrm{div}(\lambda_e \mathrm{grad}\ T_e) - B_a(T_e - T_a) - B_m(T_e - T_m) + \mathbf{W} + \sigma E^2, \tag{1.181}$$

Table 1.1: The kinetic scheme of processes in the five-component plasma.

Item	Type of reaction	Name of the forward reaction	Name of the reverse reaction
1	$N + N + P \underset{k_1^r}{\overset{k_1^f}{\rightleftarrows}} N_2 + P$	Association	Dissociation
2	$N + N \underset{k_2^r}{\overset{k_2^f}{\rightleftarrows}} N_2^+ + e$	Associative ionization	Dissociative recombination
3	$N + e \underset{k_3^r}{\overset{k_3^f}{\rightleftarrows}} N^+ + 2e$	Collisional ionization	Recombination
4	$N_2 + e \underset{k_4^r}{\overset{k_4^f}{\rightleftarrows}} N_2^+ + 2e$	Collisional ionization	Recombination
5	$N_2 + N^+ \underset{k_5^r}{\overset{k_5^f}{\rightleftarrows}} N_2^+ + N$	Exchange reaction	Exchange reaction

Note: In the given model, the processes of vibrational and electronic excitation are not considered. This is the so-called thermally equilibrium model.

where $H_e = \rho_e^{-1}(3/2n_e kT_e + p_e + n_{\mathrm{ai}}E_a + n_{\mathrm{mi}}E_m)$ is the enthalpy of electronic gas subject to the ionization energy of atoms and molecules; $p_e = n_e kT_e$ is the pressure of electronic gas; E_a, E_m are the ionization energy of atoms and molecules; $n_{\mathrm{ai}}, n_{\mathrm{mi}}$ are the volume concentration of atomic and molecular ions; E is the electric intensity module; $\mathbf{V}_e$ is the velocity of electronic gas; B_a, B_m are the efficiency factor for energy exchange at collision of electrons with atoms and molecules;

– equation of motion of electrons

$$m_e \frac{\partial \mathbf{V}_e}{\partial t} = e\mathbf{E} + e[\mathbf{V}_e \times \mathbf{B}] - \frac{\mathrm{grad} p_e}{n_e} - \sum_{k=1}^{N_k} \mu_{ek} \nu_{ek} (\mathbf{V}_e - \mathbf{V}_k), \tag{1.182}$$

where $\mathbf{V}_k$ is the averaged velocity of atoms ($k = a$), molecules ($k = m$), ions ($k = \mathrm{ai}$), and molecular ions ($k = \mathrm{ai}$); ν_{ek} is the frequency of collisions of electrons with particles of kth kind; $\mu_{ek} = m_e m_k/(m_e + m_k) \approx m_e$ is the reduced mass.

Equation (1.182) supposes essential simplification for the stationary flow of electrons without an external magnetic field. In this case, the left-hand part of this equation can be equated to zero. Let us suppose that velocities of atomic and molecular ions are close, as well as their mobilities, then

$$\mathbf{V}_e \approx \frac{\sigma}{en_e}\mathbf{E} - D_a\frac{\text{grad}n}{n} + \mathbf{V}, \tag{1.183}$$

where

$$D_a = \frac{D_e\mu_i + D_i\mu_e}{\mu_e + \mu_i};$$

D_e, D_i are the diffusion coefficients of electrons and ions; μ_e, μ_i are the mobilities of electrons and ions;

- energy conservation equation for atoms and ions

$$\text{div}(\rho_{\text{ai}}\mathbf{V}H_{\text{ai}}) = \text{div}(\lambda_{\text{ai}}\text{grad}T_a) - B_a(T_e - T_a) + A_{\tau,a}, \tag{1.184}$$

where $H_{\text{ai}} = \rho_{\text{ai}}^{-1}[(5/2)kT_a(n_a + n_i) + n_{\text{ai}}E_d]$ is the enthalpy of atomic–ionic gas subject to energy of dissociation E_d; ρ_{ai} is the density of atomic–ionic gas; λ_{ai} is the thermal conduction coefficient of atomic–ionic gas; T_a is the temperature of atomi–ionic gas; n_a, n_i are the volume concentration of atoms and ions;

- energy conservation equation for the molecular gas

$$\text{div}(\rho_m\mathbf{V}H_m) = \text{div}(\lambda_m\text{grad}T_m) + B_m(T_e - T_m) + A_{\tau,m}, \tag{1.185}$$

where $H_m = \rho_m^{-1}[(5/2)kT_m(n_m + n_{\text{mi}})]$ is the enthalpy of the molecular gas (without taking into account the energy reserved in excited vibrational and rotational energy states); ρ_m is the density of the molecular gas; λ_m is the thermal conduction coefficient of the molecular gas; T_m is the temperature of the molecular gas; $A_{\tau,m}$ is the work of viscous friction forces in the molecular gas;

- system of the kinetic equations defining a chemical composition of the plasma:

$$\text{div}(n_a\mathbf{V}) = -k_1^f n_a n_a n_p - k_2^f n_a n_a - k_3^f n_a n_e + k_5^f n_m n_{\text{ai}} + 2k_1^r n_m n_P + 2k_2^r n_{\text{mi}} n_e + k_3^r n_{\text{ai}} n_e n_e, \tag{1.186}$$

$$\text{div}(n_m\mathbf{V}) = +k_1^f n_a n_a n_p - k_4^f n_m n_e - k_5^f n_m n_{\text{ai}} - k_1^r n_m n_P + k_4^r n_{\text{mi}} n_e n_e + k_5^r n_{\text{mi}} n_a, \tag{1.187}$$

$$\text{div}(n_{\text{ai}}\mathbf{V}_{\text{ai}}) = +k_3^f n_a n_e - k_5^f n_m n_{\text{ai}} - k_3^r n_{\text{ai}} n_e n_e + k_5^r n_{\text{mi}} n_a, \tag{1.188}$$

$$\text{div}(n_{\text{mi}}\mathbf{V}_{\text{mi}}) = +k_2^f n_a n_a + k_4^f n_m n_e + k_5^f n_m n_{\text{ai}} - k_2^r n_{\text{mi}} n_e - k_4^r n_{\text{mi}} n_e n_e - k_5^r n_{\text{mi}} n_a, \tag{1.189}$$

where k_i^f, k_i^r are the rate constants of forward and reverse ith reaction; p is the index of any third particle (in the case under consideration this particle can be any of five considered species).

The rate constants, as a rule, are functions of the temperature of heavy particles and/or temperatures of electrons. However, if electrons are generally heated by electromagnetic fields, how it is observed in MW, RF, and optical PGs, the rates of ionization by electron impact can depend on the electric intensity and local pressure.

Actually, it is necessary to formulate the kinetic equation for electron concentration, similar to eqs. (1.186)–(1.189). However, it is possible to not do because one can use additional approximate condition of the quasi-neutrality:

$$n_e = n_{\rm ai} + n_{\rm mi}. \tag{1.190}$$

Nevertheless, sometimes, especially at the initial stages of the debugging of the calculation code, it is recommended to solve the kinetic equation for electronic concentrations. Then the condition (1.190) is expedient for using to check the precision of solution of the dynamic equations of atomic and molecular ions:

$$\mathbf{V}_{\rm ai} = \mu_{\rm ai}\mathbf{E} - D_{\rm ai}\frac{{\rm grad}n_{\rm ai}}{n_{\rm ai}} - D_{T,{\rm ai}}\frac{{\rm grad}(kT_a)}{kT_a} + \mathbf{V}, \tag{1.191}$$

$$\mathbf{V}_{\rm mi} = \mu_{\rm mi}\mathbf{E} - D_{\rm mi}\frac{{\rm grad}n_{\rm mi}}{n_{\rm mi}} - D_{T,{\rm mi}}\frac{{\rm grad}(kT_m)}{kT_m} + \mathbf{V}. \tag{1.192}$$

Note that eqs. (1.191) and (1.192) suppose some simplifications similar to those for eq. (1.182).

Boundary conditions at $\mathbf{V}_{\rm mi} = \mu_{\rm mi}\mathbf{E} - D_{\rm mi}({\rm grad}n_{\rm mi}/n_{\rm mi}) - D_{T,{\rm mi}}({\rm grad}(kT_m)/kT_m) + \mathbf{V}$. are formulated in usual form

$$\frac{\partial T_e}{\partial r} = \frac{\partial T_a}{\partial r} = \frac{\partial T_m}{\partial r} = 0,$$

$$\frac{\partial n_e}{\partial r} = \frac{\partial n_a}{\partial r} = \frac{\partial n_{\rm ai}}{\partial r} = \frac{\partial n_m}{\partial r} = \frac{\partial n_{\rm mi}}{\partial r} = 0.$$

Similar boundary conditions can be used for all axial velocities, while all the radial velocities are equal to zero on the symmetry axis.

Conditions of sticking are applied to heavy particles on walls of the PG channel. Sometimes the sticking condition can be replaced by conditions of gas blowing through the walls of the PG. Temperature balance with the channel wall (or with the blown gas) also should be taken into account.

Formulation of the boundary conditions for electronic gas is less obvious, and it demands the attraction of the kinetic approach. The following assumption can be additionally analyzed:

1) For the definition of electronic temperature near to the wall of the PG chamber the assumption is used that the energy of the external electric field near to the wall is spent for heating of heavy particles.
2) The model of collisionless screening layer is used close to the wall where the polarization field is created by more mobile electrons. This field promotes alignment of the ionic and electronic current toward the wall (it brakes electrons in the wall direction and accelerates ions).

3) Zero total electric current toward the wall is supposed. The requirement of a quasi-neutrality is used and the balance of energy brought by particles to the wall and energy released on the wall at their recombination is calculated.

1.2.2 Model of the three-component RF plasma generator

This model is a special case of the model presented above. The essence of the simplification of the previous model consists in assuming the plasma consisting only from three components: atoms, electrons, and ions. Despite so essential simplification, the present model is the most well-developed one among other nonequilibrium models because it corresponds to the induction PG on argon, widely used in the scientific researches.

The system of solved equations of the model is formulated in the following form:

- continuity equation

$$\operatorname{div} \rho \mathbf{V} = 0; \tag{1.193}$$

- momentum equation (the equations of motion with reference to two-dimensional cylindrical geometry)

$$\operatorname{div}(\rho u \mathbf{V}) = -\frac{\partial p}{\partial x} + F_x + S_x + \rho g_x, \tag{1.194}$$

$$\operatorname{div}(\rho v \mathbf{V}) = -\frac{\partial p}{\partial r} + F_r + S_r; \tag{1.195}$$

- energy conservation equation for electronic gas

$$\operatorname{div}(\rho \mathbf{V}_e H_e) = \operatorname{div}(\lambda_e \operatorname{grad} T_e) - B_a(T_e - T_a) + \operatorname{div} \mathbf{W} + \sigma E^2, \tag{1.196}$$

where

$$H_a = \frac{1}{\rho}\left(\frac{5}{2}kT(n_a + n_i)\right);$$

E_a is the ionization energy of atoms;

- energy conservation equation for atomic–ionic gas

$$\operatorname{div}(\rho \mathbf{V} H_a) = \operatorname{div}(\lambda_a \operatorname{grad} T) + B_a(T_e - T_a), \tag{1.197}$$

where

$$H_a = \frac{1}{\rho}\left[\frac{5}{2}kT(n_a + n_i)\right];$$

- equation of ionization equilibrium

$$\mathrm{div}(n_e\mathbf{V}) = -k_i n_a n_e - k_r n_e^2 n_i, \tag{1.198}$$

where k_i, k_r are the rate constants of atom-electronic ionization (due to their collisions) and ion-electronic recombination related to ionization by means of the condition of detailed equilibrium.

System of equations of electromagnetic part of the problem is the same as in Section 1.1.

In the case of the optically thin argon plasma, energy losses by heat radiation can be calculated under the formula (Wilberts A.T.M., et al., 1991):

$$\mathrm{div}\,\mathbf{W} = 1.19\times10^{-40}\sqrt{T_e}\exp\left(-\frac{\Delta E_i}{kT_e}\right)\times n_e\left[n_i\exp\left(\frac{h\nu}{kT_e}\right) + 4n_{2i}\exp\left(\frac{h\nu^+}{kT_e}\right)\right], \tag{1.199}$$

where $h\nu = 2.85$ eV, $h\nu^+ = 8.2$ eV; ν, ν^+ are the limit frequencies of photoionization thresholds from the ground states of neutral argon and its ion (here the second term in square brackets considers the contribution of double ions to the emissive power).

1.2.3 Two-temperature model of RF plasma under ionization equilibrium

The model is based on the following assumptions:

1. The Maxwellian distributions of particles with temperatures T_e and T assumed for electrons and heavy particles (atoms, molecules, and their ions).
2. Processes of heavy particles and electrons transfer in inhomogeneous plasma do not disturb its local ionization–recombination equilibrium. That is the plasma composition is defined by local values of thermodynamic parameters of electronic gas and heavy particles gas. In other words, velocities of the kinetic processes are supposed many times greater than the transfer processes rates.
3. It is supposed that processes of ionization of particles and their distributions on the excited electronic states are defined by the temperature of electronic gas. It allows using the Saha equation of ionization equilibrium for a definition of a plasma composition (Bittencourt J.A., 2004). For example, for three-component plasma (atoms, ions, electrons) we have

$$\frac{n_e n_i}{n_a} = \frac{2g_i}{g_a}\left(\frac{2\pi k m_e T_e}{h^2}\right)^{3/2}\exp\left(-\frac{E_a}{kT_e}\right), \tag{1.200}$$

where g_a, g_i are the statistical weights of atoms and ions; E_a is the atomic ionization potential.

Together with eq. (1.200) it is necessary to use the following correlations:

– the requirement of a quasi-neutrality

$$n_i = n_e; \tag{1.201}$$

– the Dalton law

$$(n_a + n_i)kT + n_e kT_e = p, \tag{1.202}$$

where p is the total pressure.

Relations (1.200)–(1.202) result in a quadratic equation for electron concentration, which is defined now by the two temperatures T and T_e, and also by the full pressure.

In spite of introduction two temperatures T and T_e, the given model allows to reduce the labor of the calculations due to preliminary computations of thermodynamic, thermophysical, and transport properties of the low-temperature plasma.

For argon plasma, the following correlations can be used:

– the electroconductivity and electronic thermal conduction

$$\sigma = \frac{e^2 n_e K_\sigma}{m_e v_e \sum_{k=1}^{N_k} n_k Q_k}, \quad \lambda_e = \frac{n_e k T_e^2 K_\lambda}{m_e v_e \sum_{k=1}^{N_k} n_k Q_k}, \tag{1.203}$$

where N_k is the number of particle kinds; K_σ and K_λ are the corrections to the electron and heavy particle temperatures; Q_k is the cross section of interparticle interaction (electron–electron, electron–atom, electron–ion, atom–atom, atom–ion, ion–ion). Some useful approximations are:

$$Q_{\text{ei}} = \frac{8.78 \times 10^{-10}}{T_e^2} \ln \frac{2.79 \times 10^4 T_e}{n_e^{1/3} [1 + (T_e/T)]^{1/3}}, \quad \text{m}^{-2}, \tag{1.204}$$

$$Q_{\text{ea}} = \left(3.6 \times 10^{-4} T_e - 0.1\right) \times 10^{-20}, \quad \text{m}^{-2}; \tag{1.205}$$

– thermal conduction of atomic–ionic gas is defined under the formula

$$\lambda = \frac{75k}{64} \sqrt{\frac{nkT}{M_a}} \left(\frac{n_a}{n_a Q_{\text{aa}} + n_e Q_{\text{ai}}} + \frac{n_e}{n_a Q_{\text{ia}} + n_e Q_{\text{ii}}} \right), \tag{1.206}$$

which is simplified at $T \leq 17{,}000$ K as follows:

$$\lambda = 2.5 \times 10^{-4} T^{0.75} \frac{n_a}{n_a + 5n_e}, \text{W}/(\text{mK}); \tag{1.207}$$

– viscosity of the plasma

$$\mu = 6 \times 10^{-25} T^{0.5} \sum_{k=1}^{N_k} \frac{n_k}{\sum_{l=1}^{N_k} n_l Q_{lk}}, \text{kg}/(\text{ms}); \tag{1.208}$$

- heat capacity of heavy particles gas

$$c_p = \frac{5}{2}k(n_a + n_i), \tag{1.209}$$

and of electronic gas

$$c_{p,e} = \frac{5}{2}kn_e + \left(\frac{5}{2}kT_e + E_i\right)\left(\frac{\partial n_e}{\partial T_e}\right)_p, \tag{1.210}$$

where the sum in parenthesis corresponds to the electronic energy taking into account an ionization energy.

Now we can formulate closed system of the governing equations of the two-temperature model:

- continuity equation

$$\text{div}\,\rho\mathbf{V} = 0; \tag{1.211}$$

- momentum equations (with reference to the two-dimensional cylindrical geometry)

$$\text{div}(\rho u\mathbf{V}) = -\frac{\partial p}{\partial x} + F_x + S_x + \rho g_x, \tag{1.212}$$

$$\text{div}(\rho v\mathbf{V}) = -\frac{\partial p}{\partial r} + F_r + S_r; \tag{1.213}$$

- energy conservation equation of electronic gas

$$\text{div}(\rho\mathbf{V}H_e) = \text{div}(\lambda_e \text{grad}\ T_e) - B_a(T_e - T_a) + \text{div}\ \mathbf{W} + \sigma E^2, \tag{1.214}$$

where

$$H_e = \frac{1}{\rho_e}\left(\frac{5}{2}n_e kT_e + n_e E_a\right),$$

E_a is the ionization energy of atoms;

- energy conservation equation of atomic–ionic gas

$$\text{div}(\rho\mathbf{V}H_a) = \text{div}(\lambda_a \text{grad} T) + B_a(T_e - T_a), \tag{1.215}$$

where

$$H_a = \frac{1}{\rho}\left[\frac{5}{2}kT(n_a + n_i)\right];$$

- equations of ionization equilibrium

$$n_e^2 + n_e\left(1 + \frac{T_e}{T}\right)F_{ei} - \frac{p}{kT}F_{ei} = 0, \tag{1.216}$$

where

$$F_{\rm ei} = \frac{2g_i}{g_a}\left(\frac{2\pi k m_e T_e}{h^2}\right)^{3/2} \exp\left(-\frac{E_a}{kT_e}\right),$$

p is the total pressure.

Other equations of the model (system of electrodynamic equations and the equation of radiation heat transfer) are the same, as in the previous models.

1.2.4 One-liquid two-temperature model of laser-supported plasma

This model is formulated in a two-dimensional axisymmetric geometry with reference to the stationary flow of the hydrogen plasma. This model includes the Navier–Stokes and continuity equations for the description of viscous gas dynamics, and also two energy conservation equations for electronic gas (e^-) and for heavy particle gas (H_2, H, H^+):

- continuity equation

$$\frac{\partial \rho u}{\partial x} + \frac{1}{r}\frac{\partial r\rho v}{\partial r} = 0; \tag{1.217}$$

- the Navier–Stokes equations

$$\frac{\partial}{\partial x}(\rho uu) + \frac{1}{r}\frac{\partial}{\partial r}(r\rho uv) = -\frac{\partial p}{\partial x} + S_x - \rho g, \tag{1.218}$$

$$\frac{\partial}{\partial x}(\rho uv) + \frac{1}{r}\frac{\partial}{\partial r}(r\rho vv) = -\frac{\partial p}{\partial r} + S_r, \tag{1.219}$$

where

$$S_x = -\frac{2}{3}\frac{\partial}{\partial x}(\mu \operatorname{div} \mathbf{V}) + 2\frac{\partial}{\partial x}\left(\mu\frac{\partial u}{\partial x}\right) + \frac{1}{r}\frac{\partial}{\partial r}\left[r\mu\left(\frac{\partial u}{\partial r} + \frac{\partial v}{\partial x}\right)\right],$$

$$S_r = -\frac{2}{3}\frac{\partial}{\partial r}(\mu \operatorname{div} \mathbf{V}) + \frac{\partial}{\partial x}\left[\mu\left(\frac{\partial u}{\partial r} + \frac{\partial v}{\partial x}\right)\right] + 2\frac{\partial}{\partial r}\left(\mu\frac{\partial v}{\partial r}\right) + 2\mu\frac{\partial}{\partial r}\left(\frac{v}{r}\right);$$

- energy conservation equation for heavy particle gas

$$\frac{\partial}{\partial x}\left[\left(\frac{3}{2}knu\right)T\right] + \frac{1}{r}\frac{\partial}{\partial r}\left[\left(\frac{3}{2}rknv\right)T\right] = \frac{\partial}{\partial x}\left(\lambda\frac{\partial T}{\partial x}\right) + + \frac{1}{r}\frac{\partial}{\partial r}\left(\lambda\frac{\partial T}{\partial r}\right)$$
$$-(p+P)\left[\frac{\partial u}{\partial x} + \frac{1}{r}\frac{\partial}{\partial r}(rv)\right] + A_\tau^h - Q_{\rm R}; \tag{1.220}$$

- energy conservation equation for electrons

$$\frac{\partial}{\partial x}\left[\left(\frac{3}{2}kn_eu\right)T_e\right]+\frac{1}{r}\frac{\partial}{\partial r}\left[\left(\frac{3}{2}rkn_ev\right)T_e\right]=\frac{\partial}{\partial x}\left(\lambda_e\frac{\partial T_e}{\partial x}\right)+\frac{1}{r}\frac{\partial}{\partial r}\left(r\lambda_e\frac{\partial T_e}{\partial r}\right)-A_\tau^e+Q_\mathrm{L},\tag{1.221}$$

where x, r are the axial and radial coordinates; u, v are the axial and radial components of velocity; T_e, T are the temperature of electrons and heavy particles; μ is the coefficient of dynamic viscosity; λ_e, λ are the thermal conductivities of electronic gas and gas of heavy particles; ρ is the total density; p is the dynamic component of the pressure; P is the static pressure defined by the following equation of state:

$$P=(n_ekT_e+nkT)\left(1-P_c^\mathrm{DH}\right),$$

where

$$P_c^\mathrm{DH}=\frac{1}{24\pi(n_e+n)}\left[\frac{4\pi e^2k_c}{k}\left(\frac{n_e}{T_e}+\frac{n_{\mathrm{H}^+}}{T}\right)\right]^{1.5}$$

is the correction of the Debye–Huckel; h_c is the Coulomb constant; Q_L is the power of a heat release caused by the radiation of laser source; Q_R is the power of radiative energy losses; A_τ^e, A_τ^h is the work of friction caused by interaction of electronic gas with gas of heavy particles

$$A_\tau^e=A_\tau^h=3k(T_e-T)n_em_e\sum_j^{N_h}\frac{\nu_{ej}}{m_j};\tag{1.222}$$

ν_{ej} is the frequency of collisions of electrons with jth heavy particle; N_h is the number of kinds of heavy particles (in the given case $N_h=3$, $(H_2,\ H,\ H^+)$); n_e is the volume concentration of electrons; n is the volume concentration of heavy particles $\left(n=n_{\mathrm{H}_2}+n_\mathrm{H}+n_{\mathrm{H}^+}\right)$; m_e, m_j are the masses of an electron and jth heavy particle; k is the Boltzmann constant.

For calculation of a chemical composition of nonequilibrium plasma, the system of equations is used

$$\frac{(n_\mathrm{H})^2}{n_{\mathrm{H}_2}}=\exp\left(-\frac{E_{\mathrm{H}_2}}{kT}\right)\left(\frac{\pi m_\mathrm{H}kT}{h^2}\right)^{3/2}\left(2\frac{T_e}{T}\right)\frac{Q_{\mathrm{H,ex}}}{Q_{\mathrm{H}_2,\mathrm{rot}}\cdot Q_{\mathrm{H}_2,\mathrm{vib}}},\tag{1.223}$$

$$n_e\left(\frac{n_{\mathrm{H}^+}}{n_\mathrm{H}}\right)^{T/Te}=\frac{2\exp(-E_\mathrm{H}/kT_e)}{Q_{\mathrm{H,ex}}}\left(\frac{2\pi m_ekT_e}{h^2}\right)^{3/2},\tag{1.224}$$

where $E_{\mathrm{H}_2}, E_\mathrm{H}$ are the potentials of dissociation of H_2 and ionization of H; $Q_{\mathrm{H,ex}}, Q_{\mathrm{H}_2,\mathrm{vib}}, Q_{\mathrm{H}_2,\mathrm{rot}}$ are the partition functions of the electronic states of atom H and vibrational and rotational states of molecule H_2.

Collision frequencies of electrons and heavy particles are estimated under the formula

$$\nu_{ej} = v_e n_j Q_{ej}, \tag{1.225}$$

where $v_e = \sqrt{(8k/\pi m_e)T_e}$ is the average thermal velocity of electrons; Q_{ej} is the cross section of electron–particle collision.

Depending on the type of a heavy particle, the three types of collisions are defined:

- the electron–atom collision (in the given case: $e - H$);
- the electron–molecule collisions ($e - H_2$);
- the electron–ion collisions ($e - H^+$).

In the study (Metrogul A.E., et al., 1994), the tabular data were used gained by Itakawa Y. (1974, 1978). For the Coulomb interaction of charged particles, the cross section is estimated as follows:

$$Q_{ei} = \frac{\pi e^4 k_c^2 \ln \Lambda_e}{2(kT_e)^2}, \tag{1.226}$$

where

$$\Lambda_e = \frac{1.5}{e^3 k_c^{1.5}} \sqrt{\frac{k^3 T_e^2}{\pi n_e}}. \tag{1.227}$$

The electronic thermal conduction is calculated under the formula (Metrogul A.E., et al., 1994)

$$\lambda_e = 5 \times 10^{-25} n_e \sqrt{T_e}, \tag{1.228}$$

derived before for argon plasma.

Summing up the description of the equilibrium and nonequilibrium computing models of the low-temperature plasma, we shall note that it is important to provide the models with well-tested transport and optical properties. The use of incorrect models of the properties, significant advantages of detailed nonequilibrium models of gas discharge plasma dynamics can be lost.

2 Application of numerical simulation models for investigation of laser-supported combustion waves (LSW)

2.1 Air laser-supported plasma generator

Schematic diagram of a laser-supported plasma generator (LSPG), which will be considered in this chapter, is presented in Figure 1.5. The subsonic flow model used in the present section (small pressure variations are neglected) makes it possible to reduce the continuity and Navier–Stokes equations in the form (1.95)–(1.97) to the following system of equations in time-dependent dynamic variables (for details, see Section 1.1.6):

$$\frac{\partial \omega}{\partial t} + \operatorname{div}(\omega \mathbf{V}) + \frac{\partial}{\partial x}\left(\rho v \frac{\partial u}{\partial x} - \rho u \frac{\partial u}{\partial r}\right) + \frac{\partial}{\partial r}\left(\rho v \frac{\partial v}{\partial x} - \rho u \frac{\partial v}{\partial r}\right)$$
$$+ \rho v \frac{\partial (v r^{-1})}{\partial x} - \rho u \frac{\partial (v r^{-1})}{\partial r} = \frac{1}{\mathrm{Re}}\left(\frac{\partial S_v}{\partial x} - \frac{\partial S_u}{\partial r}\right), \tag{2.1}$$

$$\operatorname{div}\left[\frac{\exp(-\varepsilon)}{r^2}\operatorname{grad}\Psi\right] = -\frac{\omega}{r}, \tag{2.2}$$

$$\rho u \frac{\partial \varepsilon}{\partial x} + \rho v \frac{\partial \varepsilon}{\partial r} = \frac{\partial \rho}{\partial t}, \tag{2.3}$$

$$\frac{\partial \Psi}{\partial r} = r \rho u \exp(\varepsilon), \quad \frac{\partial \Psi}{\partial x} = -r \rho v \exp(\varepsilon), \tag{2.4}$$

$$\omega = \frac{\partial \rho v}{\partial x} - \frac{\partial \rho u}{\partial r}, \tag{2.5}$$

and the function $\varepsilon = \ln(E+1)$ introduced using the hypothesis

$$\operatorname{div}(\mathrm{E}\rho \mathbf{V}) = \frac{\partial \rho}{\partial t} \tag{2.6}$$

makes it possible to represent the continuity equation in the form (2.3).

The use of hypothesis (2.6) is based on the fact that the time variation of the gas density is primarily initiated by internal heat sources (in our case by absorption of the laser radiation energy) under subsonic flow conditions. This variation can be assumed to be a known function to be determined from the energy conservation equation and the equation of state and not from the gas dynamic equations. On the other hand, the continuity equation, which nevertheless must unconditionally be satisfied, dictates the divergent form of the partial derivative of the density with respect to time. In order to conserve the formulation of the problem of finding the

https://doi.org/10.1515/9783110648836-003

unknown vector function we propose the following hypothesis: let us assume that there is a scalar function E satisfying eq. (2.6).

Thus, by introducing a new scalar function ε (or E) we can formulate the system of equations of a viscous compressible fluid in the dynamic variables $\varepsilon - \Psi - \omega$, which makes it possible to solve time-dependent problems of subsonic radiative gas dynamics.

The boundary conditions for functions ω and Ψ are specified as for the ordinary dynamic variables. It is also easy to write the boundary conditions for the function ε

$$x = 0, \quad \varepsilon(r) = 0, \tag{2.7}$$

$$r = R_c, \quad \varepsilon(x) = 0, \tag{2.8}$$

$$r = 0, \quad \frac{\partial \varepsilon}{\partial r} = 0, \tag{2.9}$$

$$x = L, \quad \frac{\partial \varepsilon}{\partial x} = 0. \tag{2.10}$$

Conditions (2.7) and (2.8) are the consequences of the constant temperatures on the corresponding surfaces (R_c is the radius of the cylindrical channel). Condition (2.9) determines the axial symmetry and condition (2.10) corresponds to the free boundary $x = L$ of the computational domain.

In order to determine the thermal radiation transfer to the inner surface of the plasma generator, let us consider the ray tracing method. In our calculations, we imposed the condition of axial symmetry. The algorithm of the ray tracing method is as follows. In order to find the density of the thermal radiation flux to a surface element, we introduce a local spherical coordinate system with the normal n. In this coordinate system, each ray Ω is determined by two angular coordinates, namely, the latitude angle $\theta \in [0, \pi/2]$ and the azimuthal angle $\varphi \in [0, 2\pi]$.

On the inner surface of the plasma generator, the densities of the spectral and integral radiation fluxes can be calculated from the formulas

$$W_\omega(\mathbf{r}_j) = \int_0^{2\pi} \mathrm{d}\varphi \int_0^{\pi/2} J_\omega(\mathbf{r}_j, \Omega) \cos\theta \sin\theta \, \mathrm{d}\theta,$$

$$W_\mathrm{R}(\mathbf{r}_j) = \int_0^\infty W_\omega(\mathbf{r}_j) \mathrm{d}\omega, \tag{2.11}$$

where $\mathbf{r}_j = (r_j, x_j)$ is the radius vector of the jth point on the surface in the laboratory coordinate system, r_j, x_j are the radial and axial coordinates of the jth point in the laboratory coordinate system, and $J_\omega(\mathbf{r}_j, \Omega)$ is the spectral intensity of the radiation.

Introducing a computational grid of angular directions makes it possible to integrate the spectral intensity of the radiation on the plasma generator surface over the space of the angular variables and find the total spectral density of the flux (2.11):

$$W_\omega(\mathbf{r}_j) = \frac{1}{2}\sum_{m=1}^{N_\varphi - 1}(\varphi_{m+1} - \varphi_m)\sum_{n=1}^{N_\theta} J_\omega(\mathbf{r}_j, \Omega_{m,n})(\sin\theta_{n+1}\cos\theta_{n+1} + \sin\theta_n\cos\theta_n)(\theta_{n+1} - \theta_n)$$

or

$$W_\omega(\mathbf{r}_j) = -\sum_{m=1}^{N_\varphi - 1}(\varphi_{m+1} - \varphi_m)\sum_{n=1}^{N_\theta - 1}\frac{1}{2}J_\omega(\mathbf{r}_j, \Omega_{m,n})(\cos^2\theta_{n+1} - \cos^2\theta_n) \tag{2.12}$$

The direction cosines of the vector

$$\Omega_{m,n} = (\varpi_x)_{m,n}\mathbf{i} + (\varpi_y)_{m,n}\mathbf{j} + (\varpi_z)_{m,n}\mathbf{k}$$

can be calculated from the formulas

$$\begin{aligned}(\varpi_x)_{m,n} &= \sqrt{1 - \bar{\mu}_n^2}\cos\bar{\varphi}_m,\\ (\varpi_y)_{m,n} &= \sqrt{1 - \bar{\mu}_n^2}\sin\bar{\varphi}_m,\\ (\varpi_z)_{m.n} &= \bar{\mu}_n,\\ \bar{\mu}_n = \frac{1}{2}(\mu_n + \mu_{n+1}),\ \bar{\varphi}_n &= \frac{1}{2}(\varphi_n + \varphi_{n+1}),\ \mu = \cos\theta.\end{aligned} \tag{2.13}$$

In order to determine the quantity $J_\omega(\mathbf{r}_j, \Omega_{m,n})$, it is necessary to integrate the radiation transport equation along an inhomogeneous optical ray. For these purposes, we can use a formal solution of the transport equation of the following form for a nonscattering medium:

$$J_\omega(\mathbf{r}_j, \Omega_{m,n}) = \int_{s=0}^{s=L} J_{b,\omega}[T(s')]\kappa_\omega(s')\exp\left[-\int_0^{s'}\kappa_\omega(s'')ds''\right]ds', \tag{2.14}$$

where $s = 0$ and L are the initial (on the plasma generator surface at $\mathbf{r} = \mathbf{r}_j$) and end (on the opposite generator surface along the ray or in the exit cross section) coordinates of the interval of the ray $\Omega_{m,n}$ along which integration is performed, and $J_{b,\omega}(T)$ is the radiation intensity of an absolutely black body (the Planck function).

A finite-difference grid in the spatial variable s, by means of which we carry out the numerical integration, can be found for each ray $\Omega_{m,n}$. For this purpose in the laboratory coordinate system, we find the coordinates of the intersection of the ray $\Omega_{m,n}$ with all the surfaces of the finite-difference grid $\xi = \{h_i = r_i - r_{i-1},\ h_j = x_j - x_{j-1},\ i = 1, \ldots, NI,\ j = 1, \ldots, NJ\}$ it encounters.

Figure 2.1 illustrates this algorithm graphically (see ray a). We note that in order to determine the coordinates of intersection it is necessary to use the relations of

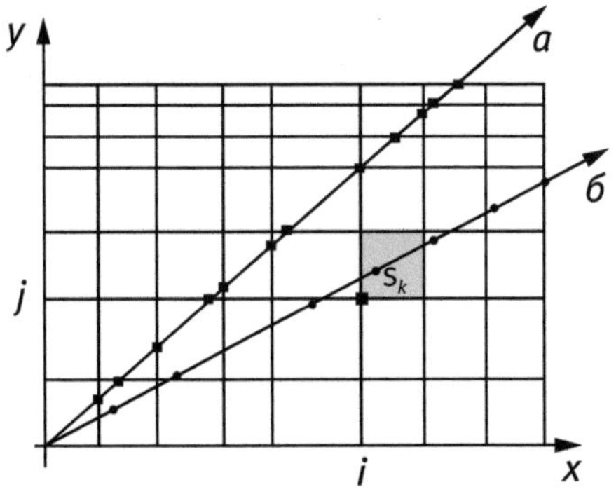

Figure 2.1: Methods of calculating selective thermal radiation transport: calculation of the coordinates of intersection of a ray with all the surfaces along it (a) and quasi-random sampling (b).

analytic geometry. However, this algorithm turns out to be uneconomical with respect to curvilinear, in particular nonstructured, computational grids. Therefore, it is recommended to use a highly efficient algorithm of quasi-random sampling of the computational grid coordinates.

This algorithm consists of the following. A segment of a ray with the directional vector $\Omega_{m,n}$ between the first ($s = 0$) and the end ($s = L$) points is divided into N_s identical fragments as shown for a ray in Figure 2.1 (in fact, the requirement of the uniformity of the computational grid on the segment $s \in [0, L]$ is not mandatory for the algorithm used). Then for each grid point along the ray $\Omega_{m,n}$ one can find the nearest grid point of the spatial computational grid (or the cell to which this grid point belongs) and on which the thermophysical and optical properties of the medium are specified. The temperature and the spectral absorption coefficient of the grid point (or cell) thus found are assigned to the current grid point of the computational grid along the ray. Thus, the temperature and the optical properties of the gas are known at each grid point of the computational grid along the ray $\Omega_{m,n}$.

We note that this algorithm of the ray tracing method (2.11)–(2.14) is far from being the most efficient. An alternative is the more efficient discrete ordinate method (Filipskii M.V., et al., 2007). However, only the ray tracing method can be developed further in order to take the specifics of radiation transport in atomic lines into account by means of statistical models of their spectrum. In this case, the algorithm outlined corresponds to the approximation based on the molecular spectrum averaged over the rotational structure and the atomic absorption spectrum averaged over the spectral line structure.

Now we will consider the results of calculating the radiative gas dynamics of an air generator obtained for three laser radiation powers: $P_L = 30$, 50, and 200 kW. Single-mode CO_2 laser radiation (see parameter $n = 2$ in eq. (1.99)) with a wavelength of $\lambda = 10.6$ μm was specified. The radiation was focused at a distance of $x_p = 3$ cm from the inlet cross section of the cylindrical channel. The divergence of the laser radiation was assumed to be equal to $\theta_L = 0.02$, 0.04, and 0.08 mrad, respectively, for the three increasing powers. On the one hand, this corresponds to the physics of the process and, on the other hand, made it possible to obtain a stable laser plasma in the similar gas flows with a velocity $u_0 = 30$ m/s.

The minimum radius of the caustic of the focused radiation was calculated using the approximate relation $R_c \approx x_p \theta_L$.

The transition from the heat conducting regime of laser plasma existence (for low power) to the radiating regime (for high power) was simulated by the input of three laser radiation powers.

Figures 2.2–2.4 show the temperature fields inside the laser generator for various laser radiation powers under different assumptions concerning the turbulent mixing and the radiative heat transfer mechanism. The radiative heat transfer calculations were carried out for the 37-group model (see details in Section 1.1.4).

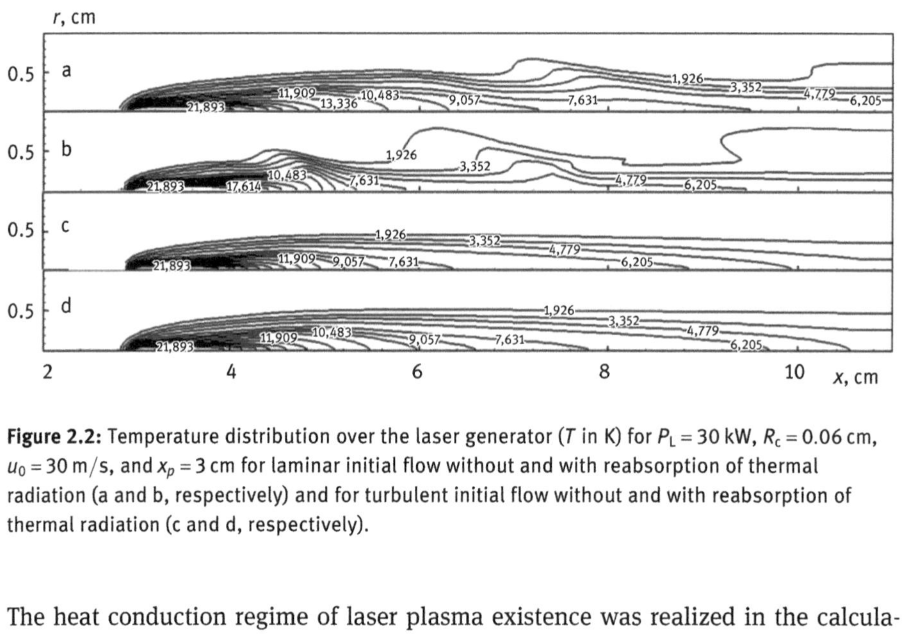

Figure 2.2: Temperature distribution over the laser generator (T in K) for $P_L = 30$ kW, $R_c = 0.06$ cm, $u_0 = 30$ m/s, and $x_p = 3$ cm for laminar initial flow without and with reabsorption of thermal radiation (a and b, respectively) and for turbulent initial flow without and with reabsorption of thermal radiation (c and d, respectively).

The heat conduction regime of laser plasma existence was realized in the calculations with the minimum laser radiation power $P_L = 30$ kW. In this case, the caustic radius $R_c = 0.06$ cm. The temperature fields reproduced in Figure 2.2a and b correspond to the assumption of a laminar flow regime in the inlet cross section of the generator. In the first case (Figure 2.2a), the reabsorption of thermal radiation was neglected, that is, the volume emission model was used in accordance with which the thermal radiation leaves the computational domain without absorption in the gas. In the second case (Figure 2.2b), the reabsorption of thermal radiation was taken into account. Clearly, the differences between the temperature fields calculated with and without taking the reabsorption into account are insignificant.

This again confirms the initial assumption that in this case the heat conduction regime was realized. As mentioned earlier, the principal reason for this is the fact that the volume of the high-temperature laser plasma region is so small that the power of the emitted thermal radiation is insufficient for intense heating of the adjacent gas layers, which can be more effectively realized by means of the heat conduction heating mechanism.

The self-oscillations of the gas flow behind the region of heat release associated with the absorption of laser radiation are extremely important for the variant of low laser radiation power in question. Figure 2.2a and b demonstrates the instantaneous structures of the temperature field. In fact, the calculations indicate that in this case there exist undamped self-oscillations of the flow. A pulsating plasma wake, in which the velocity and the temperature vary with a period of the order of $0.1 \div 0.5$ ms, is formed behind the heat release region. This period corresponds to the characteristic drift time of the vortices formed behind the heat release region.

The turbulent mixing in the gas flow and the initial flow turbulization at the inlet to the generator channel into account qualitatively changes the flow pattern (Figure 2.2c and d). In the calculations, the complete stabilization of the flow field is observed and the laser plasma becomes more stable to various perturbations of the gas flow introduced into the numerical solution for research purposes.

In the variants of the calculations presented, the initial turbulence level (boundary conditions for eqs. (1.105) and (1.106)) was simulated by specifying the boundary conditions for the functions k and ε in the form: $x = 0$ and $k = \varepsilon = 0.1$. This corresponds to approximately 15–20% of the intensity of the turbulent pulsations at the inlet to the generator channel. A decrease in the initial turbulence level led to a corresponding decrease in the effect of turbulent dissipation on the gas dynamic structure of the laser plasma and the gas flowing past it. Hence, we may conclude that in this case, it is precisely the initial flow turbulence that has the main effect on the gas flow stabilization. This is attributable to the fact that the internal turbulence of the gas flow manifests itself only at certain distances downstream from the high-temperature laser radiation absorption region, where this effect is no longer so significant.

The second series of calculations corresponds to the case of the transition from the heat conducting to the radiation regime of plasma existence. In this case, the caustic radius $R_C = 0.12$ cm. As in the first series of calculations, the temperature fields in the laser generator are obtained for a laminar inlet flow (Figure 2.3a and b) and with allowance for turbulent mixing (Figure 2.2c and d).

It is noteworthy that in this case the reabsorption of radiation already has an appreciable effect on the gas dynamic structure of the plasma generator. In particular, in this case, taking the reabsorption of radiation into account leads to gas flow stabilization. This is attributable to the decrease in the temperature gradients in the gas flow as a result of the absorption of thermal radiation by the cold gas layers. The effect of reabsorption can also be seen clearly in the temperature of the plasma jet flowing out of the generator. When the reabsorption is taken into

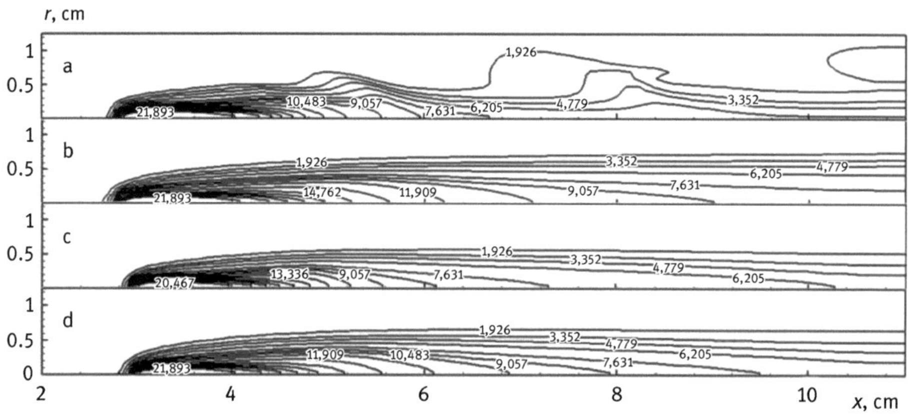

Figure 2.3: The same as in Figure 2.2 for $P_L = 50$ kW, $R_c = 0.12$ cm, $u_0 = 30$ m/s, and $x_p = 3$ cm.

account, the jet temperature increases appreciably, the increase being more marked for laminar flow. These facts indicate a decrease in the total energy losses from the plasma and the plasma wake due to the reabsorption of radiation and the absence of turbulent dissipation. However, in this case, it should again be noted that the numerical simulation results obtained are stabilized when the initial gas flow turbulence is taken into account.

In Figure 2.4, the results of the calculations for the radiative regime of the existence of the laser plasma in the plasma generator channel are presented. In this case, the caustic radius R_c is equal to 0.24 cm. As in the previous case, the reabsorption of radiation favors the stabilization of the gas flow. Attention is drawn to the large length of the high-temperature zone of the laser plasma and its appreciably increased transverse dimensions. In Figure 2.4a–d, one can clearly see that the

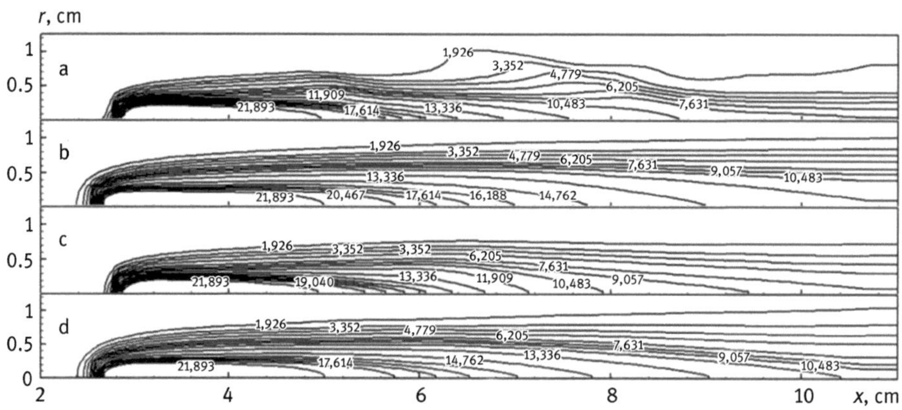

Figure 2.4: The same as in Figure 2.2 for $P_L = 200$ kW, $R_c = 0.24$ cm, $u_0 = 30$ m/s, and $x_p = 3$ cm.

reabsorption of thermal radiation leads to the laser plasma being displaced upstream toward the laser beam by more than 0.2 cm. In the case considered, the reabsorption of radiation is so significant that the initial flow turbulence has almost no effect on the gas dynamic structure of the laser plasma in the neighborhood of the highest temperature absorption zone, although, of course, in the resulting plasma jet this effect remains strong with respect to flow stabilization. As in the second calculated case, taking the reabsorption of radiation into account leads to an appreciable increase in the plasma jet temperature. Corresponding changes are also observed for the plasma jet velocity. For example, when the reabsorption of radiation was taken into account for an initial turbulent flow, the outlet cross-sectional plasma jet velocity u on the axis of symmetry was 1,860 m/s (the Mach number $M = 0.58$), while in the absence of reabsorption $u = 780$ m/s ($M = 0.32$).

The spectral and integral radiative heat fluxes on the inner generator surface were calculated by means of the ray tracing method using computational grids in the angular coordinates $N_\theta = N_\varphi = 11$, 21, and 41. The number of grid points N_s along the rays issuing from the surface varied from 10 to 40. The calculations showed that the results are most sensitive to the specified number of grid points N_s. This is easily attributable to the high degree of inhomogeneity of the laser plasma and the dependence of the solutions on the accuracy of the description of the plasma properties along with the issuing rays. The error introduced into the calculations due to grid coarseness is well known in computation physics as the "ray effect" (Filipskii et al., 2007). In the case considered, this effect was observed when N_θ and N_φ were less than 11.

Figure 2.5 shows the results of calculating the integral-over-the-spectrum radiative heat fluxes on the inner generator surface for the maximum laser radiation power. In this case, the 37-group optical model was used. These distributions completely reflect the above-mentioned features of radiation reabsorption and initial gas flow turbulence. When only the volume emission is taken into account, the decrease in the dimensions of the high-temperature laser plasma region manifests itself in an approximately two-fold decrease in the radiative heat fluxes W_R on the wall (curves *a* and *c*). The maximum fluxes W_R are observed in the case of the laminar flow with allowance for the reabsorption of thermal radiation (curve *b*). The initial gas flow turbulence leads to a regular but small reduction in the radiation fluxes associated with an insignificant decrease in the dimensions of the high-temperature laser plasma region.

In the case shown in Figure 2.6, the initial flow was assumed to be turbulent and the reabsorption of thermal radiation was taken into account. These data indicate, on the one hand, the ambiguous effect of the choice of group model on the results obtained and, on the other, the acceptable accuracy of the recommended 37-group model as compared with other group models. In this case, we note that different group models were used not only for calculating the radiative heat fluxes but also for solving the entire problem.

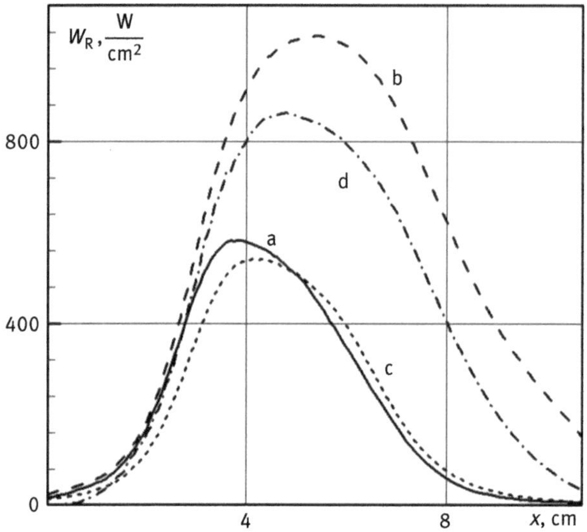

Figure 2.5: Integral radiation flux distribution along the inner generator surface for the 37-group model and $P_L = 200$ kW, $R_c = 0.24$ cm, $u_0 = 30$ m/s, and $x_p = 3$ cm: (a)–(d) – the same as in Figure 2.2.

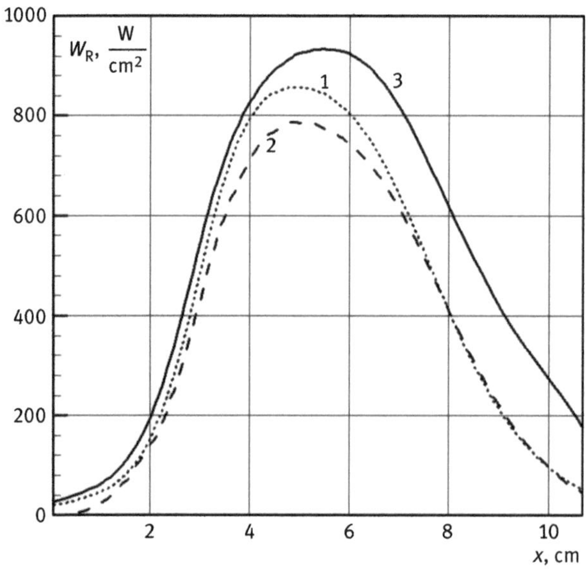

Figure 2.6: The same as in Figure 2.5 for different group models: curve 1 corresponds to 37 groups, curve 2 to 74 groups, and curve 3 to 148 groups; the input data are the same as in Figure 2.5.

In conclusion, we will consider the group radiation flux distributions at three points on the inner generator surface obtained for three group models. Figure 2.7 shows the radiation fluxes, integral over the spectral groups, which reach the inner generator surface. We note a qualitative similarity between the radiation flux spectra predicted by the three group models at each of the chosen surface points ($x = 0.4$, 4.0, and 10.7 cm). Even the roughest of these group models makes it possible to obtain a fairly complete representation of the structure of the radiation flux spectrum. Naturally, the models

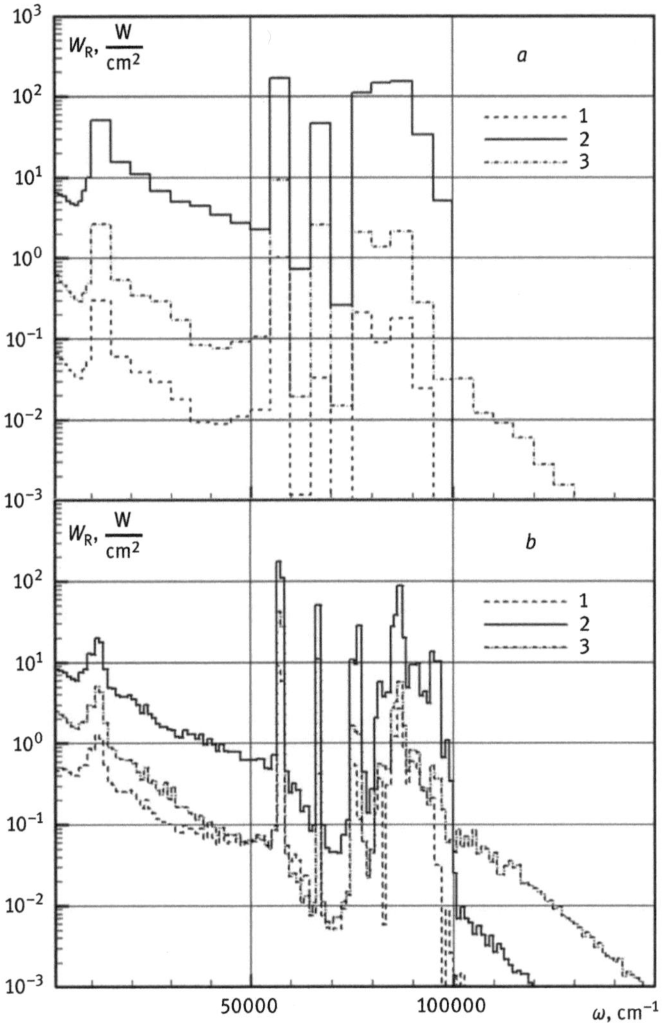

Figure 2.7: Radiation fluxes integral over the spectral groups to three parts of the generator surface: curves 1 correspond to $x = 0.4$, 4.0, and 10.7 cm for 37 (a) and 148 (b) group models; turbulent initial flow with reabsorption of thermal radiation, $P_L = 200$ kW, $R_c = 0.24$ cm, $u_0 = 30$ m/s, and $x_p = 3$ cm.

with a larger number of groups make it possible to obtain a more detailed description of these structural features.

An analysis of these group fluxes suggests that a significant fraction of the radiation emitted by the laser plasma in the ultraviolet spectrum range, primarily, by intense atomic lines reaches the inner generator surface. The radiation of the visible and near-infrared spectral regions reaching the surface is also due to atomic lines formed in quantum transitions from excited states. Therefore, as the group models solve the problem of radiation transport in atomic lines very approximately, we note the need for further development of the corresponding models.

So, the role of reabsorption of thermal radiation in the formation of the gas dynamic structure of laser plasma generators in the radiative regime was demonstrated here by comparing the thermogas dynamic parameters of generators operating in the heat conduction and radiation regimes.

Numerical experiments show that there are generator operating regimes with laminar and oscillatory gas and plasma flow regimes associated with self-oscillations of the gas flow behind the heat release region localized in the laser plasma. It is shown that initial free stream turbulence at the working chamber inlet and the reabsorption of thermal radiation suppress the large-scale flow fluctuations induced by flow past the local heat release region.

A comparative analysis of various group hot air models with reference to radiation–gas dynamic generator calculations showed that the 37-group optical model can be used for these purposes.

2.2 Hydrogen laser-supported plasma generator

Radiative gas dynamics of hydrogen laser plasma generator (see Figure 1.5) is considered in this section for the following initial conditions: the power of CW CO_2 laser is $P_L = 200$ kW, the focal length of the focusing lens is $x_f = 3$ cm, the divergence of the laser radiation is $\theta = 0.1$ mrad, the initial radius of the laser beam is $R_b = 1.0$ cm, the pressure in the LSPG channel is $p_0 = 1$ atm, the gas velocities in the input section were changed from $u_0 = 2$ to 60 m/s, and the length and radius of the LSPG channel are $L_c = 11$ cm and $R_c = 1.3$ cm.

As mentioned earlier, the investigated process develops at approximately constant pressure, which causes significant problems for a numerical solution of the task, because available disturbances of the pressure are hundreds of times less than the background pressure. However, in this case, it is quite sufficient to take into account only the temperature dependencies of the thermophysical and optical properties, which are included in the mathematical models (1.95)–(1.101). Temperature dependencies of the mentioned properties $\rho, c_p, \mu, \lambda, \mu_\omega$ and κ_g can be calculated only for the background pressure, and for the main calculations an interpolation procedure for temperature can be used. In the calculations presented here, the thermophysical and transport properties of low-temperature hydrogen plasma were borrowed from Capitelli M., et al. (2001).

Spectral and group (Figure 2.8) absorption coefficients of hydrogen plasma were calculated using the computing program ASTEROID (Surzhikov S.T., 2000). The group optical model was created by averaging the spectral absorption coefficients over 37 spectral groups. This number of spectral groups was determined by performing numerical experiments with different numbers of spectral groups (from 2 up to 300). There is a problem of optimization of the number: too low of a number provides physically unrealistic results [e.g., one can observe extinction of laser-supported waves (LSW) for some calculation cases with a low number of spectral groups], while too large a number of the spectral groups results in large computational times. Therefore, from one side the boundaries of the spectral groups (actually the spectral grid) must take into account the general peculiarity of the spectra of heat emissivity (processes of photorecombination, electronic bands of diatomic molecules, strong atomic lines, and continuous radiation processes), from the other side the spectral group model with a low number of spectral groups is favored over the spectral group model with a larger number of spectral groups. It was shown in the numerical experiments that the 37th group model is quite acceptable for the problem under consideration. Just this group model was used for calculation of radiative gas dynamics and radiation fluxes on an internal surface of the LSPG.

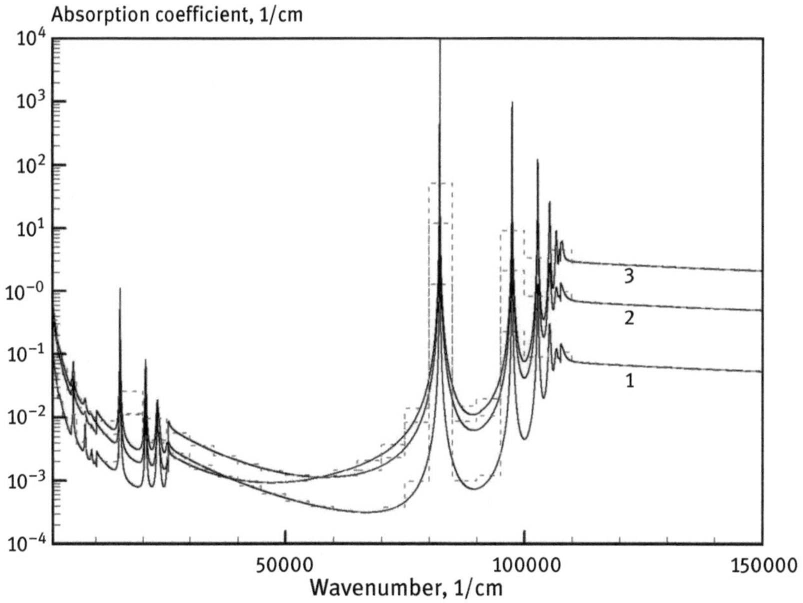

Figure 2.8: Spectral (solid line) and group absorption coefficients of a hydrogen plasma at $p = 1$ atm, $T = 20,000$ K (1), 16,000 K (2), and 12,000 K (3).

The numerical simulation results for different nonhomogeneous finite-difference meshes with compress around the high-temperature zone of the LSW and near to the wall of the cylindrical channel will be considered below. Numerical simulation results are presented for the following number of nodes along the axial and radial directions, respectively: 60 × 30, 80 × 40, 120 × 50, 240 × 100.

Let us consider the results of the following two series of calculations. First oneused the model of radiative emission (i.e., the optically thin approximation was used in this case, or the so-called volume emission approach). For an entrance velocity of $u_0 = 2$ m/s, the laser plasma was moved toward the laser and its forward front was fixed at a distance of ~2.3 cm from the channel entrance (Figures 2.9a and 2.10a). In these cases, the distributions of all thermal and gas dynamic functions quickly stabilized and did not vary with time, that is, the steady-state solution for the task was obtained. Temperature inside the LSW achieved ~24,500 K, but the plasma jet temperature was less than ~3,200 K. The temperature behavior inside the laser plasma can be explained by using the volume emission approach. Practically full absorption of the laser radiation by the plasma is observed.

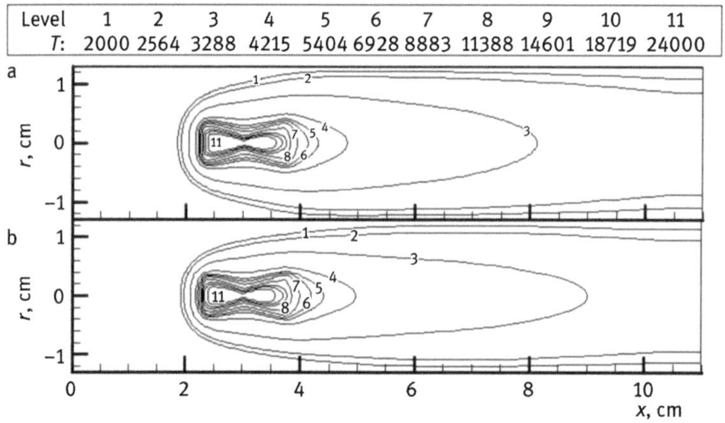

Figure 2.9: Temperature distributions in hydrogen LSPG at $u_0 = 2$ m/s (a) and $u_0 = 5$ m/s (b). The volume emission approach.

The next case studied was performed for the entrance velocity of $u_0 = 5$ m/s. Actually, this is an extreme velocity case at which the LSW can exist in a volume emission mode. A gas flow with $u_0 = 6$ m/s blows off the LSW, and the laser plasma cannot exist as the numerical solution predicts. Figures 2.9b and 2.10b show that, in this case, the high-temperature plasma is more localized than in the previous case. In this case, the maximum temperature in the LSW is also about 24,000 K and the temperature in the laser plasma jet is about 3,000 K.

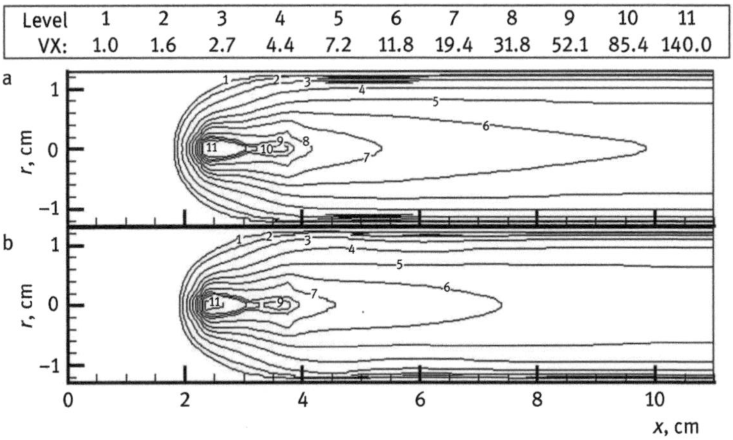

Figure 2.10: Axial velocity ($VX = u/u_0$) distributions in hydrogen LSPG at $u_0 = 2\,m/s$ (a) and $u_0 = 5\,m/s$ (b). The volume emission approach.

The numerical simulation results show extremely high cooling of the laser plasma in the case of the volume emission model. In this case, a configuration of the numerically calculated LSW follows the laser beam from near to its focusing point. It is worth noting that the volume emission approach predicts the extinction of the LSW at low entrance velocities.

The second series of calculations were performed using the full radiation heat transfer model, that is, all radiation characteristics were calculated by solving the full system of eqs. (1.95)–(1.101) by using of the 37th group model. In this series, the entrance velocity varied for significantly wide regions. Figures 2.11 and 2.12 show distributions of temperature and axial velocity for entrance velocities of $u_0 = 5$ and 50 m/s. Comparing the data presented in Figures 2.9–2.12, one can conclude that radiation heat transfer processes in the given type of LSPG are very significant, especially for the formation of the high-temperature region and for prediction of the plasma jet velocity. For the full radiation heat transfer mode, the high-temperature region of the LSW has a noticeably larger dimension. Its front is located at distance of ~1.5 cm from the entrance cross section of the LSPG channel for $u_0 = 5\,m/s$. The greatest temperature inside the LSW runs up to ~ 24,000 K. This temperature is slightly smaller than the one obtained for the volume emission mode; but due to the appreciably larger dimension of the heated region and due to reduction of total energy losses, the temperature of the plasma jet and its velocity are essentially larger. For $u_0 = 5\,m/s$, the plasma jet temperature at the exit section of the LSPG channel reaches a value of 10,800 K and the axial velocity is 250 m/s.

The maximal temperature in the plasma jet slightly decreases up to ~9,000 K by increasing the entrance velocity (see Figure 2.12). For maximum entrance velocity $u_0 = 50\,m/s$ at which the LSW exists in the conditions under consideration, the

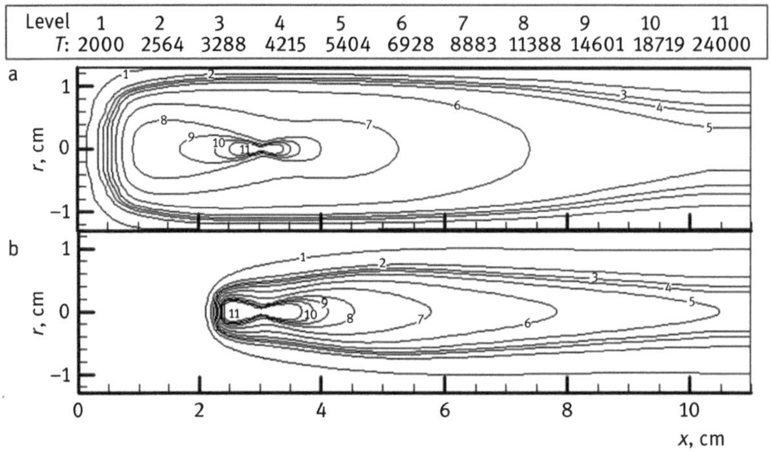

Figure 2.11: Temperature (T in K) distributions (a) in hydrogen LSPG at $u_0 = 5$ m/s (b) and 50 m/s. Full radiation heat transfer approach.

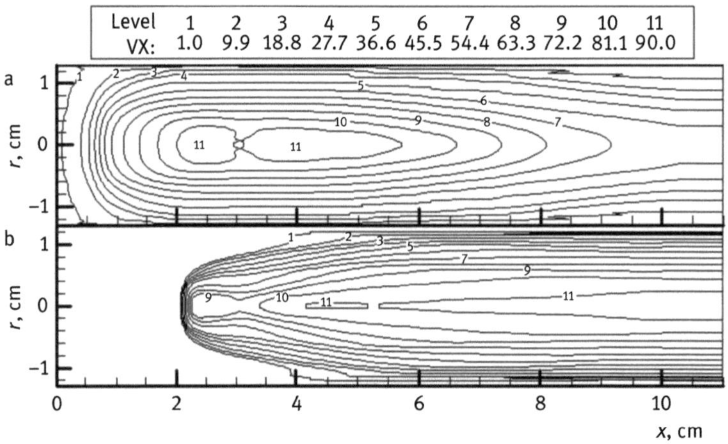

Figure 2.12: Axial velocity (VX $= u/u_0$) distributions in hydrogen LSPG at $u_0 = 5$ m/s (a) and $u_0 = 50$ m/s (b). Full radiation heat transfer approach.

plasma jet velocity achieves the value of 1,075 m/s. At velocity $u_0 = 60$ m/s the LSW is blown off by entrance hydrogen flow.

Results of the methodical study of the finite-difference method accuracy are shown in Figure 2.13, where two temperature fields obtained for two kinds of grids are presented. As mentioned earlier, to obtain a steady-state solution for the temperature field the $k-\varepsilon$ turbulent model was used for effective viscosity of gas flow inside the LSPG chamber. One can see that variation of calculation grids in the wide region

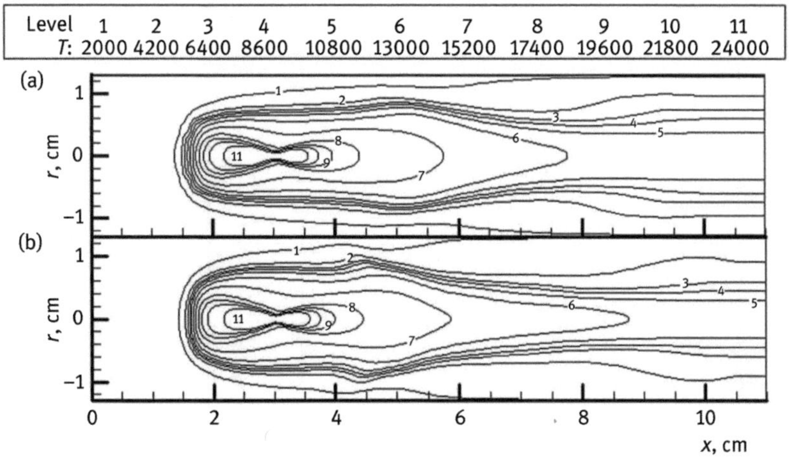

Figure 2.13: Temperature (in K) distribution with an entrance velocity of 30 m/s: (a) grid 60 × 30 and (b) grid 120 × 50.

results in some change in temperature profiles, but the general peculiarities of the thermal structure of LSW were left unchanged.

Investigation of the effect of spectral group models on numerical simulation results has shown that two spectral group models ($N_g = 37$ and 77) provide an acceptable agreement between numerical simulation results. At a lower number of spectral groups, large differences were observed in the temperature fields.

It is worth noting that the spectral group model has an especially large influence on the temperature distribution for the calculation cases when the heat radiation plays a dominant role in the energy balance inside the LSW. Sometimes, the low number of groups result in the extinction of the LSW.

Calculations of radiative heating of the internal surface of an LSPG were performed for each calculation case. The ray tracing method was used with an angular mesh of $N_\theta = N_\varphi = 21$ and $N_s = 20$, where N_s is the number of nodes along a ray. Preliminary calculations have confirmed that the radiative heat fluxes predicted by the RTM (Ray-tracing method) strongly depend on the parameters of the spatial finite-difference mesh. Numerical simulation results presented in Figures 2.14–2.16 correspond to the minimal acceptable mesh. The angular mesh was established with numerical experiments in which the number of nodes in each angular direction varied over a wide region (10–100). Figure 2.14 illustrates the dependence of the integral radiation flux distribution on the internal surface obtained by using different numbers of spectral groups. One can see that there is satisfactory convergence of the radiation flux distributions depending on the number of spectral groups. Not only the numbers of spectral group are significant here; the distribution of the wave numbers of the spectral grid is also important. Figure 2.15 gives

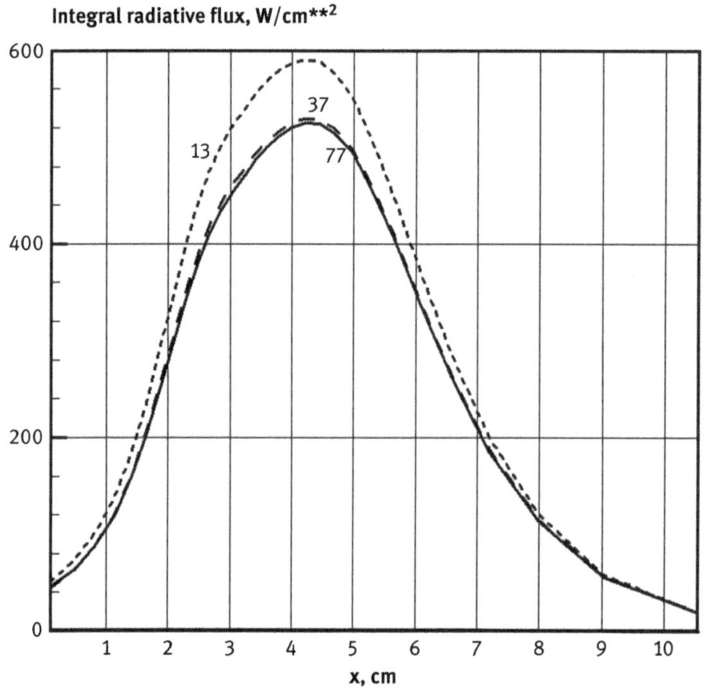

Figure 2.14: Integral radiation heat flux along the internal cylindrical surface of the hydrogen LSPG for different spectral group models (numbers near the curves designate the number of spectral groups).

representation about the used spectral grid. This figure presents spectral group radiation heat fluxes at two axial points along the internal surface of the LSPG chamber at $x = 0$ and at $x = 3.6$ cm. The spectral distribution shows that the maximum radiative heating lies in the range of the visible spectrum around ~20,000 cm^{-1}.

The tendency of decreasing radiative heating of the internal surface with increase in the entrance velocity is demonstrated in Figure 2.16. The high-temperature region of the LSW in the plasma generator channel, as well as the relatively low-temperature region, becomes smaller for increasing entrance velocities. One can also see that the region of heated gas is constricted (compare, e.g., Figures 2.11 and 2.12). These two facts lead to a decrease in the radiative heating of the internal surface. The numerical simulation results were also obtained for a power of the CW CO_2 laser of $P_L = 100$ kW and a radius of the LSPG channel of $R_c = 2.0$ cm. Temperature distributions were obtained by solving the radiative gas dynamic equations at values of entrance velocities of hydrogen flow of $u_0 = 20 - 70$ m/s. Note that for the larger channel radii, the LSW exists for higher input velocities. This can be explained by the larger mass flow, which is blowing out through the region of higher temperature of the LSW in the case of a decreased radius of the channel. It is obvious that, in this case, the conditions of

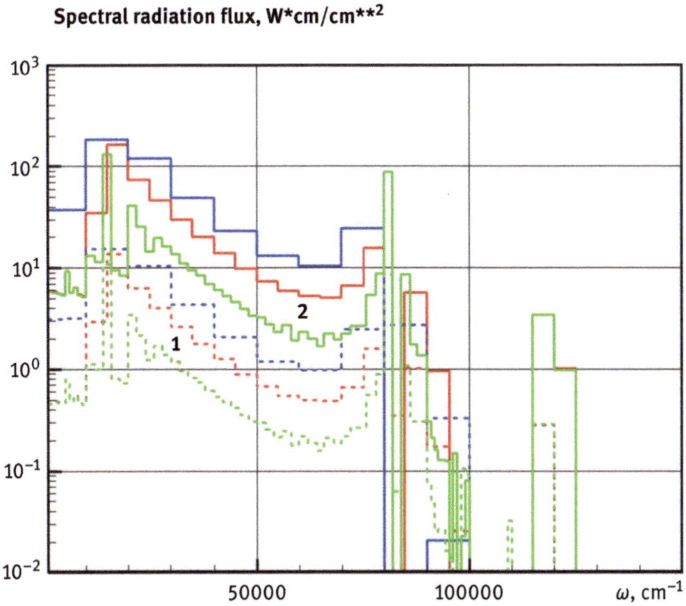

Figure 2.15: Spectral group heat flux at $x = 0$ cm (1), and $x = 3.6$ cm (2) for $N_g = 13$, 37, and 77 groups.

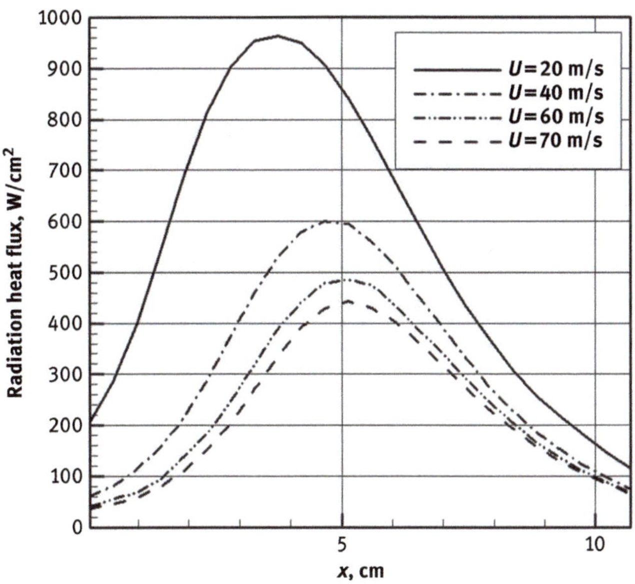

Figure 2.16: Integral radiation heat flux along internal cylindrical surface of the hydrogen LSPG, W/cm^2, at different entrance velocities. The ray tracing method.

the laser plasma cooling are much better. Therefore, the limit of the existence of the laser plasma in variables of laser power and input velocity is shifted in the direction of smaller velocities.

So, the computational model presented in this section allows the study of the hydrogen LSPG functioning in the radiative mode, that is, when there are two general mechanisms of energy exchange: absorption of laser radiation in hydrogen plasma and radiation heat transfer of the high-temperature plasma region with the ambient gas.

A numerical study has been performed for regimes of strong radiative-convection interaction. It has been shown that the distribution of the thermic and gas dynamic parameters is generally formed by the heat radiation transfer.

2.3 Bifurcation of subsonic gas flows in the vicinity of localized heat release regions

This section is dedicated to one of the little studied phenomena of subsonic gas flows. The necessity for studying the gas dynamic structure of subsonic gas flows through localized heat release areas arises in connection with the research of LSW. Certain kinds of energy devices are based on the specified phenomenon: the LSPG (Surzhikov S.T., 2005) and the laser-supported rocket engines (Mirabo L., et al., 1995; Surzhikov S.T., 1995). The interest to problems of gas flow interactions with localized heated areas has increased noticeably in physical gas dynamics recently, first of all in connection with the probable applications in practical aerodynamics.

This study concentrates basically on physical regularities of subsonic gas motion through localized heat release areas. The problem considered in the following is characterized by a number of distinctive features:

1. The study of the unsteady subsonic motion of viscous heat conducting and radiating gas in an area of heat release is based on the full system of the Navier–Stokes equations. Thus, specified nonsteady-state modes can be caused not only by the external reasons but also by own properties of gas flows through the heat release area.
2. The temperature in the heat release areas can achieve $\sim 5,000 - 20,000\,\mathrm{K}$, that is, the gas becomes completely dissociated or ionized. Differences in gas densities in the region under consideration may reach ~ 200 times because the pressure in this area only marginally differs from the atmospheric one.
3. The radiative heat transfer and real thermophysical and transport properties are also taken into account.

A heat release volume is considered in the present section as being fixed in space, that is, any changes of gas dynamic parameters do not influence on the heat release emission. In reality when the gas is heated by the LSW, the situation is slightly different: the heat release capacity depends rather strongly on the distribution of gas dynamic parameters. However, in the latter case, it is difficult to specify the reason for the occurrence of unsteady movements: whether on account of its internal properties, or whether on account of periodic change of the heat release configuration. In other words, the fixed heat release volume allows exclude the influence of variability in the heat release area and, thus, to study the regularity of occurrence of the unsteadiness caused by own properties of the gas flow.

2.3.1 Statement of the task

The problem schematic is shown in Figure 2.17.

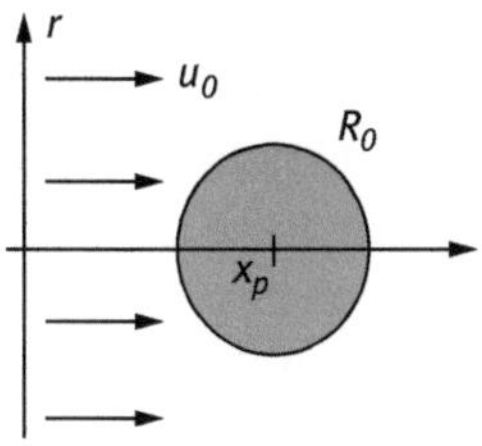

Figure 2.17: Schematic of the problem. The heat release region has a spherical shape. Typical velocities are $u_0 = 10 - 200$ m/s; typical temperatures inside the heat release region are $T = 1,000 - 20,000$ K.

On the symmetry axis ($r = 0$) in a point x_p there is a hot area with the given distribution of heat release energy:

$$Q_V = \frac{3Q_0}{4\pi R_0^3} \exp\left[- \left(\frac{r}{R_0}\right)^4 - \left(\frac{x - x_p}{R_0}\right)^4 \right], \tag{2.15}$$

where R_0 is the radius of the heat release region, x_p is the axial coordinate of its center, Q_0 is the heat release power. Note that such form practically coincides with heat release distributions received in a self-consistent model of LSW (see, e.g., Sections 2.1 and 2.2). Again, it should be stressed that in the case under consideration the distribution of heat release capacity (eq. (2.15)) is determined only by spatial variables and depends neither on time nor from gas dynamic processes.

Parameters of undisturbed gas were used at the entrance of the area under consideration: velocity u_0 and temperature T_0.

For numerical simulation of subsonic gas flows through heat release regions, the following system of the Navier–Stokes equations, mass, and energy conservation equations, and also the equation of radiative heat transfer in the form of the P_1-multigroup approximation is used

$$\frac{\partial \rho}{\partial t} + \text{div}(\rho \mathbf{V}) = 0, \tag{2.16}$$

$$\frac{\partial \rho u}{\partial t} + \text{div}(\rho u \mathbf{V}) = -\frac{\partial p}{\partial x} - \frac{2}{3}\frac{\partial}{\partial x}(\mu \,\text{div}\mathbf{V}) + \frac{1}{r}\frac{\partial}{\partial r}\left[r\mu\left(\frac{\partial u}{\partial r} + \frac{\partial v}{\partial x}\right)\right] + 2\frac{\partial}{\partial x}\left(\mu\frac{\partial u}{\partial x}\right), \tag{2.17}$$

$$\frac{\partial \rho v}{\partial t} + \text{div}(\rho v \mathbf{V}) = -\frac{\partial p}{\partial r} - \frac{2}{3}\frac{\partial}{\partial r}(\mu \,\text{div}\mathbf{V}) + \frac{\partial}{\partial x}\left[\mu\left(\frac{\partial u}{\partial r} + \frac{\partial v}{\partial x}\right)\right] + 2\frac{\partial}{\partial r}\left(\mu\frac{\partial v}{\partial r}\right) + \mu\frac{\partial}{\partial r}\left(\frac{v}{r}\right), \tag{2.18}$$

$$\rho c_p \frac{\partial T}{\partial t} + \rho c_p \mathbf{V}\text{grad}T = \text{div}(\lambda\,\text{grad}T) - Q_{\text{HR}} + Q_{\text{V}}, \tag{2.19}$$

$$Q_{\text{HR}} = \sum_{g=1}^{N_g} k_g\left(U_{b,g} - U_g\right)\Delta\omega_g, \tag{2.20}$$

$$\text{div}\left(3k_g^{-1}\text{grad}\,U_g\right) = -k_g\left(U_{b,g} - U_g\right), \quad g = 1, 2, \ldots, N_g, \tag{2.21}$$

where x, r are the radial and axial coordinates; ρ, c_p, T are the density, specific heat capacity at constant pressure and temperature; u, v are the axial and radial components of the flow velocity $\mathbf{V}$; p is the pressure; μ, λ are the coefficients of viscosity and thermal conductivity; Q_{HR} is the volume capacity due to radiation heat transfer; k, U, U_b are the absorption coefficient and radiation volume density of the medium and absolute blackbody. Subscript g indicates group properties as obtained by averaging the appropriate spectral characteristics within each of N_g spectral ranges of wave numbers $\Delta\omega_g$.

The validity of the local thermodynamic equilibrium (LTE) is assumed. The gas composition (air in this case) is considered at chemical equilibrium in each point of the calculation area at given temperature and pressure. Because of small speeds of the gas, the energy conservation equation does not contain the term representing heat release due to gas compressibility. Temperature dependence of the thermophysical (ρ, c_p), transport (μ, λ), and optical (k_g) properties of air are used only at atmospheric pressure, as their change dependence on pressure is insignificant.

At the initial time instant, a Gaussian temperature distribution with maximal temperature 2,000 K is set.

The following boundary conditions are used:

at $x = 0$: $V = (u = u_0,\ v = 0)$, $T = T_0$;

at $x = L$ $(x \to \infty)$: $\partial f/\partial x = 0$ or $\partial^2 f/\partial x^2 = 0$;

at $r = 0$: $\partial f/\partial r = 0$;

at $r = R_c$ $(r \to \infty)$: $f = f_0$ or $\partial f/\partial r = 0$;

where $f = \{u, v, T, U_g\}$.

The values x_p, L, R_c are chosen in numerical experiments from the conditions of weak influence of the boundaries of the considered area on the calculation results in the vicinity to the heat release area.

We need not formulate boundary conditions for pressure, because one is excluded from the consideration. The method of the unsteady dynamical variables was applied to the solution of the problem. Thermophysical and group optical properties of low-temperature air plasma were calculated using ASTEROID computing system (Surzhikov S.T., 2000a).

2.3.2 Qualitative analysis of the phenomenon

Essentially, the problem is as follows: if at the fixed undisturbed gas velocity, for example at $u_0 = 100\,\mathrm{m/s}$, we gradually increase heat release power Q_V, each time using the just obtained solution for thermogas dynamic functions, then a vortical gas motion (stationary or nonstationary solutions) may arise behind the heat release area at some value of Q_V^*. If then from some value $Q_V > Q_V^*$ we start to reduce gradually the value Q_V, then one can find that the vortical gas dynamic structure is preserved at $Q_V < Q_V^*$ down to certain value $Q_V^{**} < Q_V^*$, at which the gas flow once again becomes laminar. In other words, we can say about such well-known phenomenon as hysteresis. But from the point of view of computational fluid dynamics one can say about the very significant fact that there is a certain range $Q_V^{**} < Q_V < Q_V^*$, where two qualitatively different gas dynamic structures correspond to the same given data (u_0, Q_V, and other invariable entrance data).

The specified gas dynamic structures with vortical movement can be observed in a wide range of velocities $u_0 \sim 30 - 200\,\mathrm{m/s}$.

One more fact may be of interest too. One of the two obtained solutions, namely one with vortical motion, is steadier. The numerical experiments have shown that if the laminar solution is chosen as the initial one, and some indignation is introduced into the flow (e.g., change of Q_V by a value ΔQ_V within the limits $[Q_V^{**}, Q_V^*]$), then the following two results may be obtained:

- at small values ΔQ_V, the numerical solution does not leave the initial branch of the solutions, that is, the flow remains laminar;
- at large enough ΔQ_V, the numerical solution always converges to the alternative configuration, that is, becomes vortical.

If the vortical solution is chosen as the initial one, then indignations of gas dynamic functions do not result in a change in the gas dynamic pattern. It is also necessary to take into account three additional features of the obtained results:

- At speeds $u_0 \leq 20\,\mathrm{m/s}$, the bifurcation of the gas dynamic structure was not revealed. However, it should be stressed that it is impossible to interpret this conclusion as an absolute because the calculations were carried out at fixed entrance parameters (pressure, geometry, size of the heat release region, etc.).

- Near the border of the conditional stability, that is, at $Q_V \approx Q_V^*$ (the boundary of the transition «laminar flow – vortical flow»), the unsteady solutions are as a rule observed.
- With increasing ΔQ_V, reduction in the range $[Q_V^{**}, Q_V^*]$ in computing experiments is observed.

2.3.3 Quantitative results of numerical simulation

Calculations were carried out for the following entrance speeds $u_0 = 10 - 200\,\text{m/s}$, and for fixed radial and axial coordinates of the heat release region: $R_0 = 0.4\,\text{cm}$, $x_p = 3\,\text{cm}$. Maximal value of the heat release power Q_V varied in the range 2–20 kW/cm^3. The greatest value Q_V was limited by the maximal temperature inside the heated area ($T \sim 25{,}000\,\text{K}$).

First, we will consider numerical simulation results obtained step by step by increasing the entrance velocity u_0. The computed results obtained for conditions corresponding to stability boundaries of the gas flows will then be considered in detail.

Temperature and axial velocity distributions at $u_0 = 10\,\text{m/s}$ and $Q_V = 14.6\,\text{kW/cm}^3$ are shown in Figure 2.18.

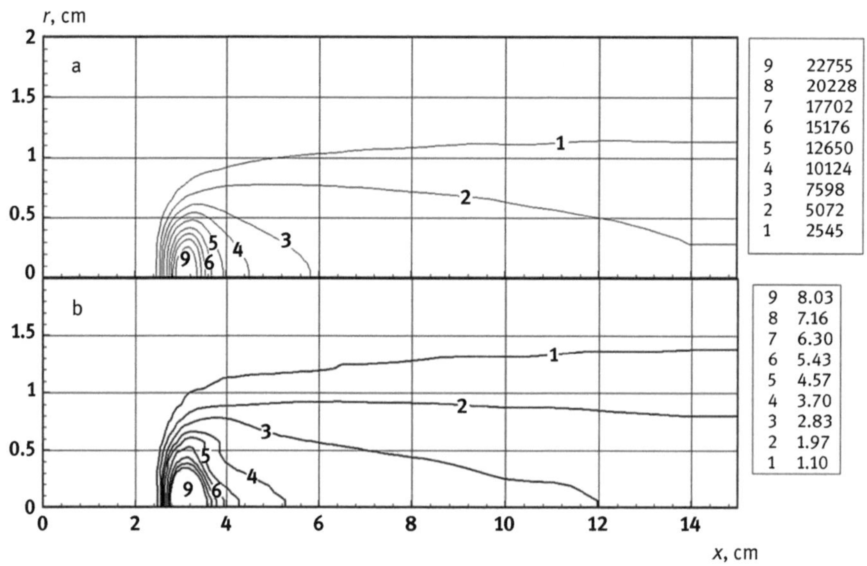

Figure 2.18: Temperature in K (a) and axial velocity u/u_0 (b) distributions at $u_0 = 10\,\text{m/s}$ and $Q_V = 14.6\,\text{kW/cm}^3$. Set of calculations with *increasing* heat release.

As mentioned earlier, the transition from laminar to a vortical mode of a gas dynamic structure is not detected in this case. A maximal temperature inside the heated region achieves approximately 24,000 K. It should be emphasized that the numerical solution, in this case, is not completely steady state, and small periodic fluctuations of the velocity and temperature fields are observed. The steady-state solution was obtained at essentially smaller heat release power $Q_V \sim 2-5\ \mathrm{kW/cm^3}$.

At entrance velocity higher than $u_0 = 30\ \mathrm{m/s}$, the phenomena of bifurcation in the flow were detected, but numerical solutions obtained at this velocity were not stable. Detailed discussion of these results will be presented slightly later. First, we will consider calculated data obtained at velocities $u_0 = 40-200\ \mathrm{m/s}$, because these calculated data are stable enough. Temperature and velocity fields for the case of stationary laminar ($Q_V = 2.5\ \mathrm{kW/cm^3}$) and quasi-steady-state vortical flow ($Q_V = 3.0\ \mathrm{kW/cm^3}$) at $u_0 = 40\ \mathrm{m/s}$ are shown in Figures 2.19 and 2.20, respectively. The name "quasi-steady state" is used here to emphasize that the solution obtained is not perfectly stationary, but rather contains moderate velocity oscillations (Figure 2.21).

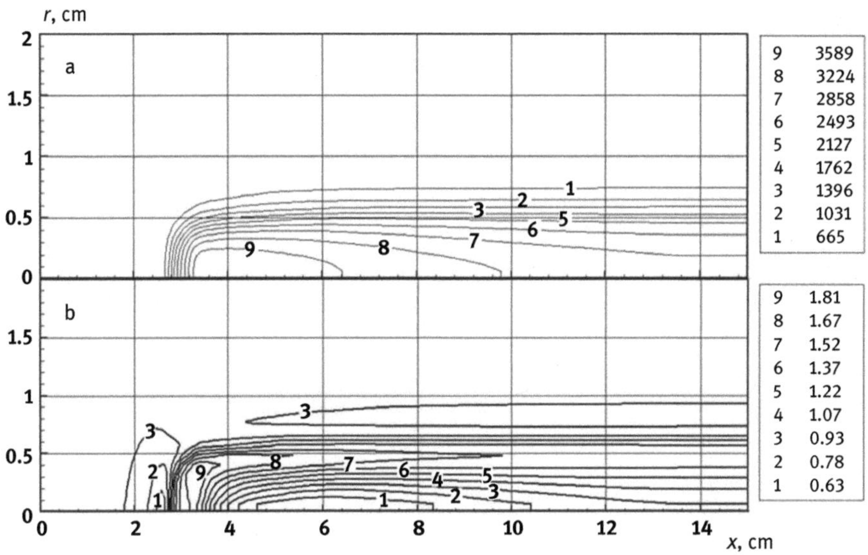

Figure 2.19: Temperature in K (a) and axial velocity u/u_0(b) distributions at $u_0 = 40\ \mathrm{m/s}$ and $Q_V = 2.5\ \mathrm{kW/cm^3}$. Set of calculations with *increasing* heat release.

The specified solutions were obtained in the calculations, where Q_V was increased with step $\Delta Q_V = +0.5\ \mathrm{kW/cm^3}$. The maximal temperature inside the heat release area achieves 5,200 K.

Numerical simulation results shown in Figures 2.22 and 2.23 were obtained for the case with gradual reduction of Q_V with the step $\Delta Q_V = -0.5\ \mathrm{kW/cm^3}$, when the vortical solution was taken as the initial one. Figure 2.22 shows temperature and velocity distributions at $Q_V = 2.25\ \mathrm{kW/cm^3}$, and Figure 2.23 shows the data at

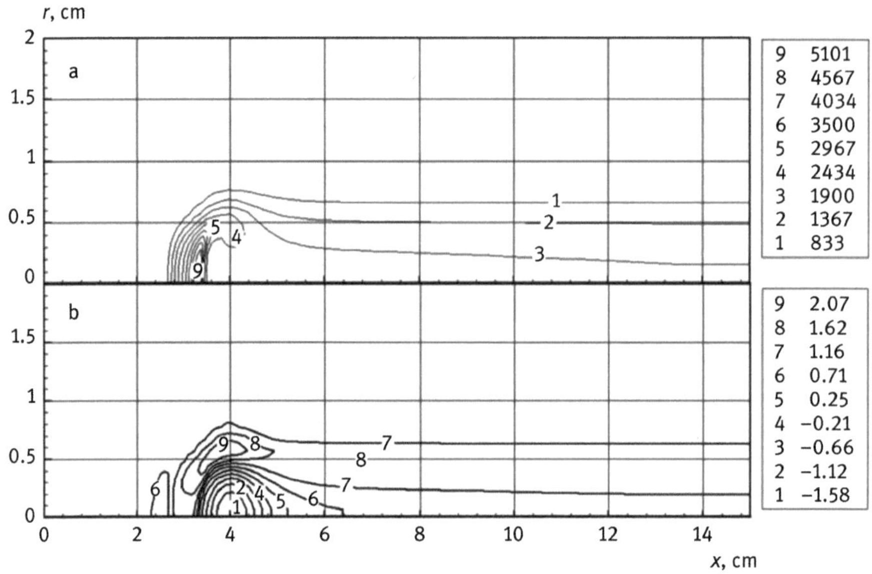

Figure 2.20: Temperature in K (a) and axial velocity distributions u/u_0 (b) at $u_0 = 40$ m/s and $Q_V = 3.0$ kW/cm^3. Set of calculations with *increasing* heat release.

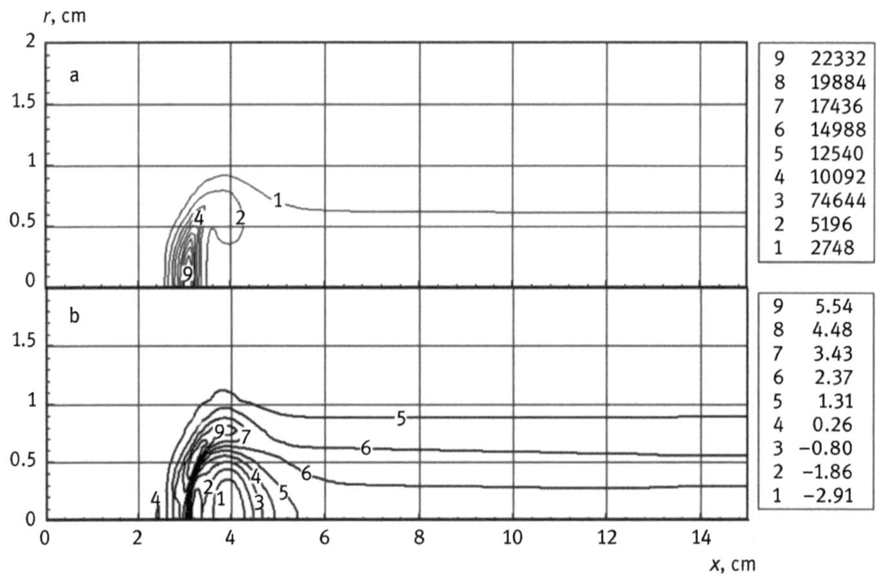

Figure 2.21: Temperature in K (a) and axial velocity distributions u/u_0(b) at $u_0 = 40$ m/s and $Q_V = 21.0$ kW/cm^3. Set of calculations with *increasing* heat release.

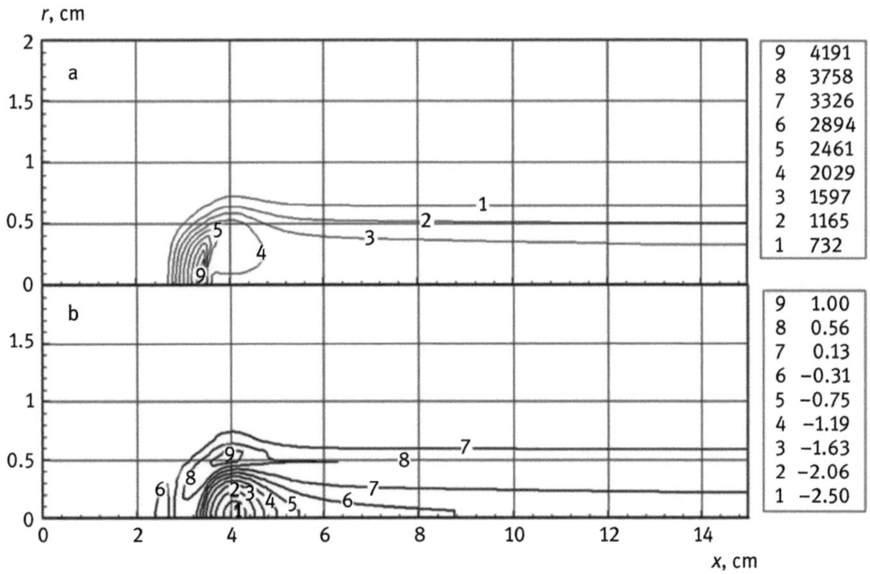

Figure 2.22: Temperature in K (a) and axial velocity distributions u/u_0(b) at $u_0 = 40$ m/s and $Q_V = 2.25$ kW/cm^3. Set of calculations with *decreasing* heat release.

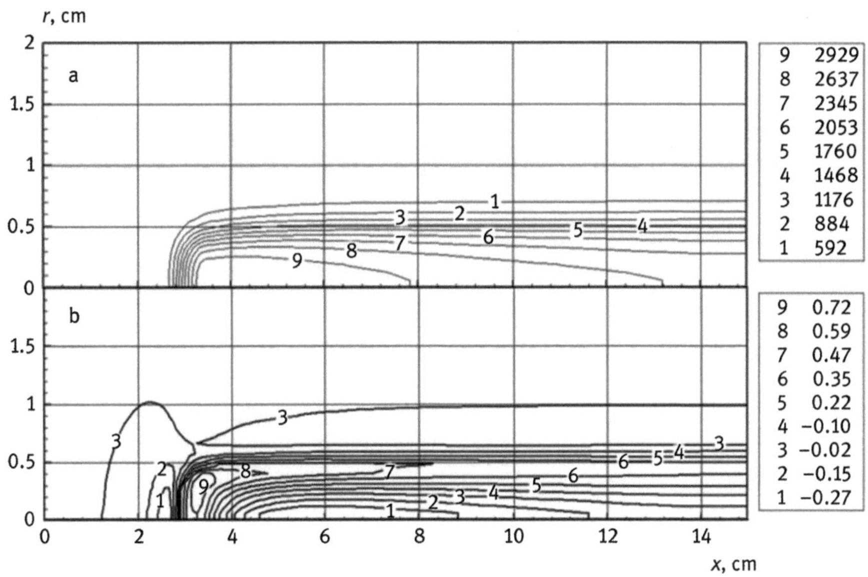

Figure 2.23: Temperature in K (a) and axial velocity distributions u/u_0(b) at $u_0 = 40$ m/s and $Q_V = 2.0$ kW/cm^3. Set of calculations with *decreasing* heat release.

$Q_V = 2.0$ kW/cm^3. Note that in the set of calculations with increasing Q_V the laminar numerical solution was obtained at $Q_V = 2.5$ kW/cm^3. Therefore, at the considered conditions it is possible to specify the range of values of Q_V inside which there is the bifurcation of the solution: $u_0 = 40$ m/s, $Q_V^{**} = 2.25 < Q_V < Q_V^{*} = 2.5$ kW/cm^3. At other speeds, these bifurcational ranges are found to be:

a) $u_0 = 50$ m/s, $Q_V^{**} = 2 < Q_V < Q_V^{*} = 3$ kW/cm^3;
b) $u_0 = 100$ m/s, $Q_V^{**} = 3.5 < Q_V < Q_V^{*} = 5.5$ kW/cm^3;
c) $u_0 = 150$ m/s, $Q_V^{**} = 5.5 < Q_V < Q_V^{*} = 7$ kW/cm^3,
d) $u_0 = 200$ m/s, $Q_V^{**} = 7.5 < Q_V < Q_V^{*} = 9$ kW/cm^3.

Numerical simulation results for $u_0 = 30$ m/s are of particular interest for the analysis of unsteady gas dynamic structures. At smaller speeds, the phenomena of a flow bifurcation were not revealed. The specified speed $u_0 = 30$ m/s is near the bottom border of the range of speeds, within which the bifurcation was found. This case is characterized by significant instability in the calculated data. Therefore, the calculations at this specified speed were performed with different values of ΔQ_V.

In the first group of calculations, the step of increasing the heat release power was $\Delta Q_V = 0.5$ kW/cm^3; in the second group, it was$\Delta Q_V^{II} = 0.1$ kW/cm^3; and in the third group, it was $\Delta Q_V^{III} = 0.025$ kW/cm^3.

The transition from laminar to the vortical motion was observed in the range $Q_V = 1.5 - 2$ kW/cm^3 in the first case and in the range $-Q_V = 3.7 - 3.8$ kW/cm^3 in the second case. In the third case, the transition from laminar to the vortical movement was found in the range $Q_V = 5.925 - 5.95$ kW/cm^3. Figure 2.24 shows temperature and velocity distributions for laminar mode at $u_0 = 30$ m/s and $Q_V = 3.7$ kW/cm^3, and Figure 2.25 shows the same data for the vortical mode at $u_0 = 30$ m/s and $Q_V = 3.8$ kW/cm^3.

Figure 2.26 shows the axial velocity distribution for laminar and vortical numerical solutions obtained with increasing heat release power. It indicates that reducing the level of change value of heat release disturbance ΔQ_V to numerical solution allows getting the transition from laminar to vortical flow at essentially larger values of the heat release powerful. It is reasonable that the gas is heated up to the large temperatures. For example, at the greatest heat release energy $Q_V = 5.925$ kW/cm^3, at which the laminar numerical solution was observed, the temperature inside the heated area achieves $T = 10,000$ K. The distributions of temperature and axial velocity u for this case are shown in Figure 2.24. Appropriate distributions of the functions after the transition to the vortical motion (at $Q_V = 5.95$ kW/cm^3) are shown in Figure 2.25. Maximal temperature in this case reaches 12,000 K.

The discussed case of calculations is also remarkable since the steady-state laminar solution does not exist at $Q_V = 5.925$ kW/cm^3. The self-oscillatory process with periodic variations of velocity components and the temperature is observed in this case. Partly, it can be seen from distributions of temperature and speed (see Figure 2.24), where an instant photo is shown.

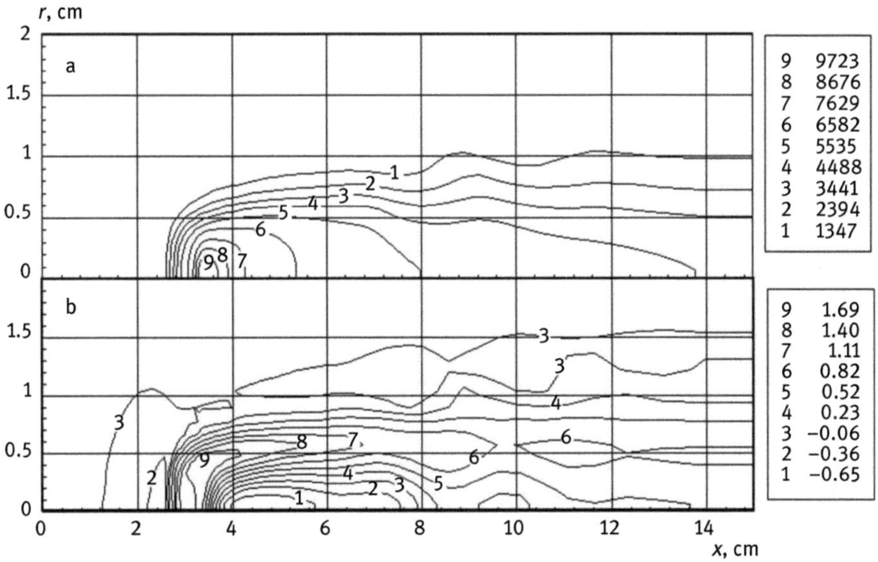

Figure 2.24: Temperature in K (a) and axial velocity distributions u/u_0 (b) at $u_0 = 30$ m/s and $Q_V = 5.925$ kW/cm^3. Set of calculations with *increasing* heat release ($\Delta Q_V = +0.025$ kW/cm^3).

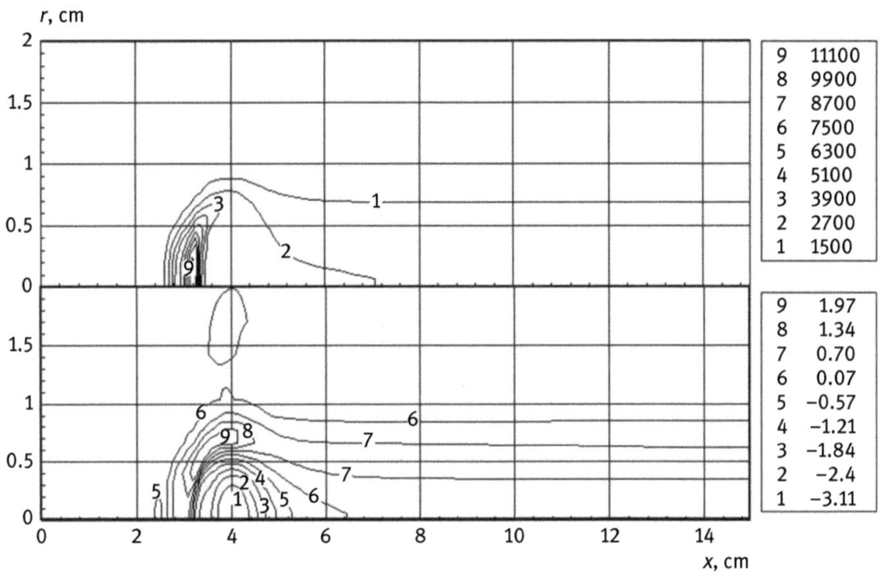

Figure 2.25: Temperature in K (a) and axial velocity distributions at u/u_0(b) $u_0 = 30$ m/s and $Q_V = 5.95$ kW/cm^3. Set of calculations with *increasing* heat release ($\Delta Q_V = +0.025$ kW/cm^3).

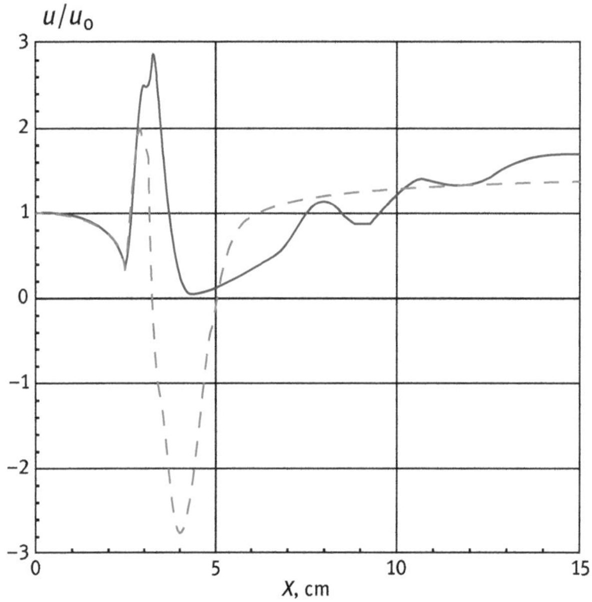

Figure 2.26: Axial velocity distributions at $r = 0$; $u_0 = 30$ m/s and $Q_V = 5.925$ (solid line) and 5.95 (dashed line) kW/cm^3. Calculations with *increasing* heat release.

So, the numerical investigation of subsonic flows with localized heat release regions showed that at certain conditions in gas flows, it is possible to detect a bifurcation of gas dynamic structure, namely, two different quasi-steady-state gas dynamic configurations at the same initial data do exist.

2.4 LSW in view of the gravity

This part deals with the late stage of optical breakdown of gas by laser radiation in the vicinity of a solid surface, where the breakdown-induced shockwave departed from the region of almost instantaneous energy release at a significant distance, where a heated plasma exists practically at atmospheric pressure. The scheme of the problem is given in Figure 2.27. It is assumed that the gas is quiescent at the site of optical breakdown. The approximation of LTE is applied, because the process develops at atmospheric pressure, and the characteristic times of development of the process (milliseconds) significantly exceed the times of relaxation to local equilibrium. The gas is taken to be vicious, heat-conducting, and selectively emitting and absorbing the radiation. As discussed before in Part 1, the LTE approximation enables one to pretabulate the data on the thermal, transport, and spectral optical properties of laser plasma in the entire temperature range of interest to us (300–20,000 K). The gas is assumed to

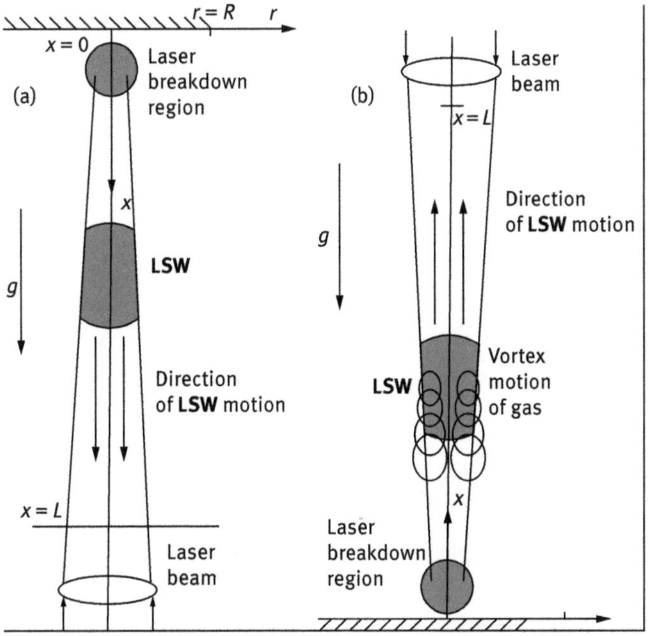

Figure 2.27: Schematic of laser-supported waves in gravity field.

be under normal conditions and quiescent at a significant distance from the site of formation of initially heated plasma cloud. Numerical analysis of regularities of propagation or steady existence of laser-supported combustion waves for the experimental conditions of Klosterman E.L., et al. (1974) was performed in Jackson J.P., et al. (1974) and Surzhikov S.T. (2000b). The distinctive element of the problem being solved in the present study is the inclusion of the direction of gravitational acceleration with respect to the direction of the laser beam. Because the problem is solved in a two-dimensional axisymmetric formulation, only two combinations shown in Figure 2.27 are possible, namely, the vector of gravitational acceleration and the laser beam have the same or opposite directions. It is their relative orientation that will define the special features of convective motion arising behind the laser-supported combustion wave, including gas dynamic instabilities (conventionally shown in Figure 2.27).

The following radiative-gas dynamic model is employed for solving the problem on motion of slow laser-supported combustion waves:

$$\frac{\partial \rho}{\partial t} + \operatorname{div}(\rho \mathbf{V}) = 0, \tag{2.22}$$

$$\frac{\partial \rho u}{\partial t}+\operatorname{div}(\rho u\mathbf{V})=-\frac{\partial p}{\partial x}-\frac{2}{3}\frac{\partial}{\partial x}(\mu\operatorname{div}\mathbf{V})+\frac{1}{r}\frac{\partial}{\partial r}\left[\mu r\left(\frac{\partial u}{\partial r}+\frac{\partial v}{\partial x}\right)\right]$$

$$+2\frac{\partial}{\partial x}\left(\mu\frac{\partial u}{\partial x}\right)+2\frac{\partial}{\partial x}\left(\mu\frac{\partial u}{\partial x}\right)+\rho\mathbf{g}, \tag{2.23}$$

$$\frac{\partial \rho \mathrm{v}}{\partial t}+\operatorname{div}(\rho v\mathbf{V})=-\frac{\partial p}{\partial r}-\frac{2}{3}\frac{\partial}{\partial r}(\mu\operatorname{div}\mathbf{V})+\frac{\partial}{\partial x}\left[\mu\left(\frac{\partial u}{\partial r}+\frac{\partial v}{\partial x}\right)\right]$$

$$+2\frac{\partial}{\partial r}\left(\mu\frac{\partial v}{\partial r}\right)+\frac{2\mu}{r}\frac{\partial v}{\partial r}-2\mu\frac{v}{r^2}, \tag{2.24}$$

$$\frac{\partial \rho H}{\partial t}+\operatorname{div}(\rho\,\mathbf{V}H)=\operatorname{div}\left(\frac{\lambda_t}{c_p}\operatorname{grad}H\right)+Q_L-Q_R, \tag{2.25}$$

$$Q_R=\int_0^\infty k_\omega\left(U_{b,\omega}-U_\omega\right)\mathrm{d}\omega\approx\sum_{g=1}^{N_g}k_g\left(U_{b,g}-U_g\right)\Delta\omega_g, \tag{2.26}$$

$$\operatorname{div}\left(\frac{1}{3k_g}\operatorname{grad}U_g\right)=k_g\left(U_g-U_{b,g}\right),\quad g=1,2,\ldots,N_g, \tag{2.27}$$

$$Q_\mathrm{L}=k_{\mathrm{L},\omega}P_\mathrm{L}\exp\left(-\frac{r^2}{R_\mathrm{L}^2}\right)\exp\left(-\int_0^x k_{\mathrm{L},\omega}\mathrm{d}x\right)\frac{1}{\pi R_\mathrm{L}^2}, \tag{2.28}$$

where u and v are projections of the velocity vector $\mathbf{V}$ onto the x- and r-axes of cylindrical coordinates; p is the pressure; ρ and c_p denote the density and specific heat capacity at constant pressure, respectively; μ is the dynamic viscosity; λ_t is the thermal conductivity coefficient; k, U, and U_b denote the volume absorption coefficient, the energy density of the medium, and the energy density of blackbody, respectively; the subscript ω indicates the spectral characteristics, and the subscript g indicates the group spectral characteristics obtained by averaging of the spectral optical properties in each of $N_g = 37$ spectral ranges $\Delta\omega_g$ into which the spectral range of thermal radiation 1,000–500,000 cm^{-1} is divided; $k_{\mathrm{L},\omega}$ is the volume absorption coefficient of laser radiation; $H = \int c_p \mathrm{d}T$ is the enthalpy; T is the temperature; and $\mathbf{g}$ is the vector of acceleration of gravity.

The presence of Gaussian distribution of radiation intensity along the beam radius R_L is assumed in eq. (2.28). The following relation

$$W_\mathrm{L}=\frac{P_\mathrm{L}}{\pi R_\mathrm{L}}$$

will characterize, in what follows, the flux density of laser radiation of power P_L.

Because the approximation of LTE in laser-supported combustion wave is assumed to be valid, it is thereby assumed that the zone of nonequilibrium ionization, which exists in the immediate vicinity of the high-temperature region because of intense ultraviolet radiation of the LSW proper, does not make a strong impact on the regularities of development of gas dynamic processes. This assumption may be partly validated by the fact that the previous calculations of the velocity of propagation of laser-supported combustion waves in the LTE approximation were indicative of the adequate description of experimental data. Nevertheless, this assumption calls for thorough verification in the future.

The boundary conditions are formulated as follows: on the free boundaries of the computational domain, the gas is taken to be immovable under atmospheric conditions (in what follows, the parameters under normal conditions are marked by subscript 0); the conditions of zero velocity and constant temperature are used on the surfaces marked by hatching (see Figure 2.27); the conditions of symmetry are employed at $r = 0$. Boundary conditions that introduce no disturbances in the flow field are applied downstream from the computational domain. Special computational experiments were performed for confirming this.

The mathematical formulation of the boundary conditions is as follows:

$$r=0:\ \frac{\partial u}{\partial r}=0,\ v=0,\ \frac{\partial \rho}{\partial r}=0,\ \frac{\partial T}{\partial r}=0;$$

$$r\to\infty:\ u=0,\ v=0,\ \rho=\rho_0,\ T=T_0;$$

$$x=0\,(\text{scheme}''\text{A}''):\ \frac{\partial u}{\partial x}=\frac{\partial v}{\partial x}=\frac{\partial T}{\partial x}=0;$$

$$x=H\,(\text{scheme}''\text{A}''):\ u=v=0,\ T=T_0;$$

$$x=0\,(\text{scheme}''\text{B}''):\ u=v=0,\ T=T_0;$$

$$x=H\,(\text{scheme}''\text{B}''):\ \frac{\partial u}{\partial x}=\frac{\partial v}{\partial x}=\frac{\partial T}{\partial x}=0.$$

Atmospheric air is considered in the calculations. The thermal, transport, and group optical properties of high-temperature air ($\rho, c_p, \mu, \lambda_t, k_g$, and k_L, ω) are given in tabulated form at atmospheric pressure in the temperature range from 300 to 20,000 K. The ASTEROID computer code is used for the calculation of spectral optical characteristics. The thermodynamic and transport properties are borrowed from Giordano D., et al. (1994).

It should be emphasized that the modes of motion of slow LSW under consideration correspond to strong radiative-gas dynamic interaction, and the thermal radiation of laser plasma has a decisive impact on the propagation of LSW. This is the so-called radiative mode of LSW motion.

The calculations were performed for CO_2 laser with wavelength $\lambda = 10.6\ \mu$m. In this case, the principal mechanism of absorption of laser radiation is by transferring the energy of electromagnetic radiation to electrons. This heating turns out to be very significant so that the electrons transmit the energy to heavy particles by colliding with the latter and heating them up to dissociation and ionization, at which new electrons are produced. In accordance with the mechanism inverse to bremsstrahlung, the spectral coefficient of absorption of laser radiation at wavelength $\lambda = 10.6\,\mu\text{m}$ is calculated by the formula

$$\mu_\omega = 6.435 p_e^2 \left(\frac{T}{10^4}\right)^{-7/2} \cdot \ln\left[27\left(\frac{T}{10^4}\right)^{4/3} p_e^{-1/3}\right]$$

with electron pressure p_e (in atm). For calculating this pressure (in the employed assumptions of LTE), one must use the Saha equation

$$\frac{x_e^2}{1 - x_e} = 6700 \frac{g_i}{g_e p} \left(\frac{T}{10^4}\right)^{5/2} \exp\left(-\frac{I}{kT}\right), \quad x_e = \frac{p_e}{p},$$

where x_e is the mole fraction of electrons; g_i and g_e denote the statistical weight of ions and electrons, respectively; $I$is the ionization potential; p is the gas pressure in atm.

The following initial numerical values were used: $u_0 = \sqrt{981L} = 31.3\,\text{cm/s}$, $L = 1.0\,\text{cm}$, $\rho_0 = 1.177 \times 10^{-3}\text{g/cm}^3$, $\lambda_0 = 1.983 \times 10^{-4}\text{W/(cm K)}$, $c_{p,0} = \lambda_0/\mu_0$.

In the present section, numerical simulation is used to investigate the regularities of the emergence of gas motion in the region of motion of laser-supported combustion wave, which is defined by the presence of the force of gravity. Two versions of relative orientation of laser beam and acceleration of gravity, which were realized in the calculations, demonstrate a fundamentally different gas dynamic structure of LSW. But the main feature is that this phenomenon is observed only at certain values of density of laser radiation.

The calculations were performed in the power density range of laser radiation $W_L = 0.06 - 0.29\ \text{MW/cm}^2$ with a fixed radius of laser beam $R_L = 1.05$ cm. This is the maximal transverse dimension of the beam in the experiments (Klosterman E.L., et al., 1974), which guarantees the radiative mode of propagation of LSW.

At the flux density of laser radiation $W_L < 0.0155\ \text{MW/cm}^2$, the laser plasma in the calculations was extinguished. At $W_L > 0.29\ \text{MW/cm}^2$, the plasma temperature exceeded 20,000 K, and the computational model employed became incorrect (first of all, because of the need for inclusion of double ionization of gas and for correction of the employed thermal and optical properties).

We will consider the results of calculation of the dynamics of LSW in a laser beam of radius $R_L = 1.05$ cm for different values of power of laser radiation $P_L = 0.06$, 0.08, 0.12, 0.207, 0.346, 0.5, and 1.0 MW. Only two of the foregoing values ($P_L = 0.207$ and 0.346 MW) correspond to the region investigated in the experiments (Klosterman E.L., et al., 1974). The calculation data are analyzed in the following sequence. First, the results of calculation by scheme A (see Figure 2.27) are given for the entire set of values of power of laser radiation, that is, for the case where the LSW moves downward toward the laser beam. Then the results are given for calculations of all values of power by scheme B, where the LSW moves upward (in opposition to the direction of vector **g**).

We will start analyzing the results of the calculations with the version corresponding to the least power of laser radiation $P_L = 0.06\,\text{MW}$, which corresponds to the flux density of laser radiation $W_L = 0.017\,\text{MW/cm}^2$. For a lower flux density of radiation, the calculation revealed the decay of LSW.

Figure 2.28 gives the temperature field in LSW at time $t = 0.54\,\text{s}$ in calculations by scheme A. The LSW moves downward at an average velocity $\langle V \rangle = 0.69\,\text{m/s}$ and leaves behind a heated region in which the gas moves at a velocity of about $u = -7.8\,\text{m/s}$. The distribution of axial velocity at the instant of time $t = 0.54\,\text{s}$ is given in Figure 2.29. Here and in what follows, the axial velocity of gas is related to the characteristic velocity of convective motion, where $u_0 = 0.313\,\text{m/s}$ is the characteristic velocity of convective motion. The negative values of axial velocity in Figure 2.29 indicate that the gas in LSW moves in the opposite direction relative to the direction of LSW motion (gas moves upward, and LSW downward).

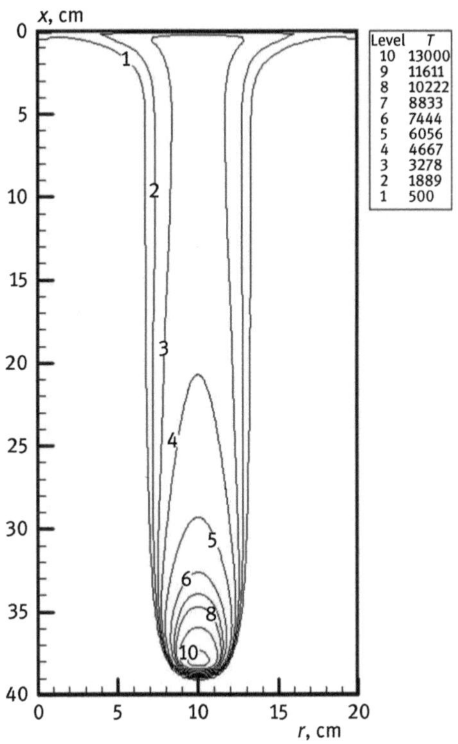

Figure 2.28: The temperature field in LSW at $P_L = 0.06$ MW at the instants of time$t = 0.54$ s. Computational scheme A.

The average velocity of LSW motion increases with the flux density of laser radiation: at $P_L = 0.08\,\text{MW}$, it is $\langle V \rangle = 2.44\,\text{m/s}$, and at $P_L = 0.12\,\text{MW}$it is nonmonotonic

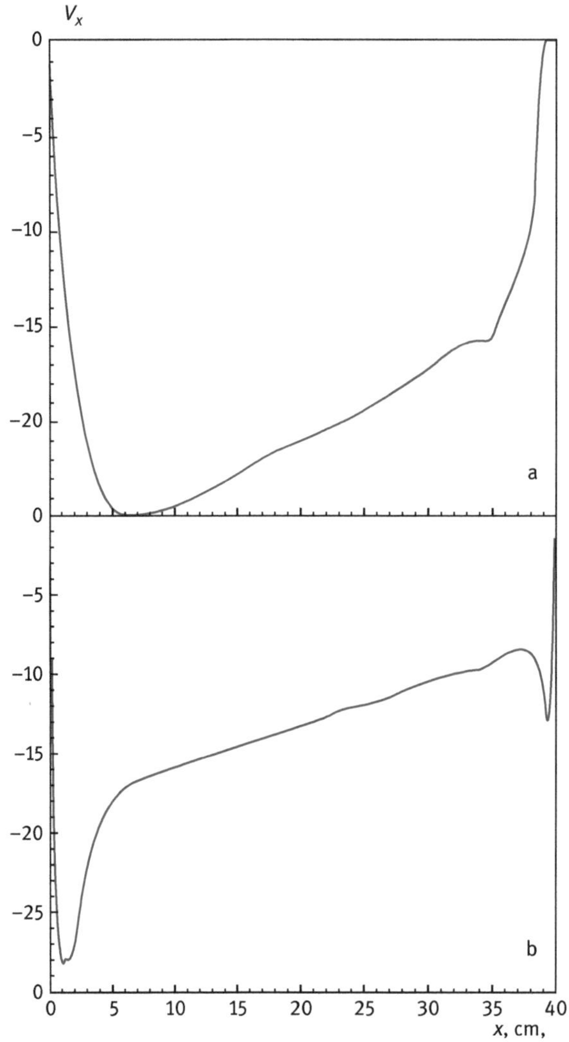

Figure 2.29: The axial velocity of gas in LSW at$P_L = 0.06$ MW (*a*) and at $P_L = 0.12$ MW (*b*).

with respect to two regions of maximal velocity, namely, in the high-temperature region of LSW and in the vicinity of the surface (Figure 2.29b).

The subsequent two versions ($P_L = 0.21$and 0.35 MW) correspond to the region of flux density of laser radiation, which was studied in the experiments (Klosterman, Byron, 1974). The average velocity of LSW at $P_L = 0.21$ MW (Figure 2.30) increased to $\langle V \rangle = 13.6$ m/s, and the maximal velocity of gas within the LSW remained unchanged: $u = -8.74$ m/s. Therefore, the LSW, in this case, moves faster than the gas accelerated within the heated region owing to the Archimedes force. Note further that the maximal temperature in the LSW likewise increases with the flux density of laser radiation.

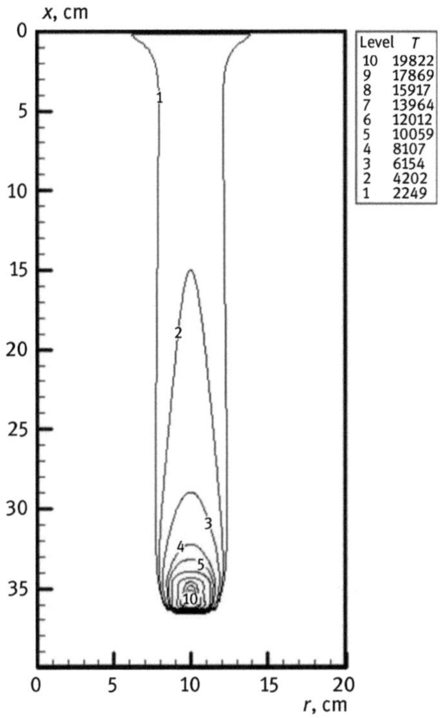

Figure 2.30: The temperature in LSW (in K) at$P_L = 0.21$ MW at the instant of time $t = 0.025$ s. Computational scheme A. The modulus of gas velocity in the vicinity of the symmetry axis is 8.74 m/s.

The degree of heating of plasma in the LSW and the velocity of its motion make an impact on the gas dynamic structure. On the one hand, the heating of plasma to a higher temperature (with increasing PL) causes an increase in the rate of thermal gravitational convection. On the other hand, the velocity of motion of the leading front of LSW increases, so that the heated gas does not have enough time to accelerate while the high-temperature region is located at the site of initial acceleration of gas. The gas in the wake of LSW has enough time for cooling down and is accelerated to a lower velocity.

The latter effect shows up at $P_L = 0.35$ MW, where the average velocity of LSW motion continues to increase ($\langle V \rangle = 19.6$ m/s), and the maximal velocity of gas decreases ($u = -5.9$ m/s). With further increase in the power of laser radiation, this tendency persists. The average velocity of LSW motion for PL = 0.5 and 1.0 MW is $\langle V \rangle = 25.8$ and 38.9 m/s, and the highest axial velocity of gas decreases from $u = -4.7$ m/s to $u = -3.8$ m/s.

In summing up the results of the analysis of regularity of LSW motion in the direction of the vector of acceleration of gravity (downward), we make the following inferences:

- At the given flux density of laser radiation, the LSW moves at an approximately unchanged velocity.
- The average velocity of LSW motion increases with flux density of radiation.
- When W_L increases in the range from 0.017 to 0.29 MW/cm^2, the average velocity of LSW increases from $\langle V \rangle = 0.69$ to 38.9 m/s.
- The dependence of the maximal velocity of gas within the LSW with increasing flux density of laser radiation is nonlinear. For relatively low values of flux density of radiation ($W_L < 0.08\,\mathrm{MW/cm^2}$), the maximal velocity somewhat increases in magnitude from $u = -7.8\,\mathrm{m/s}$ to $u = -8.7\,\mathrm{m/s}$. With further increase in W_L, the maximal velocity of gas within the LSW decreases to $u = -3.8\,\mathrm{m/s}$.
- The radial dimensions of LSW somewhat decrease with increasing flux density of radiation.
- No gas dynamic instabilities are present in the LSW wake: the gas flow is laminar.

The second series of calculations were performed for computational scheme B (see Figure 2.27). In this case, the LSW moves upward, that is, in opposition to the direction of the vector of acceleration of gravity. The gas heated by LSW likewise moves upward under the effect of the Archimedes force. The results of calculations with the power of laser radiation $P_L = 0.06\,\mathrm{MW}$ ($W_L = 0.0173\,\mathrm{MW/cm^2}$) are given in Figure 2.31a, b. The average velocity of LSW motion is, in this case, $\langle V \rangle = 1.9\,\mathrm{m/s}$. The velocity of gas within the LSW reaches a 10 value of $u{\sim}4.8\,\mathrm{m/s}$. However, a note must be made to the significant nonmonotonicity of axial distribution of velocity in the LSW. The two-dimensional distributions of temperature and velocity in the LSW, given in Figure 2.31, are indicative of the formation of vorticities in the heated gas region during the LSW motion along the beam. The data given in these figures show the periodic emergence of vortex zones. Zones of vortex motion periodically arise on the boundary of the heated region. With developing vortex motion, conditions are developed on the boundary, which promote the emergence of heated gas jets in the axial region directed toward the LSW motion. This pattern recurs with fixed periodicity so that the velocity of gas periodically changes its direction from the standpoint of observer located in the frame of reference associated with the leading front of LSW.

With the power of laser radiation increased to $P_L = 0.08\,\mathrm{MW}$ ($W_L = 0.023\,\mathrm{MW/cm^2}$), the gas dynamic structure of LSW wake becomes much smoother, although some nonmonotonicity in the distribution of axial velocity is still observed. The average velocity of LSW motion in the beam is, in this case, $\langle V \rangle = 5.97\,\mathrm{m/s}$. The axial distribution of hot gas velocity within the LSW wake has a maximum at 19 cm away from the surface. The maximal value of velocity is $u = 6.24\,\mathrm{m/s}$; one can see that this is somewhat higher than the velocity of motion of LSW front.

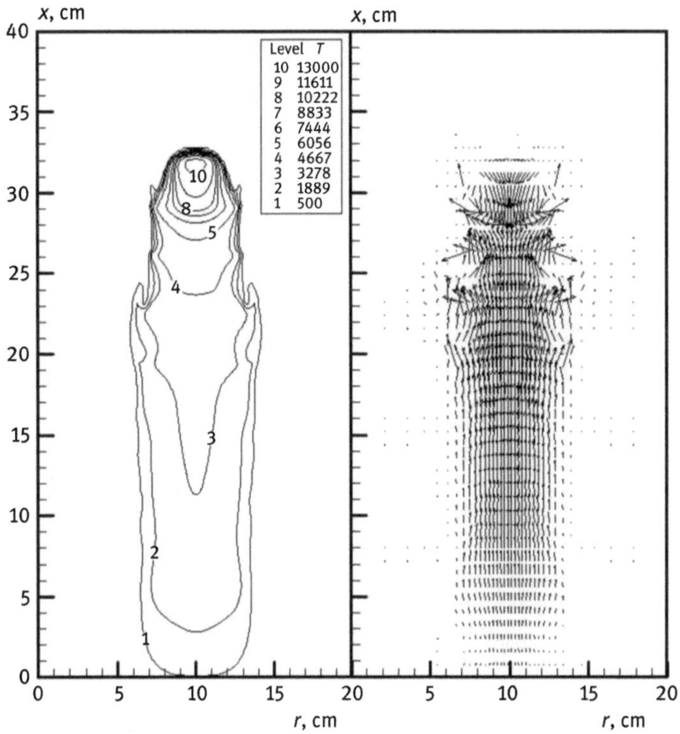

Figure 2.31: The temperature (a) and the vector velocity (b) fields in LSW at $P_L = 0.06$ MW at $t = 0.0186$ s. Computational scheme B.

For the power $P_L = 0.12$ MW, the average velocity of LSW is $\langle V \rangle = 15.7$ m/s, and the maximal velocity of hot gas within the LSW wake is only $u = 3$ m/s. The vector velocity field (Figure 2.32b) and the temperature distribution (Figure 2.32a) point to the laminar pattern of flow of gas within the LSW and its wake.

The subsequent four series of calculations, performed at $P_L = 0.21$, 0.35, 0.5, and 1.0 MW, fully confirm these regularities. At $P_L = 0.21$ MW ($W_L = 0.06$ MW/cm^2), the average velocity of LSW motion is $\langle V \rangle = 28.5$ m/s, and the maximal velocity of gas within the LSW wake is $u = 2.1$ m/s; at $P_L = 0.35$ MW ($W_L = 0.1$ MW/cm^2), we have $\langle V \rangle = 40.0$ m/s and $u = 1.9$ m/s; at $P_L = 0.5$ MW ($W_L = 0.145$ MW/cm^2), the velocity of LSW increases to $\langle V \rangle = 52.0$ m/s; and, at $P_L = 1.0$ MW ($W_L = 0.289$ MW/cm^2), the LSW velocity is $\langle V \rangle = 76.0$ m/s. In the latter two cases, the maximal velocity of gas was $u = 1.7$ and 1.56 m/s, respectively; in doing so, the maximal values of velocity in the front and tail of the wake become virtually the same.

In summing up the results of the analysis of regularities of upward LSW motion in the field of gravity, one can make the following conclusions:

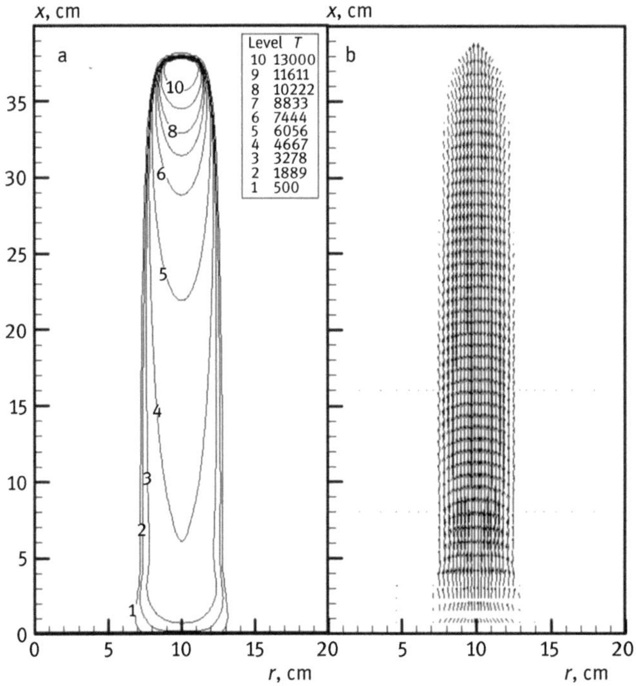

Figure 2.32: The temperature (a) and the vector velocity (b) fields in LCW at $P_L = 0.12$ MW at $t = 0.023$ s. Computational scheme B.

- At the given flux density of laser radiation, the LSW moves approximately uniformly, with a possible exception for the case of extremely low values of flux density of laser radiation.
- The average velocity of LSW motion increases with increasing flux density of laser radiation. With $W_L$increasing in the range from 0.017 to 0.29 MW/cm^2, the average velocity increases from $\langle V \rangle = 19$ to 76.4 m/s.
- At relatively low values of flux density of laser radiation ($W_L < 0.020$ MW/cm^2), the velocity of gas within the LSW exceeds the velocity of motion of LSW proper. In these cases, the emergence of vortex motion is observed: the heated gas under the effect of the Archimedes force develops a higher pressure on the LSW front, which is the principal reason for vortex formation.
- At high values of flux density of laser radiation ($W_L \geq 0.030$ MW/cm^2), the velocity of convective motion of gas within the LSW is lower than the velocity of motion of the LSW front. In this case, the convective motion of gas within the LSW is laminar, because the floating up gas falls behind the leading front of LSW motion.
- The vortex motion of the gas, which develops at a low velocity of LSW motion, causes a periodic variation in the velocity of the gas in the LSW front.

So, as a result of studying the LSW motion in the field of gravity, it was found that the thermal gravitational convection makes an appreciable impact on the velocity of this motion. The LSW moves upward (in opposition to the field of gravity) approximately twice as fast as it moves downward. It is important that the former coincides. The calculated dependences of the velocity of LSW motion on the flux density of laser radiation are given in Figure 2.33. Also given in this figure is the predicted velocity of LSW motion disregarding the thermal gravitational convection.

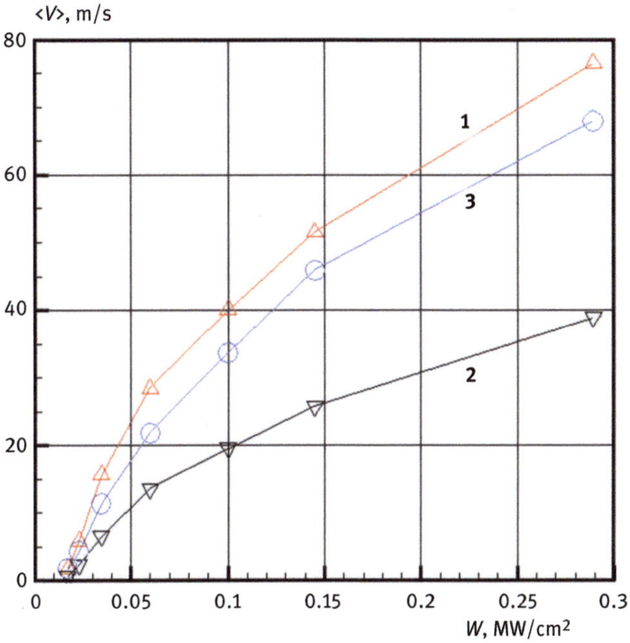

Figure 2.33: The average velocity of LCW motion: (1) in opposition to the force of gravity (upward), (2) in the direction of the force of gravity (downward), and (3) disregarding the thermal gravitational convection.

Two mechanisms of emergence of convection are identified during the upward motion of LSW. In the case of low velocity of movement of LSW front, a vortex turbulent motion of hot gas arises in the LSW wake. In doing so, a periodic variation of the direction of gas velocity is observed in the LSW front. In fact, one can speak of a new hybrid form of thermal-convective instability of the gas dynamic structure of the flow. This instability is characterized by classical gravitational convection and by the velocity of movement in space of the source of heating the gas (in this case, laser-supported combustion wave). At relatively high velocities of movement of LSW front, a laminar flow of hot gas is formed in the LSW wake.

3 Computational models of magnetohydrodynamic processes

Calculation models of homogeneous chemically equilibrium plasma considered in Section 1.1 include as the important constituent the account of a magnetic field action on plasma flows. Actually, it was the one-liquid one-temperature magnetohydrodynamic (MHD) model. Such computing models are widely applied in plasma dynamics and gas discharge physics. The set of MHD equations is also an object of interest for many basic directions of aerospace sciences, as well as the computational fluid dynamics (CFD).

However, at the numerical realization of the MHD equations, a series of problems are discussed in this chapter. The singularity of the CFD problems appears in the MHD to an even greater degree, which relates to a plurality of a statement of physically equivalent equations expressing conservation laws of mass, impulse, and energy (Hirsch C., 1990; Anderson D.A., et al., 1997). In the MHD theory, the fluid dynamic equations are added not only by the Maxwellian equations but also by the momentum conservation equation, and energy conservation equations have to be modified. The outcome is a complication of a complete system of the equations and an increase in the number of variations of the equation formulations. From the physical point of view, these various formulations of the equations are an equivalent form of the same conservation laws. From the point of view of the computing analysis, the various formulations of the equations lead to consequences of the principal character.

In this chapter, the following systems of variables will be considered:

$$\mathbf{U} = \left| \rho \quad \rho u \quad \rho \upsilon \quad \rho w \quad \rho E \quad B_x \quad B_y \quad B_z \right|^{\mathrm{T}}, \tag{3.1}$$

$$\mathbf{W}_e = \left| \rho \quad u \quad \upsilon \quad w \quad e \quad B_x \quad B_y \quad B_z \right|^{\mathrm{T}}, \tag{3.2}$$

$$\mathbf{W}_p = \left| \rho \quad u \quad \upsilon \quad w \quad p \quad B_x \quad B_y \quad B_z \right|^{\mathrm{T}}, \tag{3.3}$$

where ρ, p are the density and pressure; $\rho u, \rho v, \rho w$ are the projections of a mass flux vector to coordinate axes x, y, z; u, v, w are the velocity projections to coordinate axes; ρE is the total energy of volume unit; e is the specific internal energy (note that unlike Chapter 1 for a specific internal energy, the symbol "e" is chosen here); B_x, B_y, B_z are the projections of a magnetic induction vector to coordinate axes.

Variables (3.1) are known as *conservative variables*, while variables (3.2) and (3.3) are referred to as *natural* variables. In addition, a vector of characteristic variables will be formulated below with corresponding comments and explanations.

Conversion from one type of variables to another, as a rule, means a modification of the type of the governing equations. However, it is necessary to mean that it is not always possible to pass from one type of notation to another without using

https://doi.org/10.1515/9783110648836-004

some additional relations. The problem of using such relations with reference to three-dimensional (3D) MHD problems is the general goal of this chapter.

3.1 The governing equations

Initially, we shall choose the system of the MHD equations formulated and substantiated in the form of conservation laws of continuum mechanics together with the Maxwellian set of equations in Hirsch C. (1990), Anderson D.A., et al. (1984), and Kulikovskiy A.G., et al. (2001):

$$\frac{\partial \rho}{\partial t} + \mathrm{div}\, \rho \mathbf{V} = 0, \text{[1]} \tag{3.4}$$

$$\frac{\partial \rho \mathbf{V}}{\partial t} + \mathrm{div}\, (\rho \mathbf{V} \cdot \mathbf{V}) = -\,\mathrm{grad}\, \hat{P} + [\mathbf{J} \times \mathbf{B}], \tag{3.5}$$

$$\frac{\partial \rho e}{\partial t} + \mathrm{div}\, (\rho e \cdot \mathbf{V}) = -\,\mathrm{div}\mathbf{q} - \hat{P} : \mathrm{div}\, \mathbf{V} + (\mathbf{E}' \cdot \mathbf{J}), \tag{3.6}$$

$$\frac{\partial \mathbf{B}}{\partial t} = -\,\mathrm{rot}\mathbf{E}, \tag{3.7}$$

$$\mathbf{J} = \frac{1}{\mu_0} \mathrm{rot}\mathbf{B}, \tag{3.8}$$

$$\mathrm{div}\, \mathbf{B} = 0, \tag{3.9}$$

$$\mathrm{div}\, \mathbf{J} = 0, \tag{3.10}$$

where ρ, $\mathbf{V}$ are the density and velocity with components u, v, w along u, v, w coordinate axes; e is the specific internal energy; $\hat{P}$ is the stress tensor; $\mathbf{J}$, $\mathbf{B}$, $\mathbf{E}$ are the current density, induction of a magnetic field, electric field strength; $\mathbf{q}$ is the vector of a thermal flux; μ_0 is the magnetic permeability of free space (further the system of SI units is used; therefore $\mu_0 = 4\pi \times 10^{-7} \mathrm{kg} \cdot \mathrm{m/C}^2$ or H/m).

In the energy conservation equation (3.6), the electric field strength $\mathbf{E}'$ is calculated in a plasma moving with an average velocity $\mathbf{V}$, that is, it is related to an electric field strength in a laboratory coordinate system with the following relation:

$$\mathbf{E}' = \mathbf{E} + [\mathbf{V} \times \mathbf{B}]. \tag{3.11}$$

Components of a stress tensor can be written as follows:

$$P_{\alpha\beta} = p\,\delta_{\alpha\beta} - \tau_{\beta\alpha},$$

1 Further in equations and formulas, the letter symbols for differential operators and also inverted delta operator ∇, used in the field theory, will be used in equal measure.

where p is the pressure; $\delta_{\alpha\beta}$ is the Kronecker delta; $\tau_{\beta\alpha}$ are the components of viscous stress tensor

$$\tau_{\beta\alpha} = \mu\left[\left(\frac{\partial u_\beta}{\partial x_\alpha} + \frac{\partial u_\alpha}{\partial x_\beta}\right) - \frac{2}{3}\delta_{\beta\alpha}\,\mathrm{div}\,\vec{V}\right], \tag{3.12}$$

where μ is the coefficient of dynamic viscosity.

3.2 The vector form of Navier–Stokes equations

The compressible Navier–Stokes equations in Cartesian coordinates can be written as follows:

$$\frac{\partial \mathbf{U}}{\partial t} + \vec{\nabla}\cdot\left(\mathbf{F}^{\mathrm{Eu}} + \mathbf{F}^{\mathrm{NS}}\right) = \mathbf{Q}, \tag{3.13}$$

where $\mathbf{U}$ is the vector function of the conservative variables; $\mathbf{F}^{\mathrm{Eu}}$ is the vector function of flows in the set of Euler equations; $\mathbf{F}^{\mathrm{NS}}$ is the vector function, which includes components of a viscous stress tensor and a thermal conduction flux:

$$\vec{\mathbf{F}}^{\mathrm{Eu}} = \vec{i}\mathbf{f} + \vec{j}\mathbf{g} + \vec{k}\mathbf{h},\quad \vec{\mathbf{F}}^{\mathrm{NS}} = \vec{i}\mathbf{f}^{\mathrm{NS}} + \vec{j}\mathbf{g}^{\mathrm{NS}} + \vec{k}\mathbf{h},$$

$$\mathbf{U} = \begin{vmatrix} \rho \\ \rho u \\ \rho v \\ \rho w \\ \rho E \end{vmatrix},\quad \mathbf{f} = \begin{vmatrix} \rho u \\ \rho u^2 + p \\ \rho u v \\ \rho u w \\ \rho u[E + (p/\rho)] \end{vmatrix},\quad \mathbf{g} = \begin{vmatrix} \rho v \\ \rho v u \\ \rho v^2 + p \\ \rho v w \\ \rho v[E + (p/\rho)] \end{vmatrix},\quad \mathbf{h} = \begin{vmatrix} \rho w \\ \rho w u \\ \rho w v \\ \rho w^2 + p \\ \rho w[E + (p/\rho)] \end{vmatrix} \tag{3.14}$$

$$\mathbf{f}^{\mathrm{NS}} = -\begin{vmatrix} 0 \\ \tau_{xx} \\ \tau_{xy} \\ \tau_{xz} \\ u\tau_{xx} + v\tau_{xy} + w\tau_{xz} + \lambda\frac{\partial T}{\partial x} \end{vmatrix},\quad \mathbf{g}^{\mathrm{NS}} = -\begin{vmatrix} 0 \\ \tau_{yx} \\ \tau_{yy} \\ \tau_{yz} \\ u\tau_{yx} + v\tau_{yy} + w\tau_{yz} + \lambda\frac{\partial T}{\partial y} \end{vmatrix},$$

$$\mathbf{h}^{\mathrm{NS}} = -\begin{vmatrix} 0 \\ \tau_{zx} \\ \tau_{zy} \\ \tau_{zz} \\ u\tau_{zx} + v\tau_{zy} + w\tau_{zz} + \lambda\frac{\partial T}{\partial z} \end{vmatrix}, \tag{3.15}$$

$E = e + (\mathbf{V}^2/2)$ is the total specific energy; T is the temperature; λ is the thermal conductivity coefficient; $\mathbf{Q}$ is the source of mass, impulse, or energy in a full set of the Navier–Stokes equations.

Viscous stress tensor components have the following form:

$$\tau_{xx} = 2\mu\frac{\partial u}{\partial x} - \frac{2}{3}\mu\mathrm{div}\mathbf{V},\ \tau_{yy} = 2\mu\frac{\partial \upsilon}{\partial y} - \frac{2}{3}\mu\mathrm{div}\mathbf{V},$$

$$\tau_{zz} = 2\mu\frac{\partial \upsilon}{\partial z} - \frac{2}{3}\mu\mathrm{div}\mathbf{V};\ \tau_{xy} = \tau_{yx} = \mu\left(\frac{\partial u}{\partial y} + \frac{\partial v}{\partial x}\right),$$

$$\tau_{xz} = \tau_{zx} = \mu\left(\frac{\partial u}{\partial z} + \frac{\partial w}{\partial x}\right),\ \tau_{yz} = \tau_{zy} = \mu\left(\frac{\partial v}{\partial z} + \frac{\partial w}{\partial y}\right),$$

where μ is the coefficient of dynamic viscosity.

The perfect gas equation of state will be used further in the chapter:

$$e = \frac{p}{\rho(\gamma - 1)}, \tag{3.16}$$

where $\gamma = c_p/c_V$ is the ratio of specific thermal capacities at constant pressure and volume. We shall give also some useful relations that will be used further:

$$e = c_V T,\ \frac{p}{\rho} = \tilde{R}T = c_V(\gamma - 1)T,\ c_p = \frac{\gamma\tilde{R}}{\gamma - 1},\ c_V = \frac{\tilde{R}}{\gamma - 1},$$

$$\tilde{R} = c_V(\gamma - 1),\ p = \left(E - 0.5\vec{V}^2\right)\rho(\gamma - 1). \tag{3.17}$$

Under air normal conditions ($p = 10^5\,\mathrm{Pa}$, $T = 293$ K) one can get $\gamma = 1.4$, $\tilde{R} = 287\,\mathrm{J/(kg \times K)}$.

3.3 System of equations of magnetic induction

The basis of the equations of perfect gas MHD (further, the MHD equations) are the Faraday and the Ampere laws, which are formulated in the following form (Landau L.D., et al., 1960; Bittencourt J.A., 2004; Mitcher M., et al., 1992; Krall N.A., et al., 1973; Stratton J.A., 1948):

$$\int \mathbf{E}\mathrm{d}\mathbf{l} = -\int\limits_S \frac{\partial \mathbf{B}}{\partial t}\mathrm{d}\mathbf{S}, \tag{3.18}$$

$$\int \mathbf{H}\mathrm{d}\mathbf{l} = \int s\left(\mathbf{J} + \frac{\partial \mathbf{D}}{\partial t}\right)\mathrm{d}\mathbf{S}. \tag{3.19}$$

The Faraday low relates an electric field strength along a contour **l** enveloping an area **S** to the time change of magnetic flux through this area. The Ampere law describes the scalar magnetic potential induced in the contour **l** due to currents through this area enveloped by this contour.

Differential consequences of these conservation laws (at neglecting displacement currents) enter into a complete set of MHD equations in the form of

$$\frac{\partial \mathbf{B}}{\partial t} = -\,\mathrm{rot}\mathbf{E}, \tag{3.20}$$

$$\mathrm{rot}\mathbf{B} = \mu_0 \mathbf{J}, \tag{3.21}$$

The set of eqs. (3.20) and (3.21) is not closed. To integrate the equations, there is a need to define additional closing conditions. The best known technique is the use of generalized Ohm's law

$$\mathbf{J} = \sigma\{\mathbf{E} + [\mathbf{V} \times \mathbf{B}]\}, \tag{3.22}$$

where σ is the electroconductivity.

Basically, the presented three equations are enough to describe all variety of electromagnetic processes within the limits of ideal MHD. However, the set of Maxwellian equations contains one more equation

$$(\vec{\nabla} \cdot \mathbf{B}) = 0, \tag{3.23}$$

which should be fulfilled in the whole investigated area (including initial conditions). Besides, the above-mentioned assumption about neglecting of displacement currents imposes the following condition on total current:

$$(\vec{\nabla} \cdot \mathbf{J}) = 0, \tag{3.24}$$

which also should be fulfilled in the whole investigated area.

It has been established (Landau L.D., et al., 1960; Hughes W.F., et al., 1989; Bittencourt J.A., 2004; Mitcher M., et al., 1992; Krall N.A., et al., 1973) that the optimal procedure to complete the set of MHD equations is adding to equations of the Euler or Navier–Stokes equations for components of a magnetic induction vector obtained by a combination of eqs. (3.20–3.22). However, before starting the realization of this procedure, it is necessary to solve the basic problem, namely how to use additional, but the unconditional equations (3.23) and (3.24). We will consider two methods of obtaining the required equations. In first of them, during transformations of the initial set of eqs. (3.20–3.22) to the final form eq. (3.23) will be used, and in the second case will not.

Note that from eq. (3.20), the divergence of a magnetic field at any moment automatically should be zero if it was equal to zero in initial. However, from the mathematical point of view, the received set of equations will possess various properties, and these differences will rather be significant for the subsequent numerical realization.

First of all, let us derive equation for electric field strength with the use of eqs. (3.21) and (3.22):

$$\mathbf{E} = \frac{1}{\sigma}\mathbf{J} - [\mathbf{V}\times\mathbf{B}] = \frac{1}{\sigma\mu_0}\left[\vec{\nabla}\times\mathbf{B}\right] - [\mathbf{V}\times\mathbf{B}]. \tag{3.25}$$

Then, from eq. (3.20)

$$\frac{\partial \mathbf{B}}{\partial t} = -\left[\vec{\nabla}\times\frac{1}{\sigma}\mathbf{J}\right] + \left[\vec{\nabla}\times[\mathbf{V}\times\mathbf{B}]\right] = -\left[\vec{\nabla}\times\left\{\frac{1}{\sigma\mu_0}\left[\vec{\nabla}\times\mathbf{B}\right] - [\mathbf{V}\times\mathbf{B}]\right\}\right] =$$

$$= -\frac{1}{\sigma\mu_0}\left\{\left(\vec{\nabla}\cdot\mathbf{B}\right) - \mathbf{V}^2\mathbf{B}\right\} + \mathbf{V}\left(\vec{\nabla}\cdot\mathbf{B}\right) + \left(\mathbf{B}\cdot\vec{\nabla}\right)\mathbf{V} - \mathbf{B}\left(\vec{\nabla}\cdot\mathbf{V}\right) - \left(\mathbf{V}\cdot\vec{\nabla}\right)\mathbf{B}. \tag{3.26}$$

The first part of eq. (3.26) can be presented in the form of the sum of two functions:

$$\frac{\partial \mathbf{B}}{\partial t} = \mathbf{F}^{md} + \mathbf{F}^{B}, \tag{3.27}$$

where

$$\mathbf{F}^{md} = -\frac{1}{\sigma}\left[\vec{\nabla}\times\mathbf{J}\right] = -\frac{1}{\sigma\mu_0}\left[\vec{\nabla}\times\left[\vec{\nabla}\times\mathbf{B}\right]\right] = -\frac{1}{\sigma\mu_0}\left\{\vec{\nabla}\left(\vec{\nabla}\cdot\mathbf{B}\right) - \left(\vec{\nabla}\cdot\vec{\nabla}\right)\mathbf{B}\right\},$$

$$\mathbf{F}^{B} = \left[\vec{\nabla}\times\left[\vec{V}\times\mathbf{B}\right]\right] = \mathbf{V}\left(\vec{\nabla}\cdot\mathbf{B}\right) + \left(\mathbf{B}\cdot\vec{\nabla}\right)\mathbf{V} - \mathbf{B}\left(\vec{\nabla}\cdot\mathbf{V}\right) - \left(\mathbf{V}\cdot\vec{\nabla}\right)\mathbf{B}.$$

First of the introduced functions $\mathbf{F}^{md}$ corresponds to a change of a magnetic field induction as a result of currents in a medium with finite conductivity, and the second function $\mathbf{F}^{B}$ corresponds to a change of a magnetic induction due to electric field generation in a moving medium.

Note that the approximation of infinite electrical conductivity ($\sigma \to \infty$) which is rather widespread in MHD applications leads to the following inequality $\mathbf{F}^{md} \ll \mathbf{F}^{B}$.

Now one can write projections of vector functions $\mathbf{F}^{B}$ and $\mathbf{F}^{md}$ with and without including of the above-mentioned absolute condition $\left(\vec{\nabla}\cdot\mathbf{B}\right) = 0$.

1. Components of function $\mathbf{F}^{B}$ at an account of the equation $\left(\vec{\nabla}\cdot\mathbf{B}\right) = 0$:

$$F_x^B = -B_x\left(\frac{\partial v}{\partial y} + \frac{\partial w}{\partial z}\right) + B_y\frac{\partial u}{\partial y} + B_z\frac{\partial u}{\partial z} - u\frac{\partial B_x}{\partial x} - v\frac{\partial B_x}{\partial y} - w\frac{\partial B_x}{\partial z},$$

$$F_y^B = B_x\frac{\partial v}{\partial x} - B_y\left(\frac{\partial u}{\partial x} + \frac{\partial w}{\partial z}\right) + B_z\frac{\partial v}{\partial z} - u\frac{\partial B_y}{\partial x} - v\frac{\partial B_y}{\partial y} - w\frac{\partial B_y}{\partial z},$$

$$F_z^B = B_x\frac{\partial w}{\partial x} + B_y\frac{\partial w}{\partial y} - B_z\left(\frac{\partial u}{\partial x} + \frac{\partial v}{\partial y}\right) - u\frac{\partial B_z}{\partial x} - v\frac{\partial B_z}{\partial y} - w\frac{\partial B_z}{\partial z}. \tag{3.28}$$

2. Components of function $\mathbf{F}^B$ without taking into account the equation $(\vec{\nabla}\cdot\mathbf{B})=0$:

$$F_x^B=\frac{\partial}{\partial y}\left(uB_y-vB_x\right)+\frac{\partial}{\partial z}\left(uB_z-wB_x\right),$$

$$F_y^B=\frac{\partial}{\partial x}\left(vB_x-uB_y\right)+\frac{\partial}{\partial z}\left(vB_z-wB_y\right),$$

$$F_z^B=\frac{\partial}{\partial x}\left(vB_x-uB_z\right)+\frac{\partial}{\partial y}\left(wB_y-vB_z\right). \tag{3.29}$$

Let's pay attention to the fact that at the account of the equation $(\vec{\nabla}\cdot\mathbf{B})=0$, the set of eqs. (3.28) is obtained for definition of magnetic field induction components in the quasilinear form, in which differential operators are applied only to natural variables (u, v, w, B_x, B_y, B). If $(\vec{\nabla}\cdot\mathbf{B})=0$ has not been taken into consideration, the set of eqs. (3.29) is formulated in a conservative form. The consequences of this fact will be discussed below.

3. Components of function $\mathbf{F}^{md}$ on account of the equation $(\vec{\nabla}\cdot\mathbf{B})=0$:

$$F_x^{md}=\frac{1}{\sigma\mu_0}\left(\frac{\partial^2 B_x}{\partial x^2}+\frac{\partial^2 B_x}{\partial y^2}+\frac{\partial^2 B_x}{\partial z^2}\right),$$

$$F_y^{md}=\frac{1}{\sigma\mu_0}\left(\frac{\partial^2 B_y}{\partial x^2}+\frac{\partial^2 B_y}{\partial y^2}+\frac{\partial^2 B_y}{\partial z^2}\right),$$

$$F_z^{md}=\frac{1}{\sigma\mu_0}\left(\frac{\partial^2 B_z}{\partial x^2}+\frac{\partial^2 B_z}{\partial y^2}+\frac{\partial^2 B_z}{\partial z^2}\right). \tag{3.30}$$

4. Components of function $\mathbf{F}^{md}$ without taking into account the equation $(\vec{\nabla}\cdot\mathbf{B})=0$:

$$F_x^{md}=\frac{1}{\sigma\mu_0}\left[\left(\frac{\partial^2 B_x}{\partial x^2}+\frac{\partial^2 B_x}{\partial y^2}+\frac{\partial^2 B_x}{\partial z^2}\right)-\left(\frac{\partial^2 B_x}{\partial x^2}+\frac{\partial^2 B_y}{\partial x\partial y}+\frac{\partial^2 B_z}{\partial x\partial z}\right)\right],$$

$$F_y^{md}=\frac{1}{\sigma\mu_0}\left[\left(\frac{\partial^2 B_y}{\partial x^2}+\frac{\partial^2 B_y}{\partial y^2}+\frac{\partial^2 B_y}{\partial z^2}\right)-\left(\frac{\partial^2 B_x}{\partial x\partial y}+\frac{\partial^2 B_y}{\partial y^2}+\frac{\partial^2 B_z}{\partial y\partial z}\right)\right],$$

$$F_z^{md}=\frac{1}{\sigma\mu_0}\left[\left(\frac{\partial^2 B_z}{\partial x^2}+\frac{\partial^2 B_z}{\partial y^2}+\frac{\partial^2 B_z}{\partial z^2}\right)-\left(\frac{\partial^2 B_x}{\partial x\partial z}+\frac{\partial^2 B_y}{\partial y\partial z}+\frac{\partial^2 B_z}{\partial z^2}\right)\right]. \tag{3.31}$$

So, two physically equivalent forms of a set of magnetic induction equations have been received.

In summary of this section, we shall give also component-wise expressions of $\mathbf{F}^{md}$ with projections of a current density

$$F_x^{md}=-\left(\frac{\partial J_z}{\partial y}-\frac{\partial J_y}{\partial z}\right),\ F_y^{md}=-\left(\frac{\partial J_x}{\partial z}-\frac{\partial J_z}{\partial x}\right),\ F_z^{md}=-\left(\frac{\partial J_y}{\partial x}-\frac{\partial J_x}{\partial y}\right). \tag{3.32}$$

3.4 Force acting on ionized gas from an electric and magnetic field

The full force acting on a gas volume unit is

$$\mathbf{F}^{em} = \rho_c \mathbf{E}' + [\mathbf{J} \times \mathbf{B}], \tag{3.33}$$

where ρ_c is the density of charges. In the theory of ideal MHD $\rho_c = 0$; hence,

$$\mathbf{F}^{em} = [\mathbf{J} \times \mathbf{B}]. \tag{3.34}$$

This force is also presented in the right-hand side of eq. (3.5). To write expressions for projections of this force to the coordinate axes, we shall substitute in expression (3.34) for a current density in the form (3.21):

$$\mathbf{F}^{em} = \frac{1}{\mu_0}\left[\mathbf{B} \times \left[\vec{\nabla} \times \mathbf{B}\right]\right] = -\frac{1}{\mu_0}\left[\vec{\nabla}(\mathbf{B} \cdot \mathbf{B}) - (\mathbf{B} \cdot \vec{\nabla})\mathbf{B}\right]. \tag{3.35}$$

Then

$$F_x^{em} = \frac{1}{\mu_0}\left(-\frac{1}{2}\frac{\partial B^2}{\partial x} + B_x \frac{\partial B_x}{\partial x} + B_y \frac{\partial B_x}{\partial y} + B_z \frac{\partial B_x}{\partial z}\right), \tag{3.36}$$

$$F_y^{em} = \frac{1}{\mu_0}\left(-\frac{1}{2}\frac{\partial B^2}{\partial y} + B_x \frac{\partial B_y}{\partial x} + B_y \frac{\partial B_y}{\partial y} + B_z \frac{\partial B_y}{\partial z}\right), \tag{3.37}$$

$$F_z^{em} = \frac{1}{\mu_0}\left(-\frac{1}{2}\frac{\partial B^2}{\partial z} + B_x \frac{\partial B_z}{\partial x} + B_y \frac{\partial B_z}{\partial y} + B_z \frac{\partial B_z}{\partial z}\right), \tag{3.38}$$

where $B^2 = B_x^2 + B_y^2 + B_z^2$.

Equations (3.36–3.38) are written in the hybrid conservative and nonconservative form. To rewrite these equations in the homogeneous form, we shall use some formal transformations.

Let us add and subtract from a right-hand side of formula (3.36) an item of $(1/\mu_0)B_x(\partial B_x/\partial x)$, then

$$F_x^{em} = \frac{1}{\mu_0}\left[-\frac{1}{2}\frac{\partial B^2}{\partial x} + \frac{\partial B_x B_x}{\partial x} + \frac{\partial B_x B_y}{\partial y} + \frac{\partial B_x B_z}{\partial z} - B_x\left(\frac{\partial B_x}{\partial x} + \frac{\partial B_y}{\partial y} + \frac{\partial B_z}{\partial z}\right)\right] \tag{3.39}$$

or, with the use of $(\vec{\nabla} \cdot \mathbf{B}) = 0$,

$$F_x^{em} = \frac{1}{\mu_0}\left[\frac{1}{2}\frac{\partial}{\partial x}\left(B_x^2 - B_y^2 - B_z^2\right) + \frac{\partial B_x B_y}{\partial y} + \frac{\partial B_x B_z}{\partial z}\right]. \tag{3.40}$$

By analogy

$$F_y^{em} = \frac{1}{\mu_0}\left[\frac{1}{2}\frac{\partial}{\partial y}\left(B_y^2 - B_x^2 - B_z^2\right) + \frac{\partial B_x B_y}{\partial x} + \frac{\partial B_y B_z}{\partial z}\right], \tag{3.41}$$

$$F_z^{em} = \frac{1}{\mu_0}\left[\frac{1}{2}\frac{\partial}{\partial z}\left(B_z^2 - B_x^2 - B_y^2\right) + \frac{\partial B_x B_z}{\partial x} + \frac{\partial B_y B_z}{\partial y}\right]. \tag{3.42}$$

Thus, identical transformations of formulas (3.36)–(3.38) with the use of the equation $(\vec{\nabla} \cdot \mathbf{B}) = 0$ allow to receive the locally conservative form (3.40)–(3.42).

Otherwise, the nonconservation form for the force projections can be obtained:

$$F_x^{em} = \frac{1}{\mu_0}\left[-\frac{1}{2}\frac{\partial}{\partial x}\left(B_y^2 + B_z^2\right) + B_y\frac{\partial B_x}{\partial y} + B_z\frac{\partial B_x}{\partial z}\right], \tag{3.43}$$

$$F_y^{em} = \frac{1}{\mu_0}\left[-\frac{1}{2}\frac{\partial}{\partial y}\left(B_x^2 + B_z^2\right) + B_x\frac{\partial B_y}{\partial x} + B_z\frac{\partial B_y}{\partial z}\right], \tag{3.44}$$

$$F_z^{em} = \frac{1}{\mu_0}\left[-\frac{1}{2}\frac{\partial}{\partial z}\left(B_x^2 + B_y^2\right) + B_x\frac{\partial B_z}{\partial x} + B_y\frac{\partial B_z}{\partial y}\right]. \tag{3.45}$$

It will be shown below that exactly these formulas are most convenient for deriving the quasilinear form of the MHD equations.

3.5 A heat emission caused by the action of electromagnetic forces

Let's consider transformations of Joule heat $(\mathbf{E}' \cdot \mathbf{J})$ entering into the energy conservation equation (3.6). According to the MHD theory, the primed electric field strength $\mathbf{E}'$ considers its difference in a moving continuum from an electric field strength $\mathbf{E}$ in the same point of space of a laboratory frame (Hughes W.F., et al., 1989; Bittencourt J.A., 2004; Mitcher M., et al., 1992; Krall N.A., et al., 1973):

$$\mathbf{E}' = \mathbf{E} + [\mathbf{V} \times \mathbf{B}]. \tag{3.46}$$

Thus, Joule heat can be written in the following form:

$$\begin{aligned} Q_J &= (\mathbf{E} + [\mathbf{V}\times\mathbf{B}])\frac{1}{\mu_0}\left[\vec{\nabla}\times\mathbf{B}\right] = \frac{1}{\mu_0}\left\{\mathbf{E}\cdot\left[\vec{\nabla}\times\mathbf{B}\right] + [\mathbf{V}\times\mathbf{B}]\cdot\left[\vec{\nabla}\times\mathbf{B}\right]\right\} = \\ &= \frac{1}{\mu_0}\left\{\mathbf{B}\times\left[\vec{\nabla}\times\mathbf{E}\right] - \vec{\nabla}\cdot[\mathbf{E}\times\mathbf{B}]\right\}, \end{aligned} \tag{3.47}$$

where the following vectorial identity was used:

$$\mathbf{E}\cdot\left[\vec{\nabla}\times\mathbf{H}\right] = \mathbf{H}\cdot\left[\vec{\nabla}\times\mathbf{E}\right] - \vec{\nabla}\cdot[\mathbf{E}\times\mathbf{H}],$$

and the item $\left[\vec{\nabla}\times\mathbf{B}\right]\cdot\left[\vec{\nabla}\times\mathbf{B}\right]$ is equal to zero as a scalar product of parallel vectors.

Designed expression for Joule heat can be derived from eq. (3.47) with the use of the Maxwell equation (3.7) after calculating a vector product $[\mathbf{E}\times\mathbf{B}]$ divergence:

$$Q_J = -\frac{1}{2\mu_0}\frac{\partial B^2}{\partial t} - \frac{1}{\mu_0}\left[\frac{\partial}{\partial x}\left(E_yB_z - E_zB_y\right) + \frac{\partial}{\partial y}\left(E_zB_x - E_xB_z\right) + \frac{\partial}{\partial z}\left(E_xB_y - E_yB_x\right)\right]. \quad (3.48)$$

Projections of electric field strength $\mathbf{E}$ are expressed by projections of current density, velocity, and induction of magnetic field to coordinate axes with the use of the generalized Ohm's law

$$\mathbf{J} = \sigma\{\mathbf{E} + [\mathbf{V}\times\mathbf{B}]\} \quad (3.49)$$

as follows:

$$E_x = \frac{1}{\sigma}J_x - \left(vB_z - wB_y\right) = \frac{1}{\sigma\mu_0}\left(\frac{\partial B_z}{\partial y} - \frac{\partial B_y}{\partial z}\right) - \left(vB_z - wB_y\right), \quad (3.50)$$

$$E_y = \frac{1}{\sigma}J_y - \left(wB_x - uB_z\right) = \frac{1}{\sigma\mu_0}\left(\frac{\partial B_x}{\partial z} - \frac{\partial B_z}{\partial x}\right) - \left(wB_x - uB_z\right), \quad (3.51)$$

$$E_z = \frac{1}{\sigma}J_z - \left(uB_y - vB_x\right) = \frac{1}{\sigma\mu_0}\left(\frac{\partial B_y}{\partial x} - \frac{\partial B_x}{\partial y}\right) - \left(uB_y - vB_x\right). \quad (3.52)$$

Further transformations of derivatives in eq. (3.48) are identical; therefore, we shall consider only the first one (let's pay attention to adding of summand equal to zero):

$$Q_J^x = -\frac{1}{\mu_0}\frac{\partial}{\partial x}\left(E_yB_z - E_zB_y\right) = -\frac{1}{\mu_0}\frac{\partial}{\partial x}\left[\frac{B_z}{\sigma\mu_0}\left(\frac{\partial B_x}{\partial z} - \frac{\partial B_z}{\partial x}\right) - wB_xB_z + uB_zB_z\right] -$$
$$-\left[\frac{B_y}{\sigma\mu_0}\left(\frac{\partial B_y}{\partial x} - \frac{\partial B_x}{\partial y}\right) + uB_yB_y - vB_xB_y + uB_xB_x - uB_xB_x\right] = Q_{J,\sigma}^x + Q_{J,B}^x, \quad (3.53)$$

where

$$Q_{J,\sigma}^x = -\frac{1}{\mu_0^2\sigma}\frac{\partial}{\partial x}\left[B_z\left(\frac{\partial B_x}{\partial z} - \frac{\partial B_z}{\partial x}\right) - B_y\left(\frac{\partial B_y}{\partial x} - \frac{\partial B_x}{\partial y}\right)\right],$$

$$Q_{J,B}^x = -\frac{1}{\mu_0}\frac{\partial}{\partial x}\left[uB^2 - B_x\left(uB_x + vB_y + wB_z\right)\right].$$

Making similar calculations with two other derivatives and introducing the magnetic pressure $p_m = B^2/2\mu_0$, we receive expression for Joule thermal emission:

$$Q_J = -\frac{\partial p_m}{\partial t} - \frac{\partial}{\partial x}\left[2up_m - \frac{1}{\mu_0}B_x\left(uB_x + vB_y + wB_z\right)\right] -$$
$$-\frac{\partial}{\partial y}\left[2vp_m - \frac{1}{\mu_0}B_y\left(uB_x + vB_y + wB_z\right)\right] - \frac{\partial}{\partial z}\left[2vp_m - \frac{1}{\mu_0}B_y\left(uB_x + vB_y + wB_z\right)\right] -$$

$$-\frac{1}{\mu_0^2\sigma}\cdot\frac{\partial}{\partial x}\left[B_y\left(\frac{\partial B_x}{\partial y}-\frac{\partial B_y}{\partial x}\right)+B_z\left(\frac{\partial B_x}{\partial z}-\frac{\partial B_z}{\partial x}\right)\right]-$$
$$-\frac{1}{\mu_0^2\sigma}\cdot\frac{\partial}{\partial y}\left[B_z\left(\frac{\partial B_y}{\partial z}-\frac{\partial B_z}{\partial y}\right)+B_x\left(\frac{\partial B_y}{\partial x}-\frac{\partial B_x}{\partial y}\right)\right]-$$
$$-\frac{1}{\mu_0^2\sigma}\cdot\frac{\partial}{\partial z}\left[B_x\left(\frac{\partial B_z}{\partial x}-\frac{\partial B_x}{\partial z}\right)+B_y\left(\frac{\partial B_z}{\partial y}-\frac{\partial B_y}{\partial z}\right)\right]. \tag{3.54}$$

This expression is written in a locally conservative form which will be applied further. The time derivative from magnetic pressure is usually included in a time derivative from total volume energy, so that the energy conservation equation can be solved concerning the sum of total energy and magnetic pressure

$$\frac{\partial}{\partial t}(\rho E + p_{\mathrm{m}}) = \cdots .$$

3.6 Complete set of the MHD equations in a flux form

The generalized form of the MHD equations in 3D rectangular Cartesian coordinates is presented as follows:

$$\frac{\partial \mathbf{U}}{\partial t} + \vec{\nabla}\cdot\mathbf{F}_\Sigma = \mathbf{Q} \text{ or } \frac{\partial \mathbf{U}}{\partial t} + \frac{\partial \mathbf{f}}{\partial x} + \frac{\partial \mathbf{g}}{\partial y} + \frac{\partial \mathbf{h}}{\partial z} = \mathbf{Q}, \tag{3.55}$$

where

$$\mathbf{F}_\Sigma = \mathbf{F}^{\mathrm{Eu}} + \mathbf{F}^{\mathrm{NS}} + \mathbf{F}^{\mathrm{MGD}} + \mathbf{F}^{\mathrm{VMGD}}, \quad \mathbf{F}_\Sigma = \vec{i}\,\mathbf{f} + \vec{j}\,\mathbf{g} + \vec{k}\,\mathbf{h},$$
$$\mathbf{f} = \mathbf{f}^{\mathrm{Eu}} + \mathbf{f}^{\mathrm{NS}} + \mathbf{f}^{\mathrm{MGD}} + \mathbf{f}^{\mathrm{VMGD}},$$
$$\mathbf{g} = \mathbf{g}^{\mathrm{Eu}} + \mathbf{g}^{\mathrm{NS}} + \mathbf{g}^{\mathrm{MGD}} + \mathbf{g}^{\mathrm{VMGD}}, \quad \mathbf{h} = \mathbf{h}^{\mathrm{Eu}} + \mathbf{h}^{\mathrm{NS}} + \mathbf{h}^{\mathrm{MGD}} + \mathbf{h}^{\mathrm{VMGD}},$$

$\mathbf{f}, \mathbf{g}, \mathbf{h}$ are the projections of vector function of flows $\mathbf{F}_\Sigma$ to coordinate axes.

3.6.1 The MHD equations in projections

First, we shall give the flux form of the MHD equations at which the electric field strength and current density are present in explicit form[2]:

2 For this reason, it is impossible to divide here components of flows related only to magnetic induction and electrical conduction.

$$\mathbf{U}=\begin{vmatrix}\rho\\ \rho u\\ \rho \upsilon\\ \rho w\\ \rho E+p_m\\ B_x\\ B_y\\ B_z\end{vmatrix},\ \mathbf{f}^{\mathrm{Eu}}=\begin{vmatrix}\rho u\\ \rho u^2+p\\ \rho u\upsilon\\ \rho uw\\ \rho u\left(E+\dfrac{p}{\rho}\right)\\ 0\\ 0\\ 0\end{vmatrix},\ \mathbf{f}^{\mathrm{NS}}=\begin{vmatrix}0\\ -\tau_{xx}\\ -\tau_{xy}\\ -\tau_{xz}\\ -\left(u\tau_{xx}+\upsilon\tau_{xy}+w\tau_{xz}+\dfrac{\lambda}{c_{\mathrm{v}}}\dfrac{\partial e}{\partial x}\right)\\ 0\\ 0\\ 0\end{vmatrix},$$

$$\mathbf{g}^{\mathrm{Eu}}=\begin{vmatrix}\rho \upsilon\\ \rho \upsilon u\\ \rho \upsilon^2+p\\ \rho \upsilon w\\ \rho v\left(E+\dfrac{p}{\rho}\right)\\ 0\\ 0\\ 0\end{vmatrix},\ \mathbf{g}^{\mathrm{NS}}=\begin{vmatrix}0\\ -\tau_{yx}\\ -\tau_{yy}\\ -\tau_{yz}\\ -\left(u\tau_{yx}+\upsilon\tau_{yy}+w\tau_{yz}+\dfrac{\lambda}{c_{\mathrm{v}}}\dfrac{\partial e}{\partial y}\right)\\ 0\\ 0\\ 0\end{vmatrix},\ \mathbf{h}^{\mathrm{Eu}}=\begin{vmatrix}\rho w\\ \rho wu\\ \rho w\upsilon\\ \rho w^2+p\\ \rho w\left(E+\dfrac{p}{\rho}\right)\\ 0\\ 0\\ 0\end{vmatrix},$$

$$\mathbf{h}^{\mathrm{NS}}=\begin{vmatrix}0\\ -\tau_{zx}\\ -\tau_{zy}\\ -\tau_{zz}\\ -\left(u\tau_{zx}+\upsilon\tau_{zy}+w\tau_{zz}+\dfrac{\lambda}{c_{\mathrm{v}}}\dfrac{\partial e}{\partial z}\right)\\ 0\\ 0\\ 0\end{vmatrix},\quad \mathbf{f}^{\mathrm{MGD}}+\mathbf{f}^{\mathrm{VMGD}}=\begin{vmatrix}0\\ \dfrac{1}{\mu_0}\left(\dfrac{1}{2}B^2-B_x^2\right)\\ -\dfrac{1}{\mu_0}B_xB_y\\ -\dfrac{1}{\mu_0}B_xB_z\\ \dfrac{1}{\mu_0}(E_yB_z-E_zB_y)\\ 0\\ +uB_y-\upsilon B_x+\dfrac{1}{\sigma}J_z\\ +uB_z-wB_x-\dfrac{1}{\sigma}J_y\end{vmatrix},$$

$$\mathbf{g}^{\mathrm{MGD}}+\mathbf{g}^{\mathrm{VMGD}}=\left|\begin{array}{c}0\\ -\dfrac{1}{\mu_0}B_yB_x\\ \dfrac{1}{\mu_0}\left(\dfrac{1}{2}B^2-B_y^2\right)\\ -\dfrac{1}{\mu_0}B_yB_z\\ \dfrac{1}{\mu_0}\left(E_zB_x-E_xB_z\right)\\ \upsilon B_x-uB_y-\dfrac{1}{\sigma}J_z\\ 0\\ \upsilon B_z-wB_y+\dfrac{1}{\sigma}J_x\end{array}\right|,\quad \mathbf{h}^{\mathrm{MGD}}+\mathbf{h}^{\mathrm{VMGD}}=\left|\begin{array}{c}0\\ -\dfrac{1}{\mu_0}B_zB_x\\ -\dfrac{1}{\mu_0}B_zB_y\\ \dfrac{1}{\mu_0}\left(\dfrac{1}{2}B^2-B_z^2\right)\\ \dfrac{1}{\mu_0}\left(E_xB_y-E_yB_x\right)\\ wB_x-uB_z+\dfrac{1}{\sigma}J_y\\ wB_y-\upsilon B_z-\dfrac{1}{\sigma}J_x\\ 0\end{array}\right|, \tag{3.56}$$

where

$$E_x=\sigma^{-1}J_x-\left(\upsilon B_z-wB_y\right),\ E_y=\sigma^{-1}J_y-\left(\upsilon B_x-uB_z\right);\ E_z=\sigma^{-1}J_z-\left(uB_y-vB_x\right);$$

$$J_x=\frac{1}{\mu_0}\left(\frac{\partial B_z}{\partial y}-\frac{\partial B_y}{\partial z}\right),\ J_y=\frac{1}{\mu_0}\left(\frac{\partial B_x}{\partial z}-\frac{\partial B_z}{\partial x}\right),\ J_z=\frac{1}{\mu_0}\left(\frac{\partial B_y}{\partial x}-\frac{\partial B_x}{\partial y}\right).$$

3.6.2 Completely conservative form of the MHD equations

In this section, we rewrite eqs. (3.56) in the form containing results of Sections 3.3 and 3.4:

$$\mathbf{U}=\left|\begin{array}{c}\rho\\ \rho u\\ \rho\upsilon\\ \rho w\\ \rho E+p_m\\ B_x\\ B_y\\ B_z\end{array}\right|,\ \mathbf{f}^{\mathrm{Eu}}=\left|\begin{array}{c}\rho u\\ \rho u^2+p\\ \rho u\upsilon\\ \rho uw\\ \rho u\left(E+p/\rho\right)\\ 0\\ 0\\ 0\end{array}\right|,\ \mathbf{f}^{\mathrm{NS}}=\left|\begin{array}{c}0\\ -\tau_{xx}\\ -\tau_{xy}\\ -\tau_{xz}\\ -\left(u\tau_{xx}+\upsilon\tau_{xy}+w\tau_{xz}+\dfrac{\lambda}{c_{\mathrm{v}}}\dfrac{\partial e}{\partial x}\right)\\ 0\\ 0\\ 0\end{array}\right|,$$

$$\mathbf{g}^{\mathrm{Eu}} = \begin{vmatrix} \rho v \\ \rho v u \\ \rho v^2 + p \\ \rho v w \\ \rho v\left(E + \dfrac{p}{\rho}\right) \\ 0 \\ 0 \\ 0 \end{vmatrix}, \quad \mathbf{g}^{\mathrm{NS}} = \begin{vmatrix} 0 \\ -\tau_{yx} \\ -\tau_{yy} \\ -\tau_{yz} \\ -\left(u\tau_{yx} + v\tau_{yy} + w\tau_{yz} + \dfrac{\lambda}{c_{\mathrm{v}}}\dfrac{\partial e}{\partial y}\right) \\ 0 \\ 0 \\ 0 \end{vmatrix},$$

$$\mathbf{h}^{\mathrm{Eu}} = \begin{vmatrix} \rho w \\ \rho w u \\ \rho w v \\ \rho w^2 + p \\ \rho w\left(E + \dfrac{p}{\rho}\right) \\ 0 \\ 0 \\ 0 \end{vmatrix}, \quad \mathbf{g}^{\mathrm{NS}} = \begin{vmatrix} 0 \\ -\tau_{yx} \\ -\tau_{yy} \\ -\tau_{yz} \\ -\left(u\tau_{yx} + v\tau_{yy} + w\tau_{yz} + \dfrac{\lambda}{c_{\mathrm{v}}}\dfrac{\partial e}{\partial y}\right) \\ 0 \\ 0 \\ 0 \end{vmatrix},$$

$$\mathbf{f}^{\mathrm{MGD}} = \begin{vmatrix} 0 \\ \dfrac{1}{\mu_0}\left(\dfrac{1}{2}B^2 - B_x^2\right) \\ -\dfrac{1}{\mu_0}B_xB_y \\ -\dfrac{1}{\mu_0}B_xB_z \\ \dfrac{1}{\mu_0}uB^2 - \dfrac{1}{\mu_0}B_x\left(uB_x + vB_y + wB_z\right) \\ 0 \\ -vB_x + uB_y \\ -wB_x + uB_z \end{vmatrix} = \begin{vmatrix} 0 \\ -\dfrac{1}{2\mu_0}\left(B_x^2 - B_y^2 - B_z^2\right) \\ -\dfrac{1}{\mu_0}B_xB_y \\ -\dfrac{1}{\mu_0}B_xB_z \\ 2up_m - \dfrac{1}{\mu_0}B_x\left(uB_x + vB_y + wB_z\right) \\ 0 \\ -vB_x + uB_y \\ -wB_x + uB_z \end{vmatrix},$$

$$\mathbf{f}^{\text{VMGD}} = \left| \begin{array}{c} 0 \\ 0 \\ 0 \\ 0 \\ \dfrac{1}{\mu_0^2 \sigma} \left[B_y \left(\dfrac{\partial B_x}{\partial y} - \dfrac{\partial B_y}{\partial x} \right) + B_z \left(\dfrac{\partial B_x}{\partial z} - \dfrac{\partial B_z}{\partial x} \right) \right] \\ 0 \\ \dfrac{1}{\mu_0 \sigma} \left(\dfrac{\partial B_y}{\partial x} - \dfrac{\partial B_x}{\partial y} \right) \\ -\dfrac{1}{\mu_0 \sigma} \left(\dfrac{\partial B_x}{\partial z} - \dfrac{\partial B_z}{\partial x} \right) \end{array} \right|,$$

$$\mathbf{g}^{\text{MGD}} = \left| \begin{array}{c} 0 \\ -\dfrac{1}{\mu_0} B_y B_x \\ \dfrac{1}{\mu_0} \left(\dfrac{1}{2} B^2 - B_y^2 \right) \\ -\dfrac{1}{\mu_0} B_y B_z \\ \dfrac{1}{\mu_0} \upsilon B^2 - \dfrac{1}{\mu_0} B_y \left(u B_x + \upsilon B_y + w B_z \right) \\ -u B_y + \upsilon B_x \\ 0 \\ -w B_y + \upsilon B_z \end{array} \right| = \left| \begin{array}{c} 0 \\ -\left(\dfrac{1}{\mu_0} \right) B_y B_x \\ -\dfrac{1}{2\mu_0} \left(B_y^2 - B_x^2 - B_z^2 \right) \\ -\dfrac{1}{\mu_0} B_y B_z \\ 2 \upsilon p_m - \dfrac{1}{\mu_0} B_y \left(u B_x + \upsilon B_y + w B_z \right) \\ -u B_y + \upsilon B_x \\ 0 \\ -w B_y + \upsilon B_z \end{array} \right|,$$

$$\mathbf{g}^{\text{VMGD}} = \left| \begin{array}{c} 0 \\ 0 \\ 0 \\ 0 \\ \dfrac{1}{\mu_0^2 \sigma} \left[B_z \left(\dfrac{\partial B_y}{\partial z} - \dfrac{\partial B_z}{\partial y} \right) + B_x \left(\dfrac{\partial B_y}{\partial x} - \dfrac{\partial B_x}{\partial y} \right) \right] \\ -\dfrac{1}{\mu_0 \sigma} \left(\dfrac{\partial B_y}{\partial x} - \dfrac{\partial B_x}{\partial y} \right) \\ 0 \\ \dfrac{1}{\mu_0 \sigma} \left(\dfrac{\partial B_z}{\partial y} - \dfrac{\partial B_y}{\partial z} \right) \end{array} \right|,$$

$$\mathbf{h}^{\mathrm{MGD}} = \left| \begin{array}{c} 0 \\ -\dfrac{1}{\mu_0} B_z B_x \\ -\dfrac{1}{\mu_0} B_z B_y \\ \dfrac{1}{\mu_0}\left(\dfrac{1}{2}B^2 - B_z^2\right) \\ \dfrac{1}{\mu_0} w B^2 - \dfrac{1}{\mu_0} B_z\left(uB_x + vB_y + wB_z\right) \\ -uB_z + wB_x \\ -vB_z + wB_y \\ 0 \end{array} \right| = \left| \begin{array}{c} 0 \\ -\dfrac{1}{\mu_0} B_z B_x \\ -\dfrac{1}{\mu_0} B_z B_y \\ -\dfrac{1}{2\mu_0}\left(B_z^2 - B_x^2 - B_y^2\right) \\ 2wp_m - \dfrac{1}{\mu_0} B_z\left(uB_x + vB_y + wB_z\right) \\ -uB_z + wB_x \\ -vB_z + wB_y \\ 0 \end{array} \right|,$$

$$\mathbf{h}^{\mathrm{VMGD}} = \left| \begin{array}{c} 0 \\ 0 \\ 0 \\ 0 \\ \dfrac{1}{\mu_0^2 \sigma}\left[B_x\left(\dfrac{\partial B_z}{\partial x} - \dfrac{\partial B_x}{\partial z}\right) + B_y\left(\dfrac{\partial B_z}{\partial y} - \dfrac{\partial B_y}{\partial z}\right)\right] \\ \dfrac{1}{\mu_0 \sigma}\left(\dfrac{\partial B_x}{\partial z} - \dfrac{\partial B_z}{\partial x}\right) \\ -\dfrac{1}{\mu_0 \sigma}\left(\dfrac{\partial B_z}{\partial y} - \dfrac{\partial B_y}{\partial z}\right) \\ 0 \end{array} \right|. \tag{3.57}$$

Set of eqs. (3.55)–(3.57) can be computed with the use of any computing flux method. However, for the sake of convenience, it is recommended to normalize the system. Possible variants of such a renormalization are discussed in the following section.

3.7 The flux form of MHD equations in a dimensionless form

3.7.1 Definition of the normalizing parameters

Let us choose the following normalizing parameters in an undisturbed medium

$$\rho_*,\ c_{p^*},\ c_{V^*},\ \lambda_*,\ \sigma_*,\ \mu_*,\ B_*,\ L_*,$$

where ρ_* is the density of the undisturbed medium; c_{p^*}, c_{V^*} are the specific thermal capacities at constant pressure and volume; λ_* is the thermal conduction; σ_* is the electroconductivity; μ_* is the coefficient of dynamic viscosity; B_* is the induction of magnetic field; L_* is the characteristic size, dictated by the physical formulation of the task.

One can see that pressure p_* is not used in an explicit form as the normalizing parameter. However, the definition of this value is necessary for the definition of characteristic velocity and estimation of the temperature of the undisturbed medium.

So, the choice of density and pressure allows to introduce a characteristic velocity $V_* = \sqrt{\gamma p_*/\rho_*}$, which is the sound velocity in the medium, where $\gamma = c_{p^*}/c_{v^*}$.

For example in Table 3.1, typical parameters of the Earth's ionosphere at different heights are presented.

Table 3.1: Parameters of the Earth's ionosphere.

H, km	p_*, Pa	ρ_*, kg/m³	V_*, m/s
150	4.49×10^{-4}	2.00×10^{-9}	560
200	2.53×10^{-5}	2.50×10^{-10}	691
300	2.72×10^{-6}	1.90×10^{-11}	802
400	1.45×10^{-6}	2.80×10^{-12}	851
500	3.02×10^{-7}	5.20×10^{-13}	902

The velocity of a sound V_* has been calculated at $\gamma = 1.4$. The magnetic induction in a terrestrial ionosphere varies over a wide range, but it is possible to choose characteristic value $B_* = 3 \times 10^{-5} - 5 \times 10^{-5}$ T.

If to mark with a tilde all dimensional parameters and functions, and to rewrite the system of MHD equations in a dimensionless form, there will be the following functions in the system (dimensionless parameters and functions are not marked in anyway):

$$(x, y, z) = \frac{(\tilde{x}, \tilde{y}, \tilde{z})}{L_*}, \quad \rho = \frac{\tilde{\rho}}{\rho_*}, \quad p = \frac{\tilde{p}}{\rho V_*^2}, \quad (u, v, w) = \frac{(\tilde{u}, \tilde{v}, \tilde{w})}{V_*},$$

$$E = \frac{\tilde{E}}{V_*^2}, \quad e = \frac{\tilde{e}}{V_*^2}, \quad \lambda = \frac{\tilde{\lambda}}{\lambda_*}, \quad \mu = \frac{\tilde{\mu}}{\mu_*}, \quad \sigma = \frac{\tilde{\sigma}}{\sigma_*}, \quad c_p = \frac{\tilde{c}_p}{c_{p^*}}, \quad c_v = \frac{\tilde{c}_v}{c_{v^*}},$$

$$(B_x, B_y, B_z) = \frac{(\tilde{B}_x, \tilde{B}_y, \tilde{B}_z)}{B_*}, \quad (E_x, E_y, E_z) = \frac{(\tilde{E}_x, \tilde{E}_y, \tilde{E}_z)}{E_*},$$

$$(J_x, J_y, J_z) = \frac{(\tilde{J}_x, \tilde{J}_y, \tilde{J}_z)}{J_*}, \quad J_* = \frac{\mu_0 B_*}{L_*}, \quad E_* = V_* B_*.$$

The renormalization procedure gives the well-known dimensionless complexes. These are:

$\mathrm{Re}_* = \dfrac{\rho_* V_* L_*}{\mu_*}$ is the Reynolds number;

$\mathrm{Pr}_* = \dfrac{\mu_* c_{p^*}}{\lambda_*}$ is the Prandtl number;

$\gamma = \gamma_* = \dfrac{c_{p^*}}{c_{v^*}}$ is the adiabatic index;

$\mathrm{Re}_{\mathrm{M}^*} = \mu_0 \sigma_* V_* L_*$ is the magnetic Reynolds number;

$\mathrm{M}_{\mathrm{A}^*} = \dfrac{V_*}{B_*}\sqrt{\mu_0 \rho_*} = \dfrac{V_*}{V_{\mathrm{A}^*}}$ is the Alfvén Mach number;

$V_{\mathrm{A}^*} = \dfrac{B_*}{\sqrt{\mu_0 \rho_*}}$ is the Alfvén velocity.

If there is a need to introduce the Mach number as one of the dimensionless complexes, it is necessary to use a characteristic velocity of the investigated process (e.g.,V_0) instead of a velocity V_*. In this case, instead of criteria $\mathrm{Re}_*, \mathrm{Re}_{\mathrm{M}^*}, \mathrm{M}_{\mathrm{A}^*}$ it is necessary to write $\mathrm{Re}\,\mathrm{M}^{-1}$, $\mathrm{Re}_\mathrm{M}\,\mathrm{M}^{-1}$, $\mathrm{M}_\mathrm{A}\,\mathrm{M}^{-1}$, where $\mathrm{Re} = \rho_* V_0 L_*/\mu_*$, $\mathrm{Re}_\mathrm{M} = \mu_* \sigma_* V_0 L_*$, $\mathrm{M}_\mathrm{A} = (V_0/B_*)\sqrt{\mu_* \rho_*}$, $\mathrm{M} = V_0/V_*$ is the Mach number.

And, at last, if there is a need to introduce dimensionless pressure related to any characteristic pressure, for example, p_*, a factor of $1/\gamma$ will be appeared before pressure gradient components in momentum equations.

3.7.2 Nondimensional system of the MHD equations in flux form

$$\frac{\partial \mathbf{U}}{\partial t} + \frac{\partial \mathbf{f}}{\partial x} + \frac{\partial \mathbf{g}}{\partial y} + \frac{\partial \mathbf{h}}{\partial z} = 0,$$

$$U = \left| \rho \quad \rho u \quad \rho v \quad \rho w \quad \rho E + p_\mathrm{m} \quad B_x \quad B_y \quad B_z \right|^\mathrm{T},$$

$$\mathbf{f}^{\mathrm{Eu}} = \left| \begin{matrix} \rho u \\ \rho u^2 + p \\ \rho u v \\ \rho u w \\ \rho u \left(E + \dfrac{p}{\rho} \right) \\ 0 \\ 0 \\ 0 \end{matrix} \right|, \mathbf{f}^{\mathrm{NS}} = \frac{1}{\mathrm{Re}_*} \left| \begin{matrix} 0 \\ -\tau_{xx} \\ -\tau_{xy} \\ -\tau_{xz} \\ -\left(u\tau_{xx} + v\tau_{xy} + w\tau_{xz} + \dfrac{\gamma}{\mathrm{Pr}_*} \dfrac{\lambda}{c_v} \dfrac{\partial e}{\partial x} \right) \\ 0 \\ 0 \\ 0 \end{matrix} \right|,$$

$$\mathbf{g}^{\mathrm{Eu}} = \left| \begin{array}{c} \rho v \\ \rho v u \\ \rho v^2 + p \\ \rho v w \\ \rho v \left(E + \dfrac{p}{\rho} \right) \\ 0 \\ 0 \\ 0 \end{array} \right|, \mathbf{g}^{\mathrm{NS}} = \frac{1}{\mathrm{Re}_*} \left| \begin{array}{c} 0 \\ -\tau_{yx} \\ -\tau_{yy} \\ -\tau_{yz} \\ -\left(u\tau_{yx} + v\tau_{yy} + w\tau_{yz} + \dfrac{\gamma}{\mathrm{Pr}_*} \dfrac{\lambda}{c_v} \dfrac{\partial e}{\partial y} \right) \\ 0 \\ 0 \\ 0 \end{array} \right|,$$

$$\mathbf{h}^{\mathrm{Eu}} = \left| \begin{array}{c} \rho w \\ \rho w u \\ \rho w v \\ \rho w^2 + p \\ \rho w \left(E + \dfrac{p}{\rho} \right) \\ 0 \\ 0 \\ 0 \end{array} \right|, \mathbf{h}^{\mathrm{NS}} = \frac{1}{\mathrm{Re}_*} \left| \begin{array}{c} 0 \\ -\tau_{zx} \\ -\tau_{zy} \\ -\tau_{zz} \\ -\left(u\tau_{zx} + v\tau_{zy} + w\tau_{zz} + \dfrac{\gamma}{\mathrm{Pr}_*} \dfrac{\lambda}{c_v} \dfrac{\partial e}{\partial z} \right) \\ 0 \\ 0 \\ 0 \end{array} \right|,$$

$$\mathbf{f}^{\mathrm{MGD}} = \left| \begin{array}{c} 0 \\ \dfrac{1}{\mathrm{M}_{\mathrm{A}*}^2} \left(0.5B^2 - B_x^2 \right) \\ -\dfrac{1}{\mathrm{M}_{\mathrm{A}*}^2} B_x B_y \\ -\dfrac{1}{\mathrm{M}_{\mathrm{A}*}^2} B_x B_z \\ \dfrac{1}{\mathrm{M}_{\mathrm{A}*}^2} \left[uB^2 - B_x \left(uB_x + vB_y + wB_z \right) \right] \\ 0 \\ -vB_x + uB_y \\ -wB_x + uB_z \end{array} \right| = \left| \begin{array}{c} 0 \\ -\dfrac{1}{2\mathrm{M}_{\mathrm{A}*}^2} \left(B_x^2 - B_y^2 - B_z^2 \right) \\ -\dfrac{1}{\mathrm{M}_{\mathrm{A}*}^2} B_x B_y \\ -\dfrac{1}{\mathrm{M}_{\mathrm{A}*}^2} B_x B_z \\ 2up_{\mathrm{m}} - \dfrac{1}{\mathrm{M}_{\mathrm{A}*}^2} B_x \left(uB_x + vB_y + wB_z \right) \\ 0 \\ -vB_x + uB_y \\ -wB_x + uB_z \end{array} \right|,$$

$$\mathbf{f}^{\mathrm{VMGD}} = \left| \begin{array}{c} 0 \\ 0 \\ 0 \\ 0 \\ \frac{1}{\mathrm{Re}_{\mathrm{M}^*}\mathrm{M}_{\mathrm{A}^*}^2\sigma}\left[B_y\left(\frac{\partial B_x}{\partial y} - \frac{\partial B_y}{\partial x}\right) + B_z\left(\frac{\partial B_x}{\partial z} - \frac{\partial B_z}{\partial x}\right)\right] \\ 0 \\ \frac{1}{\mathrm{Re}_{\mathrm{M}^*}\sigma}\left(\frac{\partial B_y}{\partial x} - \frac{\partial B_x}{\partial y}\right) \\ -\frac{1}{\mathrm{Re}_{\mathrm{M}^*}\sigma}\left(\frac{\partial B_x}{\partial z} - \frac{\partial B_z}{\partial x}\right) \end{array} \right|,$$

$$\mathbf{g}^{\mathrm{MGD}} = \left| \begin{array}{c} 0 \\ -\frac{1}{\mathrm{M}_{\mathrm{A}^*}^2} B_y B_x \\ \frac{1}{\mathrm{M}_{\mathrm{A}^*}^2}\left(0.5B^2 - B_y^2\right) \\ -\frac{1}{\mathrm{M}_{\mathrm{A}^*}^2} B_y B_z \\ \frac{1}{\mathrm{M}_{\mathrm{A}^*}^2}\left[\upsilon B^2 - B_y\left(uB_x + \upsilon B_y + wB_z\right)\right] \\ -uB_y + \upsilon B_x \\ 0 \\ -wB_y + \nu B_z \end{array} \right| = \left| \begin{array}{c} 0 \\ -\frac{1}{\mathrm{M}_{\mathrm{A}^*}^2} B_y B_x \\ -\frac{1}{\mathrm{M}_{\mathrm{A}^*}^2}\left(B_y^2 - B_x^2 - B_z^2\right) \\ -\frac{1}{\mathrm{M}_{\mathrm{A}^*}^2} B_y B_z \\ 2\upsilon p_{\mathrm{m}} - \frac{1}{\mathrm{M}_{\mathrm{A}^*}^2} B_y\left(uB_x + \upsilon B_y + wB_z\right) \\ -uB_y + \upsilon B_x \\ 0 \\ -wB_y + \upsilon B_z \end{array} \right|,$$

$$\mathbf{g}^{\mathrm{VMGD}} = \left| \begin{array}{c} 0 \\ 0 \\ 0 \\ 0 \\ \frac{1}{\mathrm{Re}_{\mathrm{M}^*}\mathrm{M}_{\mathrm{A}^*}^2\sigma}\left[B_z\left(\frac{\partial B_y}{\partial z} - \frac{\partial B_z}{\partial y}\right) + B_x\left(\frac{\partial B_y}{\partial x} - \frac{\partial B_x}{\partial y}\right)\right] \\ -\frac{1}{\mathrm{Re}_{\mathrm{M}^*}\sigma}\left(\frac{\partial B_y}{\partial x} - \frac{\partial B_x}{\partial y}\right) \\ 0 \\ \frac{1}{\mathrm{Re}_{\mathrm{M}^*}\sigma}\left(\frac{\partial B_z}{\partial y} - \frac{\partial B_y}{\partial z}\right) \end{array} \right|,$$

$$\mathbf{h}^{\mathrm{MGD}} = \left|\begin{array}{c} 0 \\ -\dfrac{1}{\mathrm{M}_{\mathrm{A}^*}^2} B_z B_x \\ -\dfrac{1}{\mathrm{M}_{\mathrm{A}^*}^2} B_z B_y \\ \dfrac{1}{\mathrm{M}_{\mathrm{A}^*}^2}\left(0.5B^2 - B_z^2\right) \\ \dfrac{1}{\mathrm{M}_{\mathrm{A}^*}^2}\left[wB^2 - B_z\left(uB_x + vB_y + wB_z\right)\right] \\ -uB_z + wB_x \\ -vB_z + wB_y \\ 0 \end{array}\right| = \left|\begin{array}{c} 0 \\ -\dfrac{1}{\mathrm{M}_{\mathrm{A}^*}^2} B_z B_x \\ -\dfrac{1}{\mathrm{M}_{\mathrm{A}^*}^2} B_z B_y \\ -\dfrac{1}{\mathrm{M}_{\mathrm{A}^*}^2}\left(B_z^2 - B_x^2 - B_y^2\right) \\ 2wp_{\mathrm{m}} - \dfrac{1}{\mathrm{M}_{\mathrm{A}^*}^2} B_z\left(uB_x + vB_y + wB_z\right) \\ -uB_z + wB_x \\ -vB_z + wB_y \\ 0 \end{array}\right|,$$

$$\mathbf{h}^{\mathrm{VMGD}} = \left|\begin{array}{c} 0 \\ 0 \\ 0 \\ 0 \\ \dfrac{1}{\mathrm{Re}_{\mathrm{M}^*}\mathrm{M}_{\mathrm{A}^*}^2\sigma}\left[B_x\left(\dfrac{\partial B_z}{\partial x} - \dfrac{\partial B_x}{\partial z}\right) + B_y\left(\dfrac{\partial B_z}{\partial y} - \dfrac{\partial B_y}{\partial z}\right)\right] \\ \dfrac{1}{\mathrm{Re}_{\mathrm{M}^*}\sigma}\left(\dfrac{\partial B_x}{\partial z} - \dfrac{\partial B_z}{\partial x}\right) \\ -\dfrac{1}{\mathrm{Re}_{\mathrm{M}^*}\sigma}\left(\dfrac{\partial B_z}{\partial y} - \dfrac{\partial B_y}{\partial z}\right) \\ 0 \end{array}\right|. \tag{3.58}$$

In the presented equations, dimensionless magnetic pressure is calculated under the formula $p_{\mathrm{m}} = B^2/2\mathrm{M}_{\mathrm{A}^*}^2$, where B^2 is the squared dimensionless magnetic induction.

Let's consider further modification of vectors $\mathbf{f}^{\mathrm{VMGD}}$, $\mathbf{g}^{\mathrm{VMGD}}$, and $\mathbf{h}^{\mathrm{VMGD}}$ with the purpose of deriving more convenient calculation relations. Only the vector components corresponding to the energy conservation equation (the fifth line of each vector) are subject to transformation. Writing

$$\mathbf{f}_{\mathrm{mod}}^{\mathrm{VMGD}} = \mathbf{f}^{\mathrm{VMGD}} + \left|0 \quad 0 \quad 0 \quad 0 \quad B_x\left(\vec{\nabla}\cdot\vec{B}\right) \quad 0 \quad 0 \quad 0\right|^{\mathrm{T}},$$

$$\mathbf{g}_{\mathrm{mod}}^{\mathrm{VMGD}} = \mathbf{g}^{\mathrm{VMGD}} + \left|0 \quad 0 \quad 0 \quad 0 \quad B_y\left(\vec{\nabla}\cdot\vec{B}\right) \quad 0 \quad 0 \quad 0\right|^{\mathrm{T}},$$

$$\mathbf{h}_{\mathrm{mod}}^{\mathrm{VMGD}} = \mathbf{h}^{\mathrm{VMGD}} + \left|0 \quad 0 \quad 0 \quad 0 \quad B_z\left(\vec{\nabla}\cdot\vec{B}\right) \quad 0 \quad 0 \quad 0\right|^{\mathrm{T}},$$

one can receive

$$\mathbf{f}_{\mathrm{mod}}^{\mathrm{VMGD}} = \left| \begin{array}{c} 0 \\ 0 \\ 0 \\ 0 \\ \dfrac{1}{\mathrm{Re}_{\mathrm{M}^*}\mathrm{M}_{\mathrm{A}^*}^2\sigma} \left[\dfrac{\partial}{\partial x} \left(\dfrac{B_x^2 - B_y^2 - B_z^2}{2} \right) + \dfrac{\partial B_x \partial B_y}{\partial y} + \dfrac{\partial B_x \partial B_z}{\partial z} \right] \\ 0 \\ \dfrac{1}{\mathrm{Re}_{\mathrm{M}^*}\sigma} \left(\dfrac{\partial B_y}{\partial x} - \dfrac{\partial B_x}{\partial y} \right) \\ \dfrac{1}{\mathrm{Re}_{\mathrm{M}^*}\sigma} \left(\dfrac{\partial B_z}{\partial x} - \dfrac{\partial B_x}{\partial z} \right) \end{array} \right|, \tag{3.59}$$

$$\mathbf{g}_{\mathrm{mod}}^{\mathrm{VMGD}} = \left| \begin{array}{c} 0 \\ 0 \\ 0 \\ 0 \\ \dfrac{1}{\mathrm{Re}_{\mathrm{M}^*}\mathrm{M}_{\mathrm{A}^*}^2\sigma} \left[\dfrac{\partial B_x B_y}{\partial x} + \dfrac{\partial}{\partial y} \left(\dfrac{B_y^2 - B_x^2 - B_z^2}{2} \right) + \dfrac{\partial B_y B_z}{\partial z} \right] \\ \dfrac{1}{\mathrm{Re}_{\mathrm{M}^*}\sigma} \left(\dfrac{\partial B_x}{\partial y} - \dfrac{\partial B_y}{\partial x} \right) \\ 0 \\ \dfrac{1}{\mathrm{Re}_{\mathrm{M}^*}\sigma} \left(\dfrac{\partial B_z}{\partial y} - \dfrac{\partial B_y}{\partial z} \right) \end{array} \right|, \tag{3.60}$$

$$\mathbf{h}_{\mathrm{mod}}^{\mathrm{VMGD}} = \left| \begin{array}{c} 0 \\ 0 \\ 0 \\ 0 \\ \dfrac{1}{\mathrm{Re}_{\mathrm{M}^*}\mathrm{M}_{\mathrm{A}^*}^2\sigma} \left[\dfrac{\partial B_x B_z}{\partial x} + \dfrac{\partial B_y B_z}{\partial y} + \dfrac{\partial}{\partial z} \left(\dfrac{B_z^2 - B_x^2 - B_y^2}{2} \right) \right] \\ \dfrac{1}{\mathrm{Re}_{\mathrm{M}^*}\sigma} \left(\dfrac{\partial B_x}{\partial z} - \dfrac{\partial B_z}{\partial x} \right) \\ \dfrac{1}{\mathrm{Re}_{\mathrm{M}^*}\sigma} \left(\dfrac{\partial B_y}{\partial z} - \dfrac{\partial B_z}{\partial y} \right) \\ 0 \end{array} \right|. \tag{3.61}$$

3.8 The MHD equations in the flux form: the use of pressure instead of specific internal energy

One of effective computational methods for integrating the MHD equations is based on the quasilinear form of the equations. Transfer from one form of the equations to another is carried out with the transformation of a Jacobian from the conservative flux form

$$\frac{\partial \mathbf{U}}{\partial t}+\frac{\partial \mathbf{f}}{\partial x}+\frac{\partial \mathbf{g}}{\partial y}+\frac{\partial \mathbf{h}}{\partial z}=0 \tag{3.62}$$

to the quasilinear form

$$\frac{\partial \mathbf{W}}{\partial t}+A_x\frac{\partial \mathbf{W}}{\partial x}+A_y\frac{\partial \mathbf{W}}{\partial y}+A_z\frac{\partial \mathbf{W}}{\partial z}=0, \tag{3.63}$$

where **W** is the vector of physical primitive variables, for example:

$$\mathbf{W}=\left|\rho \quad u \quad v \quad w \quad e \quad B_x \quad B_y \quad B_z\right|^{\mathrm{T}}, \tag{3.64}$$

or

$$\mathbf{W}=\left|\rho \quad u \quad v \quad w \quad p \quad B_x \quad B_y \quad B_z\right|^{\mathrm{T}}, \tag{3.65}$$

where A_x, A_y, A_z are the Jacobian matrixes, generated by transfer from a set of eqs. (3.62) and (3.63). Such transfer procedure will be considered explicitly in Section 3.9. Here we shall pay attention to the fact that will be used further: *transfer from* eq. (3.62) *to* eq. (3.63) *is fulfilled essentially easier if as a base variable of the energy conservation equation; not the specific internal energy is used but the pressure.*

However, unfortunately, it considerably restricts a class of tasks subject to a solution with this method. For example, problems of viscous heat-conducting flow (the nonisentropic process) with thermal emission sources cannot be solved yet in such a way (without the introduction of any special methods of the linearization or the introduction of some frozen functions, etc.). Together with it, the wide class of problems of MHD satisfies to this condition; therefore, this basic possibility of simplification of a solution should be considered.

Let us rewrite the MHD equations to the conservative form concerning the pressure:

$$\frac{\partial \mathbf{U}}{\partial t}+\frac{\partial \mathbf{f}}{\partial x}+\frac{\partial \mathbf{g}}{\partial y}+\frac{\partial \mathbf{h}}{\partial z}=\mathbf{E}, \tag{3.66}$$

where

$$\mathbf{U}=\begin{vmatrix}\rho\\ \rho u\\ \rho v\\ \rho w\\ \dfrac{p}{\gamma-1}+\dfrac{\rho V^2}{2}+\dfrac{\left(B_x^2+B_y^2+B_z^2\right)}{2\mathrm{M}_{\mathrm{A}*}^2}\\ B_x\\ B_y\\ B_z\end{vmatrix}=\begin{vmatrix}\rho\\ \rho u\\ \rho v\\ \rho w\\ \dfrac{p}{\gamma-1}+\dfrac{\rho}{2}\left(u^2+v^2+w^2\right)+\dfrac{\left(B_x^2+B_y^2+B_z^2\right)}{2\mathrm{M}_{\mathrm{A}*}^2}\\ B_x\\ B_y\\ B_z\end{vmatrix},$$

$$\mathbf{f}^{\mathrm{Eu}}=\begin{vmatrix}\rho u\\ \rho u^2+p\\ \rho uv\\ \rho uw\\ u\dfrac{\gamma p}{\gamma-1}+\rho u\dfrac{1}{2}\left(u^2+v^2+w^2\right)\\ 0\\ 0\\ 0\end{vmatrix},\quad \mathbf{g}^{\mathrm{Eu}}=\begin{vmatrix}\rho v\\ \rho v u\\ \rho v^2+p\\ \rho vw\\ v\dfrac{\gamma p}{\gamma-1}+\rho v\dfrac{1}{2}\left(u^2+v^2+w^2\right)\\ 0\\ 0\\ 0\end{vmatrix},$$

$$\mathbf{h}^{\mathrm{Eu}}=\begin{vmatrix}\rho w\\ \rho wu\\ \rho wv\\ \rho w^2+p\\ w\dfrac{\gamma p}{\gamma-1}+\rho w\dfrac{1}{2}\left(u^2+v^2+w^2\right)\\ 0\\ 0\\ 0\end{vmatrix}.$$

where $\mathbf{E}$ is the vector function of the source terms that are not considered in further transformation operations. Functions $\mathbf{f}^{\mathrm{VMGD}}$, $\mathbf{g}^{\mathrm{VMGD}}$, and $\mathbf{h}^{\mathrm{VMGD}}$ remain invariable. They can be used in the form of (3.58). As to functions $\mathbf{f}^{\mathrm{NS}}$, $\mathbf{g}^{\mathrm{NS}}$ and $\mathbf{h}^{\mathrm{NS}}$ they as breaking an isentropic condition should be either excluded from reviewing, or included in the right-hand side of the equations as some source terms (as it was made in eq. (3.66)). Such an approach to a solution of the Navier–Stokes equations demands special research for each specific case, as in whole it is incorrect. However, it is admissible far from bounding surfaces and also outside of areas of a sharp changing of functions.

3.9 Eigenvectors and eigenvalues of Jacobian matrixes for the transformation of the MHD equations from conservative to the quasilinear form: statement of nonstationary boundary conditions

Formulated in Section 3.7, system of the MHD equations in the flux form possesses the property of conservatism and is rather effective for numerical solutions with the use of a wide class of finite-difference schemes. First of all, among such schemes, it is necessary to note the numerical schemes that are based on the characteristic properties of the equations (Kulikovskiy A.G., et al., 2001). Effectiveness of such schemes is defined by the simplicity of calculation of eigenvalues and eigenvectors of Jacobian matrixes for transformation of the system of equations from the conservative form to quasilinear one, and then to the characteristic form of the equations.

Unfortunately, the system of equations in the conservative form has the serious shortage connected to the definition of boundary conditions on a free surface, in particular for subsonic flow conditions. In a series of publications (Thompson K.W., 1987; 1990; Vanajakshi J.C., et al., 1989; Powell K.G., 1994; Sun M.T., et al., 1995) devoted to this problem, the possibility of use of the characteristic form of boundary conditions with subsequent formulation of corresponding conditions for conservative variables is investigated.

Thus, the problem of determination of eigenvectors and eigenvalues of Jacobian matrixes for transformation from a conservative form to quasilinear is important for further development of methods of MHD equation solution.

3.9.1 Jacobian matrixes of the passage from the conservative form to the quasilinear form of the equations

Let's take for a basis the conservative flux form of MHD equations

$$\frac{\partial \mathbf{U}}{\partial t}+\frac{\partial \mathbf{f}}{\partial x}+\frac{\partial \mathbf{g}}{\partial y}+\frac{\partial \mathbf{h}}{\partial z}=\mathbf{E}, \tag{3.67}$$

where vector functions of conservative variables $\mathbf{U}, \mathbf{f}, \mathbf{g}, \mathbf{h}$ are defined in Sections 3.6 and 3.7. The vectorial function $\mathbf{E}$ contains the variables, which are breaking hyperbolical properties of the equations and are considered further as some source terms of the equations. Our first task will be a reformulation of the set of eq. (3.67) written in the quasilinear form. As a vector of required functions, we shall choose $\mathbf{W} = \mathbf{W_p}$ (see eq. (3.3)).

Then the equation concerning variables $\mathbf{W}$ will have the following preliminary appearance:

$$P\frac{\partial \mathbf{W}}{\partial t}+Q_x\frac{\partial \mathbf{W}}{\partial x}+Q_y\frac{\partial \mathbf{W}}{\partial y}+Q_z\frac{\partial \mathbf{W}}{\partial z}=\mathbf{E}, \tag{3.68}$$

where $P,\ Q_x,\ Q_y,\ Q_z$ are the Jacobian matrixes of transformation

$$P=\frac{\partial \mathbf{U}}{\partial \mathbf{W}}, Q_x=\frac{\partial \mathbf{f}}{\partial \mathbf{W}}, Q_y=\frac{\partial \mathbf{g}}{\partial \mathbf{W}}, Q_z=\frac{\partial \mathbf{h}}{\partial \mathbf{W}}, \tag{3.69}$$

which components are defined under formulas

$$P_{ij}=\frac{\partial U_i}{\partial W_j},\ (Q_x)_{ij}=\frac{\partial f_i}{\partial W_j},\ \left(Q_y\right)_{ij}=\frac{\partial g_i}{\partial W_j},\ (Q_z)_{ij}=\frac{\partial h_i}{\partial W_j}, \tag{3.70}$$

where i is the index of the matrixes line (and simultaneously the serial number of the line in **U**, **f**, **g**, **h** vectors); j is the number of the column (corresponds to the serial number of the line in a vector **W**). Numbers of lines in vectors **U** and **W** are established under the following scheme:

$$\mathbf{U}=\begin{vmatrix} U_1\\ U_2\\ U_3\\ U_4\\ U_5\\ U_6\\ U_7\\ U_8 \end{vmatrix}=\begin{vmatrix} \rho\\ \rho u\\ \rho v\\ \rho w\\ \dfrac{p}{\gamma-1}+\dfrac{\rho}{2}\left(u^2+v^2+w^2\right)+\dfrac{1}{2\mathrm{M}_{\mathrm{A}*}^2}\left(B_x^2+B_y^2+B_z^2\right)\\ B_x\\ B_y\\ B_z \end{vmatrix},$$

$$\mathbf{W}=\begin{vmatrix} W_1 & W_2 & W_3 & W_4 & W_5 & W_6 & W_7 & W_8 \end{vmatrix}^{\mathrm{T}}=\begin{vmatrix} \rho & u & v & w & p & B_x & B_y & B_z \end{vmatrix}^{\mathrm{T}}. \tag{3.71}$$

Further, we shall name the set of variables (3.71) as primitive variables. For deriving the final form of the equation concerning vector function **W** it is necessary to multiply eq. (3.68) at the left by a matrix P^{-1}, which is inverse in relation to matrix P

$$\frac{\partial \mathbf{W}}{\partial t}+A_x\frac{\partial \mathbf{W}}{\partial x}+A_y\frac{\partial \mathbf{W}}{\partial y}+A_z\frac{\partial \mathbf{W}}{\partial z}=\left(P^{-1}\mathbf{E}\right), \tag{3.72}$$

where

$$A_x=P^{-1}Q_x,\ A_y=P^{-1}Q_y,\ A_z=P^{-1}Q_z. \tag{3.73}$$

Now the task is a determination of eigenvectors and eigenvalues of matrixes A_x, A_y, A_z. We shall consider a method of their search on an example of a matrix A_x. Equation (3.72) shall be rewritten in the form

$$\frac{\partial \mathbf{W}}{\partial t}+A_x\frac{\partial \mathbf{W}}{\partial x}+\mathbf{C}=\left(P^{-1}\mathbf{E}\right),\mathbf{C}=A_y\frac{\partial \mathbf{W}}{\partial y}+A_z\frac{\partial \mathbf{W}}{\partial z}. \tag{3.74}$$

Let's define the left eigenvectors of a matrix A_x under the formula

$$\vec{l}_i^{\,\mathrm{T}}A_x=\lambda_i^x\vec{l}_i^{\,\mathrm{T}}, \qquad i=1,2,\ldots,8, \tag{3.75}$$

where $\vec{l}_i^{\,\mathrm{T}}=(l_{i,1},\ l_{i,2},\ l_{i,3},\ \ldots,\ l_{i,8})$ is the transposed matrix of the left vectors; $l_{i,j}$ is the jth eigenvector corresponding to the ith eigenvalue.

Then eigenvalues λ_i of a matrix A_x will be defined from solution of the following equation:

$$\det\left(A_x-\lambda^x I\right)=0, \tag{3.76}$$

where I is the unit matrix. All the eigenvalues are real due to the hyperbolic properties of the system of equations. So the problem of calculation of eigenvalues of a Jacobian matrix is solved. It allows to formulate effective finite-difference schemes of Roe's type (Roe P.L., 1981).

However, as it was already marked, having the defined eigenvalues of transformation matrixes, it is possible to construct nonstationary boundary conditions using characteristic equations. For this purpose, the eigenvectors of a matrix should be determined, which allows defining the following two matrixes:

$$S_x=\begin{pmatrix} l_{1,1} & l_{2,1} & \cdots & l_{i,1} & \cdots & l_{8,1}\\ l_{1,2} & l_{2,2} & \cdots & l_{i,2} & \cdots & l_{8,2}\\ \vdots & & & & & \vdots\\ l_{1,8} & l_{2,8} & \cdots & l_{i,8} & \cdots & l_{8,8}\end{pmatrix},\quad S_x^{-1}=\begin{pmatrix} l_{1,1} & l_{1,2} & \cdots & \cdots & \cdots & l_{1,8}\\ l_{2,1} & l_{2,2} & \cdots & \cdots & \cdots & l_{2,8}\\ \vdots & & & & & \vdots\\ l_{i,1} & l_{i,2} & \cdots & \cdots & \cdots & l_{i,8}\\ \vdots & & & & & \vdots\\ l_{8,1} & l_{8,2} & \cdots & \cdots & \cdots & l_{8,8}\end{pmatrix}. \tag{3.77}$$

Rows of matrix S_x^{-1} are the left eigenvectors, and the matrix S_x has the columns, which are the left eigenvectors. Introduction of these matrixes allows to fulfill sufficiently easy procedure of A_x matrix diagonalization, that is, a determination of the diagonal matrix Λ_x corresponding to A_x:

$$\Lambda_x=S_x^{-1}A_xS_x, \tag{3.78}$$

where diagonal components of a matrix Λ_x are eigenvalues of a matrix A_x:

$$\Lambda_x = \begin{pmatrix} \lambda_1^x & 0 & \dots & 0 \\ 0 & \lambda_2^x & \dots & 0 \\ \vdots & & & \vdots \\ 0 & 0 & \dots & \lambda_8^x \end{pmatrix}. \tag{3.79}$$

The further transformations are rather formal, but laborious. We shall multiply eq. eq. (3.74) by a matrix S_x^{-1} at the left

$$S_x^{-1}\frac{\partial \mathbf{W}}{\partial t} + S_x^{-1}A_x\frac{\partial \mathbf{W}}{\partial x} + S_x^{-1}\mathbf{C} = S_x^{-1}\left(P^{-1}\mathbf{E}\right), \tag{3.80}$$

so that a component-wise notation looks as follows:

$$\vec{l}_j^{\,\mathrm{T}}\frac{\partial \mathbf{W}}{\partial t} + \lambda_j\vec{l}_j^{\,\mathrm{T}}\frac{\partial \mathbf{W}}{\partial x} + \vec{l}_j^{\,\mathrm{T}}\mathbf{C} = \vec{l}_j^{\,\mathrm{T}}\left(P^{-1}\mathbf{E}\right). \tag{3.81}$$

Let's designate

$$L_j^x = \lambda_j\vec{l}_j^{\,\mathrm{T}}\frac{\partial \mathbf{W}}{\partial x} \text{ or } \mathbf{L}^{\mathbf{x}} = S_x^{-1}A_x\frac{\partial \mathbf{W}}{\partial x}. \tag{3.82}$$

Substitute eq. (3.82) into eq. (3.80) and then multiply all terms of the equation from left by S_x:

$$\frac{\partial \mathbf{W}}{\partial t} + S_x\mathbf{L}^{\mathbf{x}} + \mathbf{C} = \left(P^{-1}\mathbf{E}\right). \tag{3.83}$$

Using designation $\mathbf{d_x} = S_x\mathbf{L}^{\mathbf{x}}$ we can finally write

$$\frac{\partial \mathbf{W}}{\partial t} + \mathbf{d_x} + \mathbf{C} = \left(P^{-1}\mathbf{E}\right). \tag{3.84}$$

To use a vector $\mathbf{d_x}$ in the given equation, it is necessary to determine its components by means of a solution of the following set of equations:

$$S_x^{-1}\mathbf{d_x} = \mathbf{L}^{\mathbf{x}}. \tag{3.85}$$

Vectors $\mathbf{d_y}$ and $\mathbf{d_z}$ are introduced similarly, so we have obtained the following set of equations for time derivatives from primitive variables:

$$\frac{\partial \mathbf{W}}{\partial t} = -\,\mathbf{d_x} - \mathbf{d_y} - \mathbf{d_z} - \mathbf{C} + \left(P^{-1}\mathbf{E}\right). \tag{3.86}$$

These equations are also used for determination of the boundary conditions for conservative variables.

3.9.2 Nonstationary boundary conditions

A method of calculation of functions $\mathbf{d_x}$, $\mathbf{d_y}$, $\mathbf{d_z}$ for supersonic and subsonic flows on boundaries is considered in Thompson K.W. (1987). It uses the fact that in the case of a supersonic flow on boundary, all perturbations that have generated in calculation domain are going out from it. But in the case of subsonic flow on the boundary, there will be at least one perturbation coming to the calculated area outside.

Presence of waves, bringing perturbations from outside, can render appreciable influence on a solution of the problem. To exclude their influence on a solution inside the area, Thomson K.W. (1987, 1990) has suggested calculating vectors $\mathbf{L^x}$, $\mathbf{L^y}$, $\mathbf{L^z}$ as follows:

$$\mathbf{L_j} = \lambda_j l_j \frac{\partial \mathbf{W}}{\partial n} \quad \text{for waves going out,} \tag{3.87}$$

$$\mathbf{L_j} = 0 \qquad \text{for entering waves } j = 1, 2, \ldots, 8,$$

where n represents one of the coordinates. As the jth eigenvalue has a physical sense of velocity of a perturbation propagation, then if the right boundary n_R of the calculated domain value $\lambda_j > 0$, the perturbation goes out from the domain, and if $\lambda_j < 0$ it enters in. On the contrary, if on the left boundary n_L: $\lambda_j < 0$, the perturbation goes out from the domain, and if $\lambda_j > 0$ it enters the domain.

Hence, the procedure of determination of eigenvalues of Jacobian matrixes for transfer from one system of functions to another was considered in the given section. It is shown how with the use of these vectors possible to construct nonstationary boundary conditions for primitive variables.

However, it is not always possible to realize this procedure in practice. But before, we indicate some calculation problems connected to the singularity of the Jacobian matrixes of transformation.

3.10 A singularity of Jacobian matrixes used for transformation of the equations formulated in the conservative form

Let's consider the transformation of the set of equations concerning conservative variables

$$\frac{\partial \mathbf{U}}{\partial t} + \frac{\partial \mathbf{f}}{\partial x} + \frac{\partial \mathbf{g}}{\partial y} + \frac{\partial \mathbf{h}}{\partial z} = \mathbf{E}, \tag{3.88}$$

where

$$\mathbf{U} = \begin{vmatrix} \rho \\ \rho u \\ \rho v \\ \rho w \\ \frac{p}{\gamma-1} + \frac{\rho}{2}\left(u^2+v^2+w^2\right) + \frac{1}{2m}\left(B_x^2+B_y^2+B_z^2\right) \\ B_x \\ B_y \\ B_z \end{vmatrix}, \tag{3.89}$$

$$\mathbf{f} = \begin{vmatrix} \rho u & 0 \\ \rho u^2 + p & -\frac{1}{2m}\left(B_x^2 - B_y^2 - B_z^2\right) \\ \rho u v & -\frac{1}{m}B_xB_y \\ \rho u w & -\frac{1}{m}B_xB_z \\ u\frac{\gamma p}{\gamma-1} + \rho u\frac{u^2+v^2+w^2}{2} & +\frac{1}{m}\left[u\left(B_x^2+B_y^2+B_z^2\right) - B_x\left(uB_x+vB_y+wB_z\right)\right] \\ 0 & \\ -vB_x + uB_y & \\ -wB_x + uB_z & \end{vmatrix} \tag{3.90}$$

$$\mathbf{g} = \begin{vmatrix} \rho v & 0 \\ \rho v u & -\frac{1}{m}B_yB_x \\ \rho v^2 + p & -\frac{1}{2m}\left(B_y^2 - B_x^2 - B_z^2\right) \\ \rho v w & -\frac{1}{m}B_yB_z \\ v\frac{\gamma p}{\gamma-1} + \rho v\frac{u^2+v^2+w^2}{2} & +\frac{\left[v\left(B_x^2+B_y^2+B_z^2\right) - B_y\left(uB_x+vB_y+wB_z\right)\right]}{m} \\ -uB_y + vB_x & \\ 0 & \\ -wB_y + vB_z & \end{vmatrix}, \tag{3.91}$$

$$
\mathbf{h} = \begin{vmatrix}
\rho w & 0 \\
\rho w u & -\frac{1}{m} B_z B_x \\
\rho w \upsilon & -\frac{1}{m} B_z B_y \\
\rho w^2 + p & -\frac{1}{2m}\left(B_z^2 - B_x^2 - B_y^2\right) \\
w\frac{\gamma p}{\gamma - 1} + \rho w \frac{u^2+\upsilon^2+w^2}{2} & +\frac{1}{m}\left[w\left(B_x^2+B_y^2+B_z^2\right) - B_z\left(uB_x + \upsilon B_y + wB_z\right)\right] \\
-uB_z + wB_x & \\
-\upsilon B_z + wB_y & \\
0 &
\end{vmatrix}. \quad (3.92)
$$

here $m = \mathrm{M}_{\mathrm{A}^*}^2$, and vector functions $\mathbf{f}^{\mathrm{NS}}$, $\mathbf{g}^{\mathrm{NS}}$, $\mathbf{h}^{\mathrm{NS}}$ and $\mathbf{f}^{\mathrm{VMGD}}$, $\mathbf{g}^{\mathrm{VMGD}}$, $\mathbf{h}^{\mathrm{VMGD}}$ are taken on the right-hand side of the equation and included in vector $\mathbf{E}$.

For transfer from the initial set of eq. (3.88) to the system of an intermediate form:

$$
P\frac{\partial \mathbf{W}}{\partial t} + Q_x\frac{\partial \mathbf{W}}{\partial x} + Q_y\frac{\partial \mathbf{W}}{\partial y} + Q_z\frac{\partial \mathbf{W}}{\partial z} = \mathbf{E}, \quad (3.93)
$$

Jacobian matrixes of the following form are turned out:

$$
P = \begin{vmatrix}
1 & 0 & 0 & 0 & 0 & 0 & 0 & 0 \\
u & \rho & 0 & 0 & 0 & 0 & 0 & 0 \\
\upsilon & 0 & \rho & 0 & 0 & 0 & 0 & 0 \\
w & 0 & 0 & \rho & 0 & 0 & 0 & 0 \\
\frac{u^2+\upsilon^2+w^2}{2} & \rho u & \rho v & \rho w & \frac{1}{\gamma-1} & \frac{B_x}{m} & \frac{B_y}{m} & \frac{B_z}{m} \\
0 & 0 & 0 & 0 & 0 & 1 & 0 & 0 \\
0 & 0 & 0 & 0 & 0 & 0 & 1 & 0 \\
0 & 0 & 0 & 0 & 0 & 0 & 0 & 1
\end{vmatrix}, \quad (3.94)
$$

$$Q_x = \begin{vmatrix} u & \rho & 0 & 0 & 0 & 0 & 0 & 0 \\ u^2 & 2\rho u & 0 & 0 & 1 & -\frac{B_x}{m} & \frac{B_y}{m} & \frac{B_z}{m} \\ uv & \rho v & \rho u & 0 & 0 & -\frac{B_y}{m} & -\frac{B_x}{m} & 0 \\ v \quad uw & \rho w & 0 & \rho u & 0 & -\frac{B_z}{m} & 0 & -\frac{B_x}{m} \\ \frac{u\left(u^2+v^2+w^2\right)}{2} & Q^x_{5,2} & Q^x_{5,3} & Q^x_{5,4} & \frac{\gamma u}{\gamma-1} & Q^x_{5,6} & Q^x_{5,7} & Q^x_{5,8} \\ 0 & 0 & 0 & 0 & 0 & 0 & 0 & 0 \\ 0 & -B_y & B_x & 0 & 0 & -v & u & 0 \\ 0 & -B_z & 0 & B_x & 0 & -w & 0 & u \end{vmatrix}, \tag{3.95}$$

where

$$Q^x_{5,2} = \frac{\gamma p}{\gamma-1} + \frac{1}{2}\rho\left(u^2+v^2+w^2\right) + \rho u^2 + \frac{B_y^2+B_z^2}{m};$$

$$Q^x_{5,3} = \rho uv - \frac{B_xB_y}{m};\ Q^x_{5,4} = \rho uw - \frac{B_xB_z}{m};\ Q^x_{5,6} = \frac{vB_y+wB_z}{m};$$

$$Q^x_{5,7} = \frac{2uB_y - vB_x}{m};\ Q^x_{5,8} = \frac{2uB_z - wB_x}{m};$$

$$Q_y = \begin{vmatrix} v & 0 & \rho & 0 & 0 & 0 & 0 & 0 \\ uv & \rho v & \rho u & 0 & 0 & -\frac{B_y}{m} & -\frac{B_x}{m} & 0 \\ v^2 & 0 & 2\rho v & 0 & 1 & \frac{B_x}{m} & -\frac{B_y}{m} & \frac{B_z}{m} \\ vw & 0 & \rho w & \rho v & 0 & 0 & -\frac{B_z}{m} & -\frac{B_y}{m} \\ \frac{v(u^2+v^2+w^2)}{2} & Q^y_{5,2} & Q^y_{5,3} & Q^y_{5,4} & \frac{\gamma v}{\gamma-1} & Q^y_{5,6} & Q^y_{5,7} & Q^y_{5,8} \\ 0 & -B_y & B_x & 0 & 0 & v & -u & 0 \\ 0 & 0 & 0 & 0 & 0 & 0 & 0 & 0 \\ 0 & 0 & B_z & -B_y & 0 & 0 & -w & v \end{vmatrix}, \tag{3.96}$$

where

$$Q^y_{5,2} = \rho\, uv - \frac{B_xB_y}{m};$$

$$Q^y_{5,3} = \frac{\gamma p}{\gamma-1} + \frac{1}{2}\rho\left(u^2+v^2+w^2\right) + \rho v^2 + \frac{B_x^2+B_z^2}{m};\ Q^y_{5,4} = \rho\, vw - \frac{B_yB_z}{m};$$

$$Q^y_{5,6} = \frac{2vB_x - uB_y}{m};\ Q^y_{5,7} = \frac{uB_x + wB_z}{m};\ Q^x_{5,8} = \frac{2vB_z - wB_y}{m};$$

$$Q_z = \begin{vmatrix} w & 0 & \rho & 0 & 0 & 0 & 0 & 0 \\ uw & \rho w & 0 & \rho u & 0 & -\frac{B_z}{m} & 0 & -\frac{B_x}{m} \\ vw & 0 & \rho w & \rho v & 0 & 0 & -\frac{B_z}{m} & -\frac{B_y}{m} \\ w^2 & 0 & 0 & 2\rho w & 1 & \frac{B_x}{m} & \frac{B_y}{m} & -\frac{B_z}{m} \\ \frac{w(u^2+v^2+w^2)}{2} & Q^z_{5,2} & Q^z_{5,3} & Q^z_{5,4} & \frac{\gamma w}{\gamma - 1} & Q^z_{5,6} & Q^z_{5,7} & Q^z_{5,8} \\ 0 & -B_z & 0 & B_x & 0 & w & 0 & -u \\ 0 & 0 & -B_z & B_y & 0 & 0 & w & -v \\ 0 & 0 & 0 & 0 & 0 & 0 & 0 & 0 \end{vmatrix}, \tag{3.97}$$

where

$$Q^z_{5,2} = \rho uw - \frac{B_x B_z}{m};\ Q^z_{5,3} = \rho vw - \frac{B_y B_z}{m};$$

$$Q^z_{5,4} = \frac{\gamma p}{\gamma - 1} + \frac{\rho(u^2+v^2+w^2)}{2} + \rho w^2 + \frac{B_x^2 + B_y^2}{m};$$

$$Q^z_{5,6} = \frac{2wB_x - uB_z}{m};\ Q^z_{5,7} = \frac{2wB_y - vB_z}{m};\ Q^z_{5,8} = \frac{uB_x + vB_y}{m}.$$

The analysis of the presented Jacobian matrixes allows drawing a conclusion on presence in matrixes Q_x, Q_y, and Q_z *the zero* lines corresponding to components of a magnetic induction B_x, B_y and B_z accordingly. In other words, the Jacobian matrixes have degenerated in relation to the corresponding projections of magnetic induction.

To receive the system of MHD equations in a quasilinear form

$$\frac{\partial \mathbf{W}}{\partial t} + A_x \frac{\partial \mathbf{W}}{\partial x} + A_y \frac{\partial \mathbf{W}}{\partial y} + A_z \frac{\partial \mathbf{W}}{\partial z} = \tilde{E}, \tag{3.98}$$

it is necessary to multiply eq. (3.93) from left by a matrix, which is the inverse matrix of P, then $A_x = P^{-1}Q_x$, $A_y = P^{-1}Q_y$, $A_z = P^{-1}Q_z$, $\tilde{\mathbf{E}} = P^{-1}\mathbf{E}$. A structure of the matrix P^{-1} looks like:

$$P=\begin{vmatrix}
1 & 0 & 0 & 0 & 0 & 0 & 0 & 0\\
-\frac{u}{\rho} & \frac{1}{\rho} & 0 & 0 & 0 & 0 & 0 & 0\\
-\frac{v}{\rho} & 0 & \frac{1}{\rho} & 0 & 0 & 0 & 0 & 0\\
-\frac{w}{\rho} & 0 & 0 & \frac{1}{\rho} & 0 & 0 & 0 & 0\\
\frac{\gamma-1}{2}(u^2+v^2+w^2) & -u(\gamma-1) & -v(\gamma-1) & -w(\gamma-1) & (\gamma-1) & -\frac{B(\gamma-1)}{m} & -\frac{B(\gamma-1)}{m} & -\frac{B(\gamma-1)}{m}\\
0 & 0 & 0 & 0 & 0 & 1 & 0 & 0\\
0 & 0 & 0 & 0 & 0 & 0 & 1 & 0\\
0 & 0 & 0 & 0 & 0 & 0 & 0 & 1
\end{vmatrix}$$

Matrix A_x is obtained by multiplication of a matrix Q_x from left by a matrix P^{-1}:

$$A_x=\begin{vmatrix}
u & \rho & 0 & 0 & 0 & 0 & 0 & 0\\
0 & u & 0 & 0 & \frac{1}{\rho} & -\frac{B_x}{\rho m} & \frac{B_y}{\rho m} & \frac{B_z}{\rho m}\\
0 & 0 & u & 0 & 0 & -\frac{B_y}{\rho m} & -\frac{B_x}{\rho m} & 0\\
0 & 0 & 0 & u & 0 & -\frac{B_z}{\rho m} & 0 & -\frac{B_x}{\rho m}\\
0 & \gamma p & 0 & 0 & u & \frac{(\gamma-1)(uB_x+vB_y+wB_z)}{m} & 0 & 0\\
0 & 0 & 0 & 0 & 0 & 0 & 0 & 0\\
0 & B_y & -B_x & 0 & 0 & -v & u & 0\\
0 & B_z & 0 & -B_x & 0 & -w & 0 & u
\end{vmatrix}. \tag{3.99}$$

Eigenvalues of the given matrix have the following appearance:

$$\lambda_1=0,\ \lambda_2=u,\ \lambda_3=u+V_{A,x},\ \lambda_4=u-V_{A,x},\ \lambda_5=u+C_f,$$
$$\lambda_6=u-C_f,\ \lambda_7=u+C_s,\ \lambda_8=u-C_s, \tag{3.100}$$

where

$$C_f=\sqrt{\frac{1}{2}\left[a^2+V_A^2+\sqrt{(a^2+V_A^2)^2-4a^2V_{A,x}^2}\right]}, \tag{3.101}$$

$$C_s=\sqrt{\frac{1}{2}\left[a^2+V_A^2-\sqrt{(a^2+V_A^2)^2-4a^2V_{A,x}^2}\right]}; \tag{3.102}$$

$V_{A,x}=B_x/\sqrt{\rho m}$ is the Alfven velocity of disturbances propagation along the xth component of a magnetic induction; $V_A=\sqrt{B_x^2+B_y^2+B_z^2}/\sqrt{\rho m}$ is the Alfven velocity; C_s, C_f are the velocities of the slow and the fast magnetosonic waves.

It is seen that the singularity of matrix Q_x causes to zero row in matrix A_x. It leads to certain difficulties in calculation of eigenvectors of matrix A_x.

It is necessary to underline that the determined eigenvalues completely correspond to available representations about the presence of the two Alfven waves, four magnetosonic waves, and two simple waves within the bounds of the MHD theory. However, from the received data nothing is possible to tell certain concerning one of those waves.

Let's consider the procedure of eigenvector definition for a matrix A_x. With this purpose we shall take eq. (3.76), having rewritten it in the following form:

$$l_1(u-\lambda)=0,\ l_1\rho+l_2(u-\lambda)+l_5\gamma p+l_7B_y+l_8B_z=0, \tag{3.103}$$

$$l_3(u-\lambda)-l_7B_z=0,\ l_4(u-\lambda)-l_8B_x=0,\ l_2\frac{1}{\rho}+l_5(u-\lambda)=0,$$

$$-l_2\frac{B_x}{\rho m}-l_3\frac{B_y}{\rho m}-l_4\frac{B_z}{\rho m}+l_5(\gamma-1)\frac{uB_z+vB_y+wB_z}{m}-l_7v-l_8w=0,$$

$$l_2\frac{B_y}{\rho m}-l_3\frac{B_x}{\rho m}+l_7(u-\lambda)=0,\ l_2\frac{B_z}{\rho m}-l_4\frac{B_x}{\rho m}+l_8(u-\lambda)=0.$$

Due to the degeneracy of a matrix A_x, the set of eq. (3.103) is redefined. To omit one of the equations, namely the sixth, it is possible to find a matrix of the left eigenvectors for the set of eigenvalues (3.103). As the normalization of a missing eigenvector $(\vec{l}_1)$ can be arbitrary, and simultaneously it should be independent linearly in relation to other vectors, it can be presented in the following form: $l_1=(0,0,0,0,0,0,1,0,0)$.

As a result one can receive a matrix S_x^{-1}, whose lines are the left eigenvectors of a matrix A_x:

$$S_x^{-1}=\begin{vmatrix} 0 & 0 & 0 & 0 & 0 & 1 & 0 & 0 \\ 1 & 0 & 0 & 0 & -\frac{1}{a^2} & 0 & 0 & 0 \\ 0 & 0 & -B_z & B_y & 0 & 0 & \frac{B_z}{\sqrt{\rho m}} & -\frac{B_y}{\sqrt{\rho m}} \\ 0 & 0 & -B_z & B_y & 0 & 0 & -\frac{B_z}{\sqrt{\rho m}} & \frac{B_y}{\sqrt{\rho m}} \\ 0 & \rho C_f & -\frac{\rho C_f B_x B_y}{z_f} & -\frac{\rho C_f B_x B_z}{z_f} & 1 & 0 & \frac{\rho C_f^2 B_y}{z_f} & \frac{\rho C_f^2 B_z}{z_f} \\ 0 & -\rho C_f & \frac{\rho C_f B_x B_y}{z_f} & \frac{\rho C_f B_x B_z}{z_f} & 1 & 0 & \frac{\rho C_f^2 B_y}{z_f} & \frac{\rho C_f^2 B_z}{z_f} \\ 0 & \rho C_s & -\frac{\rho C_s B_x B_y}{z_s} & -\frac{\rho C_s B_x B_z}{z_s} & 1 & 0 & \frac{\rho C_s^2 B_y}{z_s} & \frac{\rho C_s^2 B_z}{z_s} \\ 0 & -\rho C_s & \frac{\rho C_s B_x B_y}{z_s} & \frac{\rho C_s B_x B_z}{z_s} & 1 & 0 & \frac{\rho C_s^2 B_y}{z_s} & \frac{\rho C_s^2 B_z}{z_s} \end{vmatrix}, \tag{3.104}$$

where

$$z_f = \rho m C_f^2 - B_x^2;\ z_s = \rho m C_s^2 - B_x^2;\ a^2 = \gamma p/\rho.$$

We shall pay attention once again that serial numbers of rows in a matrix (3.104) correspond to numbers of eigenvalues (3.100).

For simplification of further calculations we shall introduce the following designations:

$$a_f = \rho C_f \frac{B_x B_y}{z_f},\ a_s = \rho C_s \frac{B_x B_y}{z_s},\ b_f = \rho C_f^2 \frac{B_y}{z_f},\ b_s = \rho C_s^2 \frac{B_y}{z_s},$$

$$p_f = \rho C_f^2 \frac{B_z}{z_f},\ p_s = \rho C_s^2 \frac{B_z}{z_s},\ g_f = \rho C_f,\ g_s = \rho C_s,$$

$$e = \frac{B_y}{\sqrt{\rho m}},\ f = \frac{B_z}{\sqrt{\rho m}},\ h_f = \rho C_f \frac{B_x B_z}{z_f},\ h_s = \rho C_s \frac{B_x B_z}{z_s}.$$

Then the set of equations for determination of the vector function $\mathbf{d_x}$ used in boundary conditions looks like the following:

$$S_x^{-1} \times \begin{pmatrix} d_1 \\ d_2 \\ \vdots \\ d_8 \end{pmatrix}_x = \begin{pmatrix} L_1^x \\ L_2^x \\ \vdots \\ L_8^x \end{pmatrix}, \tag{3.105}$$

where

$$S_x^{-1} = \begin{vmatrix} 0 & 0 & 0 & 0 & 0 & 1 & 0 & 0 \\ 1 & 0 & 0 & 0 & -1/a^2 & 0 & 0 & 0 \\ 0 & 0 & B_y & B_z & 0 & 0 & -e & -f \\ 0 & 0 & B_y & B_z & 0 & 0 & e & f \\ 0 & g_f & -a_f & -h_f & 1 & 0 & b_f & p_f \\ 0 & g_s & -a_s & -h_s & 1 & 0 & b_s & p_s \\ 0 & -g_f & a_f & h_f & 1 & 0 & b_f & p_f \\ 0 & -g_s & a_s & h_s & 1 & 0 & b_s & p_s \end{vmatrix}. \tag{3.106}$$

The set of eq. (3.105) with matrix (3.106) can be solved fairly simply owing to a special kind of the matrix. However, we shall not give this solution here, as it is not used further.

Similarly, components of vector functions $\mathbf{d_y}$ and $\mathbf{d_z}$ are determined. For reference, we shall give the structure of transfer matrixes A_y and A_z:

$$A_y = \begin{vmatrix} v & 0 & \rho & 0 & 0 & 0 & 0 & 0 \\ 0 & v & 0 & 0 & 0 & -\dfrac{B_y}{\rho m} & -\dfrac{B_x}{\rho m} & 0 \\ 0 & 0 & v & 0 & \dfrac{1}{\rho} & \dfrac{B_x}{\rho m} & -\dfrac{B_y}{\rho m} & \dfrac{B_z}{\rho m} \\ 0 & 0 & 0 & v & 0 & 0 & -\dfrac{B_z}{\rho m} & -\dfrac{B_y}{\rho m} \\ 0 & 0 & \gamma p & 0 & v & 0 & \dfrac{(\gamma-1)\left(uB_x + 1vB_y + wB_z\right)}{m} & 0 \\ 0 & -B_y & B_x & 0 & 0 & v & -u & 0 \\ 0 & 0 & 0 & 0 & 0 & 0 & 0 & 0 \\ 0 & 0 & B_z & -B_y & 0 & 0 & -w & v \end{vmatrix}, \qquad (3.107)$$

$$A_z = \begin{vmatrix} w & 0 & 0 & \rho & 0 & 0 & 0 & 0 \\ 0 & w & 0 & 0 & 0 & -\dfrac{B_z}{\rho m} & 0 & -\dfrac{B_x}{\rho m} \\ 0 & 0 & w & 0 & 0 & 0 & -\dfrac{B_z}{\rho m} & -\dfrac{B_y}{\rho m} \\ 0 & 0 & 0 & w & \dfrac{1}{\rho} & \dfrac{B_x}{\rho m} & \dfrac{B_y}{\rho m} & -\dfrac{B_z}{\rho m} \\ 0 & 0 & 0 & \gamma p & w & 0 & 0 & \dfrac{(\gamma-1)\left(uB_x + 1vB_y + wB_z\right)}{m} \\ 0 & -B_z & 0 & B_x & 0 & w & 0 & -u \\ 0 & 0 & -B_z & B_y & 0 & 0 & w & -v \\ 0 & 0 & 0 & 0 & 0 & 0 & 0 & 0 \end{vmatrix}. \qquad (3.108)$$

Thus, it is possible to say that the problem of determination of eigenvectors and eigenvalues for Jacobian transfer matrixes has been solved. But we did not yet consider significant problems that have arisen in connection with the singular character of these matrixes.

In connection with the problem of the singularity of matrixes, we shall mention two works that immediately concern this problem. In Powell K.G. (1994), an attempt to bypass this problem was made by adding to the MHD equations system the vector function of a special kind $\mathbf{F}'\left(\vec{\nabla}\cdot\vec{B}\right)$, where $\mathbf{F}'$ is a vector of functions corresponding to each equation of the complete set:

$$\vec{F}' = \begin{vmatrix} 0 & B_x & B_y & B_z & \left(\vec{V}\cdot B\right) & u & v & w \end{vmatrix}^{\mathrm{T}}.$$

It was shown in Powell K.G. (1994) that such reception eliminates a singularity of transfer matrixes, not breaking their spectral characteristics.

In Sun M.T., et al. (1995), the problem of nonstationary boundary conditions for MHD problems was solved on the basis of the use of transition matrixes that have no singularity.

How singular matrixes have been received on the basis of completely conservative MHD equations is shown in the given section. In the following section, the system of the MHD equations that were not leading to a singularity of transition matrixes will be formulated. Besides, the use of transformations will be shown there, which has helped to receive transfer matrixes used in Sun M.T., et al. (1995).

3.11 System of MHD equations without singularity in transition matrixes

Transformation of MHD equations from completely conservative form (3.88) to the quasilinear form (3.98) has shown that singularities in matrixes A_x, A_y, A_z are caused by singularities in Jacobian matrixes in those lines that correspond to the equations of magnetic induction. The reason for the specified singularity is clear: in the equation, for an xth component of a magnetic induction there is no association with a change in the magnetic field components along the x-axis, and it is similar for the other two components. Its physical reason is also clear: in the ideal magnetogasdynamics the following vectorial connection between components of a magnetic field and a velocity of flow is postulated:

$$\frac{\partial \mathbf{B}}{\partial t} = -\left[\vec{\nabla} \times [\mathbf{V} \times \mathbf{B}]\right].$$

From this equation, it follows that the variation in time of any magnetic field component depends only on the spatial variation of magnetic field components, perpendicular to the given direction.

At the record of the magnetic induction equations in the conservative form, integral conservation laws are satisfied unconditionally, which naturally results in lack of association of the given magnetic field induction component from a variation of the field in the given direction.

If to formulate the equations so that there was an explicit correlation in all directions in them, it is necessary to superimpose additional connection on components of the magnetic field in this case. This connection is the equation $(\vec{\nabla} \cdot \mathbf{B}) = 0$ which should certainly be satisfied. It is also the reason for the necessity of the given equation use at deriving a set of equations of magnetic induction in the nonconservative form:

$$\frac{\partial B_x}{\partial t} = -B_x\left(\frac{\partial v}{\partial y} + \frac{\partial w}{\partial z}\right) + B_y\frac{\partial u}{\partial y} + B_z\frac{\partial u}{\partial z} - u\frac{\partial B_x}{\partial x} - v\frac{\partial B_x}{\partial y} - w\frac{\partial B_x}{\partial z}, \tag{3.109}$$

$$\frac{\partial B_y}{\partial t} = B_x \frac{\partial v}{\partial x} - B_y \left(\frac{\partial u}{\partial x} + \frac{\partial w}{\partial z} \right) + B_z \frac{\partial v}{\partial z} - u \frac{\partial B_y}{\partial x} - v \frac{\partial B_y}{\partial y} - w \frac{\partial B_y}{\partial z}, \tag{3.110}$$

$$\frac{\partial B_z}{\partial t} = B_x \frac{\partial w}{\partial x} + B_y \frac{\partial w}{\partial y} + B_z \left(\frac{\partial u}{\partial x} + \frac{\partial v}{\partial y} \right) - u \frac{\partial B_z}{\partial x} - v \frac{\partial B_z}{\partial y} - w \frac{\partial B_z}{\partial z}. \tag{3.111}$$

This is enough to write down the system of MHD equations in quasilinear form, whose matrixes A_x, A_y, A_z *do not contain singularities* (further we shall mark these matrixes with tildes to distinguish them from singular matrixes):

$$\tilde{A}_x = \left| \begin{matrix} u & \rho & 0 & 0 & 0 & 0 & 0 & 0 \\ 0 & u & 0 & 0 & 1/\rho & 0 & B_y/\rho m & B_z/\rho m \\ 0 & 0 & u & 0 & 0 & -B_y/\rho m & -B_x/\rho m & 0 \\ 0 & 0 & 0 & u & 0 & -B_z/\rho m & 0 & -B_x/\rho m \\ 0 & \gamma p & 0 & 0 & u & 0 & 0 & 0 \\ 0 & 0 & 0 & 0 & 0 & u & 0 & 0 \\ 0 & B_y & -B_x & 0 & 0 & 0 & u & 0 \\ 0 & B_z & 0 & -B_x & 0 & 0 & 0 & u \end{matrix} \right|,$$

$$\tilde{A}_y = \left| \begin{matrix} v & 0 & \rho & 0 & 0 & 0 & 0 & 0 \\ 0 & v & 0 & 0 & 0 & -B_y/\rho m & -B_x/\rho m & 0 \\ 0 & 0 & v & 0 & 1/\rho & B_x/\rho m & 0 & B_z/\rho m \\ 0 & 0 & 0 & v & 0 & 0 & -B_z/\rho m & -B_y/\rho m \\ 0 & 0 & \gamma p & 0 & v & 0 & 0 & 0 \\ 0 & -B_y & B_x & 0 & 0 & v & 0 & 0 \\ 0 & 0 & 0 & 0 & 0 & 0 & v & 0 \\ 0 & 0 & B_z & -B_y & 0 & 0 & 0 & v \end{matrix} \right|,$$

$$\tilde{A}_z = \left| \begin{matrix} w & 0 & 0 & \rho & 0 & 0 & 0 & 0 \\ 0 & w & 0 & 0 & 0 & -B_z/\rho m & 0 & -B_x/\rho m \\ 0 & 0 & w & 0 & 0 & 0 & -B_z/\rho m & -B_y/\rho m \\ 0 & 0 & 0 & w & 1/\rho & B_x/\rho m & B_y/\rho m & 0 \\ 0 & 0 & 0 & \gamma p & w & 0 & 0 & 0 \\ 0 & -B_z & 0 & B_x & 0 & w & 0 & 0 \\ 0 & 0 & -B_z & B_y & 0 & 0 & w & 0 \\ 0 & 0 & 0 & 0 & 0 & 0 & 0 & w \end{matrix} \right|. \tag{3.112}$$

At the representation of matrixes $\tilde{A}_x, \tilde{A}_y, \tilde{A}_z$ in the form of (3.112), some transformations of the MHD equations have been made:

1) In the energy conservation equation, the term connected with Joule heat has been transferred to the right-hand part as a source summand. It does not change the spectral properties of matrixes.
2) In the momentum conservation equations, the structure of the terms connected with the magnetic field is a little bit changed.

So far in the momentum conservation equations, the following terms that correspond to the magnetic field were used (see system (3.88); unnecessary for the transformation is omitted):

$$\frac{\partial \rho u}{\partial t} + \cdots - \frac{1}{2m}\frac{\partial}{\partial x}\left(B_x^2 - B_y^2 - B_z^2\right) - \frac{1}{m}\frac{\partial B_y B_x}{\partial y} - \frac{1}{m}\frac{\partial B_z B_x}{\partial z} = 0,$$

$$\frac{\partial \rho v}{\partial t} + \cdots - \frac{1}{m}\frac{\partial B_x B_y}{\partial x} - \frac{1}{2m}\frac{\partial}{\partial y}\left(B_y^2 - B_x^2 - B_z^2\right) - \frac{1}{m}\frac{\partial B_z B_y}{\partial z} = 0,$$

$$\frac{\partial \rho w}{\partial t} + \cdots - \frac{1}{m}\frac{\partial B_x B_z}{\partial x} - \frac{1}{m}\frac{\partial B_y B_z}{\partial y} - \frac{1}{2m}\frac{\partial}{\partial z}\left(B_z^2 - B_x^2 - B_y^2\right) = 0. \qquad (3.113)$$

Let's select from parentheses in (3.113) derivative of squared components of the magnetic field on corresponding coordinates, and the remaining two terms of each equation we shall differentiate by parts. After that, we shall take an identical condition $\left(\vec{\nabla}\cdot\mathbf{B}\right) = 0$. As a result, we shall receive

$$\frac{\partial \rho u}{\partial t} + \cdots - \frac{1}{2m}\frac{\partial}{\partial x}\left(B_y^2 + B_z^2\right) - \frac{1}{m}B_y\frac{\partial B_x}{\partial y} - \frac{1}{m}B_z\frac{\partial B_x}{\partial z} = 0,$$

$$\frac{\partial \rho v}{\partial t} + \cdots - \frac{1}{m}B_x\frac{\partial B_y}{\partial x} - \frac{1}{2m}\frac{\partial}{\partial y}\left(B_x^2 + B_z^2\right) - \frac{1}{m}B_z\frac{\partial B_y}{\partial z} = 0,$$

$$\frac{\partial \rho w}{\partial t} + \cdots - \frac{1}{m}B_x\frac{\partial B_z}{\partial x} - \frac{1}{m}B_y\frac{\partial B_z}{\partial y} - \frac{1}{2m}\frac{\partial}{\partial z}\left(B_x^2 + B_y^2\right) = 0. \qquad (3.114)$$

However, the set of eq. (3.114) and, accordingly, matrixes (3.112) can be simplified even more regarding the terms that consider force action of a magnetic field. For this purpose, it is enough to pass to the quasilinear form of the equations by such a transformation, at which there will be only required functions under a differential sign. After differentiation of the terms in parentheses by parts, we'll receive the following set of equations:

$$\frac{\partial \rho u}{\partial t} + \cdots + \frac{1}{m}B_y\frac{\partial B_y}{\partial x} + \frac{1}{m}B_z\frac{\partial B_z}{\partial x} - \frac{1}{m}B_y\frac{\partial B_x}{\partial y} - \frac{1}{m}B_z\frac{\partial B_x}{\partial z} = 0,$$

$$\frac{\partial \rho v}{\partial t} + \cdots - \frac{1}{m}B_x\frac{\partial B_y}{\partial x} + \frac{1}{m}B_x\frac{\partial B_x}{\partial y} + \frac{1}{m}B_z\frac{\partial B_z}{\partial y} - \frac{1}{m}B_z\frac{\partial B_y}{\partial z} = 0,$$

$$\frac{\partial \rho w}{\partial t} + \cdots - \frac{1}{m} B_x \frac{\partial B_z}{\partial x} - \frac{1}{m} B_y \frac{\partial B_z}{\partial y} + \frac{1}{m} B_x \frac{\partial B_x}{\partial z} + \frac{1}{m} B_y \frac{\partial B_y}{\partial z} = 0. \tag{3.115}$$

In view of eq. (3.115) matrices in the quasilinear equations take the following form:

$$\tilde{A}_x = \begin{vmatrix} u & \rho & 0 & 0 & 0 & 0 & 0 & 0 \\ 0 & u & 0 & 0 & 1/\rho & 0 & B_y/\rho m & B_z/\rho m \\ 0 & 0 & u & 0 & 0 & 0 & -B_x/\rho m & 0 \\ 0 & 0 & 0 & u & 0 & 0 & 0 & -B_x/\rho m \\ 0 & \gamma p & 0 & 0 & u & 0 & 0 & 0 \\ 0 & 0 & 0 & 0 & 0 & u & 0 & 0 \\ 0 & B_y & -B_x & 0 & 0 & 0 & u & 0 \\ 0 & B_z & 0 & -B_x & 0 & 0 & 0 & u \end{vmatrix},$$

$$\tilde{A}_y = \begin{vmatrix} v & 0 & \rho & 0 & 0 & 0 & 0 & 0 \\ 0 & v & 0 & 0 & 0 & -B_y/\rho m & 0 & 0 \\ 0 & 0 & v & 0 & 1/\rho & B_x/\rho m & 0 & B_z/\rho m \\ 0 & 0 & 0 & v & 0 & 0 & 0 & -B_y/\rho m \\ 0 & 0 & \gamma p & 0 & v & 0 & 0 & 0 \\ 0 & -B_y & B_x & 0 & 0 & v & 0 & 0 \\ 0 & 0 & 0 & 0 & 0 & 0 & v & 0 \\ 0 & 0 & B_z & -B_y & 0 & 0 & 0 & v \end{vmatrix},$$

$$\tilde{A}_z = \begin{vmatrix} w & 0 & 0 & \rho & 0 & 0 & 0 & 0 \\ 0 & w & 0 & 0 & 0 & -B_z/\rho m & 0 & 0 \\ 0 & 0 & w & 0 & 0 & 0 & -B_z/\rho m & 0 \\ 0 & 0 & 0 & w & 1/\rho & B_x/\rho m & B_y/\rho m & 0 \\ 0 & 0 & 0 & \gamma p & w & 0 & 0 & 0 \\ 0 & -B_z & 0 & B_x & 0 & w & 0 & 0 \\ 0 & 0 & -B_z & B_y & 0 & 0 & w & 0 \\ 0 & 0 & 0 & 0 & 0 & 0 & 0 & w \end{vmatrix}. \tag{3.116}$$

It is remarkable that such MHD equations were used in Powell K.G. (1994) exactly in the same form.

Thus, in the given section, the MHD equations were formulated in the form which is physically equivalent to the set of eq. (3.88), however, not containing singular matrixes in the quasilinear form. In the following section, spectral characteristics of matrixes $\tilde{A}_x$, $\tilde{A}_y$, and $\tilde{A}_z$ will be studied, and also relations for the formulation of nonstationary boundary conditions will be received.

3.12 Eigenvalues and eigenvectors of nonsingular matrixes of the quasilinear system of the MHD equations

Let's investigate spectral properties of matrixes $\tilde{A}_x$, $\tilde{A}_y$, and $\tilde{A}_z$.

3.12.1 Matrix $\tilde{A}_x$

Eigenvalues of this matrix have the following expression:

$$\lambda_1 = u,\ \lambda_2 = u,\ \lambda_3 = u + V_{Ax},\ \lambda_4 = u + V_{Ax}, \tag{3.117}$$
$$\lambda_5 = u + C_f,\ \lambda_6 = u - C_f,\ \lambda_7 = u + C_s,\ \lambda_8 = u - C_s,$$

where

$$V_{A,x} = \frac{B_x}{\sqrt{\rho m}} = \frac{B_x}{\sqrt{\rho \mathrm{M}_{\mathrm{A}^*}}};\ a^2 = \frac{\gamma p}{\rho};$$

$$V_A^2 = \frac{B_x^2 + B_y^2 + B_z^2}{\rho m} = \frac{B_x^2 + B_y^2 + B_z^2}{\rho \mathrm{M}_{\mathrm{A}^*}^2};$$

$$C_f = \sqrt{\frac{1}{2}\left[a^2 + V_{\mathrm{A}}^2 + \sqrt{\left(a^2 + V_{\mathrm{A}}^2\right)^2 - 4a^2 V_{\mathrm{A},x}^2}\right]};$$

$$C_s = \sqrt{\frac{1}{2}\left[a^2 + V_A^2 - \sqrt{\left(a^2 + V_{\mathrm{A}}^2\right)^2 - 4a^2 V_{\mathrm{A},x}^2}\right]}.$$

For the determination of the left eigenvectors, it is necessary to solve the set of eq. (3.76). These equations can be written in view of a matrix $\tilde{A}_x$ structure as follows:

$$l_1(u-\lambda) = 0,\ l_1\rho + l_2(u-\lambda) + l_5\gamma p + l_7 B_y + l_8 B_z = 0, \tag{3.118}$$
$$l_3(u-\lambda) - l_7 B_x = 0, l_4(u-\lambda) - l_8 B_x = 0, l_2\rho^{-1} + l_5(u-\lambda) = 0,$$
$$l_6(u-\lambda) = 0,\ l_2\frac{B_y}{\rho m} - l_3\frac{B_x}{\rho m} + l_7(u-\lambda) = 0,$$
$$l_2\frac{B_z}{\rho m} - l_4\frac{B_x}{\rho m} + l_8(u-\lambda) = 0.$$

The procedure for determination of left eigenvectors consists of four stages.

The first stage. Let us suppose $(u-\lambda) = 0$. It means that $l_{1,1}$ is any number, including 0; $\lambda = \lambda_1 = u$. Let $l_{1,1} \neq 0$, also we shall suppose $l_{1,1} = 1$, then from system (3.118), it is followed $l_{1,2} = 0$, $l_{1,3} = 0$, $l_{1,4} = 0$, $l_{1,5} = -(1/a^2)$, $l_{1,6}$ is number, $l_{1,6}$, $l_{1,8} = 0$.

It is necessary to suppose $l_{1,6} = 0$, because having chosen a normalization for $l_{1,6} = 0$, we have no condition for normalization of the $l_{1.6}$ value. So

$$\vec{l}_1^{\mathrm{T}} = \left(1, 0, 0, 0, -\frac{1}{a^2}, 0, 0, 0\right). \tag{3.119}$$

The second stage. Let us suppose $(u - \lambda) = 0$, $\lambda = \lambda_2 = u$, and $l_{2,1} = 0$.

Then we receive all elements $l_{2,i} = 0$, except for $l_{2,6}$. Normalizing $l_{2,6}$ per unit, we receive

$$\vec{l}_2^{\mathrm{T}} = (0, 0, 0, 0, 0, 1, 0, 0). \tag{3.120}$$

Let's notice that the case of $l_{2,6} = 0$ is excluded, as the zero vector will turn out.

The third stage. Let us suppose that $(u - \lambda) \neq 0$, $\lambda = \lambda_{3,4} = u \pm V_{Ax}$, then $\xi^2 = (u - \lambda_{3,4})^2 = V_{Ax}^2$, $\xi = \pm V_{Ax}$.

From the set of eq. (3.118) $l_2 = l_5 = 0$, and remaining components of vectors are connected among themselves by relations

$$\frac{l_3}{l_8} = -\frac{B_x B_z}{\xi B_y}, \quad \frac{l_4}{l_8} = \frac{B_x}{\xi}, \quad \frac{l_7}{l_8} = -\frac{B_z}{B_y}.$$

Supposing $l_8 = 1$, we obtain

$$\vec{l}_{3,4}^{T} = \left(0, 0, -\frac{B_x B_z}{\xi B_y}, \frac{B_x}{\xi}, 0, 0, -\frac{B_z}{B_y}, 1\right).$$

Or, choosing more obvious normalization, we have:

When $\lambda_3 = u + V_{Ax}$, $\xi = -V_{Ax}$

$$\vec{l}_3^{T} = \left(0, 0, -B_z, B_y, 0, 0, +\frac{B_z}{\sqrt{\rho m}}, -\frac{B_y}{\sqrt{\rho m}}\right), \tag{3.121}$$

and at $\lambda_4 = u - V_{Ax}$, $\xi = +V_{Ax}$

$$\vec{l}_4^{T} = \left(0, 0, -B_z, B_y, 0, 0, -\frac{B_z}{\sqrt{\rho m}}, +\frac{B_y}{\sqrt{\rho m}}\right). \tag{3.122}$$

The fourth stage. Let us suppose $(u - \lambda_k) \neq 0$, $\xi^2 \neq V_{Ax}^2$, $k = 5, 6, 7, 8$.

From the set of eq. (3.118), the following relations can be received:

$l_{k,1} = 0$, $l_{k,2} = 1$ (here we use unconditioned normalization),

$$l_{k,3} = \frac{B_y B_x}{B_x^2 - \rho m \xi^2},$$

$$l_{k,4} = \frac{B_z B_x}{B_x^2 - \rho m \xi^2}, \quad l_{k,5} = -\frac{1}{\rho \xi}, \quad l_{k,7} = \frac{\xi B_y}{B_x^2 - \rho m \xi^2}, \quad l_{k,8} = \frac{\xi B_z}{B_x^2 - \rho m \xi^2}.$$

Choosing more obvious normalization, we receive

$$\vec{l}_k^T = \left(0, -\rho\xi, \frac{\xi\rho B_x B_y}{\rho m\xi^2 - B_x^2}, \frac{\xi\rho B_x B_z}{\rho m\xi^2 - B_x^2}, 1, 0, \frac{\xi^2\rho B_y}{\rho m\xi^2 - B_x^2}, \frac{\xi^2\rho B_z}{\rho m\xi^2 - B_x^2}\right). \tag{3.123}$$

In view of the received formulas (3.119)–(3.123), it is possible to generate the matrix $S_x^{-1}(\lambda_5 = u + C_f, \lambda_6 = u - C_f, \lambda_7 = u + C_s, \lambda_8 = u - C_s)$:

$$S_x^{-1} = \left|\begin{array}{cccccccc} 1 & 0 & 0 & 0 & -\frac{1}{a^2} & 0 & 0 & 0 \\ 0 & 0 & 0 & 0 & 0 & 1 & 0 & 0 \\ 0 & 0 & -B_z & B_y & 0 & 0 & \frac{B_z}{\sqrt{\rho m}} & -\frac{B_y}{\sqrt{\rho m}} \\ 0 & 0 & -B_z & B_y & 0 & 0 & -\frac{B_z}{\sqrt{\rho m}} & \frac{B_y}{\sqrt{\rho m}} \\ 0 & \rho C_f & -\frac{\rho C_f B_x B_y}{z_f} & -\frac{\rho C_f B_x B_z}{z_f} & 1 & 0 & \frac{\rho C_f^2 B_y}{z_f} & \frac{\rho C_f^2 B_z}{z_f} \\ 0 & -\rho C_f & \frac{\rho C_f B_x B_y}{z_f} & \frac{\rho C_f B_x B_z}{z_f} & 1 & 0 & \frac{\rho C_f^2 B_y}{z_f} & \frac{\rho C_f^2 B_y}{z_f} \\ 0 & \rho C_s & -\frac{\rho C_s B_x B_y}{z_s} & -\frac{\rho C_s B_x B_z}{z_s} & 1 & 0 & \frac{\rho C_s^2 B_y}{z_s} & \frac{\rho C_s^2 B_y}{z_s} \\ 0 & -\rho C_s & \frac{\rho C_s B_x B_y}{z_s} & \frac{\rho C_s B_x B_z}{z_s} & 1 & 0 & \frac{\rho C_s^2 B_y}{z_s} & \frac{\rho C_s^2 B_y}{z_s} \end{array}\right|, \tag{3.124}$$

where $z_f = \rho m C_f^2 - B_x^2$, $z_s = \rho m C_s^2 - B_x^2$.

For determination of vector function $\mathbf{d_x}$, it is necessary to solve the set of equations

$$S_x^{-1} \times \begin{pmatrix} d_1 \\ d_2 \\ \vdots \\ d_8 \end{pmatrix}_x = \begin{pmatrix} L_1^x \\ L_2^x \\ \vdots \\ L_8^x \end{pmatrix}, \tag{3.125}$$

where $\mathbf{L_j^x} = \lambda_j \vec{l}_j^T (\partial \mathbf{W}/\partial x)$, $j = 1, 2, \ldots, 8$, where j is the number of the eigenvalue; $\vec{l}_j^T$ is the left eigenvector corresponding to the jth eigenvalue.

Using procedure of Gaussian elimination, the following triangular matrix can be obtained:

$$\tilde{S}_x^{-1} = \begin{vmatrix} 1 & 0 & 0 & 0 & -1/-a^2 & 0 & 0 & 0 \\ 0 & \rho C_f & -E^f_{xy} & -E^f_{xz} & 1 & 0 & E^f_y & E^f_z \\ 0 & 0 & -B_z & B_y & 0 & 0 & \tilde{B}_z & -\tilde{B}_y \\ 0 & 0 & 0 & S_{44} & -\dfrac{-C_f + C_s}{C_f} & 0 & S_{47} & S_{48} \\ 0 & 0 & 0 & 0 & 2 & 0 & 2F^f_y & 2F^f_z \\ 0 & 0 & 0 & 0 & 0 & 1 & 0 & 0 \\ 0 & 0 & 0 & 0 & 0 & 0 & -2\tilde{B}_z & 2\tilde{B}_y \\ 0 & 0 & 0 & 0 & 0 & 0 & 0 & S_{88} \end{vmatrix}, \tag{3.126}$$

where

$$F^f_y = \frac{\rho C_f^2 B_y}{z_f}; \quad F^f_z = \frac{\rho C_f^2 B_z}{z_f}; \quad \tilde{B}_y = \frac{B_z}{\sqrt{\rho m}}; \quad F^s_y = \frac{\rho C_s^2 B_y}{z_s};$$

$$F^s_z = \frac{\rho C_s^2 B_z}{z_s}; \quad \tilde{B}_y = \frac{B_y}{\sqrt{\rho m}}; \quad F^f_{xy} = \frac{\rho C_f B_x B_y}{z_f}; \quad F^s_{xy} = \frac{\rho C_s B_x B_y}{z_s};$$

$$F^f_{xz} = \frac{\rho C_f B_x B_z}{z_f}; \; F^s_{xz} = \frac{\rho C_s B_x B_z}{z_s};$$

$$z_f = \rho m C_f^2 - B_x^2; \; z_s = \rho m C_s^2 - B_x^2;$$

$$S_{44} = \frac{B_z\left(-C_f E^s_{xz} + C_s E^f_{xz}\right) + B_y\left(-C_f E^s_{xy} + C_s E^f_{xy}\right)}{C_f B_z};$$

$$S_{47} = -\frac{B_z\left(-F^s_y C_f + F^f_y C_s\right) + \tilde{B}_z\left(E^s_{xy} C_f - E^f_{xy} C_s\right)}{C_f B_z};$$

$$S_{48} = -\frac{B_z\left(-F^s_z C_f + F^f_z C_s\right) - \tilde{B}_y\left(E^s_{xy} C_f - E^f_{xy} C_s\right)}{C_f B_z};$$

$$S_{88} = 2\frac{F^s_z \tilde{B}_z - F^f_s \tilde{B}_z + \tilde{B}_y F^s_y - \tilde{B}_y F^f_y}{\tilde{B}_z}.$$

3.12.2 Matrix $\tilde{A}_y$

Eigenvalues of this matrix have the following form:

$$\lambda_1 = \upsilon, \; \lambda_2 = \upsilon, \; \lambda_3 = \upsilon + V_{Ay}, \; \lambda_4 = \upsilon + V_{Ay},$$

$$\lambda_5 = \upsilon + C_f,\ \lambda_6 = \upsilon - C_f,\ \lambda_7 = \upsilon + C_s,\ \lambda_8 = \upsilon - C_s, \tag{3.127}$$

where

$$V_{Ay} = \frac{B_y}{\sqrt{\rho m}} = \frac{B_y}{\sqrt{\rho}\mathrm{M}_{A^*}};\quad C_f = \sqrt{\frac{1}{2}\left[a^2 + V_A^2 + \sqrt{\left(a^2 + V_A^2\right)^2 - 4a^2V_{Ay}^2}\right]};$$

$$C_s = \sqrt{\frac{1}{2}\left[a^2 + V_A^2 - \sqrt{\left(a^2 + V_A^2\right)^2 - 4a^2V_{Ay}^2}\right]};\ a^2 = \frac{\gamma p}{\rho};$$

$$V_A^2 = \frac{B_x^2 + B_y^2 + B_z^2}{\rho m} = \frac{B_x^2 + B_y^2 + B_z^2}{\rho \mathrm{M}_{A^*}^2}.$$

The set of equations for determination of left eigenvector components of a matrix $\tilde{A}_y$ has the following form:

$$l_1(\upsilon - \lambda) = 0,\ \ l_2(\upsilon - \lambda) - l_6 B_y = 0,$$

$$l_1\rho + l_2(\upsilon - \lambda) + l_5\gamma p + l_6 B_x + l_8 B_z = 0, l_4(\upsilon - \lambda) - l_8 B_y = 0,$$

$$l_3\frac{1}{\rho} + l_5(\upsilon - \lambda) = 0,\ \ -l_2\frac{B_y}{\rho m} + l_3\frac{B_x}{\rho m} + l_6(\upsilon - \lambda) = 0,$$

$$l_7(\upsilon - \lambda) = 0,\ l_3\frac{B_z}{\rho m} - l_4\frac{B_y}{\rho m} + l_8(\upsilon - \lambda) = 0. \tag{3.128}$$

Procedure for determination of the left eigenvectors is the same, as for the matrix $\tilde{A}_x$. We shall give the final result in the form of matrix S_y^{-1}:

$$S_y^{-1} = \begin{vmatrix} 1 & 0 & 0 & 0 & -\frac{1}{a^2} & 0 & 0 & 0 \\ 0 & 0 & 0 & 0 & 0 & 0 & 1 & 0 \\ 0 & -B_z & 0 & B_x & 0 & \frac{B_z}{\sqrt{\rho m}} & 0 & -\frac{B_x}{\sqrt{\rho m}} \\ 0 & -B_z & 0 & B_x & 0 & -\frac{B_z}{\sqrt{\rho m}} & 0 & \frac{B_x}{\sqrt{\rho m}} \\ 0 & -\frac{\rho C_f B_x B_y}{z_f} & \rho C_f & -\frac{\rho C_f B_z B_y}{z_f} & 1 & \frac{\rho C_f^2 B_x}{z_f} & 0 & \frac{\rho C_f^2 B_z}{z_f} \\ 0 & \frac{\rho C_f B_x B_y}{z_f} & -\rho C_f & \frac{\rho C_f B_z B_y}{z_f} & 1 & \frac{\rho C_f^2 B_x}{z_f} & 0 & \frac{\rho C_f^2 B_z}{z_f} \\ 0 & -\frac{\rho C_s B_x B_y}{z_s} & \rho C_s & -\frac{\rho C_s B_z B_y}{z_s} & 1 & \frac{\rho C_s^2 B_x}{z_s} & 0 & \frac{\rho C_s^2 B_z}{z_s} \\ 0 & \frac{\rho C_s B_x B_y}{z_s} & -\rho C_s & \frac{\rho C_s B_z B_y}{z_s} & 1 & \frac{\rho C_s^2 B_x}{z_s} & 0 & \frac{\rho C_s^2 B_z}{z_s} \end{vmatrix}, \tag{3.129}$$

where $z_f = \rho m C_f^2 - B_y^2$; $z_s = \rho m C_s^2 - B_y^2$.

For determination of vector function $\mathbf{d_y}$, it is necessary to solve the set of the following equations:

$$S_y^{-1} \times \begin{pmatrix} d_1 \\ d_2 \\ \vdots \\ d_8 \end{pmatrix}_x = \begin{pmatrix} L_1^y \\ L_2^y \\ \vdots \\ L_8^y \end{pmatrix}, \tag{3.130}$$

where $\mathbf{L_j^x} = \lambda_j \vec{l}_j^{\,T} (\partial \mathbf{W}/\partial x)$, $j = 1, 2, \ldots, 8$, where j is the number of the eigenvalue; $\vec{l}_j^{\,T}$ is the left eigenvector corresponding to the jth eigenvalue.

Triangular matrix $\tilde{S}_y^{-1}$ can be obtained using the procedure of Gaussian elimination:

$$\tilde{S}_y^{-1} = \begin{vmatrix} 1 & 0 & 0 & 0 & -\dfrac{1}{a^2} & 0 & 0 & 0 \\ 0 & E_{xy}^f & -\rho C_f & E_{yz}^f & 1 & E_x^f & 0 & E_z^f \\ 0 & 0 & S_{33} & S_{34} & \dfrac{B_z}{E_{xy}^f} & S_{36} & 0 & S_{38} \\ 0 & 0 & 0 & S_{44} & \dfrac{C_f + C_s}{C_f} & S_{46} & 0 & S_{48} \\ 0 & 0 & 0 & 0 & 2 & 2F_x^f & 0 & 2F_z^f \\ 0 & 0 & 0 & 0 & 0 & -2\tilde{B}_z & 0 & 2\tilde{B}_x \\ 0 & 0 & 0 & 0 & 0 & 0 & 1 & 0 \\ 0 & 0 & 0 & 0 & 0 & 0 & 0 & S_{88} \end{vmatrix}, \tag{3.131}$$

where

$$F_x^f = \frac{\rho C_f^2 B_x}{z_f}; \quad F_z^f = \frac{\rho C_f^2 B_z}{z_f}; \quad \tilde{B}_z = \frac{B_z}{\sqrt{\rho m}}; \quad F_x^s = \frac{\rho C_s^2 B_x}{z_s};$$

$$F_z^s = \frac{\rho C_s^2 B_z}{z_s}; \quad \tilde{B}_x = \frac{B_x}{\sqrt{\rho m}}; \quad F_{xy}^f = \frac{\rho C_f B_x B_y}{z_f} \quad ;F_{xy}^s = \frac{\rho C_s B_x B_y}{z_s};$$

$$F_{yz}^f = \frac{\rho C_f B_y B_z}{z_f}; F_{yz}^s = \frac{\rho C_s B_y B_z}{z_s}; S_{33} = -\frac{B_z \rho C_f}{E_{xy}^f};$$

$$S_{34} = \frac{B_x E_{xy}^f + B_z E_{yz}^f}{E_{xy}^f}; S_{36} = \frac{\tilde{B}_z E_{xy}^f + B_z F_x^f}{E_{xy}^f}; S_{38} = -\frac{\tilde{B}_x E_{xy}^f - B_z F_z^f}{E_{xy}^f};$$

$$S_{44} = \frac{-B_zC_fE^s_{yz} + C_sE^f_{xy}B_x + C_sB_yE^f_{yz} + E^s_{xy}C_fB_x}{B_zC_f};$$

$$S_{46} = \frac{B_zC_fF^s_x + C_sE^f_{xy}\tilde{B}_z + C_sB_zF^f_x - E^s_{xy}C_f\tilde{B}_z}{B_zC_f};$$

$$S_{48} = -\frac{-B_zC_fF^s_z + C_sE^f_{xy}\tilde{B}_x - C_sB_zF^f_z - E^s_{xy}C_f\tilde{B}_x}{B_zC_f};$$

$$S_{88} = 2\frac{F^s_z\tilde{B}_z + F^f_z\tilde{B}_z - \tilde{B}_xF^s_x + \tilde{B}_xF^f_x}{\tilde{B}_z}.$$

3.12.3 Matrix $\tilde{A}_z$

Eigenvalues of this matrix have the following form:

$$\lambda_1 = w,\ \lambda_2 = w,\ \lambda_3 = w + V_{\mathrm{A}z},\ \lambda_4 = w - V_{\mathrm{A}z},$$
$$\lambda_5 = w + C_f,\ \lambda_6 = w - C_f,\ \lambda_7 = w + C_s,\ \lambda_8 = w - C_s, \tag{3.132}$$

where

$$V_{\mathrm{A}z} = \frac{B_z}{\sqrt{\rho m}} = \frac{B_z}{\sqrt{\rho \mathrm{M}_{\mathrm{A}*}}};\ C_f = \sqrt{\frac{1}{2}\left[a^2 + V^2_{\mathrm{A}} + \sqrt{\left(a^2 + V^2_{\mathrm{A}}\right)^2 - 4a^2V^2_{\mathrm{A}z}}\right]};$$

$$C_s = \sqrt{\frac{1}{2}\left[a^2 + V^2_{\mathrm{A}} - \sqrt{\left(a^2 + V^2_{\mathrm{A}}\right)^2 - 4a^2V^2_{\mathrm{A}z}}\right]};\ a^2 = \frac{\gamma p}{\rho};$$

$$V^2_{\mathrm{A}} = \frac{B^2_x + B^2_y + B^2_z}{\rho m} = \frac{B^2_x + B^2_y + B^2_z}{\rho \mathrm{M}^2_{\mathrm{A}*}}.$$

The set of equations for determination of the left eigenvector components of the matrix $\tilde{A}_z$ has the following form:

$$l_1(w-\lambda) = 0,\ l_2(w-\lambda) - l_3B_z = 0,\ l_3(w-\lambda) - l_7B_z = 0,$$
$$l_1\rho + l_4(w-\lambda) + l_5\gamma p + l_6B_x + l_7B_y = 0,\ l_4\rho^{-1} + l_5(w-\lambda) = 0,$$
$$-l_2\frac{B_z}{\rho m} + l_4\frac{B_x}{\rho m} + l_6(w-\lambda) = 0, l_3\frac{B_z}{\rho m} + l_4\frac{B_y}{\rho m} + l_7(w-\lambda) = 0,$$
$$l_8(w-\lambda) = 0. \tag{3.133}$$

Using the same procedure for determination of components of eigenvectors, as in the case of matrix S_x^{-1}, one can obtain the matrix S_z^{-1}:

$$S_z^{-1} = \begin{vmatrix} 1 & 0 & 0 & 0 & -\dfrac{1}{a^2} & 0 & 0 & 0 \\ 0 & 0 & 0 & 0 & 0 & 0 & 0 & 1 \\ 0 & -B_y & B_x & 0 & 0 & \dfrac{B_y}{\sqrt{\rho m}} & -\dfrac{B_x}{\sqrt{\rho m}} & 0 \\ 0 & -B_y & B_x & 0 & 0 & -\dfrac{B_y}{\sqrt{\rho m}} & \dfrac{B_x}{\sqrt{\rho m}} & 0 \\ 0 & -\dfrac{\rho C_f B_x B_z}{z_f} & -\dfrac{\rho C_f B_y B_z}{z_f} & -\rho C_f & 1 & \dfrac{\rho C_f^2 B_x}{z_f} & \dfrac{\rho C_f^2 B_y}{z_f} & 0 \\ 0 & \dfrac{\rho C_f B_x B_z}{z_f} & \dfrac{\rho C_f B_y B_z}{z_f} & \rho C_f & 1 & \dfrac{\rho C_f^2 B_x}{z_f} & \dfrac{\rho C_f^2 B_y}{z_f} & 0 \\ 0 & -\dfrac{\rho C_s B_x B_z}{z_s} & -\dfrac{\rho C_s B_y B_z}{z_s} & -\rho C_s & 1 & \dfrac{\rho C_s^2 B_x}{z_s} & \dfrac{\rho C_s^2 B_y}{z_s} & 0 \\ 0 & \dfrac{\rho C_s B_x B_z}{z_s} & \dfrac{\rho C_s B_y B_z}{z_s} & \rho C_s & 1 & \dfrac{\rho C_s^2 B_x}{z_s} & \dfrac{\rho C_s^2 B_y}{z_s} & 0 \end{vmatrix}, \quad (3.134)$$

where

$$z_f = \rho m C_f^2 - B_z^2, \; z_s = \rho m C_s^2 - B_z^2.$$

For determination of vector function $\mathbf{d_z}$, it is necessary to solve the set of equations

$$S_z^{-1} \times \begin{pmatrix} d_1 \\ d_2 \\ \vdots \\ d_8 \end{pmatrix}_x = \begin{pmatrix} L_1^z \\ L_2^z \\ \vdots \\ L_8^z \end{pmatrix}, \quad (3.135)$$

where $\mathbf{L_j^z} = \lambda_j \vec{l}_j^{\mathrm{T}} (\partial \mathbf{W}/\partial z)$, $j = 1, 2, \ldots, 8$, where j is the number of an eigenvalue, $\vec{l}_j^{\mathrm{T}}$ is the characteristic left vector corresponding to the jth eigenvalue.

Triangular matrix $\tilde{S}_z^{-1}$ is obtained with the use of Gaussian elimination procedure:

$$\tilde{S}_z^{-1} = \begin{vmatrix} 1 & 0 & 0 & 0 & -\frac{1}{a^2} & 0 & 0 & 0 \\ 0 & E^f_{xz} & E^f_{yz} & \rho C_f & 1 & E^f_x & E^f_y & 0 \\ 0 & 0 & S_{33} & S_{34} & \frac{B_y}{E^f_{xz}} & S_{36} & S_{37} & 0 \\ 0 & 0 & 0 & S_{44} & S_{45} & S_{46} & S_{47} & 0 \\ 0 & 0 & 0 & 0 & 2 & 2F^f_x & 2F^f_y & 0 \\ 0 & 0 & 0 & 0 & 0 & -2\tilde{B}_y & 2\tilde{B}_x & 0 \\ 0 & 0 & 0 & 0 & 0 & 0 & S_{77} & 0 \\ 0 & 0 & 0 & 0 & 0 & 0 & 0 & 1 \end{vmatrix}, \tag{3.136}$$

where

$$F^f_x = \frac{\rho C_f^2 B_x}{z_f};\quad F^f_y = \frac{\rho C_f^2 B_y}{z_f};\quad \tilde{B}_x = \frac{B_x}{\sqrt{\rho m}};\quad F^s_x = \frac{\rho C_s^2 B_x}{z_s};\quad F^s_y = \frac{\rho C_s^2 B_y}{z_s};$$

$$\tilde{B}_y = \frac{B_y}{\sqrt{\rho m}}; F^f_{yz} = \frac{\rho C_f B_y B_z}{z_f}; F^s_{xz} = \frac{\rho C_s B_x B_z}{z_s}; F^f_{yz} = \frac{\rho C_f B_y B_z}{z_f};$$

$$F^s_{yz} = \frac{\rho C_s B_y B_z}{z_s}; S_{33} = \frac{B_x E^f_{xz} + B_y E^f_{yz}}{E^f_{xz}}; S_{34} = \frac{\rho C_f B_y}{E^f_{xz}};$$

$$S_{36} = \frac{\tilde{B}_y E^f_{xz} + B_y F^f_x}{E^f_{xz}}; S_{37} = -\frac{\tilde{B}_x E^f_{xz} - B_y F^f_y}{E^f_{xz}};$$

$$S_{44} = \rho\left(C_s - C_f \frac{B_x E^s_{xz} + B_y E^s_{yz}}{B_x E^f_{xz} + B_y E^f_{yz}}\right);\quad S_{45} = 1 - \frac{B_x E^s_{xz} + B_y E^s_{yz}}{B_x E^f_{xz} + B_y E^f_{yz}};$$

$$S_{46} = F^s_x - F^f_x \frac{B_x E^s_{xz} + B_y E^s_{yz}}{B_x E^f_{xz} + B_y E^f_{yz}} + \tilde{B}_y \frac{E^s_{xz} E^f_{yz} - E^f_{xz} E^s_{yz}}{B_x E^f_{xz} + B_y E^f_{yz}};$$

$$S_{47} = -F^s_y + F^f_y \frac{B_x E^s_{xz} + B_y E^s_{yz}}{B_x E^f_{xz} + B_y E^f_{yz}} + \tilde{B}_x \frac{E^s_{xz} E^f_{yz} - E^s_{yz} E^f_{xz}}{B_x E^f_{xz} + B_y E^f_{yz}};$$

$$S_{77} = 2\left[F^s_y - F^f_y + \frac{\tilde{B}_x}{\tilde{B}_y}\left(F^s_x - F^f_x\right)\right].$$

In conclusion of the given section, we shall pay attention to the fact that spectral characteristics of a matrix $\tilde{A}_x$, namely, eigenvalues (3.117) and left eigenvectors (3.124) practically coincide completely with spectral characteristics of complete (without simplifications) matrix A_x (3.99), which has one singular line. The unique difference is the equality to zero of the first eigenvalue of the specified singular matrix (see eq. (3.100)).

3.13 A method of splitting for three-dimensional MHD equations

In the previous sections, it has been shown how to formulate nonstationary boundary conditions for 3D MHD equations (Thompson K.W., 1987, 1990; Vanajakshi J.C., et al., 1989; Sun M.T., et al., 1995). More simple nonstationary boundary conditions for 3D MHD problems are considered in the given section. This approach is based on the method of splitting (Surzhikov S.T., et al., 2001).

As before we shall consider the system of MHD equation in the quasilinear form:

$$\frac{\partial \mathbf{W}}{\partial t} + \tilde{A}_x \frac{\partial \mathbf{W}}{\partial x} + \tilde{A}_y \frac{\partial \mathbf{W}}{\partial y} + \tilde{A}_z \frac{\partial \mathbf{W}}{\partial z} = \mathbf{G}, \tag{3.137}$$

where matrixes $\tilde{A}_x, \tilde{A}_y, \tilde{A}_z$ are taken in the form of (3.116), and vector **G** includes components of the complete set of the equations, corresponding to viscous dissipation and Joule heat. Let us rewrite matrixes $\tilde{A}_x, \tilde{A}_y, \tilde{A}_z$, having added the separation of these matrixes on blocks:

$$\tilde{A}_x = \begin{pmatrix} u & \rho & 0 & 0 & 0 & 0 & 0 & 0 \\ 0 & u & 0 & 0 & 1/\rho & 0 & B_y/\rho m & B_z/\rho m \\ 0 & 0 & u & 0 & 0 & -B_y/\rho m & -B_x/\rho m & 0 \\ 0 & 0 & 0 & u & 0 & -B_z/\rho m & 0 & -B_x/\rho m \\ 0 & \gamma p & 0 & 0 & u & 0 & 0 & 0 \\ 0 & 0 & 0 & 0 & 0 & u & 0 & 0 \\ 0 & B_y & -B_x & 0 & 0 & 0 & u & 0 \\ 0 & B_y & 0 & -B_x & 0 & 0 & 0 & u \end{pmatrix},$$

$$\tilde{A}y = \begin{pmatrix} \upsilon & 0 & \rho & 0 & 0 & 0 & 0 & 0 \\ 0 & \upsilon & 0 & 0 & 0 & -B_y/\rho m & -B_x/\rho m & 0 \\ 0 & 0 & \upsilon & 0 & 1/\rho & B_x/\rho m & 0 & B_z/\rho m \\ 0 & 0 & 0 & \upsilon & 0 & 0 & -B_z/\rho m & -B_y/\rho m \\ 0 & 0 & \gamma p & 0 & \upsilon & 0 & 0 & 0 \\ 0 & -B_y & B_x & 0 & 0 & \upsilon & 0 & 0 \\ 0 & 0 & 0 & 0 & 0 & 0 & \upsilon & 0 \\ 0 & 0 & B_z & -B_y & 0 & 0 & 0 & \upsilon \end{pmatrix},$$

$$\tilde{A}_z = \begin{pmatrix} w & 0 & 0 & \rho & 0 & 0 & 0 & 0 \\ 0 & w & 0 & 0 & 0 & -B_z/\rho m & 0 & -B_x/\rho m \\ 0 & 0 & w & 0 & 0 & 0 & -B_z/\rho m & -B_y/\rho m \\ 0 & 0 & 0 & w & 1/\rho & B_x/\rho m & B_y/\rho m & 0 \\ 0 & 0 & 0 & \gamma p & w & 0 & 0 & 0 \\ 0 & -B_z & 0 & B_x & 0 & w & 0 & 0 \\ 0 & 0 & -B_z & B_y & 0 & 0 & w & 0 \\ 0 & 0 & 0 & 0 & 0 & 0 & 0 & w \end{pmatrix}.$$

The matrixes have been divided into four blocks in such a manner that gas dynamic and magnetic parts of the problem are explicitly split. The essence of the method of decomposition is that for the solution of the set of eq. (3.137) on the current time step, the two sequential stages will be used.

At the first stage, the following set of gas dynamic equations will be solved (further, we shall suppose $\mathbf{G} = 0$):

$$\frac{\partial \mathbf{W_1}}{\partial t} + \tilde{A}_x^G \frac{\partial \mathbf{W_1}}{\partial x} + \tilde{A}_y^G \frac{\partial \mathbf{W_1}}{\partial y} + \tilde{A}_z^G \frac{\partial \mathbf{W_1}}{\partial z} = \mathbf{T}, \tag{3.139}$$

where

$$\mathrm{W}_1 = \begin{pmatrix} \rho \\ u \\ v \\ w \\ p \end{pmatrix}; \tilde{A}_x^G = \begin{pmatrix} u & \rho & 0 & 0 & 0 \\ 0 & u & 0 & 0 & 1/\rho \\ 0 & 0 & u & 0 & 0 \\ 0 & 0 & 0 & u & 0 \\ 0 & \gamma p & 0 & 0 & u \end{pmatrix}$$

$$\tilde{A}_z^G = \begin{pmatrix} w & 0 & 0 & \rho & 0 \\ 0 & w & 0 & 0 & 0 \\ 0 & 0 & w & 0 & 0 \\ 0 & 0 & 0 & w & 1/\rho \\ 0 & 0 & 0 & \gamma p & w \end{pmatrix}; \tag{3.140}$$

$$\mathbf{T} = \begin{pmatrix} T_\rho \\ T_u \\ T_v \\ T_w \\ T_p \end{pmatrix} = \begin{pmatrix} 0 \\ \frac{B_y}{\rho m}\frac{\partial B_y}{\partial x} + \frac{B_z}{\rho m}\frac{\partial B_z}{\partial x} - \frac{B_y}{\rho m}\frac{\partial B_x}{\partial y} - \frac{B_x}{\rho m}\frac{\partial B_y}{\partial y} - \frac{B_z}{\rho m}\frac{\partial B_x}{\partial y} - \frac{B_x}{\rho m}\frac{\partial B_z}{\partial z} \\ -\frac{B_y}{\rho m}\frac{\partial B_x}{\partial x} - \frac{B_x}{\rho m}\frac{\partial B_y}{\partial x} + \frac{B_x}{\rho m}\frac{\partial B_x}{\partial y} + \frac{B_z}{\rho m}\frac{\partial B_z}{\partial y} - \frac{B_z}{\rho m}\frac{\partial B_y}{\partial y} - \frac{B_y}{\rho m}\frac{\partial B_z}{\partial z} \\ -\frac{B_z}{\rho m}\frac{\partial B_x}{\partial x} - \frac{B_x}{\rho m}\frac{\partial B_z}{\partial x} - \frac{B_z}{\rho m}\frac{\partial B_y}{\partial y} - \frac{B_y}{\rho m}\frac{\partial B_z}{\partial y} + \frac{B_x}{\rho m}\frac{\partial B_x}{\partial z} + \frac{B_y}{\rho m}\frac{\partial B_y}{\partial z} \\ 0 \end{pmatrix}. \tag{3.141}$$

Components of the vector $\mathbf{W_1}$ can be calculated at the first intermediate step after diagonalization of matrixes $\tilde{A}_x^G$, $\tilde{A}_y^G$, $\tilde{A}_z^G$. We shall mark the determined values with index $(p+1)$ unlike all other functions known on the pth time layer.

At the second stage, the set of equations for magnetic induction is to be solved:

$$\frac{\partial \mathbf{W_2}}{\partial t} + \tilde{A}_x^B \frac{\partial \mathbf{W_2}}{\partial x} + \tilde{A}_y^B \frac{\partial \mathbf{W_2}}{\partial y} + \tilde{A}_z^B \frac{\partial \mathbf{W_2}}{\partial z} = \mathbf{K}, \tag{3.142}$$

where

$$\mathbf{W_2} = \begin{pmatrix} B_x \\ B_y \\ B_z \end{pmatrix}; \tilde{A}_x^B = \begin{pmatrix} u & 0 & 0 \\ 0 & u & 0 \\ 0 & 0 & u \end{pmatrix}; \tilde{A}_y^B = \begin{pmatrix} v & 0 & 0 \\ 0 & v & 0 \\ 0 & 0 & v \end{pmatrix}; \tilde{A}_z^B = \begin{pmatrix} w & 0 & 0 \\ 0 & w & 0 \\ 0 & 0 & w \end{pmatrix}; \tag{3.143}$$

$$\mathbf{K} = -\begin{pmatrix} -B_y \dfrac{\partial u^{p+1}}{\partial y} & +B_x \dfrac{\partial v^{p+1}}{\partial y} & -B_z \dfrac{\partial u^{p+1}}{\partial z} & +B_x \dfrac{\partial w^{p+1}}{\partial z} \\ B_y \dfrac{\partial u^{p+1}}{\partial x} & -B_x \dfrac{\partial v^{p+1}}{\partial x} & -B_z \dfrac{\partial v^{p+1}}{\partial z} & +B_y \dfrac{\partial w^{p+1}}{\partial z} \\ B_z \dfrac{\partial u^{p+1}}{\partial x} & -B_x \dfrac{\partial w^{p+1}}{\partial x} & +B_z \dfrac{\partial v^{p+1}}{\partial y} & -B_y \dfrac{\partial w^{p+1}}{\partial y} \end{pmatrix}. \tag{3.144}$$

Results of solution of this system will be functions B_x, B_y, B_z at the upper time layer $(p+1)$.

Let's note principal properties of the method.

1. Matrixes A_x^G, A_y^G, A_z^G are formed from the left upper blocks of matrixes A_x, A_y, A_z accordingly, while the right upper blocks form the vector of right-hand side terms $\mathbf{T}$.
2. Matrixes A_x^B, A_y^B, A_z^B are formed from the left upper blocks of matrixes A_x, A_y, A_z accordingly, while the left lower blocks form vector $\mathbf{K}$.
3. At the second stage, the system of equations in the characteristic form is solved, owing to diagonal character of matrixes A_x^B, A_y^B, A_z^B, that is, the equations are represented at once in the most simple form of possible ones.

Below the calculation relations for the first stage of the decomposition are given.

Eigenvalues of matrix $\tilde{A}_x^G$

$$\lambda_1^x = u,\ \lambda_2^x = u,\ \lambda_3^x = u,\ \lambda_4^x = u + a,\ \lambda_5^x = u - a. \tag{3.145}$$

The following matrix of left eigenvectors (rows of the matrix) corresponds to these eigenvalues:

$$\left(S_x^{-1}\right)_G = \begin{pmatrix} 1 & 0 & 0 & 0 & 0 \\ 0 & 0 & 0 & 1 & 0 \\ 0 & 0 & 1 & 0 & 0 \\ \frac{1}{a^2} & \frac{a}{\gamma p} & 0 & 0 & 1 \\ \frac{1}{a^2} & -\frac{a}{\gamma p} & 0 & 0 & 1 \end{pmatrix}, \tag{3.146}$$

where the serial number of the matrix row corresponds to the serial number of eigenvalues (3.145).

The solution of the set of equations in the form (3.85) gives the following expressions for vector $\mathbf{d}_x$ components:

$$d_{x,1} = \tilde{L}_1^x,\ d_{x,2} = \frac{\gamma p}{a}\left(\tilde{L}_2^x - \frac{1}{2}\tilde{L}_5^x\right),\ d_{x,3} = \tilde{L}_3^x$$

$$d_{x,4} = \tilde{L}_4^x,\ d_{x,5} = \frac{1}{2}\tilde{L}_5^x, \tag{3.147}$$

where functions $\tilde{L}_j^x$ are calculated by the Thompson method (Thompson K.W., 1987). Namely, if on the left boundary of the calculation domain ($x = x_0$) an eigenvalue $\lambda_j^x < 0$, then $\tilde{L}_j^x = L_j^x$; otherwise $\tilde{L}_j^x = 0$. And, on the contrary, if on the right boundary of the calculation domain ($x = x_L > x_0$) an eigenvalue $\lambda_j^x > 0$, then $\tilde{L}_j^x = L_j^x$; otherwise $\tilde{L}_j^x = 0$. These conditions mean that disturbance going out from the area is to be calculated with use of functions $\tilde{L}_j^x$, and disturbances entering into the area are considered equal to zero.

Functions $\mathbf{L}_\mathbf{j}^\mathbf{x}$ are calculated under the formula

$$\mathbf{L}_\mathbf{j}^\mathbf{x} = \lambda_j^x \vec{l}_{x,j}^{\,T} \frac{\partial \mathbf{W}_1}{\partial x} \tag{3.148}$$

or in a component-wise notation

$$L_1^x = u(1,0,0,0,0)\frac{\partial}{\partial x}\begin{pmatrix} \rho \\ u \\ v \\ w \\ p \end{pmatrix} = u\frac{\partial \rho}{\partial x},\ L_2^x = u(0,0,0,1,0)\frac{\partial}{\partial x}\begin{pmatrix} \rho \\ u \\ v \\ w \\ p \end{pmatrix} = u\frac{\partial w}{\partial x},$$

$$L_3^x = u(0,0,1,0,0)\frac{\partial}{\partial x}\begin{pmatrix} \rho \\ u \\ v \\ w \\ p \end{pmatrix} = u\frac{\partial v}{\partial x},$$

$$L_4^x = (u+a)\left(\frac{1}{a^2}, \frac{a}{\gamma p}, 0, 0, 1\right)\frac{\partial}{\partial x}\begin{pmatrix}\rho\\ u\\ v\\ w\\ p\end{pmatrix} = (u+a)\left(\frac{1}{a^2}\frac{\partial\rho}{\partial x} + \frac{a}{\gamma p}\frac{\partial u}{\partial x}\right),$$

$$L_5^x = (u-a)\left(\frac{1}{a^2}, -\frac{a}{\gamma p}, 0, 0, 1\right)\frac{\partial}{\partial x}\begin{pmatrix}\rho\\ u\\ v\\ w\\ p\end{pmatrix} = (u-a)\left(\frac{1}{a^2}\frac{\partial\rho}{\partial x} - \frac{a}{\gamma p}\frac{\partial u}{\partial x}\right).$$

Eigenvalues of matrix $\tilde{A}_y^G$

$$\lambda_1^y = v,\ \lambda_2^y = v,\ \lambda_3^y = v,\ \lambda_4^y = v + a,\ \lambda_5^y = v - a. \tag{3.149}$$

With the use of given eigenvalues, the following matrix of the left eigenvectors (row of the matrix) turns out:

$$\left(S_y^{-1}\right)_G = \begin{pmatrix} 1 & 0 & 0 & 0 & 0\\ 0 & 1 & 0 & 0 & 0\\ 0 & 0 & 0 & 1 & 0\\ 1/a^2 & 0 & a/\gamma p & 0 & 1\\ 1/a^2 & 0 & -a/\gamma p & 0 & 1\end{pmatrix}, \tag{3.150}$$

where the serial number of the matrix row corresponds to the serial number of eigenvalues (3.149). The solution of the set of equations in the form (3.85) gives the following expressions for components of the vector $\mathbf{d}_y$:

$$d_{y,1} = \tilde{L}_1^y,\ d_{y,2} = \tilde{L}_2^y,\ d_{y,3} = \frac{\gamma p}{a}\left(\tilde{L}_2^y - \frac{1}{2}\tilde{L}_5^y\right), d_{y,4} = \tilde{L}_4^y,\ d_{y,5} = \frac{1}{2}\tilde{L}_5^y,$$

where functions $\tilde{L}_j^y$ are calculated with use of the following conditions:

$y = y_{\min}$: if $\lambda_j^y < 0$, then $\tilde{L}_j^y = L_j^y$; if $\lambda_j^y > 0$, then $\tilde{L}_j^y = 0$;

$y = y_{\max}$: if $\lambda_j^y > 0$, then $\tilde{L}_j^y = L_j^y$; if $\lambda_j^y < 0$, then $\tilde{L}_j^y = 0$.

Functions $\mathbf{L}_\mathbf{j}^\mathbf{y}$ are calculated under the formula

$$\mathbf{L}_\mathbf{j}^\mathbf{y} = \lambda_j^y \vec{l}_{y,j}^T \frac{\partial \mathbf{W}_1}{\partial y} \tag{3.151}$$

or in a component-wise notation

$$L_1^y = v(1,0,0,0,0)\frac{\partial}{\partial y}\begin{pmatrix}\rho\\ u\\ v\\ w\\ p\end{pmatrix} = v\frac{\partial\rho}{\partial y},\quad L_2^y = v(0,1,0,0,0)\frac{\partial}{\partial y}\begin{pmatrix}\rho\\ u\\ v\\ w\\ p\end{pmatrix} = v\frac{\partial u}{\partial y},$$

$$L_3^y = v(0,0,0,1,0)\frac{\partial}{\partial y}\begin{pmatrix}\rho\\ u\\ v\\ w\\ p\end{pmatrix} = v\frac{\partial w}{\partial y},$$

$$L_4^y = (v+a)\left(\frac{1}{a^2},0,\frac{a}{\gamma p},0,1\right)\frac{\partial}{\partial y}\begin{pmatrix}\rho\\ u\\ v\\ w\\ p\end{pmatrix} = (v+a)\left(\frac{1}{a^2}\frac{\partial\rho}{\partial y}+\frac{a}{\gamma p}\frac{\partial v}{\partial y}\right),$$

$$L_5^y = (v+a)\left(\frac{1}{a^2},0,-\frac{a}{\gamma p},0,1\right)\frac{\partial}{\partial y}\begin{pmatrix}\rho\\ u\\ v\\ w\\ p\end{pmatrix} = (v+a)\left(\frac{1}{a^2}\frac{\partial\rho}{\partial y}-\frac{a}{\gamma p}\frac{\partial v}{\partial y}\right).$$

Eigenvalues of matrix $\tilde{A}_z^G$

$$\lambda_1^z = w,\ \lambda_2^z = w,\ \lambda_3^z = w,\ \lambda_4^z = w+a,\ \lambda_5^z = w-a. \tag{3.152}$$

With the use of given eigenvalues, the following matrix of the left eigenvectors (the matrix rows) turns out:

$$(S_z^{-1})_G = \begin{pmatrix} 1 & 0 & 0 & 0 & 0\\ 0 & 1 & 0 & 0 & 0\\ 0 & 0 & 1 & 0 & 0\\ \frac{1}{a^2} & 0 & 0 & -\frac{a}{\gamma p} & 1\\ \frac{1}{a^2} & 0 & 0 & \frac{a}{\gamma p} & 1 \end{pmatrix}, \tag{3.153}$$

where the serial number of the matrix row corresponds to the serial number of eigenvalues (3.152). The solution of the set of equations in the form (3.85) gives the following expressions for components of the vector $\mathbf{d}_z$:

$$d_{z,1} = \tilde{L}_1^z = \tilde{L}_2^z,\ d_{z,3} = \tilde{L}_3^z,$$

$$d_{z,4} = \frac{\gamma p}{a}\left(\tilde{L}_4^z - \frac{1}{2}\tilde{L}_5^z\right), d_{z,5} = \frac{1}{2}\tilde{L}_5^z, \tag{3.154}$$

where functions $\tilde{L}_j^z$ are calculated with use of the following conditions [14]:

$$z = z_{\min}:\quad \text{if } \lambda_j^z < 0, \text{ then } \tilde{L}_j^z = L_j^z;\ \text{if } \lambda_j^z > 0, \text{ then } \tilde{L}_j^z = 0,$$

$$z = z_{\max}:\quad \text{if } \lambda_j^z > 0, \text{ then } \tilde{L}_j^z = L_j^z;\ \text{if } \lambda_j^z < 0, \text{ then } \tilde{L}_j^z = 0.$$

Functions $\tilde{L}_j^z$ are calculated under the formula

$$L_j^z = \lambda_j^z \vec{l}_{z,j}^{\,T} \frac{\partial W_1}{\partial z} \tag{3.155}$$

or in a component-wise notation

$$L_1^z = w(1,0,0,0,0)\frac{\partial}{\partial z}\begin{pmatrix}\rho\\ u\\ \upsilon\\ w\\ p\end{pmatrix} = w\frac{\partial \rho}{\partial y},\quad L_2^z = w(0,1,0,0,0)\frac{\partial}{\partial z}\begin{pmatrix}\rho\\ u\\ \upsilon\\ w\\ p\end{pmatrix} = w\frac{\partial u}{\partial z},$$

$$L_3^z = w(0,0,1,0,0)\frac{\partial}{\partial y}\begin{pmatrix}\rho\\ u\\ \upsilon\\ w\\ p\end{pmatrix} = w\frac{\partial \upsilon}{\partial z},,$$

$$L_4^z = (w+a)\left(\frac{1}{a^2}, 0, 0, -\frac{a}{\gamma p}, 1\right)\frac{\partial}{\partial z}\begin{pmatrix}\rho\\ u\\ v\\ w\\ p\end{pmatrix} = (w+a)\left(\frac{1}{a^2}\frac{\partial \rho}{\partial z} - \frac{a}{\gamma p}\frac{\partial w}{\partial z}\right),$$

$$L_5^z = (w-a)\left(\frac{1}{a^2}, 0, 0, \frac{a}{\gamma p}, 1\right)\frac{\partial}{\partial z}\begin{pmatrix}\rho\\ u\\ \upsilon\\ w\\ p\end{pmatrix} = (w-a)\left(\frac{1}{a^2}\frac{\partial \rho}{\partial z} + \frac{a}{\gamma p}\frac{\partial w}{\partial z}\right).$$

Nonstationary boundary conditions at the first stage of the problem solution

Multiplying eq. (3.139) from left on matrixes $(S_x^{-1})_G$, $\left(S_y^{-1}\right)_G$, $(S_z^{-1})_G$ sequentially and then in the same sequence on reciprocal matrixes (matrixes of the right eigenvectors), we obtain:

$$\frac{\partial \mathbf{W}_1}{\partial t} = -\mathbf{d}_x - \mathbf{d}_y - \mathbf{d}_z + \mathbf{T}. \tag{3.156}$$

Equation (3.156) represents the boundary condition for vector function $\mathbf{W}_1$ on any of boundaries of the calculation domain.

The structure of vector functions $\mathbf{d}_x, \mathbf{d}_y, \mathbf{d}_z$ is that they contain only derivatives in corresponding directions x, y, z. In other words, if the boundary condition is calculated on the surface $x = \text{const}$, the normal modification of functions in relation to this boundary is considered only in vector function $\mathbf{d}_x$, while vector functions $\mathbf{d}_y, \mathbf{d}_z$ contain derivatives in tangential directions.

Vector function $\mathbf{T}$ contains derivatives in all directions; therefore, to make calculation of boundary conditions more obviously, we shall present it in the form of two terms:

$$\mathbf{T} = \mathbf{T}_n + \mathbf{T}_\tau, \tag{3.157}$$

where

$$\mathbf{T}_n = \begin{pmatrix} 0 \\ \dfrac{\partial\left(B_y^2 + B_z^2\right)}{\partial x} \\ \dfrac{\partial\left(B_x^2 + B_z^2\right)}{\partial y} \\ \dfrac{\partial\left(B_x^2 + B_y^2\right)}{\partial z} \\ 0 \end{pmatrix}, \quad \mathbf{T}_\tau = \frac{1}{\rho_m} \begin{pmatrix} 0 \\ B_y \dfrac{\partial B_x}{\partial y} + B_x \dfrac{\partial B_y}{\partial y} + B_z \dfrac{\partial B_x}{\partial z} + B_x \dfrac{\partial B_z}{\partial z} \\ B_y \dfrac{\partial B_x}{\partial x} + B_x \dfrac{\partial B_y}{\partial x} + B_z \dfrac{\partial B_y}{\partial z} + B_y \dfrac{\partial B_z}{\partial z} \\ B_z \dfrac{\partial B_x}{\partial x} + B_x \dfrac{\partial B_z}{\partial y} + B_z \dfrac{\partial B_y}{\partial y} + B_y \dfrac{\partial B_z}{\partial y} \\ 0 \end{pmatrix}.$$

Thus, as a result of the first stage calculations, the values of functions ρ^{p+1}, $u^{p+1}, v^{p+1}, w^{p+1}, p^{p+1}$ on the upper time layer are determined in view of the action of magnetic field forces. At the second stage of calculations, these values will be used for determination of boundary values of components of a magnetic field.

Nonstationary boundary conditions at the second stage of the problem solution

For determination of values of functions B_x, B_y, B_z on $(p+1)$th time layer, it is necessary to solve the following set of equations:

$$\frac{\partial B_x}{\partial t} + u^{p+1}\frac{\partial B_x}{\partial x} + v^{p+1}\frac{\partial B_x}{\partial y} + w^{p+1}\frac{\partial B_x}{\partial z} = B_y \frac{\partial u}{\partial y} - B_x \frac{\partial v}{\partial y} + B_z \frac{\partial u}{\partial z} - B_x \frac{\partial w}{\partial z}, \tag{3.158}$$

$$\frac{\partial B_y}{\partial t} + u^{p+1}\frac{\partial B_y}{\partial x} + v^{p+1}\frac{\partial B_y}{\partial y} + w^{p+1}\frac{\partial B_y}{\partial z} = -B_y\frac{\partial u}{\partial x} + B_x\frac{\partial v}{\partial x} + B_z\frac{\partial v}{\partial z} - B_y\frac{\partial w}{\partial z}, \tag{3.159}$$

$$\frac{\partial B_z}{\partial t} + w^{p+1}\frac{\partial B_z}{\partial x} + w^{p+1}\frac{\partial B_z}{\partial y} + w^{p+1}\frac{\partial B_z}{\partial z} = -B_z\frac{\partial u}{\partial x} + B_x\frac{\partial w}{\partial x} - B_z\frac{\partial v}{\partial y} + B_y\frac{\partial w}{\partial y}, \tag{3.160}$$

where superscripts $(p+1)$ specify that the corresponding values of velocity components are taken by results of the first stage. Derivative of velocities in right parts of the equations are calculated with the use of values from the lower time layer.

Thus, in the present section, the method, simple enough to realize, is formulated for calculation of nonstationary boundary conditions. Note that the question on the effectiveness of the given approach in problems of strong MHD interaction is still left open.

3.14 Application of a splitting method for nonstationary 3D MHD flow field generated by plasma plume in ionosphere

A 3D computational model and the results of the calculation for the dynamics of a plasma plume induced by a pulsed plasma thruster (PPT) in a magnetized flow of the ambient plasma are considered in this section. A nonstationary expansion process of a PPT-induced plasma plume is studied in the ionospheric plasma over 100 μs after the start of a 6 μs plasma pulse with the use of a splitting method described in the previous section. The parameters of the plasma plume in the PPT outlet section simulate the performance characteristics of a PPT-4 (Gatsonis N.A., 2001), utilizing solid Teflon propellant. The MHD model of the process makes it possible to predict the dynamics of a plasma bunch after the pulse termination during its interaction with the ambient medium and the magnetic field as well as the induced electric fields and currents in the vicinity of the expanding plasma.

Development of small spacecraft intended for optronic environmental monitoring, communications, and various defense purposes has attracted special attention to modeling and experimental investigation in electric jet propulsion. The high-performance index, simple design, and reliability of PPTs utilizing solid Teflon propellant make these thrusters highly promising for smaller space vehicles.

The theory and practice of electrojet engines were intensively developed in the 1960s–1970s. It was in that period that prototype PPTs were designed, which proved their efficiency in the real conditions of operation in space. The developed theoretical and computational models provided a comprehensive understanding of the processes in these engines and helped to significantly enhance their efficiency (Gastonis N.A., et al., 2001; Boyd I.D., et al., 2000).

The physical processes of PPT operation are studied by physical mechanics (plasma dynamics), investigating the laws of interaction of plasma flows in various conditions. In this case, one is dealing with the interaction of an originally dense plasma plume with a homogeneous flow of ionospheric plasma incident on it at an arbitrary angle. In the course of plasma expansion, its density becomes comparable with that of the rarefied ambient medium so that the reciprocal influence of these two flows becomes inevitable. Raizer Yu.P. and Surzhikov S.T. (1995) compared computational data obtained for the problem of the spread of a plasma cloud in the ionosphere by an MHD model and a collisionless plasma model. It was shown that the calculated large-scale flow characteristics (shock wave velocity, disturbance range of gas dynamic parameters, and magnetic cavity parameters) are close to both models. This is a very important conclusion because it shows that resource sparing MHD models can be used to describe large-scale phenomena accompanying the plasma cloud dynamics. The analysis of conditions for the dynamics of a plasma plume induced by a PPT shows that although the employment of the collisionless plasma model seems more justifiable for the late phase of the process, at the initial period of plasma plume dynamics the MHD model can be quite acceptable.

The scheme of the problem being solved is shown in Figure 3.1. The symmetry axis of a cylindrical nozzle and a velocity vector of the PPT plasma plume are directed along the z-axis. The center of the symmetry of the PPT outlet section has coordinates x_0, y_0, and z_0. The plasma velocity at the outlet section of the nozzle is assumed constant and equal to V_0. For the simplicity of the representation of the data, two planes (δ and γ), possessing the designed symmetry properties, are shown in the calculation domain. In the plane γ (the plane coordinate being $x = x_0$), the vectors of the unperturbed magnetic field induction $\mathbf{B}_0$ and the velocity $\mathbf{V}_\infty$ of the incident flow of the ionospheric plasma are specified. The slopes of these

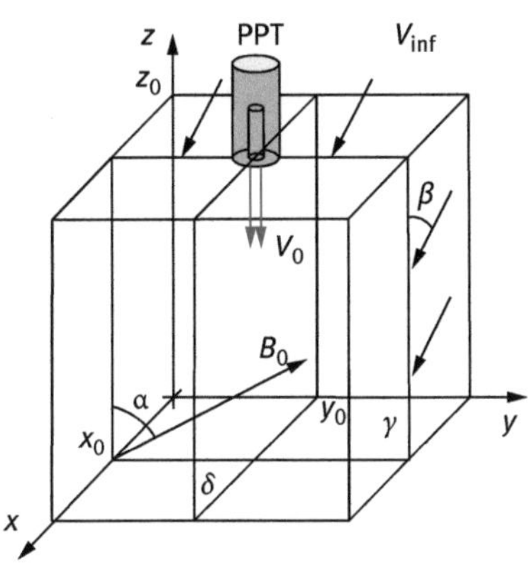

Figure 3.1: MHD problem scheme.

vectors are measured from the z-axis: the angle α determines the slope of the unperturbed magnetic field and the angle β determines the vector of the plasma's incident flow. The plane δ (the plane's coordinate being $y = y_0$) is perpendicular to the plane γ.

The following input data were used in the computations:

- The PPT pulse duration was $t = 6\,\mu s$; the time dependence of the PPT mass flow ratio during this pulse was considered constant.
- The mass of the Teflon used over a single pulse was $m = 12 \times 10^{-6}$ g.
- The radius of the outlet section of the PPT nozzle was $R_0 = 0.03$m.
- The average density of the plasma plume at the PPT outlet section was $\rho_0 = 2.83 \times 10^{-5}\,\text{kg/m}^3$.
- The average plasma pressure at the PPT outlet section was $p_0 = 250\,\text{N/m}^2$. This pressure corresponds to the plasma molecular weight $M_\Sigma = 19\,\text{kg/kmol}$ and the average temperature is $T_0 = 17,400$ K.
- The plasma velocity at the PPT outlet section was $V_0 = 25\,\text{km/s}$.
- The specified input data correspond to $M_{A^*} = 25$.

It is assumed that at the PPT outlet section, the plasma is homogeneous with respect to the radius.

Numerical modeling is performed for two orientations of the PPT symmetry axis with respect to the incident flow: for $\beta = 30°$ and $60°$ (see Figure 3.1). The slope of the magnetic field is $\alpha = 60°$. The parameters of the ambient (ionospheric) plasma correspond to the altitude of 150 km: $\rho_* = 2.0 \times 10^{-9}\,\text{kg/m}^3$, $p_* = 4.5 \times 10^{-4}\,\text{N/m}^2$, and $B_0 = 5.0 \times 10^{-5}$ T; the velocity of the incident flow of the ionospheric plasma was assumed to be equal to $V_\infty = 6$ km/s. The calculation domain had the following size $V_\infty = 6\,\text{km/s}$. The calculation domain had the following size $L_x = 2.75$ m, $L_y = 3.25$ m, and $L_z = 4$ m. In the calculations, an inhomogeneous grid, measuring $100 \times 100 \times 100$, was used.

The calculation results for the dynamics of the plasma cloud for the angle $\beta = 60°$(between the PPT axis and the incident flow of ionospheric plasma) shown in Figures 3.2–3.4 represent data for some successive time moments, whose values are mentioned in figure captions. These figures show the density distribution of the plasma bunch and the ambient medium, and the gas dynamic and magnetic pressure against the background of the vector velocity field in the plane y. In order to estimate the numerical concentration of the plasma particles (in cubic meters), the density values shown in figures must be multiplied by the coefficient 6.34×10^{16}.

It is clearly seen from the presented data that once the pulsed thruster stops operation ($t = 6\,\mu s$), the plasma plume moves across space in the form of a toroid shape plasma bunch. By the maximum density value in the expanding plasma bunch (correspondingly $\rho = 1,000$, 360, 44, 16 for $t = 6$, 25, 50, 100 μs), it is possible

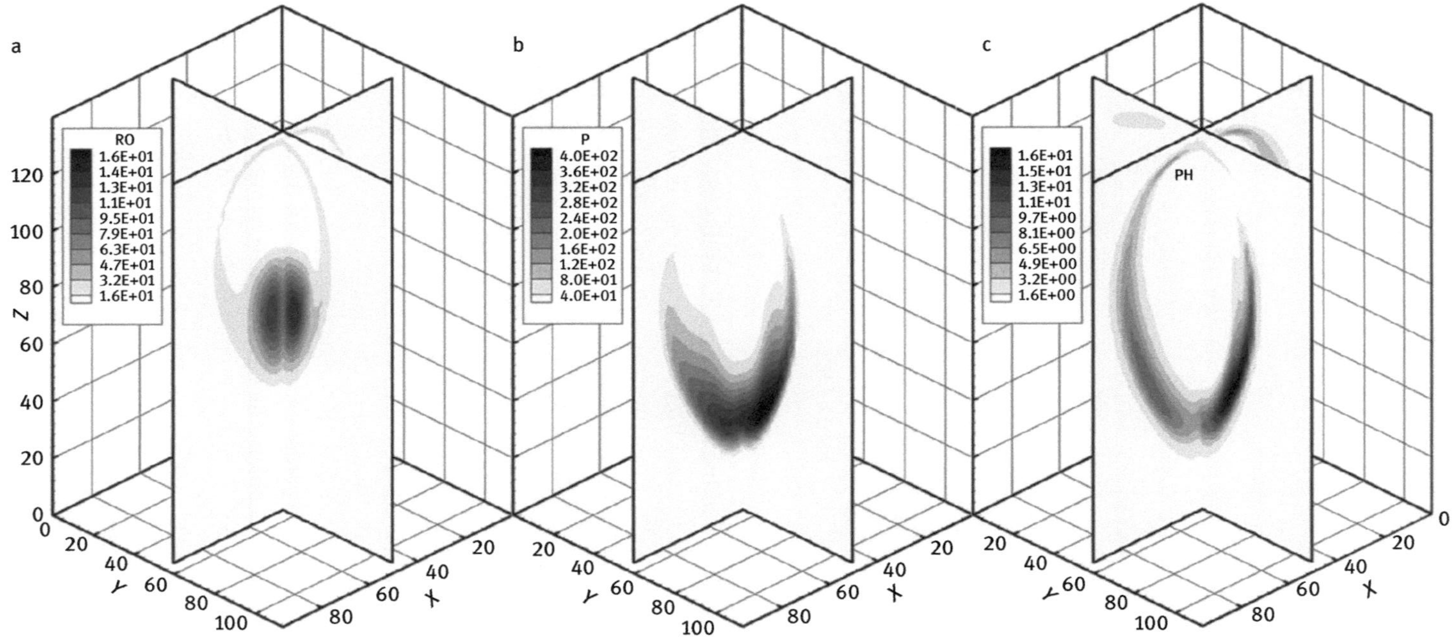

Figure 3.2: Total density distribution of the plasma plume and of the ionospheric plasma (a), gas dynamic (b), and magnetic (c) pressure at $t = 100\ \mu s$. The angle $\beta = 60°$. The tables give the density related to the ambient medium density. Gas dynamic pressure is related to $p = \tilde{p}/\rho_* V_*^2$, while the magnetic pressure is related to those in the ambient ionospheric plasma.

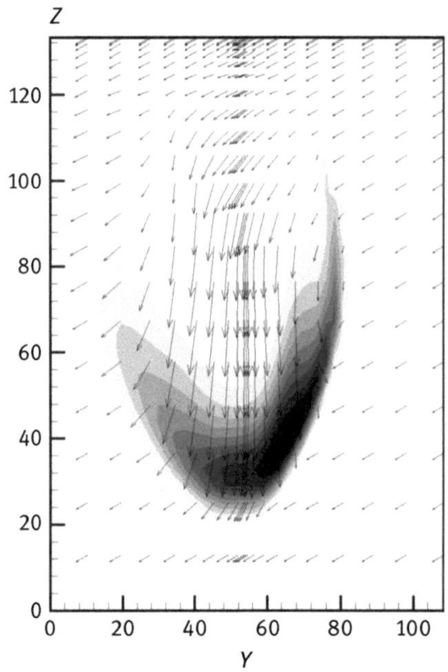

Figure 3.3: Distribution of the gas dynamic pressure and velocity field at $t = 100\,\mu s$. The angle $\beta = 60°$. The pressure in the plasma bunch reaches 401.

to judge the rate of the rarefication of the plasma bunch. At the later time stages of the expansion process ($t = 75$ and 100 μs), flow density decrease is observed in the wake of the moving bunch as well as the anisotropy in the spatial distribution of the density, which in the considered configuration points to the influence of the incident ionospheric flow.

It is obvious from the presented distributions of gas dynamic (Figure 3.2b) and magnetic (Figure 3.2c) pressures as early as time $t = 100\,\mu s$, a pressure increase is observed on the side of the incident ionospheric flow due to the collision of the plasma bunch with the ambient medium. The asymmetry of the compression wave is clearly identified from the side of the incident ionospheric flow (Figure 3.3).

It should be recalled that the distribution of the magnetic pressure (Figure 3.2c) characterizes the perturbation distribution of the magnetic field module in the vicinity of the plasma spread, and the components of the rotor of the magnetic field induction vector make it possible to obtain the electric currents induced by the expanding plasma. In this connection, the intensity distribution of the induced electric field in the vicinity of the plasma bunch is of great practical interest.

Figure 3.4 shows the projections of the vector of the intensity of the electric field intensity on the y- and z-axes, respectively, at time $t = 100\,\mu s$.

Figures 3.5–3.7 show the density, gas dynamics, and magnetic pressure at $\beta = 30°$, as well as flow field and the intensity of the electric field at time $t = 100\,\mu s$. The plume asymmetry, which is lower than it was before, clearly demonstrates the

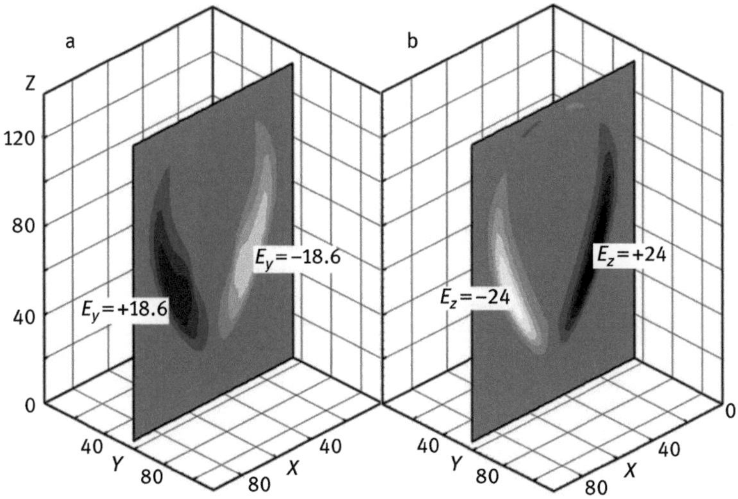

Figure 3.4: Intensity distribution of the electric field E_y (a) and E_z(b) at $t = 100\,\mu s$. The angle $\beta = 60°$.

reduction in the intensity of this interaction (compare Figures 3.3 and 3.7). This points to the fact that in the considered simulation cases, the interaction of plasma flows between each other appears to be stronger than the interaction of the plasma bunch with the ambient magnetic field.

If the characteristic scales of the expanding plasma deceleration are calculated by the estimated relations, the radius of the magnetic deceleration to be about 10 m, and the radius of the hydrodynamic braking to be about 1 m, which shows that it is the hydrodynamic braking that will be manifested first.

Therefore, both approximate theoretical estimates and the results of the 3D numerical calculations of this section show that the motion of the plasma bunch in the considered conditions is primarily influenced by the incident flow of the ionospheric plasma. Duration of the active phase of the interaction can be estimated by about 20–40 μs. It is also evident that the intensity of the interaction between the plasma plume with the ionospheric flow depends on their mutual orientation. Therefore, at $\beta = 30°$, it is easier to identify the asymmetry of the plasma plume due to the interaction of the spread plasma with the geophysical magnetic field.

Although the employed physical model was approximate, the numerical simulation succeeded in predicting practically all important parameters of the plume induced by a PPT, such as the velocity of plasma expansion, the intensity of the induced electric and magnetic fields, and the induced currents. The fact that these functions are characterized by a spatial distribution enables the researcher to analyze the pattern of interaction of the spread of the plasma with the flow of the ionospheric plasma, and, which is of special significance in providing the

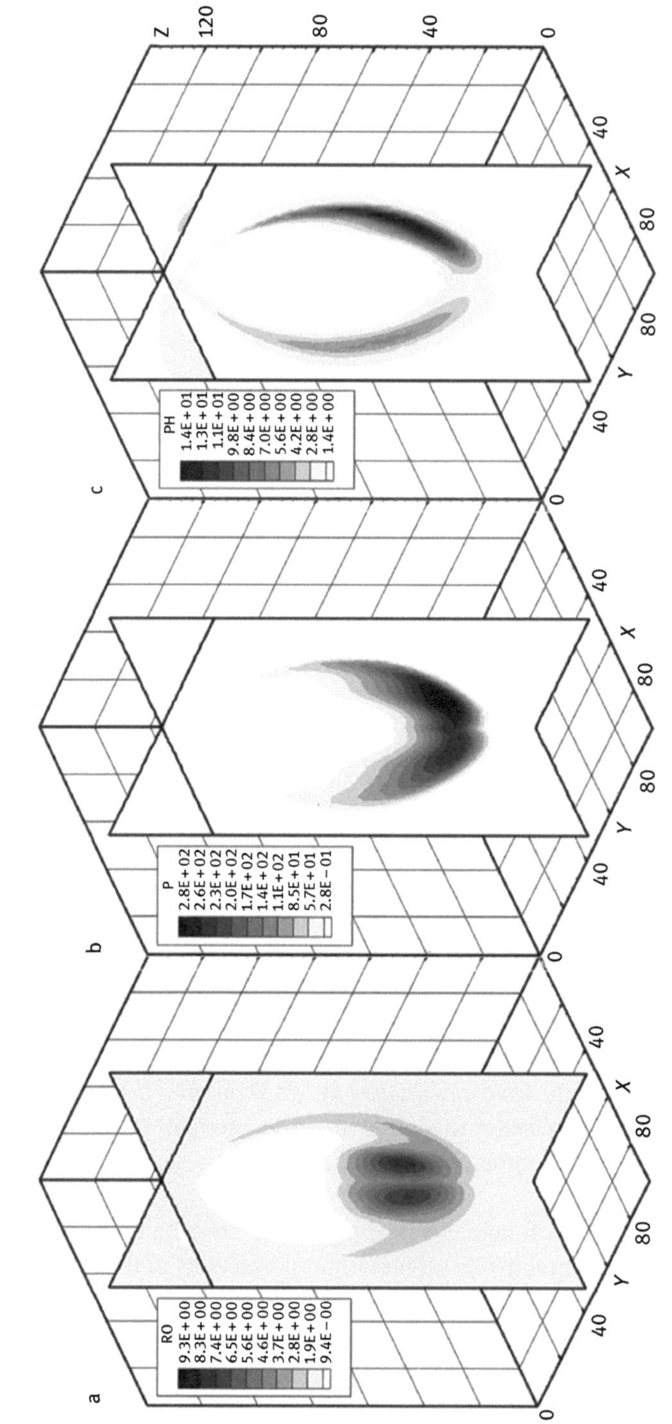

Figure 3.5: Spatial distribution of the total density of the plasma (a), the gas dynamic (b) and magnetic (c) pressure at $t = 100\,\mu s$. The angle $\beta = 30^\circ$.

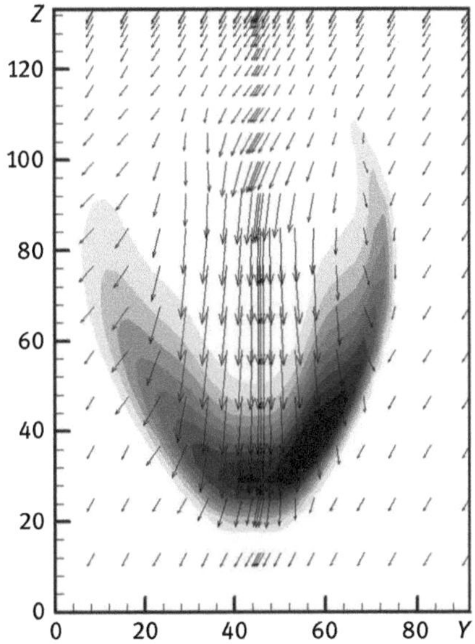

Figure 3.6: Distribution of the gas dynamic pressure and velocity field at $t = 100\,\mu s$. The angle $\beta = 30°$. The maximum pressure is 284.

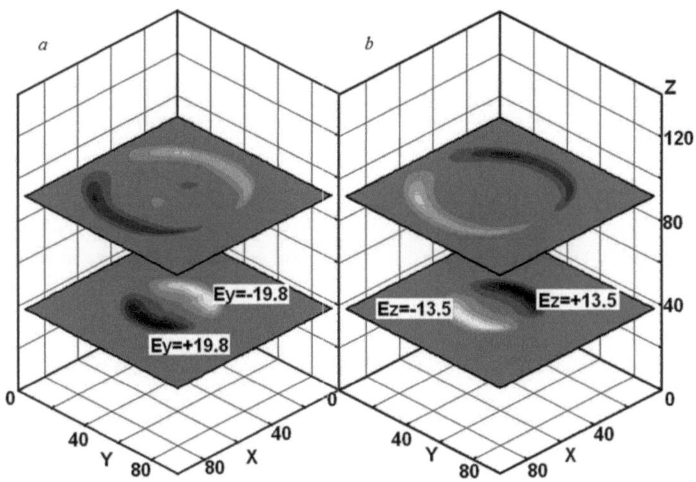

Figure 3.7: The intensity of the electric field E_y (on the left) and E_z in two planes at $t = 100\,\mu s$. The angle $\beta = 30°$.

safe operation of space vehicles, the interaction of the plasma flows with the spacecraft's surface. In the considered case, numerical modeling has shown that at a distance of 0.5 m from the outlet section of the PPT4 nozzle, the plasma plume starts its interaction with the incident ionospheric plasma.

Part II: **Numerical simulation of a glow discharge**

4 The physical mechanics of direct-current glow discharge

This chapter is devoted to computer simulation tasks of glow discharges. Physical models and numerical methods considered in this second part have various applications in the modern gas discharge physics, plasma research, and aerophysics. Chapters of this second part of the book give necessary physical models and computational physics methods for the creation of computational models of glow discharges.

Representations about the structure of glow discharge are substantially based on numerous experimental data and on the physical models describing processes in separate areas of glow discharge (Raizer Yu.P., 1991; Brown S.C., 1966). Till now the common theoretical-computational model of these phenomena is not created. Among existing models, it is necessary to note drift-diffusion models of discharge (Ward A.L., 1958; 1962; Raizer Yu.P. and Surzhikov S.T., 1987; 1988), models of quasi-neutral plasma with ambipolar diffusion (Surzhikov S.T., et al., 2005), and also the models based on the use of statistical Monte Carlo methods (Boeuf J.P. and Marode E.A., 1982). Only a few of the enumerated models are intended for research spatial (the two- and three-dimensional, 2D and 3D) structures of glow discharge. It is related, first of all, with high laboriousness of spatial problem solution. However, due to the growth of computer power and the use of parallel computing technology in computational physics, the spatial computational models including all variety of kinetic processes become more and more realistic. A necessity for this also increases in connection with a new wave of interest to electric discharge phenomena, which is generated by possible applications of these phenomena in aerophysics.

In this book, two classes of models of glow discharge are presented: the drift-diffusion model and the model of quasi-neutral plasma with ambipolar diffusion. Basically, 2D models are considered, and problems of testing of these models are solved with methods of computing physics, or by comparison with the known experimental facts. The majority of the presented results of numerical calculations correspond to the 2D glow discharge burning in conditions of a normal current density. In this case, besides the description of the axial distributions of discharge parameters (from the cathode to the anode), the significant interest is the study of regularities of 2D structures that are formed. In addition to that, it has appeared that the good test criterion for the 2D models of the normal glow discharge is the outcome of calculations under one-dimensional (1D) Engel–Steenbeck theory, created at strongly simplified suppositions for a 1D model of a cathode layer. The specified 1D model has appeared to be simple enough and rather successful; hence, it became the standard in the analysis of experimental data. Numerous verifications of the given model by experimental data allow considering its reference for the whole class of spatial computing models.

https://doi.org/10.1515/9783110648836-005

There are two well-known types of glow discharges. In a *normal glow discharge*, the current is transferred only through a part of the cathode surface (gas excited by electrons shines out only above a part of the cathode), and in the *anomalous glow discharge* the current flows through all surfaces of the cathode, and the whole cathode becomes covered by luminescence. In a normal glow discharge, the *law of a normal current density* is satisfied: the current through discharge increases with a magnification of a voltage drop on a cathode layer due to a magnification of a current spot square, thus the current density in the center of a current spot does not vary (varies slightly).

4.1 Fundamentals of the physics of direct-current glow discharge: the Engel–Steenbeck theory of a cathode layer

The classical glow discharge is characterized by the following parameters:
- The total current through discharge $I{\sim}10^{-3}-10^{-2}$ A.
- Voltage drop on discharge gap $V{\sim}200-5,000$ V.

Figure 4.1 shows the representation of glow discharge placed among other types of discharges between two electrodes. The typical values of voltage and current on a gas discharge gap of classical glow discharge are shown on the axes of this figure.

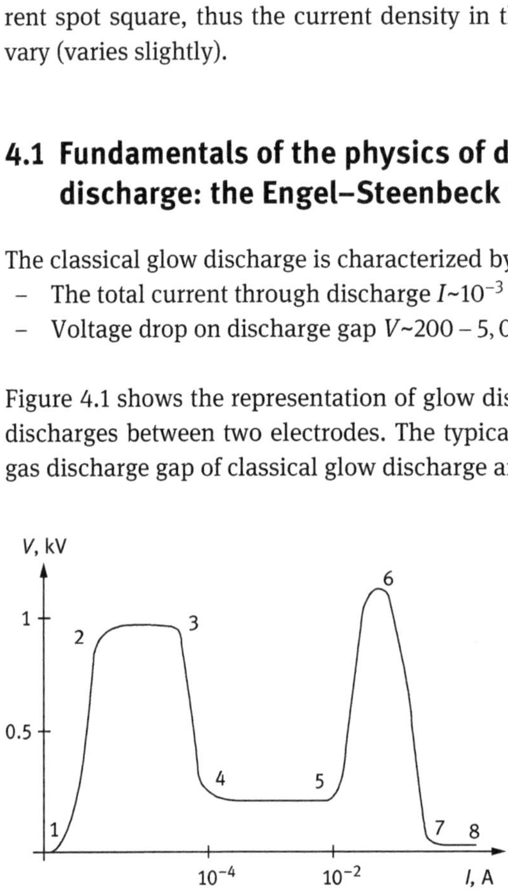

Figure 4.1: The diagrammatic representation of a voltage–current characteristic of electric discharges (boundary values of voltages and currents are approximate): 1–2, area of nonself-maintained discharge; 2–3, the dark Townsend discharge; 4–5, normal glow discharge; 5–6, anomalous glow discharge; 6–7, area of transition from glow to arc discharge; 7–8, arc discharge.

The majority of works on experimental and theoretical study of glow discharge have considered the pressure range of $p{\sim}1-50$ Torr. However, this type of discharges is used commonly enough at raised pressure though in these cases glow discharges are rather unstable.

Notwithstanding that the given type of discharges is widely applied in physical devices of various kinds, also in light sources, generators of plasma, and gas discharge optical quantum generators (lasers), it is necessary to recognize that the theoretical-computational description of glow discharges is developed in an insufficient degree. First, it is explained with a complicated structure of glow discharges, and real lack of the acceptable computational models allows describing all elements of glow discharge structure.

Basic elements of the structure of classical normal glow discharge in a gas discharge tube are shown in Figure 4.2.

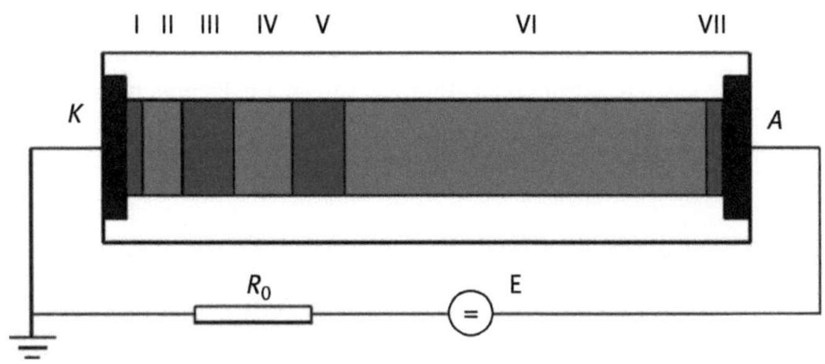

Figure 4.2: The scheme and structure of classical glow discharge: I is the region of the Aston dark space; II is the region of cathode luminescence; III is the region of cathode dark space; IV is the region of negative luminescence; V is the region of the Faraday dark space; VI is the positive column; VII is the region of anode dark space; K is the cathode; A is the anode; R_0 is the ohmic resistance of external circuit; *E* is the value of electromotive force.

Alternation of relatively light and dark areas on the scheme of discharge approximately corresponds to a real alternation of gas luminescence areas in the discharge. However, as a rule, areas *I–V* adjoin the cathode closer.

Typical distribution of electric field strength E and volumetric density of charge $\rho = e(n_i - n_e)$ are shown in Figure 4.3.

The basic physical processes ensuring the existence of the glow discharge are the secondary ion-electronic emission of electrons from the cathode, acceleration of the electrons in an applied electric field, and their collisions with neutral discharge particles, whose concentration is always by several orders more than of electrons and positive ions in the discharge. Negative ions also played an important role in some gases.

A principal source of glow discharge maintaining is the electron emission from a surface of the cathode due to the ion-electronic emission. Electrons, which abandon the cathode, have rather low energy (~1 eV) that is caused by the ion-electronic emission mechanism. These electrons cannot excite a luminescence of atoms and molecules in the gas discharge gap; therefore, near to the cathode, the dark area is

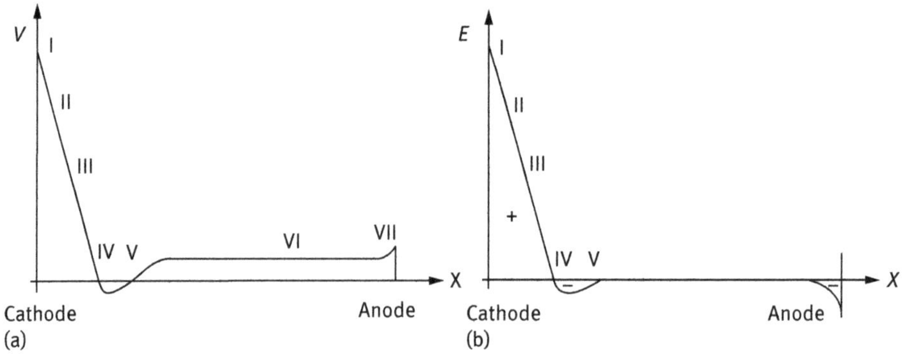

Figure 4.3: A qualitative picture of the distribution of an electric field strength (a) and a volumetric density of charge (b) along an axis of symmetry from the cathode to the anode. Figures mark areas of glow discharge structure (see Figure 4.2).

observed. This is the Aston area of glow discharge (see region I in Figure 4.2). Near to the cathode the electric field strength is great (E~$3-4\,\text{kV/cm}$), and there electrons gain energy fastly, which is sufficient to cause luminescence of gas after the Aston area, the so-called cathode luminescence (region II in Figure 4.2).

In the process of increase in energy, the electrons become capable to excite more and more high energy levels of atoms and molecules; therefore, it can be observed on a gradual modification of luminescence color (from red to violet shadowing) in the field of cathode luminescence. Finally, the energy gained by electrons becomes so high that their inelastic collisions with atoms and molecules result in ionization of the particles. It happens in the field of cathode dark space (region III in Figure 4.2), where the avalanche-like process of electron multiplication prevails. Electrons that are born as a result of neutral particles ionizing event start to gain their energy in an electric field, whose strength is essentially lower than the strength near the cathode. Nevertheless, the big number of electrons, accelerated by this small field, appears sufficient for excitation of neutral particles, and a zone of a negative luminescence arises in the discharge (region IV in Figure 4.2).

When the distance from the cathode becomes greater the electric field strength lower, and according to this the energy accumulated by electrons also decreases. This leads to that it is also possible to recognize different spectral regions (from violet to red shadowing) in the luminescence of area IV.

After the negative luminescence area, there is the Faraday dark space area (region V in Figure 4.2). The energy of electrons in this area is so low that they do not excite a luminescence of neutral particles anymore. However, the number of electrons here is already rather great; therefore, it is possible to observe an area of a spatial negative charge, and the electric field strength can change its sign to opposite here (see Figure 4.3).

The most extended area of classical glow discharge is the positive column (region VI in Figure 4.2). It is the area of quasi-neutral plasma with rather low electric field strength, and average energy of thermal motion of electrons is about 1 – 3 eV. Nevertheless, in the power spectrum of electrons, there are ones with large enough energy, and they cause excitation and luminescence of neutral particles.

In immediate proximity from the anode, the anode dark space (region VII in Figure 4.2) is observed. This is a small area of a volumetric negative charge (see Figure 4.3). The anode does not generate ions but accepts an electron flow.

The classical 1D Engel–Steenbeck model is presented below for a cathode layer of the normal glow discharge (Engel A. and Steenbeck M., 1932; Brown S.C., 1966).

Let us consider the 1D cathode layer of normal glow discharge. If we suppose that $n_{e.0}$ is the number of electrons emitted from a surface of the cathode due to ions falling on it, $n_{e.\infty}$ is the number of electrons at the end of a cathode layer with thickness d. Assume that just in this part of the cathode layer $(n_{e.\infty} - n_{e.0})$ pairs of electrons and ions are generated. We shall assume that exactly this amount of ions "beats out" electrons from the cathode, and the total number of that electrons is $n_{e.0}$:

$$n_{e,0} = \gamma(n_{e,\infty} - n_{e,0}), \tag{4.1}$$

where γ is the secondary ion-electronic emission coefficient.

It is easy to derive a formula for $n_{e.\infty}$ with the use of a definition of the first Townsend ionization coefficient $\alpha(E/p)$, being a function of field strength and pressure:

$$n_{e,\infty} = n_{e,0} \exp \int_0^d \alpha\left(\frac{E}{p}\right) \mathrm{d}x \tag{4.2}$$

and

$$n_{e,\infty} = n_{e,0} \frac{1+\gamma}{\gamma}, \tag{4.3}$$

Whence we obtain a condition of maintaining a stationary current in a cathode layer

$$\exp \int_0^d \alpha\left(\frac{E}{p}\right) \mathrm{d}x = 1 + \frac{1}{\gamma}. \tag{4.4}$$

Being based on numerous measurements of field strength in a cathode layer, Engel and Steenbeck have assumed the following linear dependence:

$$E(x) = C(d - x), \tag{4.5}$$

where C is some constant.

To define C, we shall consider the potential of an electric field inside a cathode layer

$$V(x)=\int_0^x E(x')\,\mathrm{d}x'=C\left(xd-0.5x^2\right). \tag{4.6}$$

Having set a voltage drop on a cathode layer V_d (it is possible to measure this value; the mode of its calculation will be given later), we obtain

$$V_d=C\frac{d^2}{2} \tag{4.7}$$

or

$$C=V_d\frac{2}{d^2}. \tag{4.8}$$

So, distribution of strength and a field potential in a cathode layer is set in the form of

$$E(x)=V_d(d-x)\frac{2}{d^2}, \tag{4.9}$$

$$V(x)=V_d(2d-x)\frac{x}{d^2}. \tag{4.10}$$

But from the Poisson equation, it follows that

$$\frac{\mathrm{d}^2V(x)}{\mathrm{d}x^2}=-4\pi\rho=-\frac{2}{d^2}V_d. \tag{4.11}$$

What this means is that in the cathode layer the charge density remains constant:

$$\rho=\frac{V_d}{2\pi d^2}=e(n_i-n_e). \tag{4.12}$$

Let's rewrite the condition (4.1) concerning the current density

$$j_{e,0}=\gamma j_{i,0}. \tag{4.13}$$

where $j_{e,0}, j_{i,0}$ are the current densities of electrons and ions on the cathode.

Then the total current density on the cathode is

$$j_0=j_{e,0}+j_{i,0}=j_{i,0}(1+\gamma). \tag{4.14}$$

By definition

$$j_{i,0}=n_{i,0}v_{i,\mathrm{dr},0}=n_{i,0}\mu_iE(x=0), \tag{4.15}$$

where μ_i is the ion mobility; $v_{i,\mathrm{dr},0}$ is the drift velocity of ions.

We shall determine the strength of electric field on the cathode $E(x=0)$ by assuming that $n_{e,0} < n_{i,0}$; then from formulas (4.9), (4.10), and (4.12) one can derive

$$\rho(x=0) = \frac{E(x=0)}{4\pi d} \tag{4.16}$$

and

$$j_0 = \mu_i \frac{V_d^2}{\pi d^3}(1+\gamma). \tag{4.17}$$

Voltage drop on the cathode can be calculated using the condition of self-maintained discharge (4.4). For a definition of the value $\alpha(E/p)$, we shall take an empirical relation which will be often used further:

$$\frac{\alpha(E/p)}{p} = A\exp\left[-\frac{B}{(E/p)}\right], \tag{4.18}$$

where A, B are the empirical coefficients obtained for various gases (Brown S.C., 1966; Raizer Yu.P., 1991).

Substituting (4.18) into (4.4), we get

$$\ln\left(1+\frac{1}{\gamma}\right) = Ap\int_0^d \exp\left[-\frac{Bp}{E(x)}\right]\mathrm{d}x. \tag{4.19}$$

Now it is necessary to substitute into formula (4.19) for calculating the strength of a field (4.9):

$$\ln\left(1+\frac{1}{\gamma}\right) = Ap\int_0^d \exp\left[-\frac{Bp}{\frac{2V_d}{d}\left(1-\frac{x}{d}\right)}\right]\mathrm{d}x. \tag{4.20}$$

Let's choose the value t as an integration variable

$$t = \frac{x}{d}\frac{2V_d}{Bpd}. \tag{4.21}$$

Then integration of (4.20) gives

$$\ln\left(1+\frac{1}{\gamma}\right) = \frac{AB(pd)^2}{2V_d} S\left(\frac{2V_d}{Bpd}\right), \tag{4.22}$$

where

$$S(z) = \int_0^z \exp\left(-\frac{1}{t}\right) d\,t. \tag{4.23}$$

The last integral was calculated and tabulated. In Raizer Yu.P. (1991), an assumption is used about the constancy of electric field strength in the cathode layer. It has been shown that this allows calculating integral (4.19) easily and with a sufficient exactitude for practical needs.

Relation (4.22) is possible to present in the form of the simple algebraic equation, which can be considered as a dimensionless characteristic equation for all gases (A, B) and materials of the cathode (γ):

$$\frac{(C_1 V_d)^{1/3}}{(C_2 j_0)^{2/3}} S\left[(C_1 V_d)^{1/3} (C_2 j_0)^{1/3}\right] = 1, \tag{4.24}$$

where

$$C_1 = \frac{2A}{B \ln(1 + 1/\gamma)}; \tag{4.25}$$

$$C_2 = \frac{4\pi \ln(1 + 1/\gamma)}{AB^2 p^2 (p\mu_i)(2 + \gamma)}. \tag{4.26}$$

Engel A. and Steenbeck M. (1932) have investigated the characteristic eq. (4.24) and have shown that the dependence of $C_1 V_d$ from $C_2\, j_0$ looks like a parabola, whose minimum meets normal glow discharge and corresponds to values $C_1 V_d = 6$ and $C_2\, j_0 = 0.67$. It allows receiving parameters of a cathode layer of normal glow discharge

$$V_n = \frac{3B}{A} \ln\left(1 + \frac{1}{\gamma}\right), \quad \text{V}, \tag{4.27}$$

$$\frac{j_n}{p^2} = 5.92 \times 10^{-14} \frac{AB^2 (\mu_i p)(1 + \gamma)}{\ln(1 + 1/\gamma)}, \quad \frac{\text{A}}{\text{cm}^2 \cdot \text{Torr}} \tag{4.28}$$

$$d_n p = 3.78 \frac{\ln(1 + 1/\gamma)}{A}, \quad \text{cm} \cdot \text{Torr}, \tag{4.29}$$

where the index n marks parameters of normal glow discharge.

As marked earlier, the stated theory of Engel and Steenbeck was excellently confirmed by numerous experimental researches; therefore, further, the conclusions of this theory (first, relations (4.27)–(4.29)) will be repeatedly used for interpretation of numerical simulation results.

4.2 Drift-diffusion model of a glow discharge

4.2.1 Governing equations

Equations of the drift-diffusion model of stationary glow discharge between two infinite flat electrodes (the scheme of the problem is presented in Figure 4.4) are formulated in the following form (Raizer Yu.P. and Surzhikov S.T., 1988):

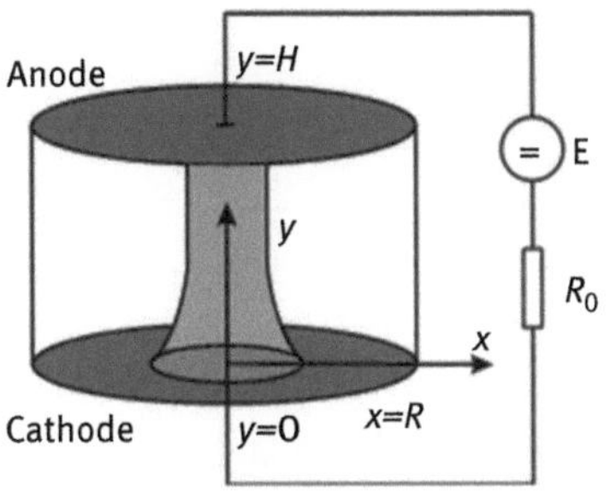

Figure 4.4: The scheme of the task.

$$\frac{\partial n_e}{\partial t} + \mathrm{div}\boldsymbol{\Gamma}_e = \alpha|\boldsymbol{\Gamma}_e| - \beta n_e n_+, \tag{4.30}$$

$$\frac{\partial n_+}{\partial t} + \mathrm{div}\boldsymbol{\Gamma}_+ = \alpha|\boldsymbol{\Gamma}_e| - \beta n_e n_+, \tag{4.31}$$

$$\mathrm{div}\mathbf{E} = 4\pi e(n_+ - n_e), \tag{4.32}$$

where

$$\boldsymbol{\Gamma}_e = -D_e \,\mathrm{grad}\, n_e - n_e \mu_e \mathbf{E}, \tag{4.33}$$

$$\boldsymbol{\Gamma}_+ = -D_+ \,\mathrm{grad}\, n_+ + n_+ \mu_+ \mathbf{E}, \tag{4.34}$$

$$\mathbf{E} = -\,\mathrm{grad}\,\varphi, \tag{4.35}$$

n_e, n_+ are the concentrations of electrons and ions in 1 cm^3 (further, in the given chapter the functions concerning ions will be designated with the index «+», as the index "i" will be used in finite-difference schemes); **E** and φ are the vector and potential of an electric field strength; D_e, D_+ are the electron and ion diffusion coefficients; μ_e, μ_+ are the electron and ion mobility; $\alpha = \alpha(E)$ is the coefficient of ionization of a molecule by the electron impact (the first Townsend coefficient); γ is the coefficient of ion–electron recombination.

In view of relation (4.35), eq. (4.32) can be rewritten in the following form:

$$\mathrm{div}[\mathrm{grad}(\varphi)] = -4\pi e(n_+ - n_e). \tag{4.36}$$

Finite-difference operators will be formulated for two orthogonal geometries, namely for the rectangular and cylindrical. Typical boundary conditions for eqs. (4.30)–(4.32) are formulated as follows:

$$x = 0: \qquad \Gamma_{e,x} = \gamma \Gamma_{+,x}, \ \varphi = 0, \tag{4.37}$$

$$x = H: \qquad n_{+} = 0, \ \varphi = V, \tag{4.38}$$

$$r = 0: \qquad \frac{\partial n_e}{\partial r} = \frac{\partial n_+}{\partial r} = \frac{\partial \varphi}{\partial r} = 0, \tag{4.39}$$

$$r = R: \qquad 1) \ n_e = n_+ = 0, \ \varphi = \frac{V}{H} x, \tag{4.40}$$

$$2) \ \frac{\partial n_e}{\partial r} = \frac{\partial n_+}{\partial r} = \frac{\partial \varphi}{\partial r} = 0, \tag{4.41}$$

where γ is the coefficient of ion–electron emission from the cathode surface; V is the voltage drop on the discharge gap.

Boundary conditions (4.40) set an undisturbed electric field and lack of charges at a great distance from the center of discharge, and condition (4.41) states the lack of transverse field gradients and particles on external boundary of the rated area. Really, at calculations, there is a recommendation to use a very small value of concentration (~10^{-6} of maximum concentration in the discharge). By methodical calculations it has been found that change of this "background" concentration by several digits do not affect parameters of the discharge.

The first of the mentioned conditions on external side boundary $r = R$ correspond to the process without radial diffusion of particles. Really, in the no-current area of discharge, there is no place for the charged particles to appear. Charges move along the electric field lines and cannot pass from one line to another. Thus, on each field line, the Townsend condition of stationary self-maintained discharge should be satisfied:

$$\int_l \alpha(l') \mathrm{d}l' > \ln\left(1 + \frac{1}{\gamma}\right). \tag{4.42}$$

If this condition is not satisfied on any field line, charges drifting along this line will abandon the discharge gap, and then this line will become no-current.

Diffusion of charged particles and their origin on boundaries of the discharge gap (on the cathode and the anode) can cause current-carrying lines without realization of condition (4.42). At the same time, it is necessary to mean that in the case of diffusion, the force and current lines do not coincide any more, and the condition of stationary self-maintained discharge has nonlocal character. In this case, there is no basic condition for charged particles origin on the periphery of the discharge area at $r \to R$. It is interpreted in the boundary condition (4.41) which assumes or the presence of a spatial symmetry in a disposition of current-carrying channels (in a cylindrical case, the discharge looks like coaxial cylinders enclosed one in another, and in a flat case it looks like channels regularly arranged on a plane), either the presence of high enough background concentration of the charged particles. If the solitary discharge is considered, this background level should be small enough to suppress the appreciable influence on the structure of the discharge.

By numerical experiments, it has been shown that in a wide range of parameters of glow discharge given a type of boundary conditions does not make an essential influence on the results of calculations. In the statement of the boundary value problem (4.30)–(4.41), diffusion of the charged particles in an axial direction is neglected.

Boundary conditions include still an undetermined value of a voltage drop on discharge gap V. For its determination it is necessary to attract conditions in an external circuit (see Figure 3.4). For stationary glow discharge, it is possible to write the obvious relation

$$\frac{E-V}{eR_0}=2\pi\int_0^R \Gamma_{e,x}(r,x=H)r^m\mathrm{d}r, \tag{4.43}$$

which specifies the equality of the sum of the voltage drop on resistance R_0 and on the discharge gap to the electromotive force ε; $\Gamma_{e,x}$ is the projection of an electron flux density vector on the x-axis. In eq. (4.43), $m=0$ corresponds to a flat case and $m=1$ corresponds to a cylindrical case.

Let us consider a glow discharge in molecular nitrogen:

$$\mu_e p=4.4\times10^5,\ \mu_+ p=1.45\times10^3,\ \left(\mathrm{Torr}\cdot\mathrm{cm}^2\right)/(\mathrm{V}\cdot\mathrm{s}), \tag{4.44}$$

$$\beta=2\times10^{-7},\ \mathrm{cm}^3/\mathrm{s},$$

$$\varepsilon^*=4\pi e=1.81\times10^{-6},\ \mathrm{V}\cdot\mathrm{cm},$$

$$\frac{\alpha}{p}=\begin{cases} A\cdot\exp\left(-\frac{B}{E/p}\right), & \frac{E}{p}>100;\\ \frac{A_1}{E/p}\exp\left(-\frac{B_1}{E/p}\right), & \frac{E}{p}<100, \end{cases}\quad (\mathrm{cm}\cdot\mathrm{Torr})^{-1}, \tag{4.45}$$

where $A=12\,(\mathrm{cm}\cdot\mathrm{Torr})^{-1}$; $B=342\,\mathrm{V}/(\mathrm{Torr}\cdot\mathrm{cm})$; $A_1=900\,\mathrm{V}/(\mathrm{Torr}\cdot\mathrm{cm})^2$; $B_1=314\,\mathrm{V}/(\mathrm{Torr}\cdot\mathrm{cm})$.

Note that similar empirical coefficients for other gases are given in Brown S.C. (1966) and Raizer Yu.P. (1991).

Diffusion coefficients were defined with the Einstein relations

$$D_e=\mu_e T_e,\ D_+=\mu_+ T_+,$$

where T_e, T_+ are the temperatures of electrons and ions, eV.

With the use of the model, it is possible to make calculations of glow discharge structure in various assumptions concerning the temperature of electrons and heavy particles:

a) $T_e = T_+ = 0$;
b) $T_e = 11,610\,\mathrm{K}\,(T_e = 1\,\mathrm{eV})$, $T_+ = 300\,\mathrm{K}\,(T_+ = 0.0258\ \mathrm{eV})$;
c) $T_e = f(x,r)$ and $T_+ = f(x,r)$.

Note that the condition (a) corresponds to lack of physical diffusion.

4.2.2 Reduction of governing equations to a form convenient for the numerical solution

Charge conservation equations (4.30) and (4.31) can be rewritten in the following dimensional form (neglecting diffusion in an axial direction):

$$\frac{\partial n_e}{\partial t} + \mu_e \frac{1}{r^m}\frac{\partial}{\partial r} n_e r^m \frac{\partial \varphi}{\partial r} + \mu_e \frac{\partial}{\partial x} n_e \frac{\partial \varphi}{\partial x} - D_e \frac{1}{r^m}\frac{\partial}{\partial r} r^m \frac{\partial n_e}{\partial r} = \alpha|\mathbf{\Gamma}_e| - \beta n_e n_+, \tag{4.46}$$

$$\frac{\partial n_+}{\partial t} - \mu_+ \frac{1}{r^m}\frac{\partial}{\partial r} n_+ r^m \frac{\partial \varphi}{\partial r} - \mu_+ \frac{\partial}{\partial x} n_+ \frac{\partial \varphi}{\partial x} - D_+ \frac{1}{r^m}\frac{\partial}{\partial r} r^m \frac{\partial n_+}{\partial r} = \alpha|\mathbf{\Gamma}_e| - \beta n_e n_+. \tag{4.47}$$

To reduce eqs. (4.46) and (4.47) to a dimensionless form, we multiply them with the factor $(H/E\mu_e)(H/N_0)$, where $N_0 = 10^9\ \mathrm{cm}^{-3}$ is the characteristic concentration of charged particles in a glow discharge. The following definitions will be used:

$$\tilde{x} = \frac{x}{H},\quad \tilde{r} = \frac{r}{H},\quad \tau = \frac{t}{t_0},\quad t_0 = \frac{H^2}{\mu_{e,0}E};$$

$$u = \left\{ N_e = \frac{n_e}{N_0},\ N_+ = \frac{n_+}{N_0},\ \Phi = \frac{\varphi}{E} \right\};$$

$$\tilde{D}_e = \frac{D_e}{\mu_{e,0}E},\quad \tilde{D}_+ = \frac{D_+}{\mu_{e,0}E},\quad Q = \frac{HN_0}{W\mu_{e,0}};$$

$$W = \frac{E}{H},\quad \tau = t\frac{W\mu_{e,0}}{H}.$$

For brevity, the sign of a tilde will not be used for nondimensional coordinates r and x; therefore,

$$\frac{\partial N_e}{\partial \tau} + \frac{1}{r^m}\frac{\partial}{\partial r} r^m N_e \frac{\partial \Phi}{\partial r} + \frac{\partial}{\partial x} N_e \frac{\partial \Phi}{\partial x} - \tilde{D}_e \frac{1}{r^m}\frac{\partial}{\partial r} r^m \frac{\partial N_e}{\partial r} =$$
$$= -\beta Q N_e N_+ + \alpha H \sqrt{\left(N_e \frac{\partial \Phi}{\partial x}\right)^2 + \left(N_e \frac{\partial \Phi}{\partial r} - \tilde{D}_e \frac{\partial N_e}{\partial r}\right)^2}, \tag{4.48}$$

$$\frac{\partial N_+}{\partial \tau} - \frac{\mu_+}{\mu_e}\frac{1}{r^m}\frac{\partial}{\partial r}r^m N_+\frac{\partial \Phi}{\partial r} - \frac{\mu_+}{\mu_e}\frac{\partial}{\partial x}N_+\frac{\partial \Phi}{\partial x} - \tilde{D}_+\frac{1}{r^m}\frac{\partial}{\partial r}r^m\frac{\partial N_+}{\partial r} =$$

$$= -\beta Q N_e N_+ + \alpha H\sqrt{\left(N_e\frac{\partial \Phi}{\partial x}\right)^2 + \left(N_e\frac{\partial \Phi}{\partial r} - \tilde{D}_e\frac{\partial N_e}{\partial r}\right)^2}. \tag{4.49}$$

Using the designations given in Table 4.1, eqs. (4.48) and (4.49) can be presented in the following canonical form:

$$\frac{\partial u}{\partial t} + \frac{1}{r^m}\frac{\partial}{\partial r}(r^m aU) + \frac{\partial}{\partial x}(bU) - c\frac{1}{r^m}\frac{\partial}{\partial r}r^m\frac{\partial u}{\partial r} = f, \tag{4.50}$$

for which a finite-difference scheme will be constructed.

Table 4.1: Coefficients in eq. (4.50).

	a	*b*	*c*	*F*
Electrons	$r^m\frac{\partial \Phi}{\partial r}$	$\frac{\partial \Phi}{\partial x}$	$\tilde{D}_e$	$\alpha H\sqrt{\left(N_e\frac{\partial \Phi}{\partial x}\right)^2 + \left(N_e\frac{\partial \Phi}{\partial r} - \tilde{D}_e\frac{\partial N_e}{\partial r}\right)^2} - \beta Q N_e N_+$
Ions	$-r^m\frac{\mu_+}{\mu_e}\frac{\partial \Phi}{\partial r}$	$-\frac{\mu_+}{\mu_e}\frac{\partial \Phi}{\partial x}$	$\tilde{D}_+$	$\alpha H\sqrt{\left(N_e\frac{\partial \Phi}{\partial x}\right)^2 + \left(N_e\frac{\partial \Phi}{\partial r} - \tilde{D}_e\frac{\partial N_e}{\partial r}\right)^2} - \beta Q N_e N_+$

The dimensionless Poisson equation for electric field potential can be obtained by analogy

$$\frac{1}{r^m}\frac{\partial}{\partial r}r^m\frac{\partial \Phi}{\partial r} + \frac{\partial^2 \Phi}{\partial x^2} = -\frac{\varepsilon^* H N_0}{W}(N_+ - N_e). \tag{4.51}$$

Boundary conditions (4.39)–(4.41) should also be transformed to the dimensionless form

$$r = 0: \quad \frac{\partial N_e}{\partial r} = \frac{\partial N_+}{\partial r} = \frac{\partial \Phi}{\partial r} = 0, \tag{4.52}$$

$$r = \frac{R}{H}: \quad 1)\ N_e = N_+ = 0, \quad \Phi = \frac{V}{E}x, \tag{4.53}$$

$$2)\ \frac{\partial N_e}{\partial r} = \frac{\partial N_+}{\partial r} = \frac{\partial \Phi}{\partial r} = 0. \tag{4.54}$$

Let us consider derivation and analysis of boundary conditions at the cathode. For this purpose, we shall consider expressions for particle flux vectors in view of radial diffusion of charged particles:

$$\mathbf{\Gamma}_e = \mathbf{e}_x\left(+\mu_e n_e \frac{\partial \varphi}{\partial x}\right) + \mathbf{e}_r\left(\mu_e n_e \frac{\partial \varphi}{\partial r} - D_e \frac{\partial n_e}{\partial r}\right), \tag{4.55}$$

$$\mathbf{\Gamma}_+ = \mathbf{e}_x\left(-\mu_+ n_+ \frac{\partial \varphi}{\partial x}\right) + \mathbf{e}_r\left(-\mu_+ n_+ \frac{\partial \varphi}{\partial r} - D_+ \frac{\partial n_+}{\partial r}\right). \tag{4.56}$$

The boundary condition (4.37) establishes the relation between modules of charged particle fluxes to cathode and anode. Note that this condition does not define their angular distributions.

For the drift-diffusion model with a constant coefficient of ion-electronic emission, the boundary condition on a cathode is formulated only for axial components of electrons and ion fluxes

$$(\mathbf{\Gamma}_e)_x = -\gamma(\mathbf{\Gamma}_+)_x. \tag{4.57}$$

If we accept condition (4.57), then it directly follows from that

$$n_e = \gamma n_+ \frac{\mu_+}{\mu_e}, \tag{4.58}$$

This condition is also used in the 1D theory of a cathode layer.

In view of (4.58), it is possible to define approximate "grid" boundary conditions for ions on the cathode.

From eqs. (4.30) and (4.31), the continuity equation is derived in the following form:

$$\frac{\partial \tilde{\rho}}{\partial t} + \operatorname{div}(\mathbf{\Gamma}_+ - \mathbf{\Gamma}_e) = 0. \tag{4.59}$$

Considering that at $x = 0$, $n_e \ll n_+$, and also introducing the designation

$$D_{\text{eff}} = D_e \gamma \frac{\mu_+}{\mu_e} - D_+ ,$$

one can obtain

$$\frac{\partial n_+}{\partial t} \approx (1+\gamma)\mu_+ \frac{\partial}{\partial x}\left(n_+ \frac{\partial \varphi}{\partial x}\right) - D_{\text{eff}} \frac{1}{r^m} \frac{\partial}{\partial r} r^m \frac{\partial n_+}{\partial r}. \tag{4.60}$$

In a dimensionless form this equation looks as follows:

$$\frac{\partial N_+}{\partial \tau} = (1+\gamma)\frac{\mu_+}{\mu_e} \frac{\partial}{\partial x} N_+ \frac{\partial \Phi}{\partial x} - \tilde{D}_{\text{eff}} \frac{1}{r^m} \frac{\partial}{\partial r} r^m \frac{\partial N_+}{\partial r}, \tag{4.61}$$

where

$$\tilde{D}_{\text{eff}} = \frac{D_{\text{eff}}}{\mathrm{E}\mu_{e,0}}.$$

Using continuity equation (4.59) and boundary condition (4.38), we will receive the equation for electrons near to the anode, which will be used for the construction of a "grid" boundary condition for electrons

$$\frac{\partial N_e}{\partial \tau} = -\frac{\partial}{\partial x} N_e \frac{\partial \Phi}{\partial x} + \tilde{D}_e \frac{1}{r^m} \frac{\partial}{\partial r} r^m \frac{\partial N_e}{\partial r}. \tag{4.62}$$

4.2.3 Initial conditions of the boundary value problem for the glow discharge

Two kinds of initial conditions are in common use at numerical simulation. The first one is the use of numerical simulation results obtained for a close boundary-value problem. The second method is based on some a priori information on a prospective solution. For example, simulating 2D normal glow discharge one can wait for a satisfactory description of the glow discharge in the 1D axial direction, whose parameters are predicted by the Engel–Steenbeck theory (see Section 4.1)

$$\frac{j_n}{p^2} = 5.92 \times 10^{-11} \frac{AB^2 \mu_+ p(1+\gamma)}{\ln(1+1/\gamma)}, \ \text{mA/(cm2} \cdot \text{Torr)} \tag{4.63}$$

$$d_n p = 3.78 A^{-1} \ln(1+1/\gamma), \ \text{cm} \cdot \text{Torr}, \tag{4.64}$$

$$V_n = 3BA^{-1} \ln(1+1/\gamma), \ \text{V}, \tag{4.65}$$

where j_n is the normal current density; d_n, V_n are the thickness of the cathode layer and the voltage drop on it.

Let us estimate a level of ion concentration in the cathode layer. For this purpose, we shall write the relationship between a current density and concentration of ions

$$j_n = n_e \mu_e E_x + n_+ \mu_+ E_x = (1+\gamma) E_x n_+ \mu_+ Q',$$

where the coefficient $Q' = 1.6 \times 10^{-16}\ N_0$ defines the dimension of j_n in mA/cm^2.

The value E_x can be estimated under the formula:

$$E_x = \left| \frac{\mathrm{d}\varphi}{\mathrm{d}x} \right| \sim \frac{V_n}{d_n},$$

whence

$$n_+ \sim \frac{j_n d_n}{(1+\gamma) V_n \mu_+ Q'} = (n_+)_0.$$

The thickness of the cathode layer d_n sets a scale of an exponential fall of ion concentration from $(n_+)_0$ to value N_0.

Initial axial distribution of electron concentration is built also under the exponential law from the value on the cathode

$$(n_e)_0 = \gamma \frac{\mu_+}{\mu_e} (n_+)_0$$

up to value N_0 on distance d_n from the cathode.

Radial distributions of ion and electron concentrations are input uniformly along the x-axis under the formula

$$n(x,r) = n(x,r=0)\exp[-(r/r_{\text{TK}})^m],$$

where $m = 2-4$; r_{ini} is the prospective radius of a current channel which is estimated by the relation

$$r_{\text{ini}} \cong \sqrt{\frac{E-2V_n}{\pi R_0 j_n}}.$$

Iterative calculation of electric field potential is performed with the use of $n_e(x,r)$ and $n_i(x,r)$. These iterations are necessary for the Poisson equation (4.32) for calculation of voltage drop V on the gas discharge gap, corresponding to given distributions of charged particles. Being obtained at each iteration, the new potential field is used for recalculated value V under formula (4.43). Due to the large nonlinearity of the equation, it is necessary to introduce a restriction on a possible change of a voltage drop, for example

$$V_n \leq V \leq E.$$

4.2.4 Direct-current glow discharge in view of gas heating

In this section, the self-consistent numerical model of glow discharge is formulated for flat and axially symmetric geometry in which heat of gas is considered. It is supposed that the discharge exists in a condition of a normal current density mode between two flat electrodes (see Figure 4.4) so the side boundary effects do not influence its structure and parameters.

As before, the electrodynamic structure of glow discharge is investigated within the limits of the drift-diffusion model formulated concerning electronic and ionic concentrations together with the Poisson equation, which defines the distribution of electric potential in the electrodischarge gap.

So, in comparison with the problem considered in the previous section, the heating of neutral gas is considered here. The Joule thermal emission is caused by the heat of neutral gas that is taken into consideration here due to collisions of molecules with electrons that receive energy from electric field (they are "warmed up" by the electric field inside the glow discharge). Not all the energy transmitted at collisions of electrons with gas molecules is used on their heat. The significant part of this energy is spent for excitation of molecules' vibrational degrees of freedom. In the given case, the solution of physical kinetic equations for determination of the electron energy distribution function, which allows predicting part of the electronmolecular collision energy transformable for the heat of gas does not realize. Instead, an approximate phenomenological effectiveness ratio f of Joule energy emission transfer to the heat of gas is introduced into the drift-diffusion model. It is known that this coefficient can vary over a wide range, and for typical laser mixtures ($N_2 + CO_2$) and conditions it makes 0.15–0.25.

The set of governing equations is formulated in the following form:

$$\frac{\partial n_e}{\partial t} + \frac{\partial \Gamma_{e,x}}{\partial x} + \frac{1}{r^m}\frac{\partial r^m \Gamma_{e,r}}{\partial r} = \alpha(E)|\mathbf{\Gamma}_e| - \beta n_+ n_e, \quad (4.66)$$

$$\frac{\partial n_+}{\partial t} + \frac{\partial \Gamma_{+,x}}{\partial x} + \frac{1}{r^m}\frac{\partial r^m \Gamma_{+,r}}{\partial r} = \alpha(E)|\mathbf{\Gamma}_e| - \beta n_+ n_e, \quad (4.67)$$

$$\frac{\partial^2 \varphi}{\partial x^2} + \frac{1}{r^m}\frac{\partial}{\partial r}\left(r^m \frac{\partial \varphi}{\partial r}\right) = 4\pi e(n_e - n_+), \quad (4.68)$$

$$\rho c_v \frac{\partial T}{\partial t} = \frac{\partial}{\partial x}\left(\lambda \frac{\partial T}{\partial x}\right) + \frac{1}{r^m}\frac{\partial}{\partial r}\left(r^m \lambda \frac{\partial T}{\partial r}\right) + Q, \quad (4.69)$$

where

$$\mathbf{\Gamma}_e = -D_e \mathrm{grad} n_e - n_e \mu_e \mathbf{E}; \; \mathbf{\Gamma}_+ = -D_+ \mathrm{grad} n_+ + n_+ \mu_+ \mathbf{E};$$

$$Q = \eta(\mathbf{jE}); \; \mathbf{j} = e(\mathbf{\Gamma}_+ - \mathbf{\Gamma}_e); \; \mathbf{E} = -\mathrm{grad}\,\varphi;$$

T is the temperature of gas; $m = 0$ is used for flat discharge; $m = 1$ is used for axially symmetric discharge. Remaining variables and coefficients of the drift-diffusion model are as mentioned earlier.

In the considered case, the energy conservation equation for neutral gas (4.69) is formulated in the form of the Fourier–Kirchhoff equation.

The important corollary of neutral gas heating is the modification of local values of neutral particles density that renders essential influence on the value of collision frequency of electrons with molecules of gas, hence, on such parameters of the drift-diffusion model as the frequency of ionization and mobility of electrons and ions.

Introducing some effective pressure, the specified variables of the drift-diffusion model can be written in the form:

$$\mu_e(p^*) = \frac{4.2 \times 10^5}{p^*}, \; \mu_i(p^*) = \frac{2280}{p^*}, \; \mathrm{cm}^2/(\mathrm{V \cdot s}),$$

$$p^* = p\frac{293}{T}, \text{ Torr},$$

$$D_e = \mu_e(p^*)T_e, \quad D_+ = \mu_+(p^*)T,$$

$$c_p = 8.314\frac{7}{2}\frac{1}{M_\Sigma}, \text{ J/(g}\cdot\text{K)}, M_\Sigma = 28 \text{ g/mol}, c_v = 0.742\text{ J/(g}\cdot\text{K)},$$

$$\rho = 1.58\times10^{-5}\frac{M_\Sigma p}{T}, \text{ g/cm}^3,$$

$$\lambda = \frac{8.334\times10^{-4}}{\sigma^2\Omega^{(2.2)*}}\sqrt{\frac{T}{M_\Sigma}}\left(0.115 + 0.354\frac{c_p M_\Sigma}{R_0}\right), \text{ W/(cm}\cdot\text{K)},$$

$$\Omega^{(2.2)*} = \frac{1.157}{(T^*)^{0.1472}}, \quad T^* = \frac{T}{(\varepsilon/k)}, \quad (\varepsilon/k) = 71.4\text{ K}, \quad \sigma = 3.68\text{ Å},$$

$$R_0 = 8.314 \text{ J/(K}\cdot\text{mol)},$$

p is the pressure of gas inside the gas discharge gap.

In this case, the ionization coefficient is calculated under the formula

$$\alpha(E) = p^* A \exp\left[-\frac{B}{(|\mathbf{E}|/p^*)}\right], \text{ cm}^{-1}. \tag{4.70}$$

4.2.5 Estimation of typical timescales of the solved problem

According to the statement of the glow discharge problem in molecular nitrogen, the following physical processes are considered:
- ionization of molecules N_2 by electron impact;
- recombination of the positive ions at collision with electrons;
- drift of ions and electrons in the external electric field;
- diffusion of charged particles;
- ambipolar diffusion of charged particles;
- heating of neutral gas.

Let us estimate characteristic timescales for the enumerated processes and compare them with a typical time of space charge relaxation. It will give the possibility to get representation concerning needed time steps for numerical simulation.

Charged particles' characteristic drift time in glow discharge

The drift velocity of electrons and ions in a constant electric field is

$$v_{\text{dr}}^k = \mu_k E, \quad (k = e, +).$$

At pressure $p = 5\,\text{Torr}$ from (4.44) one can find $\mu_e = 8.8 \times 10^4\,\text{cm}^2/(\text{s}\cdot\text{V})$ $\mu_+ = 2.9 \times 10^2\,\text{cm}^2/(\text{s}\cdot\text{V})$. There are two typical levels of an electric field in a glow discharge. These are: $E_\text{c} = 3,000\ \text{V/cm}$ for cathode layer and $E_\text{pc} = 100\,\text{V/cm}$ for the positive column. In these regions of glow discharge, we can define correspondingly:

$$\left(v_\text{dr}^e\right)_\text{c} = 2.64 \times 10^8\ \text{cm/s},\ \left(v_\text{dr}^+\right)_\text{c} = 8.7 \times 10^5\ \text{cm/s},$$

$$\left(v_\text{dr}^e\right)_\text{pc} = 8.8 \times 10^6\ \text{cm/s},\ \left(v_\text{dr}^+\right)_\text{pc} = 2.9 \times 10^4\ \text{cm/s}.$$

Each of these regions has also typical spatial scale: for a cathode layer $d_\text{c} = 0.1$ cm, and for positive column $H = 1$ cm.

Hence, the required typical times are

$$\left(t_\text{др}^e\right)_\kappa = 0.379 \times 10^{-9}\ \text{s},\ \left(t_\text{др}^+\right)_\kappa = 0.115 \times 10^{-7}\ \text{s},$$

$$\left(t_\text{др}^e\right)_\text{c} = 0.114 \times 10^{-6}\ \text{s},\ \left(t_\text{др}^+\right)_\text{c} = 0.345 \times 10^{-4}\ \text{s}.$$

Typical time of ionization

Let the characteristic time of ionization is

$$\tau_\text{i} = 1/\nu_\text{i},$$

where $\nu_\text{i}(p)$ is the frequency of ionization; p is the pressure. Frequency of ionization is connected with electron drift velocity by means of the Townsend coefficient of ionization (Brown S.C., 1966; Raizer Yu.P. and Surzhikov S.T., 1987):

$$\nu_\text{i} = \alpha(E, p) v_\text{dr}^e.$$

Let us estimate Townsend's coefficient under formula (4.45) at two characteristic values of the field strength $E_\text{c} = 3,000\,\text{V/cm}$ and $E_\text{pc} = 100\,\text{V/cm}$: $\alpha_\text{c} \approx 30\,\text{cm}^{-1}$, $\alpha_\text{pc} \approx 10^{-5}\,\text{cm}^{-1}$

It gives the estimations of ionization frequency, and typical time of ionization is

$$(\nu_\text{i})_\text{c} = 7.92 \times 10^9\,\text{s}^{-1},\ (\nu_\text{i})_\text{pc} = 3.9 \times 10^2\,\text{s}^{-1};$$

$$(\tau_\text{i})_\text{c} = 0.127 \times 10^{-9}\,\text{s},\ (\tau_\text{i})_\text{pc} = 0.256 \times 10^{-2}\,\text{s}.$$

Typical time of recombination

Frequency of ion-electronic recombination is proportional to concentration of ions

$$\nu_\text{r} = \beta n_+\,,\ \text{s}^{-1}.$$

Therefore,

$$\tau_r = 1/\beta n_+, \text{ s}.$$

Accepting $\beta = 2\times10^{-7}\ \text{cm}^3/\text{s}$ and concentration of ions equal to $n_+ = 10^{10}\text{cm}^{-3}$, one can get

$$\tau_p \approx 0.5\times10^{-3}\ \text{s}.$$

Typical time of charged particles diffusion

The average squared displacement of particles at the diffusion process is set with the Einstein formula

$$\overline{x^2} = 2Dt,$$

where D is the diffusion coefficient.

Let us consider three types of the diffusion:

- diffusion of electrons with coefficient

$$D_e = \mu_e T_e;$$

- diffusion of ions with coefficient

$$D_+ = \mu_+ T_+;$$

- ambipolar diffusion with coefficient

$$D_a = \mu_+ T_e.$$

In the given formulas, the Einstein relation for diffusion coefficients was used, where the temperature of electrons and ions were expressed in eV. Accepting $T_e = 1$ eV and $T_+ = 0.0258$ eV, one can get

$$D_e = 8.8\times10^4\ \text{cm}^2/\text{s};\ D_+ = 7.5\,\text{cm}^2/\text{s};\ D_a = 2.9\times10^2\ \text{cm}^2/\text{s}.$$

For a typical scale of diffusion transfer

$$\Lambda = \sqrt{\overline{x^2}} = 0.1\,\text{cm},$$

which corresponds approximately to the thickness of a cathode layer d_n and to an expansion of a column boundary in a radial direction, we have characteristic times of diffusion:

$$\tau_{\mathrm{d}}^{e} = \frac{\Lambda^2}{2D_e} = 0.569 \times 10^{-7}\ \mathrm{s},\ \tau_{\mathrm{d}}^{+} = \frac{\Lambda^2}{2D_+} = 0.666 \times 10^{-3}\ \mathrm{s},\ \tau_{\mathrm{d}}^{a} = \frac{\Lambda^2}{2D_a} = 0.345 \times 10^{-4}\ \mathrm{s}.$$

Typical time of relaxation of a space charge

Let there be spatially homogeneous charge $\rho = e(n_+ - n_e)$ such that

$$\operatorname{div} \mathbf{E} = 4\pi\rho = 4\pi e(n_+ - n_e) = \varepsilon^*(n_+ - n_e).$$

From the continuity equation

$$\frac{\partial \rho}{\partial t} + \operatorname{div} \mathbf{j} = 0,$$

where $\mathbf{j} = \sigma\mathbf{E}$, supposing that due to a strong difference of mobilities of electrons and ions, the current is transferred by electrons. Then

$$\frac{\partial \rho}{\partial t} + \sigma \operatorname{div} \mathbf{E} + \mathbf{E}\, \mathrm{grad}\sigma = 0.$$

For homogeneous distributions of the charged particles

$$\frac{\partial \rho}{\partial t} + \sigma \operatorname{div} \mathbf{E} = \frac{\partial \rho}{\partial t} + 4\pi\sigma\rho = 0,$$

and

$$\rho(t) = \rho(0)\ \exp(-4\pi\sigma t).$$

By definition, the time of volume charge relaxation, named as the Maxwell time, is defined as follows:

$$\tau_{\mathrm{M}} = \frac{1}{4\pi\sigma} = \frac{1}{4\pi n_e \mu_e e} = \frac{1}{\varepsilon^* n_e \mu_e}.$$

Substituting numerical values, which are characteristic for the considered problem, we define

$$\tau_{\mathrm{M}} \cong 0.63 \times 10^{-9}\ \mathrm{s}.$$

Typical time of heat conduction

Typical time of temperature wave propagation is set by the formula

$$\tau_h = \frac{L^2 \rho c_V}{\lambda}$$

This value is estimated by 10 ms for typical glow discharge.

Time steps for numerical calculations

Dimensionless time τ, entering into the system of the solved equations, looks like

$$\tau = t\frac{W\mu_{e,0}}{H} = \frac{t}{t_H}.$$

where

$$t_H = \frac{H}{W\mu_{e,0}}.$$

Obviously, this is the time necessary for an electron to fly through the discharge gap under homogeneous electric field $E_x = W = \mathrm{E}/H$ corresponding to the full magnitude of the electromotive power E.

For typical calculation cases:

1. $E = 2{,}500\ \mathrm{V}$, $W \approx 3{,}330\ \mathrm{V/cm}$, and $t_H \approx 0.256 \times 10^{-8}\ \mathrm{s}$;
2. $E = 500\ \mathrm{V}$, $W \approx 667\ \mathrm{V/cm}$, and $t_H \approx 1.28 \times 10^{-8}\ \mathrm{s}$.

One can see that the time t_H is approximately by the order of magnitude greater than the time of electron drift in a cathode layer and essentially less than the drift time in a positive column. They surpass the time of ionization in a cathode layer and time of relaxation of a volume charge, but noticeably less than diffusion and recombination times.

Relationship between timescales

Apparently from the estimations given above, the solved problem has essentially different timescales. The fastest processes are ionization of molecules by electron impact and electronic drift in the cathode layer. Their timescale is of $\sim 10^{-9}$ s. Volume charge relaxation time has the same scale. Typical time of ion drift in the cathode layer is of $\sim 10^{-7}$ s, and it makes the first basic difficulty of the numerical simulation because ions play the same significant role in forming of the discharge as electrons do.

In the problem under consideration, there are some other timescales connected to the drift of charged particles in the positive column, with the diffusion of electrons and ions, and also with ambipolar diffusion.

Typical time of charge drift in a positive column approximately by $\sim 10^2 - 10^3$ is more than the corresponding time in the cathode layer, namely for electrons it makes $\sim 10^{-7}$ s, and for ions it makes $\sim 10^{-4}$ s. The same order is for time of electron diffusion ($\sim 10^{-7}$ s) and for ambipolar diffusion $\sim 10^{-4}$ s. The greatest timescale is for processes of recombination and diffusion of ions, namely $\sim 10^{-3}$ s, and also for the process of temperature wave propagation (actually for heating of neutral gas).

So, the range of timescales is of $\sim 10^{-9} - 10^{-3}$ s. It is obvious that the direct solution of the considered problem represents excessively greater computing difficulties.

Numerical simulation of unsteady or stationary glow discharges with the use of finite-difference time-dependent methods allows performing calculations with some certain time steps, which are defined by stability conditions of used finite-difference schemes. These conditions are defined by types of solved equations, by parameters of the finite-difference schemes and by used grids. It is possible to specify two dimensionless criteria of the solved set of equation (for explicit finite-difference calculation schemes). These are the hyperbolic and parabolic CFL numbers (the Courant–Friedrichs–Lewy stability condition):

$$\mathrm{CFL}^{\mathrm{G}} = \frac{v\tau}{h} \sim 1, \quad \mathrm{CFL}^{\mathrm{P}} = \frac{2\tilde{D}\tau}{h^2} \sim 1,$$

where v is the velocity of convective transport; h, τ are the minimal step in space and maximal step in time.

The hyperbolic CFL number shows the number of grid points, which is passed by any disturbance for one calculation time step. This condition can be derived at an analysis of the stability of used finite-difference scheme for the simplest hyperbolic equation

$$\frac{\partial f}{\partial t} + v\frac{\partial f}{\partial x} = 0,$$

where v is the transport velocity.

The parabolic CFL number gives a relation of characteristic diffusion length $\sqrt{2\tilde{D}\tau}$ to the grid step h, that is, this is the number of grid points for diffusion transfer for one time step. The parabolic computational fluid dynamics (CFD) number arises at the stability analysis for the simplest equation of the form

$$\frac{\partial f}{\partial \tau} = \tilde{D}\frac{\partial^2 f}{\partial x^2},$$

where $\tilde{D}$ is the diffusion coefficient.

Using the procedure of normalization of the governing equations one can detect that there is a scale of velocity

$$\tilde{v} = \mu_{e,0}\mathrm{E}/H,$$

which corresponds to the velocity of electron motion in an electric field with strength $E_x = E/H$. For example, the velocity $\tilde{v} = 2.93 \times 10^8$ cm/s is obtained at $E = 2,500$ V and $H = 0.75$ cm.

It is obvious that characteristic dimensionless velocities for drift of electrons and ions differ extremely:

$$\overline{V_{\mathrm{dr}}^{e}} \sim \frac{V_{\mathrm{dr}}^{e}}{\tilde{v}}, \qquad \overline{V_{\mathrm{dr}}^{+}} = \frac{V_{\mathrm{dr}}^{+}}{\tilde{v}}.$$

For estimation of the greatest hyperbolic CFL numbers, it is necessary to consider subregions with the greatest velocities and the lowest steps on grid. In our case, these conditions are satisfied in cathode layer, where drift velocities are maximal and the typical grid step is equal to ~0.01 cm. For electrons, $\mathrm{CFL}_e^{\mathrm{G}} \approx 90\tau$, and for ions $\mathrm{CFL}_+^{\mathrm{G}} \approx 0.29\tau$.

Let us consider the time step τ, which is defined by the explicit finite-difference schemes from the condition $\mathrm{CFL}^{\mathrm{G}} < 1$:

$$\tau \le \frac{1}{90} = 1.11 \times 10^{-2}\,\mathrm{s},$$

which corresponds to the physical time of

$$t = \frac{\tau H}{W\mu_{e,0}} = \frac{1.11 \times 10^{-2}}{0.256 \times 10^{-8}} = 0.434 \times 10^{-9}\,\mathrm{s}.$$

It is clear that the received calculation time step corresponds to the least characteristic timescales of elementary physical processes considered above. It allows speaking that calculation of nonstationary process with $\mathrm{CFL}^{\mathrm{G}}$~1 is possible under the explicit scheme.

Numerical modeling of glow discharges by the time asymptotic method with such time steps is uneconomical; therefore, the implicit methods are in common use. These schemes allow perform numerical integration at

$$\mathrm{CFL}_e^{\mathrm{G}} \sim 9 - 90,\ \mathrm{CFL}_+^{\mathrm{G}} \sim 0.029 - 0.29.$$

It is seen that the implicitness introduced into numerical simulation algorithms allows to increase $\mathrm{CFL}^{\mathrm{G}}$ approximately 100 times in comparison with explicit ones.

The limiting value $\mathrm{CFL}_+^{\mathrm{G}}$ ~1 corresponds to the characteristic physical time of ion drift in the cathode layer $\left(t_{\mathrm{dr}}^{+}\right)_{\mathrm{c}} \sim 0.117 \times 10^{-7}$ s. Thus, it is possible to conclude that the process limiting a computing procedure is the ion drift in the cathode layer.

From here we can estimate the number of time steps ~10^4 that should be made to take into account ambipolar diffusion and ion diffusion.

For preparing estimations of the parabolic CFL numbers one can obtain:

$$\tilde{D}_e = \frac{D_e}{\mu_{e,0}E} = \frac{\mu_e T_e}{\mu_{e,0}E} = \frac{T_e}{E};\ \tilde{D}_+ = \frac{D_+}{\mu_{e,0}E} = \frac{\mu_+ T_+}{\mu_{e,0}E};\ \tilde{D}_a = \frac{D_a}{\mu_{e,0}E} = \frac{\mu_+ T_e}{\mu_{e,0}E};$$

then

$$\tilde{D}_e \approx 4 \times 10^{-4};\ \tilde{D}_+ \approx 3.4 \times 10^{-8};\ \tilde{D}_a \approx 1.3 \times 10^{-6}.$$

Setting, as before, $h \sim 10^{-2}$, we shall obtain the characteristic numerical values

$$\mathrm{CFL}_e^{\mathrm{P}} \sim 8\tau;\ \mathrm{CFL}_+^{\mathrm{P}} \sim 6.8\times10^{-4}\tau;\ \mathrm{CFL}_a^{\mathrm{P}} \sim 2.6\times10^{-2}\tau.$$

Thus, the stable conditions based on the parabolic CFL numbers in the problem under consideration are weaker than the ones based on the hyperbolic CFL numbers.

In summary, we shall note that the given estimations should be considered as approximated, as in conditions of extremely strong heterogeneity of electric field it is impossible to choose reference features.

4.3 Finite-difference methods for the drift-diffusion model

4.3.1 Finite-difference scheme for the Poisson equation

Let us introduce an inhomogeneous grid

$$\omega = \left\{ (r_i, x_j),\ 1 \le i \le NI,\ 1 \le j \le NJ \right\}$$

in calculation domain $G\{(r,x),\ 0 \le r \le R,\ 0 \le x \le H\}$, where $r_1 = 0$, $x_1 = 0$, $r_{NI} = R$, $x_{NJ} = H$ (see Figure 4.5).

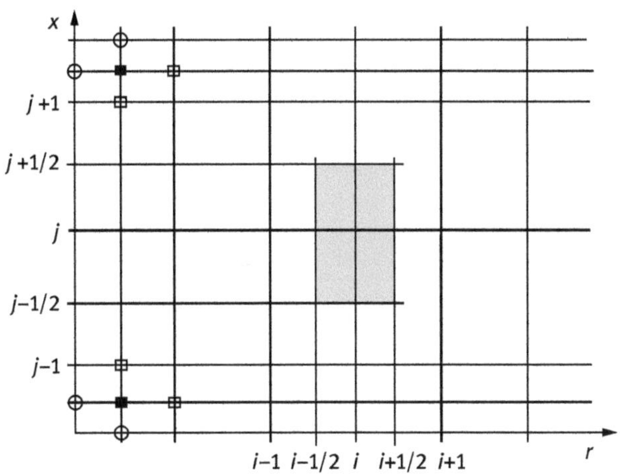

Figure 4.5: Fragment of used finite-difference grid.

Required finite-difference scheme will be created by the finite-volume method; therefore, points for determination of fluxes on boundaries of elementary volumes are introduced as follows (the so-called flux points)

$$r_{i+1/2}=\frac{1}{2}(r_i+r_{i+1}),\qquad r_{i-1/2}=\frac{1}{2}(r_i+r_{i-1});$$

$$x_{j+1/2}=\frac{1}{2}\left(x_j+x_{j+1}\right),\qquad x_{j-1/2}=\frac{1}{2}\left(x_j+x_{j-1}\right).$$

Let us consider elementary volume in the neighborhood of a point (i,j)

$$V_{i,j}=(2\pi)^m\int\limits_{r_{i+1/2}}^{r_{i+1/2}} r^m\,\mathrm{d}r\int\limits_{x_{j-1/2}}^{x_{j+1/2}}\mathrm{d}x=$$

$$=\frac{(2\pi)^m}{m+1}\frac{1}{2^{m+2}}\left(x_{j+1}-x_{j-1}\right)\left(r_{i+1}-r_{i-1}\right)\left(r_{i+1}+2r_i+r_{i-1}\right)^m, \tag{4.71}$$

where $m=0$ is used for a flat geometry and $m=1$– for axisymmetric geometry.

Let us now integrate the Poisson equation

$$\frac{H^2}{R^2}\frac{1}{r^m}\frac{\partial}{\partial r}\left(r^m\frac{\partial\Phi}{\partial r}\right)+\frac{\partial^2\Phi}{\partial x^2}=-\frac{\varepsilon^* HN_0}{W}(N_+-N_e),$$

over the elemental volume $V_{i,j}$, considering that

$$\int\limits_{r_{i-\frac{1}{2}}}^{r_{i+\frac{1}{2}}}\frac{\partial}{\partial r}r^m\frac{\partial\Phi}{\partial r}\mathrm{d}r=\frac{1}{2^m}\left[\left(r_{i+1}+r_i\right)^m\left(\frac{\partial\Phi}{\partial r}\right)_{i+\frac{1}{2}}-\left(r_i+r_{i-1}\right)^m\left(\frac{\partial\Phi}{\partial r}\right)_{i-\frac{1}{2}}\right],$$

$$\int\limits_{x_{j-1/2}}^{x_{j+1/2}}\frac{\partial^2\Phi}{\partial x^2}\mathrm{d}x=\left(\frac{\partial\Phi}{\partial x}\right)_{j+1/2}-\left(\frac{\partial\Phi}{\partial x}\right)_{j-1/2},$$

and defining derivatives in “fluxes” points under the following formulas:

$$\left(\frac{\partial\Phi}{\partial r}\right)_{i+1/2}=\frac{\Phi_{i+1}-\Phi_i}{r_{i+1}-r_i},\qquad\left(\frac{\partial\Phi}{\partial r}\right)_{i-1/2}=\frac{\Phi_i-\Phi_{i-1}}{r_i-r_{i-1}},$$

$$\left(\frac{\partial\Phi}{\partial x}\right)_{j+1/2}=\frac{\Phi_{j+1}-\Phi_j}{x_{j+1}-x_j},\qquad\left(\frac{\partial\Phi}{\partial x}\right)_{j-1/2}=\frac{\Phi_j-\Phi_{j-1}}{x_j-x_{j-1}},$$

we receive

$$\frac{2(m+1)}{\left(r_{i+1}-r_{i-1}\right)\left(r_{i-1}+2r_i+r_{i+1}\right)}\frac{H^2}{R^2}\left[\left(r_{i+1}+r_i\right)^m\frac{\Phi_{i+1}-\Phi_i}{r_{i+1}-r_i}-\left(r_i+r_{i-1}\right)^m\frac{\Phi_i-\Phi_{i-1}}{r_i-r_{i-1}}\right]+$$

$$+\frac{2}{x_{j+1}-x_{j-1}}\left(\frac{\Phi_{j+1}-\Phi_j}{x_{j+1}-x_j}-\frac{\Phi_j-\Phi_{j-1}}{x_j-x_{j-1}}\right)=-\frac{\varepsilon^* HN_0}{W}(N_+-N_e). \tag{4.72}$$

After some simple transformations, one can receive the canonical five-point finite-difference scheme:

$$A_{i,j}\Phi_{i-1,j}-C_{i,j}\Phi_{i,j}+B_{i,j}\Phi_{i+1,j}+\bar{A}_{i,j}\Phi_{i,j-1}-\bar{C}_{i,j}\Phi_{i,j}+\bar{B}_{i,j}\Phi_{i,j+1}+F_{i,j}=0, \tag{4.73}$$

$$A_{i,j}=E_iS^m_{i-1}\frac{1}{r_i-r_{i-1}},\quad B_{i,j}=E_iS^m_{i+1}\frac{1}{r_{i+1}-r_i},$$

$$C_{i,j}=E_iS^m_{i+1}\frac{1}{r_{i+1}-r_i}+E_iS^m_{i-1}\frac{1}{r_i-r_{i-1}},$$

$$\bar{A}_{i,j}=E_j\frac{1}{x_j-x_{j-1}},\quad \bar{B}_{i,j}=E_j\frac{1}{x_{j+1}-x_j},$$

$$\bar{C}_{i,j}=E_j\frac{1}{x_{j+1}-x_j}+E_j\frac{1}{x_j-x_{j-1}};$$

$$E_j=\frac{2}{x_{j+1}-x_j},\quad E_i=\frac{H^2}{R^2}\frac{2}{r_{i+1}-r_{i-1}};$$

$$S^m_{i+1}=(m+1)\frac{(r_{i+1}+r_i)^m}{(r_{i-1}+2r_i+r_{i+1})^m},\quad S^m_{i-1}=(m+1)\frac{(r_i+r_{i-1})^m}{(r_{i-1}+2r_i+r_{i+1})^m};$$

$$F_{i,j}=\left[\frac{\varepsilon^* HN_0}{W}(N_+-N_e)\right]_{i,j}. \tag{4.74}$$

This finite-difference scheme has the second-order approximation on the spatial variables.

We shall consider the conservative property of the scheme at $m=1$. The desired integral balance equation is easy for receiving from an initial equation for electric field strength

$$\operatorname{div}\mathbf{E}=\varepsilon^*(n_+-n_e)=A,$$

Having integrated it by all considered volume with the use of the Stokes theorem one can receive

$$\int_V \operatorname{div}\mathbf{E}\,\mathrm{d}V=\int_\Sigma \mathbf{E}\,\mathrm{d}\mathbf{S}=\int_{S_1}E_x(r,x=0)\mathrm{d}S_1+\int_{S_2}E_r(r=R,x)\mathrm{d}S_2-$$
$$-\int_{S_3}E_x(r,x=H)\mathrm{d}S_3=2\pi\left[-\int_0^R r\left(\frac{\partial\varphi}{\partial x}\right)_{x=0}\mathrm{d}r+\right.$$
$$\left.+\int_0^R r\left(\frac{\partial\varphi}{\partial x}\right)_{x=H}\mathrm{d}r-R\int_0^H\left(\frac{\partial\varphi}{\partial r}\right)_{r=R}\mathrm{d}x\right]=\int_V A\,\mathrm{d}V. \tag{4.75}$$

Let us multiply (4.72) by $V_{i,j}$ and sum over all elementary volumes (further $m=1$)

$$\sum_{j=2}^{NJ-1}\sum_{i=2}^{NI-1}\frac{2\pi}{4}\left(x_{j+1}-x_{j-1}\right)\frac{H^2}{R^2}\left[(r_{i+1}+r_i)\frac{\Phi_{i+1}-\Phi_i}{r_{i+1}-r_i}-(r_i+r_{i-1})\frac{\Phi_i-\Phi_{i-1}}{r_i-r_{i-1}}\right]+$$
$$+\sum_{j=2}^{NJ-1}\sum_{i=2}^{NI-1}\frac{2\pi}{2}\frac{1}{4}(r_{i+1}-r_{i-1})(r_{i-1}+2r_i+r_{i+1})\left(\frac{\Phi_{j+1}-\Phi_j}{x_{j+1}-x_j}-\frac{\Phi_j-\Phi_{j-1}}{x_j-x_{j-1}}\right)=$$
$$=-\sum_{j=2}^{NJ-1}\sum_{i=2}^{NI-1}\frac{\varepsilon^* H}{W}N_0(N_+-N_e)\frac{2\pi}{2}\frac{1}{2^3}\left(x_{j+1}-x_{j-1}\right)(r_{i+1}-r_{i-1})(r_{i-1}+2r_i+r_{i+1}).$$

The evaluation of the sum results in

$$2\pi\sum_{j=2}^{NJ-1}\frac{x_{j+1}-x_{j-1}}{2}\frac{H^2}{R^2}\left[\frac{r_{NI}+r_{NI-1}}{2}\frac{\Phi_{NI,j}-\Phi_{NI-1,j}}{r_{NI}-r_{NI-1}}-\frac{r_2+r_1}{2}\frac{\Phi_{2,j}-\Phi_{1,j}}{r_2-r_1}\right]+$$
$$+2\pi\sum_{i=2}^{NI-1}\frac{r_{i+1}-r_{i-1}}{2}\frac{r_{i-1}+2r_i+r_{i+1}}{4}\left(\frac{\Phi_{i,NJ}-\Phi_{i,NJ-1}}{x_{NJ}-x_{NJ-1}}-\frac{\Phi_{i,2}-\Phi_{i,1}}{x_2-x_1}\right)=$$
$$=-\sum_{j=2}^{NJ-1}\sum_{i=2}^{NI-1}\frac{2\pi\varepsilon^* HN_0}{W}(N_+-N_e)_{ij}\frac{x_{j+1}-x_{j-1}}{2}\frac{r_{i+1}-r_{i-1}}{2}\frac{r_{i-1}+2r_i+r_{i+1}}{4}.$$

Under the symmetry condition, it is necessary to suppose

$$\Phi_{1,j}=\Phi_{2,j},\qquad \forall j\in[1,NJ,].$$

Then we shall receive the required finite-difference analog of an integral balance equation (4.75):

$$2\pi\left[-\sum_{i=2}^{NI-1}\frac{r_{i+1}-r_{i-1}}{2}\cdot\frac{r_{i-1}+2r_i+r_{i+1}}{4}\frac{\Phi_{i,2}-\Phi_{i,1}}{x_2-x_1}\right.$$
$$\left.+\sum_{i=2}^{NI-1}\frac{r_{i+1}-r_{i-1}}{2}\cdot\frac{r_{i-1}+2r_i+r_{i+1}}{4}\frac{\Phi_{i,NJ}-\Phi_{i,NJ-1}}{x_{NJ}-x_{NJ-1}}\right]+$$
$$+2\pi\sum_{j=2}^{NJ-1}\frac{x_{j+1}-x_{j-1}}{2}\frac{H^2}{R^2}\frac{r_{NI}+r_{NI-1}}{2}\cdot\frac{\Phi_{NI,j}-\Phi_{NI-1,j}}{r_{NI}-r_{NI-1}}=$$
$$=-\sum_{j=2}^{NJ-1}\sum_{i=2}^{NI-1}2\pi\frac{\varepsilon^* H}{W}N_0(N_+-N_e)_{ij}\times\frac{x_{j+1}-x_{j-1}}{2}\frac{r_{i+1}-r_{i-1}}{2}\frac{r_{i-1}+2r_i+r_{i+1}}{4}. \tag{4.76}$$

4.3.2 Finite-difference scheme for the equation of charge motion

Let us consider the generalized form of the particle motion equation (4.50):

$$\frac{\partial u}{\partial \tau}+\frac{1}{r^m}\frac{\partial}{\partial r}(r^m au)+\frac{\partial}{\partial x}(bu)-\frac{c}{r^m}\frac{\partial}{\partial r}r^m\frac{\partial u}{\partial r}=f. \tag{4.77}$$

One can see that two physical processes are included in eq. (4.77). These are the transfer process with velocities a and b, and the dissipation process with diffusion coefficient c. Desired finite-difference scheme will be created in the following two stages. At the first step, only transfer processes of particles will be considered, while at the second step the diffusion process will be taken into account.

We will derive the conservative form for the desired finite-difference scheme. Algorithmically, the construction of conservative schemes is realized with the finite-volume method, which was already considered above.

Let us integrate the motion equation

$$\frac{\partial u}{\partial \tau}+\frac{1}{r^m}\frac{\partial}{\partial r}(r^m au)+\frac{\partial}{\partial x}(bu)=f \tag{4.78}$$

by the elementary volume V_{ij} introduced in the previous item (see Figure 4.5 and formula (4.71))

$$V_{i,j}\frac{u_{ij}^{p+1}-u_{ij}^{p}}{\tau}+(2\pi)^m(x_{j+1}-x_{j-1})\frac{1}{2}\int\limits_{r_{i-1/2}}^{r_{i+1/2}}\frac{\partial}{\partial r}(r^m au)\mathrm{d}r+(2\pi)^m\frac{1}{m+1}\times$$
$$\times\frac{1}{2^{m+1}}(r_{i+1}-r_{i-1})(r_{i-1}+2r_i+r_{i+1})\int\limits_{x_{j+1/2}}^{x_{j-1/2}}\frac{\partial}{\partial x}(bu)\mathrm{d}x=V_{ij}f_{ij}. \tag{4.79}$$

It is well known that the simplest stable finite-difference scheme for the motion equation (4.78) should be built so that derivatives have been approximated "upstream." However in the solved problem coefficients, a and b can be sign-changing; therefore, in computing algorithm, it is necessary to provide the analysis of their signs in each rated point, if necessary to change the approximating formula in an appropriate way. There are a lot of such finite-difference schemes (see, e.g., textbooks quite popular in an aerospace community: Roache P., 1980; Kulikovskii A.G., et al., 2001; Anderson D.A., et al., 1997). However, it is practically impossible to offer the uniform recipe for a solution of any specific problem. Therefore, in our case the elementary finite-difference scheme with "donor cells" is convenient (Roache P., 1980; Gentry A.A., et al., 1966). According to this scheme in the "flux" points values of velocities are defined with a special algorithm, which we will consider by way of example of the integral evaluation

$$\int\limits_{r_{i-1/2}}^{r_{i+1/2}}\frac{\partial}{\partial r}(r^m au)\mathrm{d}\,r=r_{i+1/2}^m(au)_{r_{i+1/2}}-r_{i-1/2}^m(au)_{r_{i-1/2}}.$$

Values of fluxes in points $r_{i+1/2}$ and $r_{i-1/2}$ are defined under formulas

$$(au)_{i+1/2} \equiv (au)_R = a_R u_R = u_{ij} a_R^+ + a_R^- u_{i+1,j}, \tag{4.80}$$

$$(au)_{i-1/2} \equiv (au)_L = a_L u_L = u_{ij} a_L^- + a_L^+ u_{i-1,j}, \tag{4.81}$$

where

$$a_L = \frac{1}{2}(a_i + a_{i-1}); \qquad a_R = \frac{1}{2}(a_{i+1} + a_i);$$

$$a_R^+ = \frac{1}{2}(a_R + |a_R|); \quad a_R^- = \frac{1}{2}(a_R - |a_R|);$$

$$a_L^+ = \frac{1}{2}(a_L + |a_L|); \quad a_L^- = \frac{1}{2}(a_L - |a_L|).$$

Thus,

$$r_{i+1/2}^m a_R u_R - r_{i-1/2}^m a_L u_L = r_{i+1/2}^m a_R^- u_{i+1,j} + u_{ij}\left(r_{i+1/2}^m a_R^+ - r_{i-1/2}^m a_L^-\right) - r_{i-1/2}^m a_L^+ u_{i-1,j}. \tag{4.82}$$

Under similar formulas, the second integral in (4.79) is calculated:

$$b_R u_R - b_L u_L = b_R^- u_{i,j+1} + u_{i,j}\left(b_R^+ - b_L^-\right) - b_L^+ u_{i,j-1}. \tag{4.83}$$

Substituting relations (4.80)–(4.83) in (4.79), it is simple to receive a standard canonical form of the five-point finite-difference equation similar to eq. (4.73) (see slightly below).

At the account of particle diffusion in a radial direction, there is a necessity for an approximation of the integral

$$\int_{r_{i-1/2}}^{r_{i+1/2}} \tilde{D}_k \frac{1}{r^m}\frac{\partial}{\partial r} r^m \frac{\partial u}{\partial r}(2\pi r)^m \mathrm{d}\, r \frac{x_{j+1} - x_{j-1}}{2} =$$

$$= \tilde{D}_k \frac{x_{j+1} - x_{j-1}}{2}(2\pi)^m \left[\left(r^m \frac{\partial u}{\partial r}\right)_{r_{i+1/2}} - \left(r^m \frac{\partial u}{\partial r}\right)_{r_{i-1/2}}\right] =$$

$$= \tilde{D}_k \frac{x_{j+1} - x_{j-1}}{2}(2\pi)^m \left[\left(\frac{r_{i+1} + r_i}{2}\right)^m \frac{1}{r_{i+1} - r_i} u_{i+1,j} - \left(\frac{r_{i+1} + r_i}{2}\right)^m \frac{1}{r_{i+1} - r_i} u_{ij} - \right.$$

$$\left. - \left(\frac{r_i + r_{i-1}}{2}\right)^m \frac{1}{r_i - r_{i-1}} u_{ij} + \left(\frac{r_i + r_{i-1}}{2}\right)^m \frac{1}{r_i - r_{i-1}} u_{i-1,j}\right]. \tag{4.84}$$

After the division of the received finite-difference formulas for transfer and diffusion on $V_{i,j}$, the final form of the finite-difference scheme can be presented in the following canonical five-point finite-difference equation:

$$\left(A_{i,j}+A^*_{i,j}\right)u^{p+1}_{i-1,j}+\left(B_{i,j}+B^*_{i,j}\right)u_{i+1,j}-\left(C_{i,j}+C^*_{i,j}\right)u^{p+1}_{i,j}+\bar{A}_{i,j}u^{p+1}_{i,j-1}+\bar{B}_{i,j}u^{p+1}_{i,j+1}+F_{i,j}=0, \tag{4.85}$$

$$A_{i,j}=\frac{2G^m_i}{r_{i+1}-r_{i-1}}a^+_L r^m_{i-1/-2},\quad A^*_{i,j}=-D_iS^m_{i-1}\frac{1}{r_i-r_{i-1}},$$

$$B_{ij}=-\frac{2G^m_i}{r_{i+1}-r_{i-1}}a^-_R r^m_{i+1/-2},\quad B^*_{i,j}=-D_iS^m_{i+1}\frac{1}{r_{i+1}-r_i},$$

$$C_{i,j}=\frac{1}{\tau}+\frac{2G^m_i}{r_{i+1}-r_{i-1}}\left(r^m_{i+1/2}a^+_R-r^m_{i-1/2}a^-_L\right)+\frac{2}{x_{j+1}-x_{j-1}}\left(b^+_R-b_L\right),$$

$$C^*_{i,j}=-D_iS^m_{i+1}\frac{1}{r_{i+1}-r_i}-D_iS^m_{i-1}\frac{1}{r_i-r_{i-1}},$$

$$\bar{A}_{imj}=\frac{2}{x_{j+1}-x_{j-1}}b^+_L,\ \bar{B}_{imj}=-\frac{2}{x_{j+1}-x_{j-1}}b^-_R,$$

$$F_{i,j}=\frac{u^p_{i,j}}{\tau}+f_{i,j},\ G^m_i=(m+1)\frac{2^m}{\left(r_{i-1}+2r_i+r_{i+1}\right)^m},$$

$$S^m_{i+1}=(m+1)\frac{\left(r_{i+1}+r_i\right)^m}{\left(r_{i-1}+2r_i+r_{i+1}\right)^m},\ S^m_{i-1}=(m+1)\frac{\left(r_i+r_{i-1}\right)^m}{\left(r_{i-1}+2r_i+r_{i+1}\right)^m},$$

$$D_i=-\tilde{D}_{r,k}\frac{H^2}{R^2}\frac{2}{r_{i+1}-r_i},\qquad (k=e,+).$$

In the given formula index, p marks functions on the previous (the lower) time layer.

4.3.3 Conservative properties of the finite-difference scheme for motion equation

Let us consider the motion equation in an integral form. It is completely similar to the continuity equation introduced, for example, in hydrodynamics (Landau L.D., et al., 1960):

$$\int_V\left(\frac{\partial u}{\partial t}+\mathrm{div}\boldsymbol{\Gamma}\right)\mathrm{d}V=\int_V Q\,\mathrm{d}V, \tag{4.86}$$

where Q is the source term and V is the arbitrary volume for which the finite-difference scheme is created.

Let us transform eq. (4.86) with the use of the Gauss theorem

$$\int_V\frac{\partial u}{\partial t}\mathrm{d}V+\int_\Sigma\boldsymbol{\Gamma}\mathrm{d}\mathbf{S}=\int_V Q\,\mathrm{d}V, \tag{4.87}$$

where Σ is the surface of the considered volume, which will be presented as a sum of three parts, as earlier for cylindrical geometry (see Figure 4.6):

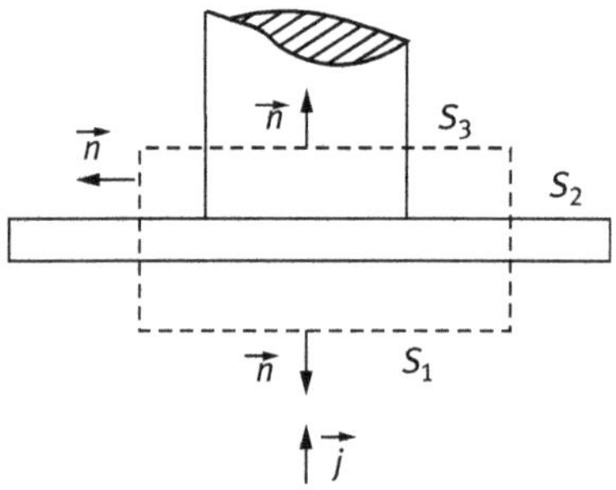

Figure 4.6: Diagram for illustrating the boundary conditions statement for a glow discharge electrode.

$$\int_{\Sigma} \mathbf{\Gamma} d\mathbf{S} = -\int_{S_1} \Gamma_x dS + \int_{S_2} \Gamma_r dS + \int_{S_3} \Gamma_x dS = (2\pi)^m \times$$

$$\times \left[-\int_0^R r^m \Gamma_x(r, x=0) dr + R^m \int_0^H \Gamma_r(r=R, x) dx + \int_0^R r^m \Gamma_x(r, x=H) dr \right]. \tag{4.88}$$

For a greater definiteness, the flux projections on coordinate axes will be presented in explicit form using formulas (4.33) and (4.34) (separately for ions and electrons)

$$\int_V \frac{\partial u}{\partial t} dV + (2\pi)^m \left[-\int_0^R r^m \left(-\mu_+ u \frac{\partial \varphi}{\partial x} \right)_{x=0} dr + R^m \int_0^H \left(-\mu_+ u \frac{\partial \varphi}{\partial r} - D_+ \frac{\partial u}{\partial r} \right)_{r=R} dr + \right.$$

$$\left. + \int_0^R r^m \left(-\mu_+ u \frac{\partial \varphi}{\partial x} \right)_{x=H} dr \right] = \int_V Q dV, \; u = n_+, \tag{4.89}$$

$$\int_V \frac{\partial u}{\partial t} dV + (2\pi)^m \left[-\int_0^R r^m \left(\mu_e u \frac{\partial \varphi}{\partial x} \right)_{x=0} dr + R \int_0^H \left(\mu_e u \frac{\partial \varphi}{\partial x} - D_e \frac{\partial u}{\partial r} \right) dx + \right.$$

$$\left. + \int_0^R r^m \left(\mu_e u \frac{\partial \varphi}{\partial x} \right)_{x=H} dr \right] = \int_V Q dV, \quad u = n_e \tag{4.90}$$

For example, let us consider boundary conditions of the second type at $r = R$:

$$\frac{\partial \varphi}{\partial r} = \frac{\partial u}{\partial r} = 0.$$

In this case, we have for ions ($u = n_+$)

$$\int_V \frac{\partial u}{\partial t} \mathrm{d}V + (2\pi)^m \int_0^R r^m \left(\mu_+ \frac{\partial \varphi}{\partial x} u\right)_{x=0} \mathrm{d}\, r - (2\pi)^m \int_0^R r^m \left(\mu u \frac{\partial \varphi}{\partial x}\right)_{x=H} \mathrm{d}\, r = \int_V Q \mathrm{d}V, \quad (4.91)$$

and for electrons ($u = n_e$)

$$\int_V \frac{\partial u}{\partial t} \mathrm{d}V - (2\pi)^m \int_0^R r^m \left(\mu_e \frac{\partial \varphi}{\partial x} u\right)_{x=0} \mathrm{d}\, r + (2\pi)^m \int_0^R r^m \left(\mu_e u \frac{\partial \varphi}{\partial x}\right)_{x=H} \mathrm{d}\, r = \int_V Q \mathrm{d}V. \quad (4.92)$$

For the undisturbed axial electric field and for a small constant concentration of particles at $r = R$, one can derive more bulky expression, which contains particle flux components on the specified boundary.

To prove the conservative properties of this finite-difference scheme we will multiply eq. (4.85) by $V_{i,j}$ and sum obtained finite-difference correlations throughout all internal points of calculation domain:

$$\sum_{i=2}^{NI-1} \sum_{j=2}^{NJ-1} V_{i,j} \Big\{ \left(A_{i,j} + A^*_{i,j}\right) u^{p+1}_{i-1,j} + \left(B_{i,j} + B^*_{i,j}\right) u^{p+1}_{i+1,j} -$$

$$- \left(C_{i,j} + C^*_{i,j}\right) u^{p+1}_{i,j} + \bar{A}_{i,j} u^{p+1}_{i,j-1} + \bar{B}_{i,j} u^{p+1}_{i,j+1} + F_{i,j} \Big\} = 0.$$

The received correlation can be simplified if we take into account the following conditions:

1) The axial symmetry condition follows

$$u^{p+1}_{2,j} = u^{p+1}_{1,j}, \quad (a^+_L)_{2,j} = (a^-_L)_{2,j} = 0;$$

2) Boundary conditions follow

$$u^{p+1}_{NI,j} = u^{p+1}_{NI-1,j}, \quad (a^+_R)_{NI-1,j} = (a^-_R)_{NI-1,j} = 0. \quad (4.93)$$

Then

$$\sum_{i=2}^{NI-1} \sum_{j=2}^{NJ-2} V_{i,j} \frac{u^{p+1}_{i,j} - u^p_{i,j}}{\tau} - (2\pi)^m \sum_{i=2}^{NI-1} \frac{r_{i+1} - r_{i-1}}{2} \frac{(r_{i-1} + 2r_i + r_{i+1})^m}{(m+1) \cdot 2^m}$$

$$\left[u^{p+1}_{i,1} (b^+_L)_{i,2} + u^{p+1}_{i,2} (b^-_L)_{i,2}\right] + (2\pi)^m \sum_{i=2}^{NI-1} \frac{r_{i+1} - r_{i-1}}{2} \frac{(r_{i-1} + 2r_i + r_{i+1})^m}{(m+1) \cdot 2^m} \times$$

$$\times \left[u^{p+1}_{i,NJ} (b^+_R)_{i,NJ-1} + u^{p+1}_{i,NJ-1} (b^-_R)_{i,NJ-1} + u^{p+1}_{i,NJ-1} (b^-_R)_{i,NJ-1}\right] = \sum_{i=2}^{NI-1} \sum_{j=2}^{NJ-1} V_{i,j} f_{i,j} \quad (4.94)$$

For the direct-current glow discharge under consideration, taking into account boundary conditions defined earlier, $\partial\varphi/\partial x > 0$, one can get for electrons $(b_L^-)_{i,2} = 0$, and for ions $(b_R^+)_{i,NJ-1} = 0$.

Normalizing the integral balance equations (4.91) and (4.92) and comparing them with the corresponding relation following from (4.94), we conclude that the constructed finite-difference scheme has the conservative property.

4.3.4 The order of accuracy of the used finite-difference approximation: inevitable diffusion due to finite-difference approximation

This problem is analyzed in many books on computational physics and CFD (Roache P., 1980; Potter D., 1973; Anderson D.A., et al., 1997). It was shown that at constant transport velocity u the donor cell finite-difference scheme has the second-order accuracy for convective transport if a joint solution of equations of motion and continuity is considered. But, at the same time, it is well known that for the transport equation, the scheme has the first-order accuracy.

To introduce the definition of the numerical (grid) diffusion, the following boundary value problem will be analyzed:

$$\frac{\partial u}{\partial t} + a\frac{\partial u}{\partial x} = 0, \quad a > 0, \tag{4.95}$$

$$u(x, t = 0) = \varphi_0(x) \tag{4.96}$$

with the use of the simplest explicit upstream finite-difference scheme (further, for short scheme)

$$\frac{u_i^{p+1} - u_i^p}{\tau} + a\frac{u_i^p - u_i^{p-1}}{h} = 0, \quad u_i^0 = \varphi_i(x_i). \tag{4.97}$$

Suppose that function $u(x, t)$ has the continuous regular limited derivatives, which are not below the second order. Let us expand this function in Taylor's series in a neighborhood points $i, i-1$ (of a spatial grid), and then p and $p+1$ (of a time grid). Then, we will receive that scheme (4.97) has the following principal part of an approximation error:

$$E = \frac{\tau}{2}\frac{\partial^2 u}{\partial t^2} - a\frac{h}{2}\frac{\partial^2 u}{\partial x^2}. \tag{4.98}$$

But proceeding from (4.95),

$$\frac{\partial^2 u}{\partial t^2} = \frac{\partial}{\partial t}\left(\frac{\partial u}{\partial t}\right) = -a\frac{\partial^2 u}{\partial x \partial t} = -a\frac{\partial}{\partial x}\left(\frac{\partial u}{\partial t}\right) = -a\frac{\partial}{\partial x}\left(-a\frac{\partial u}{\partial x}\right) = a^2\frac{\partial^2 u}{\partial x^2}$$

we have

$$E = \left(\frac{\tau a^2}{2} - \frac{ah}{2}\right)\frac{\partial^2 u}{\partial x^2} = -\frac{ah}{2}\left(1 - \frac{a\tau}{h}\right)\frac{\partial^2 u}{\partial x^2}. \tag{4.99}$$

Analyzing the form of the principal part of the approximation error, it is simple to determine that using approximation (4.97), we will actually solve the equation of the following form:

$$\frac{\partial u}{\partial t} + a\frac{\partial u}{\partial x} - D_c\frac{\partial^2 u}{\partial x^2} = 0, \tag{4.100}$$

where

$$D_m = \frac{ah}{2}\left(1 - \frac{a\tau}{h}\right). \tag{4.101}$$

Note that the effective coefficient of the numerical diffusion D_m contains the hyperbolic CFL number $\mathrm{CFL}^{\mathrm{G}} = a(\tau/h)$, which was discussed earlier.

Thus, the considered equation of convective transport describes some fictitious ("mesh") dissipative process.

With a solution to problems of heat and mass transfer, the numerical diffusion appears useful in some cases. Some effective computing algorithms are based on its use. There are also a lot of cases when the additional term $\varepsilon_c(\partial^2 u/\partial x^2)$ is introduced into the equation in an explicit form.

The value ε_c is named as "artificial viscosity (diffusion, dissipation)." Dissipative processes connected to the artificial dissipation appear also in the finite-difference equation at the use of the smoothing (regularization) of a solution, and also at some approximations of time derivatives (time-dependent viscosity, diffusion, or dissipation).

As to the numerical simulation of the drift-diffusion model of glow discharge, the occurrence of the mesh diffusion is extremely undesirable. Therefore, the mesh dissipative properties will be studied later in detail.

First, we will set the form of the mesh diffusion coefficient D_m in the case of the steady-state transport equations. It can be made by substituting $u_i^{p+1} = u_i^p$ in eq. (4.97), and then having expanded in Taylor's series the remaining part of the equation gives the following:

$$D_m^S = \frac{1}{2}ah. \tag{4.102}$$

Comparing (4.101) and (4.102), we have drawn the conclusion that a nonstationary solution of the motion equation with the numbers $\mathrm{CFL}^{\mathrm{G}}$, which strongly differ from unity, leads to a significant increase of the mesh diffusion coefficient. This implies practically important conclusion about the necessity of integration of the transport equation with different time steps for estimation of influence of the mesh diffusion. In the formulated numerical method only $\mathrm{CFL}_{\mathrm{e}}^{\mathrm{G}}$ numbers for electrons sufficiently differ from unity (generally near to the cathode), while the greatest interest is represented with their radial diffusion on edge of the current channel (at the beginning of the positive column). Numbers $\mathrm{CFL}_+^{\mathrm{G}}$ for ions remain less unity everywhere. Thus, in further estimations, we will be guided by the expression $D_m^S = (1/2)ah$.

From the viewpoint of practical needs for performing computational physical experiments, there is always a principal problem: how the mesh dissipation influences on an obtained numerical solution? More precisely: how the value of mesh diffusion is related to other significant physical processes that are taken into account in the equation?

From the structure of the dissipative term appearing in eq. (4.100), it follows that mesh diffusion will show up substantially in the region of sharp particle concentration differences and the big velocity of particle motion, and also for a coarse mesh. If to assign conditional mesh diffusion flux for the artificial dissipation process by comparing this flux with drift and physical diffusion fluxes, it is possible to obtain representation about the influence of the mesh dissipation on the desired numerical solution. Dimensional relations for determination of the drift and physical diffusion fluxes in axial and radial directions look like

$$(\Gamma_{e,\mathrm{dif}})_r = -D_e \frac{\partial n_e}{\partial r}, \quad (\Gamma_{e,\mathrm{dif}})_x = -D_e \frac{\partial n_e}{\partial x};$$

$$(\Gamma_{e,\mathrm{dr}})_r = n_e \mu_e \frac{\partial \varphi}{\partial r}, \quad (\Gamma_{e,\mathrm{dr}})_x = n_e \mu_e \frac{\partial \varphi}{\partial x};$$

$$(\Gamma_{+,\mathrm{dif}})_r = -D_+ \frac{\partial n_+}{\partial r}, \quad (\Gamma_{+,\mathrm{dif}})_x = -D_+ \frac{\partial n_+}{\partial x};$$

$$(\Gamma_{+,\mathrm{dr}})_r = -n_+ \mu_+ \frac{\partial \varphi}{\partial r}, \quad (\Gamma_{+,\mathrm{dr}})_x = -n_+ \mu_+ \frac{\partial \varphi}{\partial x}, \tag{4.103}$$

where labels “dif” and “dr” mark the diffusion and drift fluxes.

The corresponding mesh diffusion fluxes are defined as

$$(\Gamma_{e,\mathrm{c}})_r = -D_{\mathrm{c},r}^e \frac{\partial n_e}{\partial r}, \quad (\Gamma_{e,c})_x = -D_{\mathrm{c},r}^e \frac{\partial n_e}{\partial x};$$

$$(\Gamma_{+,\mathrm{c}})_r = -D_{\mathrm{c},r}^+ \frac{\partial n_+}{\partial r}, \quad (\Gamma_{+,\mathrm{c}})_x = -D_{\mathrm{c},r}^+ \frac{\partial n_+}{\partial x}. \tag{4.104}$$

Using formula (4.102), one can determine explicit expressions for coefficients of the mesh diffusion

$$D_{\mathrm{c},z}^e = \mu_e \frac{\partial \varphi}{\partial z} \frac{1}{2} \Delta z, \quad D_{\mathrm{c},z}^+ = \mu_+ \frac{\partial \varphi}{\partial z} \frac{1}{2} \Delta z,$$

$$\Delta z = (\Delta r, \Delta x).$$

Estimating $\partial\varphi/\partial z \approx \Delta\varphi/\Delta z$, one obtains

$$D^e_{c,z} \approx \frac{\mu_e \Delta\varphi}{2}, \quad D^+_{c,z} \approx \frac{\mu_+ \Delta\varphi}{2}. \tag{4.105}$$

Considering that coefficients of physical diffusion are defined by the Einstein relations, the ratio of physical diffusion fluxes to mesh diffusion fluxes look like

$$\frac{\Gamma_{e,\text{dif}}}{\Gamma_{e,\text{c}}} \approx \frac{2T_e}{\Delta\varphi}, \quad \frac{\Gamma_{+,\text{dif}}}{\Gamma_{+,\text{c}}} \approx \frac{2T_+}{\Delta\varphi}, \tag{4.106}$$

and the ratio of drift fluxes to mesh diffusion fluxes

$$\frac{\Gamma_{e,\text{dr}}}{\Gamma_{e,\text{c}}} \approx 2\frac{n_e}{\Delta n_e}, \quad \frac{\Gamma_{+,\text{dr}}}{\Gamma_{+,\text{c}}} \approx 2\frac{n_+}{\Delta n_+}. \tag{4.107}$$

It is obvious that an optimum case is realized when all given ratios are significantly superior than unity. It superimposes rather rigid requirements on a selection of the finite-difference grid. For example, at a potential drop in the cathode layer by the value of ~ 150 V per grid step, it follows from (4.106) that the number of calculation points should be so large, as it is necessary to ensure the condition $\Delta\varphi < 2T_e$ on one spatial grid step. Unfortunately, it appears inconvenient at a solution of a 2D problem. Less severe constraints are dictated by relations (4.107). Numerical simulation results show that ion concentrations in the cathode layer drop approximately 10 times, whence it follows that the number of grid points for ions should be within the limits of 10–30. About the same number of points is necessary for describing electron motion in the cathode layer. At the realization of these conditions, drift fluxes of particles will surpass mesh diffusion only in some time. It does not guarantee a lack of influence of the mesh dissipation processes on the results of numerical simulation.

The simplest way to decrees the mesh diffusion (in the frame of the considered finite-difference scheme) is improving the finite-difference grid, for example, by the use of the nonuniform grids. However, as it is known, here there are significant restrictions too, which, on the one hand, require not so sharp variation of grid steps, and, on the other hand, preservation of the certain proportion of the difference grid point disposition in various directions.

Another constructive approach to a solution of this problem is a transition to schemes of the second and higher accuracy orders with the subsequent application of regularization methods (Anderson D.A., et al., 1997; Kulikovskii A.G., et al., 2001). But unfortunately, in this case, there is a possibility of occurrence of oscillations in weak function variation areas adjoining with areas of significant drops. So, this problem should be investigated in each calculation case.

Often at an estimation of dissipative properties of finite-difference schemes, the concept of the Reynolds grid number (Roache P., 1980) is used:

$$\mathrm{Re} = \frac{uh}{\mu},$$

where u, μ are the velocity and viscosity (diffusivity). Re is the Reynolds number obtained at a local velocity of the stream, its dynamic viscosity, and a characteristic size equal to the step of the calculation grid. If instead of μ to substitute the coefficient of mesh diffusion, one can obtain the grid Reynolds number

$$\mathrm{Re_c} = \frac{uh}{D_c}.$$

It is obvious that the condition $\mathrm{Re_c} \gg \mathrm{Re}$ is necessary to provide smallness of the grid numerical diffusion in comparison with the physical one.

Formulas for calculation of different flux kinds and corresponding numerical coefficients which allow to obtain the fluxes in mA/cm^2 are given in Table 4.2. These data are useful for the analysis of the results of calculations.

$$S_1 = 1.6 \times 10^{-16} \frac{N_0 \mathrm{E}}{R} \mu_e, \; S_2 = 1.6 \times 10^{-16} \frac{N_0 \mathrm{E}}{R} \mu_+,$$
$$S_3 = 1.6 \times 10^{-16} \frac{N_0 \mathrm{E}}{H} \mu_e, \; S_4 = 1.6 \times 10^{-16} \frac{N_0 \mathrm{E}}{H} \mu_+.$$

Table 4.2: Fluxes of electrons and ions.

Fluxes along radius, mA/cm^2	Fluxes along *x*-axis, mA/cm^2
$(j_{e,\mathrm{dif}})_r = -\tilde{D}_e \frac{\partial N_e}{\partial r} S_1$	$(j_{e,\mathrm{dif}})_x = -\tilde{D}_e \frac{\partial N_e}{\partial x} S_3$
$(j_{e,\mathrm{dr}})_r = N_e \frac{\partial \Phi}{\partial r} S_1$	$(j_{e,\mathrm{dr}})_x = N_e \frac{\partial \Phi}{\partial x} S_3$
$(j_{e,\mathrm{c}})_r = \frac{\Delta r}{2} \frac{\partial \Phi}{\partial r} \frac{\partial N_e}{\partial r} S_1$	$(j_{e,\mathrm{c}})_x = \frac{\Delta x}{2} \frac{\partial \Phi}{\partial x} \frac{\partial N_e}{\partial x} S_3$
$(j_{+,\mathrm{dif}})_r = -\tilde{D}_+ \frac{\partial N_+}{\partial r} S_1$	$(j_{+,\mathrm{dif}})_x = -\tilde{D}_+ \frac{\partial N_+}{\partial x} S_3$
$(j_{+,\mathrm{dr}})_r = -N_+ \frac{\partial \Phi}{\partial r} S_2$	$(j_{+,\mathrm{dr}})_x = -N_+ \frac{\partial \Phi}{\partial x} S_4$
$(j_{+,\mathrm{c}})_r = \frac{\Delta r}{2} \frac{\partial \Phi}{\partial r} \frac{\partial N_+}{\partial r} S_2$	$(j_{+,\mathrm{c}})_x = \frac{\Delta x}{2} \frac{\partial \Phi}{\partial x} \frac{\partial N_+}{\partial x} S_4$

4.3.5 The finite-difference grids

For calculations of glow discharge, the problem of selection of acceptable finite-difference grids is not less important than in problems of CFD (Fletcher C., 1998; Anderson D.A., et al., 1997). Even in cases of simple flat or axially symmetric geometry nonhomogeneous grids are usually used. Examples of some of them are given below.

Non-uniform scales for transverse coordinate

Typical non-uniform scale on radial coordinate is built (analytically or numerically) with clustering of grid points in the neighborhood of a current column with the use of any functions of compress. If the calculation of glow discharge in an extended area is necessary, it has being reasonable to introduce a multiple block grid. For example, the two-block grid is built as follows. The grid with modification of points under the law of an arithmetical progression near an axis of symmetry is calculated under the formula

$$r_i = d_{\min}(i-1) + (i-1)(i-2)0.5H_w, \quad i = 2, 3, \ldots, N_{iw}, \quad r_1 = 0, \tag{4.108}$$

where N_{iw} is the number of grid points in internal area; $d_{\min}$ is the grid's least step near an axis. The value $d_{\min}$ is selected as the certain part of normal thickness of the cathode layer d_n (e.g., $d_{\min} = 0.1 \div 0.03 d_n$), which is known before the beginning of calculations (it is defined, for example, under the Engel–Steenbeck theory).

The arithmetical ratio H_w is determined under the formula

$$H_w = 2\frac{r_{cc} - d_{\min}(N_{iw}-1)}{(N_{iw}-1)(N_{iw}-2)},$$

where r_{cc} is the presupposed radius of the current channel; this is the boundary of the first block.

Calculation grid in region $r \in [r_{cc}, R]$ is built since a first step

$$H_1 = (r_{iw} - r_{iw-1})P,$$

where $P = 1.5 - 2$ is the scale factor. That is the first step in the external area is uniquely connected to the last step in the internal area. Grid coordinates in the external area are calculated under the formula

$$r_{N_{iw}+i} = r_{N_{iw}} + H_1 i + i(i-1)\frac{D}{2}, \tag{4.109}$$

$$D = 2\frac{(R - r_{TK}) - (NI - N_{iw})H_1}{(NI - N_{iw})(NI - N_{iw} - 1)},$$

where NI is the full number of points along the radius.

One more example of a radius-inhomogeneous grid in the second (external) block is shown later. The tangential law of grid modification by a variable r is set concerning a point of the prospective boundary of the current channel. The new variable is introduced as follows:

$$y = \mathrm{arctg}\,\beta(r - r_{cc}),$$

in which the step of the calculation grid is supposed to be constant and is equal to

$$H_y = \frac{y_{max} - y_{min}}{NI_2 - 1},$$

where $y_{max} = \mathrm{arctg}\,\beta(R_c - r_{cc})$; $y_{min} = \mathrm{arctg}(-\beta r_{cc}) = -\mathrm{arctg}\,\beta r_{cc}$; β is the grid compression factor; NI_2 is the number of points in the external area.

Calculating formulas for determination of r_i looks like

$$r_i = \frac{1}{\beta}\mathrm{tg}\,y_i + r_{TK},$$

$$y_i = y_{i-1} + H_y, \quad i = 2, 3, \ldots, NI_2,$$

$$y_1 = y_{min}, \quad r_1 = 0. \tag{4.110}$$

Note that at a numerical simulation of a glow discharge dynamics in flat channels, it is reasonable to use a homogeneous or mobile grid in the transversal direction (analog of variable r in 2D axially symmetric geometry).

Nonuniform scale on longitudinal coordinate

Unlike the previous case, inhomogeneous grids are always used in the axial direction between cathode and anode. In the cases when the anode is arranged opposite to the cathode (see Figure 4.4), as a rule, all the calculation domain between the cathode and the anode is divided into two subregions. The first one is the cathode region between the cathode and a prospective point of the cathode potential drop extremum (see Figure 4.3). The second one is the external region from the point of the cathode layer extremity to the anode. Thickness of the cathode layer d_n can be estimated by the Engel–Steenbeck theory (see, e.g., (4.29)).

In the internal area (in the cathode layer), the grid can be built, for example, under the formula

$$x_j = (j-1)d_n + (j-1)(j-2)\frac{1}{2}D_{jw}, \tag{4.111}$$

$$j = 2, 3, \ldots, N_{jw},$$

$$D_{jw} = 2\frac{d_n - (N_{jw} - 1)d_{cc}}{(N_{jw} - 1)(N_{jw} - 2)},$$

$$x_1 = 0,$$

where d_{cc} has the same value as in construction of the calculation grid along the transversal coordinate.

In the external area, the grid is built so that its greatest crowding was on the anode potential drop

$$x_{j-1} = x_j - A_j,\ j = NJ, NJ-1, \ldots, N_{jw}+1,$$

$$A_j = d_{an} + D_{je}(m-1),\ m = 1, 2, \ldots, NJ - N_{jw},$$

$$D_{je} = 2\frac{(H - d_\kappa) - (NJ - N_{jw})\,d_{an}}{(NJ - N_{jw})(NJ - N_{jw} - 1)}. \tag{4.112}$$

where NJ is the full number of points of the calculation grid in the x-direction .

The value d_{an} is the least step on axial coordinate near the anode. As a rule, $d_{an} = d_n$.

Function $y = \operatorname{arctg}\beta x$ can also be recommended for the construction of calculation grids between electrodes, because it allows receiving a crowding of grid points near the cathode and the anode.

Sometimes there is a necessity to rebuild using grids and reinterpolate arrays of calculated functions on new finite-difference grids. It is recommended to use the curve fitting method. In this case, the laws of step modification of the old and new grids can be simplified.

4.3.6 Iterative methods for solving systems of linear algebraic equations in canonical form

In the considered 2D problem of direct-current glow discharge, there is a necessity to integrate the following three finite-difference equations: the continuity equations for ions and electrons and also the Poisson equation for electric field potential. The important feature of the integration is also the algorithmic realization of these equations' solution sequence (the so-called external iterative process). The separate section of this chapter is devoted to the consideration of the external iterative process. But here we will concentrate our attention on methods of integration of finite-difference equations in canonical form.

As discussed earlier, each of the enumerated equations being written in finite-difference form can be presented in the canonical form, which is determined by using the five-point grid template:

$$A_{i,j}u^S_{i-1,j} + B_{i,j}u^S_{i+1,j} - C_{i,j}u^S_{ij} + \bar{A}_{i,j}u^S_{i,j-1} + \bar{B}_{i,j}u^S_{i,j+1} + F_{i,j} = 0, \tag{4.113}$$

$$i = 2, 3, \ldots, NI-1, \quad j = 2, 3, \ldots, NJ-1,$$

where S is the iterative index.

The boundary conditions are presented in the following general form:

$$i = 1{:}\ u_{1,j} = \varphi_{\alpha,j}u_{2,j} + \xi_{\alpha,j}, \tag{4.114}$$

$$i = NI:\ u_{NI,j} = \varphi_{\gamma,j} u_{NI-1,j} + \xi_{\gamma,j}, \tag{4.115}$$

$$j = 1:\ u_{i,1} = \varphi_{i,\bar{\alpha}} u_{i,2} + \xi_{i,\bar{\alpha}}, \tag{4.116}$$

$$j = NJ:\ u_{i,NJ} = \varphi_{i,\bar{\gamma}} u_{i,NJ-1} + \xi_{i,\bar{\gamma}}. \tag{4.117}$$

Numerous publications under the classical numerical analysis and the computing physics are devoted to methods of solution of this boundary value problem (4.113)–(4.117) (Anderson D.A., et al., 1997; Oran E.S., et al., 1987; Koonin S.E., 1986). Information on the methods presented in the quoted works, as a rule, appears enough for construction of codes for performing a numerical simulation. However, it is necessary to take into account that a solution of the glow discharge problem, as a rule, is performed by iterative methods, so the effectiveness of this or that finite-difference method should be estimated finally by analysis of the total effectiveness of the numerical simulation. In other words, the chosen finite-difference method should have a high convergence rate both on rough and on an almost exact solution, and besides it should have relatively low laboriousness. Those requirements are rather inconsistent very often. So the problem of the choice of a computing method is rather difficult.

The following three numerical methods have shown their effectiveness at a solution of various problems of glow discharge:
1) nonlinear iterative α–β algorithm (Chetverushkin B.N., 1985);
2) successive overrelaxation (SOR) method; and
3) explicit method.

Nonlinear iterative α–β-algorithm

The method has been mentioned in Chetverushkin B. N. (1985). The principal advantage of the α–β iteration scheme, which caused its effectiveness at a solution of the glow discharge problem, is lack of necessity in the knowledge of any a priori information on a spectrum of difference operator or of necessity to define its boundaries. In other words, from a practical point of view, this method is autonomous and does not demand special preliminary operations for a solution.

The solution of the finite-difference scheme (4.113) is searched in the form of

$$u_{i,j} = \alpha_{i+1,j} u_{i+1,j} + \beta_{i+1,j},$$
$$u_{i,j} = \gamma_{i-1,j} u_{i-1,j} + d_{i-1,j}, \tag{4.118}$$

$$u_{i,j} = \bar{\alpha}_{i,j+1} u_{i,j+1} + \bar{\beta}_{i,j+1},$$
$$u_{i,j} = \bar{\gamma}_{i,j-1} u_{i,j-1} + \bar{d}_{i,j-1}. \tag{4.119}$$

Sequentially substituting (4.118) and (4.119) into (4.113), it is possible to receive formulas for eight run coefficients that are connected among themselves. The situation is facilitated by the fact that it is possible to select two independent groups of those coefficients.

Computing process begins with a set of zero approximations of a matrix of coefficients $\bar{\gamma}_{i,j}^{S}$. Below examples of the special statements of the zero approximations $\bar{\gamma}_{i,j}^{S}$ considered are intended to accelerate the convergence of an iterative process. For the first time, it is possible to use the boundary conditions (4.117):

$$\bar{\gamma}_{i,j}^{S} = \varphi_{i,\bar{\gamma}}, \quad i = 1, 2, \ldots, NI; \quad j = 1, 2, \ldots, NJ.$$

In a direct cycle of the so-called α-process of calculations, the modification of an index $j = 2, 3, \ldots, NJ - 1$ is made, and the following coefficients are determined:

$$\alpha_{i+1,j}^{S+\frac{1}{2}} = \frac{B_{i,j}}{C_{i,j} - \alpha_{i,j}^{S+\frac{1}{2}} A_{i,j} - \bar{\alpha}_{i,j}^{S+\frac{1}{2}} \bar{A}_{i,j} - \bar{\gamma}_{i,j}^{S} \bar{B}_{i,j}}, \quad i = 2, 3, \ldots, NI - 1, \tag{4.120}$$

$$\gamma_{i-1,j}^{S+\frac{1}{2}} = \frac{A_{i,j}}{C_{i,j} - \gamma_{i,j}^{S+\frac{1}{2}} B_{i,j} - \bar{\alpha}_{i,j}^{S+\frac{1}{2}} \bar{A}_{i,j} - \bar{\gamma}_{i,j}^{S} \bar{B}_{i,j}}, \quad i = NI - 1, \ldots, 2, \tag{4.121}$$

$$\bar{\alpha}_{i,j+1}^{S+\frac{1}{2}} = \frac{\bar{B}_{i,j}}{C_{i,j} - \alpha_{i,j}^{S+\frac{1}{2}} A_{i,j} - \bar{\alpha}_{i,j}^{S+\frac{1}{2}} \bar{A}_{i,j} - \gamma_{i,j}^{S+\frac{1}{2}} B_{i,j}}, \quad i = 2, 3, \ldots, NI - 1. \tag{4.122}$$

In an inverse cycle of the α-process, the index j varies in the opposite direction $j = NJ - 1, NJ - 2, \ldots, 2$ and the following coefficients are calculated:

$$\alpha_{i+1,j}^{S+1} = \frac{B_{i,j}}{C_{i,j} - \alpha_{i,j}^{S+1} A_{i,j} - \bar{\alpha}_{i,j}^{S+\frac{1}{2}} \bar{A}_{i,j} - \bar{\gamma}_{i,j}^{S+1} \bar{B}_{i,j}}, \quad i = 2, \ldots, NI - 1, \tag{4.123}$$

$$\gamma_{i-1,j}^{S+1} = \frac{A_{i,j}}{C_{i,j} - \gamma_{i,j}^{S+1} B_{i,j} - \bar{\alpha}_{i,j}^{S+\frac{1}{2}} \bar{A}_{i,j} - \bar{\gamma}_{i,j}^{S+1} \bar{B}_{i,j}}, \quad i = NI - 1, \ldots, 2, \tag{4.124}$$

$$\bar{\gamma}_{i,j-1}^{S+1} = \frac{\bar{A}_{i,j}}{C_{i,j} - \gamma_{i,j}^{S+1} B_{i,j} - \alpha_{i,j}^{S+1} A_{i,j} - \bar{\gamma}_{i,j}^{S+1} \bar{B}_{i,j}}, \quad i = 2, \ldots, NI - 1. \tag{4.125}$$

For the realization of the specified runs, it is necessary to use the boundary conditions

$$\alpha_{2,j}^{S+1/2} = \varphi_{\alpha,j}, \quad \gamma_{NI-1,j}^{S+1/2} = \varphi_{\gamma,j},$$

$$\alpha_{2,j}^{S+1} = \varphi_{\alpha,j}, \quad \gamma_{NI-1,j}^{S+1} = \varphi_{\gamma,j}, \quad j = 2, \ldots, NJ - 1$$

$$\bar{\alpha}_{i,2}^{S+1/2}=\varphi_{i,\bar{\alpha}},\ \bar{\gamma}_{i,NJ-1}^{S}=\varphi_{i,\bar{\gamma}},\ \bar{\gamma}_{i,NJ-1}^{S+1}=\varphi_{i,\bar{\gamma}},\ i=2,...,NI-1. \tag{4.126}$$

After completion of the direct and inverse cycles, the estimation of α-process convergence is made, if

$$\varepsilon=\frac{\left|\bar{\gamma}_{i,j}^{S+1}-\bar{\gamma}_{i,j}^{S}\right|}{\left|\bar{\gamma}_{i,j}^{S+1}\right|}>\varepsilon_{\alpha}\left(\sim10^{-3}\right), \tag{4.127}$$

then appropriation is made

$$\bar{\gamma}_{i,j}^{S}=\bar{\gamma}_{i,j}^{S+1}\quad(i=2,...,NI-1;\quad j=2,...,NJ-1),$$

and the α-process is repeated. Otherwise, the " β-process" begins.

Prior to the beginning of the β-iterative process, the following auxiliary functions are calculated:

$$\begin{aligned}
&\psi_{\alpha,i,j}=C_{i,j}-\alpha_{i,j}A_{i,j}-\bar{\alpha}_{i,j}\bar{A}_{i,j}-\bar{\gamma}_{i,j}\bar{B}_{i,j}, \\
&\psi_{\gamma,i,j}=C_{i,j}-\gamma_{i,j}B_{i,j}-\bar{\alpha}_{i,j}\bar{A}_{i,j}-\bar{\gamma}_{i,j}\bar{B}_{i,j}, \\
&\psi_{\bar{\alpha},ij}=C_{i,j}-\alpha_{i,j}A_{i,j}-\bar{\alpha}_{i,j}\bar{A}_{i,j}-\gamma_{i,j}B_{i,j}, \\
&\psi_{\bar{\gamma},ij}=C_{i,j}-\alpha_{i,j}A_{i,j}-\gamma_{i,j}B_{i,j}-\bar{\gamma}_{i,j}\bar{B}_{i,j}, \\
&i=2,...,NI-1;\qquad j=2,...,NJ-1.
\end{aligned} \tag{4.128}$$

Zero approximations for coefficients $\bar{\alpha}_{i,j}$ are set as follows:

$$\bar{\alpha}_{i,j}^{S}=\xi_{i,\bar{\gamma}},\ i=2,...,NI;\ j=2,...,NJ. \tag{4.129}$$

Direct cycle of iterative β-process is made at modification of index $j=2,\ldots,NJ-1$:

$$\beta_{i+1,j}^{S+\frac{1}{2}}=\frac{F_{i,j}+\beta_{i,j}^{S+\frac{1}{2}}A_{i,j}+\bar{\beta}_{i,j}^{S+\frac{1}{2}}\bar{A}_{i,j}+\bar{d}_{i,j}^{S}\bar{B}_{i,j}}{\psi_{\alpha,i,j}},\ i=2,...,NI-1; \tag{4.130}$$

$$d_{i-1,j}^{S+\frac{1}{2}}=\frac{F_{i,j}+d_{i,j}^{S+\frac{1}{2}}B_{i,j}+\bar{\beta}_{i,j}^{S+\frac{1}{2}}\bar{A}_{i,j}+\bar{d}_{i,j}^{S}\bar{B}_{i,j}}{\psi_{\gamma,i,j}},\ i=NI-1,...,2; \tag{4.131}$$

$$\bar{\beta}_{i,j+1}^{S+\frac{1}{2}}=\frac{F_{i,j}+\beta_{i,j}^{S+\frac{1}{2}}A_{i,j}+\bar{\beta}_{i,j}^{S+\frac{1}{2}}\bar{A}_{i,j}+d_{i,j}^{S+\frac{1}{2}}B_{i,j}}{\psi_{\bar{\alpha},i,j}},\ i=2,...,NI-1. \tag{4.132}$$

In an inverse cycle of the β-process, the index j varies in the opposite direction $j=NJ-1,\ldots,2$

$$\beta_{i+1,j}^{S+1} = \frac{F_{i,j} + \beta_{i,j}^{S+1} A_{i,j} + \bar{\beta}_{i,j}^{S+\frac{1}{2}} \bar{A}_{i,j} + \bar{d}_{i,j}^{S+1} \bar{B}_{i,j}}{\psi_{\alpha,i,j}}, \quad i = 2, \ldots, NI-1; \tag{4.133}$$

$$d_{i-1,j}^{S+1} = \frac{F_{i,j} + d_{i,j}^{S+1} B_{i,j} + \bar{\beta}_{i,j}^{S+\frac{1}{2}} \bar{A}_{i,j} + \bar{d}_{i,j}^{S+1} \bar{B}_{i,j}}{\psi_{\gamma,i,j}}, \quad i = NI-1, \ldots, 2 \tag{4.134}$$

$$\bar{d}_{i,j-1}^{S+1} = \frac{F_{i,j} + \beta_{i,j}^{S+1} A_{i,j} + \bar{d}_{i,j}^{S+1} \bar{B}_{i,j} + d_{i,j}^{S+1} B_{i,j}}{\psi_{\bar{\gamma},i,j}}, \quad i = 2, \ldots, NI-1. \tag{4.135}$$

Boundary conditions of the following form are used in the β-process:

$$\beta_{2,j}^{S+\frac{1}{2}} = \xi_{\alpha,j}, \quad d_{NI-1,j}^{S+\frac{1}{2}} = \xi_{\gamma,j}, \tag{4.136}$$

$$\beta_{2,j}^{S+1} = \xi_{\alpha,j}, \quad d_{NI-1,j}^{S+1} = \xi_{\gamma,j}, \quad j = 2, \ldots, NJ-1;$$

$$\bar{\beta}_{i,2} = \xi_{i,\bar{\alpha}}, \quad \bar{d}_{i,NJ-1}^{S+\frac{1}{2}} = \xi_{i,\bar{\gamma}},$$

$$\bar{d}_{i,NJ-1}^{S+1} = \xi_{i,\bar{\gamma}}, \quad i = 2, \ldots, NI-1.$$

Convergence of the β-process is checked for functions $\bar{d}_{i,j}^{S+1}$. If

$$\varepsilon = \frac{\left|\bar{d}_{i,j}^{S+1} - \bar{d}_{i,j}^{S}\right|}{\left|\bar{d}_{i,j}^{S}\right|} > \varepsilon_{\beta} \quad (\sim 10^{-3}), \tag{4.137}$$

then appropriation is made $\bar{d}_{i,j}^{S} = \bar{d}_{i,j}^{S+1}$ with the subsequent recurring of the β-process. Otherwise, final calculation of required functions is made:

$$u_{i,1} = \frac{\bar{d}_{i,1}^{S+1} \cdot \varphi_{i,\bar{\alpha}} + \xi_{i,\bar{\alpha}}}{1 - \varphi_{i,\bar{\alpha}} \bar{\gamma}_{i,1}^{S+1}}, \quad i = 2, \ldots, NI-1; \tag{4.138}$$

$$u_{i,j} = \bar{\gamma}_{i,j-1}^{S+1} u_{i,j-1} + \bar{d}_{i,j-1}^{S+1}, \quad i = 2, \ldots, NI-1; \; j = 2, \ldots, NJ-1; \tag{4.139}$$

$$u_{i,NJ} = \varphi_{i,\bar{\gamma}} u_{i,NJ-1} + \xi_{i,\bar{\gamma}}, \quad i = 2, \ldots, NI-1; \tag{4.140}$$

$$u_{1,j} = \varphi_{\alpha,j} u_{2,j} + \zeta_{\alpha,j}, \quad j = 1, 2, \ldots, NJ; \tag{4.141}$$

$$u_{NI,j} = \varphi_{\gamma,j} u_{NI-1,j} + \zeta_{\gamma,j}, \quad j = 1, 2, \ldots, NJ. \tag{4.142}$$

Note that there is a possibility to calculate functions $u_{i,j}$ with the use of other formulas, for example, (4.118). Then it is necessary to change the order of use of boundary conditions.

Numerous researches of the α–β iteration algorithm have shown that α-process takes essentially less time for calculations than β-process. Therefore, the basic efforts on a diminution of calculation time are usually focused on the optimization of the β-process. For example, there is a possibility to use the overrelaxation method for the β-process.

An appreciable reduction of the number of internal α- and β-iterations can be reached due to some rise in computer RAM memory space. It is recommended to keep $\bar{\gamma}_{i,j-1}^{S+1}$ and $\bar{d}_{i,j-1}^{S+1}$ for each of the solved equations after convergence of α–β process. These remembered factors are used at the subsequent calculation as initial approximations.

Finite-difference boundary conditions for α–β-iterative method

For the Poisson equation we have

$$\begin{aligned}
&i=1: && u_{1,j}=u_{2,j}, && \varphi_{\alpha,j}=1, && \xi_{\alpha,j}=0, && j=1,\dots,NJ;\\
&i=NI: && a)\ u_{NI,j}=u_{NI-1,j}, && \varphi_{\gamma,j}=1, && \xi_{\gamma,j}=0, && j=1,\dots,NJ;\\
& && b)\ u_{NI,j}=\frac{x_jV}{E}, && \varphi_{\gamma,j}=0, && \xi_{\gamma,j}=\frac{x_jV}{E}, && j=1,\dots,NJ;\\
&j=1: && u_{i,1}=0, && \varphi_{i,\bar{\alpha}}=0, && \xi_{i,\bar{\alpha}}=0, && i=1,\dots,NI;\\
&j=NJ: && u_{i,NJ}=\frac{V}{E}, && \varphi_{i,\bar{\gamma}}=0, && \xi_{i,\bar{\gamma}}=\frac{V}{E}, && i=1,\dots,NI.
\end{aligned} \tag{4.143}$$

Boundary conditions for electrons and ions in axial directions have the first order, and in a radial direction, they can be of both first and second orders. If radial diffusion is considered, and the equations have the second order in the radial direction, the boundary conditions look like (for ions and electrons they are identical):

$$\begin{aligned}
&i=1: && u_{1,j}=u_{2,j}, && \varphi_{\alpha,j}=1, && \xi_{\alpha,j}=0, && j=1,\dots,NJ;\\
&i=NI: && a)\ u_{NI,j}=u_{NI-1,j}, && \varphi_{\gamma,j}=1, && \xi_{\gamma,j}=0, && j=1,\dots,NJ;\\
& && b)\ u_{NI,j}=0, && \varphi_{\gamma,j}=0, && \xi_{\gamma,j}=0, && j=1,\dots,NJ.
\end{aligned} \tag{4.144}$$

If particle diffusion is not considered, the equations of the first order are solved. In this case, it is necessary to formulate some "effective" boundary conditions that have no physical meaning, and as a rule, these are the direct consequence of the solved equations.

In the radial direction at a great distance from an axis of symmetry, the boundary conditions of the second kind can be used:

$$\frac{\partial n_+}{\partial r}=\frac{\partial n_e}{\partial r}=0, \tag{4.145}$$

which are the corollary of a stationary continuity equation under the condition of homogeneous background particle concentration. Boundary conditions in the axial direction have been formulated in Section 4.2.2 (see formulas (4.61) and (4.62)). In finite-difference notation, these conditions will have the following appearance:

a) At the anode

$$u_{i,NJ}^{S+1} = \varphi_{i,\bar{y}} u_{i,NJ-1}^{S+1} + \xi_{i,\bar{y}}, \tag{4.146}$$

$$\varphi_{i,\bar{y}} = \frac{1}{Z_i}\left[P_b\left(\frac{\partial\Phi}{\partial x}\right)_{i,NJ-1}\frac{1}{x_{NJ}-x_{NJ-1}}\right],$$

$$\xi_{i,\bar{y}} = \frac{1}{Z_i}\left[\frac{u_{i,NJ}^p}{\tau} - \hat{A}_{i,NJ}u_{i-1,NJ}^p - \hat{B}_{i,NJ}u_{i+1,NJ}^p\right],$$

$$Z_i = \frac{1}{\tau} + P_b\left(\frac{\partial\Phi}{\partial x}\right)_{i,NJ}\frac{1}{x_{NJ}-x_{NJ-1}} - C_{i,NJ}^*,$$

$$Z_i = \frac{1}{\tau} + P_b\left(\frac{\partial\Phi}{\partial x}\right)_{i,NJ}\frac{1}{x_{NJ}-x_{NJ-1}} - C_{i,NJ}^*,$$

$$\hat{A}_{i,NJ} = -D_i S_{i-1}^m \frac{1}{r_i - r_{i-1}}, \quad \hat{B}_{i,NJ} = -D_i S_{i+1}^m \frac{1}{r_{i+1}-r_i},$$

$$\hat{C}_{i,NJ} = \hat{A}_{i,NJ} + \hat{B}_{i,NJ}, \quad D_i = -\tilde{D}_e \frac{H^2}{R^2}\frac{2}{r_{i+1}-r_{i-1}}, \qquad P_b = 1.$$

where $\tilde{D}_e$ is the dimensionless factors of diffusion; S_{i-1}^m, S_{i+1}^m are defined earlier (see eq. (4.85)). Iterative index S used in (4.146) instead of index of a time layer p to mark the calculation of grid functions at each time layer is necessary to perform some internal iterations (S).

b) At the cathode

$$u_{i,1}^{S+1} = \varphi_{i,\bar{\alpha}} u_{i,2}^{S+1} + \xi_{i,\alpha}, \tag{4.147}$$

$$\varphi_{i,\alpha} = \frac{1}{Z_i}\left[-\frac{P_b}{x_2 - x_1}\left(\frac{\partial\Phi}{\partial x}\right)_{i,1}\right], \quad \xi_{i,\alpha} = \frac{1}{Z_i}\left[\frac{u_{i,1}^p}{\tau} + \hat{A}_{i,1}u_{i-1,1}^S + \hat{B}_{i,1}u_{i+1,1}^S\right],$$

$$Z_i = \frac{1}{\tau} - P_b\left(\frac{\partial\Phi}{\partial x}\right)_{i,1}\frac{1}{x_2 - x_1} - C_{i,1}^*,$$

$$\hat{A}_{i,1} = -D_i S_{i-1}^m \frac{1}{r_i - r_{i-1}}(1+\gamma), \quad \hat{B}_{i,1} = -D_i S_{i+1}^m \frac{1}{r_{i+1}-r_i}(1+\gamma),$$

$$\hat{C}_{i,1} = \hat{A}_{i,1} + \hat{B}_{i,1}, \quad P_b = -(1+\gamma)\frac{\mu_+}{\mu_e}, \quad D_i = -\tilde{D}_+ \frac{H^2}{R^2}\frac{2}{r_{i+1}-r_{i-1}}.$$

Note that despite neglecting diffusion in the axial direction, grid boundary conditions have included factors of radial diffusion of ions and electrons. This demonstrates the fact that we use a corollary of the solved equation as a grid boundary condition.

The Gauss–Seidel and successive relaxation methods

Computing process of this kind of relaxation methods is built as follows (Anderson D.A., et al., 1997; Peyret R., 1996):

$$u_{i,j}^{p+1} = (1-\omega)u_{i,j}^{p} + \omega\, u_{i,j}^{p+1/2}, \tag{4.148}$$

where

$$u_{i,j}^{p+1/2} = \frac{F_{i,j}^{p} + A_{i,j}u_{i-1,j}^{p+1} + B_{i,j}u_{i,j-1}^{p+1} + \bar{A}_{i,j}u_{i+1,j}^{p} + \bar{B}_{i,j}u_{i,j+1}^{p}}{C_{i,j}}, \tag{4.149}$$

p is the index of a time step or the iterative index, ω is the relaxation parameter.

It was supposed at the writing of (4.149) that the passage of calculation grid is carried out in an order of magnification of indexes i and j (see Figure 4.5).

Functions in the points marked by circles are taken on the upper time layer and in the points marked by rectangles are on the lower time layer. In our problem, such an order of tracking is convenient for calculation of properties of electrons.

If we calculate the concentration of ions, for which boundary condition is set on the anode, then the calculation area bypass with a diminution of an index j and magnification of an index i are more natural, that is

$$u_{i,j}^{p+1/2} = \frac{F_{i,j}^{p} + A_{i,j}u_{i-1,j}^{p+1} + \bar{B}_{i,j}u_{i,j+1}^{p+1} + B_{i,j}u_{i,j-1}^{p} + \bar{A}_{i,j}u_{i+1,j}^{p}}{C_{i,j}}. \tag{4.150}$$

For a solution of the Poisson equation, the direction of calculation area bypass is indifferent.

A choice of relaxation parameter ω is of great importance for the effectiveness of the considered method. The theorem is known (Roache P.J., 1980), according to which the method of a successive relaxation converges at $0 < \omega < 2$. More often, the method of relaxation at $\omega < 1$ is named as the method of successive lower relaxation, at $\omega > 1$ is named as the method of successive upper relaxation, and at $\omega = 1$ as the Gauss–Seidel method.

In the majority of practically important cases, it is possible to solve a problem of determination of optimum parameter ω. The following algorithm is in common use. First, calculation under formulas (4.149) or (4.150) at $\omega = 1$ is made. As a result of evaluations on three successive time steps, it is possible to estimate a maximum eigenvalue λ_1 of the transition operator for the Gauss–Seidel method

$$\lambda_1 \approx \sqrt{\frac{\sum_{ij}\left(u_{ij}^{p+1} - u_{ij}^{p}\right)^2}{\sum_{ij}\left(u_{ij}^{p} - u_{ij}^{p-1}\right)^2}}, \tag{4.151}$$

Whence we obtain the estimation of the relaxation parameter

$$\omega \approx \frac{2}{1-\sqrt{1-\lambda_1}}. \tag{4.152}$$

Note that the correct choice of the optimum relaxation parameter can essentially reduce the number of iterations necessary for full convergence. In Roache P. (1980) the convincing example illustrating the dependence of calculation errors on the choice of ω parameter is given. Cited data are especially useful because they show the character of a diminution of an error as approaching the optimum ω. The sharp falling of the calculation error is observed only in immediate proximity from optimum ω so that at its practical definition it is always reasonable to make numerical experiments for reaching the greatest effectiveness.

The practice of calculations of glow discharges shows that computing effectiveness of relaxation methods increases at the use of the Thomas algorithm along coordinate lines $j = \text{const}$ or $i = \text{const}$ (the SOR by lines method; the SLOR method) (Anderson D.A., et al., 1997). In particular, outcomes of calculations of the glow discharges presented in the following chapter testify more rapid convergence of the problem solution as a whole at the use of the Thomas algorithm for a solution of flow continuity equations for the charged particles in the direction from the cathode to the anode.

Boundary conditions for methods of the successive relaxation

Calculation of ion concentration

Calculation of ion concentrations is performed from the anode to the cathode. The first line $j = NJ - 1$ is nonstandard only. On the left boundary of calculation area ($i = 1$) the symmetry condition is used:

$$u_{1,NJ-1} = u_{2,NJ-2}.$$

If a boundary condition on the upper bound is known:

$$u_{i,NJ} = \varphi_{i,\bar{y}} u_{i,NJ-1} + \xi_{i,\bar{y}},$$

then in the first calculation point, it is necessary to use the formula

$$u^{p+1/2}_{2,NJ-1} = \left(F^p_{2,NJ-1} + \bar{B}_{2,NJ-1}\xi_{2,\bar{y}} + B_{2,NJ-1}u^p_{2,NJ-2} + A_{2,NJ-1}u^p_{3,NJ-1}\right)\hat{C}^{-1}_{2,NJ-1}, \tag{4.153}$$

where

$$\hat{C}_{2,NJ-1} = C_{2,NJ-1} - A_{2,NJ-1} - \bar{B}_{2,NJ-1}\varphi_{2,\bar{y}}.$$

Let us use values $A_{2,NJ-1}$ and $\bar{B}_{2,NJ-1}$ from (4.85), then one can see that $\hat{C}_{2,NJ-1} \neq 0$.

For the line $j = NJ - 1$ we can begin the calculation

$$u_{i,NJ-1}^{p+1/2} = \left(F_{i,NJ-1}^{p} + \bar{B}_{i,NJ-1}\xi_{i,\bar{y}} + B_{2,NJ-1}u_{2,NJ-2}^{p} + A_{i,NJ-1}u_{i-1,NJ-1}^{p+1/2} + \bar{A}_{i,NJ-1}u_{i+1,NJ-1}^{p}\right)\hat{C}_{i,NJ-1}^{-1},$$
$$i = 3, \ldots, NI-1, \tag{4.154}$$

where $\hat{C}_{i,NJ-1} = C_{i,NJ-1} - \bar{B}_{i,NJ-1}\varphi_{i,\bar{y}}$.

Calculations on the next line $j = NJ-2, NJ-3, \ldots, 2$ should start from the satisfaction of the symmetry boundary conditions

$$u_{2,j}^{p+1/2} = \left(F_{2,j}^{p} + \bar{B}_{2,j}u_{2,j+1}^{p+1} + B_{2,j}u_{2,j-1}^{p} + \bar{A}_{2,j}u_{3,j}^{p}\right)\hat{C}_{2,j}^{-1}, \tag{4.155}$$
$$\hat{C}_{2,j} = C_{2,j} - A_{2,j},$$

and then the subsequent calculation on the basic relation (4.150) is performed. After completion of the calculations in all internal points, the boundary values of desired functions are calculated using conditions (4.114)–(4.117).

As mentioned earlier, grid boundary conditions on the cathode are formulated in a standard form; therefore, their use does not involve any difficulties.

Calculation of electron concentrations

These calculations are performed in a reverse order being compared to the calculation of ions: from the cathode to the anode. A direction of motion along the radius is former, that is, from an axis to a periphery. Calculations along the line $i = 2$ for all j and along the line $j = 2$ for all i are performed with the nonstandard way. In the point $i = 2$, $j = 2$ grid function is calculated under the formula

$$u_{2,2}^{p+1/2} = (F_{2,2} + B_{2,2}\xi_{2,\bar{\alpha}} + \bar{B}_{2,2}u_{2,3}^{p} + \bar{A}_{2,2}u_{3,2}^{p})\hat{C}_{2,2}^{-1}, \tag{4.156}$$
$$\hat{C}_{2,2} = C_{2,2} - A_{2,2} - B_{2,2}\varphi_{2,\bar{\alpha}}.$$

It is guaranteed that $\hat{C}_{2,2} \neq 0$ due to values $A_{2,2}, B_{2,2}, \varphi_{2,\bar{\alpha}}$.

Calculations for the line $j = 2$ is performed with the formula

$$u_{i,2}^{p+1/2} = \left(F_{i,2}^{p} + B_{i,2}\xi_{2,\bar{\alpha}} + \bar{B}_{i,2}u_{i,3}^{p} + A_{i,2}u_{i-1,2}^{p+1} + \bar{A}_{i,2}u_{i+1,2}^{p}\right)\hat{C}_{i,2}^{-1}, \tag{4.157}$$
$$\hat{C}_{i,2} = C_{i,2} - B_{i,2}\varphi_{i,\bar{\alpha}}.$$

Concentrations of electrons near the symmetry axis are calculated as follows:

$$u_{2,j}^{p+1/2} = \left(F_{2,j} + \bar{B}_{2,j}u_{2,j+1}^{p} + B_{2,j}u_{2,j-1}^{p+1} + \bar{A}_{2,j}u_{3,j}^{p}\right)\hat{C}_{2,j}^{-1},$$

where $\hat{C}_{2,j} = C_{2,j} - A_{2,j}$.

Calculations of electronic concentrations in all interior points are made with the use of the basic formula (4.150). At the end of the calculation loop, the boundary values of functions are modified. Boundary conditions (4.114), (4.115), and (4.117) are applied for this purpose.

Explicit method of calculation of ion concentrations

The explicit method is reasonable for using only for calculation of ion concentrations. As shown earlier, the calculation of glow discharge parameters is performed with characteristic times corresponding to

$$\mathrm{CFL}_{+}^{\mathrm{G}} < 1, \mathrm{CFL}_{+}^{\mathrm{P}} < 1. \tag{4.158}$$

It allows the use of the explicit method for ions. Further, we will assume conditions (4.158) as fulfilled.

Let us consider eq. (4.85), and introduce the following designation:

$$C_{ij} = \tau^{-1} + E_{ij}, \tag{4.159}$$

$$E_{ij} = \frac{2G_i^m}{r_{i+1} - r_{i-1}} \left(r_{i+1/2}^m a_R^+ - r_{i-1/2}^m a_L^- \right)_{ij} + \frac{2}{x_{j+1} - x_{j-1}} \left(b_R^+ - b_L^- \right)_{ij},$$

then the explicit form of the finite-difference equation gains the following form:

$$u_{ij}^{p+1} = \tau \left(A_{ij} u_{i-1,j}^p + B_{ij} u_{i+1,j}^p - E_{ij} u_{ij}^p + \bar{A}_{ij} u_{i,j-1}^p + \bar{B}_{ij} u_{i,j+1}^p + F_{ij}^p \right). \tag{4.160}$$

The calculation for ions begins from the anode and from a symmetry axis, where boundary conditions have been fixed. Unlike an implicit method, the computing cycle ends with the use of boundary conditions. That is at the beginning, zero approximations of desired functions are used for numerical solution. Then with engaging conditions (4.114)–(4.117) the boundary values of functions are defined.

4.3.7 An iterative algorithm for the solution of a self-consistent problem

Determination of self-consistent distributions of charged particles and electric field potential is not a less important problem than the construction of a finite-difference scheme and choice of numerical methods for the solution of finite-difference equations. Basically, it is possible to use various iterative schemes, beginning from completely explicit and finishing with completely implicit. Completely explicit scheme of the iterative process is understood as the simple sequence of a solution of the Poisson equation, and continuity equations for ions and electrons. As a rule, this scheme appears overuneconomical, due to stability conditions similar to (4.158).

Improving the iterative process efficiency can be achieved by the introduction of implicitness in the algorithm. Such an algorithm is named as "explicit–implicit approximation – simple iteration." The computing cycle on the $(p+1)$ time layer begins with the determination of the electric potential field by integrating the Poisson equation. This field would not be recalculated any more and is assigned to $(p+1)$th time layer. So, in the given computing scheme the field is defined by an *explicit mode*. The *implicit part* of the iterative process consists of a sequential improvement of concentration of ions and electrons at the fixed velocity fields of drift motion (they are defined by the fixed electric field). During these so-called internal iterations, the weighting scheme for the calculation of concentrations of ions and electrons is used

$$n_e^{S+1} = n_e^S + w\left(n_e^{S+1/2} - n_e^S\right),$$
$$n_+^{S+1} = n_+^S + w\left(n_+^{S+1/2} - n_+^S\right), \tag{4.161}$$

where concentrations $n_e^{S+1/2}$ and $n_+^{S+1/2}$ are the results of numerical integration of the finite-difference equations. The weight w is defined as follows. First, the maximum relative error is determined for electron concentrations on the symmetry axis during two sequential iterations

$$\varepsilon_{\max}^e = \max_j \left\{ \frac{\left|\left(n_e^S\right)_{1,j} - \left(n_e^{S+1}\right)_{1,j}\right|}{\left(n_e^{S+1}\right)_{1,j}} \right\}. \tag{4.162}$$

Then a ratio to maximum admissible error ε_w is calculated, which is introduced as some a priori parameter

$$w_0 = \frac{\varepsilon_w}{\varepsilon_{\max}^e}. \tag{4.163}$$

The iterative weight w is selected by the following condition:

$$w = \begin{cases} 1, & w_0 \geq 1, \\ w_0, & w_0 \leq 1. \end{cases} \tag{4.164}$$

Numerous calculations have shown that the basic shortage of this scheme is the explicit mode of representation of $\alpha(E)$ function. Besides, the explicitness of such an algorithm is in the calculation method of the source summand

$$\alpha \Gamma_e^S - \beta n_+^S n_e^S.$$

Attempt to use the quasi-linearization method for calculation of this term does not bring tangible effect.

The completely implicit iterative scheme appears to be the most economic. The basic difference of this computing scheme from the previous one is the inclusion of electric potential calculation in the united iterative process. As a result, an improvement of source terms happens also in the internal iterations.

At first sight, the solution of the Poisson equation on each internal iteration is very labor-consuming. However, such an iterative process leads to a considerable increase in the integration time step. In this computing scheme, the application of a weight principle for correction of concentration fields of ions and electrons is also convenient. The greatest effectiveness is reached on the first iterations.

4.3.8 Specific properties of a solution of 2D problem about glow discharge in nonstationary statement

The stationary solution of the glow discharge problem is determined, as a rule, by the time-asymptotic method. The unique difference from a truly nonstationary problem consists in the use of relation (4.43) for an external circuit instead of real nonstationary boundary conditions.

But at the unsteady forming of glow discharge, the accumulation of charges on electrodes can render appreciable influence on a current of glow discharge, especially for fast electrodynamic processes.

Let us consider the Maxwell equation

$$\mathrm{rot}\left(\frac{1}{\mu_0}\mathbf{B}\right) = \frac{4\pi}{c}\mathbf{j} - \frac{\varepsilon_0}{c}\nabla\frac{\partial\varphi}{\partial t}, \tag{4.165}$$

where $\mathbf{B}$ is the magnetic induction vector; ε_0, μ_0 are the dielectric and magnetic permeabilities. Taking into account obvious relation

$$\mathrm{div}\,\mathrm{rot}\left(\frac{1}{\mu_0}\mathbf{B}\right) = 0,$$

and having applied the operation div to eq. (4.165), one can receive

$$4\pi\,\mathrm{div}\,\mathbf{j} = \varepsilon_0\,\mathrm{div}\,\mathrm{grad}\frac{\partial\varphi}{\partial t}$$

or

$$\mathrm{div}\left(4\pi\mathbf{j} - \varepsilon_0\frac{\partial}{\partial t}\mathrm{grad}\,\varphi\right) = 0. \tag{4.166}$$

Further, we will use the Gauss theorem

$$\int\limits_{\Sigma}\mathbf{nj}\,\mathrm{d}s - \int\limits_{\Sigma}\frac{\varepsilon_0}{4\pi}\mathbf{n}\,\mathrm{grad}\frac{\partial\varphi}{\partial t}\,\mathrm{d}s = 0. \tag{4.167}$$

The surface integrals close to an infinitely thin electrode (Figure 4.6) can be rewritten as

$$\int\limits_{S_1} \mathbf{n}\,\mathbf{j}\,\mathrm{ds} - \int\limits_{S_1} \frac{\varepsilon_0}{4\pi}\mathbf{n}\nabla\frac{\partial\varphi}{\partial t}\,\mathrm{ds} + \int\limits_{S_3} \mathbf{n}\,\mathbf{j}\,\mathrm{ds} - \int\limits_{S_3} \frac{\varepsilon_0}{4\pi}\mathbf{n}\nabla\frac{\partial\varphi}{\partial t}\,\mathrm{ds} = 0.$$

Assuming constant electric potential of the conductor, one can receive a total current

$$I = \int\limits_{S_3} \mathbf{n}\mathbf{j}\mathrm{ds} = \int\limits_{S_1} \frac{\varepsilon_0}{4\pi}\mathbf{n}\nabla\frac{\partial\varphi}{\partial t}\,\mathrm{ds} - \int\limits_{S_1} \mathbf{n}\mathbf{j}\mathrm{ds}.$$

But near the electrode $\partial\varphi/\partial r = 0$; therefore,

$$I = \int\limits_{S_1} \frac{\varepsilon_0}{4\pi}\frac{\partial}{\partial t}\frac{\partial\varphi}{\partial x}\,\mathrm{ds} - \int\limits_{S_1} \mathbf{n}\mathbf{j}\mathrm{ds}.$$

Considering the presence of an external circuit, we obtain

$$\frac{\partial Q}{\partial t} = \frac{E - V(t)}{eR_0} - \int\limits_{S_1} \mathbf{n}\mathbf{j}\mathrm{ds}, \tag{4.168}$$

where $V(t)$ is the voltage drop on the gas discharge gap; R_0 is the ohmic resistance; $\varepsilon_0 \approx 1$; Q is the charge accumulated on an electrode

$$Q = \frac{1}{4\pi e}\int\limits_{s1} \frac{\partial\varphi}{\partial x}\,\mathrm{d}\,s.$$

Equation (4.168) reflects the fact that the accumulation of charges on the electrode happens because there is a difference between velocities of their outflow in the external circuit $((\mathrm{E} - V(t))/R_0)$ and inflow from the discharge gap.

If we normalize eq. (4.168) as done earlier, we obtain

$$\frac{\mathrm{d}\tilde{Q}}{\mathrm{d}\tau} = \frac{H}{\mu_{e,0}W\tilde{R}_0}(\mathrm{E} - V) - HR^{m+1}\int\limits_0^1 \frac{\partial\Phi}{\partial x}N_e r^m \mathrm{d}\,r, \tag{4.169}$$

$$\tilde{Q} = \frac{WR^{m+1}}{N_0\varepsilon^*}\int\limits_0^1 \frac{\partial\Phi}{\partial x} r^m \mathrm{d}\,r, \tag{4.169}$$

where

$$\tilde{Q} = \frac{Q}{(2\pi)^m N_0}, \quad \tilde{R}_0 = 1.6\times 10^{-19}(2\pi)^m N_0 R_0, \quad \varepsilon^* = 4\pi e.$$

Calculation of $\tilde{Q}$ under formula (4.169) with its subsequent use in (4.169) for determination of potential V is not optimum. Insignificant errors in the definition of $\partial\Phi/\partial x$ can lead to an appreciable modification of voltage drop V, which leads, in its turn, to increasing numerical instabilities.

The following algorithm is recommended for numerical solution. If we assume that the pth time layer functions V^p, $\tilde{Q}^p$ and I^p are known, the desired solution on the $(p+1)$ th layer starts with a definition of concentration fields of ions and electrons. Then with the use of (4.169), the new value of $\tilde{Q}$ is calculated (while the iterative value, but not on the $(p+1)$ th layer)

$$\tilde{Q}^{S+1} = \tilde{Q}^p + \tau \left[\frac{H}{\mu_e W \tilde{R}_0} \left(\varepsilon - V^S \right) - HR^{m+1} \int_0^1 \frac{\partial \Phi}{\partial x} N_e r^m \mathrm{d}r \right]. \tag{4.170}$$

Now it is required to define modification of electric potential on electrodes due to received modification of charge density $(\tilde{Q}^{S+1} - \tilde{Q}^p)$. For this purpose, eq. (4.169) is used. However, from this equation, it is impossible to receive V^{p+1} in an explicit form; therefore, it is necessary to use one more internal iterative process. Setting some trial value of potential V_t the corresponding charge $\tilde{Q}$ is determined under formula (4.169). Then the estimation of potential V^{S+1} can be received with the use of linear interpolation between V^p and V_t

$$V^{S+1} = V^p + (V_t - V^p) \frac{\tilde{Q}^{S+1} - \tilde{Q}^p}{\tilde{Q}_t - \tilde{Q}^p}. \tag{4.171}$$

Now it is possible to assume $V_t = V^{S+1}$ and to repeat the process the necessary number of times up to convergence of value V. However, the desired solution of governing equations on the $(p+1)$th time step does not come to an end, because the self-consistent solution of charged particles continuity and the Poisson equations is not received yet. The iterative process should proceed until convergence of values Q^{S+1}, V^{S+1} is reached.

The iterative algorithm should contain also the block of automatic control of numerical time step. If convergence of the iterative process is not reached with $\mathrm{N_{ITER}}$ iterations ($\mathrm{N_{ITER}}$ is the initially defined number of internal iteration, e.g., $\mathrm{N_{ITER}} = 10$), the time step decreases twice, and the solution on the pth step has been chosen again as zero approximations. From another side, the automatic increasing of calculation time step can be performed when the number of iterations does not surpass some previously defined number, say 3.

4.4 Numerical simulation of the 1D glow discharge

Below the 1D numerical model of glow discharge, a nonstationary statement will be considered. Advantage of such approach is the possibility of deriving both truly nonstationary discharge and corresponding nonstationary solutions. The choice of the nonstationary variant of the glow discharge electrodynamic equations is also connected with the desire to achieve methodical unity with a technique of the 2D problem solution.

4.4.1 Governing equations and boundary conditions

Governing equations in the 1D case follow from the 2D formulation given in Section 4.2.2 and have the following form:

$$\frac{\partial u}{\partial \tau}+\frac{\partial}{\partial x}(P_b b u)+D_b\frac{\partial^2 u}{\partial x^2}=f, \tag{4.172}$$

$$f=\alpha H\sqrt{\left(N_e\frac{\partial \Phi}{\partial x}-\tilde{D}_{ex}\frac{\partial N_e}{\partial x}\right)^2}-\beta Q N_e N_+, \tag{4.173}$$

$$\frac{\partial^2 \Phi}{\partial x^2}=\varepsilon^* H\frac{N_0}{W}(N_+-N_e), \tag{4.174}$$

where

$$P_b=\begin{cases}-\mu_+/\mu_e & \text{for ions,}\\ 1 & \text{for electrons;}\end{cases}\qquad D_b=\begin{cases}-\tilde{D}_+ & \text{for ions,}\\ -\tilde{D}_e & \text{for electrons;}\end{cases}\qquad b=\frac{\partial \Phi}{\partial x}.$$

Boundary conditions are formulated as follows:

For ions ($u=N_+$)

$$x=1,\quad u\frac{\partial \Phi}{\partial x}=-\frac{\tilde{D}_+\mu_e}{2\mu_+}\frac{\partial u}{\partial x}, \tag{4.175}$$

$$x=0,\quad \frac{\partial u}{\partial x}=0; \tag{4.176}$$

For electrons ($u=N_e$)

$$x=0,\quad u\frac{\partial \Phi}{\partial x}=-\frac{1}{2}\frac{\partial u}{\partial x}+\gamma N_+\frac{\mu_+}{\mu_e}\frac{\partial \Phi}{\partial x}, \tag{4.177}$$

$$x=1,\quad \frac{\partial u}{\partial x}=0; \tag{4.178}$$

For electric potential ($u=\Phi$)

$$x=0,\quad u=0,$$

$$x=1,\quad u=V/E.$$

An important point of the problem solution is set to the fixed value of voltage drop for a gas discharge gap (further this boundary condition will be repeatedly used also in 2D calculations).

Boundary conditions for ions and electrons are obtained from the following analysis. Let us present a summarized particle flux and partial fluxes of electrons and ions in the form of the sum of two flux components in the positive and negative directions of the x-axis (Figure 4.7):

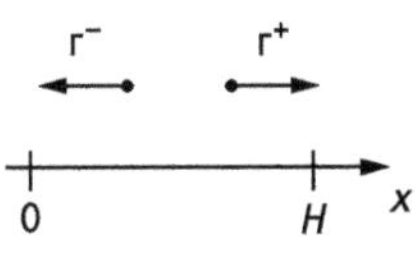

Figure 4.7: Two flux components.

$$\Gamma_\Sigma = \Gamma_\Sigma^+ - \Gamma_\Sigma^- , \tag{4.179}$$

$$\Gamma_\Sigma = \Gamma_+ - \Gamma_e, \quad \Gamma_+ = \Gamma_+^+ - \Gamma_+^-, \quad \Gamma_e = \Gamma_e^+ - \Gamma_e^- .$$

Now let us assume that on the cathode ions are not emitted in the gas discharge gap

$$x = 0, \quad \Gamma_+^+ = 0 \tag{4.180}$$

and the electron flux is connected to the flux of incident ions by the factor of a secondary electron-ionic emission γ:

$$\Gamma_e^+ = \gamma \Gamma_+^- . \tag{4.181}$$

Let us write the flux components in the following form:

$$\Gamma_e^+ = \mu_e n_e \frac{\partial \varphi}{\partial x} - \frac{1}{2} D_e \frac{\partial n_e}{\partial x} , \tag{4.182}$$

$$\Gamma_e^- = \frac{1}{2} D_e \frac{\partial n_e}{\partial x} , \tag{4.183}$$

$$\Gamma_+^+ = -\frac{1}{2} D_+ \frac{\partial n_+}{\partial x} , \tag{4.184}$$

$$\Gamma_+^- = \mu_+ n_+ \frac{\partial \varphi}{\partial x} + \frac{1}{2} D_+ \frac{\partial n_+}{\partial x} . \tag{4.185}$$

It is easy to be convinced that substituting (4.182) and (4.183) into the third condition (4.179), and relations (4.184) and (4.185) into the second relation, we will obtain usual expressions for the axial particle fluxes.

Boundary conditions (4.180) and (4.181) now can be formulated in terms of concentration and potential:

$$x = 0, \quad \frac{\partial n_+}{\partial x} = 0, \tag{4.186}$$

$$\mu_e n_e \frac{\partial \varphi}{\partial x} - \frac{1}{2} D_e \frac{\partial n_e}{\partial x} = \gamma n_+ \mu_+ \frac{\partial \varphi}{\partial x} + \gamma \frac{1}{2} D_+ \frac{\partial n_+}{\partial x}$$

or

$$\frac{\partial n_e}{\partial x} = \mu_e n_e \frac{2}{D_e} \frac{\partial \varphi}{\partial x} - \frac{2\gamma\mu_+}{D_e} n_+ \frac{\partial \varphi}{\partial x}. \tag{4.187}$$

On the anode, we set a condition of nonemission of particles

$$\Gamma_e^- = 0, \ \Gamma_+^- = 0. \tag{4.188}$$

Using (4.183) and (4.185), we obtain

$$\frac{\partial n_e}{\partial x} = 0, \tag{4.189}$$

$$\frac{\partial n_+}{\partial x} = -\frac{2\mu_+}{D_+} n_+ \frac{\partial \varphi}{\partial x}. \tag{4.190}$$

Finally, using the usual procedure of normalization, we obtain conditions (4.175)–(4.178).

4.4.2 The elementary implicit finite-difference scheme

By analogy with the 2D case, sequential application of the finite-volume method and the donor cells approximation for eq. (4.172) leads to the following finite-difference scheme:

$$\bar{A}_j u_{j-1} + \bar{B}_j u_{j+1} - C_j u_j + F_j = 0, \tag{4.191}$$

where

$$\bar{A}_j = \frac{(P_b b)_L^+}{h_j} - \bar{A}_j^*; \ \bar{B}_j = -\frac{(P_b b)_R^-}{h_j} - \bar{B}_j^*;$$

$$C_j = \frac{1}{h_j}\left[(P_b b)_R^+ - (P_b b)_L^-\right] + \frac{1}{\tau} - C_j^*; \ C_j^* = A_j^* + B_j^*;$$

$$A_j^* = D_b A_x \frac{1}{x_j - x_{j-1}}; \quad B_j^* = D_b A_x \frac{1}{x_{j+1} - x_j};$$

$$A_x = \frac{2}{x_{j+1} - x_{j-1}} = \frac{1}{h_j}; \ h_j = \frac{1}{2}(x_{j+1} - x_{j-1}).$$

With the purpose of preservation of methodological generality with a solution of the 2D problem, three-point finite-difference equations (4.191) are solved by the Thomas algorithm. For this purpose, it is necessary to formulate boundary conditions in the standard form

$$u_1 = \alpha_1 u_2 + \beta_1, \tag{4.192}$$

$$u_{NJ} = \omega u_{NJ-1} + \delta; \tag{4.193}$$

Then, for ions

$$x = 0, \qquad \alpha_1 = 1, \qquad \beta_1 = 0; \tag{4.194}$$

$$x = 1, \quad \delta = 0, \quad \omega = \frac{D_b}{2P_b h_j}\left[\left(\frac{\partial \Phi}{\partial x}\right)_{NJ} + \frac{D_b}{2P_b h_j}\right]^{-1}, \tag{4.195}$$

$$h_j = x_{NJ} - x_{NJ-1};$$

and for electrons

$$x = 0, \quad \alpha_1 = -\frac{D_b}{2h_1}\left[\left(\frac{\partial \Phi}{\partial x}\right)_1 - \frac{D_b}{2h_1}\right]^{-1}, \tag{4.196}$$

$$\beta_1 = \gamma \frac{\mu_+}{\mu_e}(N_+)_1\left(\frac{\partial \Phi}{\partial x}\right)_1\left[\left(\frac{\partial \Phi}{\partial x}\right)_1 - \frac{D_b}{2h_1}\right]^{-1}, \quad h_1 = x_2 - x_1; \tag{4.197}$$

$$x = 1, \quad \omega = 1, \quad \delta = 0. \tag{4.198}$$

Similarly, the equation for electric potential is solved. Coefficients of the Thomas algorithm and boundary conditions look like

$$\bar{A}_j = \frac{A_x}{x_j - x_{j-1}}, \qquad \bar{B}_j = \frac{A_x}{x_{j+1} - x_j}; \tag{4.199}$$

$$C_j = \bar{A}_j + \bar{B}_j, \qquad A_x = \frac{2}{x_{j+1} - x_{j-1}};$$

$$x = 0, \quad \alpha_1 = 0, \quad \beta_1 = 0;$$

$$x = 1, \quad \omega = 0, \quad \delta = V/E.$$

Note that the given finite-difference scheme allows solving problems on the structure of capacitive high frequency of glow discharges. In these cases, the typical frequency of electric field oscillations is f~0.1–100 mHz.

4.5 Diffusion of charges along a current line and effective method of grid diffusion elimination at calculations of glow discharges

Physical diffusion of charges in the direction of drift motion can appear in areas of strong longitudinal heterogeneities, in particular near the electrodes in the glow and radiofrequency capacitive discharges. Its inclusion into governing equations demands a statement of additional boundary conditions. Usually, it is done without additional physical substantiation. The account of longitudinal diffusion often appears illusory because of large grid diffusion, which is present in numerical schemes.

The sequential derivation of boundary conditions is given below, and by the example of 1D glow discharge, influence of physical and grid diffusion is shown. Diffusion effects are inspected by means of a specially developed and complicated scheme of the fourth accuracy order, where grid diffusion is actually excluded. The effective method of elimination of grid diffusion errors within the limits of standard Thomas schemes of the first and second accuracy orders is offered.

4.5.1 Governing equations for 1D case

The glow discharge is described by the standard set of equations of the drift-diffusion model for electron and ion densities n_e, n_i and for electric field E

$$\frac{\partial n_e}{\partial t}+\frac{\partial \Gamma_e}{\partial x}=\alpha\Gamma_e-\beta n_e n_+\,,\ \Gamma_e=-n_e\mu_e E-D_e\frac{\partial n_e}{\partial x}, \tag{4.200}$$

$$\frac{\partial n_+}{\partial t}+\frac{\partial \Gamma_+}{\partial x}=\alpha\Gamma_e-\beta n_e n_+\,,\ \Gamma_+=+n_+\mu_+ E-D_+\frac{\partial n_+}{\partial x}, \tag{4.201}$$

$$\frac{\partial E}{\partial x}=4\pi e(n_+-n_e),\ E=-\frac{\partial \varphi}{\partial x}, \tag{4.202}$$

where Γ are the fluxes of charges; μ is their mobility; D is the coefficients of physical diffusion; $\alpha(E)$ is the Townsend ionization coefficient; β is the coefficient of recombination; φ is the electric potential. On the cathode $(x=0)$ $\varphi=0$ and on the anode $(x=L)$ $\varphi=V$. Without taking into account diffusion $(D_{e,+}=0)$ on the cathode, it is supposed that $\Gamma_e=-\gamma\Gamma_+$, where γ is the coefficient of the secondary ion–electron emission; on the anode, it is supposed that $\Gamma_+=0$[1].

4.5.2 Boundary conditions

Expression (4.200) for Γ_e is based on the Lorentz representation of electron velocity **v** distribution

$$f(\mathbf{v}) = f_0(v) + f_1(v)\cos\vartheta,$$

where ϑ is the angle between $\mathbf{v}$ and x-axis (Raizer Yu.P. and Surzhikov S.T., 1987).

One-sided fluxes on electrodes Γ_e^+ and Γ_e^- in positive and negative directions of the x-axis ($\Gamma_e \equiv \Gamma_e^+ - \Gamma_e^-$) submit to physical real conditions. By integrating $f(\mathbf{v})$ in the limits of corresponding hemispheres, we can see

$$\Gamma_e^{\pm} = \iint v\cos\vartheta(f_0 + f_1\cos\vartheta)v^2 dv\, 2\pi\sin\vartheta\, d\vartheta =$$

$$= \frac{n_e\bar{v}_e}{4} \pm \frac{\Gamma_e}{2} = \frac{n_e\bar{v}_e}{4} \pm \frac{1}{2}\left(-\mu_e n_e E - D_e\frac{\partial n_e}{\partial x}\right), \quad E < 0. \tag{4.203}$$

The average velocity $\bar{v}_e$ is defined by temperature T_e; $D_e = \mu_e T_e = l\bar{v}_e/3$, where l is the free path of electrons.

If cathode does not absorb electrons but reflects and emits them, then $\Gamma_e^+ = \Gamma_e^- - \gamma\Gamma_i$. The condition $\Gamma_e = -\gamma\Gamma_i$ is valid. If the anode does not reflect electrons, then

$$\Gamma_e^- = 0, \quad \frac{l}{n_e}\frac{\partial n_e}{\partial x} = -\frac{3}{2} - \frac{El}{T_e}. \tag{4.204}$$

Drift velocity at the anode $v_{ed} = \mu_e|E| \ll \bar{v}_e$; therefore, $|E|l \ll T_e$. The function $n_e(x)$, being linearly continued for absorption surface of the anode, gets converted to zero on the so-called extrapolated length $l_{\text{ex}} = (2/3)l$ (in more strict theory $l_{\text{ex}} = 0.71\,l$).

The anode layer is many times larger than l, and it is assumed that error in calculation $n_e(x)$ will not be large, if we suppose that $l_{\text{ex}} = 0$. Then on the anode $n_e = 0$. Because of smallness of ionic temperature T_+ we have $v_{+d} \geq \bar{v}_+$. Therefore, there is no sense to consider longitudinal diffusion of ions in general. However, the condition $\Gamma_+(L) = 0$ is kept even, as the anode reflects ions; on the cathode $\Gamma_+^+ = 0$.

4.5.3 Numerical methods for the 1D calculation case

We consider the numerical simulation results obtained with two finite-difference methods, ensuring the first and the fourth order of accuracy. Completely implicit iterative schemes will be used. In the scheme of the first order (further refers to "the scheme 1"), particle concentrations are calculated in integer points of a computational mesh, and streams are calculated in half points.

Calculations for the fourth order of spatial accuracy are made with the use of the Petukhov's decomposition method ("the scheme 4"). In the frame of the method, eqs. (4.200) and (4.201) are formed in common form accepted for partial differential equations of the first order on t and of the third on x:

$$\left(m_0 + \sum_{i=1}^{3} u_i m_i\right)' = k_0 + \sum_{i=1}^{3} u_i k_i - p_2 \dot{u}_2 - p_3 \dot{u}_3, \tag{4.205}$$

where "prime" represents derivatives on spatial variable and "point" represents the time derivative. With reference to (4.200) and (4.201)

$$u_1 = \int_0^x n(x')\mathrm{d}x',\ u_2 = n,\ u_3 = u'_2;\ m_0, m_3, k_1, k_2, k_3 = 0;$$

$$m_1 = -D,\ m_2 = -a\varphi',\ k_0 = -F,\ p_2 = 1,\ p_3 = 0. \tag{4.206}$$

Here values are reduced to a dimensionless form: densities are measured in terms of $n_0 = 10^9\ \mathrm{cm}^{-3}$, x in L, t in $L^2/\mu_e V$, φ in V. At the same time, $a = 1$ for electrons and $(a = -\mu_+/\mu_e)$ for ions, and F is the right-hand side of the continuity equations (4.200) and (4.201)

$$F = \alpha L\left|n_e\varphi' + D_e n'_e\right| - \beta b n_e n_i,\ b = L^2 n_0/\mu_e V, \tag{4.207}$$

where α and β are dimensional.

At the integration of (4.205), the second order of time approximation accuracy is ensured. A peculiarity of application of this method is the necessity of calculation of coefficients for integer and half points of a computational mesh. In integer points, function φ', that is E, is approximated by central differences, and in half points – under the formula of the first order of accuracy. Value n in half points can be calculated as a half-sum of nodal values. Calculation of F in integer and half points is performed in (4.207).

The stationary glow discharge condition is obtained by the solution of the non-stationary problem by the time-asymptotic method. In addition to that, it is possible to enter either V and then to calculate a current density j at $t \to \infty$, or to fix $j = j_0$ on one of electrodes and to search $V(t \to \infty)$. Note that $j(x, \infty) = j_0$. The volt–ampere characteristic (VAC) of stationary discharge $V(j)$ has a minimum. Both modes give equally growing branch of VAC, but it is possible to receive falling one, if we set j only. In setting V, this branch never turns out, as it is unstable.

4.5.4 Results of 1D numerical simulation

Numerical simulation results will be analyzed for N_2 at $p = 5\ \mathrm{Torr}$, $L = 0.4\ \mathrm{cm}$, $\gamma = 1/3$. Coefficients $\alpha, \beta, \mu_{e,i}$ were borrowed from Raizer Yu.P. and Surzhikov S.T. (1988); $T_e = 0.1–5.0\ \mathrm{eV}$, $T_i = 0.026\ \mathrm{eV}$. Figure 4.8 shows the discharge structure at current density $j = 2.49\ \mathrm{mA/sm^2}$ and $T_e = 1\ \mathrm{eV}$, which is close to normal. Calculations were made under schemes 1 and 4 with average steps of $h_1 = 0.014\ \mathrm{cm}$ and $h_4 = 0.002\ \mathrm{cm}$, respectively. At approximation of gas discharge finite-difference equations and boundary conditions in the first scheme, there is greater grid diffusion, which results in additional losses of electrons because of their return to the cathode. As a result, the electric field in cathode layer, cathode voltage drop V_c, and, as a corollary, voltage drop on electrodes V increase. It is required for amplification of ionization, which should complete the specified nonphysical losses (Figure 4.8).

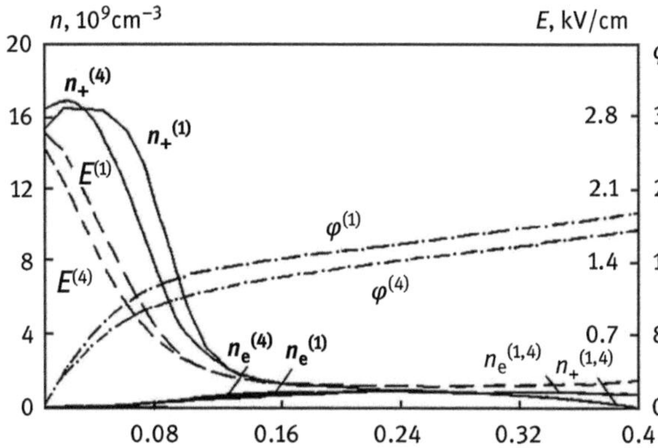

Figure 4.8: Distributions of charge densities n_e, n_+, field strength E, and potential φ, calculated under schemes of first and fourth orders (superscripts 1 and 4) (N_2, $L = 0.4$ cm, $V = 215$ V, $j = 2.5$ mA/cm^2, $T_e = 1$ eV).

As it follows from eq. (4.200) and boundary conditions, at the lack of recombination when charges disappear only on electrodes, the index of electron breeding is equal to

$$I = \int_0^L \alpha dx = \ln\left(\frac{1}{\gamma} + 1\right) = 1.39 \left(\text{for} \gamma = \frac{1}{3}\right), \tag{4.208}$$

regardless of physical diffusion, that is, the value of T_e. For compensating of the recombination losses also the breeding should be more significant. In the considered conditions, recombination losses are low, and parameter I, calculated under scheme 4, which excludes grid diffusion, differs really a little from 1.39: $I = 1.40–1.42$ at $T_e = 0.1–5.0$ eV. Small increase of I with growth of T_e, probably, is connected with redistribution of $n_{e,+}(x)$, that is, with modification of recombination and with errors of calculation. In calculations under the scheme 1, I is essentially larger ($I = 1.91$). As it was already marked, this phenomenon is connected with increasing electron breeding degree, which is caused by the necessity of compensating of their additional losses.

The overwhelming part of the integral $\int \alpha\, dx$ relates to the cathode layer. Its thickness d will be defined by the coordinate $x = d$ for which $\int_0^d \alpha\, dx$ makes 99 % from I, and a cathode drop will be defined by the condition $V_K = \varphi(d)$. VACs of the cathode layer $V_c(j)$ obtained by different schemes and different T_e are presented in Figure 4.9. It is evident as to how at the given j the value V_c steadily grows at increasing grid diffusion. For calculation under scheme 4, values V_c and d very poorly depend on T_e.

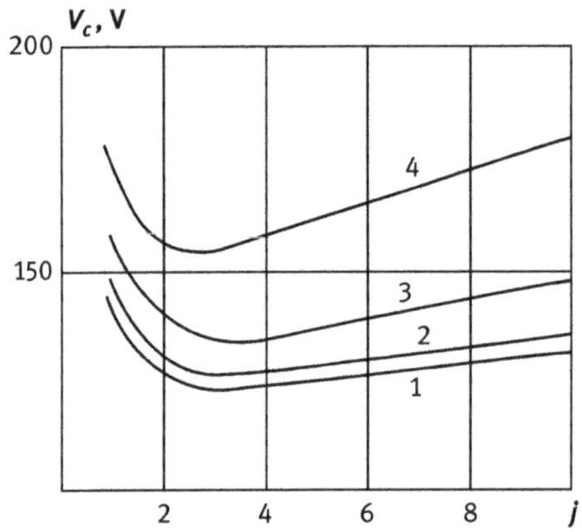

Figure 4.9: VAC in the cathode layer, derived from calculations under different schemes: 1, scheme 4, $T_e = 0.1$ eV, step $h = 0.002$ cm; 2, scheme 4, $T_e = 5$ eV, $h = 0.002$ cm; 3, scheme 1, $T_e = 1$ eV, step $h = 0.002$ cm; 4, scheme 1, $T_e = 1$ eV, step $h = 0.014$ cm.

Thus, physical diffusion in longitudinal direction renders insignificant influence on structure and parameters of glow discharge, and, apparently, it should not be taken into consideration in the problem. Much greater and rather appreciable influence is rendered with the grid diffusion, when a calculation is conducted under finite-difference schemes with grid diffusion, especially with rough meshes.

Use of schemes of the high order of accuracy (like scheme 4) is not always justified, as the accuracy is reached by large complexity and increase by the order of expenditures of computational time. However, much more simple and economic scheme 1 can be essentially improved by appreciably excluding most essential grid diffusion effects, namely losses of electrons due to their nonphysical returning to the cathode.

Let us consider numerical model of glow discharge without taking into account the longitudinal physical diffusion, as it was done in overwhelming majority of works. The boundary condition on the cathode $\Gamma_e = -\gamma \Gamma_+$ with purely drift fluxes immediately connects particle densities in calculation point $x = 0$ at the cathode $n_{e0} = \gamma(\mu_+/\mu_e)n_{+0}$. In the presence of diffusion, the boundary stream condition has the same appearance; only expression for flux Γ_e is redefined. It is offered to include grid diffusion electron flux (ionic diffusion flux is small) in a flux Γ_e in the same manner, as it would be made at the account of physical diffusion. In this case, the correct resulting balance of charges on the cathode is maintained.

In Raizer Yu.P. and Surzhikov S.T. (1988), coefficient of grid electron diffusion $D_e = 0.5\mu_e \Delta\varphi$, where $\Delta\varphi$ is the potential drop at grid step h. Note that with $D_e = 0.5\mu_e |E| h$ we can transfer from the differential boundary condition

$$-\mu_e n_e E - D_e \frac{\partial n_e}{\partial x} = -\gamma\mu_i n_i E, \ E = -|E| \tag{4.209}$$

to the corresponding finite difference condition

$$n_{e0} = \gamma \frac{\mu_i}{\mu_e} n_{i0} + \frac{1}{2}(n_{e1} - n_{e0}), \tag{4.210}$$

where n_{e0}, n_{i0} correspond to the calculation point $x = 0$ on the cathode, and n_{e1} corresponds to the next point $x = h$. Calculation under scheme 1 with a former step $h = 0.014$ cm, but with the advanced boundary condition (4.210) has given distributions of E, n_e, n_i, which are almost undistinguishable from the received on scheme 4. The integral of multiplication has decreased from $I = 1.91$ till 1.57 (against 1.42 in scheme 4). The remained discrepancy is partly connected with errors of the evaluation of integral from very sharp function $\alpha(E)$.

4.5.5 The method of the fourth order of accuracy for the solution of the drift-diffusion model equations

In the previous section, it was shown that the method of the fourth order of accuracy allows avoiding the grid diffusion effects at numerical simulation. As the specified method has appeared rather effective, below we will derive all equations of the method.

This method was developed for the solution of partial differential equations of parabolic type of the following form (Petukhov I.V., 1964):

$$\frac{dM}{d\eta} = K + \alpha P, \tag{4.211}$$

where

$$M = m_1 u_1 + m_2 u_2 + m_3 u_3 + m_4,$$

$$K = k_1 u_1 + k_2 u_2 + k_3 u_3 + k_4; \tag{4.212}$$

$$P = p_2 \frac{\partial u_2}{\partial \xi} + p_3 \frac{\partial u_3}{\partial \xi},$$

$$u_1 = \frac{\partial u_2}{\partial \eta} = u'_2, \ u_2 = \frac{\partial u_3}{\partial \eta} = u'_3, \tag{4.213}$$

where factors

$$m_i = m_i(\xi, \eta, u_1, u_2, u_3), \tag{4.214}$$

$$k_i = k_i(\xi, \eta, u_1, u_2, u_3),$$

$$p_i = p_i(\xi, \eta, u_1, u_2, u_3)$$

are the set functions of their arguments.

Equation (4.211) is solved in the calculation domain $0 \le \eta \le \eta_\delta$ and $\xi \ge 0$. The finite-difference mesh in directions η and ξ can be inhomogeneous. Here we will receive computing relations for a numerical integration of system (4.211) along a variable η (i.e., at $\alpha = 0$). Boundary conditions for the differential equation (4.211) are formulated in the following form:
at $\eta = 0$

$$\gamma_{11}u_1 + \gamma_{12}u_2 + \gamma_{13}u_3 + \gamma_{14} = 0, \tag{4.215}$$

$$\gamma_{21}u_1 + \gamma_{22}u_2 + \gamma_{23}u_3 + \gamma_{24} = 0; \tag{4.216}$$

at $\eta = \eta_\delta$

$$\gamma_{31}u_1 + \gamma_{32}u_2 + \gamma_{33}u_3 + \gamma_{34} = 0, \tag{4.217}$$

where factors γ_{ij} are known.

Connection between functions u_j^- on the left boundary of the segment η^- and functions u_j^+ on the right boundary of the segment $\eta^+ = \eta^- + 2\Delta\eta$ is set in the following form:

$$a_{i4}^+ + \sum_{j=1}^{3} a_{ij}^+ u_j^+ = a_{i4}^- + \sum_{j=1}^{3} a_{ij}^- u_j^-, \; i = 1, 2, 3, \tag{4.218}$$

where coefficients a_{i4}^+ and a_{i4}^- are subject to definition, $\Delta\eta$ is the half of range between the left and right boundaries of the segment.

To discover these coefficients, three additional conditions are used:

$$\int_{\eta^-}^{\eta^+} M'_\eta d\eta = \int_{2\Delta\eta} K d\eta, \tag{4.219}$$

$$\int_{2\Delta\eta} u_3 d\eta = (u_3^+ + u_3^-)\Delta\eta - (u_2^+ - u_2^-)\Delta\eta^2/3, \tag{4.220}$$

$$\int_{2\Delta\eta} u_2 d\eta = (u_2^+ + u_2^-)\Delta\eta - (u_1^+ - u_1^-)\Delta\eta^2/3. \tag{4.221}$$

Formulas (4.220) and (4.221) are Simpson's ones for the integration of function φ

$$\int\limits_{2\Delta\eta} \varphi d\eta = (\varphi^{+} + \varphi^{-})\Delta\eta - \left[\left(\frac{\partial\varphi}{\partial\eta}\right)^{+} - \left(\frac{\partial\varphi}{\partial\eta}\right)^{-}\right]\frac{\Delta\eta^2}{3} + \cdots. \tag{4.222}$$

For deriving the first connection between required coefficients we will take eq. (4.219) in the following form (the right term is calculated under the Simpson formula):

$$m_1^+ u_1^+ + m_2^+ u_2^+ + m_3^+ u_3^+ + m_4^+ - m_1^- u_1^- - m_2^- u_2^- - m_3^- u_3^- - m_4^- = \frac{\Delta\eta}{3}(K^+ + K^- + 4K_0), \tag{4.223}$$

where K_0 is calculated by values of functions u_j^0 in the center of the interval

$$K_0 = k_1^0 u_1^0 + k_2^0 u_2^0 + k_3^0 u_3^0 + k_4^0, \tag{4.224}$$

which, in turn, are expressed through values of functions on boundaries of the interval

$$\begin{aligned}
u_1^0 &= \frac{3}{4\Delta\eta}u_2^+ - \frac{3}{4\Delta\eta}u_2^- - \frac{1}{4}u_1^+ - \frac{1}{4}u_1^-, \\
u_2^0 &= \frac{3}{4\Delta\eta}u_3^+ - \frac{3}{4\Delta\eta}u_3^- - \frac{1}{4}u_2^+ - \frac{1}{4}u_2^-, \\
u_3^0 &= \frac{1}{2}u_3^+ + \frac{1}{2}u_3^- - \frac{\Delta\eta}{4}u_2^+ + \frac{\Delta\eta}{4}u_2^-.
\end{aligned} \tag{4.225}$$

As a result of reduction (4.223) to the form (4.218) we obtain

$$\begin{aligned}
a_{11}^+ &= m_1^+ - \frac{\Delta\eta}{3}(k_1^+ - k_1^0), \\
a_{12}^+ &= m_2^+ - \frac{\Delta\eta}{3}(k_2^+ - k_2^0) + \frac{\Delta\eta^2}{3}k_3^0 - k_1^0, \\
a_{13}^+ &= m_3^+ - \frac{\Delta\eta}{3}(k_3^+ + 2k_3^0) - k_2^0, \\
a_{14}^+ &= m_4^+ - \frac{\Delta\eta}{3}(k_4^+ + 2k_4^0)
\end{aligned} \tag{4.226}$$

and

$$\begin{aligned}
a_{11}^- &= m_1^- + \frac{\Delta\eta}{3}(k_1^- - k_1^0), \\
a_{12}^- &= m_2^- + \frac{\Delta\eta}{3}(k_2^- - k_2^0) + \frac{\Delta\eta^2}{3}k_3^0 - k_1^0,
\end{aligned}$$

$$a_{13}^{-} = m_3^{-} + \frac{\Delta\eta}{3}\left(k_3^{-} + 2k_3^{0}\right) - k_2^{0},$$

$$a_{14}^{-} = m_4^{-} + \frac{\Delta\eta}{3}\left(k_4^{-} + 2k_4^{0}\right). \tag{4.227}$$

For deriving the second connection between the coefficients we will take relation (4.220) with reference to function M:

$$\int_{2\Delta\eta} M\mathrm{d}\eta = (M^{+} + M^{-})\Delta\eta - \left(M_\eta^{+} - M_\eta^{-}\right)\Delta\eta^2/3. \tag{4.228}$$

but

$$M_\eta^{+} = K^{+},\ M_\eta^{-} = K^{-}, \tag{4.229}$$

and

$$\int_{2\Delta\eta} M\,\mathrm{d}\eta = \frac{\Delta\eta}{3}\left(M^{+} + M^{-} + 4M^{0}\right). \tag{4.230}$$

Therefore, substituting (4.229) and (4.230) into (4.228), we obtain

$$a_{21}^{+} = \frac{2}{3}\Delta\eta m_1^{+} - \frac{\Delta\eta^2}{3}k_1^{+} + \frac{\Delta\eta}{3}m_1^{0},$$

$$a_{22}^{+} = \frac{2}{3}\Delta\eta m_2^{+} - \frac{\Delta\eta^2}{3}k_2^{+} - m_1^{0} + \frac{\Delta\eta}{3}m_2^{0} + \frac{\Delta\eta^2}{3}m_3^{0},$$

$$a_{23}^{+} = \frac{2}{3}\Delta\eta m_3^{+} - \frac{\Delta\eta^2}{3}k_3^{+} - m_2^{0} - \frac{2}{3}\Delta\eta m_3^{0},$$

$$a_{24}^{+} = \frac{2}{3}\Delta\eta m_4^{+} - \frac{\Delta\eta^2}{3}k_4^{+} - \frac{2}{3}\Delta\eta m_4^{0} \tag{4.231}$$

and

$$a_{21}^{-} = -\frac{2}{3}\Delta\eta m_1^{-} - \frac{\Delta\eta^2}{3}k_1^{-} - \frac{\Delta\eta}{3}m_1^{0},$$

$$a_{22}^{-} = -\frac{2}{3}\Delta\eta m_2^{-} - \frac{\Delta\eta^2}{3}k_2^{-} - m_1^{0} - \frac{\Delta\eta}{3}m_2^{0} + \frac{\Delta\eta^2}{3}m_3^{0},$$

$$a_{23}^{-} = -\frac{2}{3}\Delta\eta m_3^{-} - \frac{\Delta\eta^2}{3}k_3^{-} - m_2^{0} + \frac{2}{3}\Delta\eta m_3^{0},$$

$$a_{24}^{-} = -\frac{2}{3}\Delta\eta m_4^{-} - \frac{\Delta\eta^2}{3}k_4^{-} + \frac{2}{3}\Delta\eta m_4^{0}. \tag{4.232}$$

The third connection between coefficients $a_{ji}^{\pm}$ is determined directly under formula (4.221):

$$a_{31}^{+}=\frac{\Delta\eta^2}{3},\ a_{32}^{+}=-\Delta\eta,\ a_{33}^{+}=1,\ a_{30}^{+}=0\,; \tag{4.233}$$

$$a_{31}^{-}=\frac{\Delta\eta^2}{3},\ a_{32}^{-}=-\Delta\eta,\ a_{33}^{-}=1,\ a_{30}^{-}=0\,. \tag{4.234}$$

Thus, in ratio (4.218), all coefficients $a_{ij}^{\pm}$ are defined under formulas (4.226), (4.227), and (4.231)–(4.234).

Now it is necessary to relate the specified coefficients with coefficients γ_{ij} in formulas for boundary conditions (4.215)–(4.217) and to receive final correlations for calculation. For boundary $\eta=0$ instead of (4.215) and (4.216) we shall write

$$\gamma_{14}+\sum_{j=1}^{3}\gamma_{1j}u_j^{-}=0\,, \tag{4.235}$$

$$\gamma_{24}+\sum_{j=1}^{3}\gamma_{2j}u_j^{-}=0\,, \tag{4.236}$$

and for boundary $\eta=\eta_\delta$ instead of (4.217) we will write

$$\gamma_{34}+\sum_{j=1}^{3}\gamma_{3j}u_j^{+}=0\,. \tag{4.237}$$

Let us assume that in each fixed point of the computational mesh, for example η, functions u_j^{-} satisfy to two equations

$$\lambda_{14}^{-}+\sum_{j=1}^{3}\lambda_{1j}^{-}u_j^{-}=0,\quad \lambda_{24}^{-}+\sum_{j=1}^{3}\lambda_{2j}^{-}u_j^{-}=0, \tag{4.238}$$

where coefficients λ_{1i}^{-} and λ_{2i}^{-} $(i=1,2,3,4)$ coincide with γ_{1i} and γ_{2i} on boundary $\eta=0$ and are subject to definition inside the calculation area. Together with (4.238) we will consider relations (4.218) and (4.234)

$$A_3^{+}=a_{31}^{+}u_1^{+}+a_{32}^{+}u_2^{+}+a_{33}^{+}u_3^{+}+a_{34}^{+}=\frac{\Delta\eta^2}{3}u_1^{-}+\Delta\eta u_2^{-}+u_3^{-}\,. \tag{4.239}$$

Whence we receive

$$u_3^{-}=A_3^{+}-\frac{\Delta\eta^2}{3}u_1^{-}-\Delta\eta u_2^{-}\,. \tag{4.240}$$

Substituting (4.240) into (4.238), we obtain

$$\left(\lambda_{i4}^{-}+\lambda_{i3}^{-}A_{3}^{+}\right)+\tilde{\lambda}_{i1}u_{1}^{-}+\tilde{\lambda}_{i2}u_{2}^{-}=0,\quad i=1,2, \tag{4.241}$$

where

$$\tilde{\lambda}_{i1}=\lambda_{i1}^{-}-\lambda_{i3}^{-}a_{31}^{-}, \tag{4.242}$$

$$\tilde{\lambda}_{i2}=\lambda_{i2}^{-}-\lambda_{i3}^{-}a_{32}^{-}. \tag{4.243}$$

The set of eq. (4.241) can be resolved concerning functions u_i^- in the following form:

$$u_i^{-}=\mu_{i4}+\mu_{i3}A_3^{+},\quad i=1,2, \tag{4.244}$$

where

$$\mu_{14}=\frac{\tilde{\lambda}_{12}\lambda_{24}^{-}-\tilde{\lambda}_{22}\lambda_{14}^{-}}{\text{Det}},\quad \mu_{13}=\frac{\tilde{\lambda}_{12}\lambda_{23}^{-}-\tilde{\lambda}_{22}\lambda_{13}^{-}}{\text{Det}},$$

$$\mu_{24}=\frac{\tilde{\lambda}_{21}\lambda_{14}^{-}-\tilde{\lambda}_{11}\lambda_{24}^{-}}{\text{Det}},\quad \mu_{23}=\frac{\tilde{\lambda}_{21}\lambda_{13}^{-}-\tilde{\lambda}_{11}\lambda_{23}^{-}}{\text{Det}};$$

$$\text{Det}=\tilde{\lambda}_{11}\tilde{\lambda}_{22}-\tilde{\lambda}_{12}\tilde{\lambda}_{21}. \tag{4.245}$$

Thus, the use of formulas (4.240) and (4.244) allows to calculate functions u_1^-, u_2^-, and u_3^-, if function A_3^+ is known.

Now we will consider the set of eq. (4.218) for $i=1,2$. Assuming that in each point of computational mesh $(\eta+2\Delta\eta)$ required functions satisfy to relations

$$\lambda_{14}^{+}+\sum_{j=1}^{3}\lambda_{1j}^{+}u_j^{+}=0,$$

$$\lambda_{24}^{+}+\sum_{j=1}^{3}\lambda_{2j}^{+}u_j^{+}=0, \tag{4.246}$$

where coefficients λ_{1i}^{+} and λ_{2i}^{+} $(i=1,2,3,4)$ are subject to definition in view of the boundary condition at $\eta=\eta_\delta$.

Let us substitute in (4.218) the function discovered above

$$u_1^{-}=\mu_{14}+\mu_{13}A_3^{+}, \tag{4.247}$$

$$u_2^{-}=\mu_{24}+\mu_{23}A_3^{+}, \tag{4.248}$$

$$u_3^{-}=\mu_{34}+\mu_{33}A_3^{+}, \tag{4.249}$$

where

$$\mu_{33}=1-\mu_{13}a_{31}^{-}-\mu_{23}a_{32}^{-},$$

$$\mu_{34} = -\mu_{14}a_{31}^{-} - \mu_{24}a_{32}^{-}.$$

Let us exclude u_i^- $(i = 1, 2, 3)$ from (4.218) at $i = 1, 2$, for this purpose we will rewrite (4.218) in the form of

$$A_i^+ = a_{i4}^+ + \sum_{j=1}^{3} a_{ij}^+ u_j^+ = a_{i4}^- + \sum_{j=1}^{3} a_{ij}^- \left(\mu_{j4} + \mu_{j3}A_3^+\right), \quad i = 1, 2 \tag{4.250}$$

or

$$A_i^+ = a_{i4}^- + \left(a_{i1}^-\mu_{14} + a_{i2}^-\mu_{24} + a_{i3}^-\mu_{34}\right) + \left(a_{i1}^-\mu_{13} + a_{i2}^-\mu_{23} + a_{i3}^-\mu_{33}\right)A_3^+, \ i = 1, 2. \tag{4.251}$$

Expressions in parentheses can be written in the form of

$$\tilde{\lambda}_{ij}^- = a_{i1}^-\mu_{1j} + a_{i2}^-\mu_{2j} + a_{i3}^-\mu_{3j}, \tag{4.252}$$

where $i = 1, 2$, $j = 3, 4$, so (4.251) gains more compact form

$$A_i^+ = a_{i4}^- + \tilde{\lambda}_{i4}^- + \tilde{\lambda}_{i3}^- A_3^+, \quad i = 1, 2. \tag{4.253}$$

Let us now pay attention to structure of formulas (4.253). Functions A_i^+ $(i = 1,2,3)$, entering in (4.253), contain coefficients a_{ij}^+ and functions u_j^+ (see formula (4.250)). At their substitution in (4.253), the system of two equations of form (4.246) will turn out, whence factors λ_{1i}^+ and λ_{2i}^+ $(i = 1, 2, 3, 4)$ will be ascertained

$$\begin{aligned}
\lambda_{11}^+ &= \tilde{\lambda}_{13}^- a_{31}^+ - a_{11}^+, \\
\lambda_{12}^+ &= \tilde{\lambda}_{13}^- a_{32}^+ - a_{12}^+, \\
\lambda_{13}^+ &= \tilde{\lambda}_{13}^- a_{33}^+ - a_{13}^+, \\
\lambda_{14}^+ &= \left(a_{14}^- - a_{14}^+\right) + \tilde{\lambda}_{14}^-;
\end{aligned} \tag{4.254}$$

$$\begin{aligned}
\lambda_{21}^+ &= \tilde{\lambda}_{23}^- a_{31}^+ - a_{21}^+, \\
\lambda_{22}^+ &= \tilde{\lambda}_{23}^- a_{32}^+ - a_{22}^+, \\
\lambda_{23}^+ &= \tilde{\lambda}_{23}^- a_{33}^+ - a_{23}^+, \\
\lambda_{24}^+ &= \left(a_{24}^- - a_{24}^+\right) + \tilde{\lambda}_{24}^-,
\end{aligned} \tag{4.255}$$

where $a_{34}^+ = 0$ is taken into account.

Now it is possible to formulate the calculation algorithm.

Step 1. In a cycle $j = 2, 3, \ldots, NJ$ auxiliary coefficients are calculated sequentially.

A. Calculation of a_{ij}^+ and a_{ij}^- under formulas (4.226), (4.227), (4.231), (4.232), (4.233), and (4.234).

B. Calculation of λ_{1j}^-, λ_{2j}^- $(j=1,2,3,4)$. At $\eta=0$ (on the left boundary of a segment), these coefficients coincide with coefficients of boundary conditions (4.215) and (4.216)

$$\lambda_{1j}^- = \gamma_{1j}, \quad \lambda_{2j}^- = \gamma_{2j}, \quad j=1,2,3,4.$$

On the right boundary of elementary segment $(\eta+2\Delta\eta)$, the specified coefficients are defined after realization of the following two stages (these factors are equated to λ_{1j}^+, λ_{2j}^+).

C. Calculation of $\mu_{13}, \mu_{14}, \mu_{23}, \mu_{24}$ under formulas (4.245). At the same time, it is reasonable to remember the values of the specified factors in each point of a computational mesh.

D. Calculation of λ_{1j}^+, λ_{2j}^+ $(j=1,2,3,4)$ under formulas (4.254) and (4.255).

Step 2. At reaching the rated point $j=NJ$ the set of equations is solved

$$\begin{aligned}
\lambda_{11}^+ u_1^+ + \lambda_{12}^+ u_2^+ + \lambda_{13}^+ u_3^+ &= -\lambda_{14}^+,\\
\lambda_{21}^+ u_1^+ + \lambda_{22}^+ u_2^+ + \lambda_{23}^+ u_3^+ &= -\lambda_{24}^+,\\
\lambda_{31}^+ u_1^+ + \lambda_{32}^+ u_2^+ + \lambda_{33}^+ u_3^+ &= -\lambda_{34}^+,
\end{aligned}$$

whence unknown functions are ascertained on an upper bound of the rated area $\eta=\eta_\delta$.

Step 3. In a cycle $j=NJ-1, NJ-2, \ldots, 1$, the following functions are calculated sequentially:

A. Functions A_3^+ under formula (4.239) (first half of equality);

B. Unknown functions u_i^+ $(j=1,2,3)$ under formulas (4.247)–(4.249).

In conclusion, it should be noted that the given method ensures accuracy of calculation of required functions proportional to $\Delta\eta^4$.

4.6 The 2D structure of glow discharge in view of neutral gas heating

The significant role in the development of our understanding of the nature of glow discharges in gas flows plays numerical modeling of these discharges in various conditions representing practical interest.

After the first works dedicated to calculations of glow discharges (Ward A.L., 1958; 1962), significant progress has been reached at the beginning of 1980s in the field of numerical model of their 2D structure (Graves D.B., et al., 1986), and the models developed afterward (Raizer Yu.P. and Surzhikov S.T., 1987, 1988) have allowed to carry out regular research of glow discharge structure by means of numerical modeling.

In order that computer models of glow discharge to be practically useful in aerophysical researches, the further development of these models is required, regarding the account of the interaction of discharges with gas flows and external magnetic field, interactions of discharges with moving gas at hypersonic velocities, and also the account of the physical and chemical transformations that occur in the region of the discharge. In the given section, the actual problem is solved about the influence of gas heating on the structure of glow discharges.

4.6.1 Statement of the 2D axially symmetric problem

The numerical model of glow discharge, in which gas heating is considered, is formulated for axially symmetric geometry. It is supposed that the discharge exists in the normal mode between two flat electrodes (Figure 4.1), so boundary effects in the radial direction do not influence its structure.

The structure of glow discharge is described within the limits of the drift-diffusion model, formulated concerning electronic and ionic concentrations, and also the Poisson equation, defining the distribution of electric potential in the electrodischarge gap (see Section 4.2.4).

It is supposed that a source of gas heating is the Joule thermal emission. As it was already marked, not the whole energy of applied electric field transmitted at electron collisions with molecules of gas goes to their heating. The significant part of this energy can be spent for excitation of molecules' vibrational degrees of freedom. For its definition, it is necessary to solve the kinetic equation for electron energy distribution function. At the same time, it is obvious that in various space zones of glow discharge a relation between the Joule thermal emission energy and the energy going on excitation of internal degrees of freedom will be different. In the offered statement, the solution of this equation is replaced with the introduction of phenomenological effectiveness ratio η of electric field energy transmission into gas heating, which does not vary by space, and this ratio is a parameter of the problem. In Petrusev A.S., et al. (2005), the phenomenological model of the account of a spatial modification of electron temperature and kinetics of excitation of vibrational conditions of molecular nitrogen was offered. With the use of this model, it was shown that in conditions similar to considered here, this ratio can reach 90%.

The calculation model of glow discharge in view of gas heating is formulated in the following form:

$$\frac{\partial n_e}{\partial t} + \mathrm{div}\boldsymbol{\Gamma}_e = \alpha(E/p)|\boldsymbol{\Gamma}_e| - \beta n_i n_e, \qquad (4.256)$$

$$\frac{\partial n_i}{\partial t} + \mathrm{div}\,\boldsymbol{\Gamma}_i = \alpha(E/p)|\boldsymbol{\Gamma}_i| - \beta n_i n_e, \qquad (4.257)$$

$$\operatorname{div}\,(\operatorname{grad}\varphi) = 4\pi e(n_e - n_i), \tag{4.258}$$

$$\rho c_V \frac{\partial T}{\partial t} = \operatorname{div}\,(\lambda \operatorname{grad} T) + Q, \tag{4.259}$$

where

$$\boldsymbol{\Gamma}_e = -D_e \operatorname{grad} n_e - n_e \mu_e \mathbf{E};\ \boldsymbol{\Gamma}_i = -D_i \operatorname{grad} n_i + n_i \mu_i \mathbf{E};$$

$$Q = \eta\,(\mathbf{j} \cdot \mathbf{E});\ \mathbf{j} = e(\boldsymbol{\Gamma}_i - \boldsymbol{\Gamma}_e);\ \mathbf{E} = -\operatorname{grad}\varphi;$$

T, p are the temperature and pressure of the gas; n_e, n_i are the concentration of electrons and ions in 1 cm^3; $\mathbf{E}$ and φ are the vector of electric field strength and its potential; $\boldsymbol{\Gamma}_e$, $\boldsymbol{\Gamma}_i$ are the vectors of fluxes of electrons and ions; D_e, D_i are the diffusivities of electrons and ions; μ_e, μ_i are the mobilities of electrons and ions; $\alpha = \alpha(E/p)$ is the coefficient of collisional ionization of molecules by electrons (the first Townsend coefficient), $E = |\mathbf{E}|$; β is the coefficient of ion–electron recombination.

At the solution of eqs. (4.256)–(4.259), an orthogonal cylindrical coordinate system is used. Boundary conditions for these equations look like

$$y = 0, \qquad \Gamma_{e,x} = \gamma \Gamma_{i,x},\ \frac{\partial n_i}{\partial y} = 0,\ \varphi = 0;$$

$$y = H, \qquad \frac{\partial n_e}{\partial y} = 0,\ n_i = 0,\ \varphi = V;$$

$$x = 0, \qquad \frac{\partial n_e}{\partial x} = \frac{\partial n_i}{\partial x} = \frac{\partial \varphi}{\partial x} = 0;$$

$$x = R, \qquad 1)\ n_e = n_i = 0,\ \varphi = \frac{V}{H} y,$$

or

$$2)\ \frac{\partial n_e}{\partial x} = \frac{\partial n_i}{\partial x} = \frac{\partial \varphi}{\partial x} = 0.$$

where γ is the coefficient of ion–electron emission from the cathode surface; V is the voltage drop on the discharge gap, $\Gamma_{e,x}$, $\Gamma_{i,x}$ are the projections of electron and ion fluxes on the x-axis; R, H are the coordinates of the calculation domain in directions x and y. Remind that the boundary conditions on electrodes for the charged particles are approximate.

Boundary conditions include the value of a voltage drop on discharge gap V that has not been determined yet. For its determination, it is necessary to attract conditions in an external circuit (see Figure 4.1). In conditions of the steady-state glow discharge, it is possible to write the obvious relation

$$\frac{E-V}{eR_0} = 2\pi \int_0^R \Gamma_e(x, y = H) x \, dx, \tag{4.260}$$

which postulates equality of the sum of the voltage drop on the resistance R_0 and voltage drop V on the discharge gap to electromotive power E.

The calculation model has intended for the study of glow discharge in molecular nitrogen at pressure $p = 1 - 20$ Torr; therefore, the following values of the coefficients entering into a mathematical statement of the problem are set:

$$\mu_e p = 4.4 \times 10^5, \quad \mu_i p = 1.45 \times 10^3 \ (\text{Torr} \cdot \text{cm}^2)/(\text{V} \cdot \text{s}),$$

$$\beta = 2 \times 10^{-7} \ \text{cm}^3/\text{s},$$

$$\frac{\alpha}{p} = A \cdot \exp\left(-\frac{B}{E/p}\right) \ (\text{cm} \cdot \text{Torr})^{-1},$$

where $A = 12 \, (\text{cm} \cdot \text{Torr})^{-1}$ and $B = 342 \, \text{V}/(\text{cm} \cdot \text{Torr})$. The empirical formula (4.11) for the first Townsend coefficient was recommended in Brown S.C. (1966) for the following range of ratios of electric field strength to a pressure:

$$100 < E/p < 600 \, \text{V}/(\text{cm} \times \text{Torr}).$$

In the calculations, an appreciable influence of used ionization coefficient approximation on integral discharge properties (total current through discharge and voltage drop across the gas-discharge gap) can be observed. Therefore, with the intention to achieve numerical proximity of calculated and experimental data, it is necessary to concern rather carefully to a choice of this approximation. The same is true for a choice of secondary ion–electron emission coefficient γ.

Diffusion coefficients can be defined by Einstein's relations:

$$D_e = \mu_e T_e, \ D_i = \mu_i T_i,$$

where T_e, T_i are the electron and ion temperatures in eV.

The energy conservation equation for neutral gas is formulated in the form of the Fourier–Kirchhoff equation (4.259) without taking into account of the convective gas motion. The stationary solution for this equation is searched in the same manner as for the concentration of the charged particles.

The important consequence of gas heating is the modification of local values of a density of neutral particles, which makes an essential influence on the value of collision frequency of electrons with gas molecules and, hence, on such parameters of the drift-diffusion model as the frequency of ionization and mobility of electrons and ions. The account of gas heating in functional dependences of the specified parameters is given in Section 4.2.4.

4.6.2 Numerical simulation results

Calculation of glow discharge is performed in the geometry shown in Figure 4.1. The axially symmetric glow discharge in molecular nitrogen (N_2) between two flat infinite electrodes in the normal current density mode or close to it is considered. Pressure in the discharge gap has been varied in the range of 1–20 Torr. Distance between electrodes is $H = 2\,\text{cm}$. Boundary conditions on an external cylindrical surface are set on distance $x = R = 2\,\text{cm}$ and 12 cm, which was quite enough to check up their influence on the structure of glow discharge. The electromotive force of the power supply varied in the range of $E = 600 - 4,000\,\text{V}$. The coefficient of secondary electron emission, the resistance of an external circuit, and electron temperature are supposed to be constant and equal according to $\gamma = 0.1$, $R_0 = 300\,\kappa\Omega$, $T_e = 11,610\,\text{K}$.

Computations were performed in such a sequence. First, the structure of glow discharge was calculated at different pressures without taking into account of its heat. In addition, a series of methodical calculations with different meshes has been fulfilled. Especially it is necessary to underline the necessity of the careful choice of computational meshes for the solution of the glow discharge task. This is due to the fact that, as discussed earlier, the mesh diffusion is caused by the use of the finite-difference schemes of the first order of accuracy or by use of various methods of correction of numerical fluxes in schemes of the second and higher orders of accuracy. This has an appreciable influence on calculation results. Therefore, in the first calculations, the meshes were defined, which do not cause appreciable influence on the desired results.

At initial data $p = 5\,\text{Torr}$, $E = 3\,\text{kV}$, the calculations have been performed on meshes $NI = 30 \times NJ = 30$, 30×59, 30×117. In these calculations, the number of points by a radial variable ($NI = 30$) did not vary. This number of points has appeared sufficient for the reliable description of glow discharge structure, as the discharge is localized near a symmetry axis, where point crowding is high enough. Besides it well known that field strength in a radial direction is noticeably lower than the field strength in an axial direction, in particular near the cathode. It was of interest to study the influence of mesh detail in an axial direction on the discharge structure. The subsequent calculations with more detailed meshes (117×201) have confirmed the validity of the specified choice of the computational mesh.

Results of calculations of glow discharge on a mesh of 30×117 at $p = 5\,\text{Torr}$ and 3 kV are shown in Figures 4.10 and 4.11.

The discharge is localized in the radialdirection near a symmetry axis, where the initial approximation has been set.

Current densities in the center of the cathode and anode spots achieve, accordingly, $j_c = 6.0\,\text{mA/cm}^2$ and $j_a = 23.3\,\text{mA/cm}^2$, and a radial size of the cathode spot is $R_c \approx 0.8\,\text{cm}$. Spatial structure of the glow discharge is shown in Figure 4.10a, where the area of spatial charges near the cathode and the anode is visible well. Axial distribution of electric field strength is shown in Figure 4.10b.

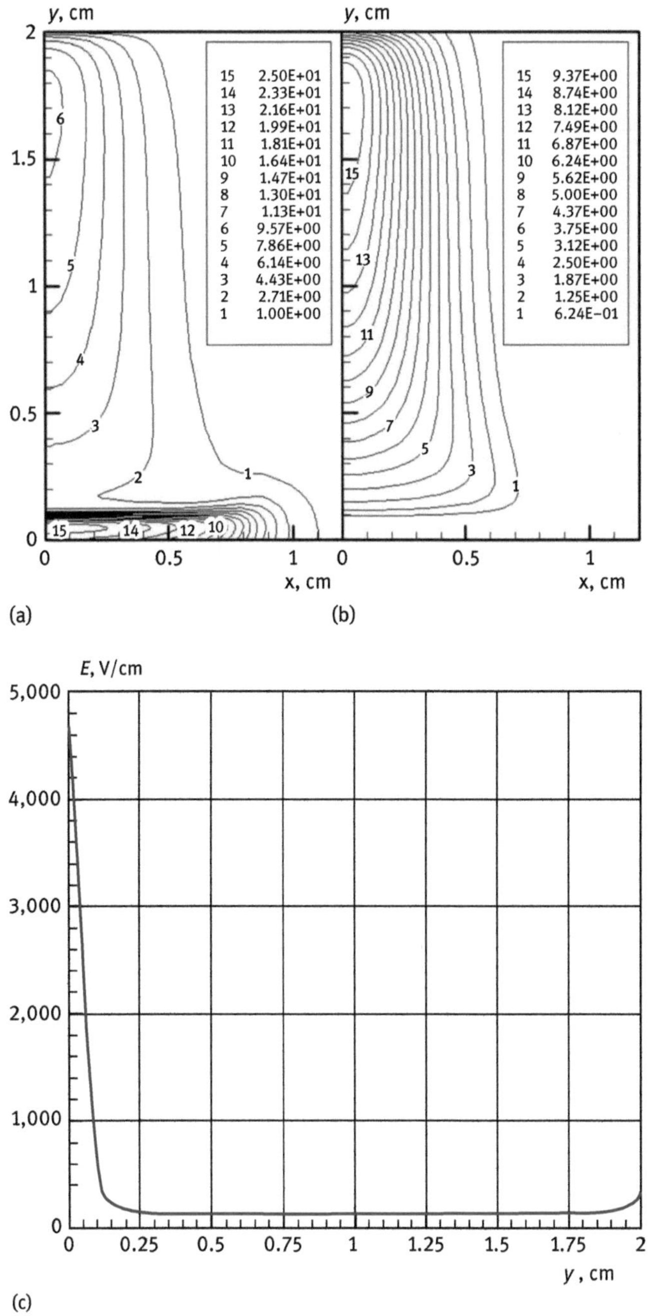

Figure 4.10: Concentration of ions (at the left) and electrons (on the right) in glow discharge at $p = 5$ Torr, $E = 3$ kV; concentration is referred to as $10^9\,cm^{-3}$ (a); distribution of electric field strength along the axis of symmetry of glow discharge at $p = 5$ Torr, $E = 3$ kV (b).

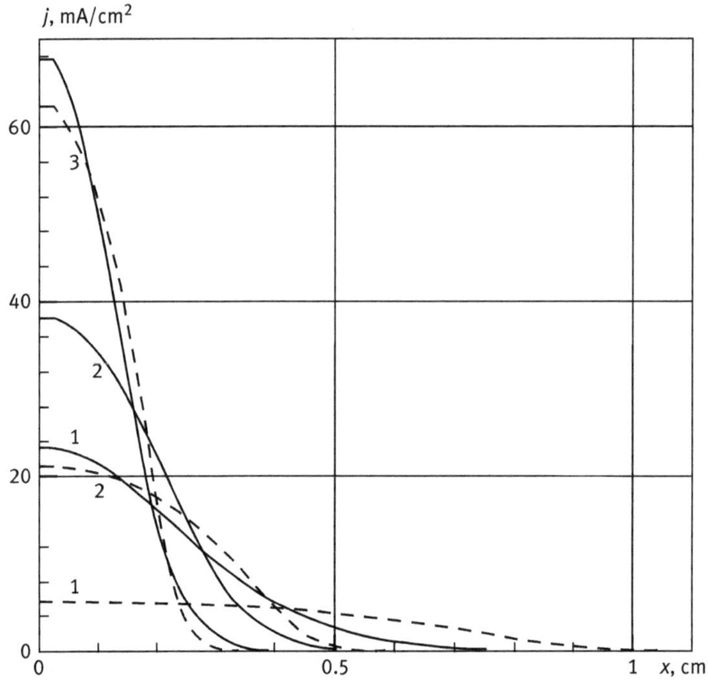

Figure 4.11: Current density on the cathode and the anode at $E = 3\,\text{kV}$: $1 - p = 5$ Torr, $2 - p = 10$ Torr, $3 - p = 20$ Torr.

The maximum value of the electric field is observed on the cathode $E_{y,\max} = 4,700\,\text{V/cm}$. There is an area of a positive column, in which the field strength is constant at a level of $E_y \approx 170\,\text{V/cm}$. In the immediate proximity from the anode, the small field strength increase is observed up to $E_y \approx 330\,\text{V/cm}$. Note that the calculations with the use of the drift-diffusion model do not allow to describe the region of the Faraday dark space.

Distributions of ion and electron concentrations in the glow discharge at $E = 3\,\text{kV}$ and $p = 20$ Torr are shown in Figure 4.12. Comparison of Figures 4.10 and 4.12 illustrate the influence of pressure on the glow discharge structure.

At the raised pressure, the discharge is strongly contracted, and the sizes of near-electrode layers decrease.

When using rough computational meshes, a nonmonotonic spatial distribution of electric potential is often observed near the cathode voltage drop. The use of rough computational meshes is also followed to the oscillations of ion concentrations near the cathode boundary of a positive column and nonmonotonic (oscillating) behavior of electron concentration distribution in the axial direction.

Common regularities in the behavior of the numerical simulation results at improving meshes that consist of the following. First, the increase of current density

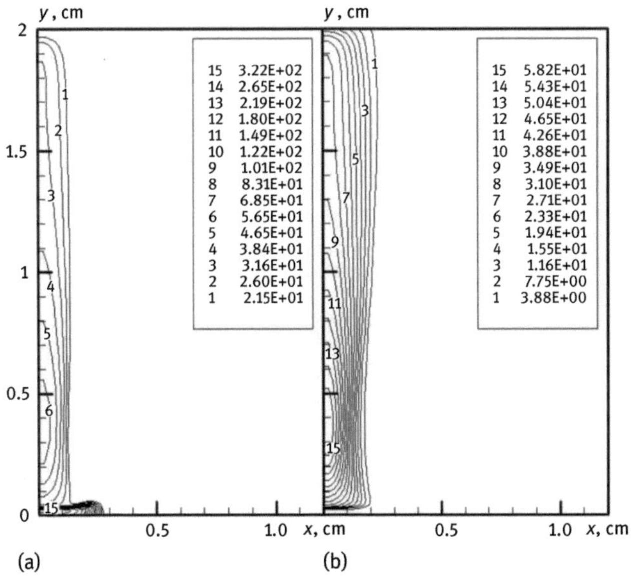

Figure 4.12: Concentration of ions (a) and electrons (b) in glow discharge at $p = 20$ Torr, $E = 3$ kV; concentration is referred to $10^9\,\text{cm}^{-3}$.

on the cathode and the anode is observed. Hence, at $E = 3$ kV and $p = 10$ Torr, the following results are observed:

for the mesh $30 \times 30 - j_c = 6.2\,\text{mA/cm}^2$ and $j_a = 11.3\,\text{mA/cm}^2$;
for the mesh $30 \times 59 - j_c = 17.95\,\text{mA/cm}^2$ and $j_a = 37.5\,\text{mA/cm}^2$;
for the mesh $30 \times 117 - j_c = 21.1\,\text{mA/cm}^2$ and $j_a = 38.0\,\text{mA/cm}^2$.

At the same time, the current in the external circuit varies slightly; therefore, the radius of the cathode and anode spots decreases noticeably. The maximum strength of the electric field E_y increases at the cathode; in the positive column and near the anode it varies slightly. Improving the computational mesh leads to the disappearance of the nonmonotony in the electric potential distribution. The important consequence of the computational mesh refinement is the increase in concentration of the charged particles in the positive column that is connected to a diminution of numerical diffusion losses.

Results of calculations of glow discharge for three pressures $p = 5, 10, 20$ Torr at $E = 3$ kV are shown in Figures 4.11 and 4.13, where the radial distributions of current densities on electrodes and axial distribution of concentration of the charged particles along the glow discharge symmetry axis are given.

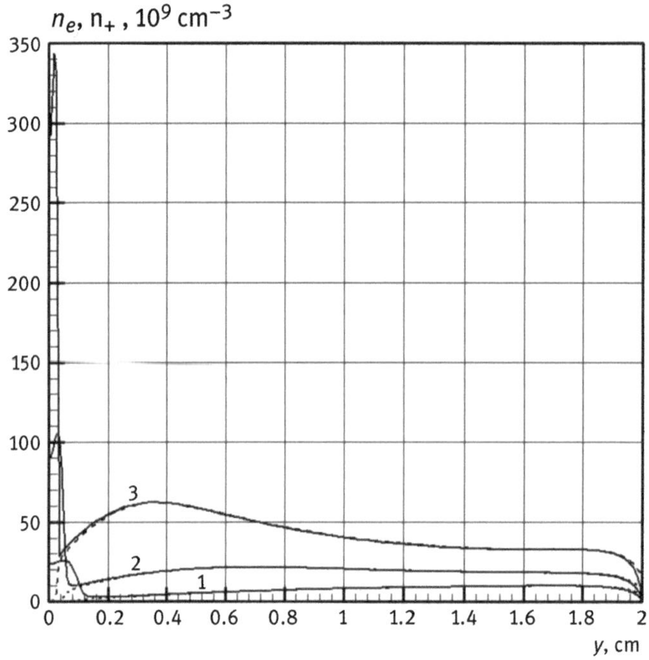

Figure 4.13: Distribution of ion (solid curves) and electron (dashed curves) concentrations along a symmetry axis of discharge at $E = 3$ kV: 1 – $p = 5$ Torr, 2 – $p = 10$ Torr, 3 – $p = 20$ Torr.

As gas pressure increases, the following modifications in glow discharge structure are observed: the current density on the cathode and on the anode increases, radial sizes of the glow discharge decrease, thickness of the near-cathode and near-anode layers decreases, electric field strength in the gas discharge increases, concentration of electrons and ions in the positive column increases (at the same time the displacement of the charged particles concentration maxima in a positive column from the anode to the cathode is observed), and concentration of ions in the cathode layer increases. Let us notice that the specified behavior of the glow discharge is in good correspondence with numerous experimental data and with the classical Engel–Steenbeck theory. It confirms the adequacy of the developed numerical model to the investigated phenomena. At the same time, it is necessary to mean that some elements of the glow discharge structure are described by the given model quite approximately (for areas of a spatial charge immediately adjoining electrodes and area of the Faraday dark space).

As it was already marked, the Joule heating in glow discharges is caused by energy transmission from electrons, heated up by the external electric field, to gas molecules at their collisions. The part of this energy is spent for excitation of molecular oscillations. It is known that molecules of N_2 that have great vibrational excitation relaxation time leads to the part of electron energy having no time to be transformed to heat and is carried away by the vibrationally excited molecules from an area of discharge, for example, at convective motion.

In the considered statement, the relaxation processes and gas motion are approximately considered by the introduction of the effectiveness ratio η of a transformation of the electron energy in heat of the gas, which varies in calculations. Despite obvious boundedness of such a statement, the specified model allows investigating the configuration of glow discharge and regularities of origin of areas with increased gas temperature in it at the account of self-consistent electrodynamic and thermal processes.

In the first series of calculations of glow discharge in view of gas heating, the pressure $p = 5$ Torr has been chosen. For two electromotive forces, $E = 2$ kV and 4 kV calculations were performed both with and without taking into account the gas heating. Effectiveness of transformation of electric field energy in gas heating W was equal to $\eta = 0.5$. In all cases, the cathode and the anode were cooled, $T_w = 300$ K. Calculations have shown that at $E = 2$ kV the thermal state of discharge is stabilized after ~1 ms, and at $E = 4$ kV after ~5 ms, which confirms the estimations of characteristic times fulfilled earlier.

Distributions of ion and electron concentrations in a glow discharge in the first series of calculations ($E = 2$ kV) taking into account both without and with gas heating are shown in Figures 4.14 and 4.15 accordingly, and the second series ($E = 4$ kV) are shown in Figures 4.16 and 4.17.

From Figure 4.18 it is evident that the account of gas heating leads to an increase in current density on the anode from $j_a = 15.5$ to 24.1 mA/cm^2 ($E = 2$ kV) and from $j_a = 14.5$ to 32.7 mA/cm^2 ($E = 4$ kV).

The current density on the cathode at gas heating falls from 5.8 to 4.0 mA/cm^2 ($E = 2$ kV) and from 7 to 4.0 mA/cm^2 ($E = 4$ kV).

The account of gas heating leads to improvement of ionization conditions by the electron impact of molecules. The thermal expansion of gas leads to a diminution of local values of particle concentration, so the parameter E/N defining the effectiveness of ionization increases.

At $E = 2$ kV, gas in the discharge gap heats approximately to 355 K. The temperature field for this case is shown in Figure 4.19.

Existence of two local maxima of axial temperature distribution near the cathode and anode attracts attention. The analysis of distributions of current density and electric field strength allows explaining such distribution of temperature. Near the cathode, the electric field strength is great, but the density of electric current is small.

The reason for that is a little electron concentration. Near the anode, the situation varies to the opposite: electron concentration is great and the electric field strength is small. Therefore, origination of nonmonotonic axial distribution of the Joule thermal emission $q_J = (\mathbf{j} \cdot \mathbf{E})$ is natural. Nevertheless, the performed numerical research has shown the possibility of realization of different configurations of gas

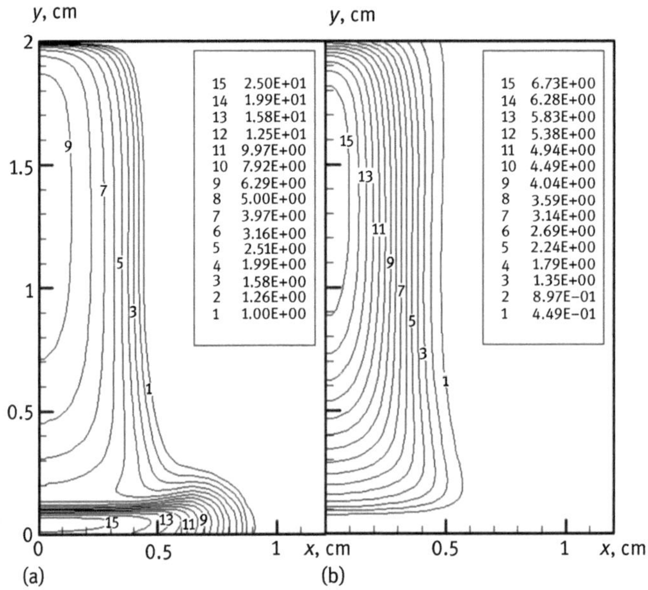

Figure 4.14: Concentration of ions (a) and electrons (b) in glow discharge at $p = 5$ Torr, $E = 2$ kV; concentration is referred to 10^9 cm^{-3}.

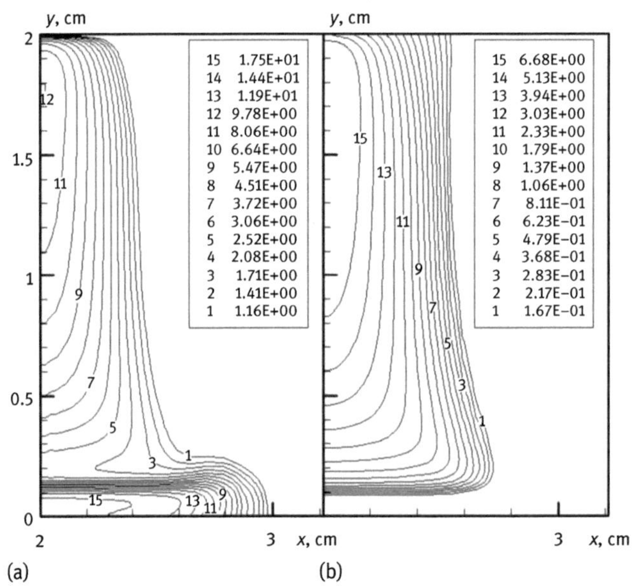

Figure 4.15: Concentration of ions (a) and electrons (b) in glow discharge at $p = 5$ Torr, $E = 2$ kV, $\eta = 0.5$; concentration is referred to 10^9 cm^{-3}.

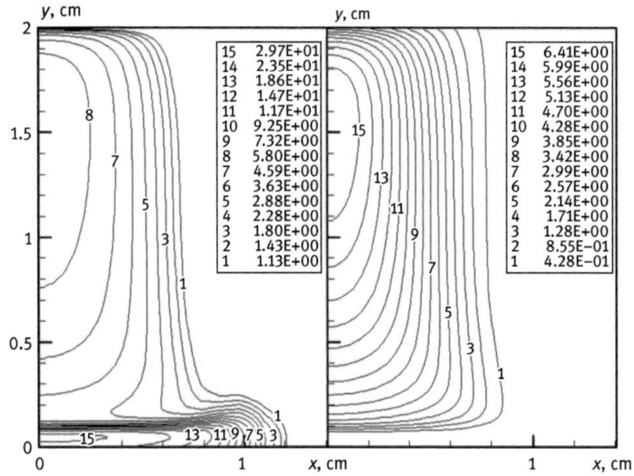

(a) (b)

Figure 4.16: Concentration of ions (a) and electrons (b) in glow discharge at $p = 5$ Torr, $E = 4$ kV; concentration is referred to 10^9 cm^{-3}.

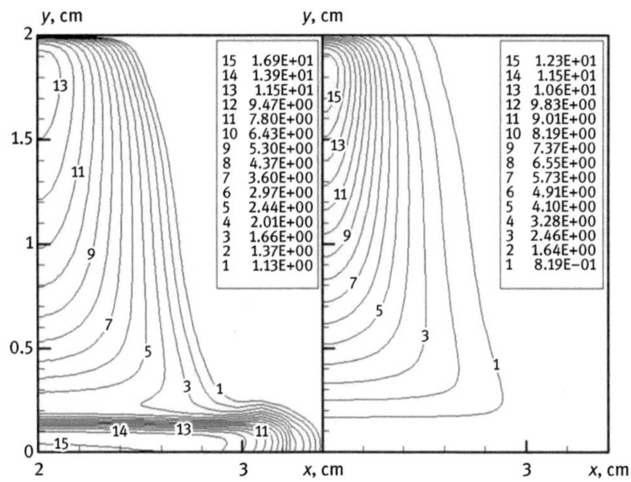

(a) (b)

Figure 4.17: Concentration of ions (a) and electrons (b) in glow discharge at $p = 5$ Torr, $E = 4$ kV, $\eta = 0.5$; concentration is referred to 10^9 cm^{-3}.

heating depending on parameters of the problem $p, \varepsilon, \gamma, \eta$ and the set boundary conditions on account of heat exchange.

Let us note one more peculiarity of the account of the gas heating. The concentration of ions in the cathode layer decreases from 2.8×10^{10} cm^{-3} to 1.85×10^{10} cm^{-3}

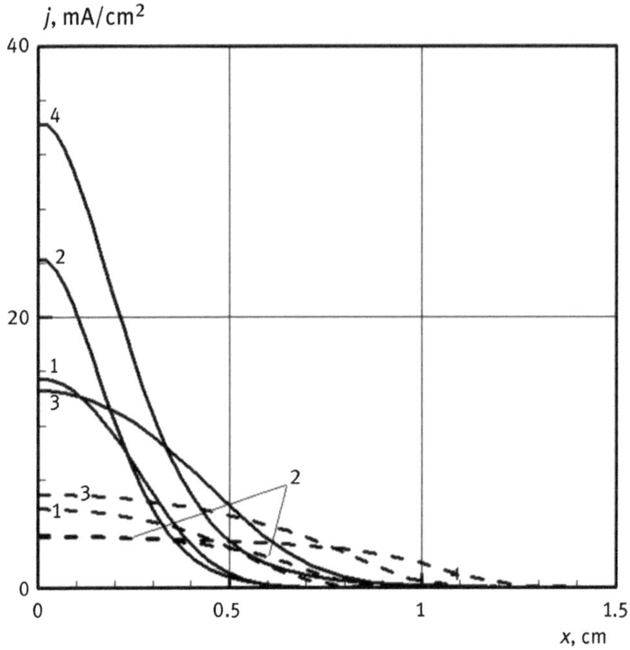

Figure 4.18: Radial distribution of a current density to the anode (continuous curves) and on the cathode (dashed curves) at $p = 5$ Torr: 1, 2, $E = 2$ kV; 3, 4, $E = 4$ kV; 1, 3, without taking into account of the gas heating; 2, 4, with taking into account the gas heating.

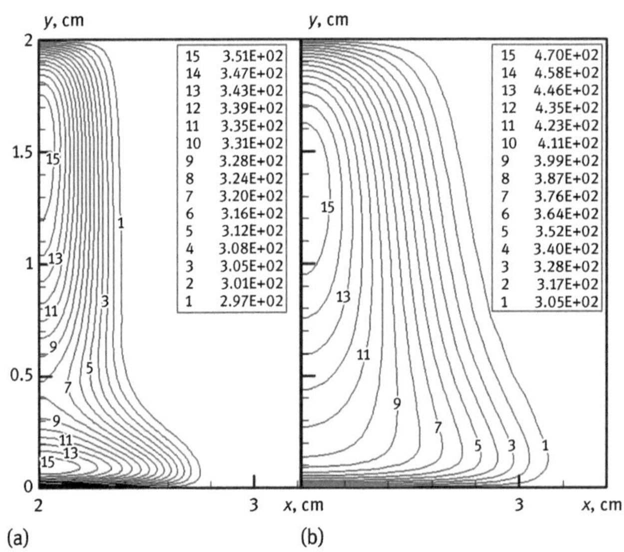

Figure 4.19: Temperature (K) in glow discharge at $p = 5$ Torr, $E = 4$ kV (a) and at $p = 5$ Torr, $E = 4$, $\varepsilon = 4$ kV (b); $\eta = 0.5$.

at $E = 2$ kV (Figure 4.20) and from 3.2×10^{10} cm^{-3} up to 1.8×10^{10} cm^{-3} at $E = 4$ kV (Figure 4.21). The concentration of the charged particles in the positive column increases. Heating of the gas leads to that the local maximum of particle concentrations in the positive column is displaced to the anode.

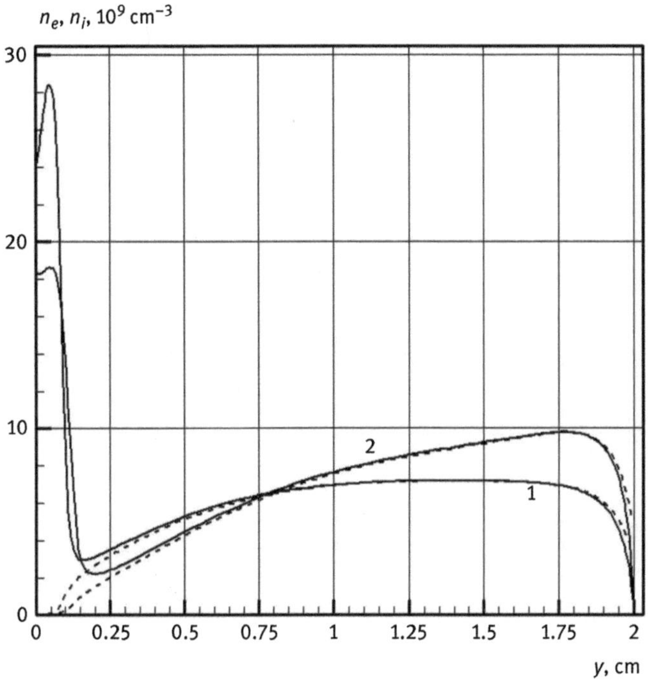

Figure 4.20: Distribution of concentration of ions (solid curves) and electrons (dashed curves) along a symmetry axis of discharge at $p = 5$ Torr, $E = 2$ kV: 1, without heating; 2, with heat release.

Now we consider the influence of pressure on heating of glow discharge. Besides the calculations with $E = 4$ kV, $p = 10$ Torr (Figures 4.22 and 4.23) the calculations with $E = 4$ kV, $p = 20$ Torr (results are not given here) have been performed.

To finish the analysis of the influence of pressure, it is necessary to add data also from Figures 4.25 and 4.26 ($E = 4$ kV, $p = 5$ Torr). With pressure growth, the gas heats more intensively, and more significant modifications of glow discharge structure are observed.

Taking into account gas heating at $p = 10$ Torr and $E = 4$ kV leads approximately to twofold drop of the cathode current density and to threefold increase of the anode current density (Figure 4.24). The electric field strength near the cathode and in the positive column falls a little: from $E_y = 9{,}000$ V/cm to $E_y = 7{,}000$ V/cm. At the same time, the radius of the current spot on the cathode increases, and on the anode, it decreases.

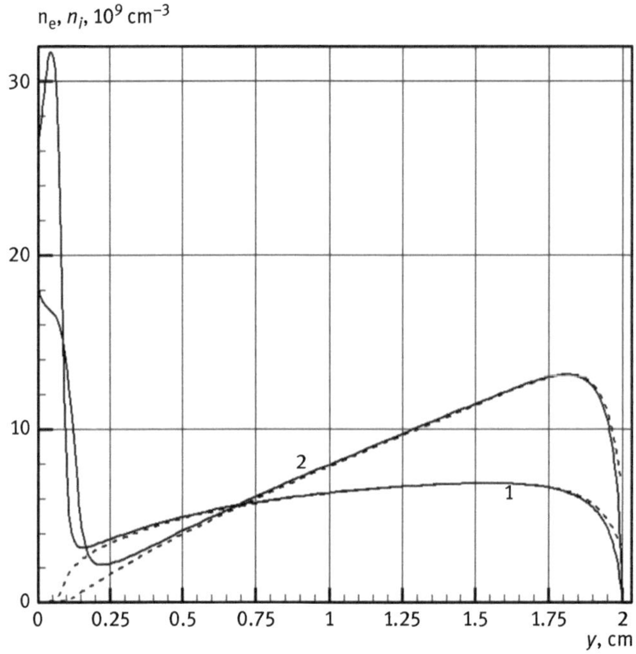

Figure 4.21: Distribution of concentration of ions (solid curves) and electrons (dashed curves) along a symmetry axis of discharge at $p = 5$ Torr, $E = 4$ kV: 1, without heating; 2, with heat release.

Distributions of particle concentrations for these two cases are shown in Figure 4.25. It is well evident that near the anode, concentration of the charged particles increases more than twice, and the concentration of ions at the cathode falls more than twice. Growth of thickness of the cathode layer at heating is remarkable too.

Let us notice that on account of heating, the calculation of discharge parameters becomes more labor-consuming not only owing to necessity of integration of the equations describing structure of discharge till thermal relaxation times but, first of all, owing to necessity of using more accurate computing meshes (the cathode spot extends and the anode spot is narrowed). Besides, with the heating of gas the maximum admissible time step of the numerical integration, ensuring the stability of the calculations, decreases.

Distributions of temperature along the symmetry axis in different conditions are shown in Figure 4.26. At pressure $p = 10$ Torr gas heated up to temperatures T~600 K while at $p = 5$ Torr the temperature does not exceed ~400 K. Such a heating leads to significant growth of charged particle concentration in the positive column and, in particular, near the anode, where $n_e \approx n_i = 4.3 \times 10^{10}$ cm^{-3} (Figure 4.25). The further growth of pressure (up to 20 Torr) does not lead to appreciable heating of glow discharge as its sizes are reduced, so the radial heat-conducting losses increase. The

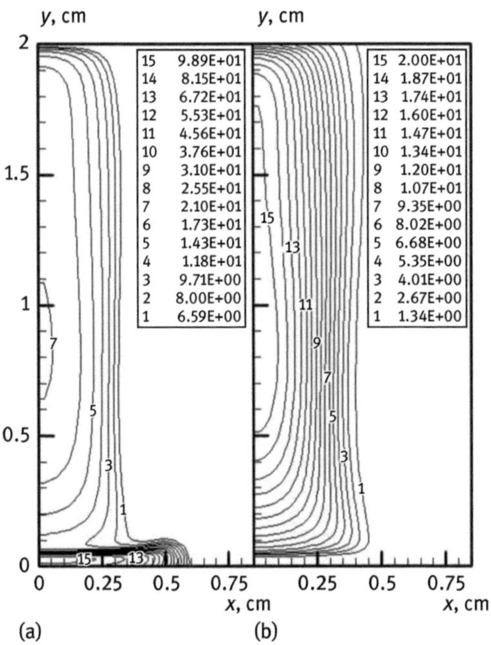

Figure 4.22: Concentration of ions (a) and electrons (b) in glow discharge at $p = 10$ Torr, $E = 4$ kV; concentration is referred to 10^9 cm^{-3}.

current density on the anode increases up to 200 mA/cm^2 (from 80 mA/cm^2 without heating), and the current density on the cathode falls to 27 mA/cm^2 (from 70 mA/cm^2 without heating). The positive column of glow discharge is narrowed in an even greater degree. The maximum temperature in discharge $\sim T = 590$ K, however, the concentration of the charged particles in the positive column reaches the value of 10^{11} cm^{-3}, and near the cathode, it is 1.4×10^{11} cm^{-3}.

In addition, a numerical investigation of the relationship of the glow discharge parameters and regularities of its heating from effectiveness ratio of the Joule heat transmission η (in the range of 0.1–0.75) and from the value of secondary ion-electron emission coefficient γ (in the range of 0.01–0.3) have shown objective modifications in its structure.

Let us now consider results of calculations of glow discharge for the least investigated pressure $p = 1$ Torr. Here gas heating appears insignificant; however, in calculations, the important modification of glow discharge structure in a radial direction is observed: the maximum of the charged particle concentrations and flux densities on the anode are displaced from a symmetry axis to a periphery. That is the gas discharge channel gains toroidal form. Distributions of ion and electron concentration at two values of electromotive force are shown in Figures 4.27 and 4.28, and distributions of current densities on electrodes are shown in Figure 4.29. Attention is

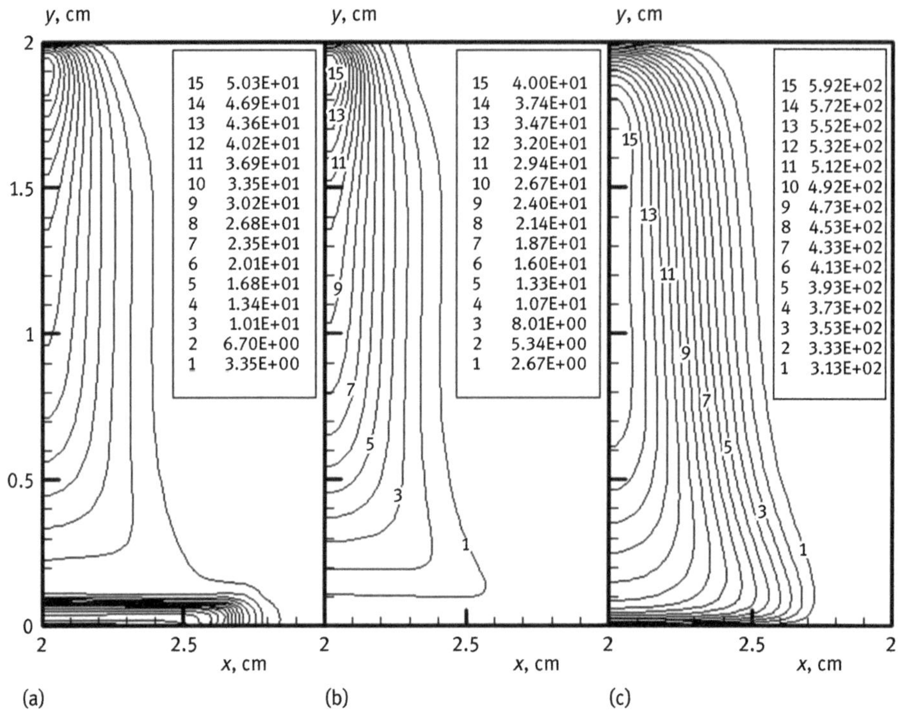

Figure 4.23: Concentration of ions (a) and electrons (b) in glow discharge at $p = 10$ Torr, $E = 4$ kV, $\eta = 0.5$; concentration is referred to 10^9 cm^{-3}; temperature (K) in glow discharge at $p = 10$ Torr, $E = 4$ kV, $\eta = 0.5$ (c).

attracted to the normal current density law: at a variation of the power supply electromotive force more than five times, the current density on the cathode practically does not vary (Figure 4.29).

One more important singularity of glow discharge predicted by the classical Engel–Steenbeck theory and confirmed by the calculations is that practically in all calculation domains, there are no areas of quasi-neutral plasma. It means that the use of quasi-neutral plasma models in aerophysical researches at small pressure is unreasonable or, at least, demanding additional substantiations.

In conclusion, we shall note that the change of glow discharge radial structure discovered by calculations is one of many effects of essentially more common regularity of structure formation in strongly nonequilibrium physical systems, in particular, in one of the classical models of such systems as a glow discharge.

The fulfilled regular research of glow discharge structure at conditions of practical interest for aerophysical applications has allowed determining a series of important singularities that are necessary for considering an analysis of experimental

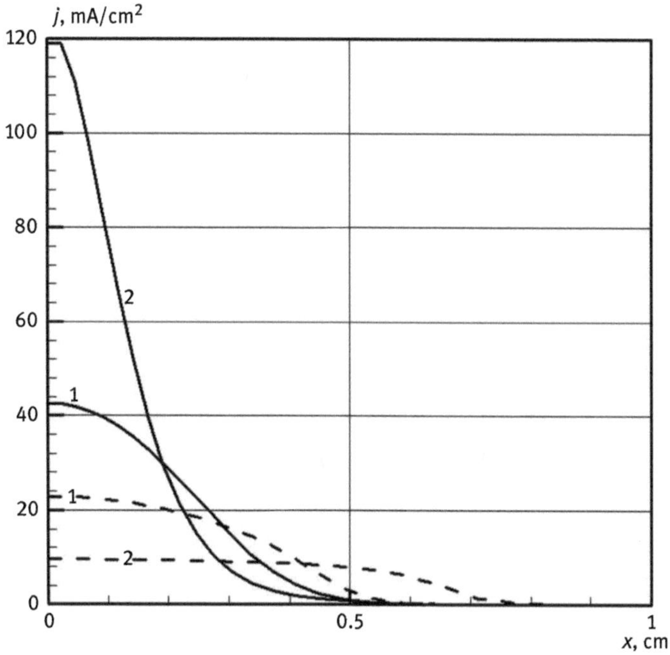

Figure 4.24: Radial distribution of a current density on the anode (solid curves) and on the cathode (dashed curves) at $p = 10$ Torr, $E = 4$ kV: 1, without heating; 2, with heat release.

data and for prediction of use of such discharges for modification of a flow field near the constructional elements of various aircraft. Namely, in a range of pressure $p < 5$ Torr, the gas heating is negligible in a wide range of electric field strength modification, and in small pressure, the glow discharge gets the torus-like form at normal current density conditions.

With a rise of pressure at more than 5 Torr and of electric field strengths, significant heating of neutral gas is observed. The specified gas heating, in turn, strongly changes the electrodynamic structure of discharges.

It should be stressed that the specified regularities are well known to that experimenters – physicists, who investigate discharges of the given type. It is important that the calculation models allow predicting the specified regularities. Nevertheless, it is obvious that the further development of the calculation models is required in view of heating of gas and regarding an account of gas motion (in particular, supersonic) and nonequilibrium physical and chemical processes.

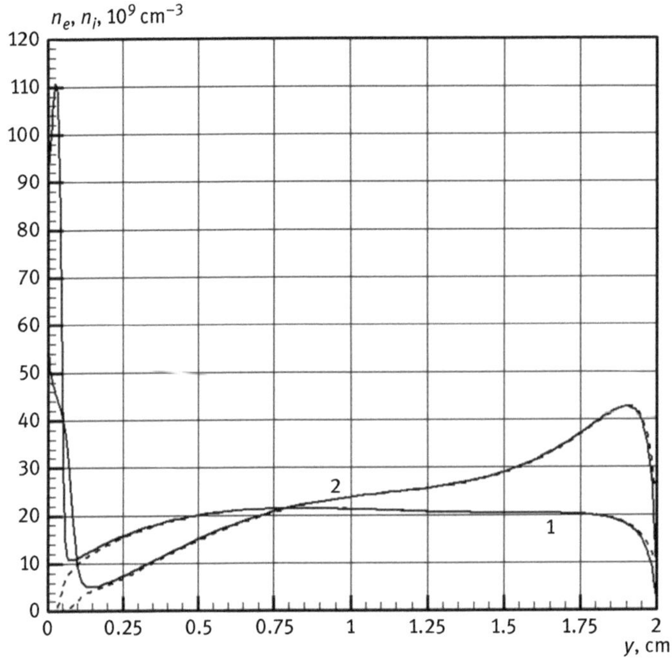

Figure 4.25: Distribution of concentration of ions (solid curves) and electrons (dashed curves) along a symmetry axis of discharge at $p = 10$ Torr, $E = 4$ kV: 1, without heating; 2, with heat release.

4.7 Normal glow discharge between curvilinear electrodes

A diffusion-drift calculation model of a normal glow discharge between curvilinear electrodes is presented in this section. The verification of the created computational model has been performed using the example of a normal glow discharge between flat electrodes. Results of numerical simulation are presented of normal glow discharge between electrodes in the form of spherical segments at pressure $p = 5$ Torr and emf of power supply in the region of 1,0–1,75 V.

A normal steady-state glow discharge is investigated in the laboratory between flat (disk) electrodes spaced a few centimeters apart. In the normal glow discharge, a column of the gas-discharge plasma should be located far from the boundaries of the electrodes. Such a discharge burns in the self-organizing mode of transverse dimensions, and the current density at the cathode is predicted with good accuracy by the 1D theory of Engel and Steenbeck (Raizer Yu.P., 1991; Engel A., et al., 1932):

$$V_n = \frac{3B}{A}\ln\left(1 + \frac{1}{\gamma}\right), \text{ V},$$

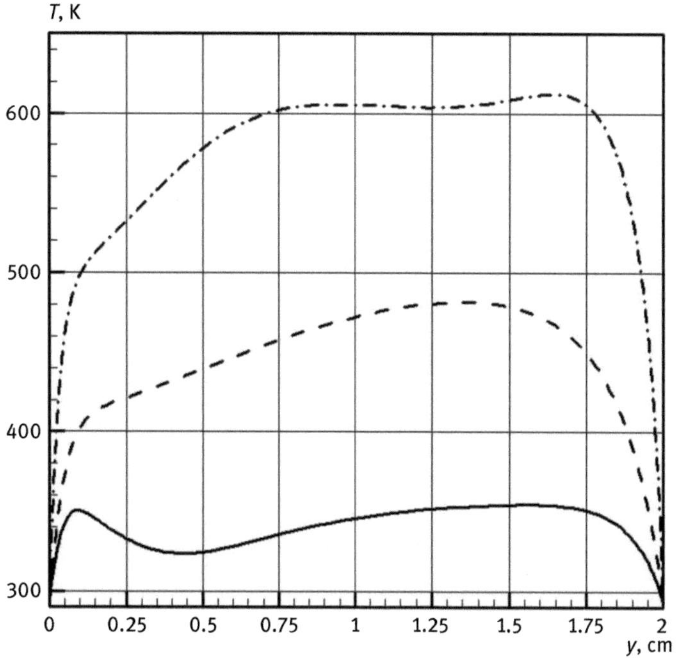

Figure 4.26: Axial distribution of temperature in glow discharge: $p = 5$ Torr, $\varepsilon = 2$ kV (solid curve), $p = 5$ Torr, $E = 4$ kV (dashed curve), $p = 10$ Torr, $E = 4$ kV (dash-dot curve).

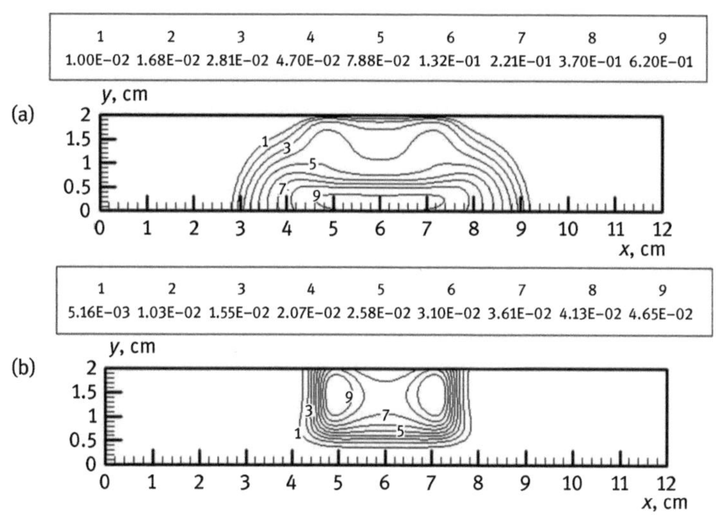

Figure 4.27: Concentration of ions (a) and electrons (b) in glow discharge at $p = 1$ Torr, $E = 0.6$ kV; concentration is referred to 10^9 cm^{-3}.

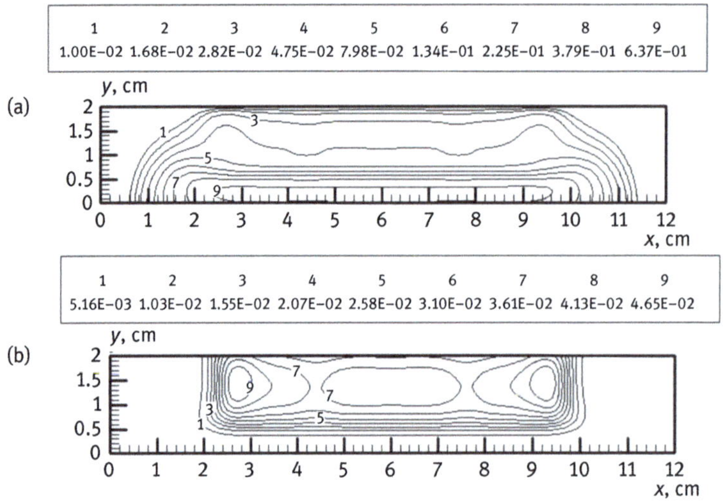

Figure 4.28: Concentration of ions (a) and electrons (b) in glow discharge at $p = 1$ Torr, $E = 2$ kV; concentration is referred to 10^9 cm^{-3}.

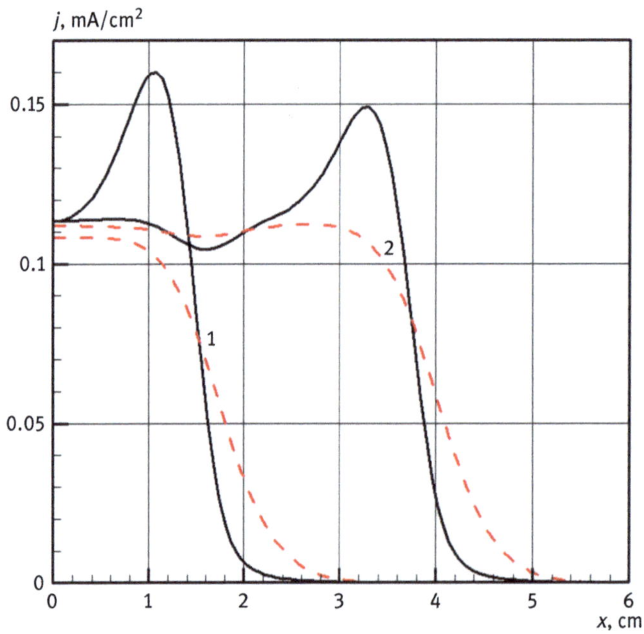

Figure 4.29: Distribution of current densities along the anode (solid curves) and the cathode (dashed curves) at $p = 1$ Torr: 1, $E = 0.6$ kV; 2, $E = 2$ kV.

$$\frac{j}{p^2} = 5.92 \times 10^{-14} \frac{AB^2 \mu_i p(1+\gamma)}{\ln(1+\gamma^{-1})}, \ \mathrm{A/(cm^2 \times Torr)},$$

$$d_n p = \frac{3.78}{A} \ln(1+\gamma^{-1}), \ \mathrm{cm \times Torr},$$

where: V_n, j_n, d_n are the voltage drop on the gas-discharge gap, current density, thickness of the cathode layer of a normal glow discharge; p is the pressure in Torr; μ_i is the mobility of ions; γ is the coefficient of secondary electron emission. It is assumed that the coefficient α of impact ionization of molecules by electrons is approximated as follows:

$$\alpha = \alpha\left(\frac{E}{p}\right) = Ap\exp\left(-\frac{B}{E/p}\right), \ \mathrm{cm}^{-1},$$

where A, B are the approximating coefficients; E is the electric field strength.

The increase in the total current through the gas-discharge gap (typical values from units to tens of mA) in such a glow discharge leads to an increase in the transverse dimensions of the current column, but the current density in the near-axis region remains practically unchanged. This phenomenon is called the law of normal current density and has been studied in classical physics of gas discharge for more than 100 years (Engel A., et al., 1932).

Since the 1980s, the symmetrical normal glow discharge is the object of study by methods of computational physics (Gladush G.G., et al., 1981; Surzhikov S.T., et al., 2004c). Numerical simulation of this type of discharge proved to be very useful for the development of computational methods and algorithms for mathematical modeling. Subsequently, the problems were investigated for calculating the heating of neutral gas, the dynamics of a glow discharge in a transverse and longitudinal magnetic field with induction up to ~ 1.0 T, taking into account the physical kinetics of excitation of N_2 vibrational states in collision with electrons (Petrusev A.S., et al., 2006a,b).

Localization of the discharge column of a normal glow discharge between flat electrodes is achieved in a computational experiment by setting the initial conditions. In a real physical experiment, it is not easy to achieve a stationary state far from the electrode boundaries (Kotov M.A., et al., 2017).

The subject of this chapter is a computational study of the structure of a normal glow discharge between convex electrodes. Schematic of the task is shown in Figure 4.30, as well as an external electric circuit consisting of a constant-current power supply with an emf ε and ohmic resistance R_0. The discharge gap is placed in a vacuum chamber. The investigated range of molecular nitrogen pressures is $p = 1$–10 Torr.

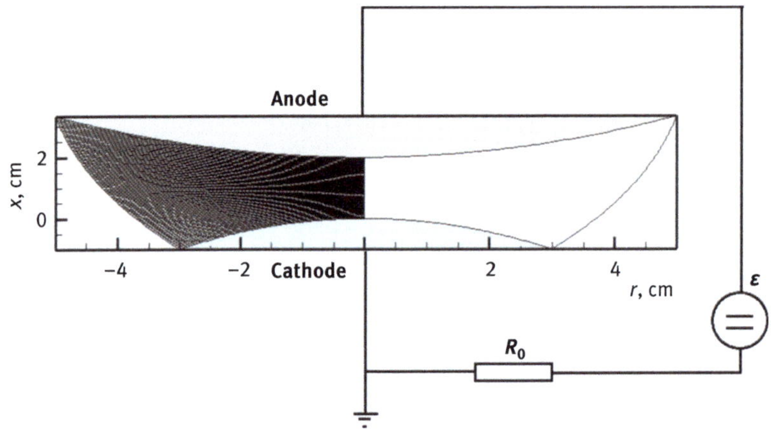

Figure 4.30: Schematic of the task.

4.7.1 Equations for the diffusion-drift model in curvilinear geometry

The equations of the diffusion-drift model used to simulate a normal glow discharge are obtained from the system of equations for a multifluid and a multitemperature partially ionized gas mixture (a detailed derivation of these equations is given in Section 4.6):

$$\mathrm{div}\mathbf{E} = -\varepsilon^{*}(n_{+} - n_e), \tag{4.261}$$

$$\frac{\partial n_e}{\partial t} + \mathrm{div}\boldsymbol{\Gamma}_e = \alpha|\boldsymbol{\Gamma}| - \beta n_{+} n_e, \tag{4.262}$$

$$\frac{\partial n_{+}}{\partial t} + \mathrm{div}\boldsymbol{\Gamma}_{+} = \alpha|\boldsymbol{\Gamma}| - \beta n_{+} n_e, \tag{4.263}$$

where n_{+}, n_e are the concentrations of ions and electrons; **E** is the vector of electric field strength with electric potential φ

$$\mathbf{E} = -\,\mathrm{grad}\,\varphi, \tag{4.264}$$

$\boldsymbol{\Gamma}_e$, $\boldsymbol{\Gamma}_{+}$ are the vectors of electron and ion flux densities,

$$\boldsymbol{\Gamma}_e = -n_e\mu_e\mathbf{E} - D_e\mathrm{grad}n_e, \quad \boldsymbol{\Gamma}_{+} = -n_{+}\mu_{+}\mathbf{E} + D_{+}\mathrm{grad}n_{+}, \tag{4.265}$$

μ_e, μ_{+} are the mobilities of electrons and ions; D_e, D_{+} are the diffusion coefficients of electrons and ions; α, β are the coefficient of impact ionization of a neutral gas by electrons (the first Townsend coefficient) and the recombination coefficient, $\varepsilon^{*} = 4\pi e = 1.81 \times 10^{-6}\ \mathrm{V \cdot cm}$.

Boundary conditions for the solution of the boundary value problem (4.261)–(4.263) are

$$\begin{aligned}
x=0:&\quad \Gamma_{e,x}=\gamma\Gamma_{+,x}, \quad \frac{\partial n_+}{\partial x}=0, \quad \varphi=0,\\
x=H:&\quad \frac{\partial n_e}{\partial x}=0, \quad n_+=0, \quad \varphi=V,\\
r=0:&\quad \frac{\partial n_e}{\partial r}=\frac{\partial n_+}{\partial r}=\frac{\partial \varphi}{\partial r}=0,\\
r=R_c:&\quad \frac{\partial n_e}{\partial r}=\frac{\partial n_+}{\partial r}=\frac{\partial \varphi}{\partial r}=0,
\end{aligned} \tag{4.266}$$

where V is the voltage drop at the electrodes, determined from the electric circuit equation

$$\frac{\varepsilon - V}{eR_0} = 2\pi \int_0^{R_a} \Gamma_{e,n_{s,a}} r\mathrm{d}r \tag{4.267}$$

For curvilinear electrodes, the integral in (4.267) should be calculated along the surface of the electrode with the projection of the electron flux density ($\Gamma_{e,n_{s,a}}$) on the local normal to the surface $n_{s,a}$.

Condition (4.267) determines the electric current through the anode. For a stationary glow discharge, the total electric current through the gas-discharge gap is a constant value and can be determined from the following equation:

$$\frac{\varepsilon - V}{eR_0} = 2\pi \int_0^{R_a} \Gamma_{+,n_{s,c}} r\mathrm{d}r,$$

where $n_{s,c}$ is the local normal to surface of cathode.

We also note that when analyzing the transient processes of establishing a stationary state, one should take into account the effect of accumulation of charges on electrodes (see Gladush G.G., et al., 1981 for more details).

The closing relations for the problem being solved are used for molecular nitrogen

$$\mu_e p = 4.4\times 10^5, \ \mu_i p = 1.45\times 10^3, \ \mathrm{Torr}\cdot\mathrm{cm}^2/\mathrm{V}\cdot\mathrm{s}, \ \beta = 2\times 10^{-7}, \ \mathrm{cm}^3/\mathrm{s},$$

$$\frac{\alpha}{p} = A\exp\left(-\frac{B}{E/p}\right), \ 1/(\mathrm{cm}\cdot\mathrm{Torr}), \ A=12, \ (\mathrm{cm}\cdot\mathrm{Torr})^{-1}, \ B=342, \ \mathrm{V}/(\mathrm{cm}\cdot\mathrm{Torr}),$$

$$D_e = \mu_e T_e, \ D_+ = \mu_i T_+, \ T_e = 1\ \mathrm{eV}, \ T_+ = 0.026\ \mathrm{eV}.$$

The construction of a computational model of a normal glow discharge in the computational domain with curvilinear boundaries requires the introduction of a curvilinear coordinate system (ξ, η) connected by single-valued transformations with the original cylindrical coordinate system (x, r). Below these transformations are given for the system of the Poisson equation and equations of the diffusion-drift model.

4.7.2 Transformation of the Poisson equation for the electric potential and the continuity equations for electrons and ions

The initial Poisson equation in an orthogonal cylindrical coordinate system is formulated as follows:

$$\frac{\partial}{\partial x}\left(\frac{\partial \varphi}{\partial x}\right)+\frac{\partial}{\partial r}\left(\frac{\partial \varphi}{\partial r}\right)+\frac{\chi}{r}\frac{\partial \varphi}{\partial r}=\Phi(r,x), \tag{4.265}$$

where $\Phi(r,x)=4\pi e(n_{+}-n_{e})$, $\chi=0$ for plane discharge, $\chi=1$ for axisymmetric discharge.

Input coordinate system (ξ, η) must be uniquely linked to the original coordinate system (x, r); therefore (4.264) can be rewritten in the form (hereinafter $y=r$)

$$\xi_x\frac{\partial}{\partial \xi}\left(\xi_x\frac{\partial \varphi}{\partial \xi}+\eta_x\frac{\partial \varphi}{\partial \eta}\right)+\eta_x\frac{\partial}{\partial \eta}\left(\xi_x\frac{\partial \varphi}{\partial \xi}+\eta_x\frac{\partial \varphi}{\partial \eta}\right)+\xi_y\frac{\partial}{\partial \xi}\left(\xi_y\frac{\partial \varphi}{\partial \xi}+\eta_y\frac{\partial \varphi}{\partial \eta}\right)+$$
$$+\eta_y\frac{\partial}{\partial \eta}\left(\xi_y\frac{\partial \varphi}{\partial \xi}+\eta_y\frac{\partial \varphi}{\partial \eta}\right)+\frac{1}{y}\left(\xi_y\frac{\partial \varphi}{\partial \xi}+\eta_y\frac{\partial \varphi}{\partial \eta}\right)=\Phi[\xi(x,y),\eta(x,y)]=\Phi$$

Division on Jacobian of the transformation J and differentiation by parts allows us to obtain the equation

$$\frac{\partial}{\partial \xi}\left(\frac{\xi_x^2}{J}\frac{\partial \varphi}{\partial \xi}\right)-\xi_x\frac{\partial \varphi}{\partial \xi}\frac{\partial}{\partial \xi}\left(\frac{\xi_x}{J}\right)+\frac{\partial}{\partial \xi}\left(\frac{\xi_x\eta_x}{J}\frac{\partial \varphi}{\partial \eta}\right)-\eta_x\frac{\partial \varphi}{\partial \eta}\frac{\partial}{\partial \xi}\left(\frac{\xi_x}{J}\right)+$$
$$+\frac{\partial}{\partial \eta}\left(\frac{\xi_x\eta_x}{J}\frac{\partial \varphi}{\partial \xi}\right)-\xi_x\frac{\partial \varphi}{\partial \xi}\frac{\partial}{\partial \eta}\left(\frac{\eta_x}{J}\right)+\frac{\partial}{\partial \eta}\left(\frac{\eta_x^2}{J}\frac{\partial \varphi}{\partial \eta}\right)-\eta_x\frac{\partial \varphi}{\partial \eta}\frac{\partial}{\partial \eta}\left(\frac{\eta_x}{J}\right)+$$
$$+\frac{\partial}{\partial \xi}\left(\frac{\xi_y^2}{J}\frac{\partial \varphi}{\partial \xi}\right)-\xi_y\frac{\partial \varphi}{\partial \xi}\frac{\partial}{\partial \xi}\left(\frac{\xi_y}{J}\right)+\frac{\partial}{\partial \xi}\left(\frac{\xi_y\eta_y}{J}\frac{\partial \varphi}{\partial \eta}\right)-\eta_y\frac{\partial \varphi}{\partial \eta}\frac{\partial}{\partial \xi}\left(\frac{\xi_y}{J}\right)+$$
$$+\frac{\partial}{\partial \eta}\left(\frac{\xi_y\eta_y}{J}\frac{\partial \varphi}{\partial \xi}\right)-\xi_y\frac{\partial \varphi}{\partial \xi}\frac{\partial}{\partial \eta}\left(\frac{\eta_y}{J}\right)+\frac{\partial}{\partial \eta}\left(\frac{\eta_y^2}{J}\frac{\partial \varphi}{\partial \eta}\right)-\eta_y\frac{\partial \varphi}{\partial \eta}\frac{\partial}{\partial \eta}\left(\frac{\eta_y}{J}\right)+$$
$$+\frac{\chi}{yJ}\left(\xi_y\frac{\partial \varphi}{\partial \xi}+\eta_y\frac{\partial \varphi}{\partial \eta}\right)=\frac{\Phi}{J}.$$

Taking into account the following relations between components of the Jacobian of the transformation

$$y_\eta = \frac{\xi_x}{J}, \quad -y_\xi = \frac{\eta_x}{J}, \quad -x_\eta = \frac{\xi_y}{J}, \quad x_\xi = \frac{\eta_y}{J},$$

note that

$$-\xi_x \frac{\partial \varphi}{\partial \xi} y_{\eta\xi} - \eta_x \frac{\partial \varphi}{\partial \eta} y_{\eta\xi} + \xi_x \frac{\partial \varphi}{\partial \xi} y_{\eta\xi} + \eta_x \frac{\partial \varphi}{\partial \eta} y_{\xi\eta} +$$

$$+\xi_y \frac{\partial \varphi}{\partial \xi} x_{\eta\xi} + \eta_y \frac{\partial \varphi}{\partial \eta} x_{\eta\xi} - \xi_y \frac{\partial \varphi}{\partial \xi} x_{\eta\xi} - \eta_y \frac{\partial \varphi}{\partial \eta} x_{\xi\eta} \equiv 0, \tag{4.269}$$

therefore,

$$\frac{\partial}{\partial \xi}\left[\frac{\left(\xi_x^2+\xi_y^2\right)}{J}\frac{\partial \varphi}{\partial \xi}\right] + \frac{\partial}{\partial \xi}\left[\frac{\left(\xi_x\eta_x+\xi_y\eta_y\right)}{J}\frac{\partial \varphi}{\partial \eta}\right] + \frac{\partial}{\partial \eta}\left[\frac{\left(\xi_x\eta_x+\xi_y\eta_y\right)}{J}\frac{\partial \varphi}{\partial \xi}\right] +$$

$$+\frac{\partial}{\partial \eta}\left[\frac{\left(\eta_x^2+\eta_y^2\right)}{J}\frac{\partial \varphi}{\partial \eta}\right] + \frac{\chi}{yJ}\left(\xi_y \frac{\partial \varphi}{\partial \xi} + \eta_y \frac{\partial \varphi}{\partial \eta}\right) = \frac{\Phi}{J}. \tag{4.270}$$

The continuity equations for the electrons and ions (4.262) and (4.263) are similar, so consider the transformation of only the equation for determining the electron concentrations. In eq. (4.262), we use the relation for the electron flux density (4.265), as well as the relation between the electric field strength and the electric potential (4.264). Rewriting (4.262) using variables (ξ, η), then dividing it by Jacobian J and using partial differentiation, we obtain

$$\frac{\partial}{\partial t}\left(\frac{n_e}{J}\right) + \frac{\partial}{\partial \xi}\left(n_e\mu_e \frac{\xi_x^2}{J}\frac{\partial \varphi}{\partial \xi}\right) - n_e\mu_e\xi_x \frac{\partial \varphi}{\partial \xi}\frac{\partial}{\partial \xi}\left(\frac{\xi_x}{J}\right) +$$

$$+\frac{\partial}{\partial \xi}\left(n_e\mu_e \frac{\xi_x\eta_x}{J}\frac{\partial \varphi}{\partial \eta}\right) - n_e\mu_e\eta_x \frac{\partial \varphi}{\partial \eta}\frac{\partial}{\partial \xi}\left(\frac{\xi_x}{J}\right) + \frac{\partial}{\partial \eta}\left(n_e\mu_e \frac{\xi_x\eta_x}{J}\frac{\partial \varphi}{\partial \xi}\right) - n_e\mu_e\xi_x \frac{\partial \varphi}{\partial \xi}\frac{\partial}{\partial \eta}\left(\frac{\eta_x}{J}\right) +$$

$$+\frac{\partial}{\partial \eta}\left(n_e\mu_e \frac{\eta_x^2}{J}\frac{\partial \varphi}{\partial \eta}\right) - n_e\mu_e\eta_x \frac{\partial \varphi}{\partial \eta}\frac{\partial}{\partial \eta}\left(\frac{\eta_x}{J}\right) + \frac{\partial}{\partial \xi}\left(n_e\mu_e \frac{\xi_y^2}{J}\frac{\partial \varphi}{\partial \xi}\right) - n_e\mu_e\xi_y \frac{\partial \varphi}{\partial \xi}\frac{\partial}{\partial \xi}\left(\frac{\xi_y}{J}\right) +$$

$$+\frac{\partial}{\partial \xi}\left(n_e\mu_e \frac{\xi_y\eta_y}{J}\frac{\partial \varphi}{\partial \eta}\right) - n_e\mu_e\eta_y \frac{\partial \varphi}{\partial \eta}\frac{\partial}{\partial \xi}\left(\frac{\xi_y}{J}\right) + \frac{\partial}{\partial \eta}\left(n_e\mu_e \frac{\xi_y\eta_y}{J}\frac{\partial \varphi}{\partial \xi}\right) - n_e\mu_e\xi_y \frac{\partial \varphi}{\partial \xi}\frac{\partial}{\partial \eta}\left(\frac{\eta_y}{J}\right) +$$

$$+\frac{\partial}{\partial \eta}\left(n_e\mu_e \frac{\eta_y^2}{J}\frac{\partial \varphi}{\partial \eta}\right) - n_e\mu_e\eta_y \frac{\partial \varphi}{\partial \eta}\frac{\partial}{\partial \eta}\left(\frac{\eta_y}{J}\right) + \frac{n_e\mu_e}{yJ}\left(\xi_y \frac{\partial \varphi}{\partial \xi} + \eta_y \frac{\partial \varphi}{\partial \eta}\right) =$$

$$= \frac{\partial}{\partial \xi}\left(D_e \frac{\xi_x^2}{J}\frac{\partial n_e}{\partial \xi} + D_e \frac{\xi_x \eta_x}{J}\frac{\partial n_e}{\partial \eta}\right) - D_e\left(\xi_x \frac{\partial n_e}{\partial \xi} + \eta_x \frac{\partial n_e}{\partial \eta}\right)\frac{\partial}{\partial \xi}\left(\frac{\xi_x}{J}\right) +$$

$$+ \frac{\partial}{\partial \eta}\left(D_e \frac{\xi_x \eta_x}{J}\frac{\partial n_e}{\partial \xi} + D_e \frac{\eta_x^2}{J}\frac{\partial n_e}{\partial \eta}\right) - D_e\left(\xi_x \frac{\partial n_e}{\partial \xi} + \eta_x \frac{\partial n_e}{\partial \eta}\right)\frac{\partial}{\partial \eta}\left(\frac{\eta_x}{J}\right) +$$

$$+ \frac{\partial}{\partial \xi}\left(D_e \frac{\xi_y^2}{J}\frac{\partial n_e}{\partial \xi} + D_e \frac{\xi_y \eta_y}{J}\frac{\partial n_e}{\partial \eta}\right) - D_e\left(\xi_y \frac{\partial n_e}{\partial \xi} + \eta_y \frac{\partial n_e}{\partial \eta}\right)\frac{\partial}{\partial \xi}\left(\frac{\xi_y}{J}\right) +$$

$$+ \frac{\partial}{\partial \eta}\left(D_e \frac{\xi_y \eta_y}{J}\frac{\partial n_e}{\partial \xi} + D_e \frac{\eta_y^2}{J}\frac{\partial n_e}{\partial \eta}\right) - D_e\left(\xi_y \frac{\partial n_e}{\partial \xi} + \eta_y \frac{\partial n_e}{\partial \eta}\right)\frac{\partial}{\partial \eta}\left(\frac{\eta_y}{J}\right) +$$

$$+ \frac{\chi D_e}{Jy}\left(\xi_x \frac{\partial n_e}{\partial \xi} + \eta_y \frac{\partial n_e}{\partial \eta}\right) + \frac{\alpha}{J}|\mathbf{\Gamma}_e| - \frac{\beta}{J} n_e n_+ .$$

By virtue of relations (4.269), this equation is simplified as follows:

$$\frac{\partial n_e}{\partial t}\left(\frac{n_e}{J}\right) + \frac{\partial}{\partial \xi}\left(n_e \mu_e \frac{\xi_x^2 + \xi_y^2}{J}\frac{\partial \varphi}{\partial \xi}\right) + \frac{\partial}{\partial \eta}\left(n_e \mu_e \frac{\eta_x^2 + \eta_y^2}{J}\frac{\partial \varphi}{\partial \eta}\right) +$$

$$+ \frac{\partial}{\partial \xi}\left(n_e \mu_e \frac{\xi_x \eta_x + \xi_y \eta_y}{J}\frac{\partial \varphi}{\partial \eta}\right) + \frac{\partial}{\partial \eta}\left(n_e \mu_e \frac{\xi_x \eta_x + \xi_y \eta_y}{J}\frac{\partial \varphi}{\partial \xi}\right) + \frac{n_e \mu_e}{yJ}\left(\xi_y \frac{\partial \varphi}{\partial \xi} + \eta_y \frac{\partial \varphi}{\partial \eta}\right) =$$

$$= \frac{\partial}{\partial \xi}\left(D_e \frac{\xi_x^2 + \xi_y^2}{J}\frac{\partial n_e}{\partial \xi}\right) + \frac{\partial}{\partial \eta}\left(D_e \frac{\eta_x^2 + \eta_y^2}{J}\frac{\partial n_e}{\partial \eta}\right) +$$

$$+ \frac{\partial}{\partial \xi}\left(D_e \frac{\xi_x \eta_x + \xi_y \eta_y}{J}\frac{\partial n_e}{\partial \eta}\right) + \frac{\partial}{\partial \eta}\left(D_e \frac{\xi_x \eta_x + \xi_y \eta_y}{J}\frac{\partial n}{\partial \xi}\right) +$$

$$+ \frac{\chi D_e}{yJ}\left(\xi_y \frac{\partial n_e}{\partial \xi} + \eta_y \frac{\partial n_e}{\partial \eta}\right) + \frac{\alpha}{J}|\mathbf{\Gamma}_e| - \frac{\beta}{J} n_e n_+ .$$

To simplify this equation, we introduce the notation

$$a = \frac{\eta_x^2 + \eta_y^2}{J},\ b = \frac{\xi_x^2 + \xi_y^2}{J},\ g = \frac{\xi_x \eta_x + \xi_y \eta_y}{J} \tag{4.271}$$

After this, the equation of continuity for electrons in the new coordinate system takes the form

$$\frac{\partial}{\partial t}\left(\frac{n_e}{J}\right) + \frac{\partial}{\partial \xi}\left(n_e \mu_e b \frac{\partial \varphi}{\partial \xi}\right) + \frac{\partial}{\partial \eta}\left(n_e \mu_e a \frac{\partial \varphi}{\partial \eta}\right) + \frac{\partial}{\partial \xi}\left(n_e \mu_e g \frac{\partial \varphi}{\partial \eta}\right) +$$

$$+ \frac{\partial}{\partial \eta}\left(n_e \mu_e g \frac{\partial \varphi}{\partial \xi}\right) + \frac{n_e \mu_e}{yJ}\left(\xi_y \frac{\partial \varphi}{\partial \xi} + \eta_y \frac{\partial \varphi}{\partial \eta}\right) =$$

$$= \frac{\partial}{\partial \xi}\left(D_e b \frac{\partial n_e}{\partial \xi}\right) + \frac{\partial}{\partial \eta}\left(D_e a \frac{\partial n_e}{\partial \eta}\right) + \frac{\partial}{\partial \xi}\left(D_e g \frac{\partial n_e}{\partial \eta}\right) + \frac{\partial}{\partial \eta}\left(D_e g \frac{\partial n_e}{\partial \xi}\right) +$$

$$+ \frac{\chi D_e}{yJ}\left(\xi_y \frac{\partial n_e}{\partial \xi} + \eta_y \frac{\partial n_e}{\partial \eta}\right) + \frac{\alpha}{J}|\mathbf{\Gamma}_e| - \frac{\beta}{J} n_e n_+ \tag{4.272}$$

We note that, in spite of the considerable complication of the equation in introduced coordinate system (ξ, η), the construction of the finite-difference scheme for (4.272) proves to be a simpler problem than for the initial equation (4.262).

For the purpose of constructing a finite-difference scheme, we use the grid template shown in Figure 4.31. To obtain the finite-difference relations by the finite-volume method the integrating of eq. (4.272) is used on the volume selected in Figure 4.31

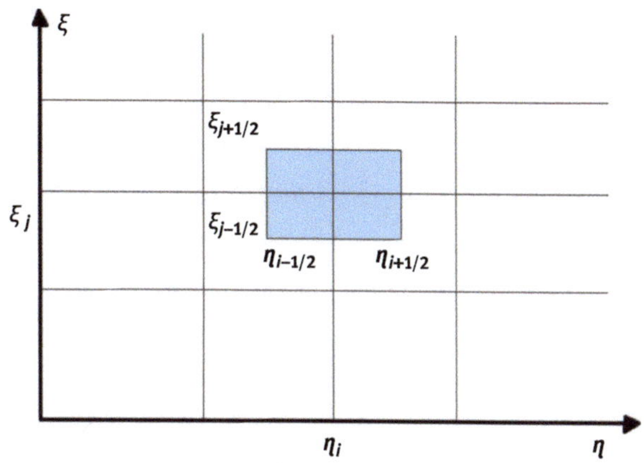

Figure 4.31: Difference grid template.

$$Vol_{i,j} = \int_{\eta_{i-1/2}}^{\eta_{i+1/2}} d\eta \int_{\xi_{j-1/2}}^{\xi_{j+1/2}} \{\} d\xi = p_i q_j, \tag{4.273}$$

where

$$p_i = \frac{1}{2}(\eta_{i+1} - \eta_{i-1}),\ q_j = \frac{1}{2}\left(\xi_{j+1} - \xi_{j-1}\right). \tag{4.274}$$

This leads to the five-point finite-difference equation of the following canonical form:

$$A_{i,j}U_{i-1,j} + B_{i,j}U_{i+1,j} - C_{i,j}U_{i,j} + \bar{A}_{i,j}U_{i,j-1} + \bar{B}_{i,j}U_{i,j+1} + F_{i,j} = 0\,, \tag{4.275}$$

where $$U_{i,j} = \varphi_{i,j},$$

$$A_{i,j} = \frac{a_{i-1/2,j} + g_{i-1/2,j}}{p_i p_i^-}, \; B_{i,j} = \frac{a_{i+1/2,j} + g_{i+1/2,j}}{p_i p_i^+} + \chi \left(\frac{\eta_y}{yJ}\right)_{i,j} \frac{1}{p_i},$$

$$\bar{A}_{i,j} = \frac{b_{i,j-1/2} + g_{i,j-1/2}}{q_j q_j^{-1}}, \; \bar{B}_{i,j} = \frac{b_{i,j+1/2} + g_{i,j+1/2}}{q_j q_j^+} + \chi \left(\frac{\xi_y}{yJ}\right)_{i,j} \frac{1}{q_j},$$

$$F_{i,j} = \frac{\varepsilon^*}{J_{i,j}} (N_+ - N_-) \frac{N_0 L^2}{\varepsilon} \tag{4.276}$$

For the convenience of numerical implementation, the following dimensionless functions and variables are introduced:

$$N_+ = \frac{n_+}{N_0}, \; N_e = \frac{n_e}{N_0}, \; U = \frac{\varphi}{\varepsilon}, \; X = \frac{x}{L}, \; Y = \frac{y}{L},$$

where N_0 is the characteristic concentration of electrons in the positive column of the glow discharge ($N_0 = 10^9$ cm^{-3}); L is the distance between the electrodes.

In (4.276) we used the variables

$$a_{i\pm 1/2,j} = \frac{1}{2}\left(a_{i,j} + a_{i\pm 1,j}\right), \; b_{i,j\pm 1/2} = \frac{1}{2}\left(b_{i,j} + b_{i,j\pm 1}\right),$$

$$g_{i\pm 1/2,j} = \frac{1}{2}\left(g_{i,j} + g_{i\pm 1,j}\right), \; g_{i,j\pm 1/2} = \frac{1}{2}\left(g_{i,j} + g_{i,j\pm 1}\right),$$

$$p_i^+ = \eta_{i+1} - \eta_i, \; p_i^- = \eta_i - \eta_{i-1}, \; q_j^+ = \xi_{j+1} - \xi_j, \; q_j^- = \xi_j - \xi_{j-1}$$

When obtaining finite-difference equations for electron concentrations, the continuity equation (4.258) is integrated according to the volume shaded in Figure 4.31 and over a time interval $\tau = t^{p+1} - t^p$

$$\mathrm{Vol}_{i,j}^{p,p+1} = \int_{t_p}^{t_{p+1}} \mathrm{d}t \int_{\eta_{i-1/2}}^{\eta_{i+1/2}} \mathrm{d}\eta \int_{\xi_{j-1/2}}^{\xi_{j+1/2}} \{\} \mathrm{d}\xi = \tau p_i q_j.$$

This leads to a five-point equation of the form (4.275), where

$$U_{i,j} = n_{e,i,j},$$

$$A_{i,j} = \frac{R_L^+}{p_i} + \frac{[D_e(a+g)]_{i-1/2,j}}{p_i p_i^-}, \; B_{i,j} = -\frac{R_R^-}{p_i} + \frac{[D_e(a+g)]_{i+1/2,j}}{p_i p_i^+} + \frac{\chi}{p_i}\left(\frac{\eta_y D_e}{yJ}\right)_{i,j},$$

$$C_{i,j} = \frac{1}{\tau J_{i,j}} + A_{i,j} + B_{i,j} + \bar{A}_{i,j} + \bar{B}_{i,j} + \left(\frac{\mu_e \eta_y}{yJ}\right)_{i,j} \frac{\varphi_{i+1,j} - \varphi_{i-1,j}}{2p_i} + \left(\frac{\mu_e \xi_y}{yJ}\right)_{i,j} \frac{\varphi_{i,j+1} - \varphi_{i,j-1}}{2q_j},$$

$$\bar{A}_{i,j} = \frac{S_L^+}{q_j} + \frac{[D_e(b+g)]_{i,j-1/2}}{q_j q_j^-}, \quad \bar{B}_{i,j} = -\frac{S_R^-}{q_j} + \frac{[D_e(b+g)]_{i,j+1/2}}{q_j q_j^+} + \frac{\chi}{q_j}\left(\frac{\xi_y D_e}{yJ}\right)_{i,j},$$

$$F_{i,j} = \frac{U_{e,i,j}^p}{\tau J_{i,j}} + \dot{\omega}_e, \quad \dot{\omega}_e = \left(\frac{\alpha}{J}|\mathbf{\Gamma}_e| - \frac{\beta}{J} n_e n_+\right)\frac{L^2}{\varepsilon \mu_e N_0}. \tag{4.277}$$

Here we use the following notation for the projection of the electron velocity in the coordinate system (ξ, η)

$$R_e = V_e + Z_e, \quad S_e = V_e + W_e, \quad V_e = \mu_e a \frac{\partial \varphi}{\partial \eta},$$

$$U_e = \mu_e b \frac{\partial \varphi}{\partial \xi}, \quad Z_e = \mu_e q \frac{\partial \varphi}{\partial \xi}, \quad W_e = \mu_e g \frac{\partial \varphi}{\partial \eta}.$$

Note that the velocity components V_e and U_e correspond to the drift velocities along the coordinate lines, and the velocity components Z_e and W_e correspond to the contribution of the corresponding cross-components. The finite-difference equation for ions is determined by the same relationships with the difference that

$$V_+ = -\mu_e a \frac{\partial \varphi}{\partial \eta}, \quad U_+ = -\mu_e b \frac{\partial \varphi}{\partial \xi}, \quad Z_+ = -\mu_e q \frac{\partial \varphi}{\partial \xi}, \quad W_+ = -\mu_e g \frac{\partial \varphi}{\partial \eta}$$

in accordance with the definition of charged particle flux densities (4.261).

Stability in the numerical integration of finite-difference equations of continuity of electrons and ions is provided by the following relationships (Figure 4.32):

$$R_{R,L}^{\pm} = \frac{1}{2}(R_{R,L} \pm |R_{R,L}|), \quad S_{R,L}^{\pm} = \frac{1}{2}(S_{R,L} \pm |S_{R,L}|)$$

and also the one-sided derivatives of the drift and diffusion terms in (4.277), proportional to $1/(yJ)$. By analogy, an equation for the concentrations of the ions could be obtained.

4.7.3 Results of numerical modeling of a normal glow discharge

Based on the calculation model described in the previous section, a computer code was created to calculate the 2D structure of a normal glow discharge. Trial test calculations were performed using the example of a well-studied normal glow discharge in molecular nitrogen between plane electrodes at $p = 5$ Torr, $\varepsilon = 1{,}000$ V, $R_0 = 300$ kΩ, $\gamma = 0.33$, $L = 2$ cm (Surzhikov S.T. and Shang J.S., 2004a). Two-dimensional fields of concentrations of ions and electrons for axisymmetric ($\chi = 1$) and flat ($\chi = 0$) glow discharges are shown in Figure 4.33. The data presented are in good agreement with the results of calculations (Surzhikov S.T. and Shang J.S., 2004a). The calculations were

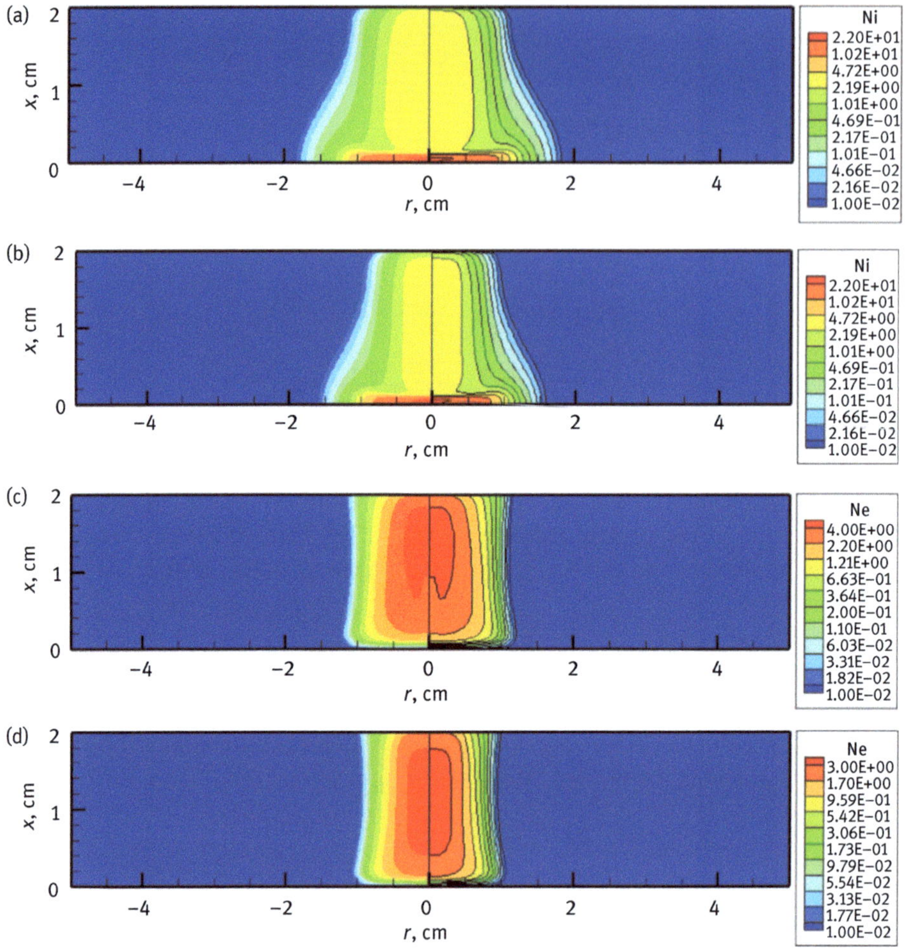

Figure 4.32: Distribution of the concentrations of ions (a, b) and electrons (c, d) in an axisymmetric (a, c) and flat (b, d) discharges between plane electrodes at $p = 5$ *Torn*, $\varepsilon = 1,000$ V, $R_0 = 300$ kΩ, $\gamma = 0.33$, $L = 2$ cm.

performed on a 71 × 71 grid. In these figures, the cathode and anodic layers are well identified, as well as the quasi-neutral positive column.

Figures 4.33 and 4.34 show the axial distributions of electron and ion concentrations along the axis of cylindrical and planar discharges, as well as ionization and recombination rates. The latter are normalized by $N_0\mu_e\varepsilon/L^2$, having the dimension 1/(cm^3×s). The concentrations of ions in the cathode layer for axisymmetric and flat discharges are close, and the concentration of charged particles in the positive column of the axisymmetric discharge is somewhat higher. It is seen in Figure 4.34

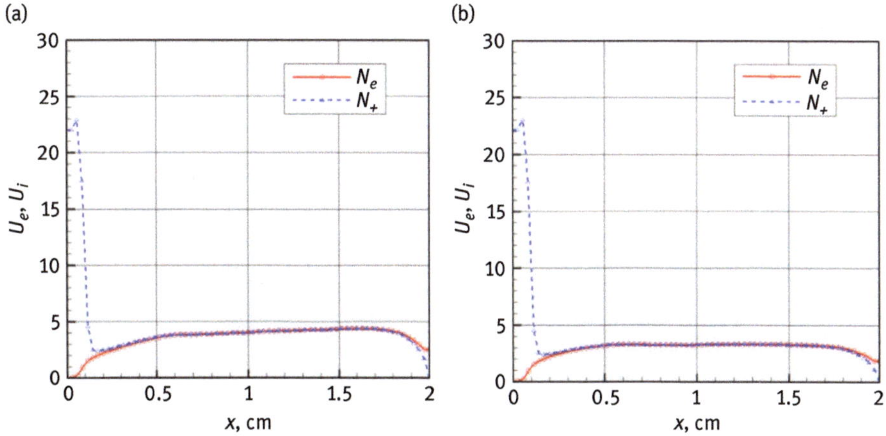

Figure 4.33: The distribution of electron and ion concentrations along the symmetry axis (a) and the symmetry plane (b) for the conditions in the discharge in Figure 4.32.

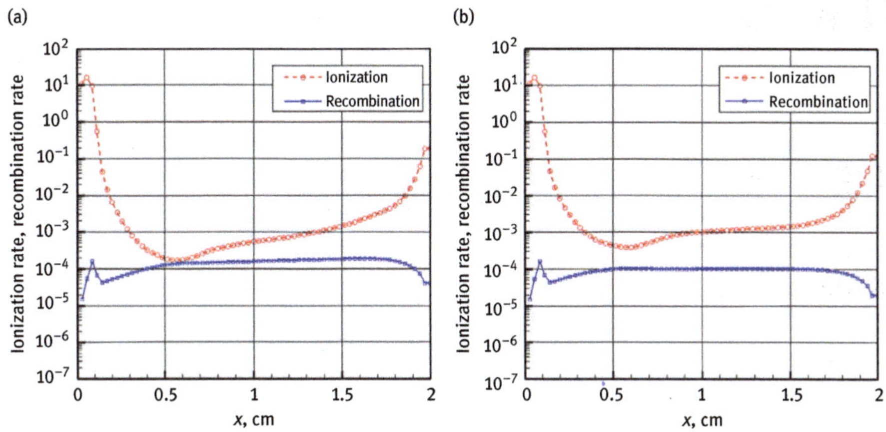

Figure 4.34: Distribution of ionization and recombination rates along the symmetry axis (a) and the symmetry plane (b) for the conditions in the discharge in Figure 4.32.

that the cathode and anode layers are the main regions of the generation of charged particles.

Calculations of an axisymmetric normal glow discharge line between curvilinear electrodes were performed for similar conditions discussed above, but in the range $\varepsilon = 1{,}0$–$1{,}75$ V. The radius of curvature of the cathode and the anode were $R_c = 5$ cm and $R_a = 10$ cm, respectively. Configuration of the computational domain with the number of nodes 141 × 141 is shown in Figure 4.30.

Figures 4.35 and 4.36 show the fields of ion and electron concentrations, electric potential and reduced electric field E/p in discharges at $\varepsilon = 1,0$ V and $\varepsilon = 1,75$ V, respectively. With increasing ε, the transverse dimensions of the glow discharge column increase.

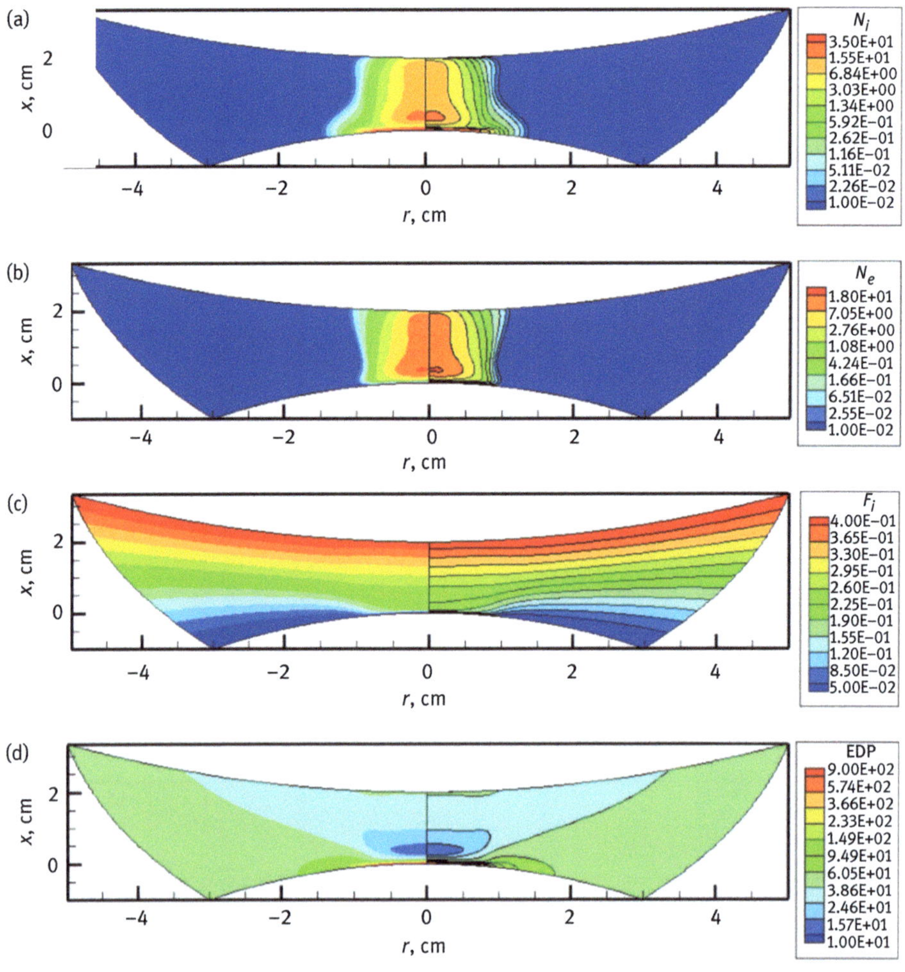

Figure 4.35: Distribution of concentrations of ions $N_i = n_+/N_0$ (a) and electrons $N_e = n_e/N_0$ (b), electric potential (c) and reduced field EDP $= E/p$ in V /cm×Torr (d) between segmental electrodes at $p = 5$ Torr, $\varepsilon = 1,000$ V, $R_0 = 300$ kΩ, $\gamma = 0.33$, $L = 2$ cm.

Also, as in the case of a discharge between plane electrodes, the cathode layer (region of increased ion concentration) is characterized by a potential well of the electric potential (Figures 4.35c and 4.36 c) and increased values of the reduced electric field (Figures 4.35d and 4.36d).

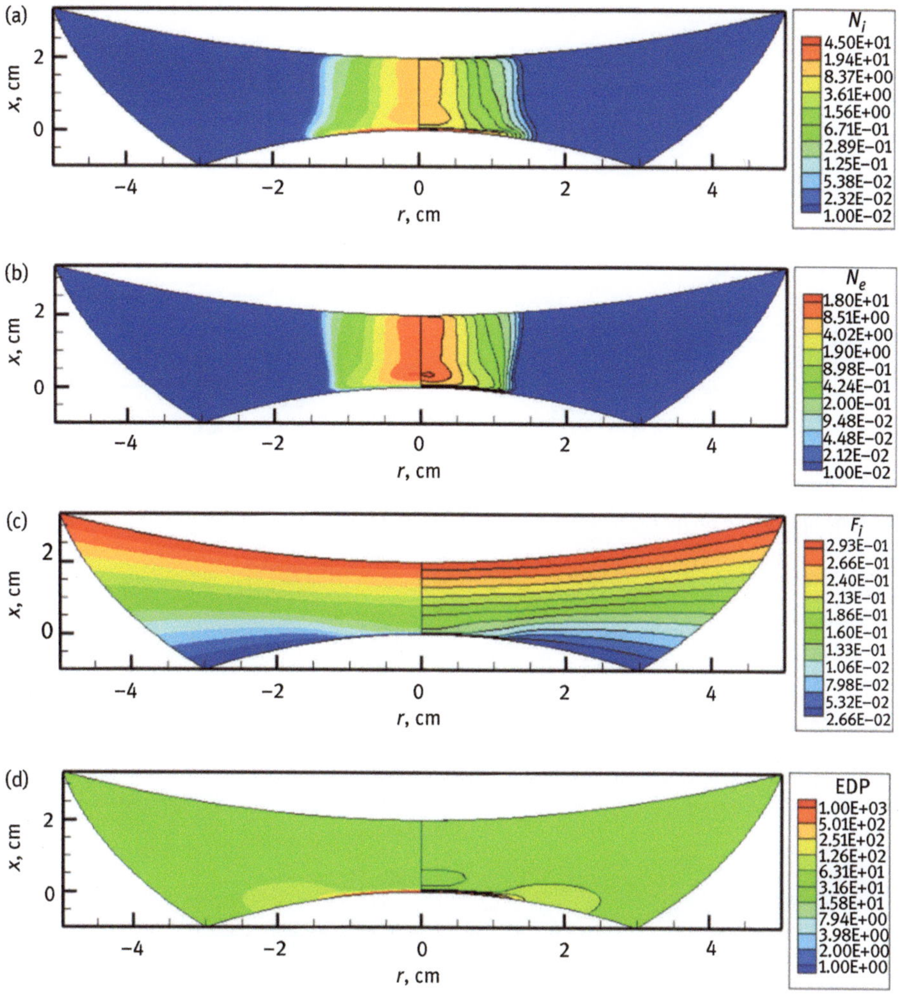

Figure 4.36: Distribution of concentrations of ions $N_i = n_+/N_0$ (a) and electrons $N_e = n_e/N_0$ (b), electric potential (c) and reduced field EDP $= E/p$ in V/cm×Torr (d) between segmental electrodes at $p = 5$ Torr, $\varepsilon = 1,750$ V, $R_0 = 300$ kΩ, $\gamma = 0.33$, $L = 2$ cm.

The axial distributions of the concentrations of charged particles in the two glow discharges, the distribution of the electric potential φ of the axial component of the electric field, and the reduced field are shown in Figures 4.37 and 4.38. Figure 4.39a, b shows the rates of ionization and recombination. We note the main differences from the normal glow discharge between flat electrodes. In the axial distribution of electron and ion concentrations, in a positive column, a local maximum appears.

An analysis of the distributions of the rates of ionization and recombination (Figure 4.39a, b) shows that recombination prevails in the vicinity of this local

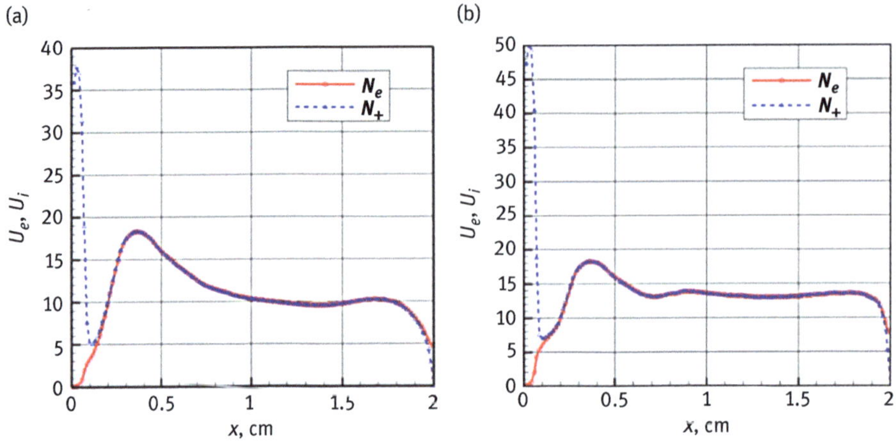

Figure 4.37: The distribution of electron and ion concentrations along the symmetry axis (a) and the symmetry plane (b) for $\varepsilon = 1,000$ V (a) and $\varepsilon = 1,750$ V (b).

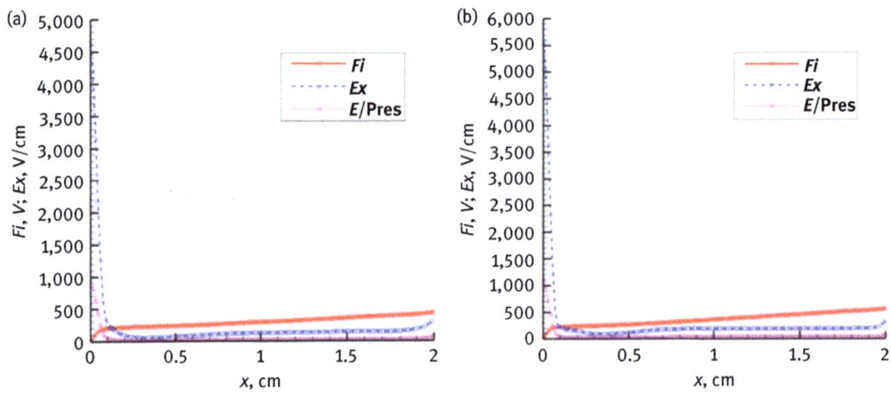

Figure 4.38: The axial distributions of the electric potential $Fi = \varphi$ (in V), the axial component of the electric field strength E_x, V/cm, and the reduced field E/p (in V/cm×Torr) along the symmetry axis at $\varepsilon = 1,000$ V (a) and $\varepsilon = 1,750$ V (b).

maximum. The rate of ionization in the cathode layer is several times greater than the corresponding value for the discharge between plane electrodes (compare Figures 4.39a and 4.34a). With increasing ε, the concentration of ions and electrons in the positive column increases (Figure 4.37a, c).

An important result of the study of the structure of a normal glow discharge between curved electrodes was the demonstration of a satisfactory fulfillment of the law of the normal current density in Figure 4.40. It can be seen that variation of ε in

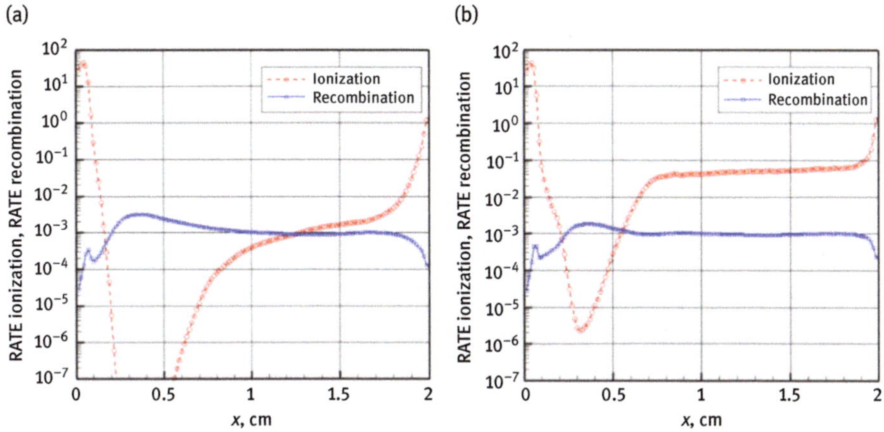

Figure 4.39: Axial distributions of the rate of ionization and recombination at $\varepsilon = 1,000$ V (a) and $\varepsilon = 1,750$ V (b), referred to the value $N_0 \mu_e \varepsilon / L^2$.

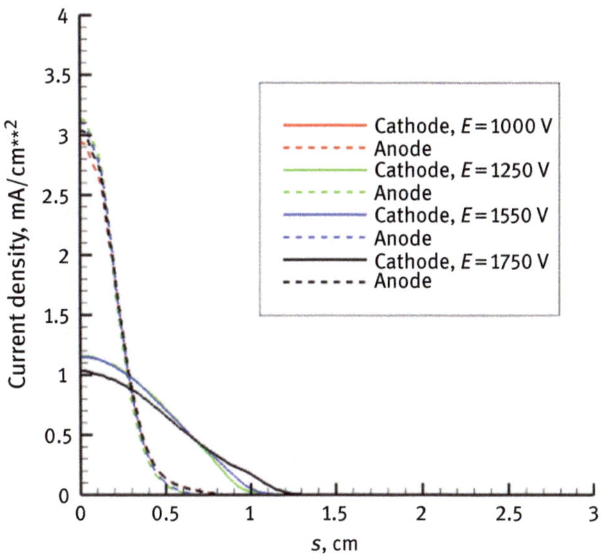

Figure 4.40: Current density at the cathode and anode as ε varies in the range $\varepsilon = 1,000 - 1,750$ V.

the range of 1,000–1,750 V resulted in an insignificant change in the current density at the cathode and anode.

So, a diffusion-drift calculation model of a normal glow discharge between curvilinear electrodes is presented. The verification of the created computational model is performed using the example of a normal glow discharge between flat electrodes.

Results of numerical simulation are presented for normal glow discharge between electrodes in the form of spherical segments with radii $R_c = 5\,\text{cm}$ and $R_a = 10$ cm at $p = 5\,\text{Torr}$, $\varepsilon = 1,000 - 1,750\,\text{V}$, $R_0 = 300\,\text{k}\Omega$, $\gamma = 0.33$, the distance between electrodes $L = 2\,\text{cm}$. Satisfactory fulfillment of the law of the normal current density for the investigated discharge is shown.

4.8 Numerical simulation of the normal glow discharge for conditions of the experimental research at low distance between plane electrodes

In this section, the 2D drift-diffusion calculation model of charged particles in a normal glow discharge between plane disk electrodes is used for description recently performed experiments (Kozlov P.V., et al., 2018). The verification of the created computational model has been performed using the experimental data on a normal glow discharge between flat electrodes with distance $H = 1\,\text{cm}$. Results of numerical simulation of the VACs and experimental data are presented for the pressure of $p = 5\,\text{Torr}$ and emf of power supply in the region of 1,00–1,75 V. The sensitivity is analyzed for this computational model relative to empirical parameters used.

Experimental studies of a normal glow discharge were started about 100 years ago in connection with observations of current structures in gas discharge devices and the behavior of cathode and anode spots on electrodes. Recently, experimental studies have been performed on a large-scale experimental device with plate electrodes specially created for this purpose (Kozlov P.V., et al., 2018). A normal glow discharge in the interelectrode gas discharge gap of height $H = 2\,\text{cm}$ at pressures $p = 1\text{–}10\,\text{Torr}$ was studied. The glow discharge, in this case, was characterized by the presence of a sufficiently long positive column (about 1 cm) and near-electrodes layers.

Kozlov P.V., et al. (2018) performed a series of experiments in a gas-discharge gap at $H = 1\,\text{cm}$ and $p = 1\text{–}10\,\text{Torr}$. In this case, at low pressures, the positive column is practically absent, and the cathode layer immediately passes into the anode layer. The experimental data on current–voltage characteristics and visual observations obtained make it possible to compare with the results of numerical modeling of the structure of a glow discharge. Figures 4.41 and 4.42 show photo of the experimental setup and schematic of the task considered.

Systematic research of the normal glow discharge showed that its structure is well described by the diffusion-drift model (Shang J.S. and Surzhikov S.T.). However, this model belongs to the class of phenomenological models, in which there are a number of physical constants that are borrowed from experiment or calculations using kinetic plasma models. These constants include the coefficient of secondary ion-electron emission from the cathode and the ionization coefficient of

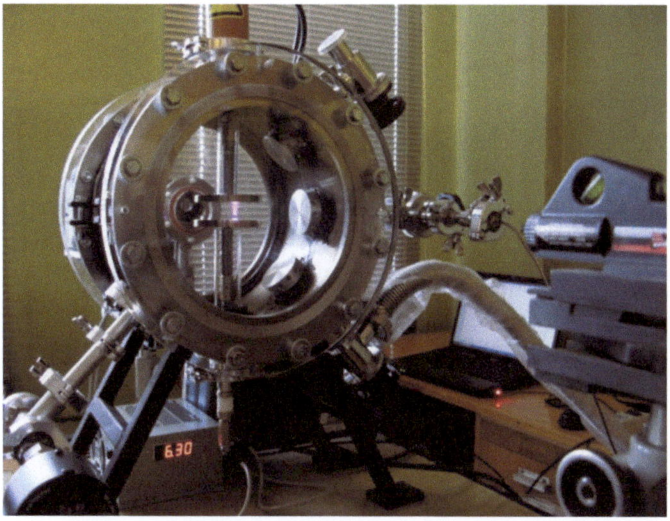

Figure 4.41: Experimental vacuum chamber and location of disk electrodes for investigation of normal glow discharge. A normal glow discharge is visible between the electrodes.

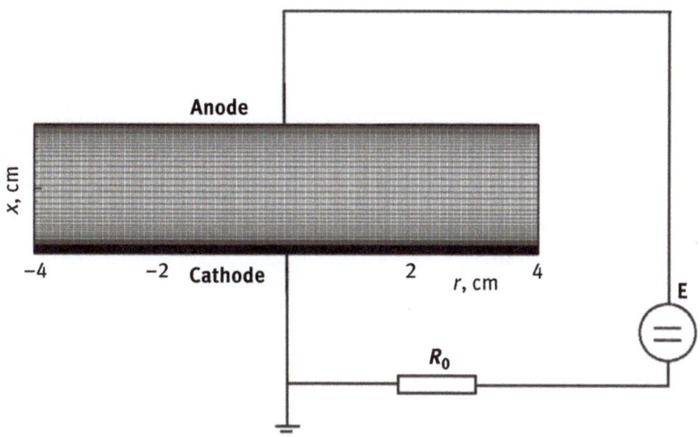

Figure 4.42: Schematic of the direct-current normal glow discharge.

the gas by electron impact. When attempting to compare the calculated and experimental data, it turned out that the successful calculated prediction of the experimental values of the current–voltage characteristic largely depends on the choice of these values.

This section presents the results of calculations of a 2D structure of a normal glow discharge in molecular nitrogen in the pressure range $p = 3$–10 Torr in a gas discharge gap of height $H = 1$ cm. Comparison with the obtained experimental data is discussed. The analysis of the structural features of a normal glow discharge was carried out with various initial data.

4.8.1 The drift-diffusion model used for the analysis of experimental data

As shown in Section 4.2, the equations of the drift-diffusion model which are used to simulate a normal glow discharge were obtained from the system of equations for a multifluid and multitemperature partially ionized gas mixture. These are

$$\mathrm{div}\mathbf{E} = -\varepsilon^{*}(n_{+} - n_e), \tag{4.278}$$

$$\frac{\partial n_e}{\partial t} + \mathrm{div}\mathbf{\Gamma}_e = \alpha|\mathbf{\Gamma}_e| - \beta n_{+} n_e, \tag{4.279}$$

$$\frac{\partial n_{+}}{\partial t} + \mathrm{div}\mathbf{\Gamma}_{+} = \alpha|\mathbf{\Gamma}_e| - \beta n_{+} n_e, \tag{4.280}$$

where $n_{+} n_e$ are the concentration of ions and electrons; $\mathbf{E}$ is the vector of electric field strength with electric potential φ

$$\mathbf{E} = -\mathrm{grad}\varphi. \tag{4.281}$$

The vectors of electron and ion flux densities $\mathbf{\Gamma}_e$, $\mathbf{\Gamma}_{+}$ are defined as follows:

$$\mathbf{\Gamma}_e = -n_e\mu_e\mathbf{E} - D_e\,\mathrm{grad}\,n_e, \mathbf{\Gamma}_{+} = -n_{+}\mu_{+}\mathbf{E} - D_{+}\,\mathrm{grad}\,n_{+}, \tag{4.282}$$

where μ_e, μ_{+} are the mobilities of electrons and ions; D_e, D_{+} are the diffusion coefficients of electrons and ions; α, β are the coefficient of impact ionization of a neutral gas by electrons (the first Townsend coefficient) and the recombination coefficient; $\varepsilon^{*} = 4\pi e = 1.81 \times 10^{-6}$ V × cm.

Boundary conditions for eqs. (4.278)–(4.280) are the following:

$$\begin{aligned}
x = 0:&\quad \Gamma_{e,x} = \gamma\Gamma_{+,T},\ \frac{\partial n_{+}}{\partial x} = 0,\ \varphi = 0;\\
x = H:&\quad \frac{\partial n_e}{\partial x} = 0,\ n_{+} = 0,\ \varphi = V;\\
r = 0:&\quad \frac{\partial n_e}{\partial r} = \frac{\partial n_{+}}{\partial r} = \frac{\partial \varphi}{\partial r} = 0;\\
r = R_c:&\quad \frac{\partial n_e}{\partial r} = \frac{\partial n_{+}}{\partial r} = \frac{\partial \varphi}{\partial r} = 0,
\end{aligned} \tag{4.283}$$

where V is the voltage drop at the electrodes, determined from the electric circuit equation

$$\frac{E - V}{eR_0} = 2\pi \int_0^{R_a} \Gamma_{e,n_S,a} r\mathrm{d}r. \tag{4.284}$$

The integral in (4.284) should be calculated along the surface of the electrode with the projection of the electron flux density ($\Gamma_{e,n_{s,a}}$) on the local normal to the surface $n_{s,a}$.

Condition (4.284) determines the electric current through the anode. For a stationary glow discharge, the total electric current through the gas-discharge gap is a constant value and can also be determined from the following equation:

$$\frac{E-V}{eR_0}=2\pi\int_0^{R_a}\Gamma_{+,n_{s,c}}\,r\mathrm{d}r,$$

where $n_{s,c}$ is the local normal to surface of cathode.

As before, it should be noted that to precisely take into account the relaxation of volume charge one should take into account the effect of the accumulation of charges on electrodes.

The closing relations for the problem being solved are for molecular nitrogen

$$\mu_e p=4.4\times10^5,\ \ \mu_i p=1.45\times10^3\ \mathrm{Torr}\cdot\mathrm{cm}^2/\mathrm{V}\cdot\mathrm{s},\ \ \beta=2\times10^{-7}\ \mathrm{cm}^3/\mathrm{s}$$

$$D_e=\mu_e T_e,\ \ D_+=\mu_i T_+\,,\ \ T_e=1\,\mathrm{eV},\ \ T_+=0.026\,\mathrm{eV}.$$

Two approximations were used for the first Townsend coefficient

$$\frac{\alpha}{p}=A\exp\left(-\frac{B}{E/p}\right),\ 1/(\mathrm{cm}\cdot\mathrm{Torr}).$$

1) In the case of the one-mode approximation, the following approximation coefficients

$$A=12\,(\mathrm{cm}\cdot\mathrm{Torr})^{-1},\ \ B=342\,\mathrm{V}/(\mathrm{cm}\cdot\mathrm{Torr}),$$

were used for any value of E/p.

2) In the case of the two-mode approximation, values of A and B were different for two regions of variation of the reduced electric field E/p:

$$\text{if } E/p>200\,\mathrm{V}/(\mathrm{cm}\cdot\mathrm{Torr}),\text{ then } A=12\,(\mathrm{cm}\cdot\mathrm{Torr})^{-1},\ B=342\,\mathrm{V}/(\mathrm{cm}\cdot\mathrm{Torr});$$
$$\text{if } E/p<200\,\mathrm{V}/(\mathrm{cm}\cdot\mathrm{Torr}),\text{ then } A=8.8\,(\mathrm{cm}\cdot\mathrm{Torr})^{-1},\ B=275\,\mathrm{V}/(\mathrm{cm}\cdot\mathrm{Torr}).$$

4.8.2 Numerical simulation results

Calculations of the normal glow discharge were performed for the electric circuit shown in Figure 4.42. The resistive resistance of the external circuit in all cases was $R_0=300\,\mathrm{k}\Omega$ and the height of the electric charging gap $H=1\,\mathrm{cm}$. The best conditions for the existence of the normal glow discharge were realized at $p=5\,\mathrm{Torr}$. At lower pressures, as will be shown below, the electric current column significantly increased its radial dimensions and the discharge approached the anomalous discharge. At high pressures, the radial dimensions of the current column decreased so much that the discharge passed into the subnormal combustion mode.

A systematic study of the normal glow discharge was also performed at $p = 3$ Torr. In addition, the calculations were performed for different values of the secondary electron emission coefficient $\gamma = 0.05$, 0.1, and 0.33, as well as for two approximation dependences of the first Townsend coefficient. Electrodynamic fields presented in Figures 4.43–4.52 correspond to the one-mode approximation of the first Townsend coefficient.

Figures 4.43–Figures 4.45 show the fields of electrodynamic functions in the normal glow discharge with $p = 5$ Torr, $\gamma = 0.33$ for three values of emf: $E = 1,000$,

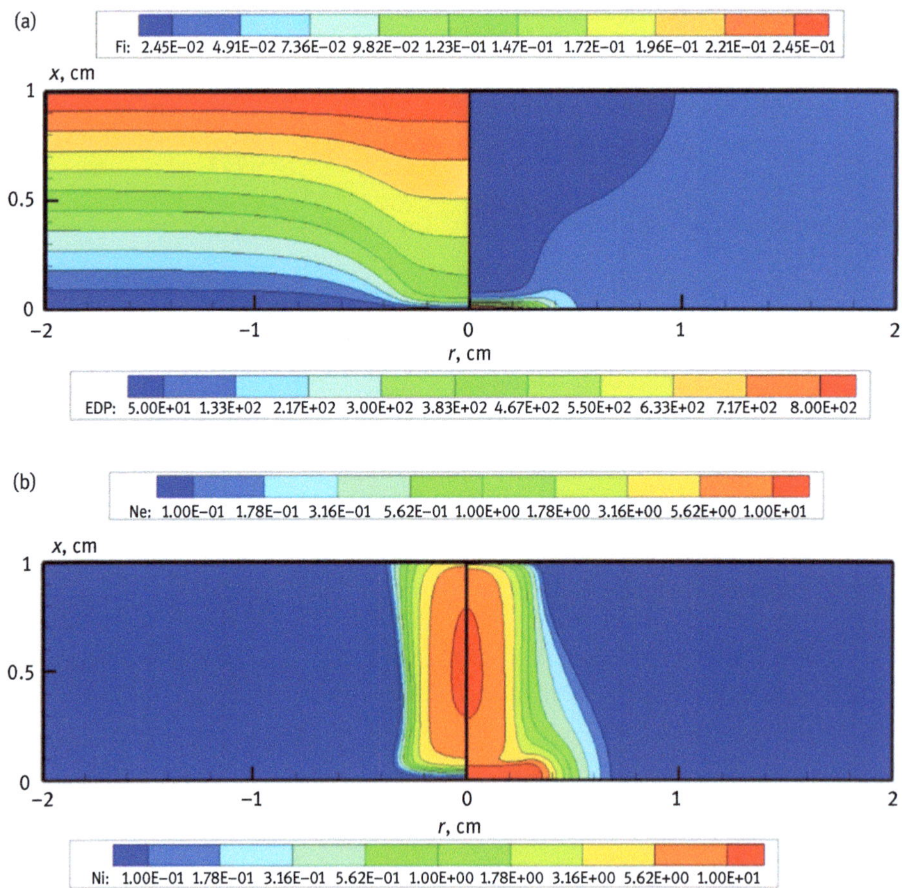

Figure 4.43: Normal glow discharge at $p = 5$ Torr, $\gamma = 0.33$, $E = 1$ kV: a) Electric potential $Fi = \varphi/E$ (to left) and reduced electric field $\mathrm{EDP} = E_x/p$, V/(cm·Torr) (to right); b) Volume concentration of electrons (to left) and ions (to right); $N_e, N_i = (n_e, n_i)/10^9\,\mathrm{cm}^{-3}$.

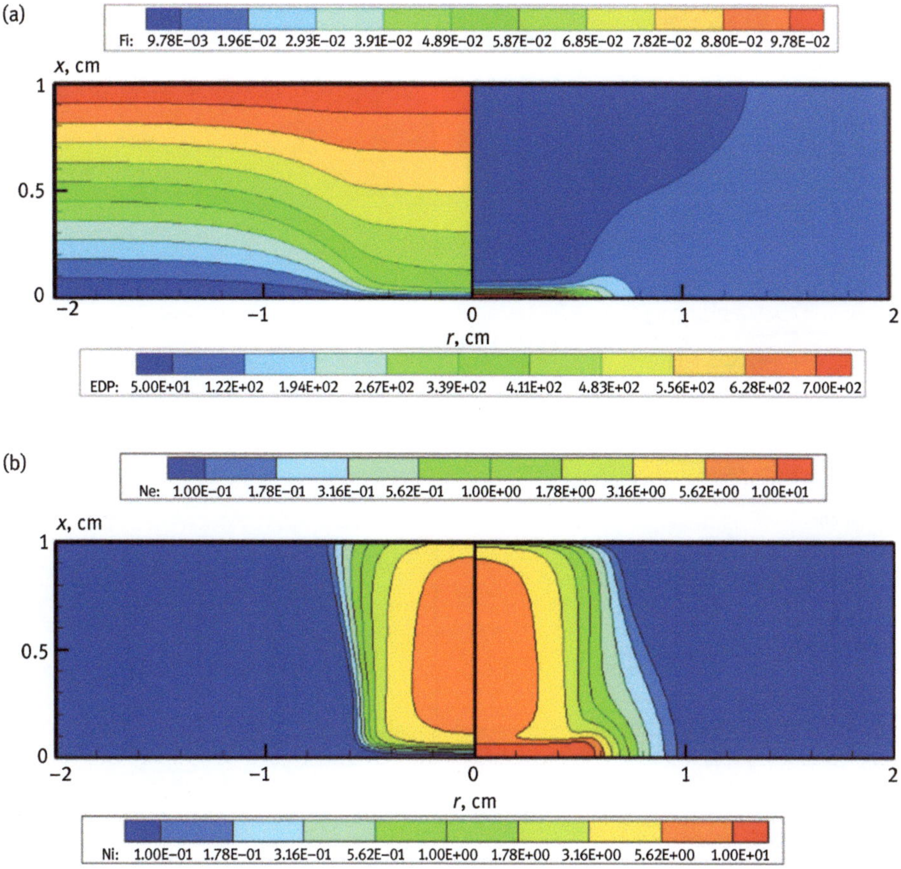

Figure 4.44: Normal glow discharge at $p = 5$ Torr, $\gamma = 0.33$, $E = 2.5$ kV: a) Electric potential $Fi = \varphi/E$ (to left) and reduced electric field $\mathrm{EDP} = E_x/p$, V/(cm·Torr) (to right); b) Volume concentration of electrons (to left) and ions (to right); N_e, $N_i = (n_e, n_i)/10^9\,\mathrm{cm}^{-3}$.

2,500, and 4,000 V. Clearly seen how the radial dimensions of the electric current column increase. However, the absolute values of the reduced field and the concentrations of ions and electrons in the vicinity of the axis of symmetry change slightly.

A comparison of the axial distributions of the concentrations of ions and electrons is given in Figure 4.46 for emf: $E = 1,000$ and 4,000 V. The distribution of current density at the cathode and anode, corresponding to different values of emf, indicates good compliance with the law of normal current density, especially for relatively large emf.

The distribution of current densities is shown in Figure 4.47a for $\gamma = 0.33$ and in Figure 4.47b for $\gamma = 0.05$. From this, it is seen that a decrease in γ leads to a noticeable decrease in current density at the cathode, but a slight decrease in the anode. This significantly increases the radial dimensions of the current column.

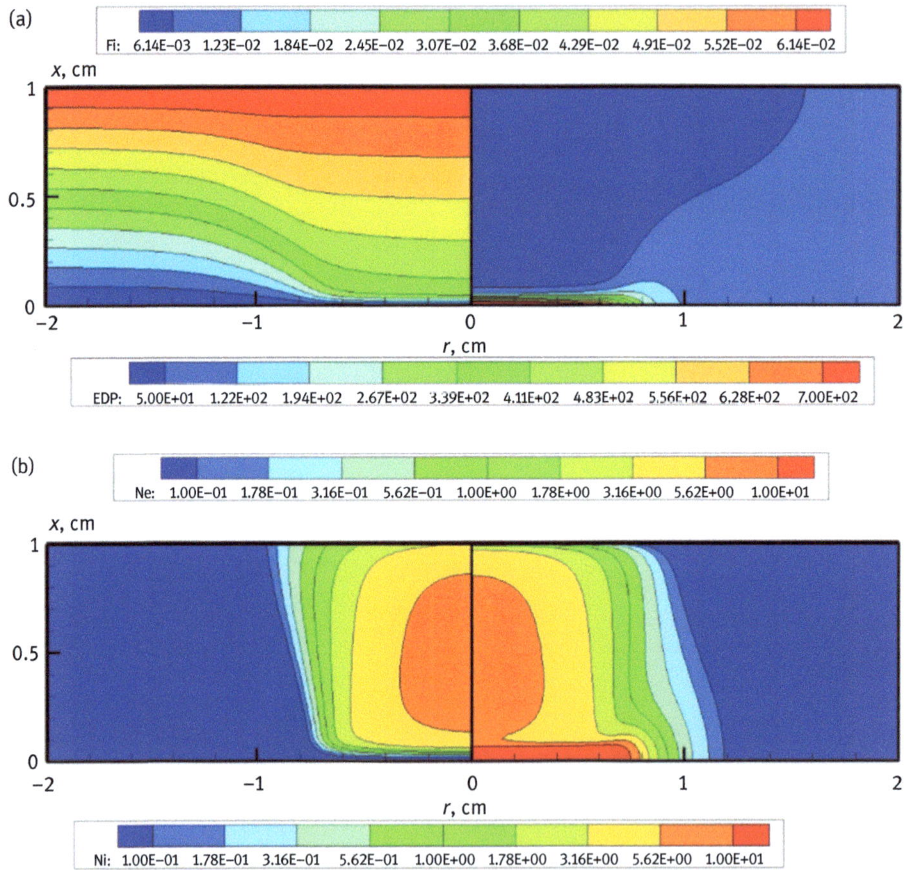

Figure 4.45: Normal glow discharge at $p = 5$ Torr, $\gamma = 0.33$, $E = 4.0$ kV: a) Electric potential $Fi = \varphi/E$ (to left) and reduced electric field $\text{EDP} = E_x/p$, V/(cm·Torr) (to right); b) Volume concentration of electrons (to left) and ions (to right); $N_e,\ N_i = (n_e,\ n_i)/10^9\,\text{cm}^{-3}$.

Figures 4.48–4.52 show the distribution of electric potential, reduced field, and concentrations of charged particles at $p = 3$ Torr; $\gamma = 0.05$ for three values of emf: $E = 1{,}000$, 2,500, and 4,000 V.

We note a significant increase in the radial dimensions of the electric current columns in comparison with the previous calculated series (Figures 4.43–4.45). In the case under consideration, the radial dimensions of the cathode and anode spots practically coincided, which follows from the spatial distributions of functions in Figures 4.48–4.50 and from the current density distributions on the cathode and anode shown in Figure 4.51. In this case, the axial distribution of the concentrations of charged particles in the axial regions changes slightly. We also pay attention to the effect of increasing the current density at the boundary of the electric current column.

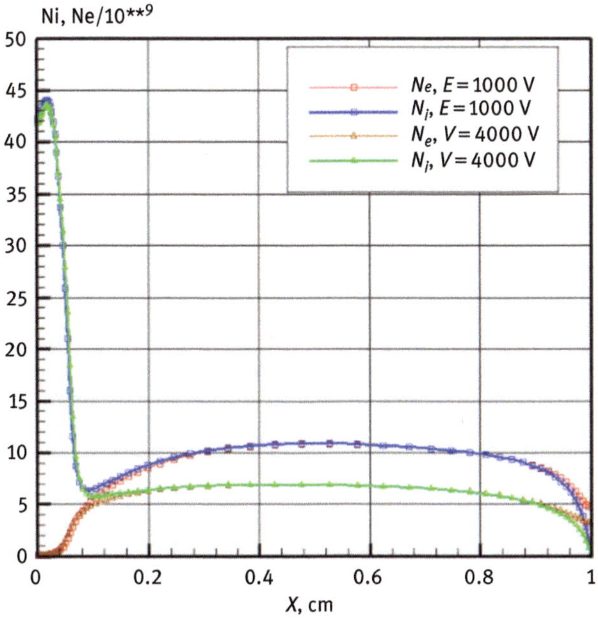

Figure 4.46: Normal glow discharge at $p = 5$ Torr, $\gamma = 0.33$, $E = 1$ and 4.0 kV. Distributions of volume concentrations of electrons and ions along the axis of symmetry.

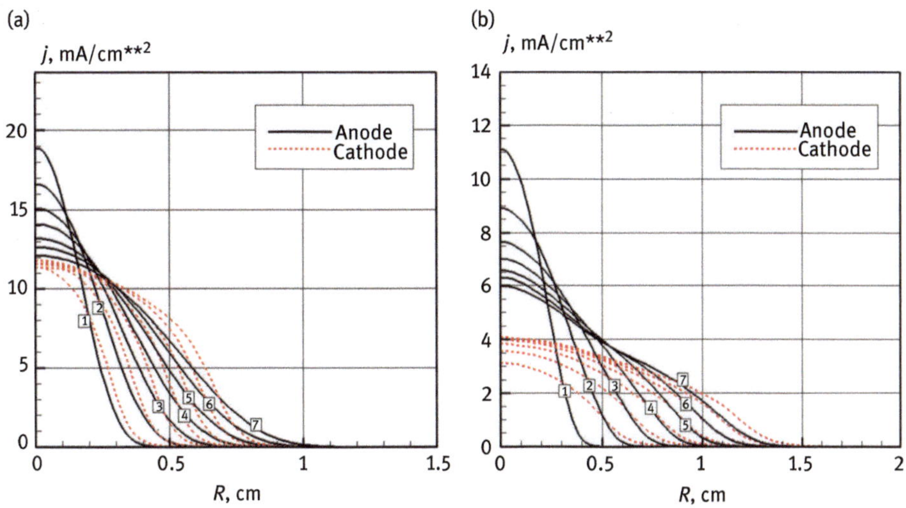

Figure 4.47: Distributions of electric current density on the cathode and anode at $E = 1.0$ kV (curves 1), $E = 1.5$ kV (curves 2), $E = 2.0$ kV (curves 3), $E = 2.5$ kV (curves 4), $E = 3.0$ kV (curves 5), $E = 3.5$ kV (curves 6), $E = 4.0$ kV (curves 7). Normal glow discharge at $p = 5$ Torr, $\gamma = 0.33$ (a) and $\gamma = 0.05$ (b).

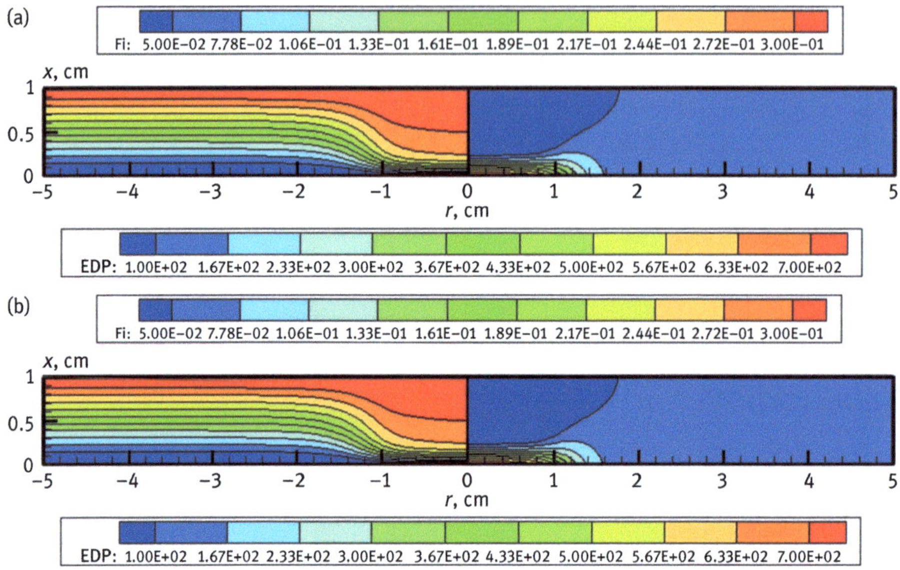

Figure 4.48: Normal glow discharge at $p = 5$ Torr, $\gamma = 0.05$, $E = 1.0$: (a) electric potential $Fi = \varphi/E$ (to left) and reduced electric field $\text{EDP} = E_x/p$, $\text{V}/(\text{cm} \times \text{Torr})$ (to right) and (b) volume concentration of electrons (to left) and ions (to right); N_e, $N_i = (n_e, n_i)/10^9$ cm^{-3}.

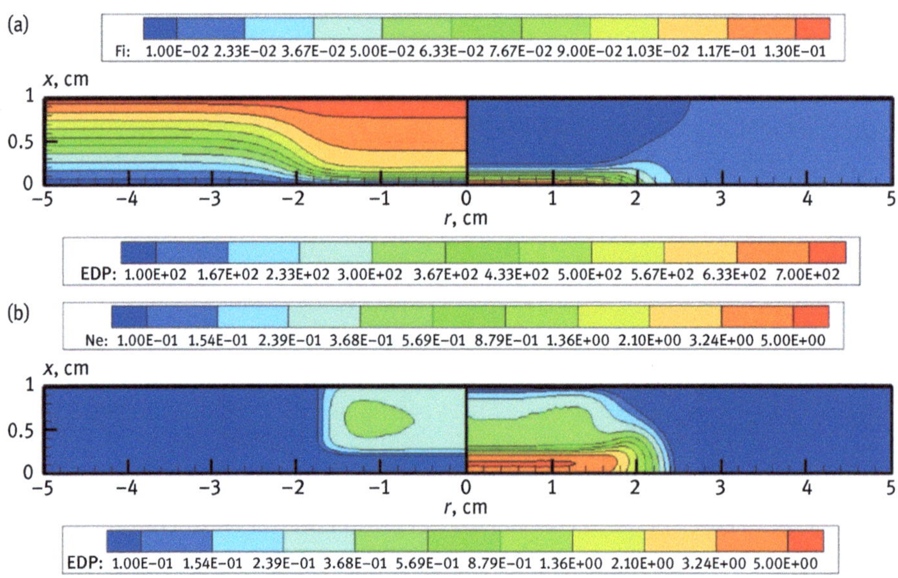

Figure 4.49: Normal glow discharge at $p = 5$ Torr, $\gamma = 0.05$, $E = 2.5$ kV: (a) electric potential $Fi = \varphi/E$ (to left) and reduced electric field $\text{EDP} = E_x/p$, $\text{V}/(\text{cm} \times \text{Torr})$ (to right) and (b) volume concentration of electrons (to left) and ions (to right); N_e, $N_i = (n_e, n_i)/10^9$ cm^{-3}.

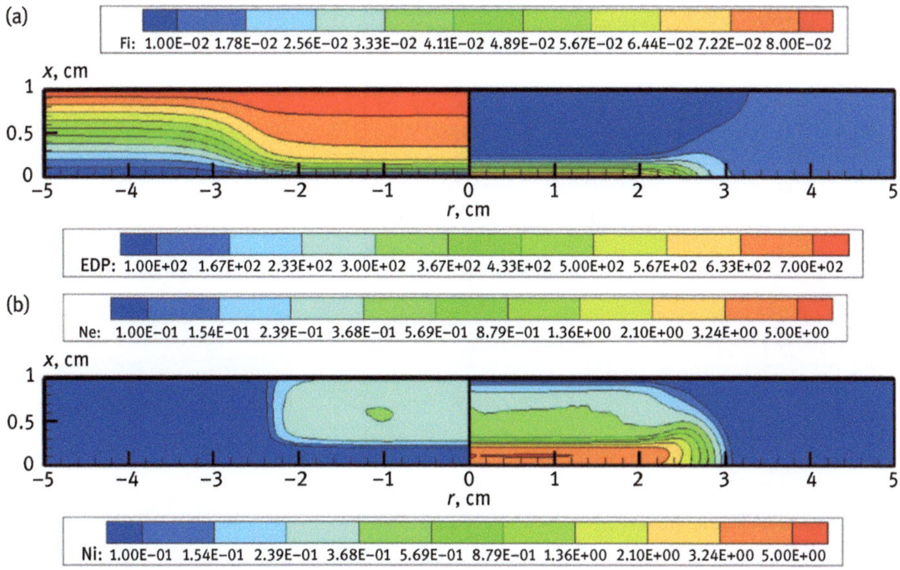

Figure 4.50: Normal glow discharge at $p = 5$ Torr, $\gamma = 0.05$, $E = 4.0$ kV: (a) electric potential $Fi = \varphi/E$ (to left) and reduced strength of the electric field $\text{EDP} = E_x/p$, V/(cm × Torr) (to right) and (b) volume concentration of electrons (to left) and ions (to right); N_e, $N_i = (n_e, n_i)/10^9$ cm^{-3}.

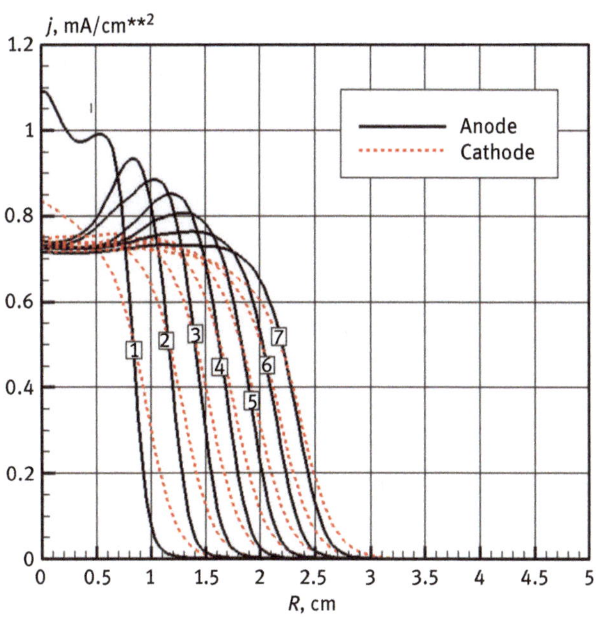

Figure 4.51: Distributions of electric current density on the cathode and anode at $E = 1.0$ kV (curves 1), $E = 1.5$ kV (curves 2), $E = 2.0$ kV (curves 3), $E = 2.5$ kV (curves 4), $E = 3.0$ kV (curves 5), $E = 3.5$ kV (curves 6), $E = 4.0$ kV (curves 7). Normal glow discharge at $p = 3$ Torr, $\gamma = 0.05$.

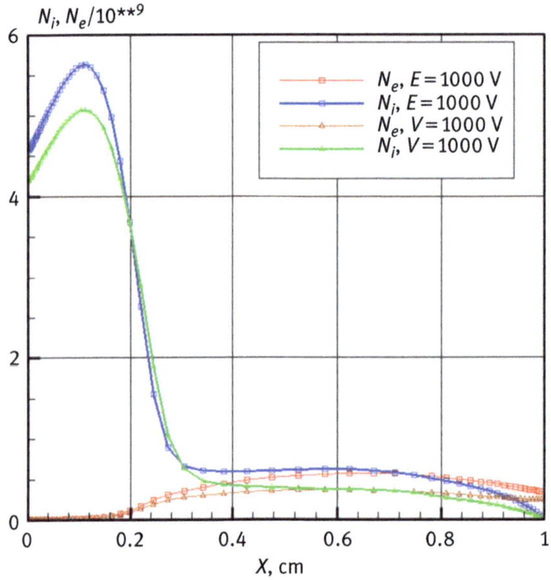

Figure 5.52: Normal glow discharge at $p = 5$ Torr, $\gamma = 0.05$, $E = 1.0$ kV and 4.0 kV. Distributions of volume concentrations of electrons and ions along the axis of symmetry.

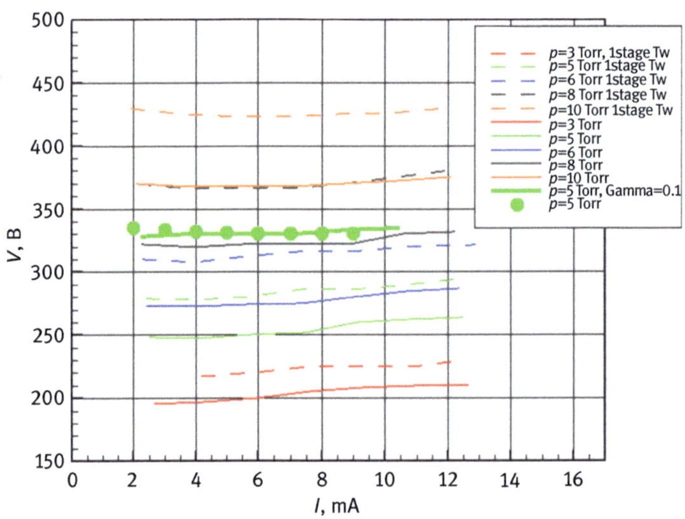

Figure 4.53: Volt–ampere characteristics of the normal glow discharge at $\gamma = 0.33$ and $\gamma = 0.1$. Solid curves correspond to the use of the two-mode approximation of the first Townsends coefficient; dotted lines correspond to the use of the one-mode approximation of the first Townsends coefficient; circles are the experimental data.

Comparison of the calculation results with the experimental data of the current–voltage characteristic is given in Figure 4.53. From this figure, there is a noticeable effect on the VACs of the method of approximation of the first Townsend coefficient and the secondary electron emission coefficient. For $\gamma = 0.1$ and $p = 5$ Torr, good agreement between the calculated and experimental data was obtained.

So, a numerical simulation of the 2D structure of a normal glow discharge in molecular nitrogen in the pressure range $p = 3$ Torr and $p = 5$ Torr was performed in this section. It is shown that two empirical coefficients (ionization coefficients and second electron emission coefficients) included in the diffusion-drift model have a significant effect on the parameters of the current–voltage characteristic.

An example of a good calculated description of experimental data is given in Figure 4.53.

5 Drift-diffusion model of glow discharge in an external magnetic field

The theory and computing model of glow discharge in two-dimensional flat geometry in view of an external magnetic field are presented in this chapter. It is supposed that the magnetic field is perpendicular to a plane, where the glow discharge is considered.

The considered problem represents significant interest for basic physics of gas discharge it as allows to investigate the behavior of glow discharges in assssn external magnetic field in view of positive columns and near-electrode layers of a spatial charge.

It was shown in the previous chapter that the drift-diffusion model of glow discharge allows predicting its performances with sufficient accuracy for practical needs in a wide range of pressures $p \sim 0.5$–50 Torr and voltages on electric discharge gap $V \sim 0.3$–10 kV. Considering that the glow discharge represents partially ionized gas with typical concentration of the charged particles of cm^3 against the background of $\sim 10^{17}\,\mathrm{cm}^{-3}$ of neutral particles and with small absolute value of currents across the discharge (tens milliamps), it is obvious that the external magnetic field $B \sim 0.01$–1 T can strongly influence the structure of the glow discharges. Furthermore, we will use the assumption that the specified magnetic field will not be distorted by the discharge. The given assumption is laid as a basis of glow discharge numerical model in an external magnetic field.

5.1 Derivation of the equations for calculation model

The flat two-dimensional discharge in molecular nitrogen between two infinite flat electrodes is considered. The configuration of the discharge and the external magnetic field is shown in Figure 5.1. It is supposed that the column of glow discharge is not limited in the direction of the z-axis.

Equations of the drift-diffusion model are formulated for concentration of electrons n_e and positive ions n_i, and for electric field potential φ, which defines a vector of an electric field strength $\mathbf{E} = -\operatorname{grad} \varphi$

$$\frac{\partial n_e}{\partial t} + \frac{\partial \Gamma_{e,x}}{\partial x} + \frac{\partial \Gamma_{e,y}}{\partial y} = \alpha(E)|\mathbf{\Gamma}_e| - \beta n_i n_e \tag{5.1}$$

$$\frac{\partial n_i}{\partial t} + \frac{\partial \Gamma_{i,x}}{\partial x} + \frac{\partial \Gamma_{i,r}}{\partial y} = \alpha(E)|\mathbf{\Gamma}_e| - \beta n_i n_e, \tag{5.2}$$

https://doi.org/10.1515/9783110648836-006

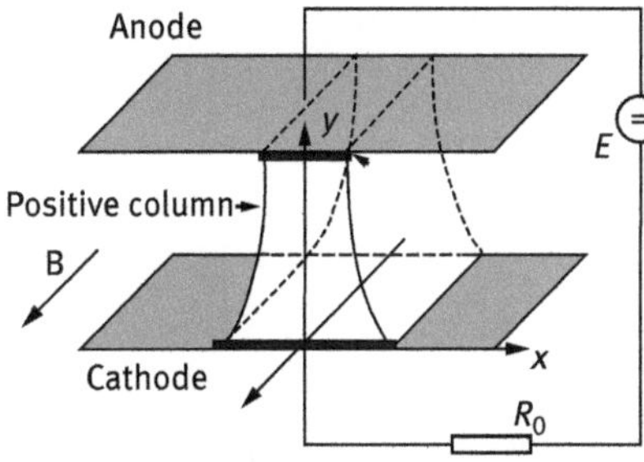

Figure 5.1: The scheme of glow discharge with an external magnetic field.

$$\frac{\partial^2\varphi}{\partial x^2} + \frac{\partial^2\varphi}{\partial y^2} = 4\pi e(n_e - n_i), \tag{5.3}$$

where

$$\mathbf{\Gamma}_e = -D_e \mathrm{grad} n_e - n_e\mu_e\mathbf{E}, \quad \mathbf{\Gamma}_i = -D_i \mathrm{grad} n_i + n_i\mu_i\mathbf{E};$$

$$Q = \eta(\mathbf{jE}), \quad \mathbf{j} = e(\mathbf{\Gamma}_i - \mathbf{\Gamma}_e),$$

$\alpha(E)$ and β are the ionization and recombination coefficients; $\mathbf{\Gamma}_e, \mathbf{\Gamma}_i$ are the densities of fluxes of electrons and positive ions; μ_e, μ_i are the mobilities of electrons and ions; D_e, D_i are the diffusion coefficients of electrons and ions.

To introduce a magnetic field into the computing model, the following equations expressing a momentum conservation law for electronic and ionic liquids are used (Bittencourt J.A., 2004; Chen F.F., 1984):

$$\begin{aligned}\rho_e\frac{\partial \mathbf{u}_e}{\partial t} + \rho_e(\mathbf{u}_e\cdot\nabla)\mathbf{u}_e = -\nabla p_e - \tau_e - en_e\left(\mathbf{E} + \frac{1}{c}[\mathbf{u}_e\mathbf{B}]\right) \\ - m_e\nu_{en}n_e(\mathbf{u}_e - \mathbf{u}_n) - m_e\nu_{ei}n_e(\mathbf{u}_e - \mathbf{u}_i),\end{aligned} \tag{5.4}$$

$$\begin{aligned}\rho_i\frac{\partial \mathbf{u}_i}{\partial t} + \rho_i(\mathbf{u}_i\cdot\nabla)\mathbf{u}_i = -\nabla p_i - \tau_i + en_i\left(\mathbf{E} + \frac{1}{c}[\mathbf{u}_i\mathbf{B}]\right) \\ - m_i\nu_{ie}n_i(\mathbf{u}_i - \mathbf{u}_e) - m_i\nu_{in}n_i(\mathbf{u}_i - \mathbf{u}_n),\end{aligned} \tag{5.5}$$

where $\mathbf{u}_e, \mathbf{u}_i$ are the velocities of electronic and ionic liquids; ρ_e, ρ_i are the densities of electronic and ionic liquids; $\rho_e = m_e n_e$, $\rho_i = m_i n_i$; m_e, m_i are the masses of electron and ion; $\mathbf{u}_n$ is the average velocity of neutral particles; p_e, p_i are the pressures of electrons and ions; τ_e, τ_i are the viscous stress tensors of electronic and ionic liquids; $\nu_{en}, \nu_{ei}, \nu_{in}$ are the collision frequencies of the electron impacts with neutral particles, with ions, and also ions with neutral particles; $\mathbf{B}$ is the magnetic field induction vector; c is the speed of light.

If we take into account that

$$m_e \ll m_i \tag{5.6}$$

and to use the condition of smallness of inertial motion of electrons

$$\rho_e(\mathbf{u}_e \cdot \nabla)\mathbf{u}_e \ll \rho_i(\mathbf{u}_i \cdot \nabla)\mathbf{u}_i, \tag{5.7}$$

that eq. (5.4) can be essentially simplified as

$$-\nabla p_e - en_e\left(\mathbf{E} + \frac{1}{c}[\mathbf{u}_e\mathbf{B}]\right) - m_e\nu_{en}n_e(\mathbf{u}_e - \mathbf{u}_n) - m_e\nu_{ei}n_e(\mathbf{u}_e - \mathbf{u}_i) = 0. \tag{5.8}$$

Even for supersonic gas flow velocities, the inequality is true

$$\mathbf{u}_e > \mathbf{u}_n, \mathbf{u}_i. \tag{5.9}$$

Therefore eq. (5.8) supposes further simplification

$$kT_e\nabla n_e + en_e\mathbf{E} + \frac{en_e}{c}[\mathbf{u}_e\mathbf{B}] + (m_e\nu_e)n_e\mathbf{u}_e = 0, \tag{5.10}$$

where $p_e = n_e kT_e$.

Having divided (5.10) on the product in parentheses, we will get

$$n_e\mathbf{u}_e = -D_e\nabla n_e - \mu_e n_e\mathbf{E} - \frac{\mu_e n_e}{c}[\mathbf{u}_e\mathbf{B}], \tag{5.11}$$

where $\mu_e = e/m_e\nu_e$ is the mobility of electrons; $D_e = (kT_e/e)\mu_e$ is the coefficient of electronic diffusion; $\nu_e = \nu_{en} + \nu_{ei}$.

In the considered formulation, the vector of electronic liquid velocity has two components $\mathbf{u}_e = \{u_{e,x}; u_{e,y}\}$, while the vector of magnetic field induction has only one component B_z; therefore

$$\Gamma_{e,x} = n_e u_{e,x} = \frac{1}{1+b_e^2}\left(-D_e\frac{\partial n_e}{\partial x} - \mu_e n_e E_x\right) - \frac{b_e}{1+b_e^2}\left(-D_e\frac{\partial n_e}{\partial y} - \mu_e n_e E_y\right), \tag{5.12}$$

$$\Gamma_{e,y} = n_e u_{e,y} = \frac{1}{1+b_e^2}\left(-D_e\frac{\partial n_e}{\partial y} - \mu_e n_e E_y\right) + \frac{b_e}{1+b_e^2}\left(-D_e\frac{\partial n_e}{\partial x} - \mu_e n_e E_x\right), \tag{5.13}$$

where

$$b_e = \frac{\mu_e}{c}B_z, \tag{5.14}$$

E_x, E_y are the components of the electric field. It is significant that cross derivatives in the considered statement have disappeared.

By analogy, assuming that in considered range of velocities it is possible to omit the left part of eq. (5.5), we obtain

$$-\nabla p_i + en_i\left(\mathbf{E} + \frac{1}{c}[\mathbf{u}_i B]\right) - m_i\nu_{ie}n_i(\mathbf{u}_i - \mathbf{u}_e) - m_i\nu_{in}n_i(\mathbf{u}_i - \mathbf{u}_n) = 0. \tag{5.15}$$

Considering that

$$m_i\nu_{ie}n_i(\mathbf{u}_i-\mathbf{u}_e)=-m_e\nu_{ei}n_e(\mathbf{u}_e-\mathbf{u}_i), \tag{5.16}$$

and also that in the considered case

$$\mathbf{u}_n=0, \tag{5.17}$$

and

$$m_e\nu_e<m_i\nu_{in}, \tag{5.18}$$

it is possible to get

$$n_i\mathbf{u}_i=-D_i\nabla n_i+\mu_i n_i\mathbf{E}+\frac{\mu_i n_i}{c}[\mathbf{u}_i\mathbf{B}], \tag{5.19}$$

where $\mu_i=e/m_i\nu_{in}$ is the mobility of ions; $D_i=(kT_i/e)\mu_i$ is the diffusivity of ions.

Also for electronic liquid, the average velocity of ions has only two nonzero components $\mathbf{u}_i=\{u_{i,x};u_{i,y}\}$, and the induction of an external magnetic field has only one component B_z; therefore,

$$\Gamma_{i,x}=n_iu_{i,x}=\frac{1}{1+b_i^2}\left(-D_i\frac{\partial n_i}{\partial x}+\mu_i n_iE_x\right)+\frac{b_i}{1+b_i^2}\left(-D_i\frac{\partial n_i}{\partial y}+\mu_i n_iE_y\right), \tag{5.20}$$

$$\Gamma_{i,y}=n_iu_{i,y}=\frac{1}{1+b_i^2}\left(-D_i\frac{\partial n_i}{\partial y}+\mu_i n_iE_y\right)+\frac{b_i}{1+b_i^2}\left(D_i\frac{\partial n_i}{\partial x}-\mu_i n_iE_x\right), \tag{5.21}$$

where $b_i=(\mu_i/c)B_z$.

Let us introduce an effective electric field for electrons and ions with the following components:

$$E_{e,x}=\frac{b_eE_y-E_x}{1+b_e^2},\quad E_{e,y}=-\frac{b_eE_x+E_y}{1+b_e^2}, \tag{5.22}$$

$$E_{i,x}=\frac{E_x+b_iE_y}{1+b_i^2},\quad E_{i,y}=\frac{E_y-b_iE_x}{1+b_i^2}. \tag{5.23}$$

Substituting values of densities of electron and ion fluxes with effective electric fields in the continuity equations of electronic and ionic liquids (5.1) and (5.2), one can obtain

$$\frac{\partial n_e}{\partial t}+\frac{\partial}{\partial x}\left(\mu_e n_eE_{e,x}-\frac{D_e}{1+b_e^2}\frac{\partial n_e}{\partial x}\right)+\frac{\partial}{\partial y}\left(\mu_e n_eE_{e,y}-\frac{D_e}{1+b_e^2}\frac{\partial n_e}{\partial y}\right)=\alpha|\Gamma_e|-\beta n_en_i, \tag{5.24}$$

$$\frac{\partial n_i}{\partial t}+\frac{\partial}{\partial x}\left(\mu_i n_iE_{i,x}-\frac{D_i}{1+b_i^2}\frac{\partial n_i}{\partial x}\right)+\frac{\partial}{\partial y}\left(\mu_i n_iE_{i,y}-\frac{D_i}{1+b_i^2}\frac{\partial n_i}{\partial y}\right)=\alpha|\Gamma_e|-\beta n_en_i, \tag{5.25}$$

where $|\Gamma_e| = \sqrt{\Gamma_{e,x}^2 + \Gamma_{e,y}^2}$.

Introduced factors b_e and b_i which take into account the magnetic field in model (5.24) and (5.25)

$$b_e = \frac{\mu_e B_z}{c} = \frac{\omega_e}{\nu_e}, \quad b_i = \frac{\mu_i B_z}{c} = \frac{\omega_i}{\nu_{in}}, \tag{5.26}$$

are well-known in plasma physics. These are the Hall parameters of electronic and ionic liquids. Factors $\omega_e = eB_z/m_e c$ and $\omega_i = eB_z/m_i c$ in (5.26) are the Larmor frequencies of electron and ion rotation in a magnetic field.

Boundary conditions for ion and electron concentrations and also for electric potential are formulated in the following form:

$$y = 0: \quad \frac{\partial n_i}{\partial y} = 0, \quad \Gamma_e = \gamma \Gamma_i, \quad \varphi = 0; \tag{5.27}$$

$$y = H: \quad n_i = 0, \quad \frac{\partial n_e}{\partial y} = 0, \quad \varphi = \frac{V}{E}; \tag{5.28}$$

$$x = 0: \quad \frac{\partial n_e}{\partial x} = \frac{\partial n_i}{\partial x} = \frac{\partial \varphi}{\partial x} = 0; \tag{5.29}$$

$$x = L: \quad \frac{\partial n_e}{\partial x} = \frac{\partial n_i}{\partial x} = \frac{\partial \varphi}{\partial x} = 0. \tag{5.30}$$

Initial conditions for the integration of equations of the drift-diffusion model of glow discharge were discussed in Chapter 4. It is important to keep in mind that wrong initial conditions can result not only in essential deceleration of the integrating procedure but also to its divergence. Therefore, even at the analytical formulation of initial conditions, it is desirable to use the correct configuration of the discharge and reasonable values of concentration of the charged particles. It is obvious that the integrating procedure could be accelerated essentially if we use the preliminarily obtained solutions for the calculation of new variants.

In the considered problem, heating up of neutral gas has not been taken into account; therefore, mobilities of electrons and ions, and also diffusivities are supposed to be constant:

$$\mu_e(p) = \frac{4.2 \times 10^5}{p}, \ \frac{\text{cm}^2}{\text{V} \cdot \text{s}}, \quad \mu_i(p) = \frac{2280}{p}, \ \frac{\text{cm}^2}{\text{V} \cdot \text{s}},$$

$$D_e = \mu_e(p) T_e, \quad D_i = \mu_i(p) T,$$

where p is the pressure in the glow discharge.

Coefficient of ion-electronic recombination and temperature of electrons are constants: $\beta = 2 \times 10^{-7} \text{cm}^3/\text{s}$, $T_e = 11610$ K.

The ionization coefficient (the first Townsend coefficient) is set in the form of

$$\alpha(E) = pA \exp\left[-\frac{B}{(|\mathbf{E}|/p)}\right], \quad \text{cm}^{-1},$$

where for molecular nitrogen $A = 12\ \frac{1}{\text{cm}\cdot\text{Torr}}$, $B = 342\ \frac{\text{V}}{\text{cm}\cdot\text{Torr}}$.

The equations are solved together with the equation for an external electric circuit, which for a direct current looks as follows:

$$E = V + IR_0,$$

where V is the voltage drop on electrodes; I is the total current through the discharge gap; $E$is the electromotive force of the power supply; R_0 is the resistance of an external circuit.

5.2 Numerical simulation results

Calculations were performed for the following input data:

gas $-N_2$; pressure $p = 2$–50 Torr; $E = 2$–9 kV; $R_0 = 300$ kOhm;
$\gamma = 0.01$–0.33; $H_c = 1$–2 cm; $R_c = 2$–4 cm; $B = 0$–0.1 T.

To show the influence of the external magnetic field on the discharge structure, first we will consider the glow discharge at $p = 5$ Torr, $E = 2\,\text{kV}$, $\gamma = 0.1$ in the absence of magnetic field. Numerical simulation results, for the glow discharge configuration shown in Figure 5.1, are presented in Figures 5.2 and 5.3.

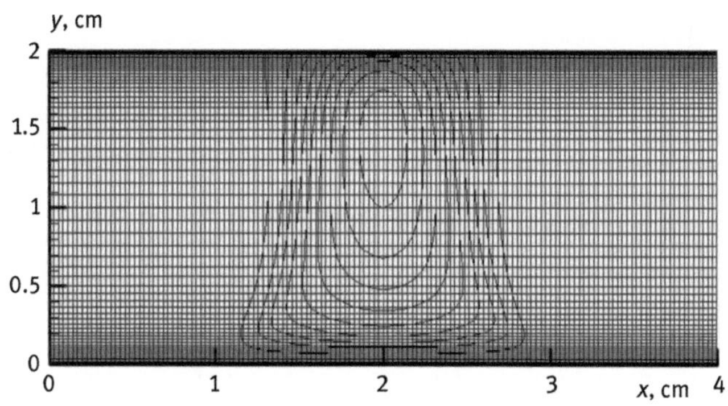

Figure 5.2: Examples of computational mesh (141 × 61).

Here concentrations of the charged particles are referred to value of $N_0 = 10^9\ \text{cm}^{-3}$. For the calculation of total current through the gas discharge, it was supposed that the length of discharge along z-direction is equal to 1 cm. Fields of electron and ion concentrations are shown in Figure 5.3a and b. Distributions of current densities on

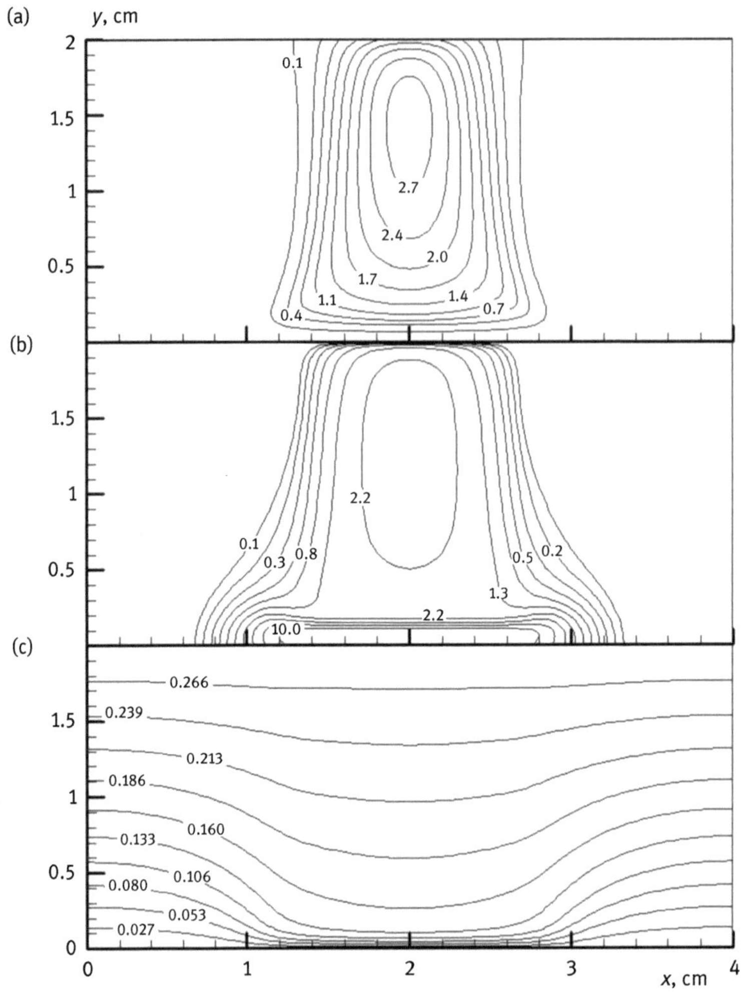

Figure 5.3: Results of numerical modeling of discharge without magnetic field at $p = 5$ Torr, $E = 2$ kV, $\gamma = 0.1$, $H = 2$ cm, $I = 4.85$ mA, $j_{c,\max} = 3.58$ mA/cm^2, $j_{a,\max} = 6.03$ mA/cm^2, $V = 533$ V: (a) electron concentration; (b) ion concentration; (c) electric potential (numbers at curves correspond to φ/E); (d) current density on the anode (solid line) and on the cathode (dashed line); (e) distribution of ion (continuous line) and electron (dashed line) concentrations along the axis of glow discharge; (f) electrical conduction (Ohm$^{-1}\cdot$cm^{-1}). Levels of concentration are referred to value of $N_0 = 10^9$ cm^{-3}. The parameters of one-dimensional normal glow discharge calculated using the Engel–Steenbeck theory: $d_n = 0.15$ cm, $V_n = 205$ V, $j_n = 1.37$ mA/cm^2.

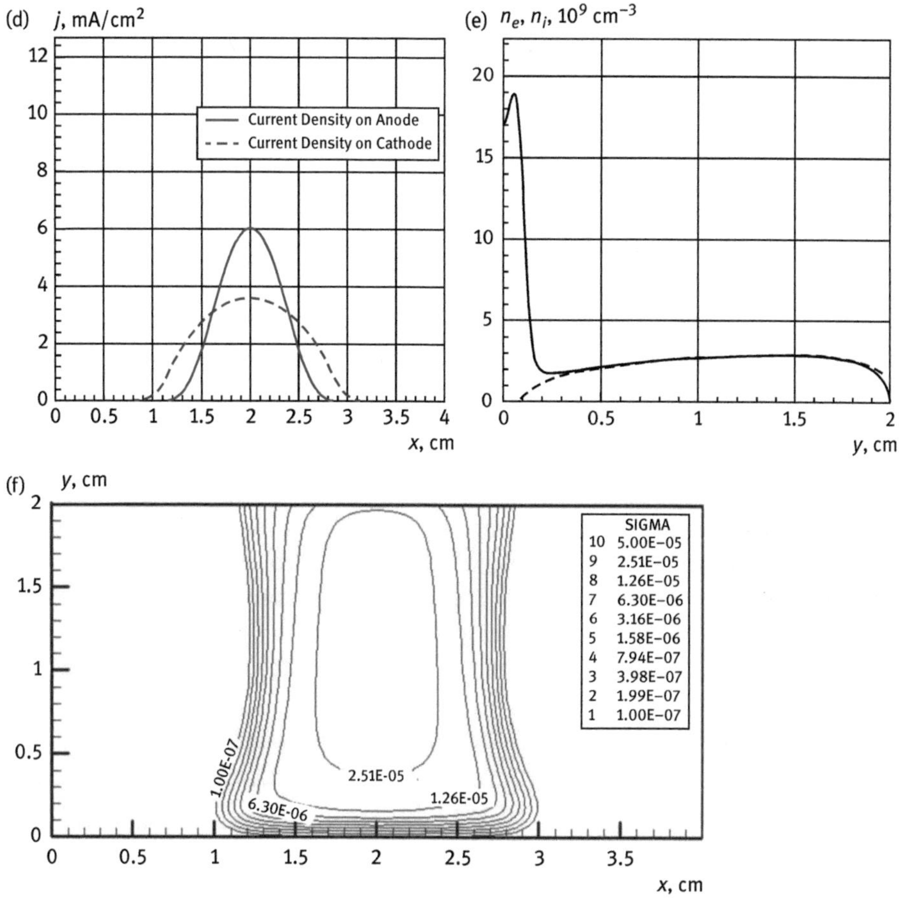

Figure 5.3 (continued)

the cathode and the anode are shown in Figure 5.3c, and distributions of electron and ion concentrations along an y-axis are shown in Figure 5.3d.

Analyzing the data presented it is possible to define three basic regions in the glow discharge:

- The area of positive volume charge near to the cathode. This is the cathode layer (Figure 5.3a, b and d).
- The area of negative volume charge near to the anode. This is the anode layer (Figure 5.3d).
- The area of quasineutral plasma. This is the positive column of glow discharge (Figure 5.3a, b and d).

The specified areas of the normal mode glow discharge are clearly visible in Figure 5.3e, where the distribution of electric potential is shown. In the cathode layer, ($y \le 0.1$ cm) sharp growth of the electric potential is observed; in the positive column ($0.1 \le y \le 0.9$ cm) distribution of the potential is almost linear; in the anode layer, a small increase of the potential is observed again.

As it was marked earlier, the important advantage of the drift-diffusion model, which has great importance for the gas-discharge physics, is the possibility of electric conductivity prediction. Results of calculations of the electric conductivity in the absence of a magnetic field, $\sigma_e = e\mu_e n_e$ $(\mathrm{Ohm \cdot cm})^{-1}$, are given in Figure 5.3f.

Results of numerical modeling of glow discharges without a magnetic field, which are represented in Figure 5.3, will be used later for the analysis of the influence of various factors, including parameters of numerical modeling. Similar researches performed in Chapter 4 relate to an axially symmetric glow discharge:

$$\sigma_e = e\mu_e n_e$$

Common representation of the influence of various input data on the structure of normal gas discharge is listed in Table 5.1, where numerical results for various pressures, electromotive forces of a power supply, and coefficients of a secondary electron emission are presented. In this table, the values of the normal current density on the cathode j_n, predicted using the Engel – Steenbeck theory, are also given.

Table 5.1: Numerical simulation results for flat glow discharge without external magnetic field.

p, Torr,	E, kV	γ	I, mA	$j_{c,\,max}$, mA/cm^2	$j_{a,\,max}$, mA/cm^2	j_n, mA/cm^2
5	2	0.1	4.85	3.58	6.03	1.37
10	2	0.1	3.90	7.03	8.89	5.49
5	2	0.3	4.85	4.97	6.70	2.65
5	3	0.1	8.19	4.34	6.22	1.37
5	4	0.1	11.5	4.79	5.01	1.37

As it was already discussed, this theory allows to calculate even a voltage drop on the cathode V_n and thickness of the cathode layer $(pd)_n$ for the glow discharge in the normal current density mode:

$$V_n = \frac{2.72B}{A}\ln\left(\frac{1}{\gamma}+1\right), \quad (pd)_n = \frac{3.78}{A}\ln\left(\frac{1}{\gamma}+1\right),$$

$$j_n = p^2\frac{(\mu_i p)V_n^2}{4\pi e(pd)_n^3}.$$

Comparing the results of numerical modeling with the data obtained using the Engel – Steenbeck theory, it is necessary to take into account that the basis of the given theory is a one-dimensional model with a cathode layer and a linear increase of electric potential in the cathode layer. Nevertheless, this has been confirmed by numerous experimental data that allow considering the correspondence of numerical simulation results to the Engel – Steenbeck theory as evidence of the computing model's adequacy.

In this table, I is the total current; $j_{c,\ \max}, j_{a,\ \max}$ are the maximal current densities on the cathode and the anode.

Numerical simulation results for other input data are shown in Figures 5.4–5.8.

Configuration of glow discharge at raised pressure $p = 10$ Torr is shown in Figure 5.4. It is evident that in comparison with the previous calculation variant ($p = 5$ Torr), the cross sizes of current columns have been noticeably reduced (Figure 5.4a); thickness of the cathode layer (Figure 5.4b) has decreased. The current across the discharge gap has decreased from $I = 4.85$ mA ($p = 5$ Torr) to $I = 3.9$ mA ($p = 10$ Torr). However, at the same time, the current density on the cathode and the anode has increased. One-dimensional Engel–Steenbeck theory specifies the same as well. An important point in that with the growth of pressure the increasing of discrepancy from the Engel – Steenbeck model should be observed owing to the amplification of influence of the two-dimensional effects.

Modifications in the structure of the glow discharge at the increasing of the secondary electronic emission coefficient are shown in Figure 5.5 ($p = 5$ Torr, $\gamma = 0.3$), comparing these data with similar data in Figure 5.3 ($p = 5$ Torr, $\gamma = 0.1$).

With an increase in γ, the current in discharge gap will not practically vary, but the current density on the cathode and on the anode will increase. It is obvious that at the same time cross sizes of the current channel will decrease. One more singularity of changing the structure of the discharge is some diminution of the thickness of the cathode layer that also corresponds to the prediction of the approximated one-dimensional theory.

Results of calculations of flat glow discharge at increased electromotive force $E = 3$ kV are shown in Figure 5.6 and at $E = 4$ kV in Figure 5.7.

Based on the given calculated data, we can specify the main regularities of glow discharge structure formation without magnetic field:

1. The positive charge area (the cathode layer) has a small expansion along y-axis. At $p = 5$ Torr, $E = 2$ kV, $\gamma = 0.1$, $H = 2$ cm (see Figure 5.3b) the height of the cathode layer reaches approximately $d_c \cong 0.15$ cm. It well coincides with thickness of a cathode layer $d_n \cong 0.151$ cm, predicted by the Engel–Steenbeck theory.
2. At the parameters specified above, the glow discharge exists in conditions of the subnormal glow discharge (NGD). Under the condition of the normal current density with a modification of a discharge current across the gas-discharge gap (e.g., at other parameters being equal, with an increase in the electromotive force E), the current density on the cathode remains practically constant, and cross sizes of the current column change. In the case considered here, cross

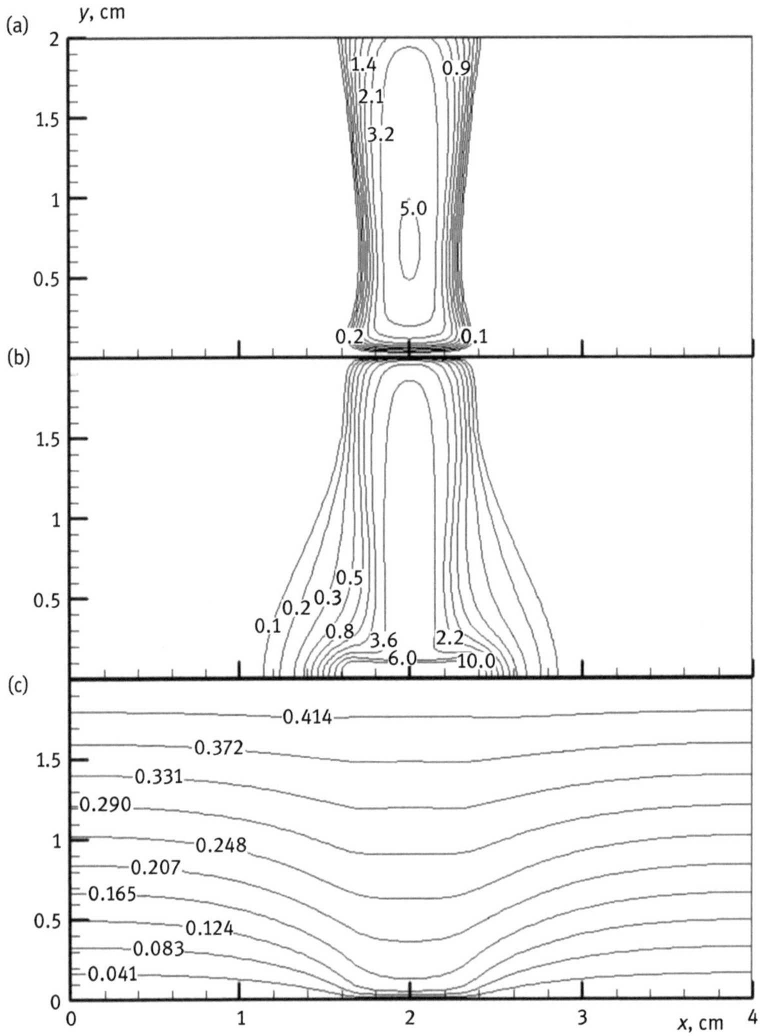

Figure 5.4: Results of numerical modeling of glow discharge without magnetic field at $p = 10$ Torr, $E = 2$ kV, $\gamma = 0.1$, $H = 2$ cm, $I = 3.9$ mA, $j_{c,\max} = 7.03$ mA/cm^2, $j_{a,\max} = 8.89$ mA/cm^2, $V = 835$ V: (a) concentration of electrons; (b) concentration of ions; (c) electric potential (numbers at curves correspond to φ/E); (d) current density on the anode (solid line) and on the cathode (dashed line); (e) distribution of electron (dashed line) and ion (solid line) concentrations along the axis of glow discharge; (f) electrical conduction in gas-discharge gap (Ohm$^{-1}\cdot$cm^{-1}). Levels of concentration are referred to value of $N_0 = 10^9$ cm^{-3}. Parameters of one-dimensional normal glow discharge (using the Engel–Steenbeck theory): $d_n = 0.0755$ cm, $V_n = 205$ V, $j_n = 5.49$ mA/cm^2.

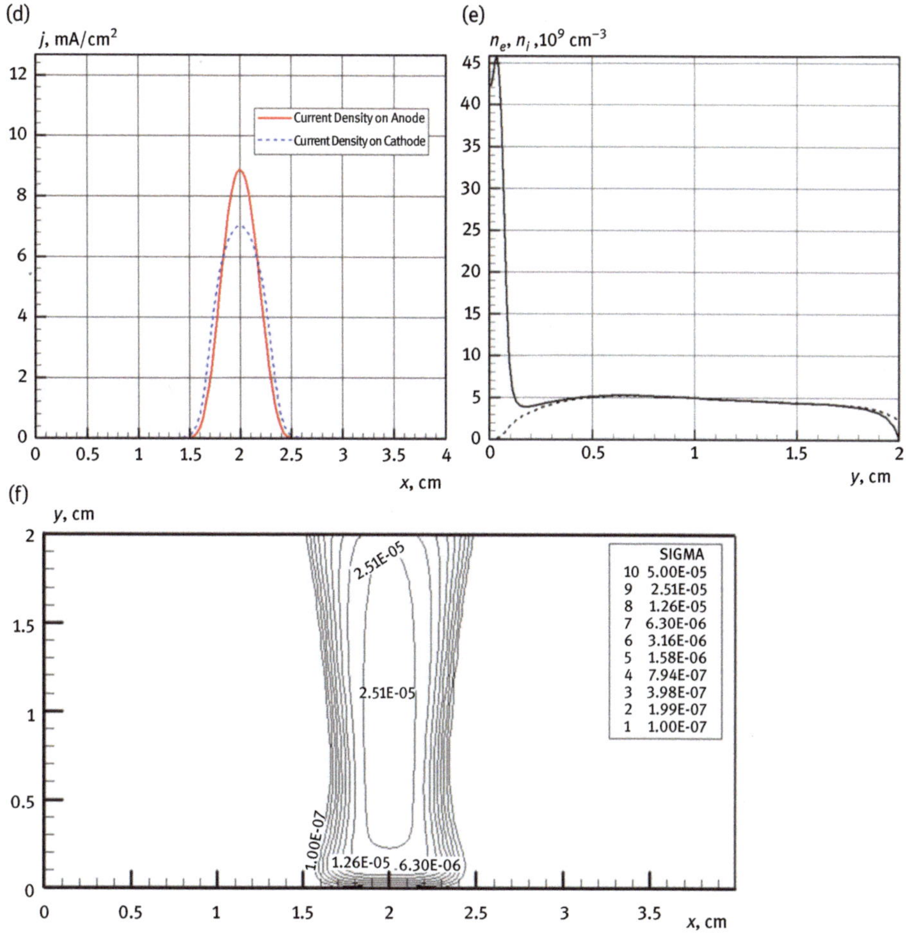

Figure 5.4 (continued)

sizes of discharge are comparable to the thickness of the cathode layer; therefore, the law of the normal current density is not exactly satisfied because charge losses in a cross direction are too high.

Nevertheless, this discharge in subnormal mode is close enough to the normal discharge. With an increase in the discharge current in a range of $I = 4.85–11.5$ mA (at a growth of electromotive force from $E = 2$ kV up to 4 kV) the greatest current density on the cathode varies only in the range $j_c = 3.58–4.79$ mA/cm^2. The Engel–Steenbeck theory for these conditions predicts $j_n = 1.37$ mA/cm^2.

3. Growth of electromotive force E of an external electric circuit leads to an increase in cross sizes of the glow discharge, both the positive column and the cathode layer. This tendency is observed for all investigated pressures $p = 2.5, 10, 50$ Torr and for various factors of a secondary electronic emission ($\gamma = 0.01–0.3$).

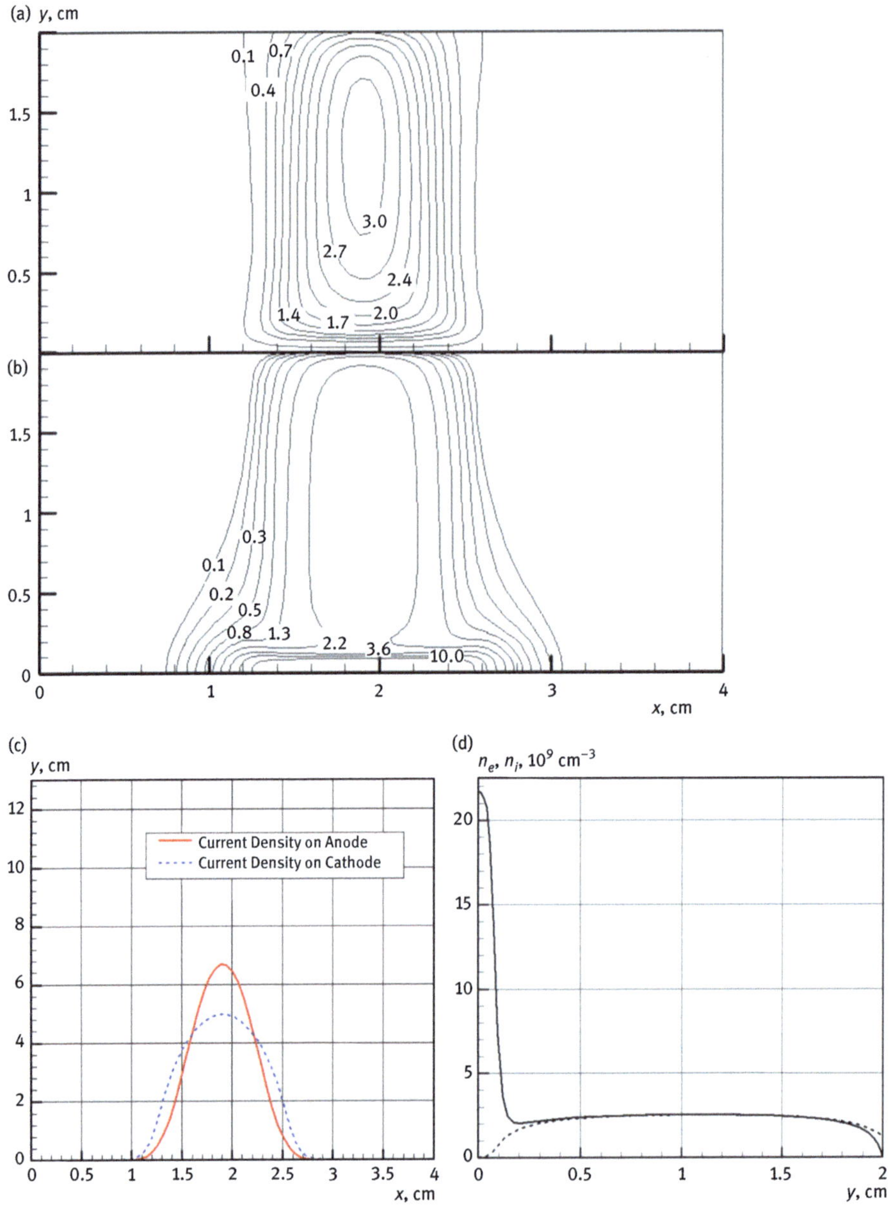

Figure 5.5: Results of numerical modeling of glow discharge without magnetic field at $p = 5$ Torr, $E = 2$ kV, $\gamma = 0.3$, $H = 2$ cm, $I = 4.85$ mA, $j_{c,\,max} = 4.97$ mA/cm^2, $j_{a,\,max} = 6.7$ mA/cm^2, $V = 454$ V: (a) electron concentration; (b) ion concentration; (c) current density on the anode (solid line) and on the cathode (dashed line); (d) distribution of electron (dashed line) and ion (solid line) concentrations along the axis of glow discharge. Parameters of one-dimensional normal glow discharge (using the Engel–Steenbeck theory): $d_n = 0.0924$ cm, $V_n = 125$ V, $j_n = 2.65$ mA/cm^2.

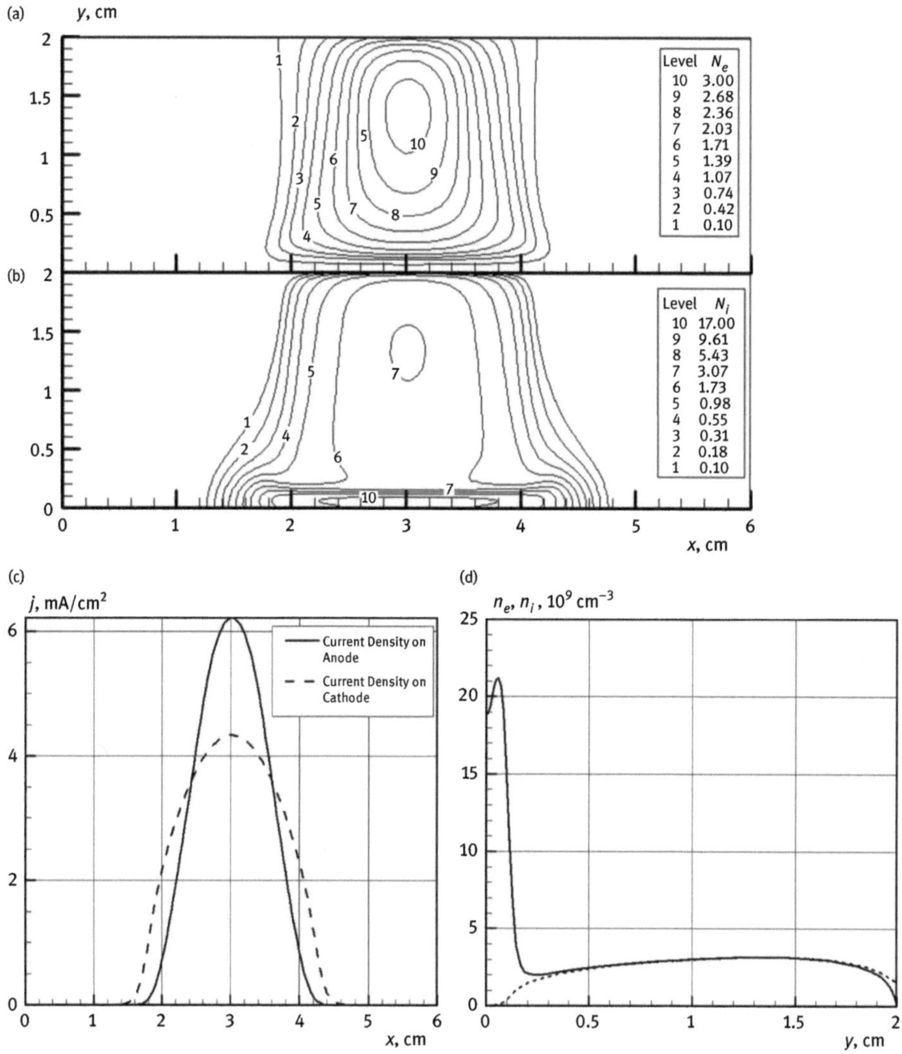

Figure 5.6: Results of numerical modeling of discharge without magnetic field at $p = 5$ Torr, $E = 3$ kV, $\gamma = 0.1$, $H = 2$ cm, $I = 8.19$ mA, $j_{c,\,\max} = 4.34$ mA/cm^2, $j_{a,\,\max} = 6.22$ mA/cm^2, $V = 555$ V: (a) concentration of electrons; (b) concentration of ions; (c) current density on the anode (solid line) and on the cathode (dashed line); (d) distribution of concentrations of electrons (dashed line) and ions (solid line) along the axis of glow discharge. Levels of concentration are referred to value $N_0 = 10^9$ cm^{-3}. Parameters of one-dimensional normal glow discharge (using the Engel–Steenbeck theory) are: $d_n = 0.15$ cm, $V_n = 205$ V, $j_n = 1.37$ mA/cm^2.

4. The diminution of pressure in gas-discharge gap leads to an appreciable increase not only in cross sizes of discharge but also in the height of the cathode layer. Numerical results of the specified research are confirmed by the Engel–Steenbeck theory.

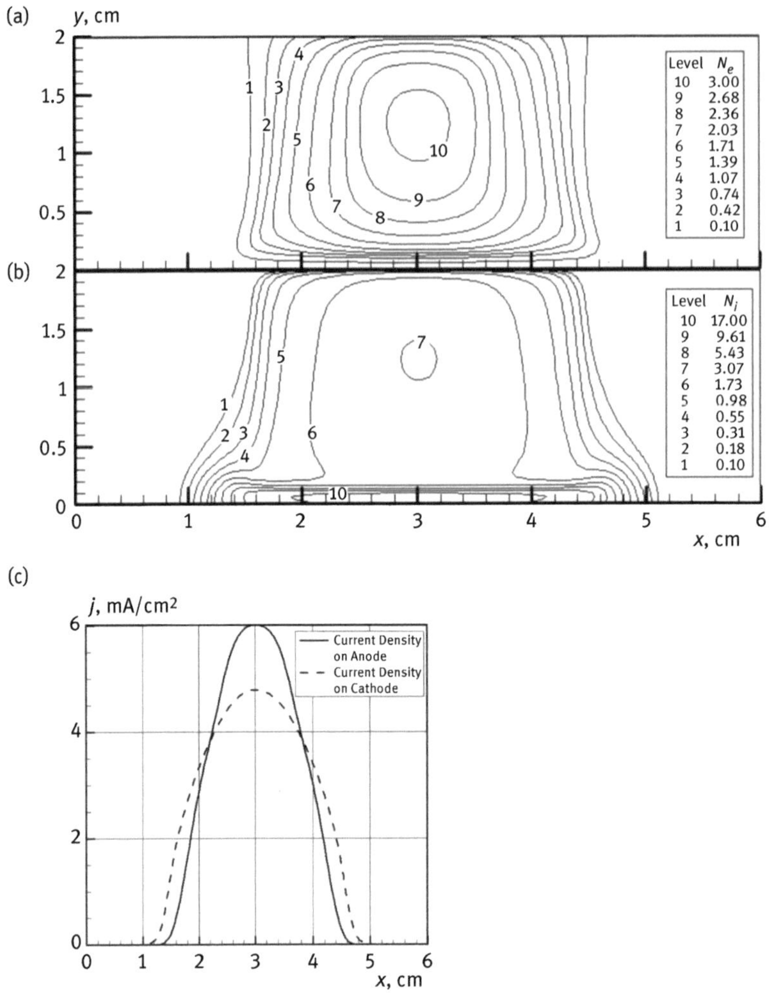

Figure 5.7: Results of numerical modeling of discharge without a magnetic field at $p = 5$ Torr, $E = 4$ kV, $\gamma = 0.1$, $H = 2$ cm, $I = 11.5$ mA, $j_{c,\,max} = 4.79$ mA/cm^2, $j_{a,\,max} = 5.01$ mA/cm^2, $V = 561$ V: (a) concentration of electrons; (b) concentration of ions; (c) current density on the anode (solid line) and on the cathode (dashed line). Levels of concentration are referred to value $N_0 = 10^9$ cm^{-3}. Parameters of one-dimensional normal glow discharge (using the Engel–Steenbeck theory): $d_n = 0.15$ cm, $V_n = 205$ V, $j_n = 1.37$ mA/cm^2.

5. Numerical calculations confirm that the secondary electronic emission coefficient plays an important role in glow discharge structure formation and is one of the defining parameters of numerical modeling. From the physical point of view, this parameter, defining the effectiveness of "breaking-away" electrons from the cathode by ion stream falling on the cathode, allows modeling of various materials that the cathode is made from. From the calculated data presented

in Table 5.1, the strong influence of γ is obvious. In particular, by comparing the numerical simulation results at $p = 5$ Torr and $E = 2$ kV, it follows that though at growth γ value from 0.1 up to 0.3 the total current practically does not vary, but a significant change of a current density on the cathode from 3.58 mA/cm^2 till 4.97 mA/cm^2 is observed. Under these conditions, the Engel–Steenbeck theory predicts an increase in the current density in the NGD cathode from 1.37 mA/cm^2 till 2.65 mA/cm^2.

It is significant that in all listed cases there was a steady-state solution for the structure of glow discharge. However, insignificant oscillations of all parameters of the glow discharge were observed. Most likely, principal causes of the specified oscillations are singularities of a numerical solution of the problem.

Influence of an external magnetic field on the structure of glow discharge has been studied for the same initial data, as without magnetic field. In calculations, the induction of the magnetic field varied in the following range: $B_z = 0.01 - 0.1$ T.

Configuration of glow discharge at $p = 5$ Torr, $E = 2$ kV, $\gamma = 0.1$ and magnetic induction of external field $B_z = -0.01$ T are shown in Figure 5.8. Let us note two singularities of the presented data:

1) The calculations were performed for negative orientation of the magnetic field induction vector $B_z = -0.01$ T (see the field configuration in Figure 5.1).

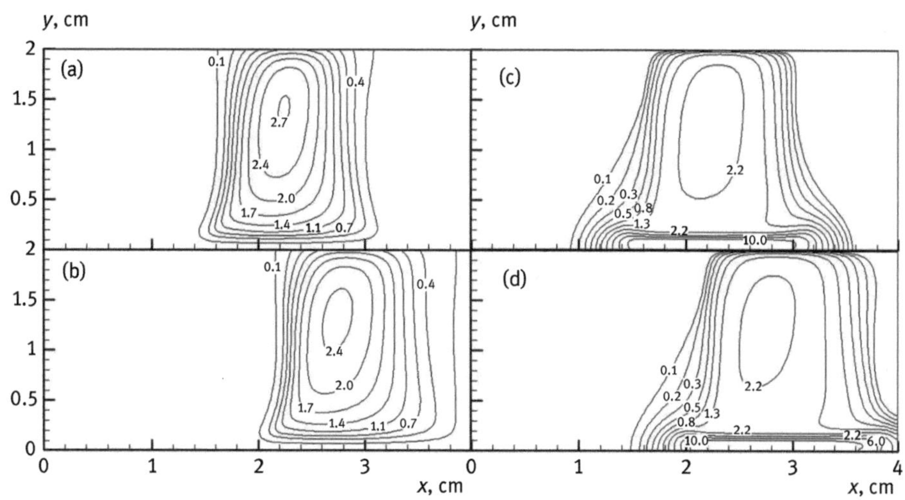

Figure 5.8: Concentration of electrons (a, b) and ions (c, d) (referred to 10^9 cm^{-3}) in glow discharge with magnetic field at $p = 5$ Torr, $E = 2$ kV, $\gamma = 0.1$, $H = 2$ cm, $B_z = -0.01$ T at sequential time instants: (a, c) $t = 50$ μs; (b, d) $t = 150$ μs.

2) Two instant configurations of the glow discharge are shown in Figure 5.8 (at instants of time $t = 50$ and 150 μs after energizing the magnetic field), moving in a positive direction of x-axis. The velocity of the specified motion will be analyzed later.

Let us concentrate on the distribution of the charged particle concentrations fields in a cross magnetic field at various initial data.

For investigating the influence of magnetic field on electrodynamic structure of glow discharge, additional calculations at $B_z = -0.05$ T have been performed. Results of the specified calculations are shown in Figure 5.9. It is evident that with an increase in the magnetic field induction the deformation of glow discharge's configuration increases and the velocity of the discharge transverse motion in the magnetic field considerably increases. To measure the current column displacement velocity of the discharge by using the electron isolines $n_e = 1.1 \times 10^9$ cm^{-3}, then at $B_z = -0.01$ T this velocity is equal to $u_x \approx 5.5 \times 10^3$ cm/s, and at $B_z = -0.05$ T it is of $u_x \approx 3.25 \times 10^4$ cm/s.

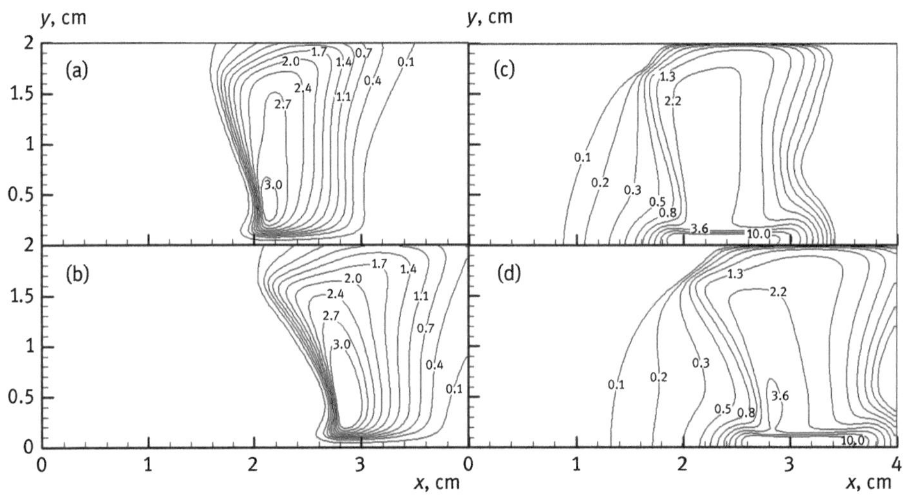

Figure 5.9: Concentration of electrons (a, b) and ions (c, d) (referred to 10^9 cm^{-3}) in glow discharge with magnetic field at $p = 5$ Torr, $E = 2$ kV, $\gamma = 0.1$, $H = 2$ cm, $B_z = -0.05$ T at sequential time instants: (a, c) $t = 10$ μs; (b, d) $t = 30$ μs.

It was observed that the change of the magnetic field induction vector direction leads to displacement of glow discharge in the opposite direction. Results of calculation of glow discharge are shown in Figure 5.10 at $p = 5$ Torr, $E = 2$ kV, $\gamma = 0.1$ in cross magnetic field $B_z = +0.05$ T.

Dynamics of glow discharge for increased pressure ($p = 10$ Torr) in cross magnetic field $B_z = -0.1$ T is shown in Figure 5.11.

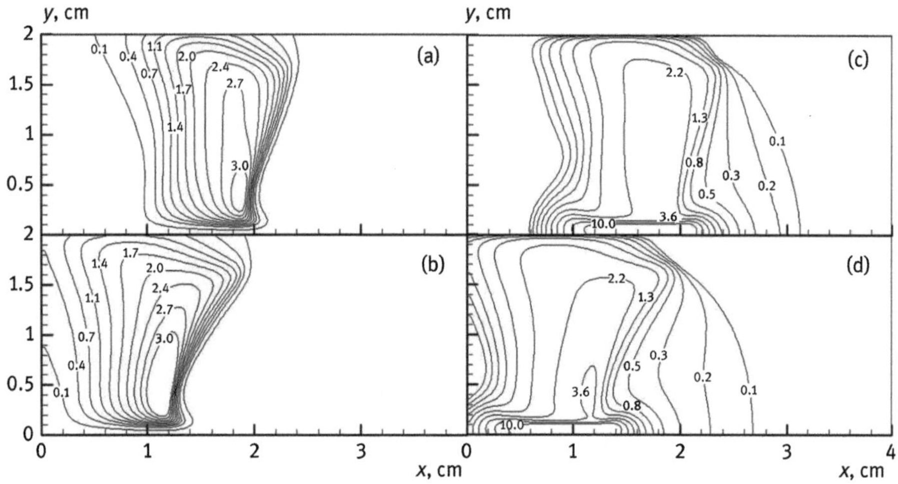

Figure 5.10: Concentration of electrons (a, b) and ions (c, d) (referred to 10^9 cm^{-3}) in glow discharge with magnetic field at $p = 5$ Torr, $E = 2$ kV, $\gamma = 0.1$, $H = 2$ cm, $B_z = +0.05$ T at sequential time instants: (a, c) $t = 10$ μs; (b, d) $t = 30$ μs.

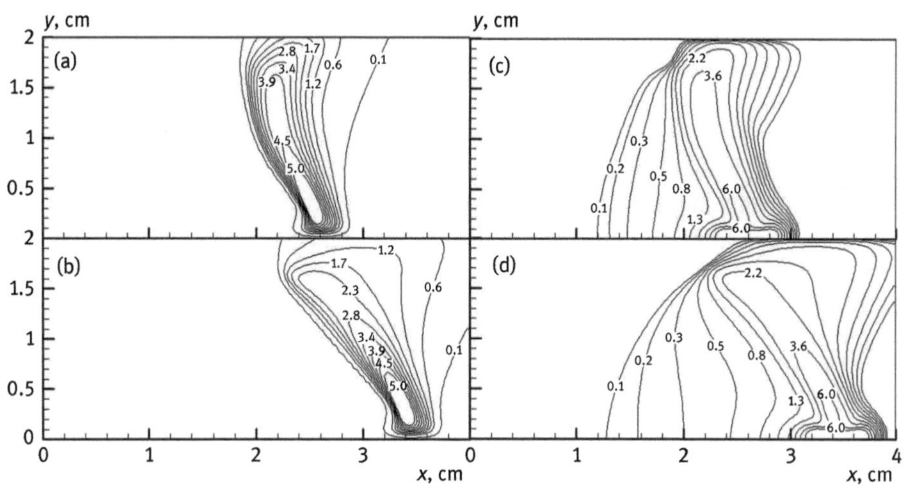

Figure 5.11: Concentration of electrons (a, b) and ions (c, d) (referred to 10^9 cm^{-3}) in glow discharge with magnetic field at $p = 10$ Torr, $E = 2$ kV, $\gamma = 0.1$, $H = 2$ cm, $B_z = -0.1$ T at sequential time instants: (a, c) $t = 10$ μs; (b, d) $t = 30$ μs.

The average velocity of discharge displacement, in this case, is estimated by value of $u_x \approx 4.5 \times 10^4$ cm/s. The discharge is displacing, virtually keeping the configuration.

Numerical simulation results on moving glow discharge in cross magnetic field $B_z = -0.05$ T, which are shown in Figures 5.12 and 5.13 show the influence of electromotive force on glow discharge structure. For this purpose an increase in electromotive force from $E = 3$ kV (Figure 5.12) up to $E = 4$ kV (Figure 5.13) was studied. It is expedient to compare the presented data on the electrodynamic structure of glow discharge with numerical simulation results shown in Figure 5.9 corresponding to electromotive force $E = 2$ kV. The specified increase in electromotive force is actually equivalent to the growth of the total current through discharge, but it does not mean proportional growth of current densities on electrodes and an appreciable change of concentration of the charged particles in near-electrode layers and in the positive column.

Comparison of instant configurations of gas discharge at various electromotive forces allows in drawing a conclusion about the appreciable deceleration of displacement of discharge with an increase in E (with an increase in total current across discharge).

The phenomenon of glow discharges motion in a cross magnetic field, which has been investigated in this section, is the well known in the electrodynamics and in plasma physics the Hall effect of current origin at the motion of charged particles in crossed electric and magnetic fields (Bittencourt J.A., 2004; Chen F.F., 1984).

The magnetic field induction, used as the input data for the calculated variants, corresponds to the Hall parameter for electrons of the order of unity.

It should be stressed that in the general case of three-dimensional geometry and at arbitrary orientation of a magnetic field induction, a computing model with magnetic field will be essentially more complicated and contain cross derivatives, which have been omitted in the considered case owing to a special choice of geometry and configuration of a magnetic field.

5.3 Glow discharge in cross magnetic field in view of heating of neutral gas

As it was mentioned earlier, the unique property of glow discharge is the non-equilibrium weakly ionized gas, in which electrons have the temperature $T_e = 10000–20000$ K, while the ions and neutral particles have the low temperature close to the room temperature. The principal reason for that is the great mass distinction between electrons and heavy particles M_n, M_i (neutrals and ions). The basic physical mechanism of the existing of glow discharge plasma is the electron emission from a surface of the cathode under impacts of incident ions and ionization of atoms and molecules at their collisions with electrons accelerated in an external electric field. The quantitative characteristic of this process, as is well known, is the

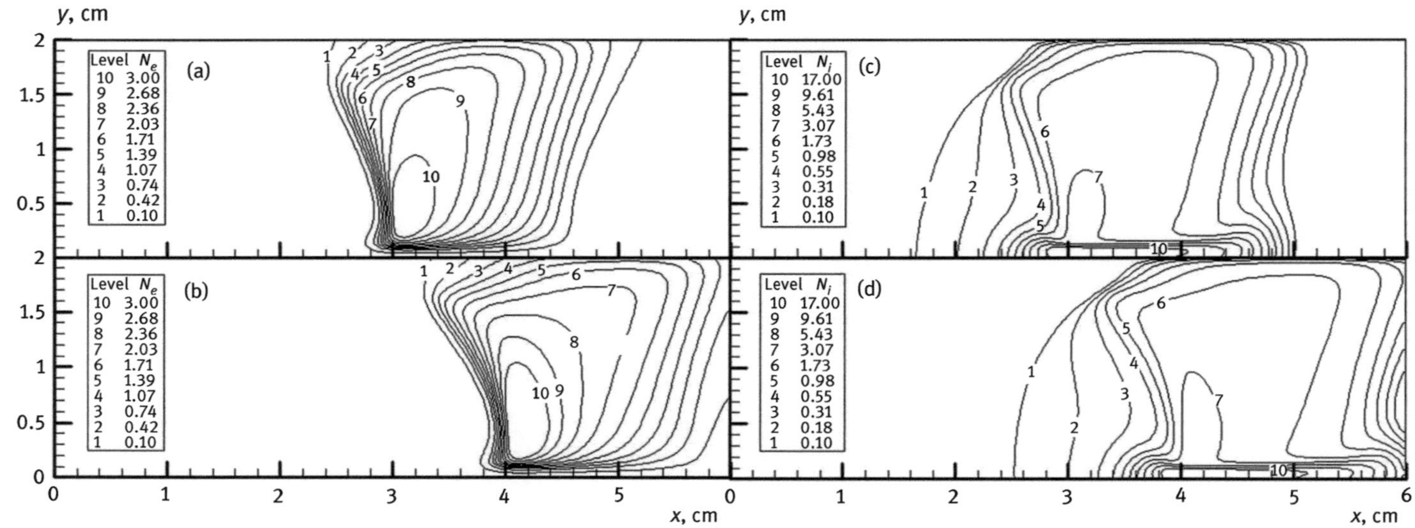

Figure 5.12: Concentration of electrons (a, b) and ions (c, d) (referred to 10^9 cm^{-3}) in glow discharge with magnetic field at $p = 5$ Torr $E = 3$ kV, $\gamma = 0.1$, $H = 2$ cm, $B_z = -0.05$ T in sequential time instants: (a, c) $t = 17$ µs; (b, d) $t = 51$ µs.

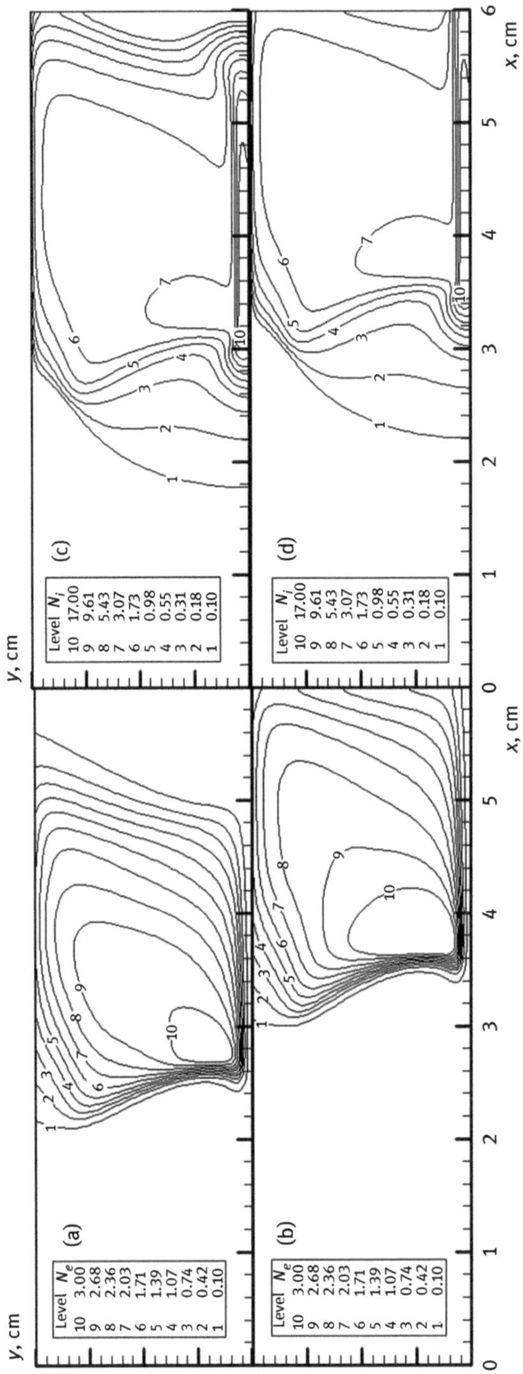

Figure 5.13: Concentration of electrons (a, b) and ions (c, d) (referred to 10^9 cm^{-3}) in glow discharge with magnetic field at $p = 5$ Torr, $E = 4$ kV, $\gamma = 0.1$, $H = 2$ cm, $B_z = -0.05$ T in sequential time instants: (a, c) $t = 17$ μs; (b, d) $t = 51$ μs.

frequency of ionization by electron impact. In the drift-diffusion theory of glow discharge, the ionization coefficient α is widely used. Its analytical approximation has been offered by Townsend and is named as the first Townsend coefficient:

$$\alpha = Ap\exp\left[-\frac{B}{(E/p)}\right]. \tag{5.31}$$

This coefficient is connected with frequency of ionization ν_i by the following formula:

$$\alpha = \frac{\nu_i}{v_{\mathrm{dr},e}} = \frac{\nu_i}{\mu_e E}, \tag{5.32}$$

where $v_{\mathrm{dr},e}$ is the electron drift velocity.

From physics of ionization process and from relation (5.32), it is obvious that the frequency of ionization ν_i depends on the concentration of neutral particles n_n which at a given pressure p is immediately connected with the temperature of neutral particles. It means that despite a huge difference of electron and neutral particle temperatures, the temperature of the neutral particles substantially defines the rate of the ionization process.

It is well known from the laser physics that in gas-discharge lasers the gas inside ones can be heated up to temperatures ~600 K. At higher temperatures, the ionization process is intensified so much that the glow discharge loses its stability. Instead of homogeneous glow discharge in such heated weakly ionized gas, the contracted discharge channels are observed.

Hence, the study of glow discharge at various levels of heating of neutral gas is of great scientific and applied interest. In the earlier part, the numerical modeling of axially symmetric glow discharge in view of heating gas was performed but without external magnetic field. It is obvious that the presence of external magnetic field complicates a common scheme of a description of glow discharges in view of heating of neutral gas. In the given section, the computing model of such discharge in two-dimensional flat geometry is considered.

5.3.1 Problem formulation

The schematic diagram of the gas-discharge gap with the external magnetic field, which will be investigated here, is shown in Figure 5.1. If we consider one of the cathode sections, then the scheme of the solved problem can be presented in the form of Figure 5.14. The glow discharge with heating will be considered between infinite and flat electrodes.

The electrodynamic structure of the glow discharge is described by using the drift-diffusion model for electron concentration n_e and positive ions n_i together with the Poisson equation for electric potential φ and electric field strength vector

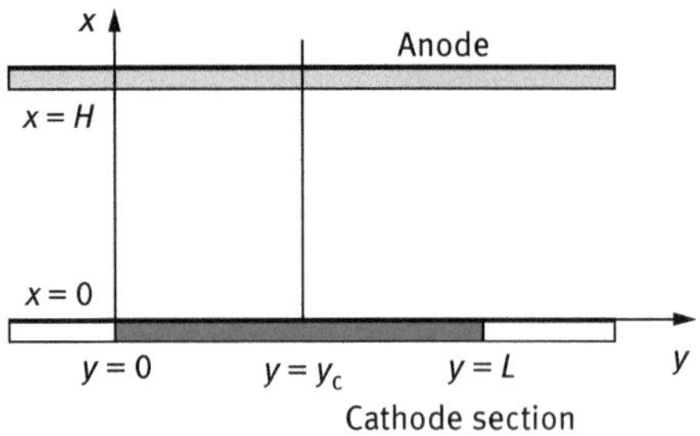

Figure 5.14: The calculation domain.

$\mathbf{E} = -\operatorname{grad} \varphi$. In addition, the Fourier–Kirchhoff equation is formulated for a temperature of neutral particles:

$$\frac{\partial n_e}{\partial t} + \frac{\partial}{\partial x}\left(\mu_e n_e E_{e,x} - \frac{D_e}{1+b_e^2}\frac{\partial n_e}{\partial x}\right) + \frac{\partial}{\partial y}\left(\mu_e n_e E_{e,y} - \frac{D_e}{1+b_e^2}\frac{\partial n_e}{\partial y}\right) = \alpha|\Gamma_e| - \beta n_e n_i, \quad (5.33)$$

$$\frac{\partial n_i}{\partial t} + \frac{\partial}{\partial x}\left(\mu_i n_i E_{i,x} - \frac{D_i}{1+b_i^2}\frac{\partial n_i}{\partial x}\right) + \frac{\partial}{\partial y}\left(\mu_i n_i E_{i,y} - \frac{D_i}{1+b_i^2}\frac{\partial n_+}{\partial y}\right) = \alpha|\Gamma_e| - \beta n_e n_i, \quad (5.34)$$

$$\frac{\partial^2 \varphi}{\partial x^2} + \frac{\partial^2 \varphi}{\partial y^2} = 4\pi e(n_e - n_i), \quad (5.35)$$

$$\rho c_V \frac{\partial T}{\partial t} = \frac{\partial}{\partial x}\left(\lambda \frac{\partial T}{\partial x}\right) + \frac{\partial}{\partial y}\left(\lambda \frac{\partial T}{\partial y}\right) + Q_J, \quad (5.36)$$

$$E_{e,x} = \frac{b_e E_y - E_x}{1+b_e^2}, \quad E_{e,y} = -\frac{b_e E_x + E_y}{1+b_e^2}, \quad (5.37)$$

$$E_{i,x} = \frac{E_x + b_i E_y}{1+b_i^2}, \quad E_{i,y} = \frac{E_y - b_i E_x}{1+b_i^2}, \quad (5.38)$$

where $\mathbf{\Gamma}_e, \mathbf{\Gamma}_i$ are the densities of electron and ion fluxes:

$$|\Gamma_e| = \sqrt{\Gamma_{e,x}^2 + \Gamma_{e,y}^2};$$

B is the vector of external magnetic field induction (this vector is shown in Figure 5.1); $Q_J = \eta(\mathbf{jE})$, $\mathbf{j} = e(\mathbf{\Gamma}_i - \mathbf{\Gamma}_e)$; $\alpha(E)$ and β are the first Townsend ionization coefficient and coefficient of ion-electron recombination; μ_e, μ_i are the mobilities of electrons and ions; D_e, D_i are the diffusivities of electrons and ions; η is the part of electron energy, gained in the electric field and intended for heating up of gas; the Hall parameters for electrons and ions are

$$b_e = \frac{\mu_e B_z}{c} = \frac{\omega_e}{\nu_e},$$

$$b_i = \frac{\mu_i B_z}{c} = \frac{\omega_i}{\nu_{in}}. \tag{5.39}$$

The Larmor frequency for electrons is

$$\omega_e = \frac{eB_z}{m_e c}; \tag{5.40}$$

and the Larmor frequency for ions is

$$\omega_i = \frac{eB_z}{m_i c}. \tag{5.41}$$

5.3.2 Thermophysical and electrophysical parameters

The discharge in molecular nitrogen is considered; therefore, we will use the following coefficients:

$$\mu_e(p^*) = \frac{4.2\times 10^5}{p^*}\frac{\text{cm}^2}{\text{V}\cdot\text{s}}, \quad \mu_i(p^*) = \frac{2280}{p^*}\frac{\text{cm}^2}{\text{V}\cdot\text{s}}, \tag{5.42}$$

$$p^* = p\frac{293}{T}\ \text{Torr}, \quad D_e = \mu_e(p^*)T_e, \quad D_i = \mu_i(p^*)T,$$

$$\rho = 1.58\times 10^{-5}\frac{M_\Sigma p}{T}\frac{\text{g}}{\text{cm}^3},$$

$$\lambda = \frac{8.334\times 10^{-4}}{\sigma^2\Omega^{(2.2)*}}\sqrt{\frac{T}{M_\Sigma}}\left(0.115 + 0.354\frac{c_p M_\Sigma}{\tilde{R}}\right)\frac{\text{W}}{\text{cm}\cdot\text{K}},$$

$$c_p = 8.314\frac{7}{2}\frac{1}{M_\Sigma}\frac{\text{J}}{\text{g}\cdot\text{K}}, \quad M_\Sigma = 28\frac{\text{g}}{\text{mole}},$$

$$\Omega^{(2.2)*} = \frac{1.157}{(T^*)^{0.1472}}, \quad T^* = \frac{T}{(\varepsilon/k)}, \quad (\varepsilon/k) = 71.4\ \text{K}, \quad \sigma = 3.68\ \text{Å}$$

$$c_V = 0.742\frac{\text{J}}{\text{g}\cdot\text{K}}, \quad \tilde{R} = 8.314\frac{\text{J}}{\text{K}\cdot\text{mole}};$$

p is the pressure; $n_n = 0.954\times 10^{19}(p/T)$ is the concentration of neutral particles.

Coefficient of ion–electron recombination β and electron temperature T_e are the chosen constants: $\beta = 2\times 10^{-7}\ \text{cm}^3/\text{s}$, $T_e = 11610$ K.

The first Townsend ionization coefficient in view of gas heating is formulated as follows:

$$\alpha(E) = p^{*} A \exp\left[-\frac{B}{(|\mathbf{E}|/p^{*})} \right] \quad (\mathrm{cm} \cdot \mathrm{Torr})^{-1}, \tag{5.43}$$

where $A = 12\,(\mathrm{cm} \cdot \mathrm{Torr})^{-1}$, $B = 342\ \mathrm{V}/(\mathrm{cm\ Torr})$.

Set of eqs. (5.33)–(5.35) is solved together with the equation for an external electric circuit:

$$E = V + IR_0, \tag{5.44}$$

where V is the voltage drop on electrodes; I is the total current; E is the electromotive force of power supply; R_0 is the Ohmic resistance of an external circuit.

5.3.3 The method of numerical integration

Continuity equations for the charged particles and the Poisson equation are formulated in the following canonical form:

$$\frac{\partial u}{\partial \tau} + \frac{\partial a u}{\partial \tilde{x}} + \frac{\partial b u}{\partial \tilde{y}} = \frac{\partial}{\partial \tilde{x}}\left(D \frac{\partial u}{\partial \tilde{x}}\right) + \frac{\partial}{\partial \tilde{y}}\left(D \frac{\partial u}{\partial \tilde{y}}\right) + f, \tag{5.45}$$

where $\tilde{x} = \frac{x}{H}$, $\tilde{y} = \frac{y}{H}$, $\tau = \frac{t}{t_0}$, $t_0 = \frac{H^2}{\mu_{e,0}\mathrm{E}}$; $u = \{n_e,\ n_i, \varphi\}$.

Coefficients $a,\ b,\ D$ and functions $u,\ f$are set by the following formulas:

- for electrons

$$u = \frac{n_e}{N_0},\ a = \tilde{\mu}_e \frac{\partial \Phi}{\partial \tilde{x}},\ b = \tilde{\mu}_e \frac{\partial \Phi}{\partial \tilde{y}},\ D = \tilde{D}_e, \tag{5.46}$$

$$f = f_e = \alpha H \sqrt{\left(u\tilde{\mu}_e \frac{\partial \Phi}{\partial \tilde{x}} - \tilde{D}_e \frac{\partial u}{\partial \tilde{x}}\right)^2 + \left(u\tilde{\mu}_e \frac{\partial \Phi}{\partial \tilde{y}} - \tilde{D}_e \frac{\partial u}{\partial \tilde{y}}\right)^2} - \beta \frac{H^2 N_0}{\mu_{e,0}\mathrm{E}} u_e u_i, \tag{5.47}$$

where $\tilde{\mu}_e = \frac{\mu_e}{\mu_{e,0}}$, $\Phi = \frac{\varphi}{E}$, $\tilde{D}_e = \frac{D_e}{\mu_{e,0}E}$, $u_e = \frac{n_e}{N_0}$, $u_i = \frac{n_i}{N_0}$;

$\mu_{e,0}$ is the characteristic value of electron mobility ($\mu_{e,0} = 8.8 \times 10^4 \mathrm{cm}/(\mathrm{s} \cdot \mathrm{V})$); N_0 is the typical concentration of electrons in positive column ($N_0 = 10^9\ \mathrm{cm}^{-3}$);

- for ions

$$u = \frac{n_i}{N_0},\ a = -\tilde{\mu}_i \frac{\partial \Phi}{\partial \tilde{x}},\ b = -\tilde{\mu}_i \frac{\partial \Phi}{\partial \tilde{y}},\ D = \tilde{D}_i,\ f = f_e, \tag{5.48}$$

where $\tilde{\mu}_i = \frac{\mu_i}{\mu_{e,0}}$, $\tilde{D}_i = \frac{D_i}{\mu_{e,0}E}$;

- for electric potential

$$a = b = 0, D = 1, f = \tilde{\varepsilon}(u_i - u_e), \tag{5.49}$$

where $\tilde{\varepsilon} = 4\pi e \frac{H^2 N_0}{\mathrm{E}}$, $4\pi e = 1.86 \times 10^{-6}\ \mathrm{V} \cdot \mathrm{cm}$

Boundary and initial conditions for the numerical integration of the set of eq. (5.45) were discussed in Section 5.1.

5.3.4 The finite-difference scheme

For the solution of the boundary-value problem (5.45)–(5.49), the following five-point finite-difference scheme was used:

$$A_{i,j}u_{i-1}^{m+1} + B_{i,j}u_{i+1,j}^{m+1} + \bar{A}_{i,j}u_{i,j-1}^{m+1} + \bar{B}_{i,j}u_{i,j+1}^{m+1} - C_{i,j}u_{i,j}^{m+1} + F_{i,j}^{m+1/2} = 0, \tag{5.50}$$

$$A_{i,j} = \frac{a_L^+}{p_i} + \frac{D_{i-1/2,j}}{p_i^- p_i}, \quad B_{i,j} = \frac{a_R^-}{p_i} + \frac{D_{i+1/2,j}}{p_i^+ p_i}, \tag{5.51}$$

$$A_{i,j} = \frac{b_L^+}{q_j} + \frac{D_{i,j-1/2}}{q_j^- q_j}, \quad B_{i,j} = \frac{b_R^-}{q_j} + \frac{D_{i,j+1/2}}{q_j^+ q_j}, \tag{5.52}$$

$$C_{i,j} = \frac{1}{\tau} + \frac{a_R^+ - a_L^-}{p_i} + \frac{b_R^+ - b_L^-}{q_j} + \left(\frac{D_{i+1/2,j}}{p_i^+} + \frac{D_{i-1/2,j}}{p_i^-}\right)\frac{1}{p_i} + \left(\frac{D_{i,j+1/2}}{q_j^+} + \frac{D_{i,j-1/2}}{q_j^-}\right)\frac{1}{q_i}, \tag{5.53}$$

$$F_{i,j} = \frac{u_{i,j}^m}{\tau} + f_{i,j}^{m+1/2}, \tag{5.54}$$

where

$$p_i = \frac{1}{2}(x_{i+1} - x_{i-1}), \quad p_i^- = x_i - x_{i-1}, \quad p_i^+ = x_{i+1} - x_i;$$

$$q_j = \frac{1}{2}(y_{j+1} - y_{j-1}), \quad q_j^- = y_j - y_{j-1}, \quad q_j^+ = y_{j+1} - y_j;$$

$$D_{i\pm1/2,j} = \frac{1}{2}(D_{i,j} + D_{i\pm1,j}), \quad D_{i,j\pm1/2} = \frac{1}{2}(D_{i,j} + D_{i,j\pm1});$$

$$a_R = \frac{1}{2}(a_{i,j} + a_{i+1,j}), \quad a_L = \frac{1}{2}(a_{i,j} + a_{i-1,j});$$

$$a_R^\pm = \frac{1}{2}(a_R \pm |a_R|), \quad a_L^\pm = \frac{1}{2}(a_L \pm |a_L|);$$

$$b_R = \frac{1}{2}(b_{i,j} + b_{i,j+1}), \quad b_L = \frac{1}{2}(b_{i,j} + b_{i,j-1});$$

$$b_R^\pm = \frac{1}{2}(b_R \pm |b_R|), \quad b_L^\pm = \frac{1}{2}(b_L \pm |b_L|);$$

m is the index of a time layer.

Boundary conditions were formulated in the following canonical form:

$$\begin{aligned} u_{i,1} &= \alpha_i u_{i,2} + \beta_i, & i &= 1, 2, \ldots, NI; \\ u_{i,NJ} &= \tilde{\alpha}_i u_{i,NJ-1} + \tilde{\beta}_i, & i &= 1, 2, \ldots, NI; \end{aligned} \tag{5.55}$$

$$u_{1,j} = \gamma_j u_{2,j} + \delta_j, \qquad j = 1, 2, \ldots, NJ;$$
$$u_{NI,j} = \tilde{\gamma}_j u_{NI-1,j} + \tilde{\delta}_j, \quad j = 1, 2, \ldots, NJ. \tag{5.56}$$

The singularity of the solved problem is the strong nonlinearity of function $F_{i,j}^{m+1/2}$, which contains the function of sources of the charged particles (see (5.54)):

$$f_{i,j}^{m+1/2} = \left[\alpha(E_{i,j})\, |\Gamma_e|_{i,j} - \beta n_{e_{i,j}} n_{+_{i,j}}\right]^{m+1/2}. \tag{5.57}$$

The stability of the used numerical algorithm is provided by the method of approximation of summand $\alpha(E)\,|\Gamma_e|$ because $\alpha(E)$ is the exponential function of electric field strength and concentration of charged particles.

The module of electron flux density is approximated in the form of

$$|\Gamma_e|_{i,j} = \sqrt{(\Gamma_{e,x})_{i,j}^2 + (\Gamma_{e,y})_{i,j}^2}, \tag{5.58}$$

where

$$(\Gamma_{e,x})_{i,j} = u_{e,i-1,j}^{m+1/2} E_{x,i,j}^{+} + u_{e,i+1,j}^{m+1/2} E_{x,i,j}^{-} + \tilde{D}_{e,i,j} \frac{u_{e,i+1,j}^{m+1/2} - u_{e,i-1,j}^{m+1/2}}{\tilde{x}_{i+1} - \tilde{x}_{i-1}};$$

$$(\Gamma_{e,y})_{i,j} = u_{e,i,j-1}^{m+1/2} E_{y,i,j}^{+} + u_{e,i,j+1}^{m+1/2} E_{y,i,j}^{-} + \tilde{D}_{e,i,j} \frac{u_{e,i,j+1}^{m+1/2} - u_{e,i,j-1}^{m+1/2}}{\tilde{y}_{j+1} - \tilde{y}_{j-1}};$$

$$E_{x,i,j}^{\pm} = \frac{1}{2}\left(E_{x,i,j} \pm |E_{x,i,j}|\right), \; E_{y,i,j}^{\pm} = \frac{1}{2}\left(E_{y,i,j} \pm |E_{y,i,j}|\right);$$

$$E_{x,i,j} = -\frac{\Phi_{i+1,j} - \Phi_{i-1,j}}{x_{i+1} - x_{i-1}}; \; E_{y,i,j} = -\frac{\Phi_{i,j+1} - \Phi_{i,j-1}}{y_{j+1} - y_{j-1}};$$

$$E_{i,j} = \sqrt{E_{x,i,j}^2 + E_{y,i,j}^2}\,; \quad u_e = n_e/N_0.$$

The superscript "$m + 1/2$" specifies the necessity of organization of additional iterations between the equations for n_e, n_i, and φ on each time layer.

The finite-difference equation (5.50) is solved on an inhomogeneous mesh:

$$\omega_h^D = \{x_i, i = 1, 2, \ldots, NI; x_1 = 0, \; x_{NI} = H; y_j, j = 1, 2, \ldots, NJ;$$
$$y_j = 0, y_{NJ} = L; \; t^{m+1} = t^m + \tau, \; m = 0, 1, \ldots\} \tag{5.59}$$

with the use of a successive under-relaxation method ($\omega = 0.5$–0.75) by runs along the x-axis. Inside each time layer, some iterations (3–5) should be performed for the coordination of n_e, n_i, φ. The specified iterations also led to local convergence of a voltage drop on electrodes V by satisfying the equation for external circuit:

$$V = E - IR_0, \tag{5.60}$$

where $I = \int_0^L en_e\mu_e E(y, x = H)\mathrm{d}y$.

5.3.5 The method of numerical integration of the heat conductive equation

First we will consider the solution of the thermal conduction equation for neutral gas in the absence of its motion:

$$\rho c_V \frac{\partial T}{\partial t} = \frac{\partial}{\partial x}\left(\lambda \frac{\partial T}{\partial x}\right) + \frac{\partial}{\partial y}\left(\lambda \frac{\partial T}{\partial y}\right) + Q_J \tag{5.61}$$

under the following boundary conditions:

$$y = 0: \quad \frac{\partial T}{\partial y} = 0; \tag{5.62}$$

$$y = L: \quad \frac{\partial T}{\partial y} = 0; \tag{5.63}$$

$$x = 0: \quad T = T_w; \tag{5.64}$$

$$x = H_c: \quad T = T_w. \tag{5.65}$$

Here ρ is the density of gas; c_V is the specific thermal capacity at constant volume; T is the temperature; λ is the coefficient of thermal conduction; $x,\ y$ are the axial and transversal variables; Q_J is the volumetric power of the Joule thermal emission; L is the transversal size of the calculated domain. We shall consider that eq. (5.61) is written for axially symmetric cylindrical geometry. In the given section, the finite-difference scheme that was used in Chapter 4 for the solution of the axially symmetric problem will be considered and used for the calculation of glow discharge in a flat geometry.

For scaling of (5.61) we shall take the following parameters:

$$\rho_0,\ c_{V,0},\ T_0,\ t_0 = \frac{H^2}{\mu_{e,0}E},$$

where H is the characteristic dimension of the problem (e.g., the height of the interelectrode gap); E is the electromotive force of the power supply; $\mu_{e,0}$ is the mobility of electrons in cold gas.

By using the dimensionless variables, the heat conduction equation will take the following form:

$$\rho c_V \frac{\partial T}{\partial \tau} = \frac{\partial}{\partial x}\left(\tilde{\lambda} \frac{\partial T}{\partial x}\right) + \frac{\partial}{\partial r}\left(\tilde{\lambda} \frac{\partial T}{\partial r}\right) + \tilde{Q}_J, \tag{5.66}$$

where $\tilde{\lambda} = \frac{\lambda}{\mu_{e,0}\mathrm{E}}$, $\tilde{Q}_J = Q_J \frac{H^2}{\mu_{e,0}\mathrm{E}}$. Remaining functions and arguments are obtained by the division of dimensional values on corresponding scaling values.

Boundary conditions are reduced to the following form:

$$y = 0: \quad \frac{\partial T}{\partial y} = 0; \tag{5.67}$$

$$y = \frac{L}{H}: \quad \frac{\partial T}{\partial y} = 0; \tag{5.68}$$

$$x = 0: \quad T = \frac{T_w}{T_0}; \tag{5.69}$$

$$x = 1: \quad T = \frac{T_w}{T_0}. \tag{5.70}$$

The finite-volume method will be used, as before, for deriving of the five-point finite-difference scheme. We shall integrate (5.66) by volume of an elementary computational mesh:

$$W_{i,j}\{*\} = \int_{\tau^m}^{\tau^{m+1}} d\tau \int_{x_{i-1/2}}^{x_{i+1/2}} dx \int_{y_{j-1/2}}^{y_{j+1/2}} \{*\} dy, \tag{5.71}$$

where $\{*\}$ means all terms in (5.66).

If $\{*\} = 1$, then $W_{i,j} = pq\tau$, where

$$q = \frac{1}{2}(x_{i+1} - x_{i-1}), \quad \tau = \tau^{m+1} - \tau^m, \quad p = \frac{1}{2}(y_{j+1} - y_{j-1}). \tag{5.72}$$

The used finite-difference mesh has the following parameters:

$$\begin{aligned} \omega_h^{\mathbb{C}} = \{p^+ = y_{j+1} - y_j,\ p^- = y_j - y_{j-1},\ j = 1, 2, 3, \ldots, NJ; \\ q^+ = x_{i+1} - x_i,\ q^- = x_i - x_{i-1},\ i = 1, 2, \ldots, NI; \tau = \tau^{m+1} - \tau^m,\ m = 0, 1, 2, \ldots\} \end{aligned} \tag{5.73}$$

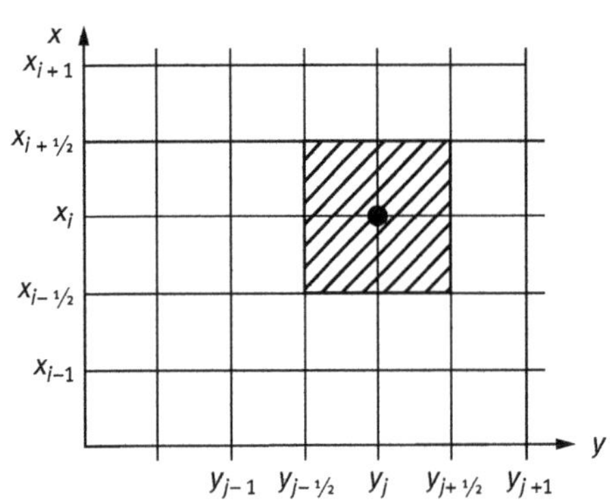

Figure 5.15: Fragment of the finite-difference mesh.

It is schematically shown in Figure 5.15.

Applying the integral operator (5.71) sequentially to all items in eq. (5.66), we will get

$$W_{i,j}\left\{\rho c_V \frac{\partial T}{\partial \tau}\right\} = \rho_{i,j}^m c_{V,i,j}^m \left(T_{i,j}^{m+1} - T_{i,j}^m\right) W_{i,j}; \tag{5.74}$$

$$W_{i,j}\left\{\frac{\partial}{\partial x}\left(\tilde{\lambda}\frac{\partial T}{\partial x}\right)\right\} = \frac{\tau p}{q^+}\tilde{\lambda}_{i+1/2,j}T_{i+1,j}^{m+1} - \tau p\left(\frac{\tilde{\lambda}_{i+1/2,j}}{q^+} + \frac{\tilde{\lambda}_{i-1/2,j}}{q^-}\right)T_{i,j}^{m+1} + \frac{\tau p}{q^-}\tilde{\lambda}_{i-1/2,j}T_{i-1,j}^{m+1}, \tag{5.75}$$

where $\tilde{\lambda}_{i\pm1/2,j} = \frac{1}{2}\left(\tilde{\lambda}_{i,j} + \tilde{\lambda}_{i\pm1,j}\right)$;

$$W_{i,j}\left\{\frac{\partial}{\partial y}\left(\tilde{\lambda}\frac{\partial T}{\partial y}\right)\right\} = \frac{\tau q}{p^+}\tilde{\lambda}_{i,j+1/2}T_{i,j+1}^{m+1} - \tau q\left(\frac{\tilde{\lambda}_{i,j+1/2}}{p^+} + \frac{\tilde{\lambda}_{i,j-1/2}}{p^-}\right)T_{i,j}^{m+1} + \frac{\tau q}{p^-}\tilde{\lambda}_{i,j-1/2}T_{i,j-1}^{m+1}, \tag{5.76}$$

where $\tilde{\lambda}_{i,j\pm1/2} = \frac{1}{2}\left(\tilde{\lambda}_{i,j} + \tilde{\lambda}_{i,j\pm1}\right)$;

$$W_{i,j}\{\tilde{Q}_J\} = qp\tau\tilde{Q}_J. \tag{5.77}$$

Combining (5.74)–(5.77) in a single equation, we will get the five-point finite-difference scheme of the following form:

$$A_{i,j}T_{i,j-1}^{m+1} + B_{i,j}T_{i,j+1}^{m+1} - \mathbb{C}_{i,j}T_{i,j}^{m+1} + \bar{A}_{i,j}T_{i-1,j}^{m+1} + \bar{B}_{i,j}T_{i+1,j}^{m+1} + F_{i,j} = 0, \tag{5.78}$$

where

$$B_{i,j} = \frac{\tilde{\lambda}_{i,j+1/2}}{pp^+},\quad \bar{A}_{i,j} = \frac{\tilde{\lambda}_{i-1/2,j}}{qq^-},\quad \bar{B}_{i,j} = \frac{\tilde{\lambda}_{i+1/2,j}}{qq^+},\quad F_{i,j} = \frac{\rho_{i,j}^m c_{V,i,j}^m}{\tau}T_{i,j}^m + \tilde{Q}_J, \tag{5.79}$$

$$\mathbb{C}_{i,j} = A_{i,j} + B_{i,j} + \bar{A}_{i,j} + \bar{B}_{i,j} + \frac{\rho_{i,j}^m c_{V,i,j}^m}{\tau}. \tag{5.80}$$

The received finite-difference scheme has the second-order approximation on space and the first order of approximation on time.

5.3.6 Numerical simulation results for glow discharge in a magnetic field in view of heating of a gas

Calculations of glow discharge were performed in the calculation domain shown in Figure 5.14. For all calculations, the following initial data were common: $H = 2$ cm is the height of electrodischarge gap, $y = y_c = 2$ cm is the coordinate of the center of the plasma column at an initial instant, and $y = L = 4$ cm is the breadth of electric discharge gap. For the estimation of initially charged particle concentrations, the theory of normal current density was used.

It should be stressed that the fixed location of discharge can be ensured by the initial conditions and by the computing scheme, which does not break the equilibrium position of glow discharge. Without external magnetic field, the glow discharge is localized in the region where it has been initiated.

Let us start analyzing the numerical simulation results obtained for gas pressure $p = 10$ Torr, electromotive force $E = 2$ kV, secondary electronic emission coefficient $\gamma = 0.1$, and effectiveness ratio of transformation of electric field energy in heating of gas $\eta = 0$–0.3. Each of the listed parameters of glow discharge strongly influence its structure; therefore, first of all the qualitative analysis of the basic tendencies of this influence will be presented. Data shown in Figure 5.16 correspond to absence of magnetic field ($B = 0$).

One-dimensional Engel–Steenbeck theory predicts an increasing in the thickness of cathode layer and a voltage drop in it, and also a diminution of normal current density with the decrease of γ. We would remind that the secondary emission coefficient is defined as the ratio of a number of secondary electrons to the total number of ions falling on the cathode surface. All this is due to a diminution of γ, which actually means a decrease in the flux of electrons abandoning the cathode, which reduces the effectiveness of ionization processes and impedes the process of burning of glow discharge. It is a reason for an increase in the voltage drop across the discharge gap ($V = 451$ V at $\gamma = 0.33$ and $V = 555$ V at $\gamma = 0.1$) and the diminution of the total current ($I = 3.5$ mA at $\gamma = 0.33$ and $I = 3.2$ mA at $\gamma = 0.1$; E = 1.5 kV, $p = 5$ Torr). Increasing the voltage drop on the gap leads to an increase in cross sizes of the cathode spot. Let us also note the twofold drop of concentration of the charged particles in the positive column ($n_e = n_i = 2.8 \times 10^9$ cm^{-3} at $\gamma = 0.33$ and $n_e = n_i = 1.4 \times 10^9$ cm^{-3} at $\gamma = 0.1$).

Growth of gas pressure leads to proportional diminution of the thickness of the cathode layer and a quite abrupt increase in the current density on the cathode. According to the Engel–Steenbeck theory, the cathode voltage drop does not depend on pressure; therefore, the current density should increase by the pressure quadratically.

Two-dimensional calculations give a good agreement with this theory, despite the fact that with an increase in pressure the voltage drop varies on the whole discharge gap as well. Results of the calculations of glow discharge at pressure Torr are shown in Figure 5.16. An appreciable diminution of cross sizes of the cathode layer and the positive column can be noted with an increase in pressure.

With an increase in pressure, the voltage drop on gas-discharge gap has increased from 544 V up to 835 V, and the current density on the cathode has increased from 3.58 mA/cm^2 to 6.94 mA/cm^2, and the current density on the anode has increased from 6.03 mA/cm^2 up to 8.89 mA/cm^2. The total current in the circuit has decreased from 4.86 mA to 3.91 mA. It is important to note that an abrupt increase in the concentration of ions in the cathode layer from 1.9×10^{10} cm^{-3} up to 4.6×10^{10} cm^{-3}.

By comparing Figures 5.16 and 5.17 it is possible to see the modification of glow discharge structure by considering gas heating. Isolines of electron and ion

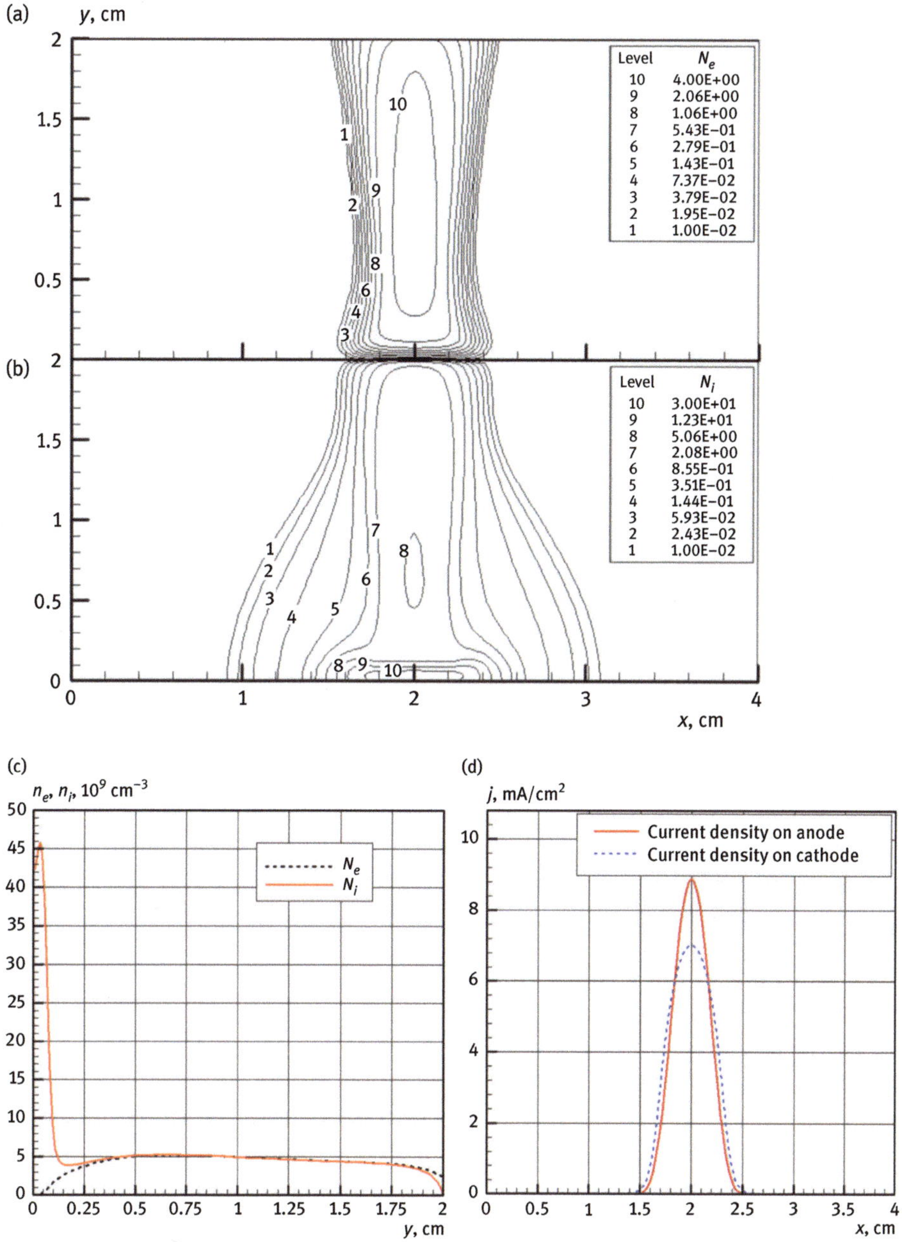

Figure 5.16: Structure of two-dimensional flat glow discharge at $p = 10$ Torr, $E = 2.0$ kV, $\gamma = 0.1$; the computational mesh is 141×61: (a) concentration of electrons; (b) concentration of ions; (c) distribution of ions and electrons along the axis of symmetry of glow discharge; (d) current density on the cathode and the anode.

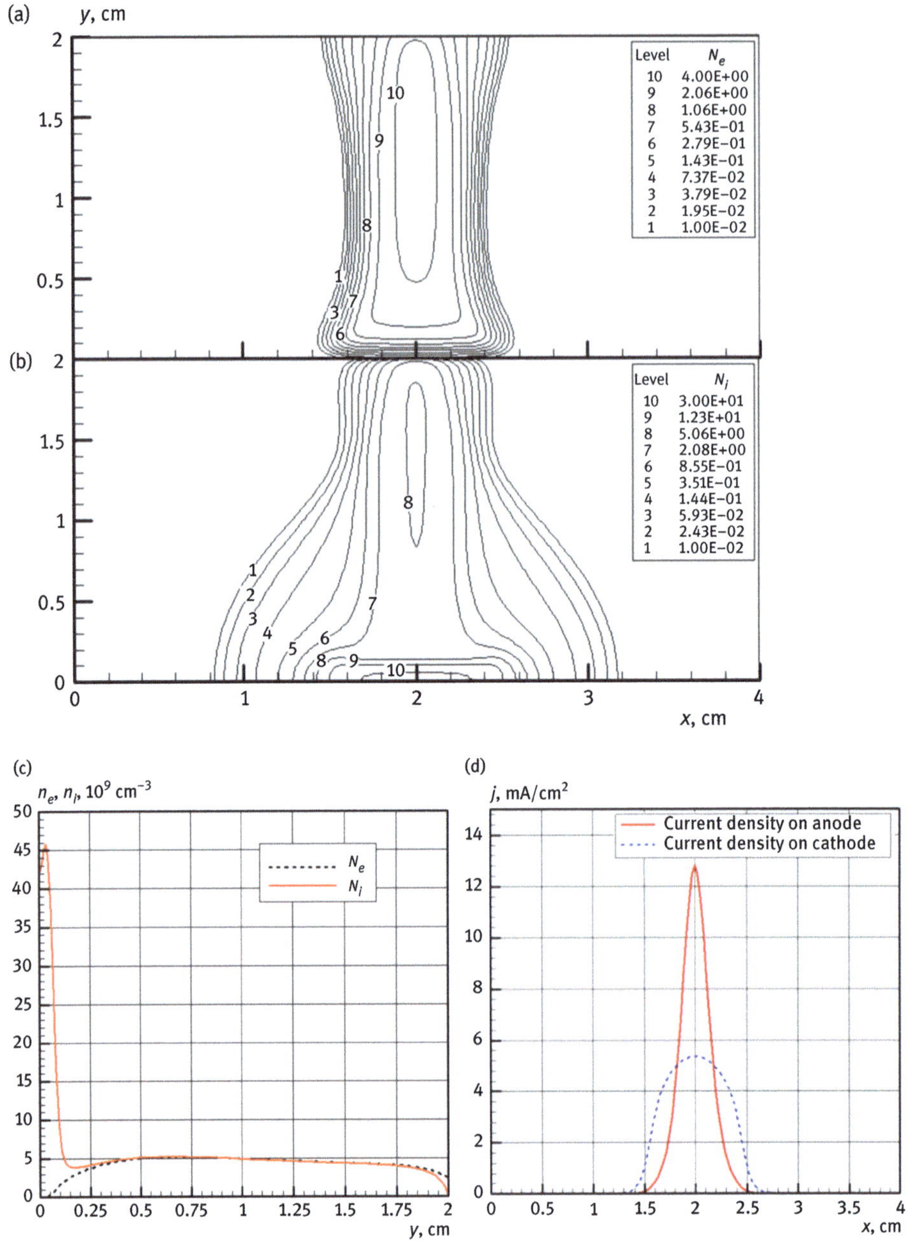

Figure 5.17: Structure of a two-dimensional flat glow discharge at $p = 10$ Torr, $E = 2.0$ kV, $\gamma = 0.1$, $\eta = 0.3$; the computational mesh is 141×61: (a) concentration of electrons; (b) concentration of ions; (c) distribution of ions and electrons along glow discharge symmetry axis; (d) current density on the cathode and the anode; e distribution of temperature along a symmetry axis of discharge.

(e)

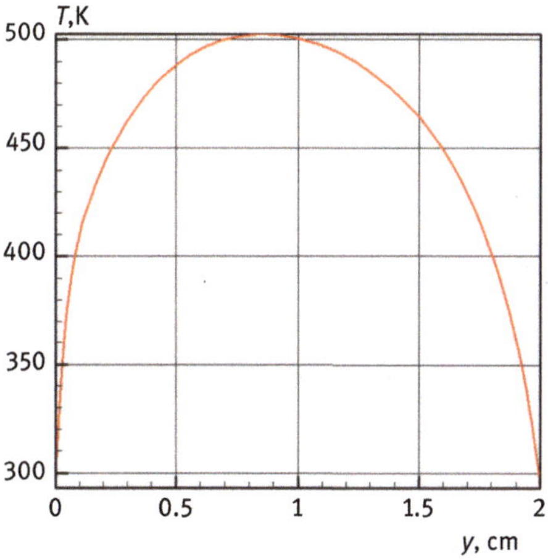

Figure 5.17 (continued)

concentrations in glow discharge at $p = 10$ Torr and $E = 2.0$ kV are shown in Figure 5.17. Because the gas heating leads to a drop of effective local pressure (see (5.42)), the discovered tendency of the discharge integral parameters diminution (voltage drop, current density) answers to some real decreasing of pressure. The voltage drop on the gap decreases to 756 V, the current in the circuit increases to 4.13 mA, and the current density on the cathode drops to 5.36 mA/cm^2.

The concentration of ions in the cathode layer also decreases (to 3.5×10^{10} cm^{-3}). However, the concentration of the charged particles in the positive column, in this case, increases up to the value of 6.0×10^9 cm^{-3}.

Results of calculations of glow discharge in external magnetic field $B_z = +0.1$ T in view of gas heating ($\eta = 0.3$) are shown in Figures 5.18, and 5.19 for successive time instants. The principal peculiarity of the considered calculation case is the discharge motion perpendicular to the directions of the magnetic field and the current. At comparatively high inductions of the magnetic field (in the considered case, at $B_z = +0.1$ T), the current channel of glow discharge is displaced so fast that the neutral gas has not time to get warm. As the initial condition for calculations in the presence of the magnetic field, the results of calculations of glow discharge in view of gas heating were used at the same values of p, E, γ; therefore in process of glow discharge motion in the magnetic field, the temperature of neutral gas inside a zone of the current column decreased. At $B_z = +0.1$ T the velocity of the current column motion in a direction of y-axis was $v_y \sim 0.3 \times 10^5$ cm/s.

Figure 5.20 represents the glow discharge structure with a relatively weak magnetic field ($B_z = +0.01$ T). These calculations were performed from initially heated gas.

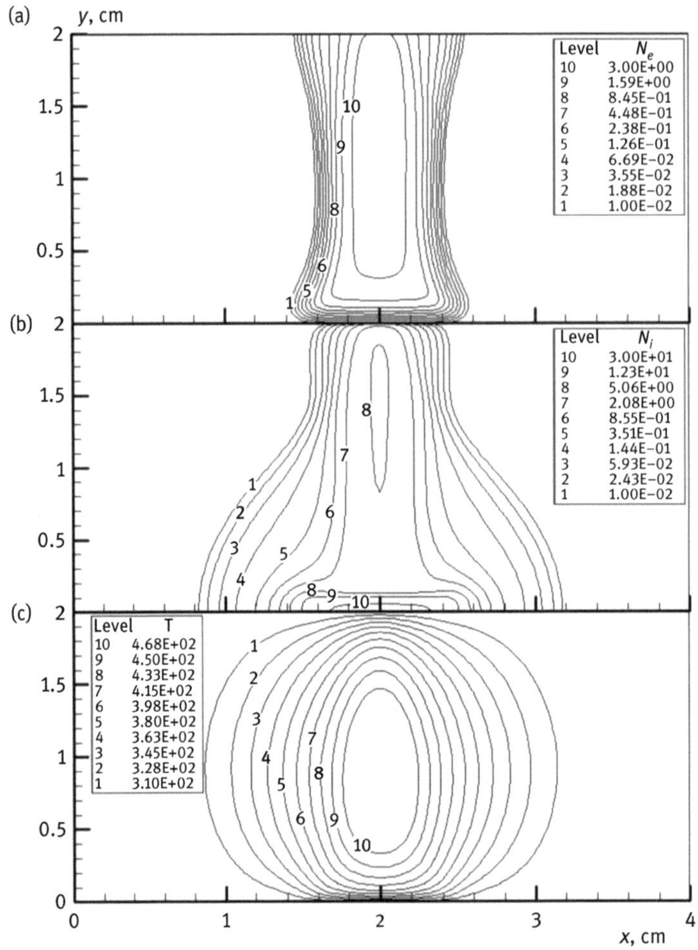

Figure 5.18: Structure of two-dimensional flat glow discharge at $t = 0$, $p = 10$ Torr, $E = 2.0$ kV, $B_z = +0.1$ T, $\gamma = 0.1$, $\eta = 0.3$; the computational mesh is 141 × 61: (a) concentration of electrons; (b) concentration of ions; (c) temperature.

At $B = 0.01$ T the current column motion in y direction is observed with a velocity $v_y \sim 0.2 \times 10^4$ cm/s.

Thus, in this section, we have considered the possibility of the two-dimensional drift-diffusion model for the calculation of glow discharge in flat geometry with a transverse magnetic field and neutral gas heating.

A distinctive peculiarity of the performed research is self-consistent with electrodynamic structure account of neutral gas heating in regions occupied by the discharges.

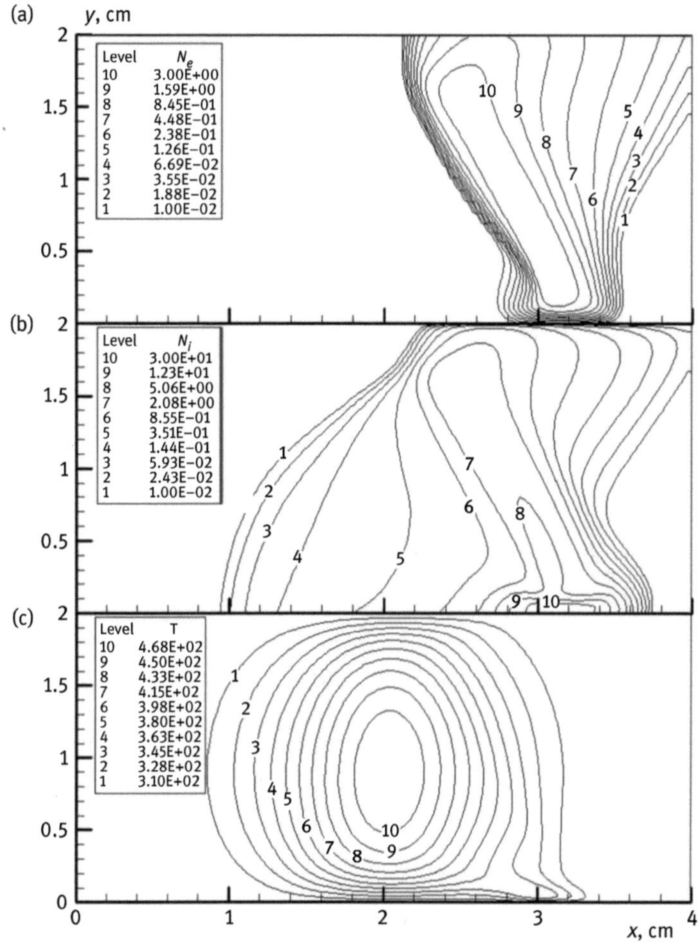

Figure 5.19: Structure of two-dimensional flat glow discharge at $t = 20$ µs, $p = 10$ Torr, E = 2.0 kV, $B_z = +0.1$ T, $\gamma = 0.1$, $\eta = 0.3$; the computational mesh is 141 × 61: (a) concentration of electrons; (b) concentration of ions; (c) temperature.

It has been established that molecular nitrogen in glow discharge at $p \sim 5$–10 Torr and $E \sim 2$–3 kV is heated up to temperatures of ~500 K, and its current column in a magnetic field $B_z = +(0.01–0.1)$ T moves with a velocity of $v_y \sim 0.2\times10^4–0.3\times10^5$ cm/s, with no time to heat up neutral gas to the specified temperatures, but leaving a trail of the warmed gas.

Note that the numerical simulation allows to improve our understanding of the processes in glow discharges. In particular, results of the calculations suggest the possibility of the use of an external magnetic field as the control parameter near to streamline surfaces.

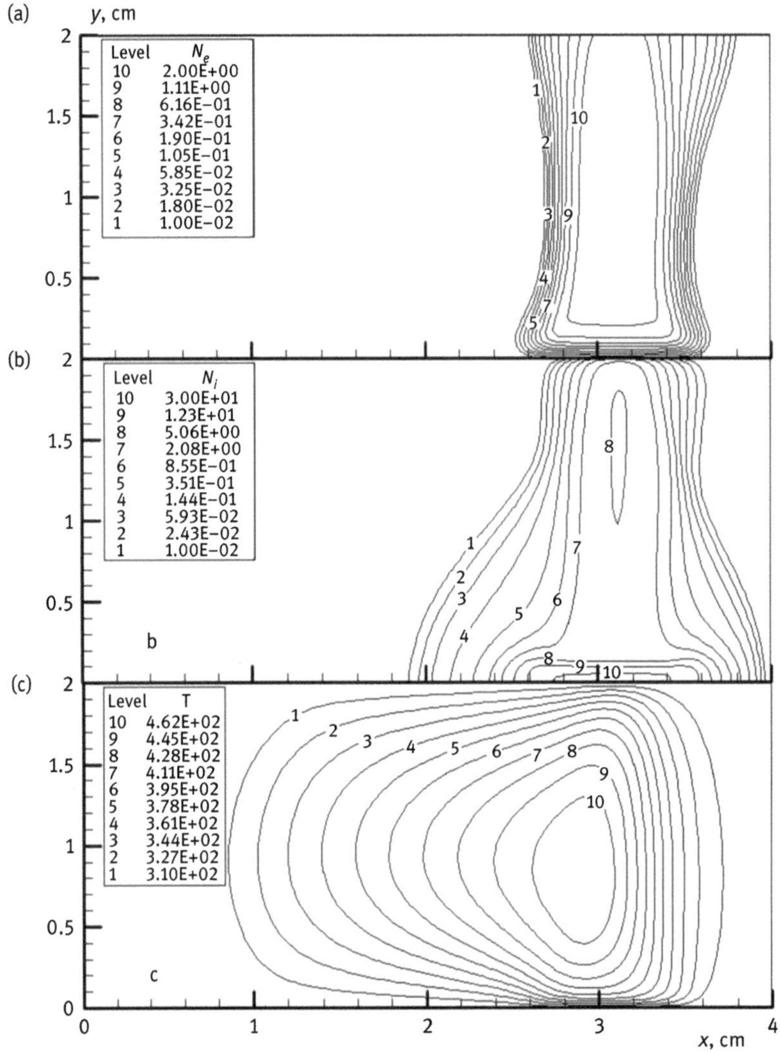

Figure 5.20: Structure of two-dimensional flat glow discharge at $t = 450$ μs, $p = 10$ Torr, $E = 2.0$ kV, $B_z = +0.01$ T, $\gamma = 0.1$, $\eta = 0.3$; initially warmed gas; the computational mesh is 141 × 61: (a) concentration of electrons; (b) concentration of ions; (c) temperature.

5.4 The glow discharge in the cross-flow of neutral gas and in the magnetic field

The analysis of characteristic time scales of elementary physical processes in the glow discharges that had been considered in the previous chapters allows to draw a conclusion that approximately in $\tau_p \sim 10^{-4}$ s after electric breakdown the basic relaxation electrodynamic processes are completed. If we take into account that the characteristic spatial scale of the discharge channel is of $L \sim 1$ cm, then we can omit considering the influence of gas flow with velocities $V \leq 0.1(L/\tau_p) \approx 10^3$ cm/s on the electrodynamic structure of gas discharges. The given estimations allow to approve that the influence of gas motion on the electrodynamic structure of glow discharge cannot be neglected at supersonic and, in particular, at hypersonic velocities of gas flow (~10^5 cm/s). It is obvious that the fact of high-speed motion of gas should be considered in computational models of glow discharge. In this chapter the computing model of two-dimensional flat glow discharge in a cross gas flow and in an exterior magnetic field is given.

5.4.1 A computational model of glow discharge with cross gas flow

Schematic of the problem is shown in Figure 5.21.

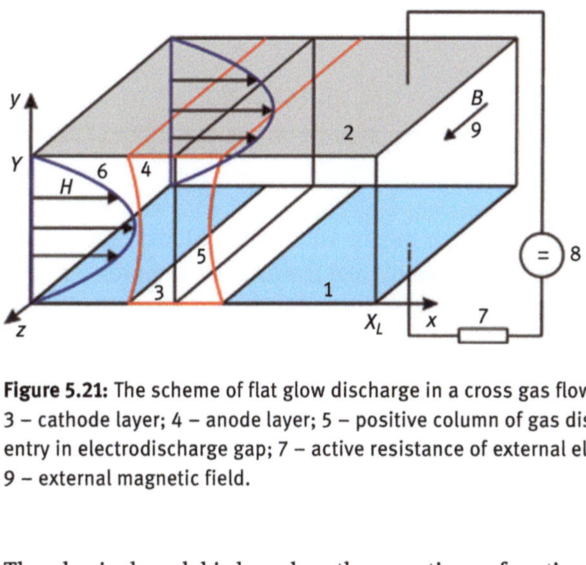

Figure 5.21: The scheme of flat glow discharge in a cross gas flow: 1 – cathode; 2 – anode; 3 – cathode layer; 4 – anode layer; 5 – positive column of gas discharge; 6 – flow of neutral gas at entry in electrodischarge gap; 7 – active resistance of external electric circuit; 8 – power supply; 9 – external magnetic field.

The physical model is based on the equations of motion of multifluid partially ionized gas mixture, which are obtained from the Boltzmann equation by using the moment procedure (Krall N.A., et al., 1973; Shkarofsky I.P., et al., 1966)

$$n_e m_e \frac{\partial \mathbf{u}_e}{\partial t} + n_e m_e (\mathbf{u}_e \nabla) \mathbf{u}_e = -\nabla p_e - \tau_e + n_e \mathbf{F}_e - e n_e \left(\mathbf{E} + \frac{1}{c} [\mathbf{u}_e \mathbf{H}] \right) - \\ - m_e \nu_{en} n_e (\mathbf{u}_e - \mathbf{u}_n) - m_e \nu_{ei} n_e (\mathbf{u}_e - \mathbf{u}_i), \quad (5.81)$$

$$n_i m_i \frac{\partial \mathbf{u}_i}{\partial t} + n_i m_i(\mathbf{u}_i \nabla)\mathbf{u}_i = -\nabla p_i - \tau_i + n_i \mathbf{F}_i + e n_i \left(\mathbf{E} + \frac{1}{c}[\mathbf{u}_i \mathbf{H}] \right) - m_i \nu_{in} n_i(\mathbf{u}_i - \mathbf{u}_e) - m_i \nu_{in} n_i(\mathbf{u}_i - \mathbf{u}_n), \tag{5.82}$$

where n_e, n_i are the volumetric concentration of electrons and ions; m_e, m_i are the masses of an electron and an ion; $\mathbf{u}_e, \mathbf{u}_i, \mathbf{u}_n$ are the average velocities of electronic and ionic liquids, and also of neutral gas; p_e, p_i are the pressure of electronic and ionic liquids; τ_e, τ_i are the components of viscous stress tensor of electronic and ionic liquids; $\mathbf{F}_e, \mathbf{F}_i$ are the volumetric forces effecting the particles of electron and ion liquids; e is the electron charge; c is the speed of light; $\mathbf{E}, \mathbf{H}$ are the intensities of electric and magnetic fields; ν_{en}, ν_{ei} are the frequencies of collisions of electrons with neutral particles and with ions; ν_{in}, ν_{ie} are the frequencies of collisions of ions with neutral particles and with electrons.

Let us fix some inequalities and introduce some assumptions for further simplification of the initial eqs. (5.81) and (5.82):

1) $m_e \ll m_i$.
2) A degree of ionization of gas $\alpha < 10^{-4}$, that is, the concentration of the charged particles is more than four orders of magnitude lower than the concentration of neutral particles $n_e \sim n_i \ll n_n$.
3) The characteristic velocities of neutral gas flow are of $V_0 = 10^4 – 3 \times 10^5$ cm/s.
4) There is no influence of the glow discharge plasma on gas dynamics of neutral particle flow.

Glow discharge in molecular nitrogen or in the air will be considered later; therefore, for numerical estimations, it is possible to take the following approximations of particle collision cross-sections:

$$\nu_{eN_2} = 2.5 \times 10^{-11} n_{N_2} T_e \left(1 + 9.3 \times 10^{-3} \sqrt{T_e}\right) \mathrm{c}^{-1}, \tag{5.83}$$

$$\nu_{eO_2} = 1.5 \times 10^{-10} n_{O_2} T_e \left(1 + 4.2 \times 10^{-2} \sqrt{T_e}\right) \mathrm{c}^{-1}, \tag{5.84}$$

$$\nu_{eO} = 2.8 \times 10^{-10} n_O \sqrt{T_e}\, \mathrm{c}^{-1}, \tag{5.85}$$

where T_e is the temperature of electrons, K.

Let us consider molecular nitrogen at $p = 5\,\text{Topp} \cong 6.58 \times 10^3$ erg/cm³, and $T = 300$ K, then

$$n_{N_2} \approx \frac{p}{kT} = 1.59 \times 10^{17}\ \mathrm{cm}^{-3}.$$

Suppose that $T_e = 11610$ K, one can estimate

$$\nu_{eN_2} \cong 9.2 \times 10^{10}\ \mathrm{s}^{-1}.$$

For comparison, we shall adduce values of collision frequencies of electrons with molecules O_2 and atoms O (with other parameters being equal):

$$\nu_{eO_2} \cong 1.42 \times 10^{10}\ \mathrm{c}^{-1},\ \nu_{eO} \cong 4.79 \times 10^{9}\ \mathrm{c}^{-1}.$$

The collision frequency of ions with neutral particles is estimated by the following formula (Raizer Yu.P., 1987):

$$\nu_{in} = \frac{2\sqrt{2}}{3}\pi a^2 \sqrt{\frac{8kT}{m_e \pi}} n_n, \tag{5.86}$$

where a is the effective radius of a neutral particle in relation with the interaction of ions with molecules and atoms. For our estimations, it is enough to use the Bohr radius under

$$a = 0.529 \times 10^{-8}\ \mathrm{cm},$$

then $\nu_{in} = 1.42 \times 10^{8}\ \mathrm{s}^{-1}$.

To estimate the frequency of collision of electrons with ions it is possible to apply the approximated relation, which considers shielding of a field of the given charged particle by other charged particles:

$$\nu_{ei} = 5.5 T_e^{-3/2} n_e \ln\left(220 T_e n_e^{-1/3}\right)\ \mathrm{s}^{-1}. \tag{5.87}$$

For discussed initial data

$$\nu_{ei} \approx 2.27 \times 10^{6}\ \mathrm{s}^{-1}.$$

Further, we will consider the basic scales of the most significant processes rates:

1) A characteristic velocity of gas flow is $V_0 = 10^4 - 3 \times 10^5$ cm/s.
2) A thermal velocity of electrons (by considering the Maxwell of their motion):

$$\mathrm{v}_{e,T} = \sqrt{\frac{8kT_e}{\pi m_e}} \approx 6.21 \times 10^{5} \sqrt{T_e} \approx 6.69 \times 10^{7}\ \mathrm{cm/s}. \tag{5.88}$$

3) Drift velocities of electrons and ions in an electric field with strength $E_{pc} = 300$ V/cm (in a positive column):

$$\mathrm{v}_{e\,dr} = \mu_e E_{pc} = 2.52 \times 10^{7}\ \mathrm{cm/s}, \tag{5.89}$$

$$\mathrm{v}_{i\,dr} = \mu_i E_{pc} = 8.64 \times 10^{4}\ \mathrm{cm/s}, \tag{5.90}$$

where for the mobilities of electrons and ions in molecular nitrogen the following values are chosen:

$$\mu_e = \frac{4.2 \times 10^5}{p} \frac{\text{cm}^2}{\text{V} \cdot \text{s}}, \tag{5.91}$$

$$\mu_i = \frac{1.44 \times 10^3}{p} \frac{\text{cm}^2}{\text{V} \cdot \text{s}} \tag{5.92}$$

where the pressure is measured in Torr.

Presented estimations allow to make important conclusion about the presence of the two scales of velocities in the problem under consideration:
1) The velocities of electronic motion (a thermal velocity, a drift velocity) have the scale of $v_e \sim 2 \times 10^7$ cm/s.
2) The velocities of ionic motion and motion of neutral particles of $v_i \approx V_0 \sim 10^4 – 10^5$ cm/s.

Let us consider the left-hand side of eq. (5.81), by introducing a characteristic time τ_p and a spatial scale $L = 1$ cm, and estimating orders of its summands:

a) $n_e m_e \dfrac{\partial u_e}{\partial t} \sim n_e m_e \dfrac{u_e}{\tau_p} \approx 2 \times 10^{-5}$;

b) $n_e m_e u_e \dfrac{\partial u_e}{\partial x} \sim n_e m_e \dfrac{u_e^2}{L} \approx 3.6 \times 10^{-2}$.

The summands in the right-hand part of eq. (5.81) can be estimated as follows:

a) $\nabla p_e \sim \dfrac{n_e k T_e}{L} \approx 0.16$;

b) $e n_e E \approx 1.44 \times 10^4$;

c) $e n_e c^{-1} u_e H \approx 3.2 \times 10^2$,

where the strength of a magnetic field is accepted to be equal to $H = 10^4$ Oe;

d) $m_e v_{en} n_e (u_e - u_n) \approx m_e v_{en} n_e u_e \approx 167.6$;

e) $m_e v_{ei} n_e (u_e - u_i) \approx m_e v_{ei} n_e u_e \approx 5.05 \times 10^{-3}$.

We shall fulfill similar estimations with the left- and right-hand sides of eq. (5.82). In the left-hand side of eq. (5.82):

a) $n_i m_i \dfrac{\partial u_i}{\partial t} \sim n_i m_i \dfrac{u_i}{\tau_p} \approx 4.65 \times 10^{-3}$;

b) $n_i m_i u_i \dfrac{\partial u_i}{\partial x} \sim n_i m_i \dfrac{u_i^2}{L} \approx 4.65 \times 10^{-2}$.

In the right-hand side of eq. (5.82):

a) $\nabla p_i \sim \frac{p_i}{L} = \frac{n_i k T_i}{L} \approx 4.14 \times 10^{-3}$;
b) $e n_i E \approx 1.44 \times 10^4$;
c) $e n_i c^{-1} u_i H \approx 1.6$;
d) $m_i \nu_{ie} n_i (u_i - u_e) \approx m_i \nu_{ie} n_i u_e \approx 1.29$;
e) $m_i \nu_{in} n_i (u_i - u_n) \approx m_i \nu_{in} n_i u_i \approx 66$.

In the considered estimations, the analysis of the possible influence of viscous stress tensor components of electron and ion liquids has been omitted. To estimate the value of the items corresponding to the viscous stress tensor components of electron and ion gases, we shall present them in the form of

$$\Gamma_e = \eta_e \Delta \mathbf{u}_e + \frac{1}{3} \eta_e \operatorname{grad}(\operatorname{div} \mathbf{u}_e), \tag{5.93}$$

$$\Gamma_i = \eta_i \Delta \mathbf{u}_i + \frac{1}{3} \eta_i \operatorname{grad}(\operatorname{div} \mathbf{u}_i), \tag{5.94}$$

where η_e, η_i are the coefficients of dynamic viscosity of electron and ion fluids.

The phenomenological set of the specified coefficients of viscosity (for the absence of a strong magnetic field) gives

$$\eta_e = n_e m_e \frac{k T_e}{m_e \nu_{ee}}, \tag{5.95}$$

$$\eta_i = n_i m_i \frac{k T_i}{m_i \nu_{ii}}, \tag{5.96}$$

or for the considered conditions

$$\eta_e \approx 5.78 \times 10^{-8} \ \frac{\text{g}}{\text{cm} \cdot \text{s}}, \quad \eta_i \approx 0.149 \times 10^{-8} \ \frac{\text{g}}{\text{cm} \cdot \text{s}}.$$

Considering the scales of electronic and ionic velocities (according to 2×10^7 cm/s and 10^4–10^5 cm/s) it is possible to conclude that in regions of large gradients of velocities of ion and electron liquids it is impossible to neglect viscous stress tensor components. Further, we will assume insignificance of the given terms in the considered problem about dynamics of glow discharge. Such an assumption is based on the fact that in the considered cases the possible change of discharge parameters is insignificant in the direction across a streamline.

Comparison of the order of all values in eqs. (5.81) and (5.82) allows to draw a conclusion that at the considered conditions the equation for the determination of the average velocity of electrons and ions can be essentially simplified. Having taken the specified estimations, we can write

$$\begin{cases} m_e \nu_{en} n_e \mathbf{u}_e = -\nabla p_e - e n_e (\mathbf{E} + c^{-1}[\mathbf{u}_e \mathbf{H}]), \\ m_i \nu_{in} n_i \mathbf{u}_i = -\nabla p_i - m_i n_i \nu_{in} \mathbf{u}_n + e n_i (\mathbf{E} + c^{-1}[\mathbf{u}_i \mathbf{H}]). \end{cases} \tag{5.97}$$

Let us suppose that

$$\nabla p_e = kT_e \nabla n_e, \tag{5.98}$$

$$\nabla p_i = kT_i \nabla n_i. \tag{5.99}$$

Then

$$\mathbf{u}_e = -\frac{kT_e}{m_e \nu_{en} n_e}\nabla n_e - \frac{e}{m_e \nu_{en}}\left(\mathbf{E} + \frac{1}{c}[\mathbf{u}_e \mathbf{H}]\right),$$

$$\mathbf{u}_i = -\frac{kT_i}{m_i n_i \nu_{in}}\nabla n_i + \mathbf{u}_n + \frac{e}{m_i \nu_{in}}\left(\mathbf{E} + \frac{1}{c}[\mathbf{u}_i \mathbf{H}]\right).$$

By definition

$$\mu_e = \frac{e}{m_e \nu_{en}}, \quad \mu_i = \frac{e}{m_i \nu_{in}}, \tag{5.100}$$

then

$$\frac{kT_e}{m_e \nu_{en}} = \frac{e}{m_e \nu_{en}} \cdot \frac{kT_e}{e} = \mu_e \frac{kT_e}{e}.$$

But $\frac{D_e}{\mu_e} = \frac{k}{e} T_e \to D_e = \mu_e \frac{k}{e} T_e$, where $\frac{k}{e} T_e$ is the electron temperature, in eV; therefore

$$\frac{kT_e}{m_e \nu_{en}} = \mu_e T_e[\text{эВ}] = D_e, \tag{5.101}$$

and by analogy,

$$\frac{kT_i e}{m_i \nu_{in} e} = \mu_i T_i[\text{эВ}] = D_i. \tag{5.102}$$

In view of (5.101), (5.102) we will obtain expressions for the averaged velocities of electrons and ions:

$$\mathbf{u}_e = -D_e \nabla n_e \frac{1}{n_e} - \mu_e\left(\mathbf{E} + \frac{1}{c}[\mathbf{u}_e \mathbf{H}]\right), \tag{5.103}$$

$$\mathbf{u}_i = -D_i \nabla n_i \frac{1}{n_i} + \mathbf{u}_n + \mu_i\left(\mathbf{E} + \frac{1}{c}[\mathbf{u}_i \mathbf{H}]\right). \tag{5.104}$$

Let us rewrite eqs. (5.24) and (5.25) in x and y components (see Figure 5.24)

$$\begin{cases} u_{e,x} = -D_e \dfrac{1}{n_e}\dfrac{\partial n_e}{\partial x} - \mu_e E_x - \mu_e \dfrac{1}{c} u_{e,y} H_z, \\ u_{e,y} = -D_e \dfrac{1}{n_e}\dfrac{\partial n_e}{\partial y} - \mu_e E_y + \mu_e \dfrac{1}{c} u_{e,x} H_z; \end{cases} \tag{5.105}$$

$$\begin{cases} u_{i,x} = u_{n,x} - D_i \dfrac{1}{n_i} \dfrac{\partial n_i}{\partial x} + \mu_i E_x + \mu_i \dfrac{1}{c} u_{i,y} H_z, \\ u_{i,y} = u_{n,y} - D_i \dfrac{1}{n_i} \dfrac{\partial n_i}{\partial y} + \mu_i E_y - \mu_i \dfrac{1}{c} u_{i,x} H_z. \end{cases} \tag{5.106}$$

From (5.105) it is possible to derive x and y components of velocity of electrons. For a velocity $u_{e,x}$

$$u_{e,x}\left(1 + \frac{\mu_e^2}{c^2} H_z^2\right) = -\frac{D_e}{n_e}\frac{\partial n_e}{\partial x} + \frac{\mu_e}{c} H_z \frac{D_e}{n_e}\frac{\partial n_e}{\partial y} + \frac{\mu_e}{c} H_z \mu_e E_y - \mu_e E_x,$$

where

$$b_e = \frac{\mu_e H_z}{c}, \tag{5.107}$$

is the Hall parameter for electrons.

Then

$$u_{e,x} = -\mu_e E_{e,x} - \frac{D_e}{1 + b_e^2} \frac{1}{n_e} \frac{\partial n_e}{\partial x} + \frac{b_e D_e}{1 + b_e^2} \frac{1}{n_e} \frac{\partial n_e}{\partial y}, \tag{5.108}$$

where

$$E_{e,x} = \frac{E_x - b_e E_y}{1 + b_e^2}. \tag{5.109}$$

By analogy for a velocity $u_{e,y}$ we shall write the expression:

$$u_{e,y} = -\mu_e E_{e,y} - \frac{D_e}{1 + b_e^2} \frac{1}{n_e} \frac{\partial n_e}{\partial y} - \frac{b_e D_e}{1 + b_e^2} \frac{1}{n_e} \frac{\partial n_e}{\partial x}, \tag{5.110}$$

where

$$E_{e,y} = \frac{E_y + b_e E_x}{1 + b_e^2}. \tag{5.111}$$

We shall adduce similar calculations for ions. The component of a velocity $u_{i,x}$

$$u_{i,x} = \frac{u_{n,x} + b_i u_{n,y}}{1 + b_i^2} + \mu_i E_{i,x} - \frac{D_i}{1 + b_i^2} \frac{1}{n_i} \frac{\partial n_i}{\partial x} - \frac{b_i D_i}{1 + b_i^2} \frac{1}{n_i} \frac{\partial n_i}{\partial y}, \tag{5.112}$$

where

$$E_{i,x} = \frac{E_x + b_i E_y}{1 + b_i^2}. \tag{5.113}$$

The component of a velocity $u_{i,y}$

$$u_{i,y} = \frac{u_{n,y} - b_i u_{n,x}}{1 + b_i^2} + \mu_i E_{i,y} - \frac{D_i}{1 + b_i^2} \frac{1}{n_i} \frac{\partial n_i}{\partial y} + \frac{b_i D_i}{1 + b_i^2} \frac{1}{n_i} \frac{\partial n_i}{\partial x}, \tag{5.114}$$

where

$$E_{i,y} = \frac{E_y - b_i E_x}{1 + b_i^2}. \tag{5.115}$$

Let us introduce densities of particle flux vectors

$$\Gamma_e = \mathbf{i}\Gamma_{e,x} + \mathbf{j}\Gamma_{e,y}, \tag{5.116}$$

$$\Gamma_i = \mathbf{i}\Gamma_{i,x} + \mathbf{j}\Gamma_{i,y}, \tag{5.117}$$

where $\Gamma_{e,x} = n_e u_{e,x}$, $\Gamma_{e,xy} = n_e u_{e,y}$, $\Gamma_{i,x} = n_i u_{i,x}$, $\Gamma_{i,y} = n_i u_{i,y}$.

Continuity equations for electron and ion liquids are formulated, as before, in view of ionization of neutral gas and an ion–electron recombination

$$\frac{\partial n_e}{\partial t} + \operatorname{div}\Gamma_e = \alpha\Gamma_e - \beta n_e n_i,$$

$$\frac{\partial n_i}{\partial t} + \operatorname{div}\Gamma_i = \alpha\Gamma_e - \beta n_e n_i.$$

Let us consider the equation of continuity of an electronic liquid

$$\frac{\partial n_e}{\partial t} + \frac{\partial \Gamma_{e,x}}{\partial x} + \frac{\partial \Gamma_{e,y}}{\partial y} = \alpha\Gamma_e - \beta n_e n_i.$$

Having substituted in it relation (5.108), we obtain

$$\begin{aligned} &\frac{\partial n_e}{\partial t} + \frac{\partial}{\partial x}\left(- n_e \mu_e E_{e,x} - \frac{D_e}{1 + b_e^2} \frac{\partial n_e}{\partial x} + \left\langle \frac{b_e D_e}{1 + b_e^2} \frac{\partial n_e}{\partial y} \right\rangle \right) \\ &\quad + \frac{\partial}{\partial y}\left(- n_e \mu_e E_{e,y} - \frac{D_e}{1 + b_e^2} \frac{\partial n_e}{\partial y} + \left\langle \frac{b_e D_e}{1 + b_e^2} \frac{\partial n_e}{\partial x} \right\rangle \right) = \alpha\Gamma_e - \beta n_e n_i. \end{aligned}$$

Let us consider a combination of summands noted by the angular brackets. We shall suppose that the value of D_e varies insignificantly, then

$$\begin{aligned} &\frac{\partial}{\partial x}\left(\frac{b_e D_e}{1 + b_e^2} \frac{\partial n_e}{\partial y} \right) - \frac{\partial}{\partial y}\left(\frac{b_e D_e}{1 + b_e^2} \frac{\partial n_e}{\partial x} \right) \\ &\quad = \frac{\partial}{\partial x}\left(\frac{b_e D_e}{1 + b_e^2} \right) \frac{\partial n_e}{\partial y} + \frac{b_e D_e}{1 + b_e^2} \frac{\partial^2 n_e}{\partial x \partial y} - \frac{\partial}{\partial y}\left(\frac{b_e D_e}{1 + b_e^2} \right) \frac{\partial n_e}{\partial x} - \frac{b_e D_e}{1 + b_e^2} \frac{\partial^2 n_e}{\partial x \partial y} \\ &\quad = \frac{\partial}{\partial x}\left(\frac{b_e D_e}{1 + b_e^2} \right) \frac{\partial n_e}{\partial y} - \frac{\partial}{\partial y}\left(\frac{b_e D_e}{1 + b_e^2} \right) \frac{\partial n_e}{\partial x} \approx 0. \end{aligned}$$

Therefore, the continuity equation for electrons will take the following form:

$$\frac{\partial n_e}{\partial t}+\frac{\partial}{\partial x}\left(-n_e\mu_e E_{e,x}-\frac{D_e}{1+b_e^2}\frac{\partial n_e}{\partial x}\right)+\frac{\partial}{\partial y}\left(-n_e\mu_e E_{e,y}-\frac{D_e}{1+b_e^2}\frac{\partial n_e}{\partial y}\right)=\alpha\Gamma_e-\beta n_e n_i. \tag{5.118}$$

By analogy, we will write the continuity equation for ions

$$\begin{aligned}\frac{\partial n_i}{\partial t}&+\frac{\partial}{\partial x}\left(\frac{u_{n,x}+b_i u_{n,y}}{1+b_i^2}n_i+n_i\mu_i E_{i,x}-\frac{D_i}{1+b_i^2}\frac{\partial n_i}{\partial x}\right)\\&+\frac{\partial}{\partial y}\left(\frac{u_{n,y}-b_i u_{n,x}}{1+b_i^2}n_i+n_i\mu_i E_{i,y}-\frac{D_i}{1+b_i^2}\frac{\partial n_i}{\partial y}\right)=\alpha\Gamma_e-\beta n_e n_i,\end{aligned} \tag{5.119}$$

Let us introduce the effective diffusivities into consideration

$$\tilde{D}_e=\frac{D_e}{1+b_e^2},\ \tilde{D}_i=\frac{D_i}{1+b_i^2}.$$

Then in the final form, the equations for the concentration of electrons and ions are formulated as follows:

$$\frac{\partial n_e}{\partial t}+\frac{\partial}{\partial x}\left(-n_e\mu_e E_{e,x}-\tilde{D}_e\frac{\partial n_e}{\partial x}\right)+\frac{\partial}{\partial y}\left(-n_e\mu_e E_{e,y}-\tilde{D}_e\frac{\partial n_e}{\partial y}\right)=\alpha\Gamma_e-\beta n_e n_i, \tag{5.120}$$

$$\begin{aligned}\frac{\partial n_i}{\partial t}&+\frac{\partial}{\partial x}\left(\frac{u_{n,x}+b_i u_{n,y}}{1+b_i^2}+n_i\mu_i E_{i,x}-\tilde{D}_i\frac{\partial n_i}{\partial x}\right)\\&+\frac{\partial}{\partial y}\left(\frac{u_{n,y}-b_i u_{n,x}}{1+b_i^2}+n_i\mu_i E_{i,y}-\tilde{D}_i\frac{\partial n_i}{\partial y}\right)=\alpha\Gamma_e-\beta n_e n_i.\end{aligned} \tag{5.121}$$

These equations testify that in the considered conditions the motion of neutral particles immediately influences the distribution of concentration of ions. However, there is no direct influence of the motion of neutrals on electron behavior. Nevertheless, such influence is due to the electric field connecting dynamics of electron and ion liquids and also through the process of recombination.

It is evident from eq. (5.121) that in the absence of a magnetic field motion of neutral particles there is an additional component to a drift velocity of ion motion in an electric field. So, for example, if neutral gas has moved in the x-direction with a constant velocity, the glow discharge would simply be blown off by the stream of gas in the same direction. However, in reality, everything is more complicated. Due to the boundary layer, there is a distribution of a longitudinal velocity by the electric discharge channel height. So at different distances from electrodes, the gas flow will influence the motion of ions differently. The influence of neutral gas motion on a configuration of glow discharge will become considerably more complicated in the presence of a magnetic field.

5.4.2 Simplified hydrodynamics of the problem under consideration: the Couette flow

The Couette flow (laminar gas flow between two flat surfaces) is characterized by following distribution of a longitudinal velocity inside of the flat channel (Anderson D.A., et al., 1997):

$$\mathrm{v} = -\frac{1}{2\mu}\frac{\partial p}{\partial x}\left(h^2 - y^2\right).$$

Integrating this equation by the height of the channel, it will turn out

$$G = \int_{-h}^{h} v dy,$$

$$G = -\frac{2h^3 b}{3\mu}\frac{\partial p}{\partial x} \rightarrow \frac{\partial p}{\partial x} = -\frac{3\mu}{2h^3 b}G,$$

$$v = -\frac{1}{2}\left(-\frac{3G}{2h^3 b}\right)\left(h^2 - y^2\right) = \frac{3G}{4h^3 b}\left(h^2 - y^2\right) = A\left(h^2 - y^2\right),$$

where G is the value of the gas flow in unit time through a rectangular area $b2h$; $h = 0.5Y_H$; b is the width of the channel.

If we suppose that the velocity along a symmetry axis equal to V_0,

$$V_0 = Ah^2 \rightarrow A = \frac{V_0}{h^2},$$

then

$$\mathrm{v} = \frac{V_0}{h^2}\left(h^2 - y^2\right) = V_0\left(1 - \frac{y^2}{h^2}\right).$$

For the qualitative analysis of gas flow dynamics in the rectangular channel, a distribution of a longitudinal velocity was set in the following form:

$$v = V_0[1 - (y/h)^m], \tag{5.122}$$

where m = 2 or 6. We shall emphasize that the basic purpose of the given section is to demonstrate the influence of neutral particle stream on the structure of glow discharge. Therefore, the distribution of an axial velocity by the height of the channel is not significant in principle. In a more complete problem statement, the distributions of gas-dynamic functions depend on parameters of discharge.

5.4.3 Numerical simulation results for the glow discharge in neutral gas flow

Calculations were performed using following initial data: gas pressure $p = 5$ Torr, the electromotive force of the power supply $E = 2$ kV, the resistance of an external electric circuit $R_0 = 300$ kOhm, the distance between flat surfaces of the cathode and the anode is $Y_H = 2$ cm, and length of the flat channel $X_L = 6$ cm (see Figure 5.24). The induction of the magnetic field directed along the z-axis varied in the range of $B = 0 - 0.05$ T. All numerical simulation results are related to 1 cm along z-axis.

In an initial instant, the plasma cloud was placed close to the cathode at $x_0 = 3$ cm with the concentration of the charged particles $n_0 = 10^{11}$ cm^{-3}. For calculations in the absence of a magnetic field and at zero velocity of the gas, the relaxation regime of formation of the glow discharge plasma in a condition of a normal current density was observed. Typical relaxation time is about ~10–20 μs after the beginning of the process.

It should be stressed that from the computational point of view this part of the problem is the most laborious. It is explained by the nonlinearity of the solved problem and by strong disturbances, introduced in distributions of required functions by the setting of arbitrary initial conditions.

The received stationary solution of the problem about glow discharge in two-dimensional flat geometry was used afterward as the initial conditions for the solution of the problem about dynamics of glow discharge in a magnetic field and in a gas flow. The specified solution is presented in Figures 5.22 and 5.23. In this case, the electron temperature was supposed to be constant and was equal to $T_e = 1$ eV, $\gamma = 0.1$.

Concentrations of electrons and ions in the given glow discharge are shown in Figure 5.24. In the figures, near-electrode areas of glow discharge (the cathode and anode layers) and also the area of the positive column are well visible. These areas are especially evident in Figure 5.22b, where the axial distribution of the electric field strength is shown. The greatest strength is reached near to the cathode $E_y = 4350$ V/cm.

According to the distribution of the electric field strength, the thickness of the cathode layer is estimated as ~0.15–0.18 cm, which is in good correspondence with the Engel–Steenbeck theory. The strength of the electric field in the positive column is practically constant and equal to $E_y = 180$ V/cm. We shall emphasize that the strength of the electric field in the positive column is prebreakdown but sufficient for compensating for the electron losses from the positive column due to diffusion and drift. We shall also pay attention to some growth of the field strength near the anode where it reaches $E_y = 290$ V/cm.

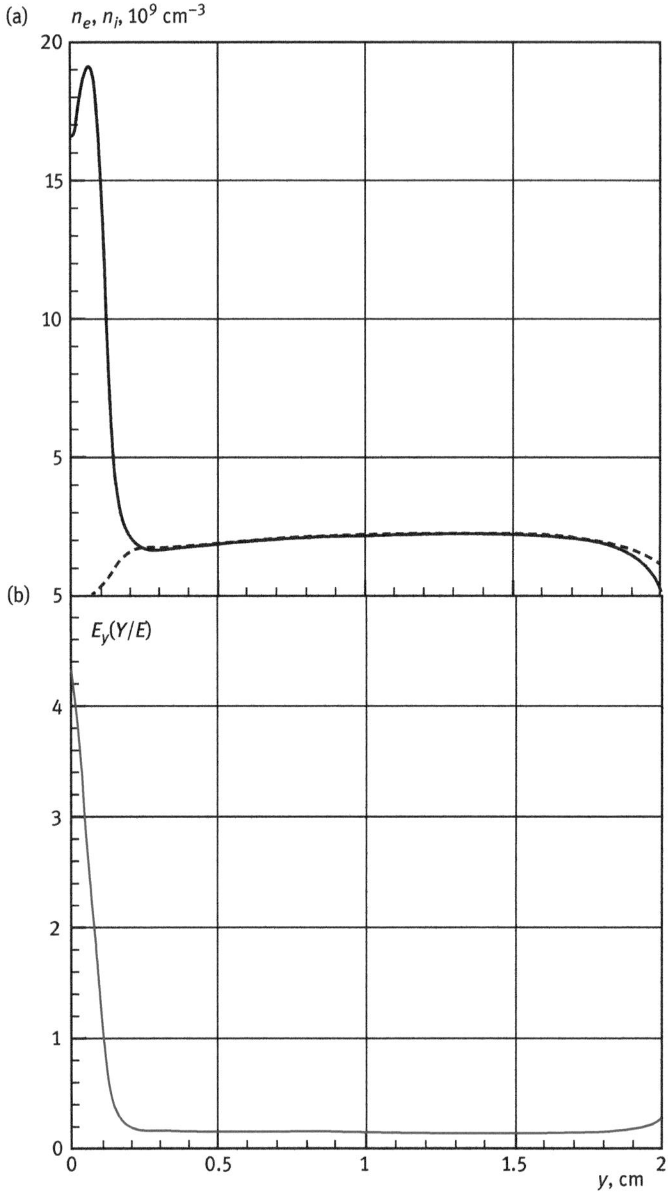

Figure 5.22: Axial distribution of electron (dashed curve) and ions (continuous curve) concentration (a), and axial distribution of electric field strength E_y (b) at $p = 5$ Torr, $E = 2$ kV.

The normal mode of the glow discharge proves to be true by Figure 5.23, where the distribution of the current density to the cathode and the anode is given. In the calculations, an additional acknowledgment of existence the normal mode of glow

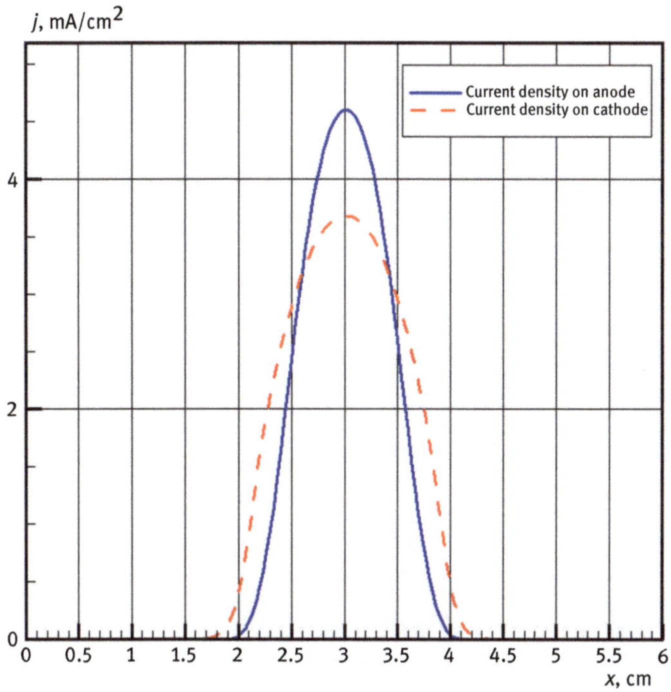

Figure 5.23: Current density on the anode (continuous curve) and the cathode (dashed curve); $p = 5$ Torr, $E = 2$ kV.

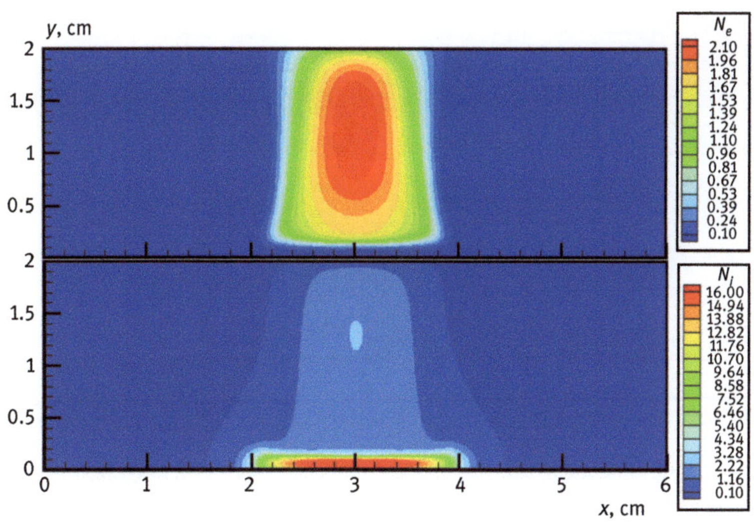

Figure 5.24: Volumetric concentration of electrons (at the top of the figure) and ions, 10^9 cm^{-3}; $p = 5$ Torr, $E = 2$ kV.

discharge is received: with the increase of electromotive force of the power supply the current density on the anode increases, and the current density on the cathode varies insignificantly. At the same time, of course, the breadth of the cathode spot increases, and so the full current through the discharge gap varies slightly.

The account of the dependence of the spectrum-averaged electron energy T_e on the electric field strength (Petrusev A.S., et al., 2005) has not led to appreciable modifications of the electrodynamic structure of the glow discharge. To be convinced of this, it is enough to compare distributions of concentration of the charged particles and current densities to the electrodes, which are shown in Figures 5.25, and 5.26, with the corresponding distributions obtained at a constant temperature of electrons (see Figures 5.24 and 5.23). Calculated field of electron temperature T_e is shown in Figure 5.25c, where it is evident that, as it follows from (5.123), the electron temperature is high enough only in the cathode layer where the concentration of electrons is still small. The approximation of electron temperature was used in the form of

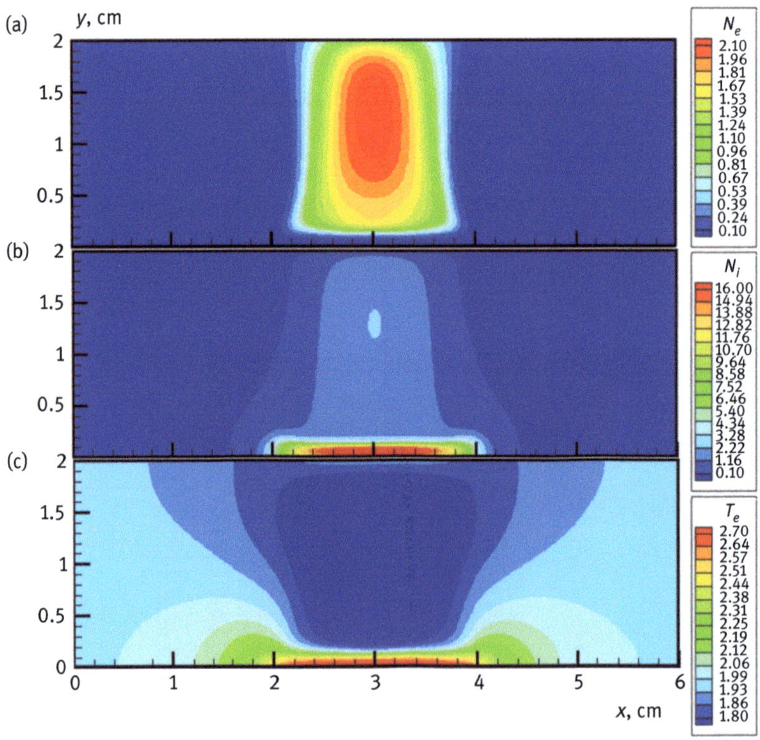

Figure 5.25: Volumetric concentration of electrons (a) and ions (b), 10^9 cm^{-3}, and electronic temperature (in eV) in gas-discharge gap. The T_e depends on $|\mathbf{E}|$; $p = 5$ Torr, $E = 2$ kV.

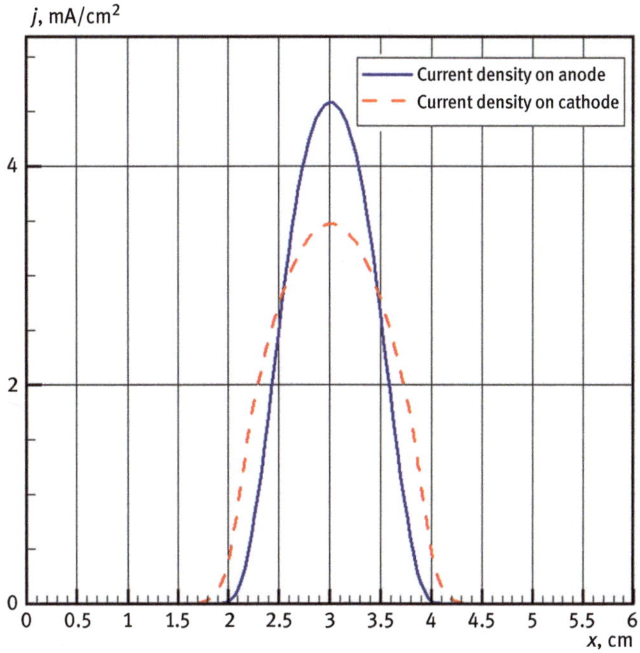

Figure 5.26: Current density on the anode (solid curve) and the cathode (dashed curve); $p = 5$ Torr, $E = 2$ kV.

$$\frac{T_e}{T} = 29.96 \ln\left(\frac{E}{p}\right) + 24.64, \tag{5.123}$$

where E/p is measured in V/(cm · Torr). This approximation has been offered in the work (Petrusev A.S., et al., 2005) on the basis of the experimental data discussed in the work (Brown S.C., 1966).

As it was already discussed earlier, the obtained distributions of concentrations and electric potential in glow discharge were used as the initial conditions for calculations in view of gas motion and in the presence of the external magnetic field. We shall notice that the use of the empirical formula (5.123) is reasonable in the drift-diffusion model for the estimation of the role of electronic diffusion. For the solution of problems of physical and chemical kinetics, the use of T_e means the acceptance of assumption about thermalization of electrons, which does not match the reality.

In the presence of gas flow in the positive x-direction, the discharge comes to motion in the same direction that is caused by conditions of the computing experiments, when the glow discharge is in a condition of an indifferent equilibrium near the area of starting initialization. However according to the calculation model, the electrons do not sense motion of neutral particles explicitly, and first the motion of the glow discharge is caused by the motion of ions. However, a velocity of their displacement is defined not only by a velocity of neutral particles but also by the drift and diffusion velocities. In addition, the distribution of a velocity of neutral particles is inhomogeneous. From here it follows that the velocity of motion of glow discharge can essentially differ from an average velocity of gas flow. In this case, the analysis becomes more complicated due to the inhomogeneous distribution of longitudinal velocity (5.122). The cross velocity was supposed to be equal to zero.

The important peculiarity of the calculations is the oscillation of total current and voltage drop in the gas-discharge gap. In this case, the oscillation amplitude of current and voltage make ~17 %. The specified singularity is inherent practically in all calculated data.

Dynamics of the glow discharge motion along the x-direction is shown in Figures 5.27 (distributions of electron concentrations), 5.28 (ions) and 5.29 (electron temperature). In the work (Surzhikov S.T., et al., 2004) it has been established that the glow discharge in an external cross magnetic field moves either in positive or in negative x-direction depending on the direction of the magnetic field.

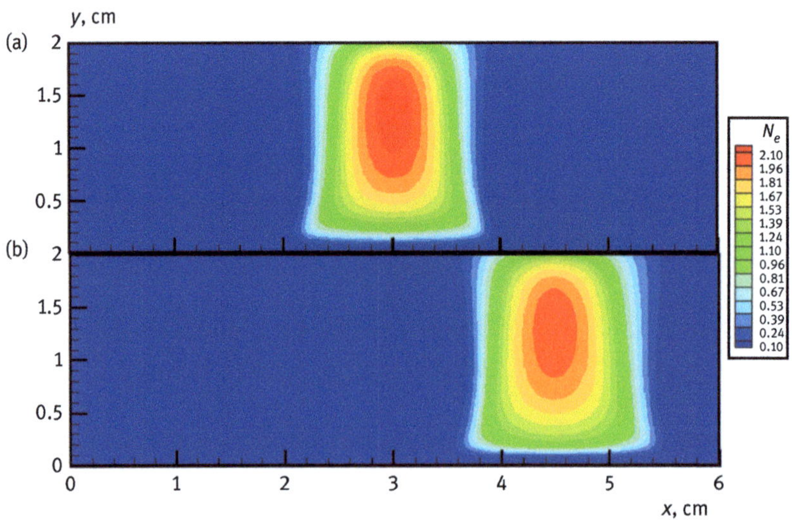

Figure 5.27: Distribution of electron concentrations, 10^9 cm^{-3}; $p = 5$ Torr, $E = 2$ kV: (a) $t = 0$ µs; (b) $t = 160$ µs.

By considering the two factors on glow discharge (the magnetic field and the gas flow) it is possible to obtain various cases both for acceleration and deceleration of motion of gas discharge, also up to inverse motion (against the gas flow). It is also realized in the following three series of calculations. Dynamics of gas discharge is shown in Figures 5.30–5.32 in gas-discharge gap with gas flow and with the cross

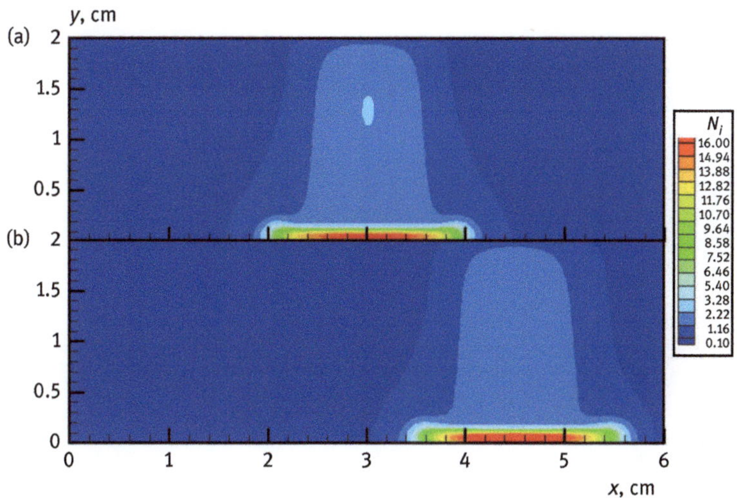

Figure 5.28: Distribution of ion concentrations, 10^9 cm^{-3}; $p = 5$ Torr, $E = 2$ kV: (a) $t = 0$ μs; (b) $t = 160$ μs.

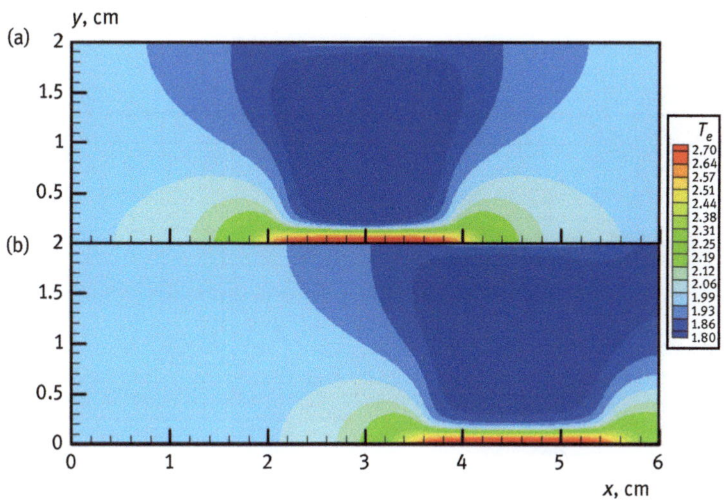

Figure 5.29: Distribution of electron temperature, eV; $p = 5$ Torr, $E = 2$ kV: (a) $t = 0$ μs; (b) $t = 160$ μs.

magnetic field $B_z = +0.05$ T. Calculations were performed also for $B_z = +0.01$ T and $B_z = +0.02$ T.

If in the case $B_z = +0.01$ T appreciable deceleration of the gas-discharge motion along the gas flow is observed, then in the case $B_z = +0.02$ T the discharge is slowly displaced toward the gas flow. At last, in the case $B_z = +0.05$ T the gas discharge is

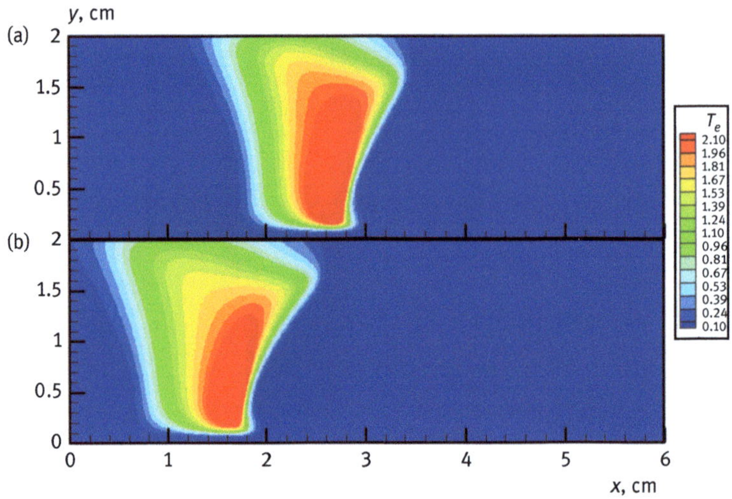

Figure 5.30: Concentration of electrons, 10^9 cm^{-3}; $p = 5$ Torr, $E = 2$ kV, $B_z = 0.05$ T: (a) $t = 20$ µs; (b) $t = 60$ µs.

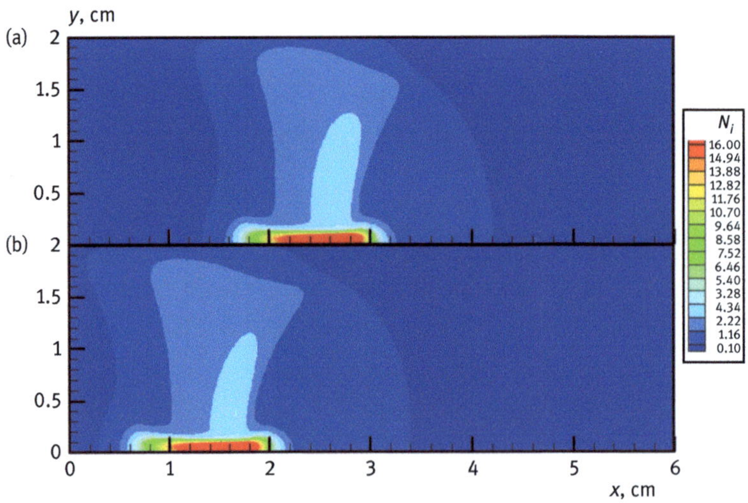

Figure 5.31: Concentration of ions, 10^9 cm^{-3}; $p = 5$ Torr, $E = 2$ kV, $B_z = 0.05$ T: (a) $t = 20$ µs; (b) $t = 60$ µs.

displaced to the left with a velocity, which many times exceeds the velocity of the gas flow.

By increasing the magnetic field induction even more strong distortion in the electrodynamic structure of the glow discharge is observed. Numerical experiments confirm that the Hall currents generated in plasma are rather inhomogeneous by the

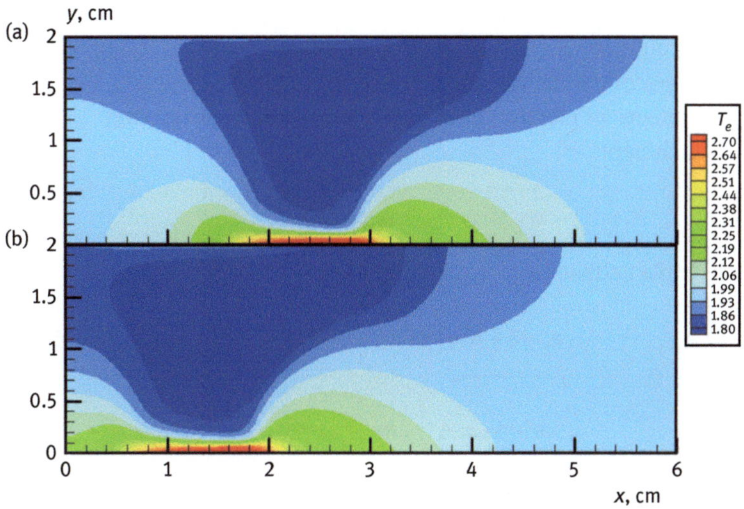

Figure 5.32: Electron temperature, eV; $p = 5$ Torr, $E = 2$ kV, $B_z = 0.05$ T: a $t = 20$ μs; b $t = 60$ μs.

thickness of the electrodischarge gap. A similar conclusion can be drawn by considering the discharge plasma in gas flow in accelerating magnetic field in the x-direction.

Thus, the two-dimensional computer model of glow discharge in gas flows and cross magnetic fields was considered in this chapter. But, let us remind that the calculation model is limited by conditions

$$|\mathbf{u}_e| >> |\mathbf{u}_i|, |\mathbf{u}_n|,$$

that is usually fulfilled with a greater margins. Nevertheless at greater hypersonic velocities, this inequality ceases to be true.

By using the given computer model, the investigation of glow discharges dynamics in a cross magnetic field and in cross gas flow can be done. Nevertheless, the braking conditions by the cross magnetic field for glow discharges in gas flows and motion of the discharges toward gas flows are estimated by the calculations in the section.

5.5 Computing model of glow discharge in electronegative gas

Till now we have considered computer models of glow discharges consisting only from positive ions and electrons. Specified models well describe glow discharges, for example, in molecular nitrogen (N_2). However, the theory and computing models of glow discharges in the air (dry and humid) and in other gases containing negative ions are of great practical interest. In the gases containing atoms and molecules, which possess *electron affinity*, steady negative ions are formed. Such gases are referred to as *electronegative*. Halogens, dry and, in particular, humid air are called as *electronegative gases*. Ions O_2^-, H_2O^-, F^-, Br^-, I^-, Cl_2^-, OH^-, and so on are formed in them. Taking into account the formation of such negative ions leads to appreciable electron losses (*to an electron attachment*).

Following mechanisms of electron attachment are usually considering:

- Dissociative attachment

$$\mathrm{e} + \mathrm{M} \Leftrightarrow \mathrm{A}_1^- + \mathrm{A}_2; \tag{5.124}$$

- Attachment in triple collisions

$$\mathrm{e} + \mathrm{M} + \mathrm{N} \Leftrightarrow \mathrm{M}^- + \mathrm{N}; \tag{5.125}$$

- Photocapture

$$\mathrm{e} + \mathrm{A} \Leftrightarrow \mathrm{A}^- + h\nu, \tag{5.126}$$

  where e is the electron; M is the molecule consisting of atoms A_1 and A_2; M is any third particle that takes part in the collision; $h\nu$ is the emitted quantum of electromagnetic radiation.

In kinetic schemes (5.124)–(5.125), reactions of the destruction of negative ions are also specified. The recombination of negative ions happens at their collisions with various particles (more effectively with excited particles), at a photodetachment, and as a result of ion-molecular reactions. Mechanisms of the formation and loss of negative ions are considered, for example, in the book (Capitelli M., et al., 2000). In the same book methods of calculation of elementary process probabilities for the formation and destruction of negative ions are given.

5.5.1 Computational model

The drift-diffusion model of glow discharge in electronegative gas is formulated using the same assumptions, as in the previous chapters. The basic difference from the previous formulation is the necessity to take into consideration the equations of motion and continuity for the negative ions. The equations of motion for three kinds of the particles (electrons, positive, and negative) are formulated by using the moment procedure applied to Boltzmann equations:

$$\begin{aligned} m_e n_e \frac{\partial \mathbf{u}_e}{\partial t} + m_e n_e (\mathbf{u}_e \cdot \nabla)\mathbf{u}_e = &- \nabla p_e + \frac{m_e n_e \mathbf{F}_e}{m_e} - \tau_e - e n_e \left(\mathbf{E} + c^{-1}[\mathbf{u}_e \mathbf{B}]\right) \\ &- m_e n_e \nu_{en}(\mathbf{u}_e - \mathbf{u}_n) - m_e n_e \nu_{e+}(\mathbf{u}_e - \mathbf{u}_+) \\ &- m_e n_e \nu_{e-}(\mathbf{u}_e - \mathbf{u}_-), \end{aligned} \tag{5.127}$$

$$m_+n_+\frac{\partial \mathbf{u}_+}{\partial t}+m_+n_+(\mathbf{u}_+\cdot\nabla)\mathbf{u}_+=-\nabla p_++\frac{m_+n_+\mathbf{F}_+}{m_+}-\tau_++en_+\left(\mathbf{E}+c^{-1}[\mathbf{u}_+\mathbf{B}]\right)$$
$$-m_+n_+\nu_{+e}(\mathbf{u}_+-\mathbf{u}_e)-m_+n_+\nu_{+n}(\mathbf{u}_+-\mathbf{u}_n)$$
$$-m_+n_+\nu_{+-}(\mathbf{u}_+-\mathbf{u}_-), \quad (5.128)$$

$$m_-n_-\frac{\partial \mathbf{u}_-}{\partial t}+m_-n_-(\mathbf{u}_-\cdot\nabla)\mathbf{u}_-=-\nabla p_-+\frac{m_-n_-\mathbf{F}_-}{m_-}-\tau_--en_-\left(\mathbf{E}+c^{-1}[\mathbf{u}_-\mathbf{B}]\right)$$
$$-m_-n_-\nu_{-e}(\mathbf{u}_--\mathbf{u}_e)-m_-n_-\nu_{-n}(\mathbf{u}_--\mathbf{u}_n)$$
$$-m_-n_-\nu_{-+}(\mathbf{u}_--\mathbf{u}_+), \quad (5.129)$$

where

$$\tau_\alpha=-\mathbf{i}\frac{\partial p_{\alpha,i,m}}{\partial x_m}-\mathbf{j}\frac{\partial p_{\alpha,j,m}}{\partial x_m}-\mathbf{k}\frac{\partial p_{\alpha,k,m}}{\partial x_m};$$

$p_{\alpha,(i,j,k),m}$ are the components of a viscous stress tensor (summation by repeating indexes is supposed); $\alpha=(e,+,-,n)$; $\mathbf{F}_\alpha$ is the mass volumetric force acting on particles α; m_e, m_α are the masses of electrons and ions; n_α is the volumetric concentration of particles of a class α; $\mathbf{u}_\alpha$ is the averaged velocity of particles of a classα; p_α is the pressure of particles of a class α; e is the electron charge; c is the velocity of light; $\mathbf{E}=\mathbf{i}E_x+\mathbf{j}E_y$ is the vector of electric field strength; $\mathbf{B}=\mathbf{k}B_z$ is the vector of magnetic field induction; $\nu_{\alpha\beta}$ is the collision frequency for particles of α and β classes.

For partially ionized gas, as before, we suppose that

$$n_n>>(n_+,n_e,n_-). \quad (5.130)$$

Let us consider glow discharge in air at $p=5$ Torr and electromotive force of external power supply $E=1$ kV. In this case $n_n\sim10^{17}$ cm^{-3}, $n_e\approx n_+=10^{11}$ cm^{-3}, $n\sim10^8$ cm^{-3}. Volumetric mass forces and viscosity are supposed negligible, that is, $\mathbf{F}_\alpha=0$, $\tau_\alpha=0$.

For the estimation of collision frequencies for electrons with ions, and ions and electrons with neutral particles we will use the following formulas:

$$\nu_{e+}=5.5T_e^{-3/2}n_*\ln\left(220T_e/n_*^{1/3}\right)\ \mathrm{s}^{-1}; \quad (5.131)$$

$$\nu_{+n}=\frac{2\sqrt{2}}{3}\pi a^2\sqrt{\frac{8kT}{m_e\pi}}n_n\ \mathrm{s}^{-1}; \quad (5.132)$$

$$\nu_{eN_2}=2.5\times10^{-11}n_{N_2}T_e\left(1+9.3\times10^{-3}\sqrt{T_e}\right)\ \mathrm{s}^{-1},$$

$$\nu_{eO_2}=1.5\times10^{-10}n_{O_2}\sqrt{T_e}\left(1+4.2\times10^{-2}\sqrt{T_e}\right)\ \mathrm{s}^{-1},$$

$$\nu_{eO}=2.8\times10^{-10}n_O\sqrt{T_e}\ \mathrm{s}^{-1}, \quad (5.133)$$

where $n_* = n_e = n_+$; a is the effective radius of neutral particles; T_e is the temperature of electrons.

Further, we will consider the physical processes averaged on a period

$$t \sim \max\{\tau_{en}, \tau_{e+}, \tau_{e-}, \tau_{+n}, \tau_{-+}\},$$

where $\tau_{\alpha\beta} = 1/\nu_{\alpha\beta}$. In this case, it is possible to neglect the left-hand side of eqs. (5.127)–(5.129). Let

$$p_e = n_e kT_e, \quad p_+ = n_+ kT, \quad p_- = n_- kT, \tag{5.134}$$

where T_e, T are the temperatures of electrons and all heavy particles. We shall suppose that these temperatures are constant. Then it is possible to rewrite (5.127)–(5.129) in the form of

$$-kT_e\nabla n_e - en_e\mathbf{E} - en_e c^{-1}[\mathbf{u}_e\mathbf{B}] - m_e n_e \nu_{en}\mathbf{u}_e - m_e n_e \nu_{e+}(\mathbf{u}_e - \mathbf{u}_n) - m_e n_e \nu_{e-}(\mathbf{u}_e - \mathbf{u}_+) = 0, \tag{5.135}$$

$$-kT\nabla n_+ + en_+\mathbf{E} + en_+ c^{-1}[\mathbf{u}_+\mathbf{B}] - m_+ n_+ \nu_{+e}(\mathbf{u}_+ - \mathbf{u}_e) - m_+ n_+ \nu_{+n}\mathbf{u}_+ - m_+ n_+ \nu_{+-}(\mathbf{u}_+ - \mathbf{u}_-) = 0, \tag{5.136}$$

$$-kT\nabla n_- - en_-\mathbf{E} - en_- c^{-1}[\mathbf{u}_-\mathbf{B}] - m_- n_- \nu_{-e}(\mathbf{u}_- - \mathbf{u}_e) - m_- n_- \nu_{-n}\mathbf{u}_- - m_- n_- \nu_{-+}(\mathbf{u}_- - \mathbf{u}_+) = 0. \tag{5.137}$$

Let us suppose that the velocity of neutral particles $\mathbf{u}_n$ is smaller than other velocities in (5.135)–(5.137), and consider estimations of values of some terms in (5.135). From (5.131)–(5.133) it follows that $\nu_{en} >> (\nu_{e+}, \nu_{e-})$; therefore instead of (5.135) it is possible to write

$$-kT_e\nabla n_e - en_e\mathbf{E} - en_e c^{-1}[\mathbf{u}_e\mathbf{B}] = m_e n_e \mathbf{u}_e \nu_{en}$$

or

$$n_e\mathbf{u}_e = -D_e\nabla n_e - \mu_e n_e\mathbf{E} - \mu_e n_e c^{-1}[\mathbf{u}_e\mathbf{B}], \tag{5.138}$$

where

$$\mu_e = \frac{e}{m_e \nu_{en}}, \tag{5.139}$$

$$D_e = \frac{kT_e}{m_e \nu_{en}} = \left(\frac{kT_e}{e}\right)\left(\frac{e}{m_e \nu_{en}}\right) = T_{e[\mathrm{eV}]}\mu_e. \tag{5.140}$$

Instead of (5.136) for positive ions, it is possible to write

$$-kT\nabla n_+ + en_+\mathbf{E} + en_+ c^{-1}[\mathbf{u}_+\mathbf{B}] - m_+ n_+ \mathbf{u}_+ \nu_+ + m_+ n_+ (\nu_{+e}\mathbf{u}_e + \nu_{+-}\mathbf{u}_-) = 0, \tag{5.141}$$

where $\nu_+ = \nu_{+e} + \nu_{+n} + \nu_{+-}$.

For $|\mathbf{u}_+| \approx |\mathbf{u}_-|$ and $\nu_{+n} \gg \nu_{+-}$ the last two terms in (5.141) are equal to $(-m_+ n_+ \mathbf{u}_+ \nu_+ + m_+ n_+ \mathbf{u}_e \nu_{+e})$. But for $\nu_+ \gg \nu_{+e}$ (even for $\mathbf{u}_e \gg \mathbf{u}_+$) it is possible to write

$$m_+ n_+ \mathbf{u}_+ \nu_+ = -kT\nabla n_+ + en_+ \mathbf{E} + en_+ c^{-1}[\mathbf{u}_+ \mathbf{B}]$$

or

$$n_+ \mathbf{u}_+ = -D_+ \nabla n_+ + \mu_+ n_+ \mathbf{E} + \mu_+ n_+ c^{-1}[\mathbf{u}_+ \mathbf{B}], \tag{5.142}$$

where

$$\mu_+ = \frac{e}{m_+ \nu_+}, \tag{5.143}$$

$$D_+ = \frac{kT}{m_+ \nu_+} = \left(\frac{kT}{e}\right)\left(\frac{e}{m_+ \nu_+}\right) = T_{[\mathrm{eV}]}\mu_+. \tag{5.144}$$

By analogy the equations for negative ions can be obtained:

$$-kT\nabla n_- - en_- \mathbf{E} - en_- c^{-1}[\mathbf{u}_- \mathbf{B}] - m_- n_- \mathbf{u}_- \nu_- +$$
$$+ m_- n_- (\nu_{-e} \mathbf{u}_e + \nu_{-+} \mathbf{u}_+) = 0, \tag{5.145}$$

where $\nu_- = \nu_{-e} + \nu_{-n} + \nu_{-+}$.

As

$$\nu_{-+}|\mathbf{u}_+| \ll \nu_-|\mathbf{u}_-| \text{ and } \nu_{-e}|\mathbf{u}_e| \ll \nu_-|\mathbf{u}_-|,$$

it is possible to write

$$m_- n_- \mathbf{u}_- \nu_- = -kT\nabla n_- - en_- \mathbf{E} - en_- c^{-1}[\mathbf{u}_- \mathbf{B}]$$

or

$$n_- \mathbf{u}_- = -D_- \nabla n_- - \mu_- n_- \mathbf{E} - \mu_- n_- c^{-1}[\mathbf{u}_- \mathbf{B}], \tag{5.146}$$

where

$$\mu_- = \frac{e}{m_- \nu_-}, \tag{5.147}$$

$$D_- = \frac{kT}{m_- \nu_-} = \left(\frac{kT}{e}\right)\left(\frac{e}{m_- \nu_-}\right) = T_{[\mathrm{eV}]}\mu_-. \tag{5.148}$$

Equations (5.138), (5.142), and (5.146) will be used for further transformations. In the two-dimensional statement (the configuration of an external magnetic field is shown in Figure 5.39)

$$[\mathbf{u}_\alpha \mathbf{B}] = \mathbf{i}\mathbf{u}_{\alpha y} B - \mathbf{j}\mathbf{u}_{\alpha x} B, \ B = B_z, \tag{5.149}$$

therefore, projecting eq. (5.138) on x- and y-axes, one can get

$$n_e u_{e,x} = -D_e \frac{\partial n_e}{\partial x} - \mu_e n_e E_x - \mu_e n_e \frac{u_{e,y} B_z}{c}, \tag{5.150}$$

$$n_e u_{e,y} = -D_e \frac{\partial n_e}{\partial y} - \mu_e n_e E_y + \mu_e n_e \frac{u_{e,x} B_z}{c}. \tag{5.151}$$

The given set of equations can be solved in relation to $u_{e,x}$ and $u_{e,y}$

$$n_e u_{e,x} = \frac{1}{1+b_e^2}\left(-D_e \frac{\partial n_e}{\partial x} - \mu_e n_e E_x\right) - \frac{b_e}{1+b_e^2}\left(-D_e \frac{\partial n_e}{\partial y} - \mu_e n_e E_y\right), \tag{5.152}$$

$$n_e u_{e,y} = \frac{1}{1+b_e^2}\left(-D_e \frac{\partial n_e}{\partial y} - \mu_e n_e E_y\right) - \frac{b_e}{1+b_e^2}\left(-D_e \frac{\partial n_e}{\partial x} - \mu_e n_e E_x\right), \tag{5.153}$$

where

$$b_e = \mu_e B/c. \tag{5.154}$$

Now we can introduce the effective electric field into relations (5.152) and (5.153) by the following way:

$$n_e u_{e,x} = -\mu_e n_e E_{e,x} - \frac{1}{1+b_e^2} D_e \frac{\partial n_e}{\partial x} + \frac{b_e}{1+b_e^2} D_e \frac{\partial n_e}{\partial y} = \Gamma_{e,x}, \tag{5.155}$$

$$n_e u_{e,y} = -\mu_e n_e E_{e,y} - \frac{1}{1+b_e^2} D_e \frac{\partial n_e}{\partial y} - \frac{b_e}{1+b_e^2} D_e \frac{\partial n_e}{\partial x} = \Gamma_{e,y}, \tag{5.156}$$

where

$$E_{e,x} = \frac{E_x - b_e E_y}{1+b_e^2}, \tag{5.157}$$

$$E_{e,y} = \frac{E_y - b_e E_x}{1+b_e^2}. \tag{5.158}$$

By analogy one can transform eq. (5.142) for the positive ions

$$n_+ u_{+,x} = -D_+ \frac{\partial n_+}{\partial x} + \mu_+ n_+ E_x + \mu_+ n_+ \frac{u_{+,y} B_z}{c}, \tag{5.159}$$

$$n_+ u_{+,y} = -D_+ \frac{\partial n_+}{\partial y} + \mu_+ n_+ E_y - \mu_+ n_+ \frac{u_{+,x} B_z}{c}, \tag{5.160}$$

or

$$n_+ u_{+,x} = +\mu_+ n_+ E_{+,x} - \frac{D_+}{1+b_+^2} \frac{\partial n_+}{\partial x} - \frac{b_+}{1+b_+^2} D_+ \frac{\partial n_+}{\partial y} = \Gamma_{+,x}, \tag{5.161}$$

$$n_+ u_{+,y} = +\mu_+ n_+ E_{+,y} - \frac{D_+}{1+b_+^2} \frac{\partial n_+}{\partial y} + \frac{b_+}{1+b_+^2} D_+ \frac{\partial n_+}{\partial x} = \Gamma_{+,y}, \tag{5.162}$$

where

$$b_+ = \frac{\mu_+}{c} B_z, \tag{5.163}$$

$$E_{+,x} = \frac{E_x + b_+ E_y}{1 + b_+^2}, \tag{5.164}$$

$$E_{+,y} = \frac{E_y - b_+ E_x}{1 + b_+^2}. \tag{5.165}$$

Equations for the negative ions will be obtained from (5.146)

$$n_- u_{-,x} = -D_- \frac{\partial n_-}{\partial x} - \mu_- n_- E_x - \mu_- n_- \frac{u_{-,y} B_z}{c}, \tag{5.166}$$

$$n_- u_{-,y} = -D_- \frac{\partial n_-}{\partial y} - \mu_- n_- E_y + \mu_- n_- \frac{u_{-,x} B_z}{c}, \tag{5.167}$$

or

$$n_- u_{-,x} = -\mu_- n_- E_{-,x} - \frac{D_-}{1+b_-^2} \frac{\partial n_-}{\partial x} + \frac{b_-}{1+b_+^2} D_- \frac{\partial n_-}{\partial y} = \Gamma_{-,x}, \tag{5.168}$$

$$n_- u_{-,y} = -\mu_- n_- E_{-,y} - \frac{D_-}{1+b_-^2} \frac{\partial n_-}{\partial y} - \frac{b_-}{1+b_-^2} D_- \frac{\partial n_-}{\partial x} = \Gamma_{-,y}, \tag{5.169}$$

where

$$b_- = \left(\frac{\mu_-}{c}\right) B_z; \tag{5.170}$$

$$E_{-,x} = \frac{E_x - b_- E_y}{1 + b_-^2}; \tag{5.171}$$

$$E_{-,y} = \frac{E_y + b_- E_x}{1 + b_-^2}. \tag{5.172}$$

Now we will consider the continuity equations for all kinds of charged particles. These equations include densities of fluxes of particles $\Gamma_{\alpha,x}$ and $\Gamma_{\alpha,y}$, and also coefficients of the kinetic processes for the charged particles:

$$\frac{\partial n_e}{\partial t} + \frac{\partial \Gamma_{e,x}}{\partial x} + \frac{\partial \Gamma_{e,y}}{\partial y} = \alpha(E)|\Gamma_e| - \beta_e n_+ n_e - \nu_a n_e + k_d n_n n_-, \tag{5.173}$$

$$\frac{\partial n_+}{\partial t} + \frac{\partial \Gamma_{+,x}}{\partial x} + \frac{\partial \Gamma_{+,y}}{\partial y} = \alpha(E)|\Gamma_e| - \beta_e n_+ n_e - \beta_- n_- n_+, \tag{5.174}$$

$$\frac{\partial n_-}{\partial t} + \frac{\partial \Gamma_{-,x}}{\partial x} + \frac{\partial \Gamma_{-,y}}{\partial y} = \nu_a n_e - k_d n_n n_- - \beta_- n_- n_+, \tag{5.175}$$

where $\alpha(E)$ is the coefficient of ionization; β_e is the coefficient of electron–ion recombination; β_- is the coefficient of recombination at collision of positive and negative ions; ν_a is the frequency of attachment; k_d is the detachment coefficient.

Using formulas for densities of electron and ion fluxes (5.155), (5.156), (5.161), (5.162), (5.168), and (5.169), we will obtain a system of equations of the drift-diffusion model of glow discharge in electronegative gas:

$$\frac{\partial n_e}{\partial t} + \frac{\partial}{\partial x}\left(-\mu_e n_e E_{e,x} - \frac{De}{1+b_e^2}\frac{\partial n_e}{\partial x}\right) + \frac{\partial}{\partial y}\left(-\mu_e n_e E_{e,y} - \frac{De}{1+b_e^2}\frac{\partial n_e}{\partial y}\right)$$
$$= \alpha(E)|\mathbf{\Gamma}_e| - \beta_e n_+ n_e - \nu_a n_e + k_d n_n n_-, \tag{5.176}$$

$$\frac{\partial n_+}{\partial t} + \frac{\partial}{\partial x}\left(+\mu_+ n_+ E_{+,x} - \frac{D_+}{1+b_+^2}\frac{\partial n_+}{\partial x}\right) + \frac{\partial}{\partial y}\left(+\mu_+ n_+ E_{+,y} - \frac{D_+}{1+b_+^2}\frac{\partial n_+}{\partial y}\right)$$
$$= \alpha(E)|\mathbf{\Gamma}_e| - \beta_e n_+ n_e - \beta_- n_- n_+, \tag{5.177}$$

$$\frac{\partial n_-}{\partial t} + \frac{\partial}{\partial x}\left(-\mu_- n_- E_{-,x} - \frac{D_-}{1+b_-^2}\frac{\partial n_-}{\partial x}\right) + \frac{\partial}{\partial y}\left(-\mu_- n_- E_{-,y} - \frac{D_-}{1+b_-^2}\frac{\partial n_-}{\partial y}\right)$$
$$= \nu_a n_e - k_d n_n n_- - \beta_- n_- n_+, \tag{5.178}$$

$$\frac{\partial^2 \varphi}{\partial x^2} + \frac{\partial^2 \varphi}{\partial y^2} = 4\pi e(n_e + n_- - n_+), \tag{5.179}$$

$$\mathbf{E} = -\,\mathrm{grad}\varphi. \tag{5.180}$$

Here the components of the effective electric field are defined under formulas (5.157), (5.158), (5.164), (5.165), (5.171), and (5.172).

To integrate eqs. (5.176)–(5.180) it is necessary to define kinetic coefficients in the right-hand side of eqs. (5.176)–(5.178).

Frequency of the attachment is defined using the formula

$$\nu_a = \left(\frac{\alpha_a}{p}\right)\nu_{\mathrm{dr}} p,$$

where α_a/p is the coefficient of attachment; ν_{dr} is the drift velocity of electrons. The following approximate relations for air are used:

$$\frac{\alpha_a}{p} = 0.005\,\frac{1}{\mathrm{cm}\cdot\mathrm{Torr}} \quad \text{at} \quad \frac{E}{p} < 40\,\frac{\mathrm{B}}{\mathrm{cm}\cdot\mathrm{Torr}};$$

$$\frac{\alpha_a}{p} = 0.005 + \left[\frac{E}{p} - 40\right]\cdot 0.35\times 10^{-3}\,\frac{1}{\mathrm{cm}\cdot\mathrm{Torr}} \quad \text{at} \quad \frac{E}{p} > 40\,\frac{\mathrm{B}}{\mathrm{cm}\cdot\mathrm{Torr}}.$$

The coefficient of ion–ion recombination is estimated for the following reaction:

$$N_2^+ + O_2^- \rightarrow N_2 + O_2$$

and is supposed to be equal to $\beta_- = 1.6 \times 10^{-7}$ cm^3/s.

The detachment coefficient is estimated by the value of

$$k_d = 10^{-14}\ \text{cm}^3/\text{s}.$$

The coefficients of ionization $\alpha(E)$ and recombination β_e are taken into account as in the form of

$$\frac{\alpha}{p} = A \exp\left[-\frac{B}{(E/p)}\right] \frac{1}{\text{cm}\cdot\text{Torr}},$$

$$\beta_e = 2 \times 10^{-7}\ \text{cm}^3/\text{s},$$

where

$$A = 12 \frac{1}{\text{cm}\cdot\text{Torr}}, \quad B = 342 \frac{\text{V}}{\text{cm}\cdot\text{Torr}}.$$

Also other approximations (Raizer, 1987; Brown, 1961) are used

$$\left(\frac{\alpha}{p}\right)_{\text{air}} = A_{\text{air}} \exp\left[-\frac{B_{\text{air}}}{(E/p)}\right] \frac{1}{\text{cm}\cdot\text{Torr}},$$

$$A_{\text{air}} = 15 \frac{1}{\text{cm}\cdot\text{Torr}}, \quad B_{\text{air}} = 365 \frac{\text{V}}{\text{cm}\cdot\text{Torr}} \quad \text{for } \frac{E}{p} \in [100 - 800] \frac{\text{V}}{\text{cm}\cdot\text{Torr}}$$

or

$$\left(\frac{\alpha}{p}\right)_{\text{air}} = 1.17 \times 10^{-4} \left(\frac{E}{p} - 32.2\right) \frac{1}{\text{cm}\cdot\text{Torr}}$$

for

$$\frac{E}{p} \in [44 - 176] \frac{\text{V}}{\text{cm}\cdot\text{Torr}}.$$

Let us notice that the selection of the electrophysical constants describing the kinetics of physical and chemical processes in partially ionized electronegative gases is an extremely important stage of construction of a computing model. A detailed analysis is necessary for the adequate account of all processes in each concrete case. It is possible to recommend the works (Kossyi I.A., et al., 1992; Macheret S.O., et al., 2002).

Boundary conditions for problem is formulated in the following form:

$$y = 0: \frac{\partial n_+}{\partial y} = \frac{\partial n_-}{\partial y} = 0, \ \Gamma_e = \gamma \Gamma_+, \ \varphi = 0,$$

$$y = H: n_+ = 0, \ \frac{\partial n_e}{\partial y} = \frac{\partial n_-}{\partial y} = 0, \ \varphi = V,$$

$$x = 0: \frac{\partial n_e}{\partial y} = \frac{\partial n_+}{\partial y} = \frac{\partial n_-}{\partial y} = \frac{\partial \varphi}{\partial y} = 0,$$

$$x = L: \frac{\partial n_e}{\partial y} = \frac{\partial n_+}{\partial y} = \frac{\partial n_-}{\partial y} = \frac{\partial \varphi}{\partial y} = 0,$$

where γ is the coefficient of the secondary ion–electron emission.

5.5.2 Numerical simulation results

Numerical modeling of glow discharge in electronegative gases was performed in the calculation domain as shown in Figure 5.33. The following initial data were used: the height of the gas-discharge gap $y = H = 2$ cm, the breadth of the calculation area $x = L = 4$ cm, and initial position of glow discharge $x = L/2 = 2$ cm. The computational mesh is shown in Figure 5.2 was used for the calculations.

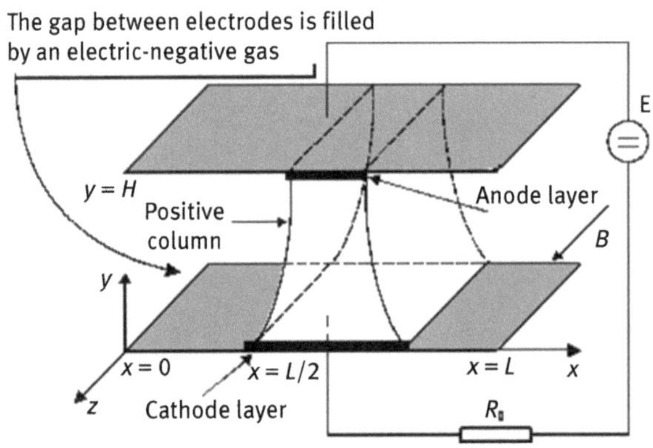

Figure 5.33: Glow discharge in external magnetic field.

Calculations have been performed for glow discharge at pressure $p = 5$–10 Torr, emf of the power supply is $E = 2$ kV, coefficient of the secondary electron emission $\gamma = 0.05$, the induction of cross external magnetic field $B = \pm 0.05$ T. Note that in the previous section numerical simulation results were considered at $\gamma = 0.1$.

First of all, let us compare the calculation results for electropositive and electronegative gases. Current densities on electrodes of glow discharge in electropositive gas (molecular nitrogen, N_2) at two pressures are shown in Figure 5.34a and b.

Figures 5.35 and 5.36 show concentrations of electrons and ions for pressures $p = 5$ and 10 Torr. Common tendency of modification of glow discharge structure with the growth of pressure is the diminution of the thickness of near-electrode layers and substantial growth of the current density on the cathode. According to the Engel–Steenbeck theory, the voltage drop on the cathode layer does not depend on pressure; therefore the current density on the cathode should increase quadratically with pressure.

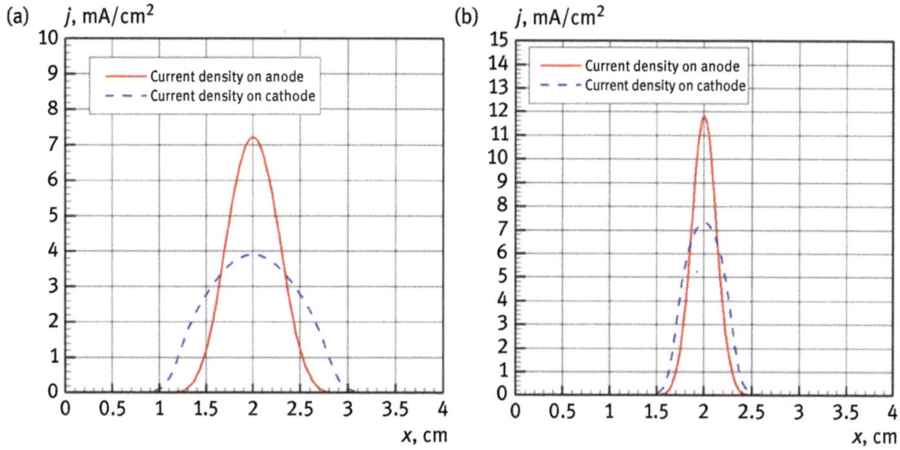

Figure 5.34: The current density on the anode (solid line) and on the cathode (dashed line) in N_2 at $E = 2$ kV, $\gamma = 0.05$: (a) $p = 5$ Torr; (b) $p = 10$ Torr.

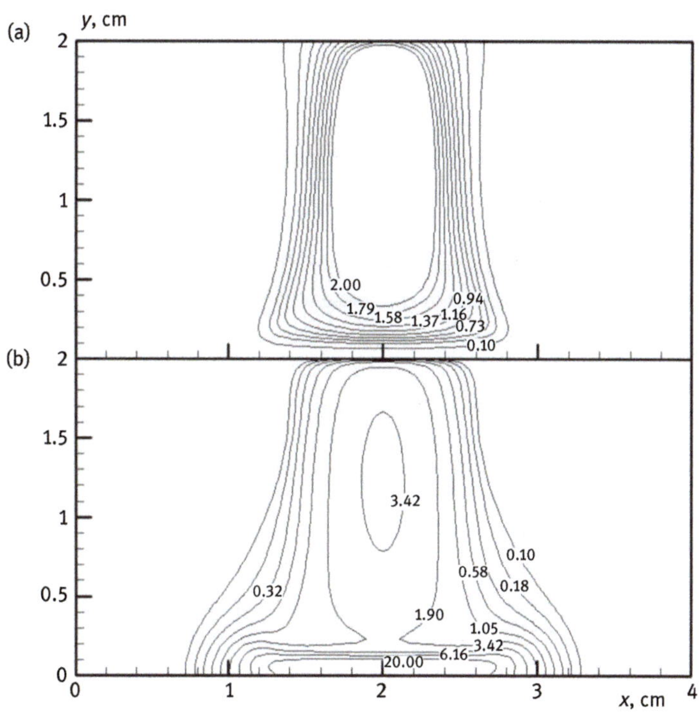

Figure 5.35: Fields of concentrations of electrons and ions (related to 10^9 cm^{-3}) in electropositive gas at $p = 5$ Torr, $E = 2$ kV, $\gamma = 0.05$: (a) electrons; (b) ions.

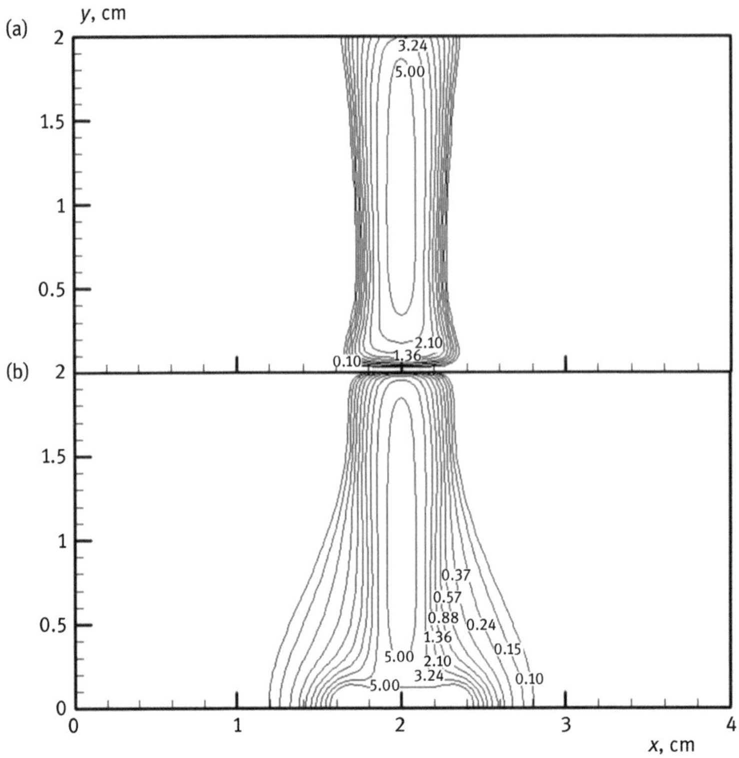

Figure 5.36: Fields of concentrations of electrons and ions (related to 10^9 cm^{-3}) in electropositive gas at $p = 10$ Torr, $E = 2$ kV, $\gamma = 0.05$: (a) electrons; (b) ions.

There is one more important singularity of the glow discharge, namely the cross sizes of the cathode spot and positive column decrease with the increase in pressure. For example, with a double increase in pressure the voltage drop between electrodes increases from 544 V up to 827 V, and the maximum current density on the cathode increases from 3.95 mA/cm^2 up to 7.3 mA/cm^2. At the same time, the total current through the discharge decreases from 4.86 mA to 3.91 mA. In addition at $\gamma = 0.1$ appreciable growth of concentration of ions in the cathode layer is observed.

Presented numerical simulation results for glow discharge in the electropositive gas were used as initial conditions for calculations of the glow discharge in the electronegative gas. These calculations were performed both without an external magnetic field and with an external cross magnetic field. The structure of such a glow discharge in air at $p = 5$ Torr, E = 2 kV and $\gamma = 0.05$ is shown in Figure 5.37.

Comparing the presented data with the results of calculations of glow discharge in molecular nitrogen, one can note that modifications in the structure of the discharge are not so essential. In the given calculation, the concentration of negative ions is rather low (see Figure 5.44c). The current densities on the cathode and

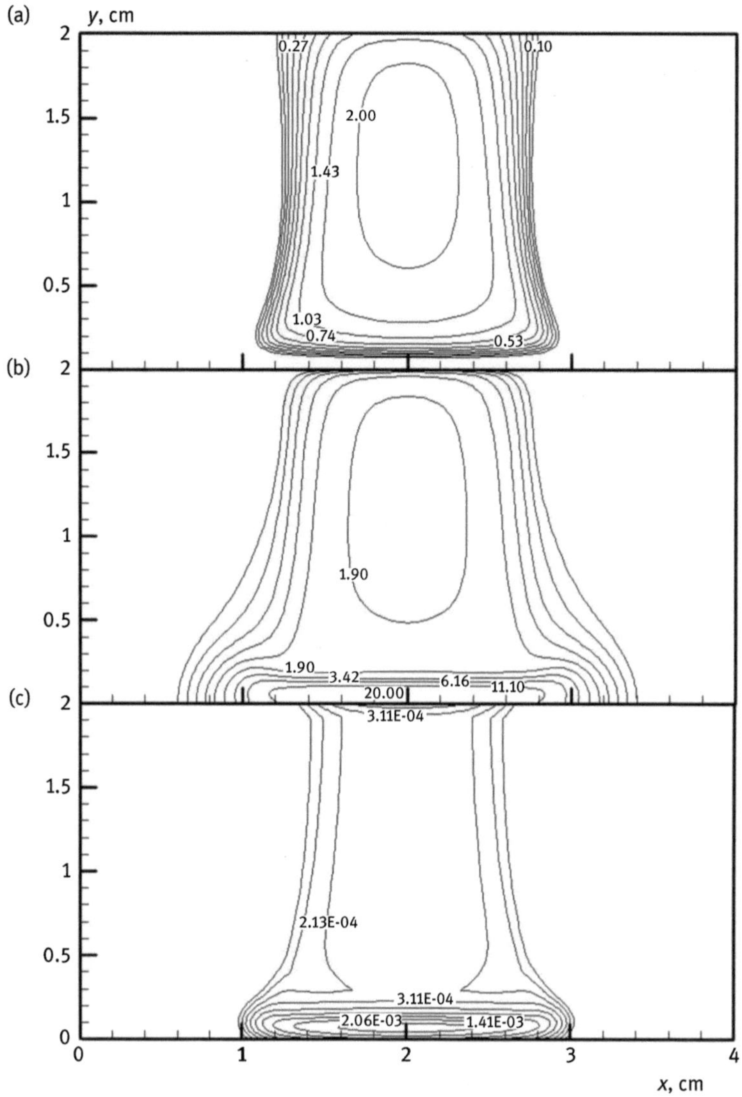

Figure 5.37: Structure of glow discharge in electronegative gas (air) in the absence of a magnetic field at $p = 5$ Torr, $E = 2$ kV: (a) concentration of electrons (10^9 cm^{-3}); (b) concentration of positive ions (109 cm^{-3}); (c) concentration of negative ions (10^9 cm^{-3}).

anode, cross sizes of the positive column, and concentrations of positive ions and electrons have decreased a little. Most likely, the principal reason for this is the use of some other approximations of the Townsend ionization coefficient.

Numerical simulation results for glow discharge in the cross magnetic field are shown in Figures 5.38–5.40.

As before, the cross magnetic field leads to the motion of glow discharge plasma along an x-axis. If the magnetic field induction is higher, the velocity will be higher as well. Figures 5.38 and 5.39 show glow discharges in electronegative gas for two values of the magnetic field inductions, $B = -0.01$ T and $B = -0.05$ T. In the first case, ($B = -0.01$ T) the instant configuration of charged particles is shown at $t = 99$ μs after energizing the magnetic field, and in the second case ($B = -0.05$ T) the

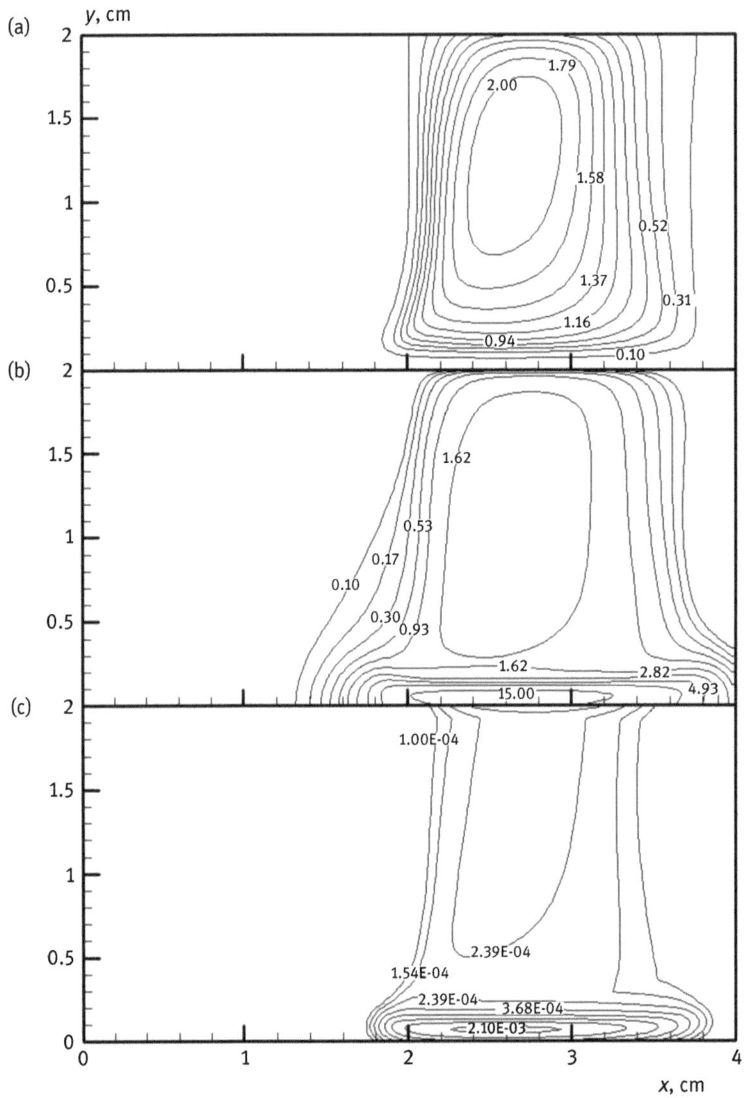

Figure 5.38: Concentration of the charged particles (109 cm^{-3}) at $p = 5$ Torr, $E = 2$ kV, $B = -0.01$ T, at the instant $t = 99$ μs: (a) for electrons; (b) for positive ions; (c) for negative ions.

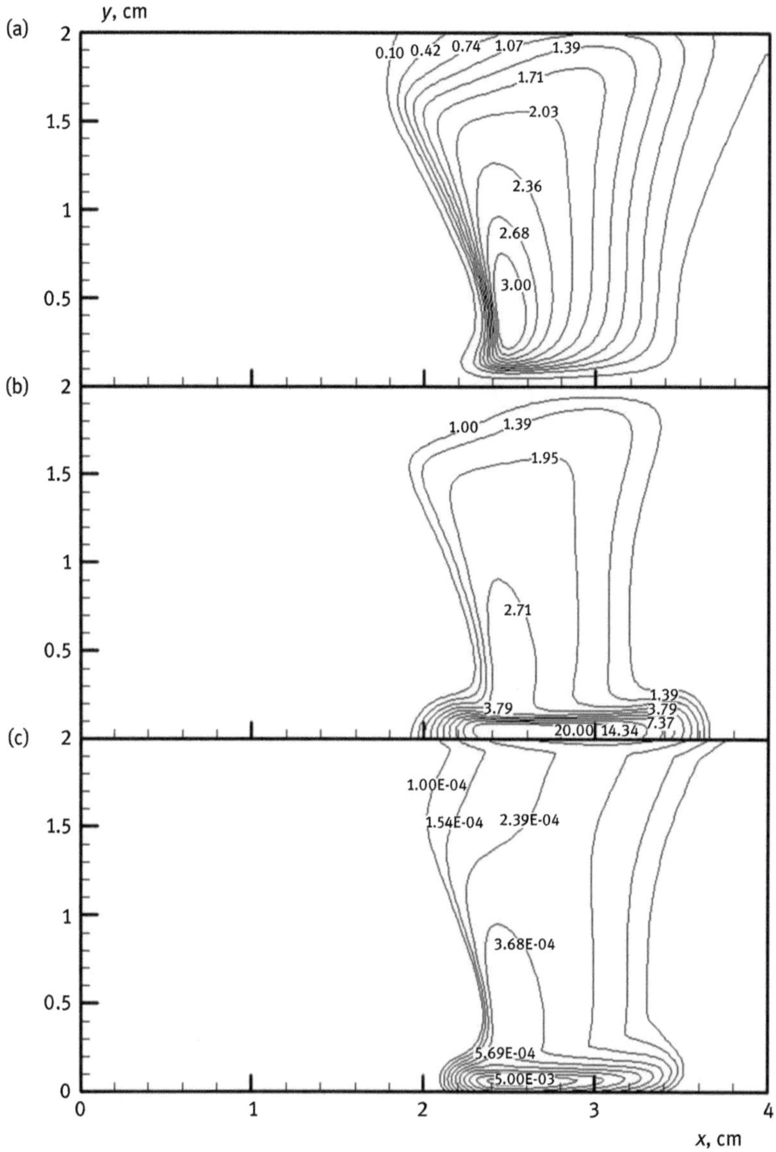

Figure 5.39: Concentration of the charged particles (109 cm^{-3}) at $p = 5$ Torr, $E = 2$ kV, $B = -0.05$ T, at the instant $t = 19$ μs: (a) for electrons; (b) for positive ions; (c) for negative ions.

structure of glow discharge is shown at $t = 19$ μs. It is obvious that in the second case the glow discharge essentially moves faster, and its structure in a greater degree is distorted by this motion. Calculations of glow discharge at $B = -0.05$ T up to $t = 19$ μs show that it is displaced practically without changing its configuration.

As it was already discussed in the case of glow discharge in molecular nitrogen, the direction of discharge motion depends on the direction of the cross magnetic field. It is natural that this also regularity is maintained for discharges in electronegative gases. For example, a configuration of the glow discharge at $B = -0.05$ T for conditions similar to the previous calculation is shown in Figure 5.40. Results of these

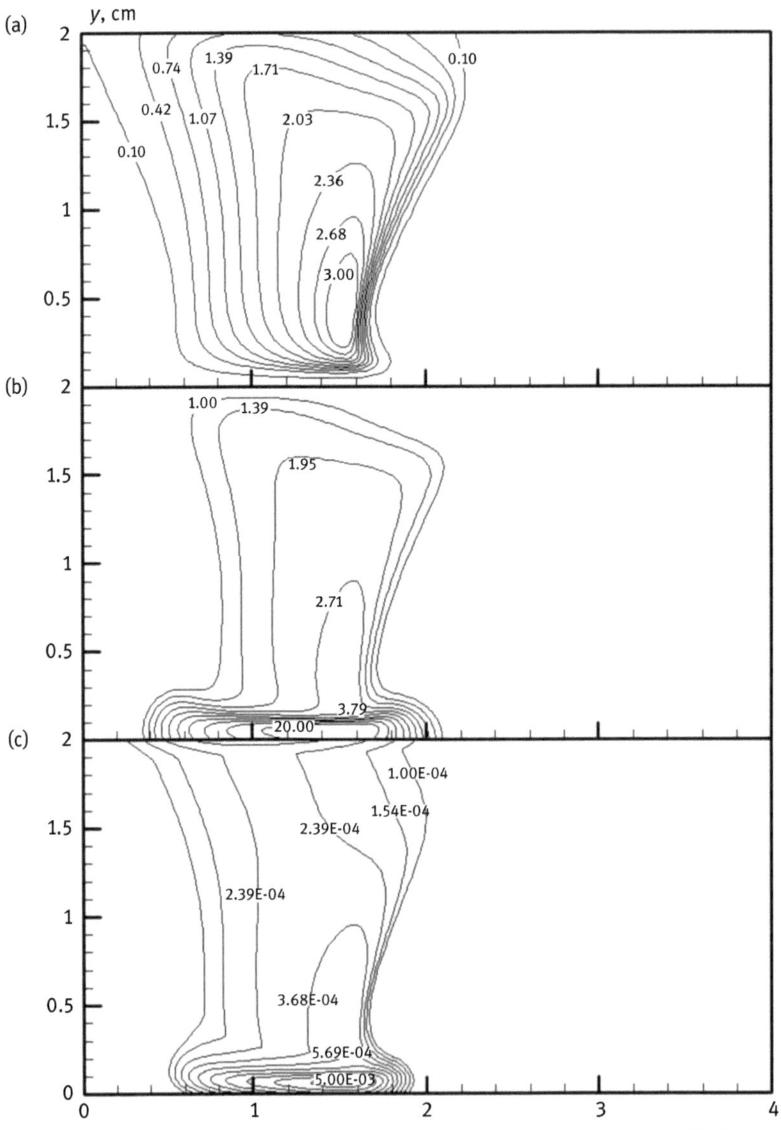

Figure 5.40: Concentration of the charged particles (109 cm^{-3}) at $p = 5$ Torr, $E = 2$ kV, $B = +0.05$ T, at the instant $t = 19$ μs: (a) for electrons; (b) for positive ions; (c) for negative ions.

calculations correspond to the instant $t = 19\ \mu s$. Note that the full symmetry of the calculation results in relation to the variant $B = -0.05$ T not only confirms physical regularities but also supports the adequacy of the developed numerical model.

5.6 Numerical modeling of glow discharge between electrodes arranged on the same surface

The drift-diffusion model has been applied to numerical modeling of glow discharge between two flat infinite electrodes (the cross scheme of discharge). Such a scheme of discharge is the most convenient one for the research of the NGDs. However, for practical applications, the longitudinal scheme, when two electrodes are arranged on the same plane, is of more importance. This configuration is shown in Figure 5.41. It is obvious that such a scheme of discharge is most acceptable for aerophysical applications of discharges on various streamline surfaces.

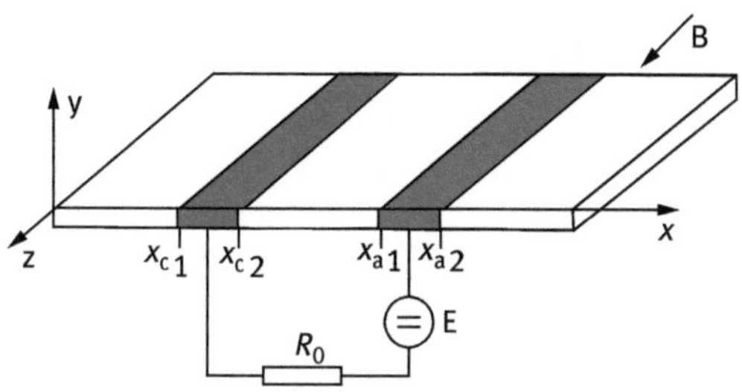

Figure 5.41: The scheme of a surface glow discharge.

The glow discharge of this kind has the following important peculiarities:

1) The discharge cannot exist in the condition of normal current density because it is limited by boundaries of the cathode and anode sections.
2) It is necessary to expect abrupt growth of electric field strength near the boundaries of the electrode sections.
3) The glow discharge region becomes nonsymmetric: one its border is located on a dielectric surface, and another in a nondisturbed gas flow.

The specified properties of the surface direct current glow discharges are manifested in the formulation of the calculation model and boundary conditions.

5.6.1 The equations of the drift-diffusion model for surface glow discharge

The surface glow discharge is considered in molecular nitrogen between two flat electrodes arranged on the same surface (see Figure 5.41). The glow discharge is described by the drift-diffusion model of the motion of electrons and ions together with the

Poisson equation for defining electric potential φ and electric field strength $\mathbf{E} = -\,\text{grad}\,\varphi$, and the energy conservation equation describing neutral gas heating-up:

$$\frac{\partial n_e}{\partial t} + \frac{\partial}{\partial x}\left(\mu_e n_e E_{e,x} - \frac{D_e}{1+b_e^2}\frac{\partial n_e}{\partial x}\right) + \frac{\partial}{\partial y}\left(\mu_e n_e E_{e,y} - \frac{D_e}{1+b_e^2}\frac{\partial n_e}{\partial y}\right) = \alpha|\Gamma_e| - \beta n_e n_i, \quad (5.181)$$

$$\frac{\partial n_i}{\partial t} + \frac{\partial}{\partial x}\left(\mu_i n_i E_{i,x} - \frac{D_i}{1+b_i^2}\frac{\partial n_i}{\partial x}\right) + \frac{\partial}{\partial y}\left(\mu_i n_i E_{i,y} - \frac{D_i}{1+b_i^2}\frac{\partial n_i}{\partial y}\right) = \alpha|\Gamma_e| - \beta n_e n_i, \quad (5.182)$$

$$\frac{\partial^2 \varphi}{\partial x^2} + \frac{\partial^2 \varphi}{\partial y^2} = 4\pi e(n_e - n_i), \quad (5.183)$$

$$\rho c_V \frac{\partial T}{\partial t} = \frac{\partial}{\partial x}\left(\lambda \frac{\partial T}{\partial x}\right) + \frac{\partial}{\partial y}\left(\lambda \frac{\partial T}{\partial y}\right) + q_J, \quad (5.184)$$

$$\Gamma_e = u_e n_e = n_e \mathbf{V} - D_e \text{grad} n_e - n_e \mu_e (\mathbf{E} + \mathbf{u}_e \times \mathbf{B}), \quad (5.185)$$

$$\Gamma_i = u_i n_i = n_i \mathbf{V} - D_i \text{grad} n_i + n_i \mu_i (\mathbf{E} + \mathbf{u}_i \times \mathbf{B}), \quad (5.186)$$

where n_e, n_i are the volumetric concentration of electrons and ions; Γ_e, Γ_i are the vectors of densities of electron and ion flows $|\Gamma_e| = \sqrt{\Gamma_{e,x}^2 + \Gamma_{e,y}^2}$; $\mathbf{B}$ is the eternal magnetic field (its configuration is shown in Figure 5.49); $q_J = \eta(\mathbf{j} \cdot \mathbf{E})$; $\mathbf{j} = e(\mathbf{\Gamma}_i - \mathbf{\Gamma}_e)$; $\alpha(E)$ and β are the coefficients of ionization and recombination; μ_e, μ_i are the mobilities of electrons and ions; $\mathbf{u}_e, \mathbf{u}_i$ are the averaged velocities of electrons and ions; D_e, D_i are the diffusivities of electrons and ions; η is the part of the energy of the electric field used for heating-up of neutral gas; V is the velocity of gas motion (e.g., along a surface);

$$b_e = \frac{\mu_e B_z}{c} = \frac{\omega_e}{\nu_e}, \quad b_i = \frac{\mu_i B_z}{c} = \frac{\omega_i}{\nu_i} \quad (5.187)$$

are the Hall parameters for electrons and ions;

$$\omega_e = \frac{eB_z}{m_e c} = \frac{eH_z}{m_e c} \quad (5.188)$$

is the Larmor frequency for electrons;

$$\omega_i = \frac{eB_z}{m_i c} = \frac{eH_z}{m_i c} \quad (5.189)$$

is the Larmor frequency for ions; T, ρ, c_V are the temperature, density, and specific thermal capacity at constant volume, respectively; λ is the coefficient of thermal conduction.

Components of an external effective electric field are expressed, as earlier, by the following relations:

$$E_{e,x} = \frac{b_e E_y - E_x}{1+b_e^2}, \quad E_{e,y} = -\frac{b_e E_x + E_y}{1+b_e^2}; \quad (5.190)$$

$$E_{i,x} = \frac{E_x + b_i E_y}{1 + b_i^2}, \quad E_{i,y} = \frac{E_y - b_i E_x}{1 + b_i^2}. \tag{5.191}$$

It is supposed that the glow discharge does not distort an external magnetic field.

The used constitutive thermodynamic relationships were discussed in the previous sections; therefore, here they are presented without additional explanation:

$$\mu_e(p^*) = \frac{4.2 \times 10^5}{p^*} \frac{\text{cm}^2}{\text{V} \cdot \text{s}}, \quad \mu_i(p^*) = \frac{2280}{p^*} \frac{\text{cm}^2}{\text{V} \cdot \text{s}}, \tag{5.192}$$

$$p^* = p \frac{293}{T} \text{ Torr}, \quad D_e = \mu_e(p^*) T_e, \quad D_i = \mu_i(p^*) T,$$

$$c_p = 8.314 \frac{7}{2} \frac{1}{M_\Sigma} \frac{\text{J}}{\text{g} \cdot \text{K}}, \quad M_\Sigma = 28 \frac{\text{g}}{\text{mole}}, \quad \rho = 1.58 \times 10^{-5} \frac{M_\Sigma p}{T} \frac{\text{g}}{\text{cm}^3},$$

$$\lambda = \frac{8.334 \times 10^{-4}}{\sigma^2 \Omega^{(2.2)*}} \sqrt{\frac{T}{M_\Sigma}} \left(0.115 + 0.354 \frac{c_p M_\Sigma}{\widetilde{R}} \right) \frac{\text{W}}{\text{cm} \cdot \text{K}},$$

$$\Omega^{(2.2)*} = \frac{1.157}{(T^*)^{0.1472}}, \quad T^* = \frac{T}{(\varepsilon/k)}, \quad \frac{\varepsilon}{k} = 71.4 \text{ K}, \quad \sigma = 3.68 \text{ Å},$$

$$c_V = 0.742 \text{ J}/(\text{g} \times \text{K}), \widetilde{R} = 8.314 \text{ J}/(\text{K} \cdot \text{mole}).$$

p is the pressure in non-disturbed gas.

Coefficients of recombination β and electronic temperature T_e are assumed to be constant: $\beta = 2 \times 10^{-7} \text{cm}^3/\text{s}$, $T_e = 11610$ K.

The coefficient of ionization is defined as follows (the first Townsend coefficient):

$$\alpha(E) = p^* A \exp\left[-\frac{B}{(|\mathbf{E}|/p^*)} \right] \frac{1}{\text{cm} \cdot \text{Torr}}, \tag{5.193}$$

where

$$A = 12 \frac{1}{\text{cm} \cdot \text{Torr}}, \quad B = 342 \frac{\text{V}}{\text{cm} \cdot \text{Torr}}.$$

Equations (5.181)–(5.183) are solved together with the equation for an external electric circuit, which in the stationary case looks like

$$E = V + IR_0, \tag{5.194}$$

where V is the voltage drop on the electrodes; I is the discharge current; E is the emf of the power supply; R_0 is the external ballast resistance.

The total current in the discharge is calculated using the formula:

$$I = \int_0^L (\mathbf{jn})_c \mathrm{d}x = \int_0^L (\mathbf{jn})_a \mathrm{d}x, \tag{5.195}$$

where **n** is the unit normal vector to the cathode (c) and anode (a) surface; **j** is the current density.

Heat emission in a gas due to the discharge current is calculated using the formula:

$$q_J = \eta(\mathbf{j}\cdot\mathbf{E}) = 1.6\times10^{-19}\eta\left[nE^2(\mu_i+\mu_e) + (D_e - D_i)\mathbf{E}\,\mathrm{grad}n\right],$$

where η is the effectiveness of transformation of the electric field energy in heating-up of neutral gas ($\eta \sim 0.1$–0.9).

For the calculation of parameters of the surface glow discharge in flat geometry, all integral parameters (e.g., the total current) should be related to the length unit in the z-direction (see Figure 5.49).

In this chapter the simplified gas-dynamic model is used. Validity of the equation $\rho u = \mathrm{const}$ along each streamline along the surface is assumed.

5.6.2 Boundary conditions for the surface discharge

Boundary conditions are formulated in the following form:
at $y=0,\ \ x\in[x_{c1}, x_{c2}]$ (surface of the cathode)

$$n_e = \gamma n_i\mu_i,\quad \frac{\partial n_i}{\partial y} = 0,\quad \varphi = 0; \tag{5.196}$$

at $y=0,\ \ x\in[x_{a1}, x_{a2}]$ (surface of the anode)

$$n_i = 0,\quad \frac{\partial n_e}{\partial y} = 0,\quad \varphi = V; \tag{5.197}$$

at $y=0,\ \ x\notin[x_{c1}, x_{c2}],\ x\notin[x_{a1}, x_{a2}]$ (surface of a dielectric)

$$n_e = n_i = n_0,\quad \frac{\partial\varphi}{\partial y} = 0; \tag{5.198}$$

$$y\to\infty:\ \frac{\partial n_e}{\partial y} = \frac{\partial n_i}{\partial y} = \frac{\partial\varphi}{\partial y} = \frac{\partial T}{\partial y} = 0; \tag{5.199}$$

$$x = 0,\ L:\ \frac{\partial n_e}{\partial x} = \frac{\partial n_i}{\partial x} = \frac{\partial\varphi}{\partial x} = 0; \tag{5.200}$$

$$x = 0:\ T = T_\infty; \tag{5.201}$$

$$x = L:\ \frac{\partial T}{\partial x} = 0. \tag{5.202}$$

Here n_0 is the typical concentration of electrons on a surface of the dielectric (n_0 is several orders lesser than that in electrodischarge plasma above electrodes, e.g., $n_0 \sim 10^3$–$10^7\ \mathrm{cm}^{-3}$); V is the potential of the anode relative to the cathode. Note that electrophysical boundary conditions on the dielectric surface are substantially defined by its catalytic ability. Following coordinates of cathode and anode sections were used in calculations: $x_{c1} = 1.476$ cm, $x_{c2} = 2.111$ cm, $x_{a1} = 5.861$ cm, $x_{a2} = 6.496$ cm.

5.6.3 Initial conditions of numerical modeling

Clouds of quasineutral plasma above the cathode and the anode were set as the initial conditions. The initial distribution of potential was obtained from the solution of the Laplace equation under given boundary conditions (5.196)–(5.200). The temperature was supposed to b a constant in all calculation domains, $T = 300$ K.

5.6.4 Numerical simulation results of surface glow discharge

The typical configuration of surface glow discharge is shown in Figure 5.42. This discharge corresponds to the following initial data: $p = 5$ Torr, $E = 500$ V, $R_0 = 12$ kOhm, $\gamma = 0.1$, $\eta = 0.9$. Concentrations of charged particles are related to value of $N_0 = 10^9$ cm^{-3}.

The calculations performed have confirmed some singularities of surface glow discharges that are well known from experimenters. Positively charged area of the discharge (the cathode layer) is well visible in Figure 5.42a. The increased concentration of positive ions is also observed near the boundaries of the anode with a dielectric, though on the anode surface (according to the boundary conditions) their concentration is equal to zero.

The concentration of electrons in the cathode layer is very low (Figure 5.42a). The greatest concentration of electrons is observed near the boundaries of the anode.

The numerical simulation results show that the area of quasineutral plasma is pushed aside from the dielectric surface at a distance of ~0.5–1 cm. Figure 5.50c shows the distribution of electroconductivity of discharge plasma and vectorial electric field. Let us consider the basic peculiarities of the presented distributions:

1) The maximum electric field strength (~3 kV/cm) is observed in the cathode layer near the cathode. This result is in correspondence with the theory of glow discharges.
2) The electric field strength in the anode layer is commensurable with the field strength in the plasma layer near the dielectric surface between electrodes.
3) The local maxima of electric field strengths (~4 kV/cm) are observed near the boundaries of the cathode with the dielectric surface. It corresponds well to the known experimental fact of a strength step on boundaries of electrodes. For this reason, in real designs, the electrodes with smoothed boundaries are used, as the specified strength steps lead to the development of breakdown and instabilities of discharge plasma.

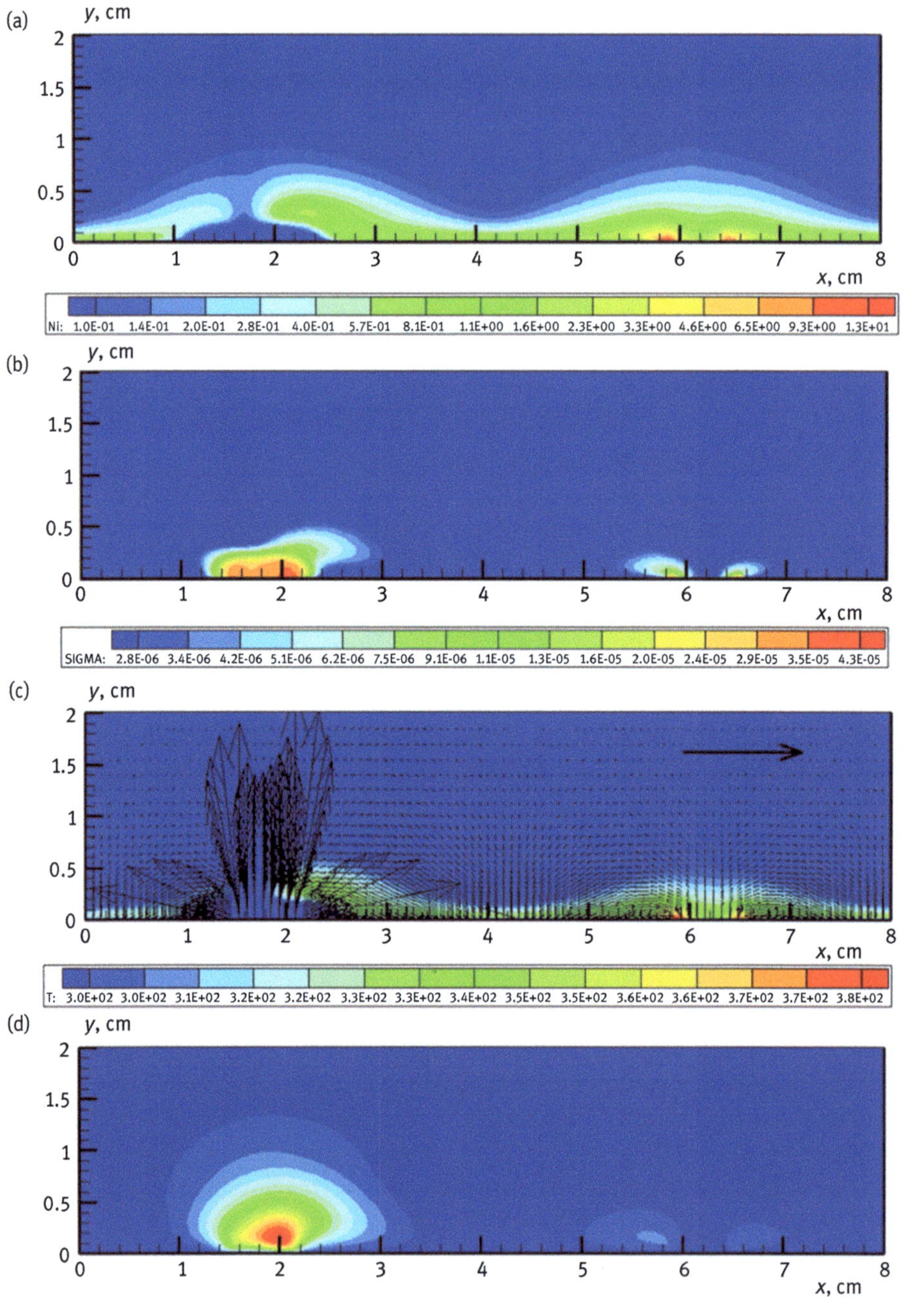

Figure 5.42: Configuration of surface glow discharge at $p = 5$ Torr, $E = 500$ V, $R_0 = 12$ kOhm, $\gamma = 0.1$, $\eta = 0.9$: (a) concentration of electrons; (b) concentration of ions; (c) electroconductivity $(\text{Ohm} \cdot \text{cm})^{-1}$ and electric field strength E (the arrow shows electrostatic intensity 3,000 V/cm); d temperature, K.

The temperature of neutral gas in the surface glow discharge is shown in Figure 5.42d. The assumption of a cooled surface was used; therefore, the maximum of the temperature field is reached at some distance from the surface (approximately at a distance of 0.2 cm). Neutral gas near the anode of the discharge is heated up to a lesser degree.

In the subsequent figures (Figure 5.43a–d) the similar function distributions of surface glow discharge in a cross gas flow with the velocity of 362 m/s are shown (the gas flow from left to right). It is obvious that the gas flow appreciably distorts the distribution of electron and ion concentrations (Figure 5.43a and b). The temperature field (compare Figures 5.42d and 5.43d) is also greatly deformed.

As it follows from the theory of glow discharges, the cross magnetic field strongly influences an electrodischarge structure at $b_e \sim 1$ (and, especially, at $b_i \sim 1$). The calculations of surface glow discharge with the cross magnetic field of $|B_z| \sim 0.03 \div 0.5$ T have confirmed this assumption. In the specified cases $|b_e| = 0.264–0.44$ and $|b_i| = 0.000864–0.00144$. A significant influence of the magnetic field induction vector direction on the discharge structure has been observed.

Figures 5.44 and 5.45 show the electrodynamic structure of the surface glow discharge at $B_z = -0.03$ and $+0.03$ T. First, we fix to significant modifications in fields of electric conductivity in immediate proximity to the surface. As it follows from the theory, superposition of a magnetic field changes a configuration of an effective electric field. It can be seen in Figures 5.44 and 5.45.

Figure 5.46 illustrates amplification of this influence at the growth of the magnetic field induction (compare Figures 5.45a and 5.46a, and also Figures 5.45b and 5.46b). Presented here are the distributions of concentrations of charged particles that correspond to the values of emf $E = 500$ V. An increase in the voltage drop on electrodes (in our case, emf) leads to an abrupt increase in all parameters of the surface glow discharge, and then to its breakdown.

For example, Figure 5.47 shows fields of electron concentration at $E = 900$ V. The increased level of concentration is can be seen.

5.7 Normal glow discharge in axial magnetic field

In this section, the theory and results of numerical simulation of a glow discharge between two parallel electrodes with an axial magnetic field are presented. The model consists of continuity equations for electron and ion fluids, as well as the Poisson equation for the self-consistent electric field.

As discussed earlier, the NGD is better in the study of the configuration of the two "infinite" flat electrodes. The term "infinite" electrodes is used to emphasize the fact that the boundary effects on the electrodes do not affect the structure of the NGD. A typical configuration of the NGD and its circuitry are shown in Figure 5.48. As a rule, such discharges permanently exist with parameters close to the minimum

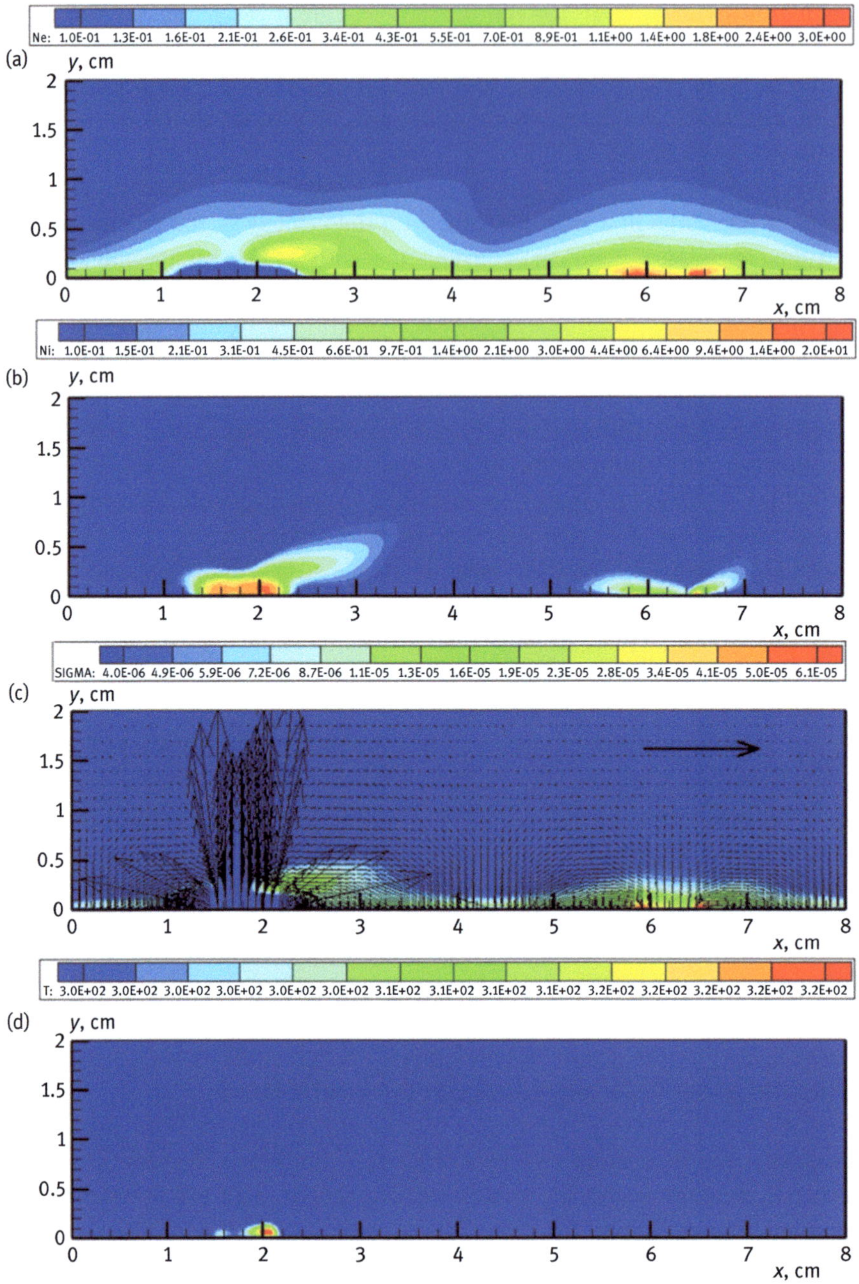

Figure 5.43: Configuration of surface glow discharge at $p = 5$ Torr, $E = 500$ V, $R_0 = 12$ kOhm, $\gamma = 0.1$, $\eta = 0.9$, $V_\infty = 365$ m/s: (a) concentration of electrons; (b) concentration of ions; (c) electroconductivity $(\text{Ohm}\cdot\text{cm})^{-1}$ and electric field strength E (the arrow shows electrostatic intensity 3,000 V/cm); (d) temperature, K.

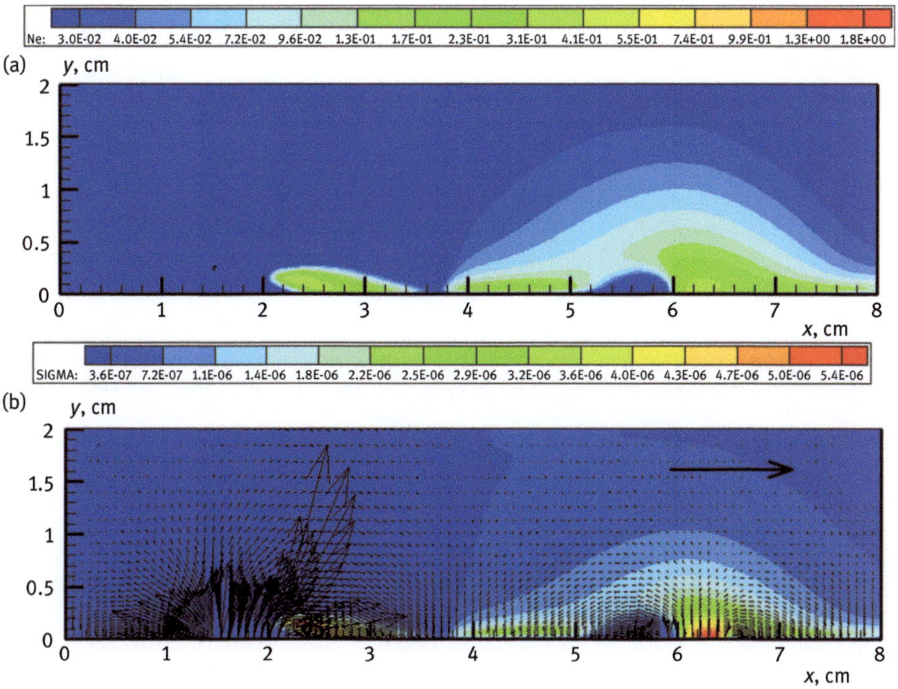

Figure 5.44: Configuration of surface glow discharge at $p = 5$ Torr, $E = 500$ V, $\gamma = 0.1$, $B = -0.03$ T: (a) concentration of electrons; (b) electroconductivity $(\text{Ohm} \cdot \text{cm})^{-1}$ and an effective electric field E_e (the arrow shows electrostatic intensity 3,000 V/cm).

of the Paschen breakdown curve. For example, for molecular nitrogen N_2, the voltage drop between electrodes reaches only a few hundred volts, and the total current through the discharge gap is measured by several milliamps.

The law of "normal current density" states that with an increase in the total current through the glow discharge the current density at the center of the current spot remains practically unchanged. This is a simple, but very effective, one-dimensional method for calculating the parameters of the NGD that was created by Engel and Steenbeck (Engel A., et al., 1932). This one-dimensional theory of the cathode layer in NGD allows to predict the current density, voltage drop, and depth of the cathode layer with an acceptable accuracy for practical needs. But unfortunately, this model assumes the linear dependence of the intensity of the electric field in a cathode layer, as well as the use of the local field approximation for the ionization source term. It means that the ionization source term is a function of the local value of the reduced electric field, E/p, where E is the magnitude of the electric field and p is the gas pressure. As a result, this model could not describe the negative glow region where the ionization source term depends on the potential distribution upstream in the cathode fall and not on the local value of

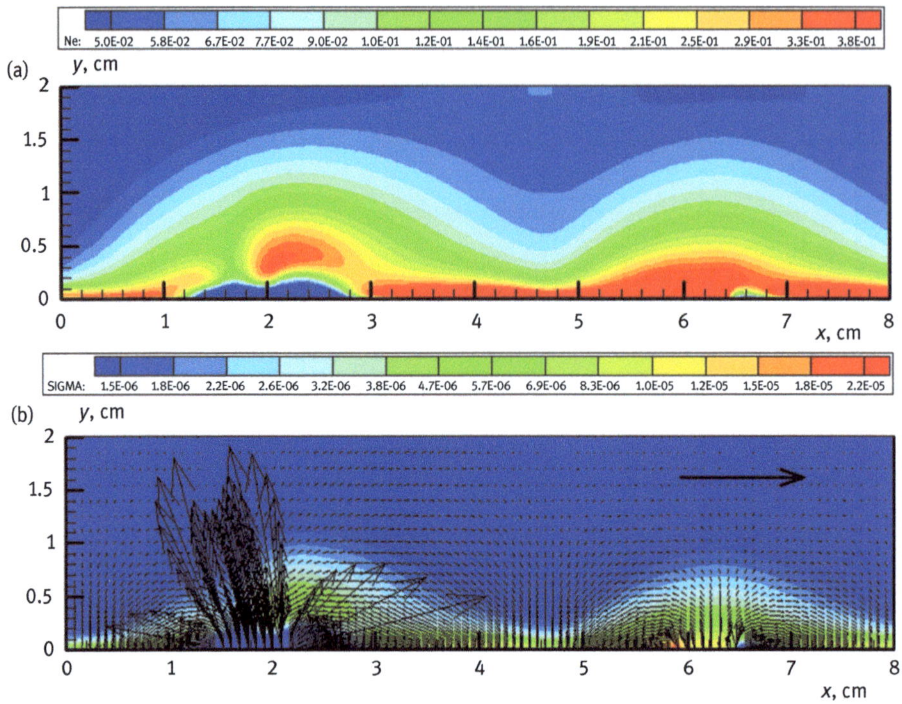

Figure 5.45: Configuration of surface glow discharge at $p = 5$ Torr, $E = 500$ V, $\gamma = 0.1$, $B = +0.03$ T: (a) concentration of electrons; (b) electroconductivity $(\text{Ohm}\cdot\text{cm})^{-1}$ and an effective electric field E_e (the arrow shows electrostatic intensity 3,000 V/cm).

the E/p. The two-dimensional hybrid fluid-dynamic model (Fiala A., et al.,1994) has demonstrated the real physics of the phenomenon.

The drawback of the local field approximation was also considered in detail in Kolobov V.I. and Tsendin L.D. (1992), where a self-consistent analytic model of the cathode region of a direct current discharge was created.Another specific peculiarity of the glow discharge that cannot be described using the Engel and Steenbeck theory is the phenomenon of the subnormal current oscillations, which were studied by Aslanbekov R.R. and Kolobov V.I. (2003), where the drift-diffusion model was supplemented with the Boltzmann equation for electron energy distribution function.

Nevertheless, the classical Engel and Steenbeck theory, as well as the local field approximation, can fairly well provide the global features of NGD, and therefore they are in common use in gas-discharge physics and computational physics of gas discharges in two-dimensional and, especially, in three-dimensional geometry (Almeida P.G.C., et al., 2011).

The NGD is the weakly ionized nonequilibrium gas, in which electrons are heated up to a few electron volts, and the ions remain almost cold and their temperature is only slightly higher than the room temperature.

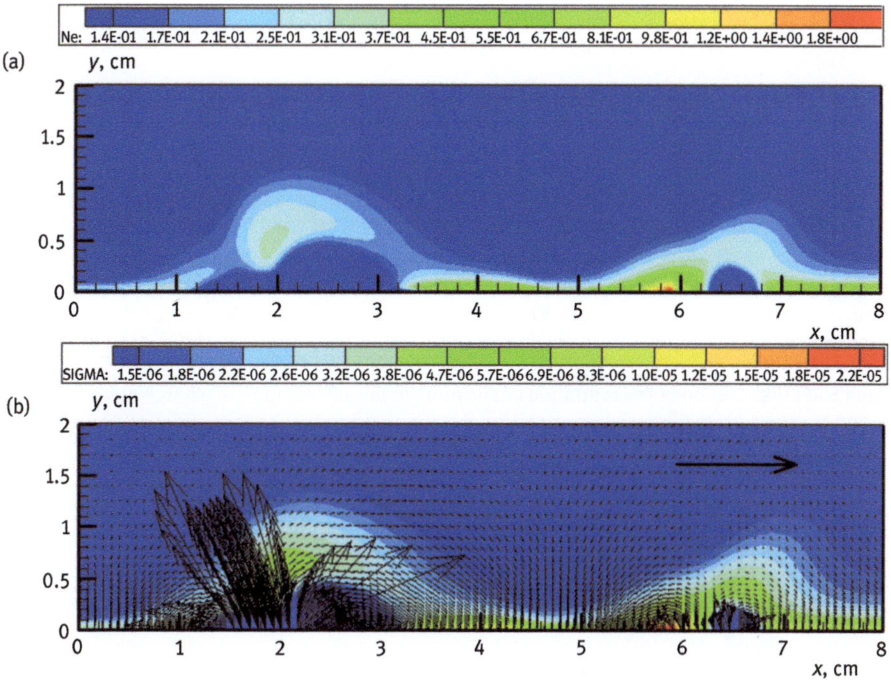

Figure 5.46: Configuration of surface glow discharge at $p = 5$ Torr, $E = 500$ V, $\gamma = 0.1$, $B = +0.05$ T: (a) concentration of electrons; (b) electroconductivity $(\text{Ohm}\cdot\text{cm})^{-1}$ and an effective electric field E_e (the arrow shows electrostatic intensity 3,000 V/cm).

Modifying the properties of glow discharges external magnetic field results in many new plasma-dynamic and gas-dynamic effects, which are traditionally considering in plasma physics. These are plasma generators and radiation sources, devices with magnetic plasma confinement, plasma chemical reactors, and microelectronics (Bittecourt J.A., 2004). Different plasma-dynamic processes are increasingly used in space technology (Morozov A.I., 2003; Smirnov A., et al., 2007).

The unique properties of the plasma in a magnetic field was studied in recent years for creating the electromagnetic actuators using aerothermodynamics' technologies (Shang J.S., et al., 2005a,b,c; Shin et al, 2007; Kumar H., et al., 2005). A specific feature of such technologies is the use of weakly ionized gas. To study the properties of electromagnetic actuators in highly rarefied hypersonic flows, it is useful to analyze the structure of normal and abnormal glow discharges without and in the presence of a transverse magnetic field (Surzhikov S.T., et al., 2004b; Shang J.S., et al., 2009; Macheret S.O., 2006). Several important for hypersonic aerodynamics the regularities of the interaction of gas flow with a glow discharge, such as the increase or decrease of the pressure of the gas on the surface (Shang J.S., et al., 2005a,b,c), and also inertialess movement of discharge along the surface in a

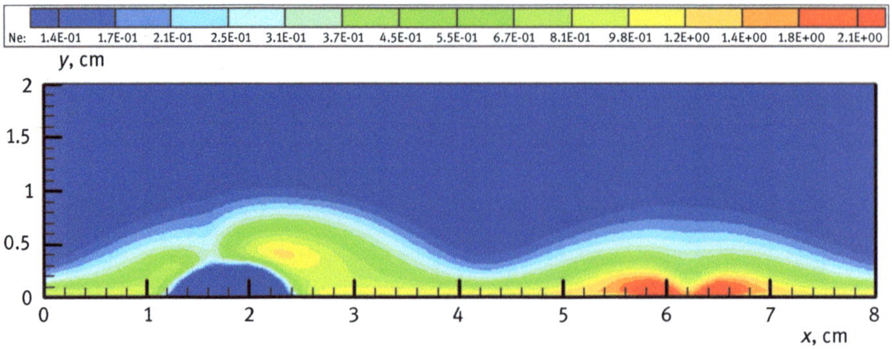

Figure 5.47: Distribution of concentration of electrons in the surface glow discharge at $p = 5$ Torr, $E = 900$ V, $\gamma = 0.1$, $\eta = 0.9$.

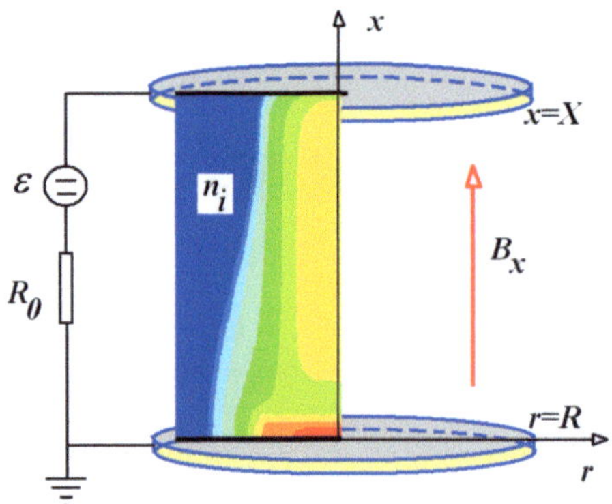

Figure 5.48: Schematic of a glow discharge with external magnetic field.

transverse magnetic field were subject of research in the field of fundamental physics of gas discharges (Shang et al., 2005; Macheret, 2006).

In this section, we will study the laws of the existence of NGD in the longitudinal (the axial) magnetic field (Figure 5.48 shows the scheme of this problem) in a theoretically "ideal" formulation when the current discharge column is located far from the borders of the electrodes.

It should be noted that the glow discharge in the longitudinal magnetic field was studied more than 100 years ago (Phillips, 1901; Penning, 1937). Despite the fact that this type of discharge is widely used in the ion accelerators and in the plasma-chemical reactors, a systematic study of the NGD in an axial magnetic field was not carried out. The main reason for this is that the aforementioned theoretical "ideal" configuration NGD is rarely realized in plasma technologies.

However, the study of the NGD in a longitudinal magnetic field is important for knowing the fundamental nature that is directly related to the problem of the formation of the current areas (spots) on the electrodes. An NGD is a good example of self-organization of the electrical current when the radial structure of the discharge is determined exclusively by its own internal processes, and is not related to the influence of boundaries in the radial direction.

As an object of study in this section NGD is selected in molecular nitrogen at a pressure of $p = 2 \div 10$ Torr at a gas-discharge gap equal to $X = 2$ cm. Parameters of the discharge geometry and external current circuit were taken from Raizer Yu.P and Surzhikov (1987), where an axial magnetic field is not taken into account.

5.7.1 Analysis

A current column of a cylindrical glow discharge in molecular nitrogen between flat electrodes in an axial magnetic field is considered (see Figure 5.48). The drift-diffusion theory is used for the description of the processes in a gas discharge.

The system of governing equations is based, as before, on equations of continuity for the concentration of electrons n_e and positive ions n_i together with the equations for the electrostatic field:

$$\frac{\partial n_e}{\partial t} + \frac{\partial \Gamma_{e,x}}{\partial x} + \frac{1}{r}\frac{\partial r\Gamma_{e,r}}{\partial r} = \alpha(E)|\mathbf{\Gamma}_e| - \beta n_i n_e, \tag{5.203}$$

$$\frac{\partial n_i}{\partial t} + \frac{\partial \Gamma_{i,x}}{\partial x} + \frac{1}{r}\frac{\partial r\Gamma_{i,r}}{\partial r} = \alpha(E)|\mathbf{\Gamma}_e| - \beta n_i n_e, \tag{5.204}$$

$$\frac{\partial^2 \varphi}{\partial x^2} + \frac{1}{r}\frac{\partial}{\partial r} r \frac{\partial \varphi}{\partial r} = 4\pi e(n_e - n_i), \tag{5.205}$$

where

$$\mathbf{\Gamma}_e = -D_e \mathrm{grad} n_e - n_e \mu_e \mathbf{E}, \quad \mathbf{\Gamma}_i = -D_i \mathrm{grad} n_i + n_i \mu_i \mathbf{E},$$

$$\mathbf{j} = e(\mathbf{\Gamma}_i - \mathbf{\Gamma}_e), \quad \mathbf{E} = -\mathrm{grad}\varphi; \tag{5.206}$$

$\alpha(E)$ and β are the ionization and recombination coefficients; Γ_e, Γ_i are the electron and ion flux densities; μ_e, μ_i are the electron and ion mobilities; D_e, D_i are the electron and ion diffusion coefficients; φ is the potential of electric field.

Now we need to include an axial magnetic field to the governing equations. To introduce a magnetic field in the model, we will use the following momentum conservation equations for electronic and ionic gases:

$$\rho_e \frac{\partial \mathbf{u}_e}{\partial t} + \rho_e(\mathbf{u}_e \cdot \nabla)\mathbf{u}_e = -\nabla p_e - \tau_e - en_e\left(\mathbf{E} + \frac{1}{c}[\mathbf{u}_e\mathbf{B}]\right)$$
$$- m_e\nu_{en}n_e(\mathbf{u}_e - \mathbf{u}_n) - m_e\nu_{ei}n_e(\mathbf{u}_e - \mathbf{u}_i), \qquad (5.207)$$

$$\rho_i \frac{\partial \mathbf{u}_i}{\partial t} + \rho_i(\mathbf{u}_i \cdot \nabla)\mathbf{u}_i = -\nabla p_i - \tau_i + en_i\left(\mathbf{E} + \frac{1}{c}[\mathbf{u}_i\mathbf{B}]\right)$$
$$- m_i\nu_{ie}n_i(\mathbf{u}_i - \mathbf{u}_e) - m_i\nu_{in}n_i(\mathbf{u}_i - \mathbf{u}_n), \qquad (5.208)$$

where $\mathbf{u}_e, \mathbf{u}_i$ are the velocities of electronic and ionic gases; ρ_e, ρ_i are the densities of electronic and ionic gases, $\rho_e = m_e n_e$, $\rho_i = m_i n_i$; m_e, m_i are the mass of electron and ion; $\mathbf{u}_n$ is the averaged mass velocity of neutral particles; p_e, p_i are the electronic and ionic pressures; τ_e, τ_i are the viscosity stress tensor components of electronic and ionic gases; $\nu_{en}, \nu_{ei}, \nu_{in}$ are the frequencies of electron-neutral, electron-ion, and ion-neutral collisions; $\mathbf{B}$ is the inductivity of magnetic field; c is the speed of light.

Let take into account that

$$m_e << m_i,$$

and introduce the following general assumption to our analysis

$$\rho_e(\mathbf{u}_e \cdot \nabla)\mathbf{u}_e << \rho_i(\mathbf{u}_i \cdot \nabla)\mathbf{u}_i.$$

In this case it is possible to simplify eq. (5.207)

$$-\nabla p_e - en_e\left(\mathbf{E} + \frac{1}{c}[\mathbf{u}_e\mathbf{B}]\right) - m_e\nu_{en}n_e(\mathbf{u}_e - \mathbf{u}_n) - m_e\nu_{ei}n_e(\mathbf{u}_e - \mathbf{u}_i) = 0. \qquad (5.209)$$

Let take into account that

$$\mathbf{u}_e >> \mathbf{u}_n, \mathbf{u}_i,$$

and $p_e = n_e kT_e$.

Then, eq. (5.207) can be simplified:

$$kT_e\nabla n_e + en_e\mathbf{E} + \frac{en_e}{c}[\mathbf{u}_e\mathbf{B}] + (m_e\nu_e)n_e\mathbf{u}_e = 0,$$

and, finally

$$\boldsymbol{\Gamma}_e = n_e\mathbf{u}_e = -D_e\nabla n_e - \mu_e n_e\mathbf{E} + \frac{\mu_e n_e}{c}[\mathbf{u}_e\mathbf{B}], \qquad (5.210)$$

where $\mu_e = e/m_e\nu_e$ is the electron mobility; $D_e = (kT_e/e)\mu_e$ is the electron diffusion coefficient; $\nu_e = \nu_{en} + \nu_{ei}$. These coefficients were introduced in the phenomenological model of glow discharge.

Taking into account that the magnetic field has only the x-component B_x, one can find that averaged electronic velocity has three components $\mathbf{u}_e = \{u_{e,x}, u_{e,r}, u_{e,\varphi}\}$ and corresponding three components of electron fluxes:

$$\Gamma_{e,x} = -D_e\frac{\partial n_e}{\partial x} - \mu_e n_e E_x,$$

$$\Gamma_{e,r} = -\frac{D_e}{1+b_e^2}\frac{\partial n_e}{\partial r} - \frac{\mu_e}{1+b_e^2}n_e E_r,$$

$$\Gamma_{e,\varphi} = -b_e\Gamma_{e,r},$$

where $b_e = (\mu_e/c)B_x$ is the Hall parameter for electrons.

Let us now consider, by analogy, momentum conservation of the ionic species:

$$-\nabla p_i + en_i\left(\mathbf{E} + \frac{1}{c}[\mathbf{u}_i\mathbf{B}]\right) - m_i\nu_{ie}n_i(\mathbf{u}_i - \mathbf{u}_e) - m_i\nu_{in}n_i(\mathbf{u}_i - \mathbf{u}_n) = 0 \tag{5.211}$$

Taking into account that

$$m_i\nu_{ie}n_i(\mathbf{u}_i - \mathbf{u}_e) = -m_e\nu_{ei}n_e(\mathbf{u}_e - \mathbf{u}_i),$$

$$\mathbf{u}_i = 0,$$

and $m_e\nu_e << m_i\nu_{in}$

one can obtain

$$\mathbf{\Gamma}_i = n_i\mathbf{u}_i = -D_i\nabla n_i + \mu_i n_i\mathbf{E} + \frac{\mu_i n_i}{c}[\mathbf{u}_i\mathbf{B}], \tag{5.212}$$

where $\mu_i = e/m_i\nu_{in}$ is the ions mobility; $D_i = (kT_i/e)\mu_i$ is the ions diffusion coefficient.

Again, taking into account that the magnetic field has only the x-component B_x, we can write:

$$\Gamma_{i,x} = -D_i\frac{\partial n_i}{\partial x} + \mu_i n_i E_x,$$

$$\Gamma_{i,r} = -\frac{D_i}{1+b_i^2}\frac{\partial n_i}{\partial r} + \frac{\mu_e}{1+b_i^2}n_i E_r,$$

$$\Gamma_{i,\varphi} = -b_i\Gamma_{i,r},$$

where $b_i = (\mu_i/c)B_x$ is the Hall parameter for ions.

Finally, the charged particles continuity equations can be written in following form:

$$\frac{\partial n_e}{\partial t} + \frac{\partial}{\partial x}\left(-\mu_e n_e E_x - D_e\frac{\partial n_e}{\partial x}\right) + \frac{1}{r}\frac{\partial}{\partial r}r\left(-\frac{\mu_e}{1+b_e^2}n_e E_r - \frac{D_e}{1+b_e^2}\frac{\partial n_e}{\partial r}\right) = \alpha|\Gamma_e| - \beta n_e n_i, \tag{5.213}$$

$$\frac{\partial n_i}{\partial t} + \frac{\partial}{\partial x}\left(\mu_i n_i E_x - \frac{D_i}{1+b_i^2}\frac{\partial n_i}{\partial x}\right) + \frac{1}{r}\frac{\partial}{\partial r}r\left(\frac{\mu_i}{1+b_i^2}n_i E_r - \frac{D_i}{1+b_i^2}\frac{\partial n_i}{\partial r}\right) = \alpha|\Gamma_e| - \beta n_e n_i, \tag{5.214}$$

where

$$|\Gamma_e| = \sqrt{\Gamma_{e,x}^2 + \Gamma_{e,r}^2 + \Gamma_{e,\varphi}^2}, \tag{5.215}$$

$$\Gamma_{e,x} = n_e u_{e,x} = -D_e \frac{\partial n_e}{\partial x} + \mu_e n_e \frac{\partial \varphi}{\partial x}, \tag{5.216}$$

$$\Gamma_{e,r} = n_e u_{e,r} = -\frac{D_e}{1+b_e^2}\frac{\partial n_e}{\partial r} + \frac{\mu_e}{1+b_e^2} n_e \frac{\partial \varphi}{\partial r}, \tag{5.217}$$

$$\Gamma_{e,\varphi} = n_e u_{e,\varphi} = -b_e \Gamma_{e,r}. \tag{5.218}$$

Note that the appearance of the azimuthal component of the electron flux density in the total electron flux density (5.215) means that now there is an additional source of ionization concerned to the azimuthal rotation of the discharge plasma in an axial magnetic field.

It should also be noted that the introduced coefficients, which are taking into account a magnetic field, can be presented in the following form:

$$b_e = \frac{\mu_e B_x}{c} = \frac{\omega_e}{\nu_e}, \quad b_i = \frac{\mu_i B_x}{c} = \frac{\omega_i}{\nu_i},$$

where

$$\omega_e = \frac{eB_x}{m_e c} \sim 1.76 \times 10^{10}\,\mathrm{s}^{-1},$$

$$\omega_i = \frac{eB_x}{m_i c} \sim 6.02 \times 10^{5}\,\mathrm{s}^{-1}$$

are the Larmor frequencies of electrons and ions, where B_x should be measured in Gs ($1\,\mathrm{T} = 10^4$ Gs); $m_e = 9.1 \times 10^{-28}$ g, $m_i = 16 \times 1.66 \times 10^{-24}$ g, $B_x = 0.1$ T, $e = 4.802 \times 10^{-10}$SGSE.

If we use the approximation dependences for the mobility of charged particles

$$\mu_e = 4.2 \times 10^5 / p_{\mathrm{Torr}}, \ \mathrm{cm}^2/(\mathrm{V \cdot s}),$$

$$\mu_i = 2280 / p_{\mathrm{Torr}}, \ \mathrm{cm}^2/(\mathrm{V \cdot s}),$$

then at $p = 5$ Torr one can get $b_e \approx 0.84$, $b_i \approx 0.45 \times 10^{-2}$.

The Larmor radii of electrons and ions have the following values:

$$R_{Le} = \frac{v_{e\perp} m_e c}{eB_x} \sim 1.42 \times 10^{-3}\ \mathrm{cm},$$

$$R_{Li} = \frac{v_{i\perp} m_i c}{eB_x} \sim 0.142\ \mathrm{cm},$$

where

$$v_{e\perp}\sim 2.5\times 10^7\ \text{cm/s},\ v_{i\perp}\sim 8.6\times 10^4\text{cm/s}.$$

5.7.2 Boundary conditions

Boundary conditions for charged particles and electrical potential are formulated as follows:

$$x=0:\ \frac{\partial n_i}{\partial x}=0,\ \ \Gamma_{e,x}=\gamma\Gamma_{i,x},\ \ \varphi=0;$$

$$x=X:\ n_i=0,\ \ \frac{\partial n_e}{\partial x}=0,\ \ \varphi=V; \tag{5.219}$$

$$r=0:\ \frac{\partial n_e}{\partial r}=\frac{\partial n_i}{\partial r}=\frac{\partial \varphi}{\partial r}=0;$$

$$r=R:\ \frac{\partial n_e}{\partial r}=\frac{\partial n_i}{\partial r}=\frac{\partial \varphi}{\partial r}=0, \tag{5.220}$$

where γ is the secondary emission coefficient; V is the voltage drop on the discharge gap; R is the radial boundary.

A quasineutral plasma cloud near to cathode is used as the initial condition.

5.7.3 Constitutive relationship

In the present investigation, the variation of neutral species temperature is not taken into consideration.

The recombination coefficient β and electron temperature are taken as constants $\beta=2\times 10^{-7}$ cm^3/s, $T_e=11610$ K. It was shown in Raizer and Surzhikov (1988) that the variation of T_e in wide region (1–5 eV) does not change cardinally the global features of NGD.

The ionization coefficient for N_2 is determined as follows:

$$\alpha(E)=pA\exp\left[-\frac{B}{(|\mathbf{E}|/p)}\right]\ \text{cm}^{-1}, \tag{5.221}$$

where

$$A=12\ (\text{cm})^{-1},\ B=342\ (\text{V/cm}\cdot\text{Torr}).$$

Equations (5.213)–(5.214) are supplemented with the equation for the external circuit (see Figure 5.48), which is written for a stationary current as

$$\varepsilon=V+IR_0, \tag{5.222}$$

where I is the total discharge current; ε is the emf of power supply, and R_0 is the external resistance, $R_0 = 300$ kΩ.

5.7.4 Numerical simulation results

A typical configuration of axisymmetric glow discharge is shown in Figure 5.49. This figure shows two-dimensional distributions of ion and electron concentrations in glow discharge without magnetic field at $\varepsilon = 2000$ V and $p = 5$ Torr. Here and further the charged particle concentrations are referred to $N_0 = 10^{10}$ cm^{-3}: $N_e = n_e/N_0$, $N_i = n_i/N_0$.

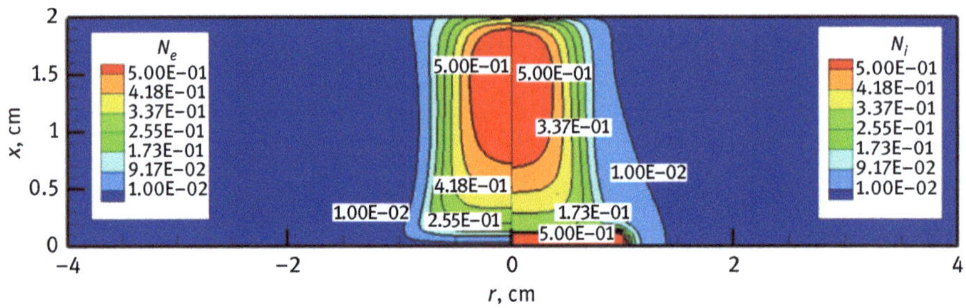

Figure 5.49: Electron (from left) and ion contours in 10^{10} cm^{-3} at $p = 5$ Torr, $\varepsilon = 2$ kV, $\gamma = 0.33$, $B_x = 0$.

Distributions of electron and ion concentrations along x-axis are shown in Figure 5.50 for the same conditions, where one can define the main areas of a glow discharge, such as:
- the area of positive charge space near to cathode (cathode layer), at $x < 0.15$ cm;
- the area of negative charge space near to anode (anode layer), at $x > 1.85$ cm;
- the area of a quasineutral plasma (positive column), at $0.2 < x < 1.85$ cm.

The same characteristic areas of the NGD can be fixed also in Figure 5.49.

As it was shown earlier, the formation of the glow discharge column occurs as a result of nonlinear interaction of the following processes occurring in the discharge gap:
- the secondary electron emission (with a factor γ), which is a consequence of bombardment of the cathode surface by ions moving from the positive column; these ions are accelerated in the cathode layer;
- birth of electrons in the discharge current column, which is localized in the potential well, as shown in Figure 5.51;
- loss of the electrons from the discharge current column due to diffusion.

As for radial diffusion of electrons, it can play a significant role in subnormal existence mode of NGD when the glow discharge is near to the minimum limit of the voltage drop, which provides its existence (Raizer, Surzhikov, 1987).

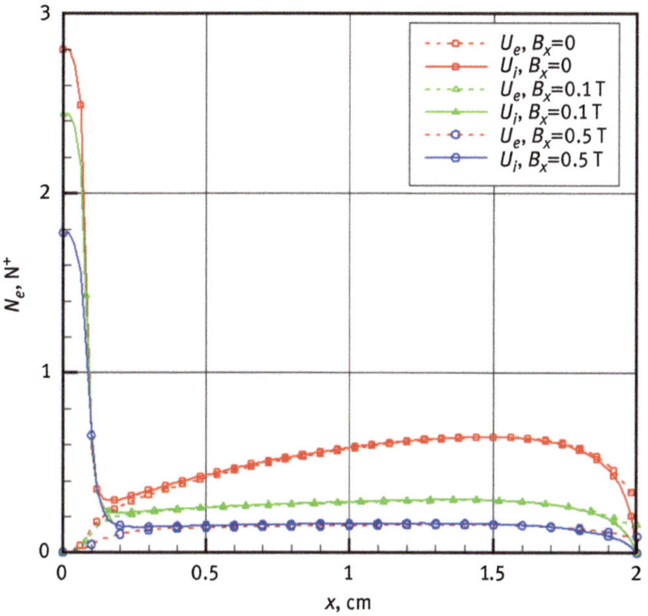

Figure 5.50: Ion and electron distributions along axis of symmetry versus magnetic field induction at $p = 5$ Torr, $\varepsilon = 2$ kV, $\gamma = 0.33$.

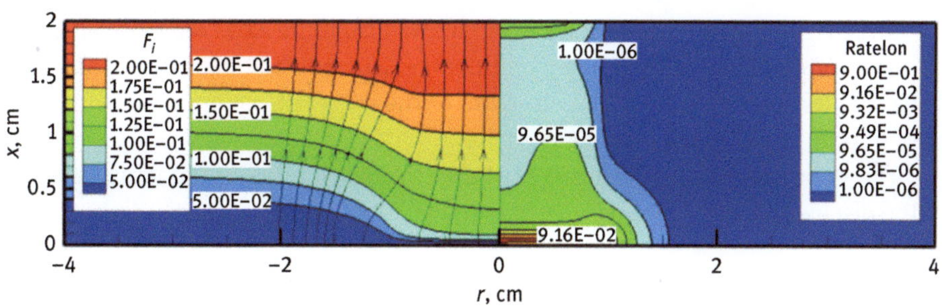

Figure 5.51: Electric potential $Fi = \varphi/V$ (from the left) and rate of ionization contours at $p = 5$ Torr, $\varepsilon = 2$ kV, $\gamma = 0.33$, $B_x = 0$.

Figure 5.51 shows the current lines in the glow discharge. The self-maintaining glow discharge can exist if there are current lines for which the Townsend condition is provided

$$\int_l \alpha(l')\,\mathrm{d}l' > \ln(1 + 1/\gamma). \tag{5.223}$$

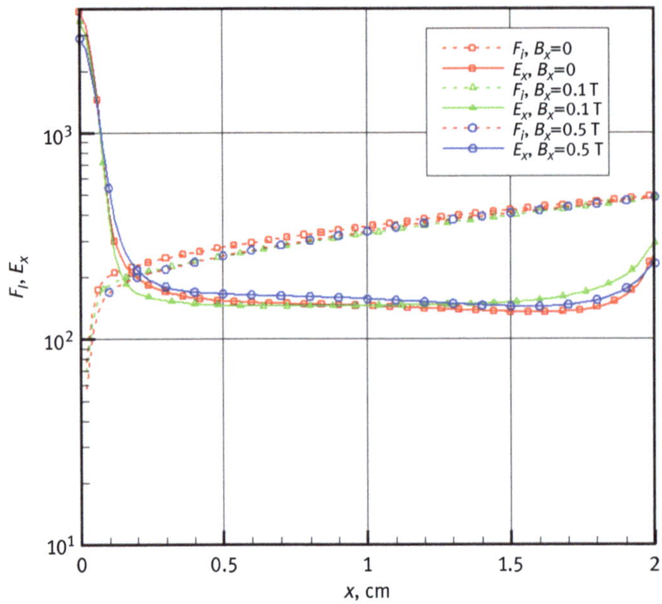

Figure 5.52: Electric potential (*Fi*, in V) and axial projection of electric field intensity (*Ex*, in V/cm) distributions along the axis of symmetry versus magnetic field induction at $p = 5$ Torr, $\varepsilon = 2$ kV, $\gamma = 0.33$.

If the current line is located far enough away from the axial region, so that the diffusion of electrons does violate condition (5.223), this line becomes currentless. Thus, the formation of the radial structure of the current is determined by the balance of births and loss of charged particles.

The right-hand side of Figure 5.51 shows the distribution of the ionization rate in the discharge gap

$$q(\alpha|\mathbf{\Gamma}| - \beta n_e n_i),$$

where $q = 2.38 \times 10^{-18}$ cm^{3}s. The normalization factor $q = t_0/N_0$ is used for dimensionless eqs. (5.213) and (5.214), $t_0 = X^2/(\mu_e \varepsilon)$ is the average time of electronic drift between cathode and anode in the electric field of ε/X.

Throughout the discharge gap, the ionization rate exceeds the rate of recombination, although in the axial region of the positive column adjacent to the anode layer the rates of ionization and recombination are quite close. Note that for the existence of a stationary discharge there cannot be the superiority of the recombination rate over the rate of ionization because there are also diffusion losses. Figure 5.51 shows that the highest rate of ionization occurs in the cathode layer.

Figure 5.52 shows distribution of electric potential along x-axis and axial projection of electric field E_x. The cathode voltage drop and quasilinear distribution of the electric potential in positive column are distinctly seen in this figure.

In the cathode layer, the electric field strength reaches 4,000 V/cm, and in the positive column, it remains approximately constant, ~180 V/cm. In the anode layer E_x increases about twice.

The data presented in Figures 5.49–5.52 will be used further for comparison with other calculation cases. Investigation of the influence of different input parameters, such as $p, \varepsilon, \gamma, R_0$, on the electrodynamic structure of a direct current glow discharge without axial magnetic field was done in Section 4.6. The main objective of this study is to identify the change in the electrodynamic structure of the NGD when the axial magnetic field is added.

Concentrations of electrons and ions in a glow discharge at $p=5$ Torr and $\varepsilon=2000$ V for two values of induction of the magnetic field B_x= 0.1 and 0.5 T are shown in Figures 5.53 and 5.54.

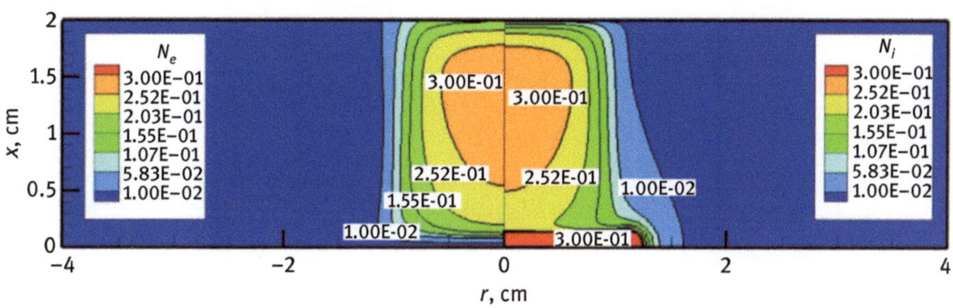

Figure 5.53: Electron (from left) and ion contours in 10^{10} cm^{-1} at $p=5$ Torr, $\varepsilon=2$ kV, $\gamma=0.33$, $B_x=0.1$ T.

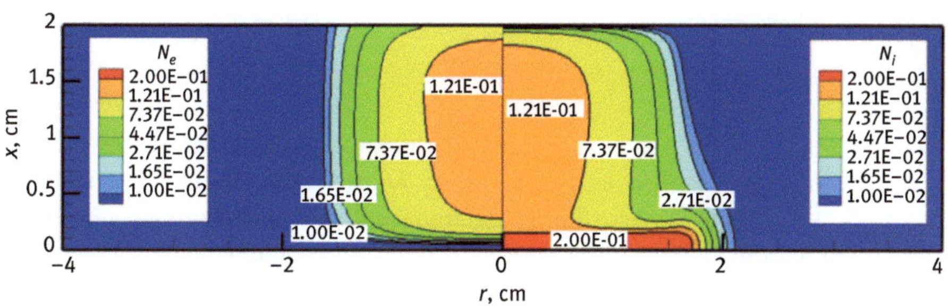

Figure 5.54: Electron (from left) and ion contours in 10^{10} cm^{-1} at $p=5$ Torr, $\varepsilon=2$ kV, $\gamma=0.33$, $B_x=0.5$ T.

Note that the increase in the radial dimensions of the current column and falling of concentrations of charged particles in the positive column of a glow discharge. This drop of concentrations of electrons and ions can be clearly seen in Figure 5.50, which shows the axial distributions of charged particle concentrations. The reduction of the concentration of ions in the cathode layer and of the maximum value of the electric field at the cathode (see Figure 5.52) is also seen.

Broadening of the discharge current column, as well as reducing of the concentrations of charged particles and the electric field at the cathode, leads to a natural reduction in the current density at the cathode; the radial distribution of which is shown in Figure 5.55.

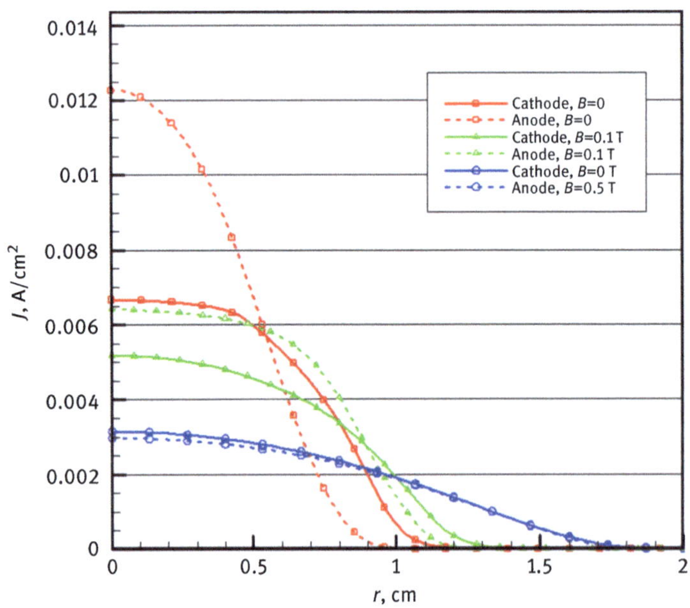

Figure 5.55: Cathode and current density versus magnetic field induction at $p = 5$ Torr, $\varepsilon = 2$ kV, $\gamma = 0.33$.

An important feature of a glow discharge in a longitudinal magnetic field is the azimuthal motion of electrons and ions, which follows from the theoretical model (5.213)–(5.218). Azimuthal velocities of electrons and ions in comparison with corresponding longitudinal velocities for two values of the magnetic field are shown in Figures 5.56–5.59. Maximal axial electron velocity, $V_{e,x} \sim 10^8$ cm/s, is achieved in the cathode layer, where the electric field is maximum.

Maximal velocity of the rotation of electronic clouds, $\sim 1.5 \times 10^7$ cm/s, is reached near the radial boundary of the cathode layer. The azimuthal velocity of electrons is several times smaller on the radial boundary of the positive column of a glow discharge. The velocity of the azimuthal motion of electrons increases proportionally to magnetic field induction (compare Figures 5.56 and 5.57).

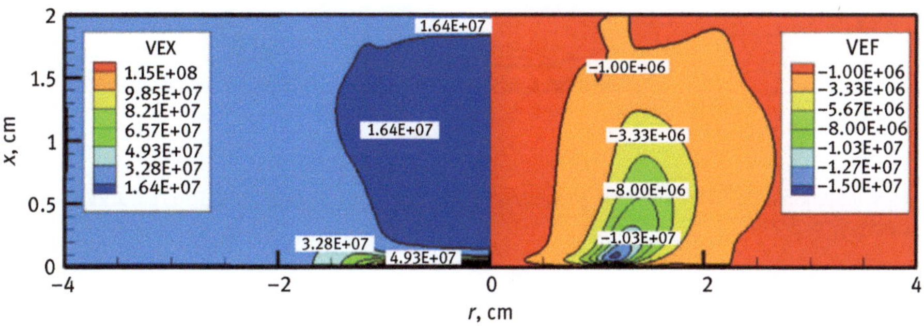

Figure 5.56: Electron axial (from the left) and azimuthal velocities in cm/s $p = 5$ Torr, $\varepsilon = 2$ kV, $\gamma = 0.33$, $B_x = 0.1$ T.

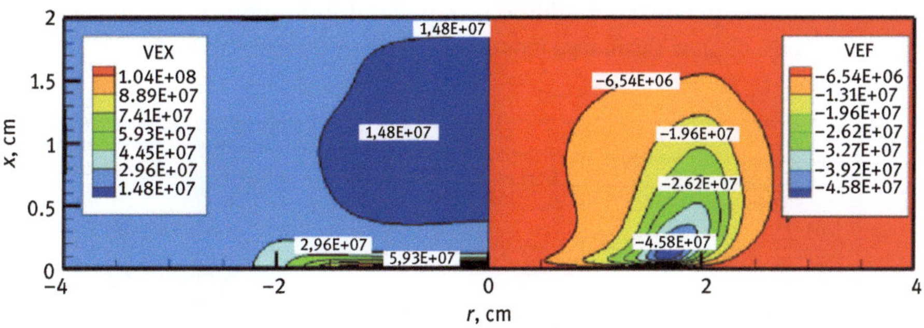

Figure 5.57: Electron axial (from the left) and azimuthal velocities in cm/s at $p = 5$ Torr, $\varepsilon = 2$ kV, $\gamma = 0.33$, $B_x = 0.5$ T.

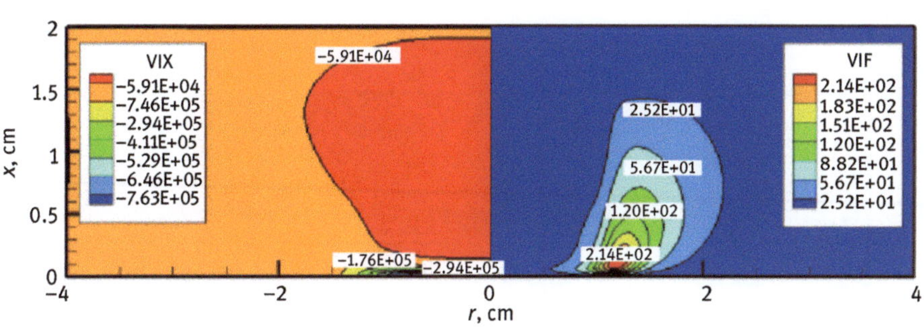

Figure 5.58: Ion axial (from the left) and azimuthal velocities in cm/s at $p = 5$ Torr, $\varepsilon = 2$ kV, $\gamma = 0.33$, $B_x = 0.1$ T.

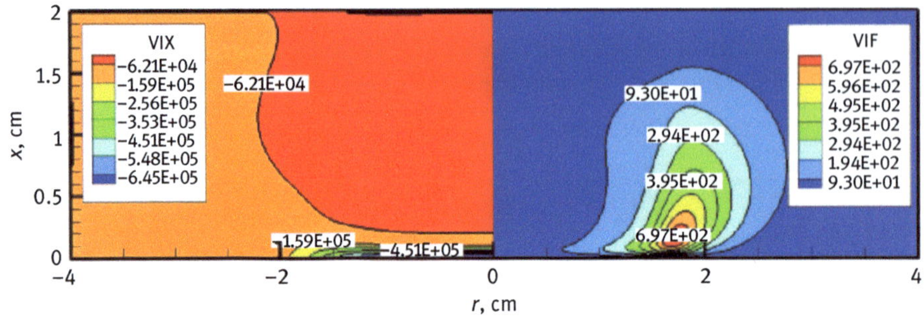

Figure 5.59: Ion axial (from the left) and azimuthal velocities in cm/s at $p = 5$ Torr, $\varepsilon = 2$ kV, $\gamma = 0.33$, $B_x = 0.5$ T.

The azimuthal motion of ionic clouds opposite to the rotational direction of electrons and their velocities vary from $V_{i,\varphi} \sim 200$ cm/s to 700 cm/s, although the speeds of their axial movement achieve (3–5) × 10^5 cm/s.

Azimuthal rotation of the electrons is a fundamental feature of a glow discharge in a longitudinal magnetic field, because in accordance with Surzhikov and Shang (2004) it actually provides an additional source of ionization on radial boundary current column of glow discharge (compare Figures 5.60 and 5.61).

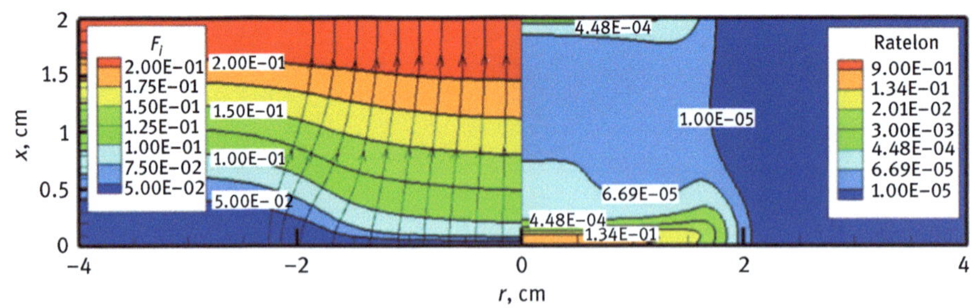

Figure 5.60: Electric potential $Fi = \varphi/V$ (from left) and rate of ionization contours at $p = 5$ Torr, $\varepsilon = 2$ kV, $\gamma = 0.33$, $B_x = 0.5$ T.

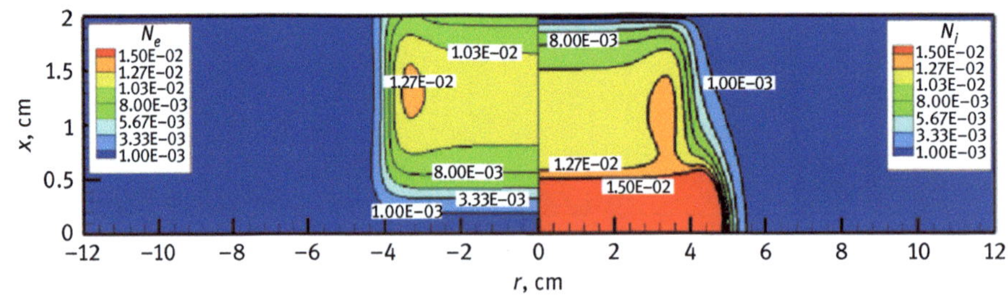

Figure 5.61: Electron (from left) and ion contours in 10^{10} cm^{-1} at $p = 1$ Torr, $\varepsilon = 2$ kV, $\gamma = 0.33$, $B_x = 0$.

This leads to a marked increase in the radial dimensions of the discharge current column. Shifting the balance between birth and loss electrons in the direction of increasing the number of charged particles leads to a decrease in the required electric field in the cathode layer (Figure 5.52) as well as the required level of concentration of electrons and ions (Figure 5.50).

With a decrease of gas pressure in the discharge gap, the degree of influence of the magnetic field on the electrodynamic structure of a glow discharge becomes larger.

Figures 5.62 and 5.63 demonstrate distributions of concentrations of electrons and ions at $p = 1$ Torr in the longitudinal magnetic field with induction T and without magnetic field. One can observe a significant increase in the radial dimensions of the gas-discharge current column. As with the higher pressure the longitudinal magnetic field decreases the concentration of electrons and ions. Note also that a toroidal structure of a glow discharge, which was previously observed at low pressures, disappears in a longitudinal magnetic field.

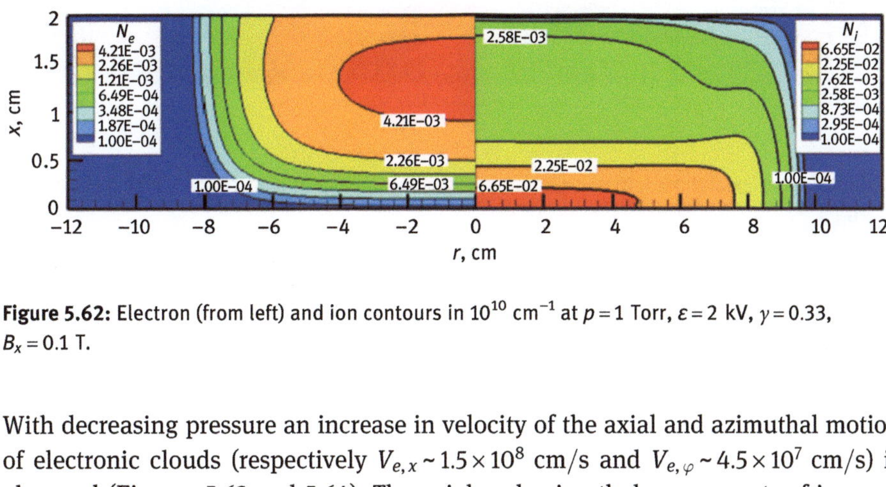

Figure 5.62: Electron (from left) and ion contours in 10^{10} cm^{-1} at $p = 1$ Torr, $\varepsilon = 2$ kV, $\gamma = 0.33$, $B_x = 0.1$ T.

With decreasing pressure an increase in velocity of the axial and azimuthal motion of electronic clouds (respectively $V_{e,x} \sim 1.5 \times 10^8$ cm/s and $V_{e,\varphi} \sim 4.5 \times 10^7$ cm/s) is observed (Figures 5.63 and 5.64). The axial and azimuthal components of ions velocity also increases slightly.

Let us analyze fulfillment of the law of normal current density as an example of calculation of the structure of a glow discharge at a pressure of $p = 1$ Torr. Figure 5.65 shows the radial distribution of the current density at the cathode and the anode for two values of emf with and without considering the longitudinal magnetic field. One can see that without a magnetic field normal current density law is well satisfied: at at increasing of the EMF value in 2 times and significant changing of the radius of the cathode spot the current density is practically remained unchanged.

In the axial magnetic field, the current density at the cathode spot is also changed insignificantly, although in these calculations the value of EMF was changed at two

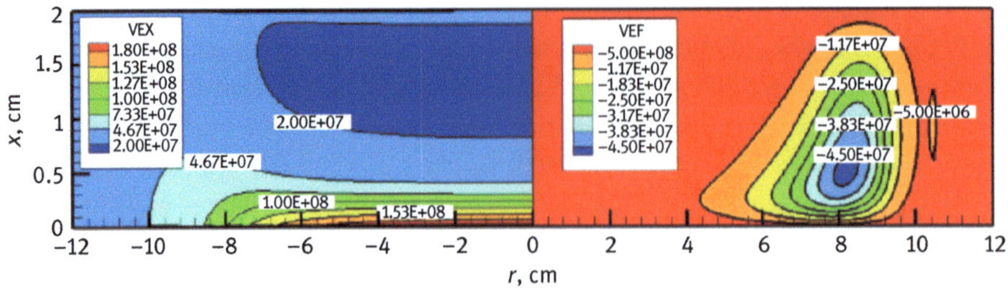

Figure 5.63: Electron axial (from left) and azimuthal velocities in cm/s at $p = 1$ Torr, $\varepsilon = 2$ kV, $\gamma = 0.33$, $B_x = 0.1$ T.

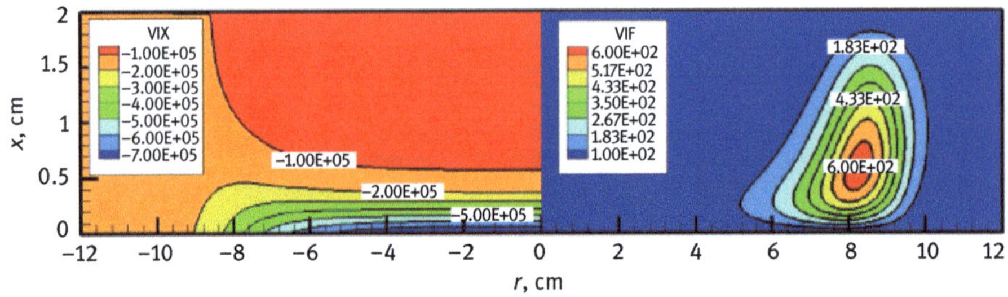

Figure 5.64: Ion axial (from left) and azimuthal velocities in cm/s at $p = 1$ Torr, $\varepsilon = 2$ kV, $\gamma = 0.33$, $B_x = 0.1$ T.

times and significantly increased radial dimensions of the current column. It is noteworthy that in this case the current density at the cathode and the anode is virtually identical. At this convergence of the radial distributions of current densities at the cathode and the anode, there is also an increase in the induction of axial magnetic field and at increased pressure (see Figure 5.55).

Thus, a theory and two-dimensional numerical simulation model intended for modeling of the electrodynamic structure of the NGDs with axial magnetic field are presented in this section.

A numerical study of the electrodynamic structure of the NGD in molecular nitrogen in the pressure range $p = 1$–5 Torr has been performed when the power supply emf $\varepsilon = 1000$–2000 V in an axial magnetic field with induction $B_x = 0$–0.5 T.

It is shown that in the investigated range of discharge parameters in an axial magnetic field the law of the normal current density is retained.

With an increasing in the magnetic field induction, the radial dimensions of the cathode and anode spots, as well as a positive column, increase.

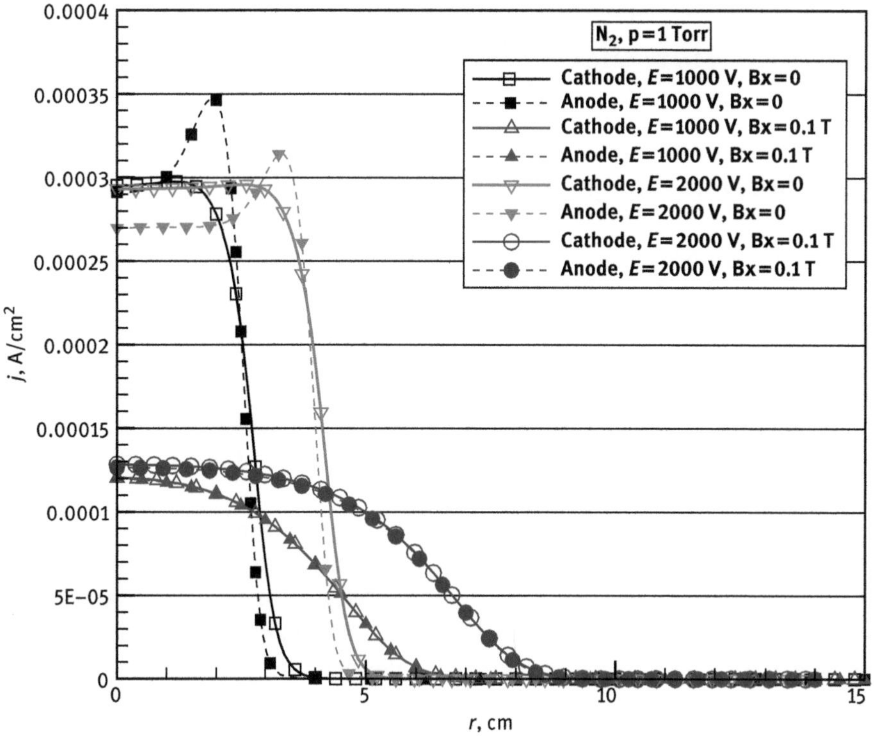

Figure 5.65: Cathode current density versus magnetic field induction and emf of power supply ε at $p = 1$ Torr, $\gamma = 0.33$.

In an axial magnetic field, there is an azimuthal rotation of electrons and ions in opposite directions with characteristic velocities $V_{e,\varphi} \sim 10^7$ cm/s and $V_{i,\varphi} \sim 10^3$ cm/s. This azimuthal rotation of electrons results in additional source of ionization in direct current discharge.

6 The radiofrequency capacitive normal glow discharge

6.1 The radiofrequency capacitive normal glow discharge in transverse magnetic field

The problems associated with computer simulation of a two-dimensional structure of a radiofrequency capacitive (RFC) glow discharge are discussed. The diffusion-drift computational model of the discharge, burning in a quasistationary mode, was used to analyze the behavior of the RFC discharge in a transverse magnetic field. The calculations were performed for a two-dimensional RFC glow discharge at a pressure of 5 – 10 Torr and an emf of 520 – 1000 V power source with a frequency of 13.59 MHz. The magnetic field vector with induction $B = 0.2\,\mathrm{T}$ is directed across the current column of the RFC discharge.

The possibility of using electromagnetic methods for modifying partially ionized gas flows in various aerospace technologies has been discussed in the literature for more than 60 years. One of the possible ways of such a modification is the use of glow gas discharge of continuous current and high-frequency discharges of induction and capacitive type, well known in physics. These types of discharges are characterized by relatively low energy costs and ease of organization.

An important feature of these types of discharges is that, when the pressure in gas flows is of the order of 1 – 10 Torr, they are fairly uniform, and the characteristic time of their formation is microseconds, which is noticeably less than the time of formation of a gas-dynamic structure.

In this section, the studies described in the previous chapters in the field of numerical simulation of glow discharges in rarefied gases and flows are continued with respect to a RFC glow discharge burning in a quasistationary mode at a pressure of $p = 5 - 10$ Torr. The discharge circuit is shown in Figure 6.1. The quasistationary mode of existence of a discharge is characterized by the fact that after its formation from some initial plasma cloud that was localized in the interelectrode gap, which is used as initial calculation data (the process of the formation of a current column takes several microseconds), the RFC glow discharge retains its configuration for a much longer period of time (more ~ 100 μs), resulting in a gas discharge gap. As a rule, in calculations, there is a slow expansion of the current column while preserving the internal electrodynamic structure of the discharge in near-electrode regions of the space charge and a positive column.

In addition to a direct current (DC) glow discharge, the RFC glow discharge under investigation is characterized by a low degree of ionization of the rarefied gas, $n_i/n_n \sim 10^{-5}$ (n_i, n_n are the numerical concentrations of ions and neutral particles), as well as by high degree of nonequilibrium at which the temperature of neutral particles and ions is close to the room temperature, while the electrons are

https://doi.org/10.1515/9783110648836-007

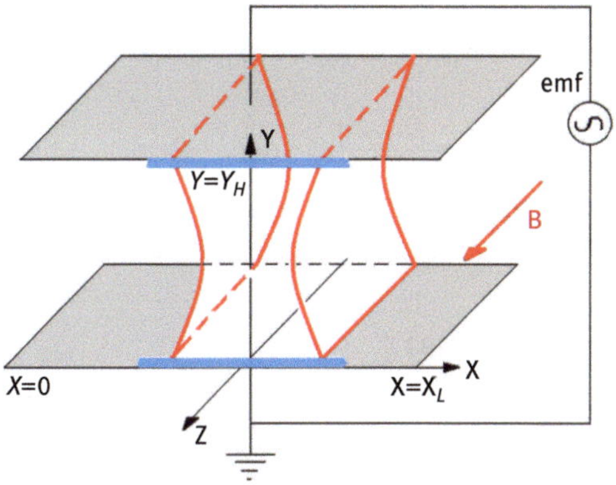

Figure 6.1: RFC discharge schematic in an external magnetic field.

heated by an external electric field up to 20,000–30,000 K. The normal current density mode is supported by a balance between the ionization, recombination and the diffusion of the charged particles in the column of electrical discharge. The results of numerical studies of the one-dimensional structure of the RFC discharge are published in Raizer Yu.P., et al. (1995).

Note that the high-frequency electric field leads to the possibility of the existence of a greater variety of current structures than in the classical DC glow discharge. In particular, two forms of the RFC glow discharge are observed in experiments (Levitskii S.M., 1957): the so-called α- and γ-forms of the discharge.

In this section, a quasistationary discharge forms are obtained by calculation, and the evolution of one of the discharge forms in a transverse magnetic field is investigated. We emphasize that this section discusses the numerical solutions obtained for the quasistationary RFC glow discharge at time intervals $< 200\ \mu s$.

6.1.1 Governing equations

A two-dimensional structure of an RFC glow discharge in molecular nitrogen, which exists between two infinite flat electrodes, is considered (see Figure 6.1). The problem is solved in a rectangular Cartesian coordinate system so that the discharge structure is an infinite layer in the direction of the z-axis.

The electrodynamic structure of the discharge is described by the continuity equations for the volume density of electrons n_e and ions n_i, together with the Poisson equation $\mathbf{E} = -\text{grad}\varphi$ for the electric field, as well as the heat equation for neutral particles. This system of equations is presented herein in a convenient form for numerical implementation in (Surzhikov S.T., et al., 2004c)

$$\frac{\partial n_e}{\partial t} + \frac{\partial}{\partial x}\left(\mu_e n_e E_{e,x} - \frac{D_e}{1+b_e^2}\frac{\partial n_e}{\partial x}\right) + \frac{\partial}{\partial y}\left(\mu_e n_e E_{e,y} - \frac{D_e}{1+b_e^2}\frac{\partial n_e}{\partial y}\right) = \alpha|\Gamma_e| - \beta n_e n_i, \quad (6.1)$$

$$\frac{\partial n_i}{\partial t} + \frac{\partial}{\partial x}\left(\mu_i n_i E_{i,x} - \frac{D_i}{1+b_i^2}\frac{\partial n_i}{\partial x}\right) + \frac{\partial}{\partial y}\left(\mu_i n_i E_{i,y} - \frac{D_i}{1+b_i^2}\frac{\partial n_i}{\partial y}\right) = \alpha|\Gamma_e| - \beta n_e n_i, \quad (6.2)$$

$$\frac{\partial^2 \varphi}{\partial x^2} + \frac{\partial^2 \varphi}{\partial y^2} = 4\pi e(n_e - n_i), \quad (6.3)$$

$$\rho c_p \frac{\partial T}{\partial t} = \frac{\partial}{\partial x}\left(\lambda \frac{\partial T}{\partial x}\right) + \frac{\partial}{\partial y}\left(\lambda \frac{\partial T}{\partial y}\right) + q_J, \quad (6.4)$$

$$\Gamma_e = n_e \mathbf{V} = -D_e \mathrm{grad} n_e - n_e \mu_e (\mathbf{E} + \mathbf{V} \times \mathbf{B}), \quad (6.5)$$

$$\Gamma_i = n_i \mathbf{V} = -D_i \mathrm{grad} n_i + n_i \mu_i (\mathbf{E} + \mathbf{V} \times \mathbf{B}), \quad (6.6)$$

where D_e, D_i are the diffusion coefficients of electrons and ions; p, T are the pressure and temperature of neutral gas; φ, e are the electric potential and charge of an electron; $\mathbf{\Gamma}_e, \mathbf{\Gamma}_i$ are the density of electron and ion fluxes; $|\Gamma_e| = \sqrt{\Gamma_{e,x}^2 + \Gamma_{e,y}^2}$; $\mathbf{B}$ is the vector of magnetic field induction (the direction of this vector is shown in Figure 6.1); $q_J = \eta(\mathbf{jE})$; $\mathbf{j} = e(\Gamma_i - \Gamma_e)$; $\mathbf{V}$ is the gas velocity; $\alpha(E)$, β are the ionization and recombination coefficients; μ_e, μ_i are the mobility of electrons and ions; η is the part of Joule heating, which goes into heat (a significant part of the energy goes into the excitation of the vibrational degrees of freedom of molecular nitrogen); $\mathbf{B} = (0,\ 0,\ B_z)$ is the magnetic field induction

$$b_e = \frac{\mu_e B_z}{c} = \frac{\omega_e}{\nu_e},\ b_i = \frac{\mu_i B_z}{c} = \frac{\omega_i}{\nu_{in}}, \quad (6.7)$$

are the Hall parameters for electrons and ions; $\omega_e = \frac{eB_z}{m_e c}$, $\omega_i = \frac{eB_z}{m_i c}$ are the Larmor radii of electrons and ions; ν_e, ν_{+n} are the collision frequencies of electrons and ions in a partially ionized gas; m_e, m_i are the mass of electrons and ions; c is the speed of light.

Equations (6.1)–(6.4) are integrated with the equation for the potential of the anode relative to the zero potential of the cathode. In the calculations, it is possible to implement two modes of maintaining RFC glow discharge. In the first case, to determine the potential of the anode with respect to the potential of the cathode, the equation of the external current circuit is used (Gladush G.G., Samokhin A.A., 1986):

$$\frac{E_{mf} - V(t)}{R_0} + \int_{-\frac{1}{2}X_L}^{+\frac{1}{2}X_L} \left(j_{y,i} - j_{y,e}\right)_{y=Y_H} Z_0 \mathrm{d}x = \frac{\mathrm{d}Q}{\mathrm{d}t}, \quad (6.8)$$

$$\varepsilon_0 Q(t) = \int_{-\frac{1}{2}X_L}^{+\frac{1}{2}X_L} \left(\frac{\partial \varphi}{\partial y}\right)_{y=Y_H} Z_0 \mathrm{d}x, \quad (6.9)$$

$$\varepsilon(t) = E_0 \mathrm{Sin}(2\pi f t), \tag{6.10}$$

where $\varepsilon_0 = 1.81 \times 10^{-6}$ B · cm; $V(t)$ is the potential drop at the discharge gap, E_{mf} is the emf power source; $Q(t)$ is the amount of excess ions at the anode, due to the current imbalance in the electrical discharge gap and the external circuit; R_0 is the Ohmic resistance of an electrical circuit; $Z_0 = 1\,\mathrm{cm}$; f is the potential change frequency; t is the time.

In the second case, it is assumed that the potential of the anode varies according to a given law, for example

$$V(t) = E_0 \mathrm{Sin}(2\pi f t). \tag{6.11}$$

This section presents examples of the implementation of condition (6.11).

The boundary conditions for particle concentrations and for electric potential are given in the following form:

$$y = 0, \quad \left.\frac{\partial \varphi}{\partial y}\right|_{y=0} > 0: \ \frac{\partial n_i}{\partial y} = 0, \ \Gamma_{y,e} = \gamma \Gamma_{y,i}, \ \varphi = 0; \tag{6.12}$$

$$y = H, \quad \left.\frac{\partial \varphi}{\partial y}\right|_{y=H} > 0: \ n_i = 0, \ \frac{\partial n_e}{\partial y} = 0, \ \varphi = E(t); \tag{6.13}$$

$$y = 0, \quad \left.\frac{\partial \varphi}{\partial y}\right|_{y=0} < 0: \ n_i = 0, \ \frac{\partial n_e}{\partial y} = 0, \ \varphi = 0; \tag{6.14}$$

$$y = H, \quad \left.\frac{\partial \varphi}{\partial y}\right|_{y=H} < 0: \ \frac{\partial n_i}{\partial y} = 0, \ \Gamma_{y,e} = \gamma \Gamma_{y,i}, \ \varphi = E(t); \tag{6.15}$$

$$x = 0: \ \frac{\partial n_e}{\partial x} = \frac{\partial n_i}{\partial x} = \frac{\partial \varphi}{\partial x} = 0; \tag{6.16}$$

$$x = L: \ \frac{\partial n_e}{\partial x} = \frac{\partial n_i}{\partial x} = \frac{\partial \varphi}{\partial x} = 0. \tag{6.17}$$

In (6.12) and (6.15) γ is the coefficient of secondary electron emission.

It is assumed that the thermophysical and transport properties of the neutral particles of the discharge depend on temperature; therefore:

$$\mu_e(p^*) = \frac{4.2 \times 10^5}{p^*}\ \mathrm{cm}^2/(\mathrm{V} \cdot \mathrm{s}), \ \ \mu_i(p^*) = \frac{1140}{p^*}\ \mathrm{cm}^2/(\mathrm{V} \cdot \mathrm{s}),$$

$$p^* = p\frac{293}{T}\ \mathrm{Torr},$$

$$D_e = \mu_e(p^*) T_e, \ D_i = \mu_i(p^*) T\ \mathrm{cm}^2/\mathrm{s},$$

$$c_p = 8.314\frac{7}{2}\frac{1}{M_\Sigma}\ \mathrm{J}/(\mathrm{g} \cdot \mathrm{K}), \ M_\Sigma = 28\ \mathrm{g/mole}, \ \rho = 1.58 \times 10^{-5}\frac{M_\Sigma p}{T}\ \mathrm{g/cm}^3,$$

$$\lambda = \frac{8.334 \times 10^{-4}}{\sigma^2 \Omega^{(2.2)*}} \sqrt{\frac{T}{M_\Sigma}} \left(0.115 + 0.354 \frac{c_p M_\Sigma}{\tilde{R}}\right) \text{W}/(\text{cm} \cdot \text{K}), \ \Omega^{(2.2)*} = \frac{1.157}{(T^*)^{0.1472}}, \ T^* = \frac{T}{(\varepsilon/k)},$$

$$(\varepsilon/k) = 71.4\text{K}, \ \sigma = 3.68 \ \text{Å}, \ \tilde{R} = 8.314 \ \text{J}/(\text{K} \cdot \text{mole});$$

p is the undisturbed pressure in the environment; $N = 0.954 \times 10^{19} (p/T)$, cm^{-3} is the volume concentration of neutral particles. The recombination coefficient β is considered to be constant, $\beta = 2 \times 10^{-7}\ \text{cm}^3/\text{s}$.

The electron temperature T_e is calculated using the empirical relationship (Surzhikov, et al., 2005):

$$\frac{T_e}{T} = 29.96 \ln\left(\frac{E}{p}\right) + 24.64, \tag{6.18}$$

where E/p is the reduced electric field (is the discharge parameter), V/(cm · Torr). Note that relation (6.18) is extrapolating for the cathode layer.

Ionization coefficient (the first Townsend coefficient) is given in the following form:

$$\alpha(E) = p^* A \exp\left[-\frac{B}{(|\mathbf{E}|/p^*)}\right], \ 1/(\text{cm} \cdot \text{Torr}), \tag{6.19}$$

where

$$A = 12\,1/(\text{cm} \cdot \text{Torr}), \ B = 342\,\text{V}/(\text{cm} \cdot \text{Torr}).$$

6.1.2 Results of numerical simulation

The electrodynamic structure of the RFC glow discharge in molecular nitrogen at a pressure of $p = 5 - 10$ Torr studied using a numerical simulation method. The following parameters of the discharge were used: the amplitude values of the emf of power supply $E_0 = 520 - 1000$ V, the distance between the flat electrodes (cm) and the width of the gas discharge gap (cm; see Figure 6.1). The magnetic field induction vector is directed along the z-axis, and its modulus is $B_z = 0.2$ T. The numerical simulation was performed per 1 cm of the length of the discharge gap in the direction of the z-axis. A nonuniform structured grid $n_x \times n_y = 101 \times 51$ was used (n_x, n_y are the number of nodes of the computational grid along the x- and y-axes).

The initial conditions in the interelectrode gap were set in two stages. At first, a quasineutral plasma cloud of the spherical shape with the concentration of charged particles of $n_0 = 10^{11}\ \text{cm}^{-3}$ was located at $x_0 = 3$ cm near the cathode. The formation of glow discharge of continuous current in the regime of normal current density in the absence of a magnetic field was observed for ~10 μs. The developed numerical method ensured the symmetry of the numerical solution in the directions of the x- and y-axes, which is important in simulating the RFC discharge. At the second stage, the obtained numerical solution for a stationary glow discharge of DC was used as the initial condition in simulating an RFC discharge with similar parameters. We draw attention to an important fact: the initial condition for simulating an RFC discharge is a normal glow discharge at the center of the computational domain far from the lateral boundaries.

The first series of calculations of the RFC discharge was performed at $p = 5$ Torr, $E_0 = 520$ V without an external magnetic field. The two-dimensional electrodynamic structure of the RFC discharge is shown in Figures 6.2 and 6.3 for the two phases of the anode potential.

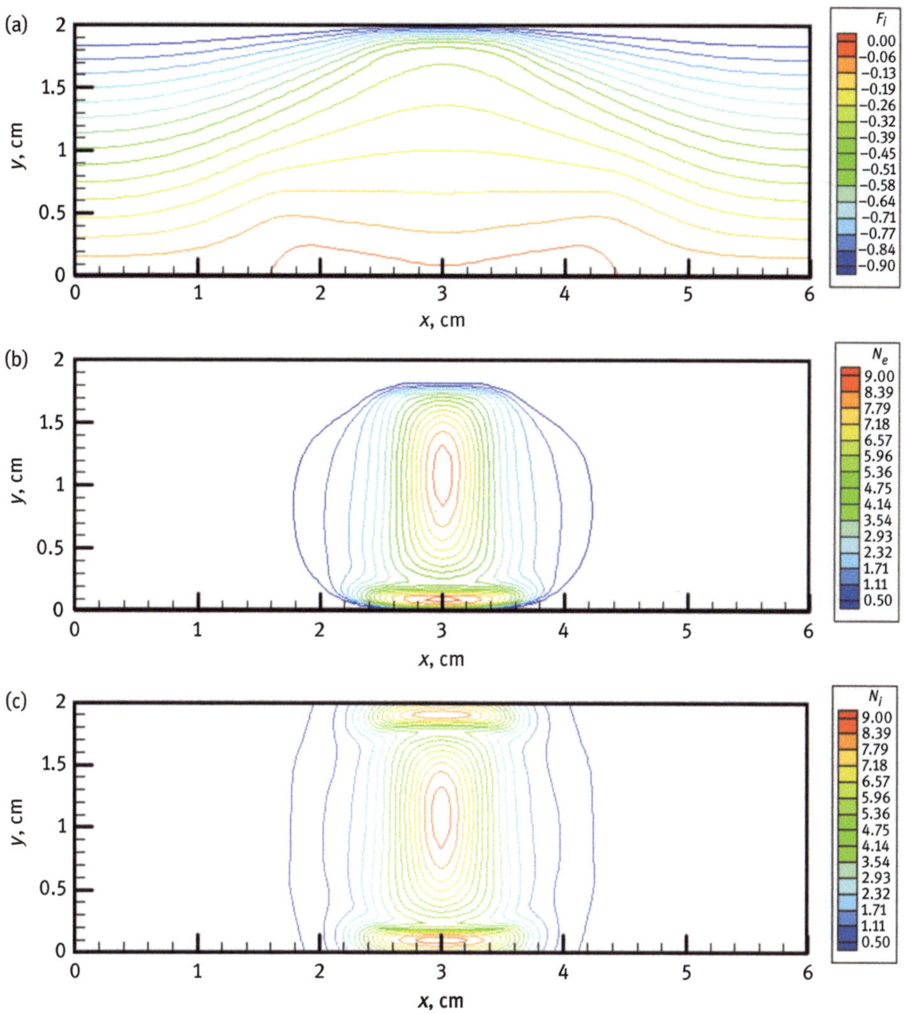

Figure 6.2: Electrodynamic structure of the RFC discharge at $p = 5$ Torr, $E_0 = 520$ V, $f = 13.59$ MHz in phase A: (a) electric potential $\Phi = \varphi/E_0$, (b) $N_e = n_e \times 10^{-9}$ cm^{-3}; (c) $N_i = n_i \times 10^{-9}$ cm^{-3}.

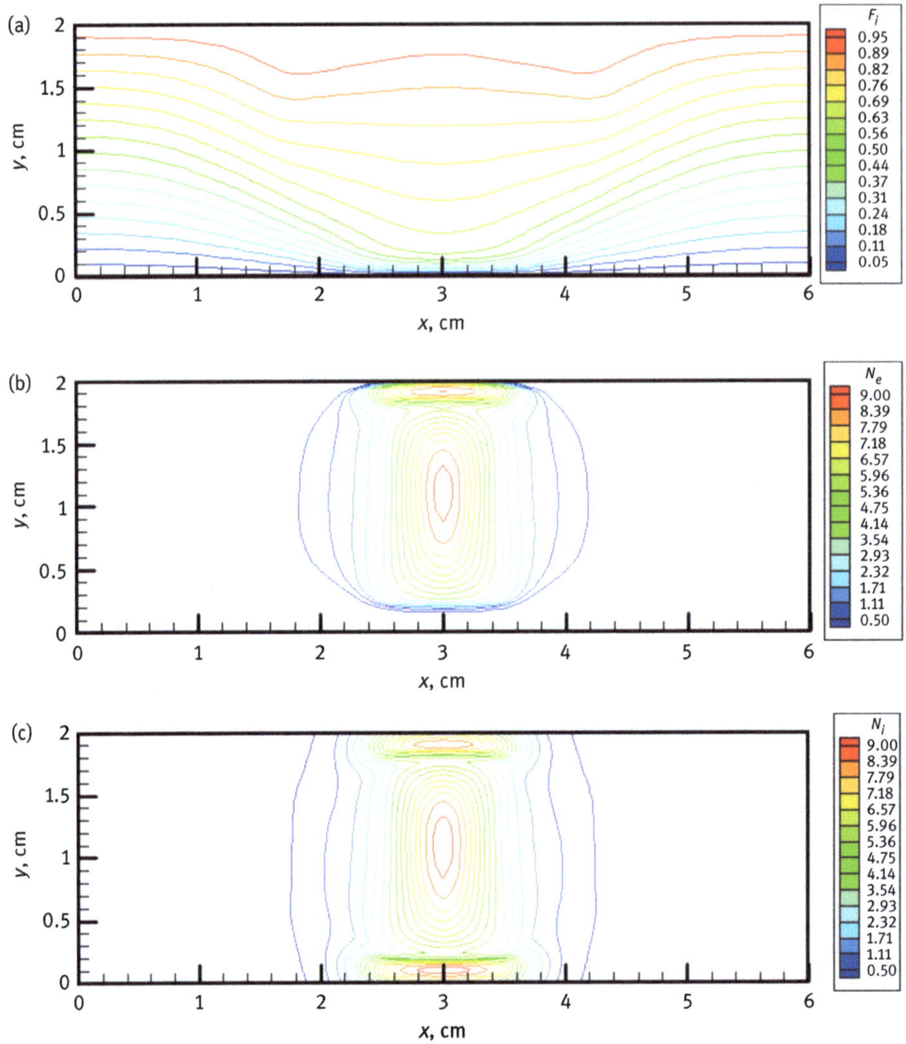

Figure 6.3: Electrodynamic structure of the RFC discharge at $p = 5$ Torr, $E_0 = 520$ V, $f = 13.59$ MHz in phase B: (a) electric potential $\Phi = \varphi / E_0$, (b) $N_e = n_e \times 10^{-9}\,\mathrm{cm}^{-3}$; (c) $N_i = n_i \times 10^{-9}\,\mathrm{CM}^{-3}$.

Phase A corresponds to the negative potential of the anode, and phase B to the positive potential. Figures 6.2a and 6.3a show the distribution of electric potential and Figures 6.2b, c and 6.3b, c show electron and ion concentrations. The configuration of the positive column of a quasineutral plasma is clearly visible, which in fact does not change during the anode polarity reversal. Electron concentrations oscillate in the electric discharge gap in accordance with the oscillations of the electric potential. Significantly heavier ions remain almost immovable.

The second series of calculations gives an idea about the change in the structure of the RFC discharge when the conditions change: when the pressure rises and the peak value of the voltage between the electrodes. Figures 6.4–6.7 show the results of numerical simulation of the RFC discharge structure at a pressure of $p = 10$ Torr and an amplitude of emf $E_0 = 1000$ V. As before, the magnetic field was not taken into account. Note that in this case, the amplitude value of the current is ~40 mA, which significantly exceeds this value in the case of previous calculation (~2 mA). As before, the distribution of electrons in the gas discharge gap oscillates in accordance with the electric potential between the electrodes. In the case under consideration, the thickness of the space charge regions (the cathode and anode layers) decreased significantly when compared with the previous case, and the concentration of ions in them markedly increased. Pronounced near-electrode layers were formed near the surfaces of the electrodes.

If we use a qualitative classification of the forms of existence of RFC discharges (Gladush G.G., Samokhin A.A., 1986), then the two categories considered can be conditionally attributed to α- and γ-forms. In the latter case, of the two considered, the discharge exists in γ-form. The calculations did not reveal the evolution of the electric current column, as in the first case. The discharge was simulated up to ~500 μs. This allows us to hope that a stationary configuration of an RFC discharge of γ-form is obtained.

The inclusion of a transverse magnetic field noticeably changes the electrodynamic structure of a glow discharge. Figures 6.6 and 6.7 show the results of numerical simulation under the following conditions: $p = 5$ Torr, $E_0 = 520$ V, $B_z = 0.2$ T. The obtained calculated data show that the fields of electron and ion concentrations stabilize in space during the transition from phase A to phase B.

From Figures 6.6 and 6.7 it is seen that the maximum concentration of charged particles decreases when compared with the case of the absence of a magnetic field. An important consequence of the imposition of an external transverse magnetic field is the disappearance of the near-electrode layers of the space charge in the form in which they are shown in Figures 6.2 and 6.3. As a result, the electric field intensity decreases in the near-electrode zone.

The plasma structure of the RFC discharge with a magnetic field is long-lived, but evolving in the direction transverse to the electric current channel. In the time interval of ~200 μs, the highest concentration in the positive column decreased by about five times.

Thus, the analysis of the results of numerical modeling of a two-dimensional structure of the RFC discharge, existing between two flat electrodes in two forms, is performed. In the calculations, two burning modes of the RFC discharge are identified. A quasistationary solution was obtained for an RFC discharge of α-modification that exists as a localized, slowly evolving plasma current column in the interelectrode gap with parameters close to the normal glow discharge. The second solution was

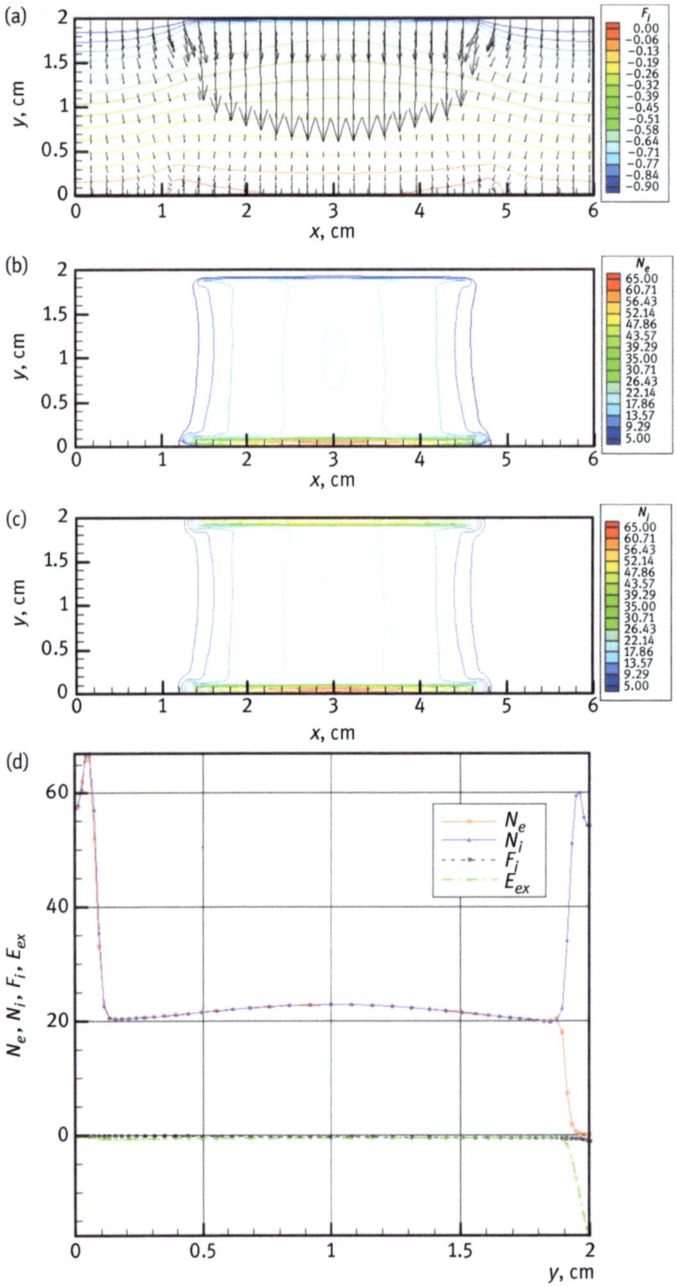

Figure 6.4: Electrodynamic structure of the RFC discharge at $p = 10$ Torr, $E_0 = 1000$ V, $f = 13.59$ MHz in phase A: (*a*) electric potential $\Phi = \varphi/E_0$, (*b*) $N_e = n_e \times 10^{-9}$ cm^{-3}, (*c*) $N_i = n_i \times 10^{-9}$cm^{-3}, (*d*) axial distributions of ions and electrons ($N_{e,i} = n_{e,i} \times 10^{-9}$CM^{-3}), electric potential φ, and electric field strength $E = (\partial\varphi/\partial y)(H/E_0)$.

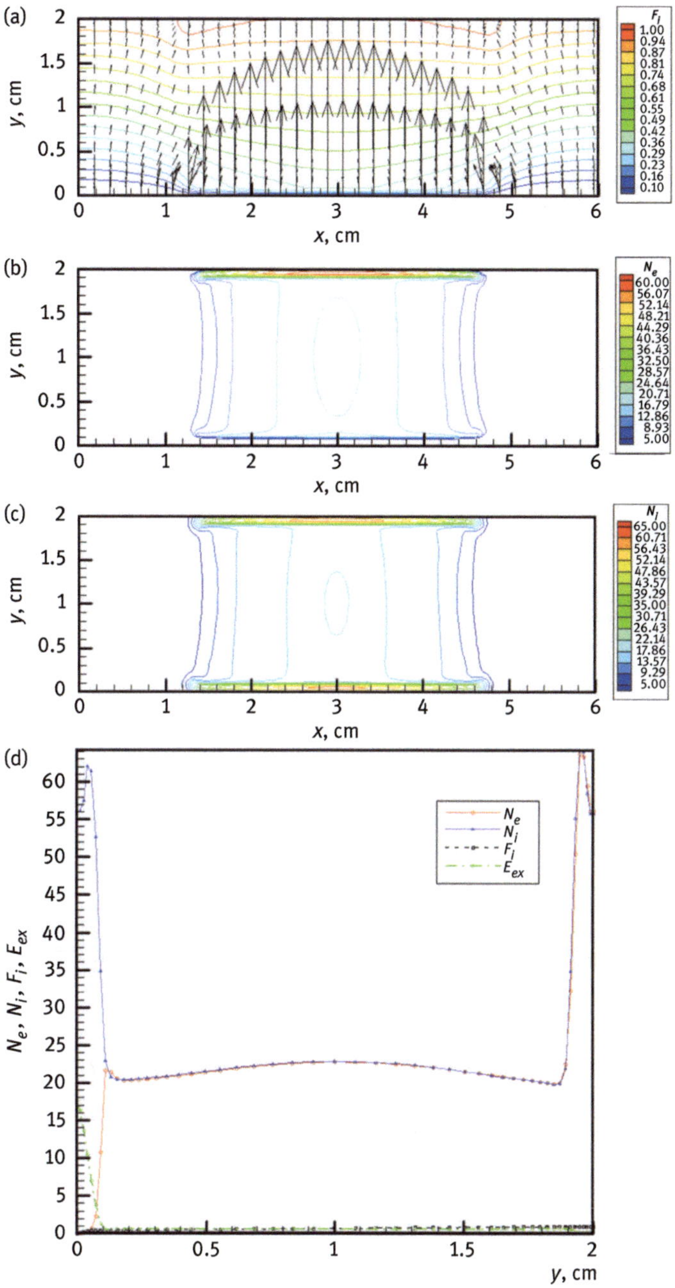

Figure 6.5: Electrodynamic structure of the RFC discharge at $p = 10$ Torr, $E_0 = 1000$ V, $f = 13.59$ MHz in phase B: (*a*) electric potential $\Phi = \varphi/E_0$, (*b*) $N_e = n_e \cdot 10^{-9}\,\text{cm}^{-3}$, (*c*) $N_i = n_i \cdot 10^{-9}\,\text{cm}^{-3}$, (d) axial distributions of ions and electrons ($N_{e,i} = n_{e,i} \times 10^{-9}\,\text{cm}^{-3}$), electric potential φ, and electric field strength $E = (\partial\varphi/\partial y)(H/E_0)$.

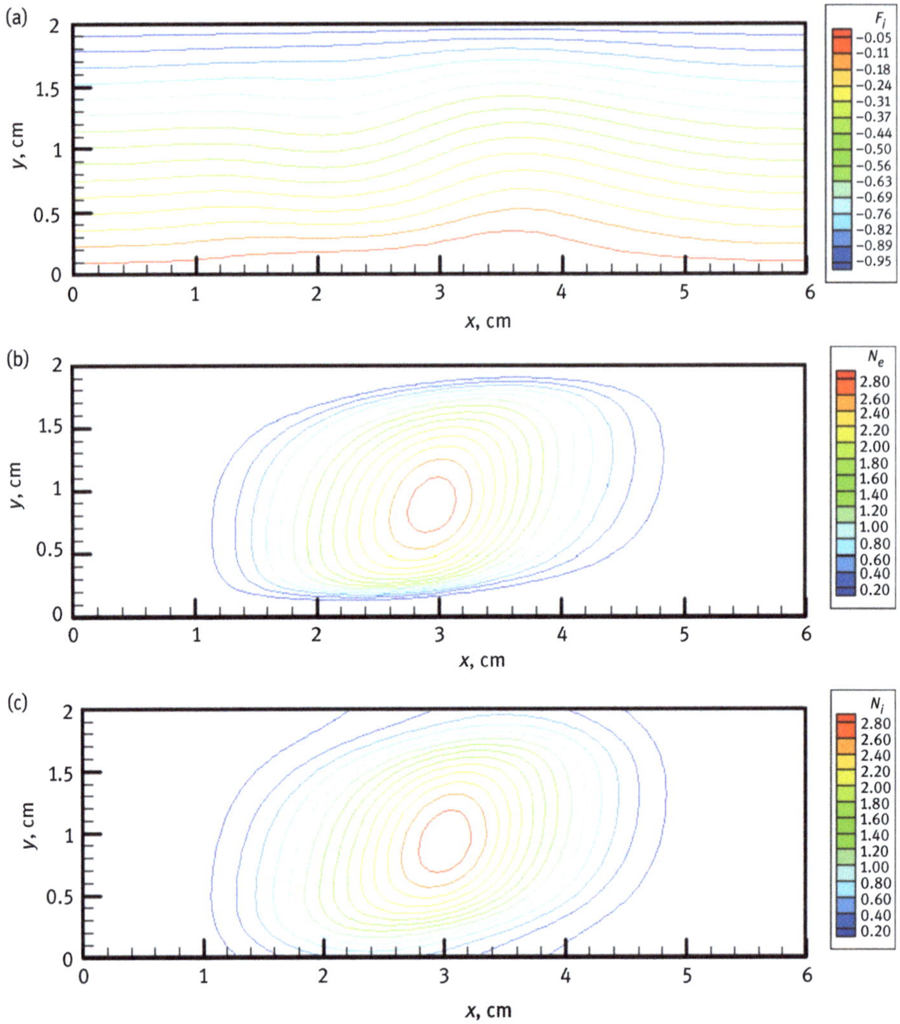

Figure 6.6: Electrodynamic structure of the RFC discharge at $p = 5$ Torr, $E_0 = 550$ V, $f = 13.59$ MHz, $B_z = 0.2$ T in phase A: (a) electric potential $\Phi = \varphi/E_0$, (b) $N_e = n_e \times 10^{-9}\,\text{cm}^{-3}$, (c) $N_i = n_i \times 10^{-9}\,\text{cm}^{-3}$.

obtained for the RFC discharge, which can be associated with the γ-form of the RFC discharge. The structure of this form of the RFC discharge remained unchanged for at least ~500 μs.

An external transverse magnetic field with induction $B_z = 0.2$ T changes the current structure of the γ-form of RFC discharge, leading to the disappearance of spatial charge regions in the near-electrode regions.

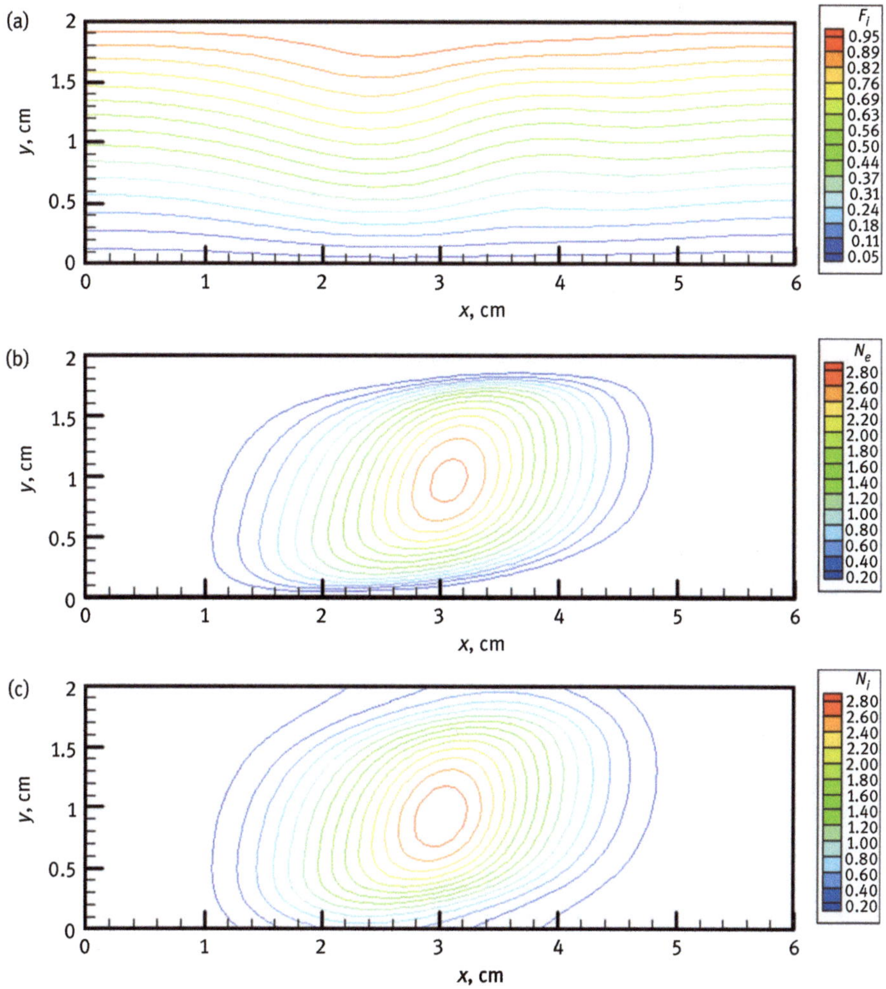

Figure 6.7: Electrodynamic structure of the RFC discharge at $p = 5$ Torr, $E_0 = 550$ V, $f = 13.59$ MHz, $B_z = 0.2$ T in phase B: (a) electric potential $\Phi = \varphi/E_0$, (b) $N_e = n_e \times 10^{-9}\,\text{cm}^{-3}$, (c) $N_i = n_i \times 10^{-9}\,\text{cm}^{-3}$.

The experience of conducting numerical studies of the two-dimensional structure of the RFC discharge in the presented formulation, as well as the numerical simulation results presented in this chapter, show that the electrodynamic structure of the RFC discharge obtained in a numerical solution is very sensitive to such source data as the pressure, the voltage amplitude on the electrodes. All this is in the correct qualitative agreement with experimental data and one-dimensional numerical calculations.

6.2 The radiofrequency capacitive normal glow discharge in axial magnetic field

A numerical model of a normal axisymmetric high-frequency capacitive glow discharge in an axial magnetic field is considered in this section. The analysis of previous investigations on α- and γ-forms of high-frequency discharges was performed. Using the authors-created numerical model, a study of the electrodynamic structure of a normal high-frequency axially symmetric glow discharge was performed with the following parameters: distance between electrodes of $H = 2\,\text{cm}$, pressure of $p = 5\,\text{Torr}$, amplitude voltage drop on the interelectrode gap of $E_0 = 460$ and 478 V, and coefficient of the secondary ion–electron coefficient emissions of $\gamma = 0.33$. The two-dimensional electrodynamic structure of the γ-form of the high-frequency glow discharge was studied without and with an axial magnetic field induction of $B_x = 0.1\,\text{T}$. A qualitative explanation is given of the reasons for the radial expansion of the current channel in a normal mode due to a high-frequency electric field oscillation.

The theory of a normal glow discharge in an axial magnetic field was discussed in Chapter 5. Using this theory and the computer program developed by the authors, the electrodynamic structure of a normal glow discharge in molecular nitrogen was studied in the pressure range $p = 1-5\,\text{Torr}$. In the above calculations, an external electrical circuit with Ohmic resistance $R_0 = 300\,\text{k}\Omega$ and emf of power supply of $E = 1-2\,\text{kV}$ was taken into account. It was found that with the induction of an axial magnetic field of $B_x = 0.1-0.5\,\text{T}$, the distinctive properties of a normal glow discharge are retained: when the current column expands due to an increase in total current through the gas discharge gap, the current density at the center of the cathode spot remains almost unchanged. The calculations showed an increase in the transverse radial dimensions of the current column with an increase in the absolute value of the magnetic field induction.

A new property of a normal glow discharge placed in an axial magnetic field, namely the rotation of ions and electrons around the axis of symmetry of the discharge, was also noted. At the same time, the question of the appearance of an additional ionization source in the discharge gap, associated with the observed rotation of charged particles around the axis of symmetry, was analyzed.

In this section, a computer model of an RFC glow discharge in an axial magnet field is constructed and analyzed.

In one of the first papers devoted to the numerical study of the two-dimensional structure of RFC glow discharge (Gladush G.G., et al. 1986), the problem of the stability of such a discharge in the regime of normal current density at $p = 5\,\text{atm}$ was studied. The external electrical circuit took into account the Ohmic resistance $R_0 = 250\,\text{k}\Omega$ and the source of alternating voltage $E = E_0\,\text{Sin}\omega t$, $E_0 = 700\,\text{V}$, $\omega = 2\pi f\,\text{MHz}$, $f = 1$.

It was mentioned earlier that the choice of parameters of the external circuit is very important for studying RFC glow discharge since there are at least two of its forms: the high-current discharge (γ-form) and the low-current discharge (α-form). These discharge forms were studied experimentally (Levitskii S.M., 1957; Yatsenko N.A., 1981).

Among the criteria for the transition of one form of discharge to another, a comparison of values of the bias current in the electrode layer with the conduction current (Smirnov A.S., 1984) or the layer breakdown condition (Raizer Yu.P., et al, 1995) is used. Due to the complex nature of RFC glow discharges these criteria are more qualitative than quantitative. The contribution to the stable existence of the α-form RFC glow discharge secondary photoemission of electrons is analyzed in (Baranov, 2002).

The numerical simulation parameters chosen in Gladush G.G. and Samokhin A.A. (1986) for RFC glow discharge correspond to the γ-form of the discharge. These authors distinguish three frequency ranges ω/p leading to three types of near-electrode volt–ampere characteristics. At low frequencies, there is only one normal current density corresponding to an Ohmic discharge (γ-form). At high frequencies, there is a normal current density corresponding to the low-current form of the discharge (α-form). In the intermediate frequency range, in principle, two values of the normal current density can coexist, corresponding to low- and high-current forms of discharge. However, the authors suggest that such a discharge mode may not be implemented in practice.

In Gladush G.G. and Samokhin A.A. (1986), a two-dimensional numerical RFC glow discharge model in flat geometry with an electrode spacing of $H = 1\,\text{cm}$ was built and implemented. That is, the glow discharge current column had the shape of a plasma "knife." Important numerical results of this paper are a result of a two-dimensional electrodynamic structure of a quasistationary RFC glow discharge and a detailed analysis of the ionization processes in the volume and near the electrodes. However, these authors (Gladush G.G., et al., 1986) paid attention to the fact that by the time $t = 50\,\mu\text{s}$ after the ignition of the discharge, stationary values of current density and total current through the gas discharge gap could not be achieved. Over the entire specified time interval, a decrease of the total current and an increase of the current density at the electrodes, as well as the field strength in quasineutral plasma, are observed.

In this section, we consider a numerical model of a normal axisymmetric RFC glow discharge between flat electrodes of radius $R = 5\,\text{cm}$ with an axial magnetic field of $B_x = 0.1\,\text{T}$. The normal mode of RFC glow discharge with parameters close to those of Gladush G.G. and Samokhin A.A. (1986) is investigated.

6.2.1 Numerical simulation model

The diffusion-drift model of an axisymmetric normal glow discharge in an axial magnetic field between plane electrodes was formulated in Surzhikov S.T. and Shang J.S. (2014). The main difference between the problem solved in this section and that of the mentioned work is the sinusoidal varying voltage drop according to the law

$E = E_0$ Sinωt for different values of E_0 and f. The principal scheme of such kind of glow discharge is shown in Figure 6.8.

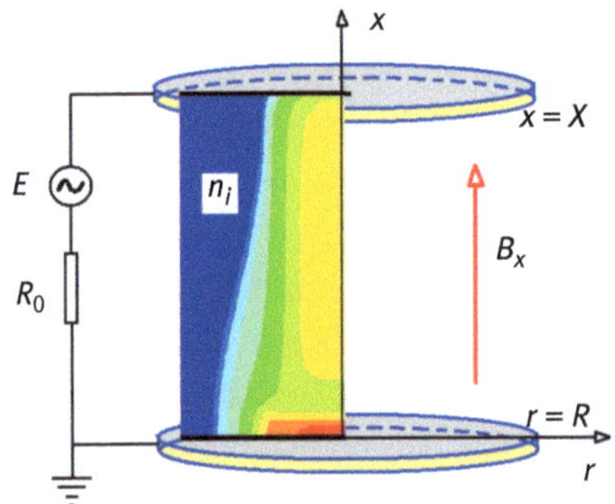

Figure 6.8: Schematic of an axial radiofrequency glow discharge with external magnetic field.

The system of governing equations has the following form:

$$\frac{\partial n_e}{\partial t} + \frac{\partial}{\partial x}\left(-\mu_e n_e E_x - D_e \frac{\partial n_e}{\partial x}\right) + \frac{1}{r}\frac{\partial}{\partial r} r\left(-\frac{\mu_e}{1+b_e^2} n_e E_r - \frac{D_e}{1+b_e^2}\frac{\partial n_e}{\partial r}\right) = \alpha|\Gamma_e| - \beta n_e n_i, \quad (6.20)$$

$$\frac{\partial n_i}{\partial t} + \frac{\partial}{\partial x}\left(\mu_i n_i E_x - \frac{D_i}{1+b_i^2}\frac{\partial n_i}{\partial x}\right) + \frac{1}{r}\frac{\partial}{\partial r} r\left(\frac{\mu_i}{1+b_i^2} n_i E_r - \frac{D_i}{1+b_i^2}\frac{\partial n_i}{\partial r}\right) = \alpha|\Gamma_e| - \beta n_e n_i, \quad (6.21)$$

$$\frac{\partial^2 \varphi}{\partial x^2} + \frac{1}{r}\frac{\partial^2 \varphi}{\partial r^2} = 4\pi e(n_e - n_i), \quad (6.22)$$

where n_e, n_i are the volume density of electrons and ions; D_e, D_i are the diffusion coefficients of electrons and ions; φ is the electric potential; e is the electron charge; $\alpha(|E|/p)$, β are the ionization and recombination coefficients ($\beta = 2\times 10^{-7}, \mathrm{cm}^3/\mathrm{s}$);

$$|\Gamma_e| = \sqrt{\Gamma_{e,x}^2 + \Gamma_{e,r}^2 + \Gamma_{e,\varphi}^2},$$

$$\Gamma_{e,\varphi} = n_e u_{e,\varphi} = -b_e \Gamma_{e,r},$$

$$\Gamma_{e,x} = -D_e \frac{\partial n_e}{\partial x} + \mu_e n_e \frac{\partial \varphi}{\partial x} = -D_e \frac{\partial n_e}{\partial x} - \mu_e n_e E_x,$$

$$\Gamma_{e,r} = -\frac{D_e}{1+b_e^2}\frac{\partial n_e}{\partial r} + \frac{\mu_e}{1+b_e^2} n_e \frac{\partial \varphi}{\partial r} = -\frac{D_e}{1+b_e^2}\frac{\partial n_e}{\partial r} - \frac{\mu_e}{1+b_e^2} n_e E_r; \quad (6.23)$$

$\Gamma_{e,x}$, $\Gamma_{e,r}$, $\Gamma_{e,\varphi}$ are the axial, radial, and azimuth components of the density of the electron flux Γ_e; E_r, E_r are the axial and radial components of the electric field strength.

Several coefficients of this model take into account an axial magnetic field $\mathbf{B} = \mathbf{i}B_x$, where $\mathbf{B}$ is the vector of magnetic field induction. These are

$$b_e = \frac{\mu_e B_x}{c} = \frac{\omega_e}{\nu_e},\ b_i = \frac{\mu_i B_x}{c} = \frac{\omega_i}{\nu_i},$$

$$\omega_e = \frac{eB_x}{m_e c} \sim 1.76 \times 10^{10}\ \mathrm{s}^{-1},\ \omega_i = \frac{eB_x}{m_i c} \sim 3.0 \times 10^5\ \mathrm{s}^{-1}, \tag{6.24}$$

where ω_e, ω_i are the Larmor frequencies of electrons and ions; ν_e, ν_i are the frequency of collisions of electrons and ions with neutral particles; note that in (6.25) B_x should be measured in Gs ($1\ \mathrm{T} = 10^4$ Gs); $m_e = 9.1 \times 10^{-28}$ g, $m_i = 32 \times 1.66 \times 10^{-24}$ g, $B_x = 0.1\ \mathrm{T}$, $e = 4.802 \times 10^{-10}$ SGSE; μ_e, μ_i are the mobilities of charged particles, $\mu_e = 4.4 \times 10^5/p$, $\mathrm{cm}^2/(\mathrm{V}\cdot\mathrm{s})$, $\mu_i = 1140/p$, $\mathrm{cm}^2/(\mathrm{V}\cdot\mathrm{s})$; p is the pressure.

At $p = 5$ Torr the Larmor radii of electrons and ions have the following values:

$$R_{Le} = \frac{v_{e\perp} m_e c}{eB_x} \sim 1.42 \times 10^{-3}\ \mathrm{cm},\ R_{Li} = \frac{v_{i\perp} m_i c}{eB_x} \sim 0.142\ \mathrm{cm},$$

where $v_{e\perp} \sim 2.5 \times 10^7\ \mathrm{cm/s}$, $v_{i\perp} \sim 8.6 \times 10^4\ \mathrm{cm/s}$ are the typical averaged velocities of electrons and ions.

For the convenience of the numerical solution, eqs. (6.20) and (6.21) were normalized to

$$Q = \frac{t_0}{n_{e,0}} = \frac{H^2}{E\mu_{e,0} n_{e,0}} = 4.5 \times 10^{-17}\ \mathrm{cm}^3 \cdot \mathrm{c}.$$

As mentioned earlier, in the case under consideration the equation for the external electric circuit was changed on the equation for voltage drop on electrodes

$$E = E_0 \mathrm{Sin}(2\pi f t).$$

Equations (6.20)–(6.22) were integrated with the following boundary conditions:

$$x = 0,\ \text{at}\ \frac{\partial \varphi}{\partial x}\Big|_{x=0} > 0:\ \frac{\partial n_i}{\partial x} = 0,\ \Gamma_{x,e} = \gamma \Gamma_{x,i},\ \varphi = 0; \tag{6.25}$$

$$x = H,\ \frac{\partial \varphi}{\partial x}\Big|_{x=H} > 0:\ n_i = 0,\ \frac{\partial n_e}{\partial x} = 0,\ \varphi = E(t); \tag{6.26}$$

$$x = 0,\ \frac{\partial \varphi}{\partial x}\Big|_{x=0} < 0:\ n_i = 0,\ \frac{\partial n_e}{\partial x} = 0,\ \varphi = 0; \tag{6.27}$$

$$x = H,\ \frac{\partial \varphi}{\partial x}\Big|_{x=H} < 0:\ \frac{\partial n_i}{\partial x} = 0,\ \Gamma_{x,e} = \gamma \Gamma_{x,i},\ \varphi = E(t); \tag{6.28}$$

$$r = 0:\ \frac{\partial n_e}{\partial r} = \frac{\partial n_i}{\partial r} = \frac{\partial \varphi}{\partial r} = 0; \tag{6.29}$$

$$r = R: \quad \frac{\partial n_e}{\partial r} = \frac{\partial n_i}{\partial r} = \frac{\partial \varphi}{\partial r} = 0. \tag{6.30}$$

In (6.25) and (6.28) γ is the coefficient of the secondary electron emission.

The first Townsend coefficient (the ionization coefficient) was used in the following form:

$$\alpha(E) = p^* A_{in} \exp\left[-\frac{B_{in}}{(|\mathbf{E}|/p^*)}\right], \; 1/(\text{cm} \cdot \text{Torr})$$

where

$$A_{in} = 12, \; 1/(\text{cm} \cdot \text{Torr}), \; B_{in} = 342, \; \text{V}/(\text{cm} \cdot \text{Torr}).$$

It was assumed that the general mechanism of ionization is as follows:

$$N_2 + e^- \rightarrow N_2^+ + 2e^-.$$

6.2.2 Results of numerical simulation

As initial data for modeling a high-frequency discharge, the time-stable solution of the problem of a normal DC glow discharge with $p = 5$ Torr, $E = 1$ kV, $R_0 = 300$ kΩ, $\gamma = 0.33$ was used.

Figures 6.9 and 6.10 show the distributions of electron and ion concentrations, the reduced field $|E_x|/p$, and the rate of ionization. This is a fully stationary solution. The calculation was carried out up to $t = 800\,\mu$s. As a result of the calculation, a voltage drop across the gas-discharge gap was obtained, which was then used (with some variations) in the calculations of the high-frequency discharge. Figure 6.4 shows the axial distribution of the concentration of the ions and the electrons in the normal DC glow discharge. The space charge regions at the cathode and the anode, as well as the positive quasineutral column, are clearly identified here. Figures 6.9 and 6.10 clearly show the spatial structure of the discharge in the cathode and anode

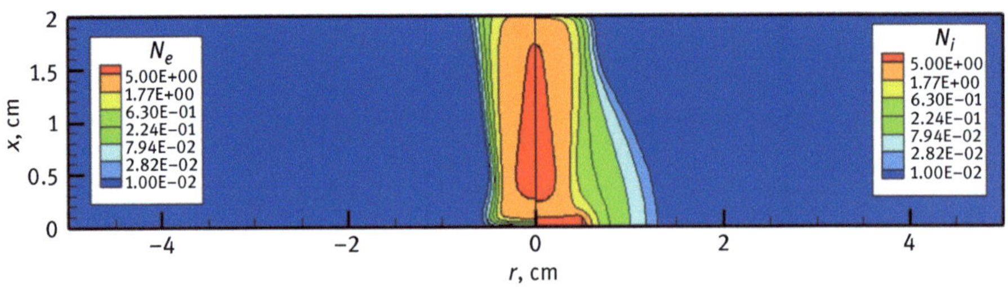

Figure 6.9: Volume concentrations of electrons (from left, $N_e = n_e \times 10^{-9}$ cm^{-3}) and ions (from right, $N_i = n_i \times 10^{-9}$ cm^{-3}) for the normal DC glow discharge at $p = 5$ Torr, $E = 1$ kV, $R_0 = 300$ kOhm, $\gamma = 0.33$.

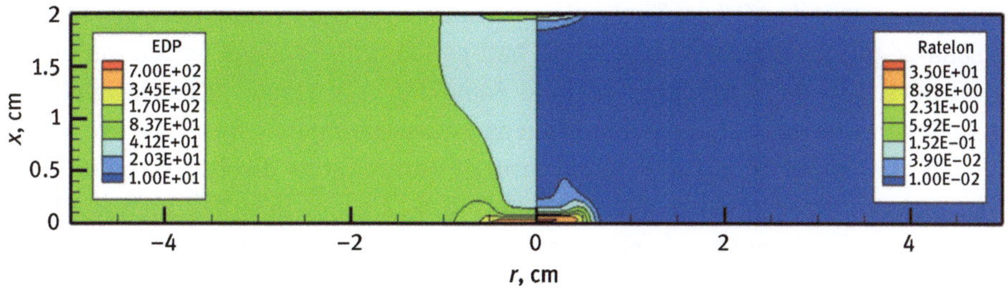

Figure 6.10: Reduced field E/p, V/(cm × Torr) (from left) and rate of ionization, 1/(cm^3 s) (from right) for the normal DC glow discharge at $p = 5$ Torr, $E = 1$ kV, $R_0 = 300$ kOhm, $\gamma = 0.33$.

layers. The highest values of the reduced field are observed in the cathode layer. Ionization occurs mainly in the cathode and a little in the anode layer.

As it was mentioned earlier, the obtained distributions of the electrodynamic parameters of the normal DC glow discharge were set as the initial data in the RFC glow discharge simulation (Figure 6.11).

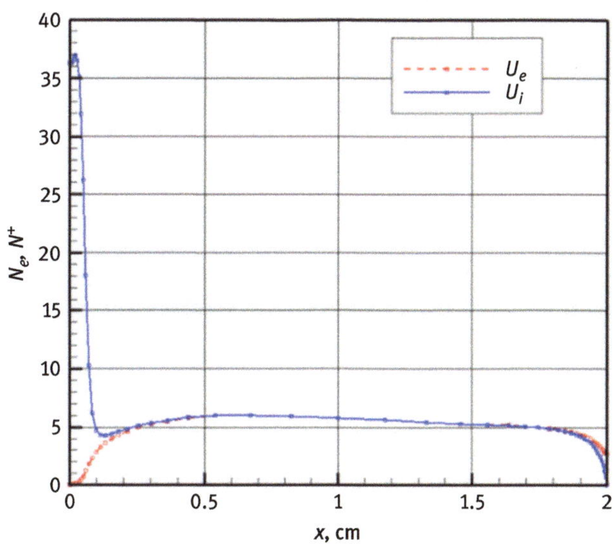

Figure 6.11: Axial distribution of the volume concentrations of electrons (from left, $N_e = n_e \times 10^{-9}$ cm^{-3}) and ions (from right, $N_i = n_i \times 10^{-9}$cm^{-3}) for the DC glow discharge at $p = 5$ Torr, $E = 1$ kV, $R_0 = 300$ kOhm, $\gamma = 0.33$.

The first series of calculations was performed at $p = 5$ Torr, $E_0 = 478$ V, and $B_x = 0$. The numerical simulation results for the fields of electron and ion concentrations, electric potential, and reduced field $|E_x|/p$ are shown in Figures 6.12 and 6.13 for the time instant $t = 40.9\,\mu s$. These electrodynamic functions are shown for the two discharge phases corresponding to the half-period of the voltage oscillation on the electrodes. Just as in Gladush G.G. and Samokhin A.A. (1986), it is reasonable to define the given discharge configuration as a quasistationary γ-form of a discharge.

The transition from the original form of normal glow discharge (Figures 6.9 and 6.10) to that shown in Figures 6.12 and 6.13 occur in a few microseconds. Then, there is a very slow expansion of the current column while preserving the structure of two near-electrode regions of the space charge with an increased ion concentration at the electrodes. Computer animation of the results obtained indicates that against the background of a quasistationary distribution of ion concentrations, a cloud of electrons oscillates with the frequency of the applied electric field between the electrodes (see the vector field in Figure 6.14). The axial distributions of electron and ion concentrations for the indicated time points are shown in Figure 6.15.

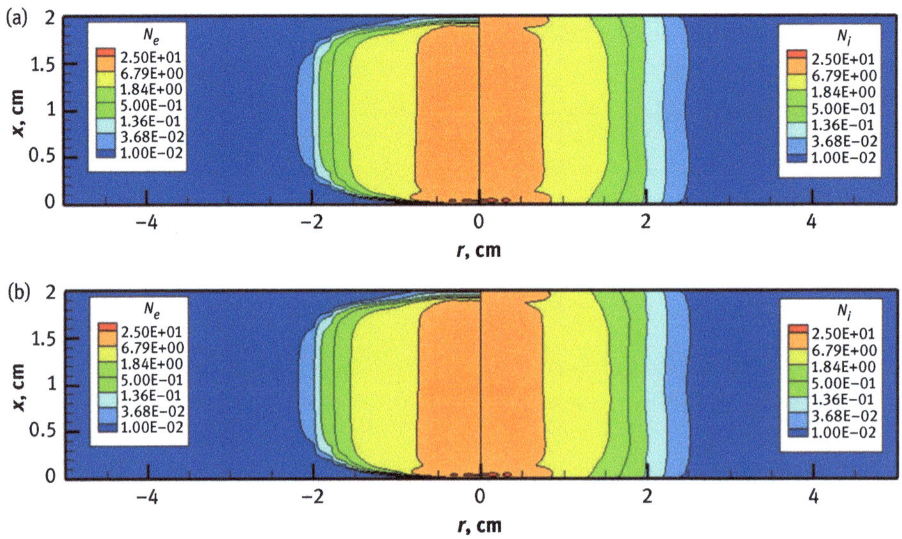

Figure 6.12: Volume concentrations of electrons (from left, $N_e = n_e \times 10^{-9}$ cm^{-3}) and ions (from right, $N_i = n_i \times 10^{-9}$ cm^{-3}) for the RFC glow discharge at $f = 1.359$ MHz, $p = 5$ Torr, $E_0 = 460$ V, $\gamma = 0.33$. The two opposite phases of the discharge are shown.

Here one should pay attention to one important property of the high-frequency discharge, which is related to the fact that, unlike a normal DC glow discharge, electrons left the current column in the radial direction have no time to leave the discharge gap while moving to the anode, because the electric field changes its polarity (see Figure 6.14). In this case, the electrons remain in the gas-discharge gap for a long time and it increases the probability of gas ionization. This leads to continuous

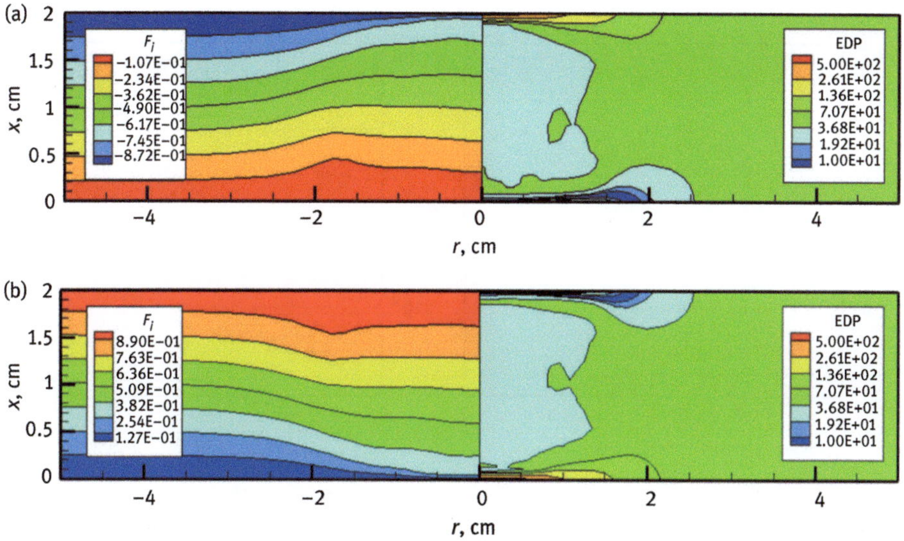

Figure 6.13: Electric potential $Fi = \varphi/E$ (from left) and reduced field E/p, V/(cm · Torr) (from right) for the RFC glow discharge at $f = 1.359$ MHz, $p = 5$ Torr, $E_0 = 460$ V, $\gamma = 0.33$. The two opposite phases of the discharge are shown.

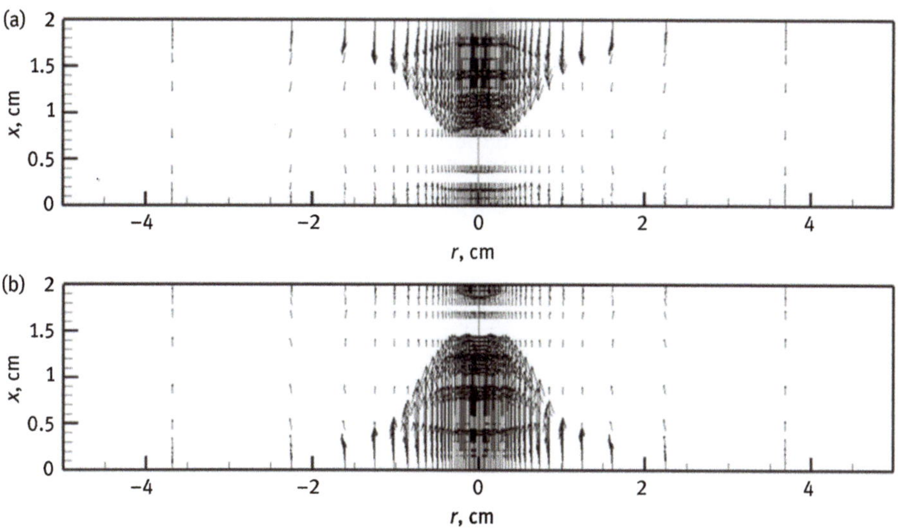

Figure 6.14: Electric field strength (V/cm) for the RFC glow discharge at $f = 1.359$MHz, $p = 5$ Torr, $E_0 = 460$ V, $\gamma = 0.33$. The two opposite phases of the discharge are shown.

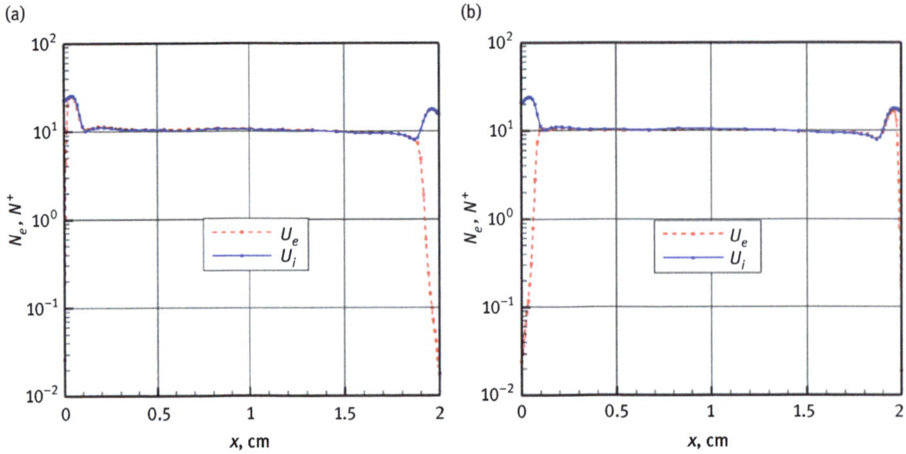

Figure 6.15: Axial distribution of the volume concentrations of electrons (from left, $N_e = n_e \times 10^{-9}$ cm^{-3}) and ions (from right, $N_i = n_i \times 10^{-9}$ cm^{-3}) for the RFC glow discharge at $f = 1.359$ MHz, $p = 5$ Torr, $E_0 = 460$ V, $\gamma = 0.33$. The two opposite phases of the discharge are shown.

expansion of the discharge. Obviously, at substantially lower frequencies, this will not be observed, since the electrons will have time to leave the discharge gap.

The calculation of RFC glow discharge with the same parameters ($p = 5$ Torr, $E_0 = 478$ V, $\gamma = 0.33$) for the frequency $f = 1.359$ was carried out up to $t = 128\,\mu$s. After the formation of the γ-form of the high-frequency discharge, within several microseconds, the current pole began to slowly come down to $t = 100\,\mu$s, demonstrating behavior of the integral characteristics (the total current and the electric field strength in the positive column), as in Gladush G.G. and Samokhin A.A. (1986). Then the discharge quickly extinguished.

The third series of calculations was performed by taking into account the axial magnetic field $B_x = 0.1$ T at $p = 5$ Torr, $E_0 = 460$ V, $\gamma = 0.33$, and $f = 1.359$. The distributions of the electrodynamic parameters of such discharge are shown in Figures 6.16–6.19. The calculation was carried out up to $t = 182.4\,\mu$s. In this case, after the classic γ-form of the discharge was quickly formed, the current column began to slowly expand. From this, it follows that the presence of a magnetic field made it possible to expand the region of existence of a high-frequency discharge (compare these results with previous).

Thus, the presented preliminary results of numerical simulation of axisymmetric RFC glow discharge taking into account the axial magnetic field allow us to conclude that there is a quasistationary γ-form of a high-frequency axisymmetric RFC glow discharge. Accounting for the external magnetic field, as in the case of the direct current glow discharge leads to the expansion of the current column in the mode of the RFC normal glow discharges.

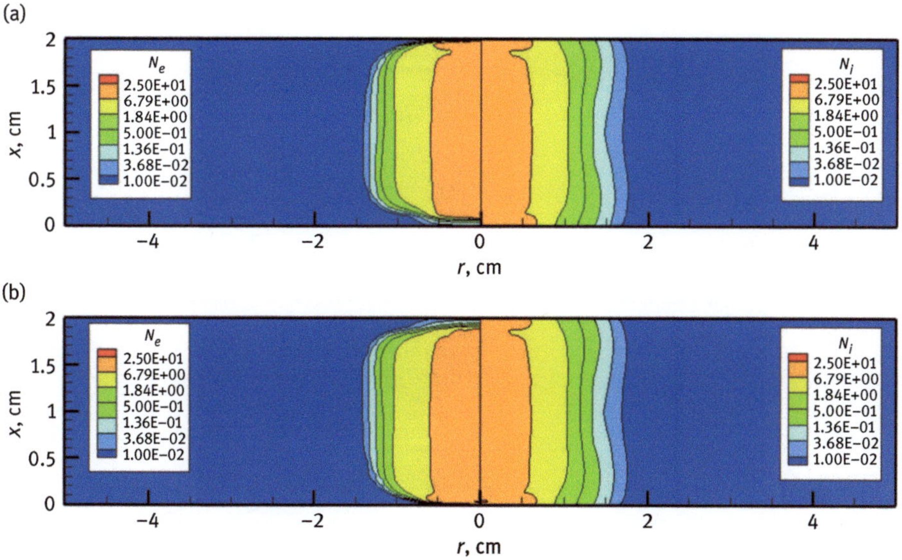

Figure 6.16: Volume concentrations of electrons (from left, $N_e = n_e \times 10^{-9}$ cm^{-3}) and ions (from right, $N_i = n_i \times 10^{-9}$ cm^{-3}) for the RFC glow discharge at $f = 13.59$ MHz, $p = 5$ Torr, $E_0 = 478$ V, $\gamma = 0.33$. The two opposite phases of the discharge are shown.

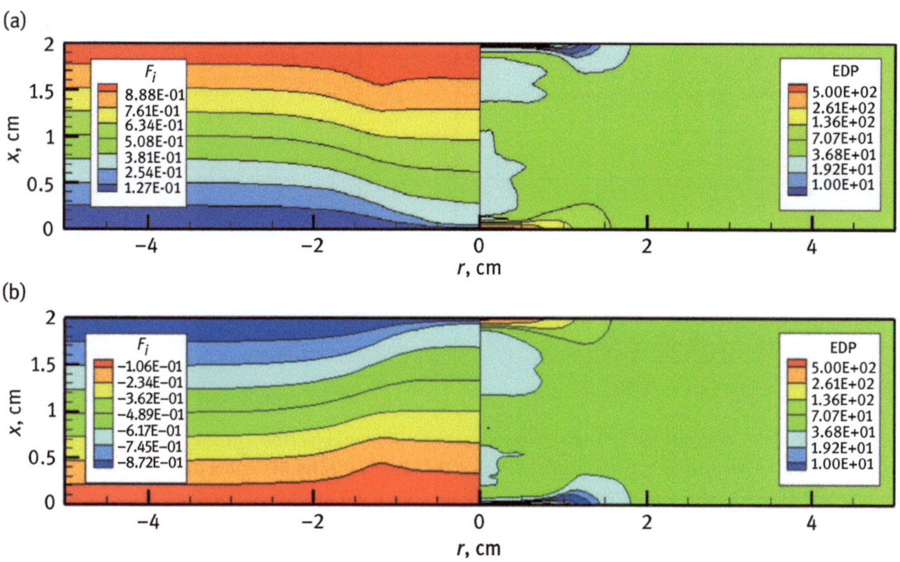

Figure 6.17: Electric potential $Fi = \varphi/E$ (from left) and reduced field E/p, V/(cm · Torr) (from right) for the RFC discharge at $f = 13.59$ MHz, $p = 5$ Torr, $E_0 = 478$ V, $\gamma = 0.33$. The two opposite phases of the discharge are shown.

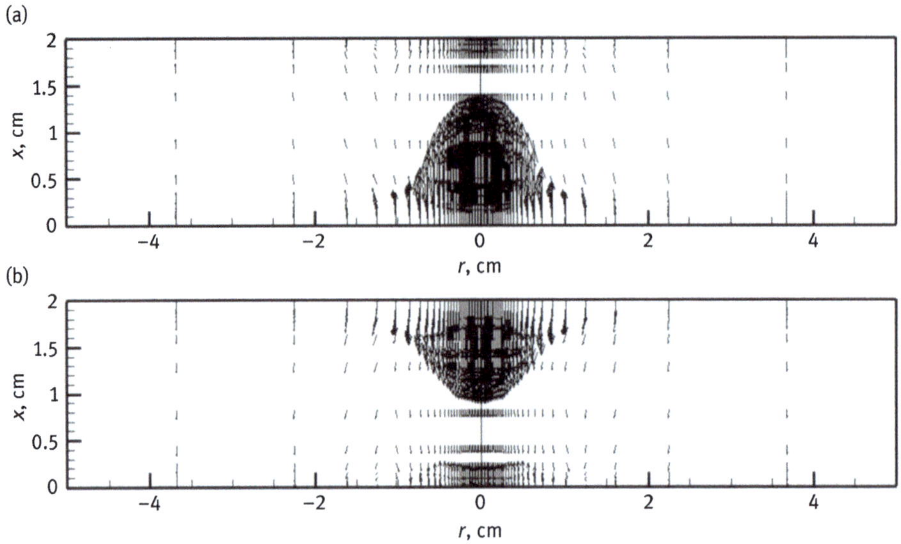

Figure 6.18: Electric field strength (V/cm) for the RFC glow discharge at $f = 13.59$ MHz, $p = 5$ Torr, $E_0 = 478$ V, $\gamma = 0.33$. The two opposite phases of the discharge are shown.

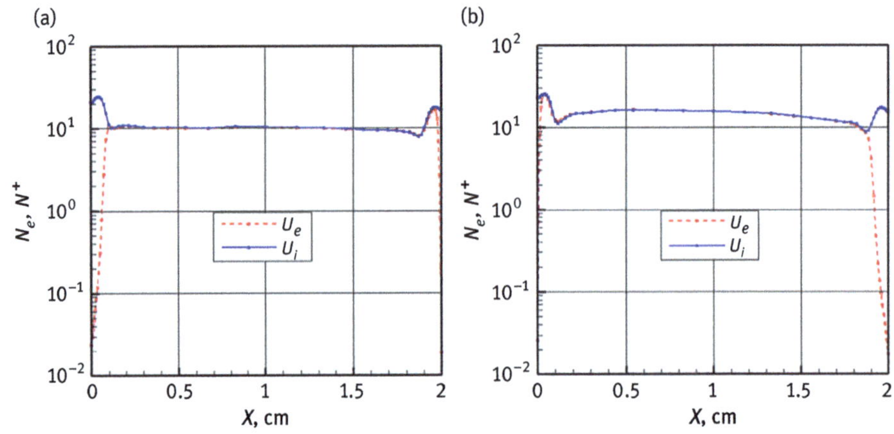

Figure 6.19: Axial distribution of the volume concentrations of electrons (from left, $N_e = n_e \times 10^{-9}$ cm^{-3}) and ions (from right, $N_i = n_i \times 10^{-9}$ cm^{-3}) for the RFC glow discharge at $f = 13.59$ MHz, $p = 5$ Torr, $E_0 = 478$ V, $\gamma = 0.33$. The two opposite phases of the discharge are shown.

7 Numerical simulation of Penning discharge

7.1 Application of the modified drift-diffusion theory for study of the two-dimensional structure of the Penning discharge

The basic configuration of the Penning discharge is defined as the discharge in a cylindrical chamber that is equipped by two plane cathodes and hollow anode (Penning F.M., 1936; Hirsch E.N., 1964; Safronov B.G., et al., 1974). An axial magnetic field is a significant peculiarity of the discharge. The typical voltage drop between electrodes in such a discharge is of $V \sim 1000$V and the magnetic field induction is of ~ 0.1T. In various applications, the most popular are the Penning discharge chambers at the pressures of $p \sim 10^{-2} - 10^{-5}$ Torr filled with the gases H_2, D_2, Ar, or Xe. At such low pressures, the magnetic field plays a significant role in maintaining a gas discharge.

In spite of the wide field of application of these discharges, scientific literature contains a very limited papers containing the description of the spatial structure of such discharges. A bounded number of fragmentary experimental data do not allow to perform validation of computational codes that were developed. Note that modern methods of mathematical modeling allow to predict a spatial structure of gas discharges of various kinds, and, unfortunately, corresponding experimental data are practically absent.

Among developments of numerical simulation methods that are intended for modeling spatial structure of gas discharges, we can stay on the methods of particles in cells (PIC). Application of the method for plasma in the magnetic field, including the simulation of the Penning discharge, has been demonstrated in Birdsall C.K. and Langdon A.B. (1985) and Surzhikov S.T. (1999).

However, the well-known advantages of the PIC methods for the simulation of rarefied plasma are not of much use with an increase in the pressure in discharge chambers up to the order of 1 Torr. In this case, the models of the magnetohydrodynamics (MHD) appear to be efficient. In turn, the MHD models lost their advantages at low pressures. It is obvious that there is a need for developing both classes of models. Two-dimensional MHD model of the Penning discharge was investigated in Surzhikov S.T. (2015).

Theory and computational model of the Penning discharge at pressure ~1 mTorr in a cylindrical cone with an axial magnetic field is presented in this section. Electrodynamic model of this kind of a low-pressure discharge is formulated. The model consists of continuity equations for electron and ion fluids and the Poisson equation for the self-consistent electric field.

Inside gas discharge chambers at so low pressures (e.g., at pressure $p \sim 1$mTorr) the reduced electric field achieves high values of order $E/p = 10^6\,\mathrm{V/cm\,Torr}$. This means that while calculating the mobilities it is necessary to take into account a

https://doi.org/10.1515/9783110648836-008

change of physical mechanisms of interaction between charged and neutral particles under transition from low to high reduced fields (Huxley L.G.H., et al., 1974; McDaniel E.W., 1964).

The general idea of the modification of the drift-diffusion model developed in this section is the introduction of the nonlinear dependencies of mobilities for electrons and ions (from the electric field intensity) and modification of ionization coefficient, which can be used in weakly ionized gas with the magnetic field at low pressure. Numerical simulation results are presented for two-dimensional Penning discharge at various initial conditions. The results are obtained for atomic hydrogen at pressure $1-4\,\mathrm{mTorr}$, emf of power supply 2.5 kV, and magnetic field induction of $B = 0 - 0.1\,\mathrm{T}$.

It will be shown that in an axial magnetic field an azimuthal oppositely directed gyration of ions and electrons in the glow discharge is observed.

Thus, the objective of this chapter is to study the two-dimensional structure of the Penning discharge in an axial magnetic field, and patterns of counter-rotating ions (protons) and electrons fluids in imposing the axial magnetic field. It also allows to study the structure of accelerated beam of protons, which is of great practical interest for aerospace applications.

7.1.1 Preliminary analysis

It was established in earlier studies that a model of glow discharge based on drift-diffusion approach for two fluids (ions and electrons) allows predicting all general characteristics of such discharges for pressures $\sim 1-10\,\mathrm{Torr}$ and voltage drop on gas discharge gaps $\sim 1-3\,\mathrm{kV}$. Taking into account that a glow discharge is a weakly ionized gas, it is possible to include the undisturbed external magnetic field, which will affect the electrodynamical structure of the glow discharge but will not be disturbed by this discharge.

A distinctive feature of this section is to study a configuration of the Penning discharge in molecular hydrogen when the external magnetic field is applied in the axial direction, that is, along the current column. In these circumstances, there is the emergence of oppositely directed flows of ions and electrons in the azimuthal direction, which significantly alters the structure of the Penning discharge electrodynamics. The principal scheme of the Penning discharge is shown in Figure 7.1. The ring cathode (at $x = X_c$) is a significant part of the cylindrical discharge chamber. An internal open part of the cathode is intended for forming proton plume, which is formed inside the gas-discharge volume.

A cylindrical two-dimensional (2D) glow discharge in atomic hydrogen between flat electrodes in the presence of an axial magnetic field is considered (see Figure 7.1). A drift-diffusion theory is used for describing gas-discharge processes. This theory is based on the continuity equations for concentration of electrons n_e and positive

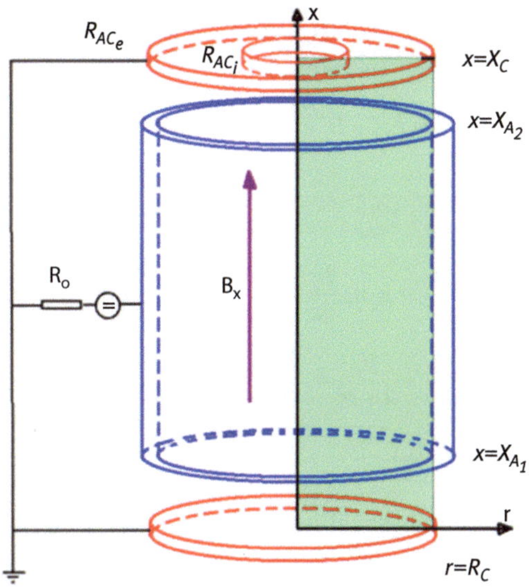

Figure 7.1: Schematic of a glow discharge with external magnetic field.

ions n_i together with the equations for the electrostatic field $\mathbf{E} = -\text{grad}\varphi$ and energy conservation equation for the neutral species. Classic equations of the drift-diffusion theory (without magnetic field) have the following form:

$$\frac{\partial n_e}{\partial t} + \frac{\partial \mathbf{\Gamma}_{e,x}}{\partial x} + \frac{1}{r}\frac{\partial r\mathbf{\Gamma}_{e,y}}{\partial r} = \alpha(E)|\mathbf{\Gamma}_e| - \beta n_i n_e \tag{7.1}$$

$$\frac{\partial n_i}{\partial t} + \frac{\partial \mathbf{\Gamma}_{i,x}}{\partial x} + \frac{1}{r}\frac{\partial r\mathbf{\Gamma}_{i,y}}{\partial r} = \alpha(E)|\mathbf{\Gamma}_e| - \beta n_i n_e, \tag{7.2}$$

$$\frac{\partial^2 \varphi}{\partial x^2} + \frac{1}{r}\frac{\partial}{\partial r} r \frac{\partial \varphi}{\partial r} = 4\pi e(n_e - n_i), \tag{7.3}$$

where $\mathbf{\Gamma}_e = -D_e \text{grad} n_e - n_e \mu_e \mathbf{E},\ \mathbf{\Gamma}_i = -D_i \text{grad} n_i + n_i \mu_i \mathbf{E},$

$$Q = \eta(\mathbf{jE}),\ \mathbf{j} = e(\mathbf{\Gamma}_i - \mathbf{\Gamma}_e), \tag{7.4}$$

$\alpha(E)$ and β are the ionization and recombination coefficients; $\mathbf{\Gamma}_e$, $\mathbf{\Gamma}_i$ are the electron and ion flux densities; μ_e, μ_i are the electron and ion mobility; D_e, D_i are the electron and ion diffusion coefficients; η is the part of Joule heating, which goes into heating the gas; φ is the potential of electric field.

The concept of a magnetic field in the model was introduced in Section 5.7. The charged particles continuity equations can be written as follows:

$$\frac{\partial n_e}{\partial t} + \frac{\partial}{\partial x}\left(\mu_e n_e E_{e,x} - D_e \frac{\partial n_e}{\partial x}\right) + \frac{1}{r}\frac{\partial}{\partial r}\left[\frac{r}{1+b_e^2}\left(\mu_e n_e E_{e,r} - D_e \frac{\partial n_e}{\partial r}\right)\right] = \alpha|\mathbf{\Gamma}_e| - \beta n_e n_i \quad (7.5)$$

$$\frac{\partial n_i}{\partial t} + \frac{\partial}{\partial x}\left(\mu_i n_i E_{i,x} - D_i \frac{\partial n_i}{\partial x}\right) + \frac{1}{r}\frac{\partial}{\partial r}\left[\frac{r}{1+b_i^2}\left(\mu_i n_i E_{i,r} - D_i \frac{\partial n_i}{\partial r}\right)\right] = \alpha|\mathbf{\Gamma}_e| - \beta n_e n_i \quad (7.6)$$

where
$$|\mathbf{\Gamma}_e| = \sqrt{\mathbf{\Gamma}_{e,x}^2 + \mathbf{\Gamma}_{e,r}^2 + \mathbf{\Gamma}_{e,\varphi}^2}$$

$$\mathbf{\Gamma}_{e,x} = n_e u_{e,x} = -D_e \frac{\partial n_e}{\partial x} + \mu_e n_e \frac{\partial \varphi}{\partial x},$$

$$\mathbf{\Gamma}_{e,r} = n_e u_{e,r} = -\frac{D_e}{1+b_e^2}\frac{\partial n_e}{\partial r} + \frac{\mu_e}{1+b_e^2} n_e \frac{\partial \varphi}{\partial r},$$

$$\mathbf{\Gamma}_{e,\varphi} = n_e u_{e,\varphi} = -b_e \mathbf{\Gamma}_{e,r}.$$

Coefficients of this computational model, which take into account a magnetic field, can be presented in the following form:

$$b_e = \frac{\mu_e B_x}{c} = \frac{\omega_e}{\nu_e}, \; b_i = \frac{\mu_i B_x}{c} = \frac{\omega_i}{\nu_i},$$

where $\omega_e = \frac{eB_x}{m_e c} = \frac{eH_x}{m_e c}$ is the Larmor frequency of electrons; $\omega_i = \frac{eB_x}{m_i c} = \frac{eH_x}{m_i c}$ is the Larmor frequency of ions.

Boundary conditions for charged particles and electrical potential are formulated as follows:

$$x = 0: \quad \frac{\partial n_i}{\partial x} = 0, \; \mathbf{\Gamma}_e = \gamma \mathbf{\Gamma}_i, \; \varphi = 0; r = 0: \frac{\partial n_e}{\partial r} = \frac{\partial n_i}{\partial r} = \frac{\partial \varphi}{\partial r} = 0;$$

$$x = X, \; r > R_{AC_i}: \quad \frac{\partial n_i}{\partial x} = 0, \; \mathbf{\Gamma}_e = \gamma \mathbf{\Gamma}_i, \; \varphi = 0; r = R: \frac{\partial n_e}{\partial r} = 0; n_i = 0; \; \varphi = V.$$

Boundary conditions for ions (H_2^+ or H^+) and electrons at $r < R_{ACi}$ (a nozzle of the chamber) should be formulated in view of electric field and concentration of charged particles outside the discharge chamber. Numerical simulation in this section was performed for the following boundary conditions:

$$x = X, \; r < R_{AC_i}: \; \frac{\partial n_i}{\partial x} = \frac{\partial n_e}{\partial x} = 0, \; \varphi = 0.$$

A quasineutral plasma cloud of a spherical form in the center of the chamber is used as the initial condition.

In the given case, the variation of neutral gas temperature is not taken into consideration. The transport and thermophysics properties are therefore do not depend on the temperature. For example, for atom H

$$\mu_e(p)=\frac{3.7\times10^5}{p}\frac{\text{cm}^2}{\text{V}\cdot\text{s}},\ \mu_i(p)=\frac{6.55\times10^3}{p}\frac{\text{cm}^2}{\text{V}\cdot\text{s}},\ D_e=\mu_e(p)T_e,\ D_i=\mu_i(p)T,$$

$$c_p=8.314\frac{7}{2}\frac{1}{M_\Sigma}\frac{\text{J}}{\text{g}\cdot\text{K}},\ M_\Sigma=2\frac{\text{g}}{\text{mol}},\ \rho=1.58\times10^{-5}\frac{M_\Sigma p}{T}\frac{\text{g}}{\text{cm}^3},$$

$$\lambda=\frac{8.334\times10^{-4}}{\sigma^2\Omega^{(2.2)*}}\sqrt{\frac{T}{M_\Sigma}}\left(0.115+0.354\frac{c_pM_\Sigma}{R_g}\right),$$

$$\Omega^{(2,2)^*}=\frac{1.157}{(T^*)^{0.1472}},\ T^*=\frac{T}{(\varepsilon/k)},\ \frac{\varepsilon}{k}=59.7\,\text{K},\ \sigma=2.827\,\mathring{\text{A}},\ R_g=8.314\frac{\text{J}}{\text{K}\cdot\text{mol}},$$

where p is the undisturbed pressure in Torr. The recombination coefficient β and electron temperature are taken as constants $\beta=2\times10^{-7}\ \text{cm}^3/\text{s}$, $T_e=11610$ K.

The ionization coefficient for H is determined as follows:

$$\alpha(E)=p^*A\exp\left[-\frac{B}{(|\mathbf{E}|/p^*)}\right],\text{cm}^{-1},$$

where

$$A=5\,\frac{1}{\text{cm}\cdot\text{Torr}},\ B=130\,\frac{\text{V}}{\text{cm}\cdot\text{Torr}}.$$

Equations (7.1)–(7.3) are supplemented with the equation for the external circuit (see Figure 7.1), which is written for a stationary current as

$$E=V+IR_0,$$

where V is the voltage on the electrodes, I is the total discharge current, E is the Emf. in power supply, and R_0 is the external resistance.

7.1.2 Modified drift-diffusion model

Let us consider methods for calculating drift velocities in relatively low and asymptotically large reduced fields.

Drift velocity for small reduced field ($E/p \lessapprox 100\text{V/Torr}$) is

$$V_{\text{dr},i}=\mu_i\mathbf{E}. \tag{7.7}$$

Reduced mobility of H^+ in H is (McDaniel E.W., et al., 1973):

$$K_0=\frac{p_{\text{Torr}}}{760_{\text{Torr}}}\frac{273.16_K}{T_K}\mu_i=15.7\,\text{cm}^2/(\text{V}\cdot\text{s})$$

From this relation one can find the mobility of ion H^+:

$$\mu_i = K_0 \left(\frac{p_{\text{Torr}}}{760_{\text{Torr}}} \frac{273.16_K}{T_K} \right)^{-1}$$

Further we will us the following correlations:

$$(\mu_i p_{\text{Torr}})_{\text{H}^+} = 1.31 \times 10^4 \text{ cm}^2\text{Torr}/(\text{V} \cdot \text{s}),$$

$$(\mu_i p_{\text{Torr}})_{\text{H}_2{}^+} \sim 6.55 \times 10^3 \text{cm}^2\text{Torr}/(\text{V} \cdot \text{s}).$$

Let us estimate the drift velocity

1) at $\frac{E}{p} = 1000 \text{ V}/10 \text{ Torr}$:

$$V_{\text{dr, H}_2^+} = (\mu_i p_{\text{Torr}})_{\text{H}_2^+} \frac{E}{p} = 6.55 \times 10^5 \text{ cm/s}$$

2) at $\frac{E}{p} = 1000\ V/0.001 \text{ Torr}$:

$$V_{\text{dr, H}_2^+} = (\mu_i p_{\text{Torr}})_{E_2^+} \frac{E}{p} = 6.55 \times 10^9 \text{cm/s!}$$

Obviously, it is impossible from basic physics point of view.

We have to use another correlation for asymptotically large reduced field

$$V_{\text{dr}, i} = \left(\frac{M_i}{M} \right)^{1/4} \left(1 + \frac{M_i}{M} \right)^{1/4} \sqrt{\frac{eEl}{M_i}}, \tag{7.8}$$

where l is the mean free path; M_i, M are the mass of ion and background gas.

One can see that there is the square root dependence for ion drift velocity on electric field:

$$V_{\text{dr}, i} = 9.2 \times 10^4 \sqrt{\frac{E_{\text{V}}}{p_{\text{Torr}}}} \text{ cm/s}, \tag{7.9}$$

as it was mentioned earlier, at low reduced field

$$V_{\text{dr, H}_2^+} = (\mu_i p_{\text{Torr}})_{\text{H}_2^+} \frac{E_{\text{V}}}{p_{\text{Torr}}}.$$

Let us save regular connection for velocity drift:

$$V_{\text{dr}, i} = \mu_i \mathbf{E},$$

then

$$\mu_i p_{\text{Torr}} = 9.2 \times 10^4 \sqrt{\frac{p_{\text{Torr}}}{E_{\text{V}}}}. \tag{7.10}$$

For any fields we can use the following approximation:

$$\mu_{i,\,eff} = \min\left\{\frac{\mu_i p}{p},\ \frac{9.2\times 10^4}{\sqrt{p_{\mathrm{Torr}} E_{\mathrm{V}}}}\right\}, \tag{7.11}$$

and consequence

$$D_{i,\,eff} = \mu_{i,\,eff} T_i \tag{7.12}$$

By the analogy for electrons

$$\mu_{e,\,eff} = \min\left\{\frac{\mu_e p}{p},\ \frac{2.4\times 10^6}{\sqrt{p_{\mathrm{Torr}} E_{\mathrm{V}}}}\right\}, \tag{7.13}$$

and

$$D_{e,\,eff} = \mu_{e,\,eff} T_e \tag{7.14}$$

Let us now consider the ionization coefficient. At low reduced fields, as before

$$\alpha = p_{\mathrm{Torr}} A_{1/\mathrm{cm}\cdot\mathrm{Torr}} \exp\left[-\frac{B_{\mathrm{V/cm}\cdot\mathrm{Torr}}}{E_{\mathrm{V}}/p_{\mathrm{Torr}}}\right]. \tag{7.15}$$

It is known that for large reduced field

$$\alpha = N_n \sigma_i, \tag{7.16}$$

and the ionization cross section could be calculated with the Tomson formula

$$\sigma_i = 4\pi a_0^2 \left(\frac{I_H}{\varepsilon}\right) \frac{\varepsilon - I}{I}\ \mathrm{cm}^2, \tag{7.17}$$

where the kinetic energy has the following form:

$$\varepsilon_{[\mathrm{eV}]} = 2.84\times 10^{-14} V^2_{e[\mathrm{cm/s}]}, \tag{7.18}$$

and for heavy particles

$$\varepsilon_{[\mathrm{eV}]} = 1.05\times 10^{-12} V^2_{A[\mathrm{cm/s}]}, \tag{7.19}$$

where A is the mass of the heavy particles.

Now we can form constitutive relations for the modified theory for electrons

$$\mathbf{\Gamma}_{e,x} = -D_{e,\,eff}\frac{\partial n_e}{\partial x} - \mu_{e,\,eff} n_e E_x, \tag{7.20}$$

$$\mathbf{\Gamma}_{e,r} = -\frac{D_{e,\,eff}}{1+b^2_{e,\,eff}}\frac{\partial n_e}{\partial r} - \frac{\mu_{e,\,eff}}{1+b^2_{e,\,eff}} n_e E_r, \tag{7.21}$$

$$\mathbf{\Gamma}_{e,\varphi} = -b_{e,eff}\mathbf{\Gamma}_{e,r}, \tag{7.22}$$

$$b_{e,eff} = \frac{\mu_{e,eff}B_x}{c}, \tag{7.23}$$

and for ions

$$\mathbf{\Gamma}_{i,x} = -D_{i,eff}\frac{\partial n_i}{\partial x} + \mu_{i,eff}n_iE_x, \tag{7.24}$$

$$\mathbf{\Gamma}_{i,r} = -\frac{D_{i,eff}}{1+b_{i,eff}^2}\frac{\partial n_i}{\partial r} + \frac{\mu_{e,eff}}{1+b_{i,eff}^2}n_iE_r, \tag{7.25}$$

$$\mathbf{\Gamma}_{i,\varphi} = -b_{i,eff}\mathbf{\Gamma}_{i,r}, \tag{7.26}$$

$$b_{i,eff} = \frac{\mu_{i,eff}B_x}{c}. \tag{7.27}$$

7.1.3 Procedures

The equations are solved by an implicit finite-difference method of second-order spatial accurate. A conservative difference approximation of the initial-boundary value problem is obtained by a finite-volume method. Time-dependent terms are retained in (7.1) and (7.2) because the steady-state asymptote is derived by the time-relaxation method.

The following inhomogeneous finite-difference grid was used:

$$\{r_i, i=1,2,\ldots,NI;\ r_i=0, r_{NI}=R; x_i, j=1,2,\ldots,NJ;\ x_i=0, x_{NJ}=H;$$
$$t^{n+1}=t^n+\tau, n=0,1,\ldots;\ NI{\sim}151, NJ=101\},$$

where τ is the time step, X is the height of the gas discharge channel, and R is the transversal dimension of the gas discharge channel. The grid was condensed near to electrodes.

The finite-difference approximation of eqs. (7.1)–(7.3) (together with appropriate boundary conditions) has resulted in the following finite-difference form:

$$A_{i,j}U_{i-1,j} + B_{i,j}U_{i+1,j} + \bar{A}_{i,j}U_{i,j-1} + \bar{B}_{i,j}U_{i,j+1} - C_{i,j}U_{i,j} + F_{i,j} = 0, \tag{7.28}$$

where U are the functions that are liable for the determination of $U_{ij} = \{n_e, n_i, \varphi\}_{ij}$; $A_{ij}, B_{ij}, C_{ij}, \bar{A}_{ij}, \bar{B}_{ij}, F_{ij}$ are the coefficients of the finite-difference scheme.

For obtaining dimensionless functions in eqs. (7.1)–(7.3) the following normalizing parameters were used: X, E, $\mu_{e,0}$, $N_0 = 10^{10}$ cm^{-3}.

A successive overrelaxation method by lines method (with the Thomas algorithm) was used to obtain the solution of finite-difference equation (7.28). The sequential algorithm of integration for these equations with additional internal iterations was adopted. At the end of each time step, the voltage Von electrodes was calculated by the following formula:

$$V = E - IR_0,$$

where

$$I = \int_0^R e n_e \mu_e E_{\text{anode}} \mathrm{d}x.$$

Then the computing procedure is repeated for the next time step until the convergence of the solution to a steady asymptote.

7.1.4 Numerical simulation results

The calculations were performed for the following initial data: pressure in molecular hydrogen H_2 was $p = 1.2\,\text{mTorr}$; emf of power supply $E = 2.5\,\text{kV}$; $R_0 = 3\,\text{kOhm}$; $\gamma = 0.33$; $B_x = 0.1\,\text{T}$, $R_C = R_{AC_e} = 0.55\,\text{cm}$, $R_{AC_i} = 0.4\,\text{cm}$, $X_{A1} = 0.3\,\text{cm}$, $X_{A2} = 0.8\,\text{cm}$, $X_C = 1.1\,\text{cm}$, $L = X_C = 1.1\,\text{cm}$ (this value was used as a special scale of the task).

Figures 7.2 and 7.3 show concentrations of electrons and ions in the discharge chamber. The main feature of these distributions is the high concentrations of charged particles in the central region of the discharge chamber. Figure 7.4 shows the radial distribution of the current density at the cathode, at the anticathode, as well as at the anticathode outlet.

It is seen that in the near-axial region the current density is higher by several orders of magnitude than that near to the rest regions of the cathode and an anticathode. This means that the radial distribution of ions emerging from the discharge chamber into accelerating optical system ($x > X_c$) has the form of the ion beam. The distributions of concentrations of electrons and ions near the cathode are shown in Figure 7.5 and the radial distribution of ion concentration in the middle section of the discharge chamber at $x = 0.55$ cm is given in Figure 7.6. Increased concentration of charged particles in the near-axis region is observed in all cross sections of the discharge.

In the drift-diffusion model, a plasma configuration is determined by several factors. These are as follows: the electric field distribution, averaged fluxes of charged particles, and distribution of the ionization rate in the discharge space. The distribution of electric potential is shown in Figure 7.7. Let us note the low potential in the near-axial region. Distribution of the electric field module (see Figure 7.8) shows the concentration of the field strength at the borders of the anode. The module of the field strength increases by more than an order of magnitude near borders of the anode.

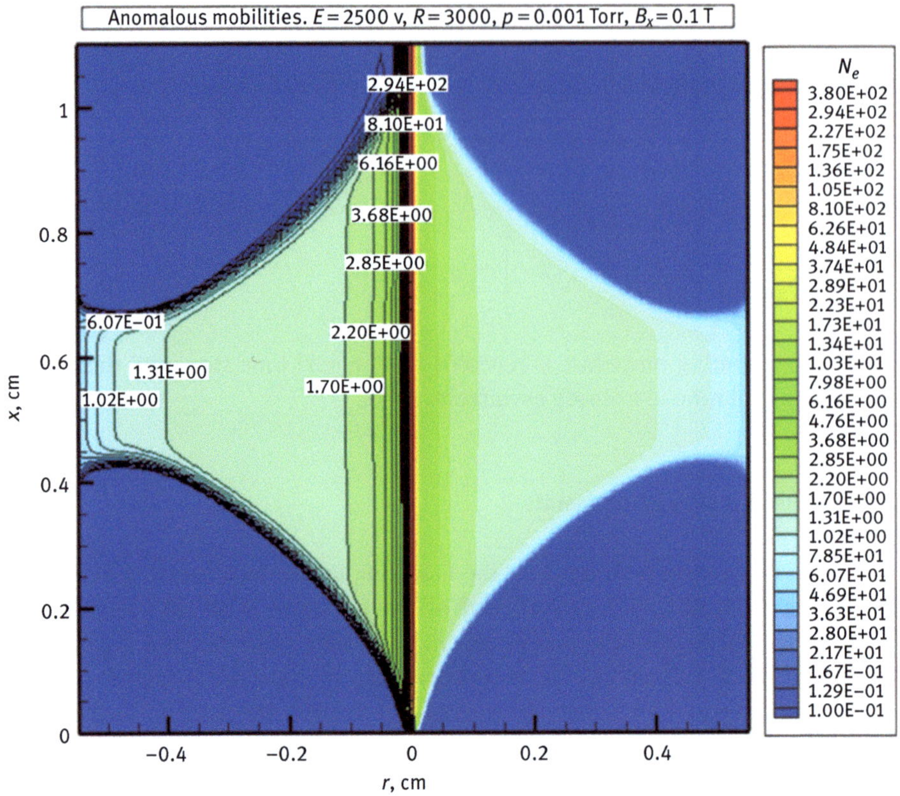

Figure 7.2: Concentration of electrons (in 10^{10} cm^{-3}).

Similar data are shown in Figure 7.9, which shows the distribution of the reduced electric field. This figure clearly shows the area with the maximum strength of the field near the boundaries of the anode, as well as the areas with the lowest electric field strength near the axis of symmetry and in the near-axial region at the cathode, as well as near the anticathode.

Figures 7.10 and 7.11 show projections of the electric field along with the axial and radial coordinates. As might be expected, a local increase in the axial and radial electric field is observed at the boundary of the anode.

Although the local nonhomogeneous of the electric field is observed near the border of the anode, unlike what is observed in the glow discharge at relatively high pressure (about 1 Torr, or higher), rates of ionization are not very large in these areas. Figures 7.12 and 7.13 show an electron impact ionization rate calculated using the averaged electron flux in the axial direction of the axis of symmetry, and in the azimuthal direction. It can be seen that the main source of ionization is the axial fluxes of electrons near the axis of symmetry of the electric discharge chamber.

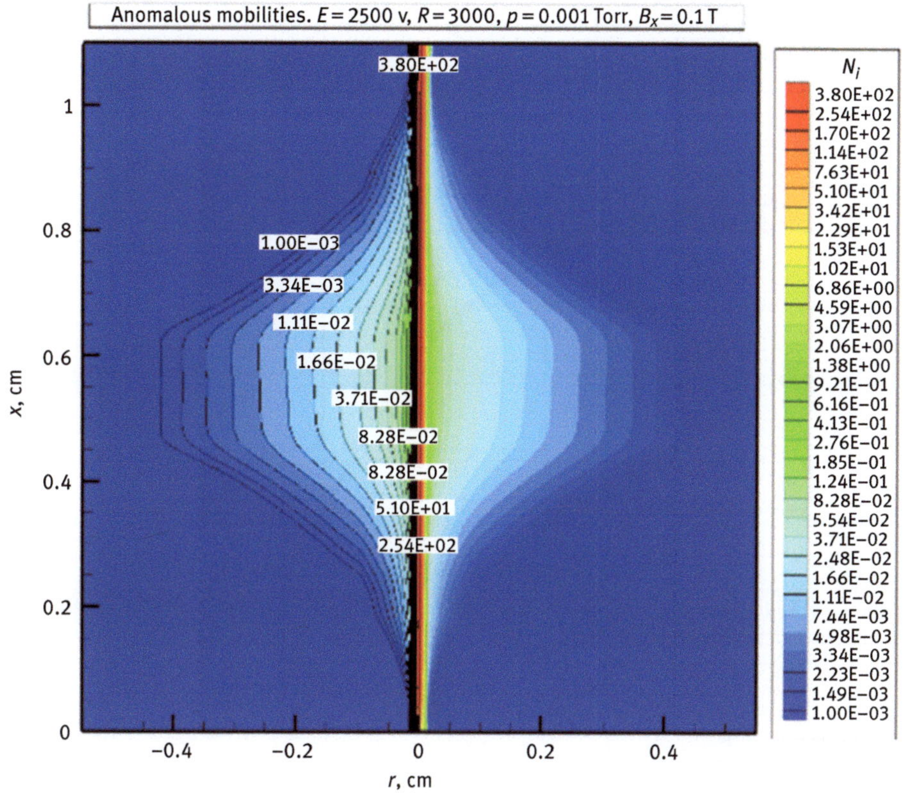

Figure 7.3: Concentration of ions (in 10^{10} cm^{-3}).

Somewhat surprisingly it turned out to be a significant contribution to the ionization rate of axial fluxes of ions in the near-axial region of the discharge (Figures 7.14–7.16). The contribution of ion fluxes in the radial and azimuthal direction is not so significant. Nevertheless, it should be noted that the azimuthal motion of ions and electrons is a significant source of ionization. But it should be borne in mind that here we are talking about the average movements, which are discussed in the drift-diffusion model. When considering the kinetic model, the motion of the electrons and ions of the gas-discharge plasma is more complicated.

Note that, in Figures 7.12–7.16, the rates of ionizations were calculated as follows:

$$\dot{\varpi}_{e,x} = \alpha_e |\mathbf{\Gamma}_{e,x}| Q_{un\,\mathrm{dim}}, \quad |\mathbf{\Gamma}_{e,x}| = n_e |V_{e,x}|, \quad Q_{un\,\mathrm{dim}} = \frac{L^2}{E\mu_{e,0} N_{n,0}},$$

$$\dot{\varpi}_{e,r} = \alpha_e |\mathbf{\Gamma}_{e,r}| Q_{un\,\mathrm{dim}}, \quad |\mathbf{\Gamma}_{e,r}| = n_e |V_{e,r}|, \; \dot{\varpi}_{e,\varphi} = \alpha_e |\mathbf{\Gamma}_{e,\varphi}| Q_{un\,\mathrm{dim}}, \quad |\mathbf{\Gamma}_{e,\varphi}| = n_e |V_{e,\varphi}|,$$

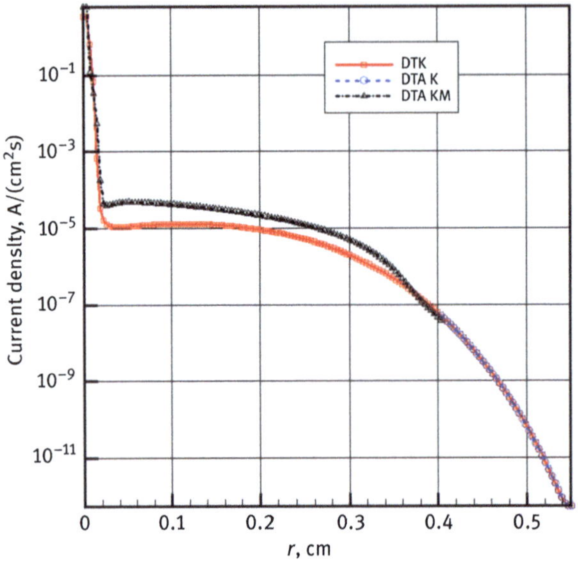

Figure 7.4: Radial distributions of current densities on cathode (DTK) and on anticathode (DTAK), as well as in anticathode outlet (DTAKM).

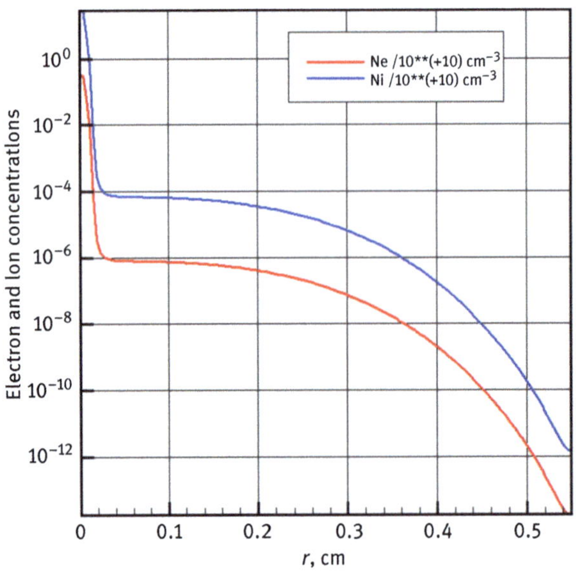

Figure 7.5: Radial distributions of concentrations of electron and ions above cathode.

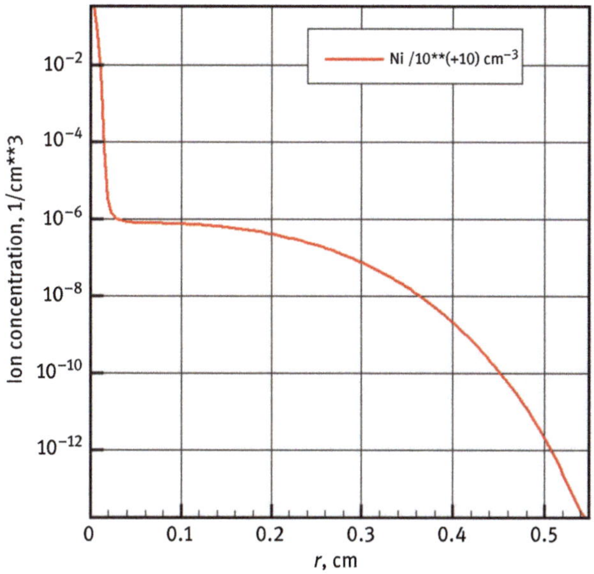

Figure 7.6: Radial distributions of concentrations of electron in cross section with axial coordinate $x = 0.55$ cm.

$$\dot{\varpi}_{i,x} = \alpha_i |\mathbf{\Gamma}_{i,x}| Q_{un\,\mathrm{dim}}, \quad |\mathbf{\Gamma}_{i,x}| = n_i |V_{i,x}|, \quad \dot{\varpi}_{i,r} = \alpha_i |\mathbf{\Gamma}_{i,r}| Q_{un\,\mathrm{dim}}, \quad |\mathbf{\Gamma}_{i,r}| = n_i |V_{i,r}|,$$
$$\dot{\varpi}_{i,\varphi} = \alpha_i \left|\mathbf{\Gamma}_{i,\varphi}\right| Q_{un\,\mathrm{dim}}, \quad \left|\mathbf{\Gamma}_{i,\varphi}\right| = n_i \left|V_{i,\varphi}\right|.$$

We emphasize that these terms give only a glimpse concerning streams of charged particles, which are the most important for the rate of ionization of neutral gas. A more reliable method for the calculation of ionization rates should be based on kinetic models.

Figures 7.17 and 7.18 give coefficients of electron impact ionization that were calculated with the axial and radial components of the electric field.

Figure 7.19 demonstrates the difference between electronic and ionic concentrations. One can see the absence of quasineutrality in the discharge volume. In this figure the space charge regions near the cathode is clearly visible, where the boundary condition of the secondary electron emission is used.

Figure 7.20 shows the axial projection of the averaged velocity of electrons. One can see a correlation between the velocity of the electrons and the electric field. A significant local increase in velocity of the electrons is observed in areas of high values of the electric field (see Figure 7.9). Note that the average velocity of the electrons and ions that is calculated with the drift-diffusion model depends on the model of the drift of charged particles used in the physical model. Near the axis of symmetry in the gas discharge chamber, the oscillations of the axial velocity of

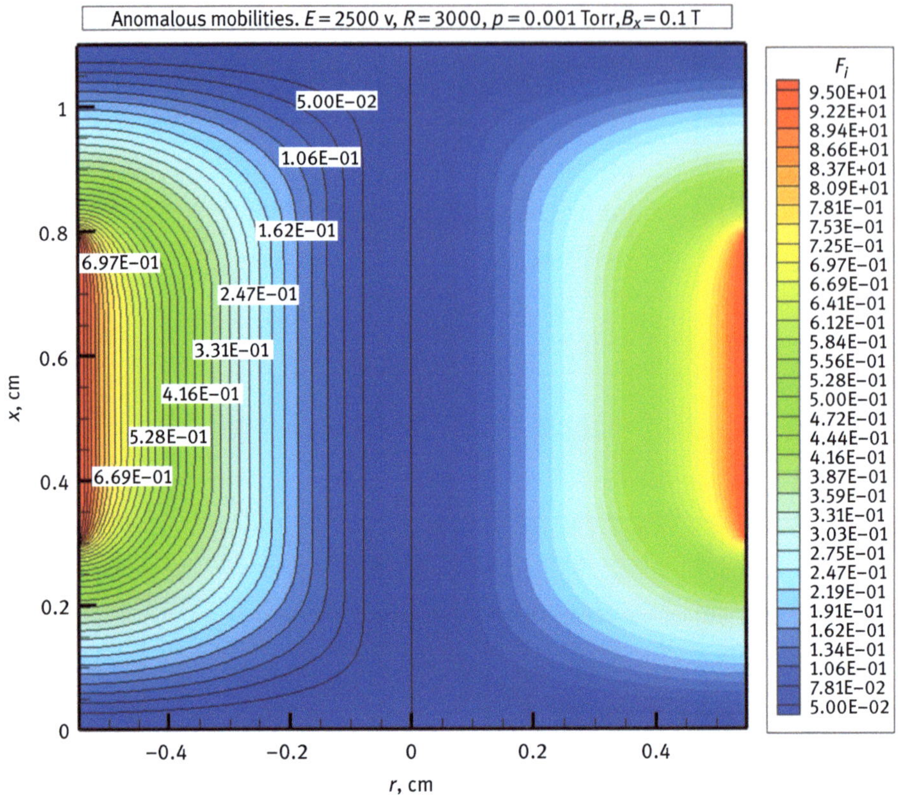

Figure 7.7: The electric potential ($Fi = \varphi/E$).

electrons are observed. These oscillations are also seen in the distributions of axial velocity (Figure 7.21) and in the distribution of electron flux (Figure 7.22).

Fields of axial, radial, and azimuthal projections of the ion velocity are shown in Figures 7.23–7.25.

By analogy with the distribution of the axial velocity of the electrons, the acceleration of ions is also observed near the borders of the anode, as well as the oscillations of ions' velocities near the axis of symmetry. The local maxima in the velocity distribution in the radial and azimuthal direction are located near the boundaries of the anode and near the axis of symmetry. The axial distribution of the velocity of ions and fluxes of ions are shown in Figures 7.26 and 7.27.

Flux densities of electrons and ions, which are shown in Figures 7.22 and 7.27, in accordance with (7.20), (7.21) and (7.24), (7.25) comprise two components. The first part corresponds to the drift of the particles, and the second part corresponds to diffusion. It can be seen that the diffusion of electrons is significant near the cathode and the anticathode. One can see that the diffusion of ions is not significant.

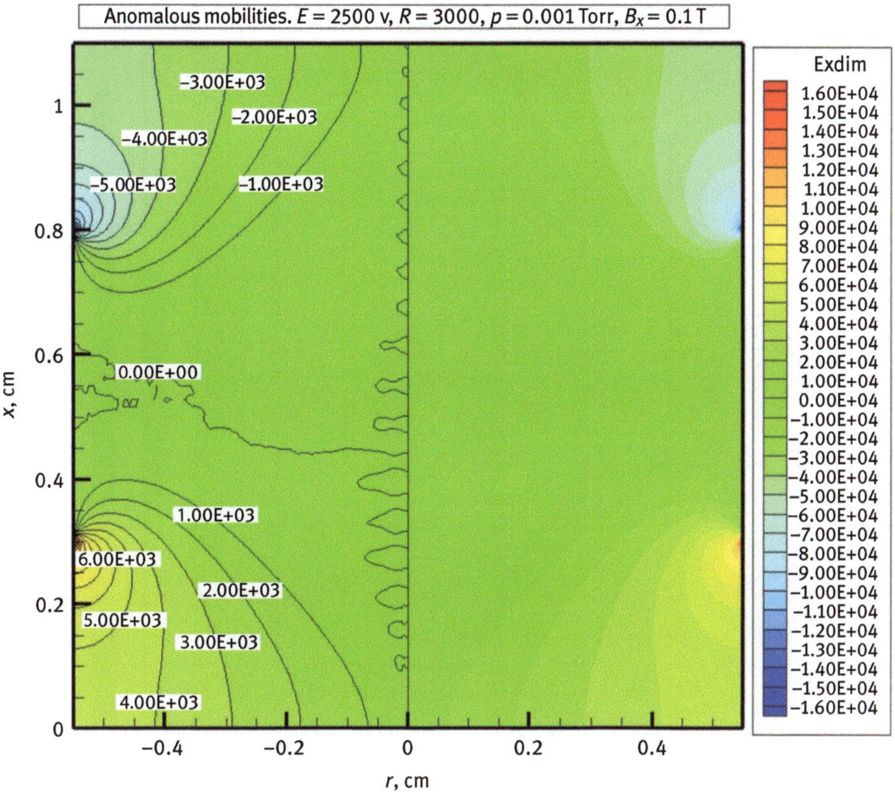

Figure 7.8: Modulus of the electric field, V/cm.

Note especially that the numerical simulation results have one feature connected with axial velocity inversion in the vicinity of the axis of symmetry in the plane of axial symmetry. Figure 7.28 shows the radial distribution of the axial component of the velocity of the electrons and ions in the plane of the anticathode. We see that in the vicinity of the axis of symmetry the ion velocity becomes negative, while the speed of the electrons becomes positive. However, the dimension of the region with the inversion of velocity is insignificant. Computational experiments have shown that the fact of the inversion of velocity strongly depends on the height of the anode.

In addition, an exceptionally strong computational model parameter is the set the boundary conditions for the electric potential in the plane of the anticathode in the meniscus region. Apparently, to get reliable solutions, it is necessary to solve the conjugate problem of discharge processes in the ionization chamber of the Penning discharge and motion of ions in the ion-optical system.

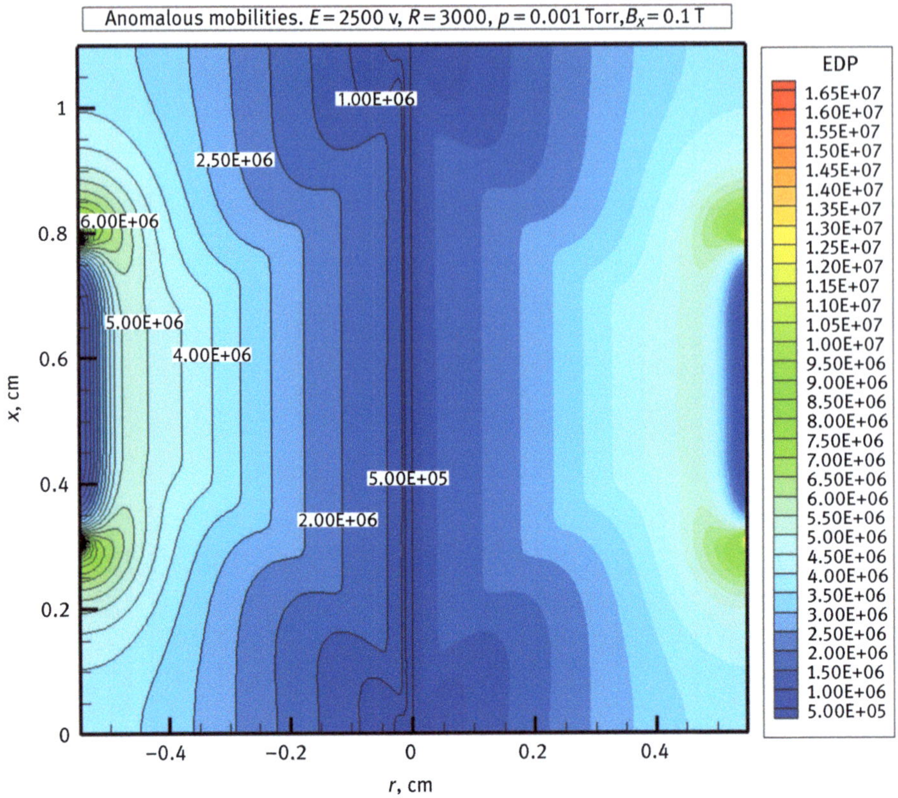

Figure 7.9: Reduced electric field $|E|/p$, V/(cm · Torr).

Figures 7.29–7.31 show radial distribution of electrodynamic parameters in the plane with coordinates $x = 0.55$ cm, that is, at the center of the anode. The radial distributions of flux of electron due to the mobility and diffusion are shown in Figure 7.29. The electronic drift prevails in the discharge volume. However, the radial diffusion of electrons becomes significant near the axis of symmetry. Similar distributions for ions are shown in Figure 7.30.

The radial distribution of electrons and ions in cross section $x = 0.55$ cm is shown in Figure 7.31. An increase in concentrations near the axis of symmetry of the discharge is clearly seen.

Finally, Figure 7.32 illustrates the radial distribution of the ionization rate in cross section at $x = 0.55$ cm, calculated taking into account the impact ionization of electrons and ions. It is evident that the axial fluxes give the main contribution into ionization of neutral particles. However, the ionization of neutral gas due to azimuthal fluxes is important in the greater part of the discharge volume.

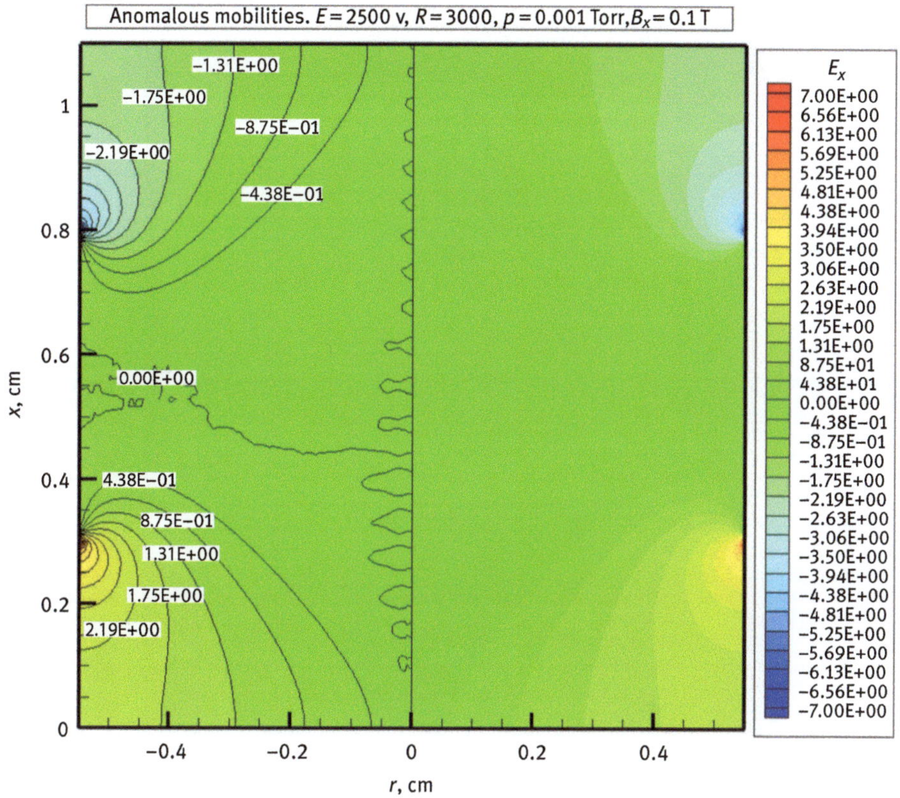

Figure 7.10: Axial electric field $\tilde{E}_x = E_x/(E/L)$.

So, in this section, the modified drift-diffusion model of the Penning discharge has been created, which is based on the nonlinear dependence of the velocity of charged particles from the electric field. This nonlinear dependence is obtained by taking into account the singularities of movement of the charged particles at high fields.

Using the modified drift-diffusion model allowed us to obtain information about the spatial structure of the Penning discharge in a gas-filled chamber at a pressure of $p = 0.0012$ Torr. It is shown that the structure of the Penning discharge is significantly nonhomogeneous in the considered conditions, and there is no quasi-neutrality in the volume of discharge. In the axial region of the discharge, the chamber is formed from an increased concentration of charged particles, and a beam of ions is observed in the meniscus of anticathode. The role of rotation of charged particles in the discharge volume is also shown, primarily for the ionization of neutral gas.

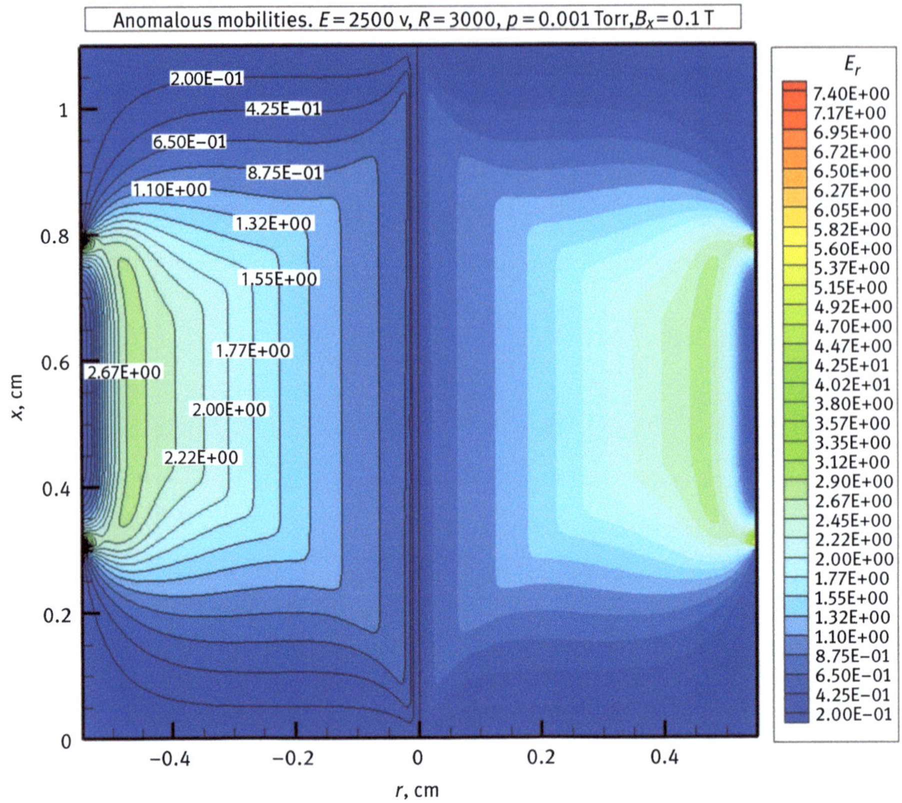

Figure 7.11: Radial electric field $\tilde{E}_r = E_r/(\mathrm{E}/L)$.

Despite the fact that the use of the developed modified drift-diffusion model has provided the possibility to obtain a two-dimensional structure of the Penning discharge, the problematic issues created by the computational and theoretical models that require special consideration should also be noted.

First, despite the presence of a magnetic field, an assessment of the mean free path for collisions and ionization shows that the use of the drift-diffusion approach is not entirely justified. Second, the model of change of the prevailing mechanism of the drift velocity of the electric field should be regarded as provisional.

Third, the ionization model requires its refinement in terms of accounting stepwise ionization, non-Maxwellian distribution functions of electrons and ions.

An important element of the further development of the model should be more detailed by taking into account the plasma-chemical processes of dissociation and ionization. Obviously, these problems can be solved by doing special experiments with diagnostics volume discharge structure, and maybe after

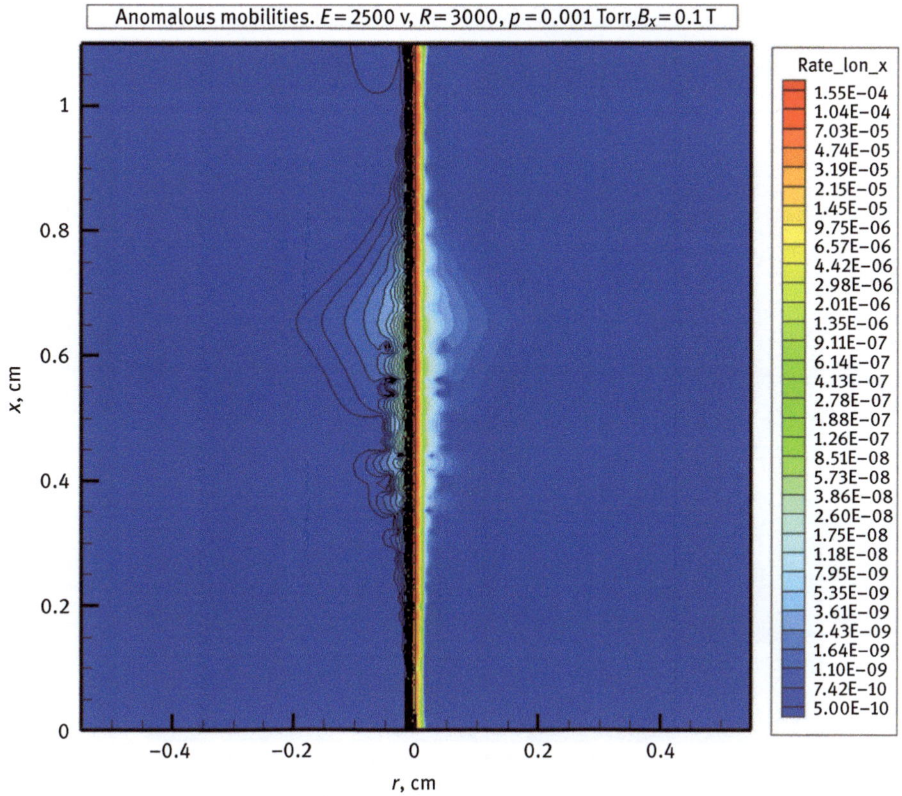

Figure 7.12: Rate of ionization by electronic collisions $\dot{\varpi}_{e,x}$ predicted for axial direction of electrons drift.

applying the study of the structure of the discharge different hybrid PIC/DSMC models or fully kinetic models.

7.2 Numerical study of effectiveness of hydrogen ions acceleration with the Penning discharge at moderate pressure

A model task about the effectiveness of ions acceleration with the use of a diminutive Penning discharge chamber of cylindrical geometry at pressures ~1 Torr is considered in this section.

The moderate pressures about p~1 Torr are considered here; therefore contrary to the model of the Penning discharge presented in the previous section, the

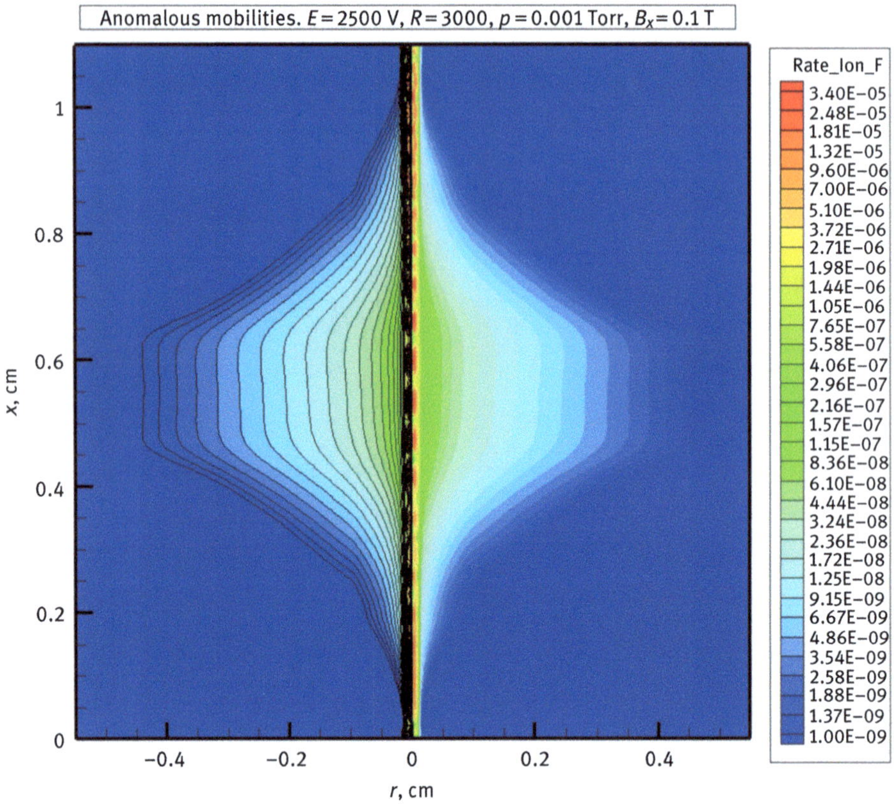

Figure 7.13: Rate of ionization by electronic collisions $\dot{\varpi}_{e,\varphi}$ predicted for azimuthal direction of electrons drift.

classical Townsend approach for ionization coefficient and the linear dependences of electrons and ions mobility from the electric field intensity are used.

Numerical simulation results are presented for two-dimensional Penning discharge at various initial conditions. The results are obtained for atomic hydrogen at pressure 1 – 5 Torr, emf of power supply 500 – 2500 kV, and magnetic field induction of $B = 0 - 0.1$ Tl.

The objective of this model task is to study the two-dimensional structure of the Penning discharge in an axial magnetic field and patterns of counter-rotating ions (protons) and electrons fluids in imposing the axial magnetic field. It also allows to study the structure of accelerated beam of protons, which is of great practical interest.

The principal scheme of the Penning discharge is shown in Figure 7.1. The ring cathode (at $x = X_c$) is the significant part of the cylindrical discharge chamber. An internal open part of the cathode is intended for formation.

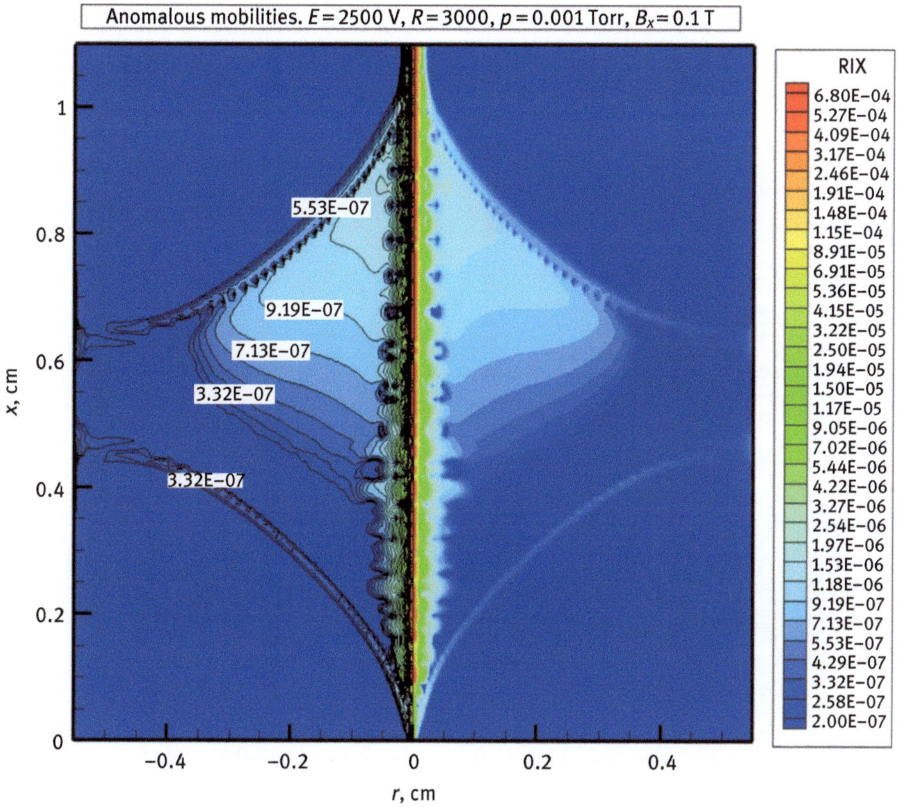

Figure 7.14: Rate of ionization by ionic collisions $\dot{\varpi}_{i,x}$ predicted for axial direction of ions drift.

7.2.1 Governing equations and boundary conditions

A cylindrical 2D glow discharge in atomic hydrogen between flat electrodes in the presence of an axial magnetic field is considered (see Figure 7.1). A drift-diffusion theory is used for the description of gas discharge processes. This theory is based on continuity equations for a concentration of electrons and positive ions together with the equations for the electrostatic field $\mathbf{E} = -\mathrm{grad}\varphi$ and energy conservation equation for the neutral species. Governing equations for the considered task were presented in the previous section. Due to relatively high pressure, the constitutive empirical coefficients were used as for the base drift-diffusion model, namely

$$\mu_e(p) = \frac{3.7\times10^5}{p}\frac{\mathrm{cm}^2}{\mathrm{V\cdot s}},\quad \mu_i(p) = \frac{6.55\times10^3}{p}\frac{\mathrm{cm}^2}{\mathrm{V\cdot s}},$$

Where p is the pressure in Torr. The recombination coefficient β and electron temperature are taken as constants $\beta = 2\times10^{-7}\ \mathrm{cm}^3/\mathrm{s}$, $T_e = 11610\ \mathrm{K}$.

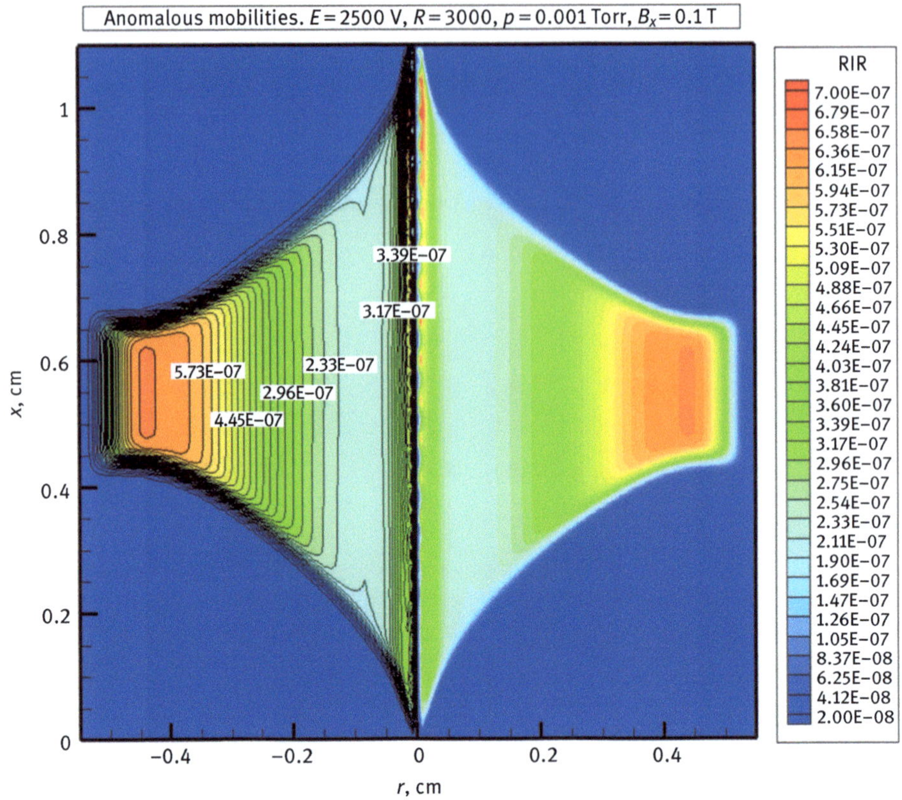

Figure 7.15: Rate of ionization by ionic collisions $\dot{\varpi}_{i,r}$ predicted for radial direction of ions drift.

The ionization coefficient for H is determined as follows:

$$\alpha(E) = pA \exp\left[-\frac{B}{(|\mathbf{E}|/p)}\right] \text{cm}^{-1},$$

where
$$A = 5 \ \frac{1}{\text{cm}\cdot\text{Torr}}, \ B = 130 \ \frac{\text{V}}{\text{cm}\cdot\text{Torr}}.$$

Boundary conditions in the case under consideration is given as follows:

$$x = 0: \qquad \frac{\partial n_i}{\partial x} = 0, \ \mathbf{\Gamma}_e = \gamma \mathbf{\Gamma}_i, \ \varphi = 0;$$

$$x = X, \ r > R_{AC_i}: \qquad \frac{\partial n_i}{\partial x} = 0, \ \mathbf{\Gamma}_e = \gamma \mathbf{\Gamma}_i, \ \varphi = 0;$$

$$r = 0: \qquad \frac{\partial n_e}{\partial r} = \frac{\partial n_i}{\partial r} = \frac{\partial \varphi}{\partial r} = 0;$$

$$r = R: \qquad \frac{\partial n_e}{\partial r} = 0, \ n_i = 0, \ \varphi = V.$$

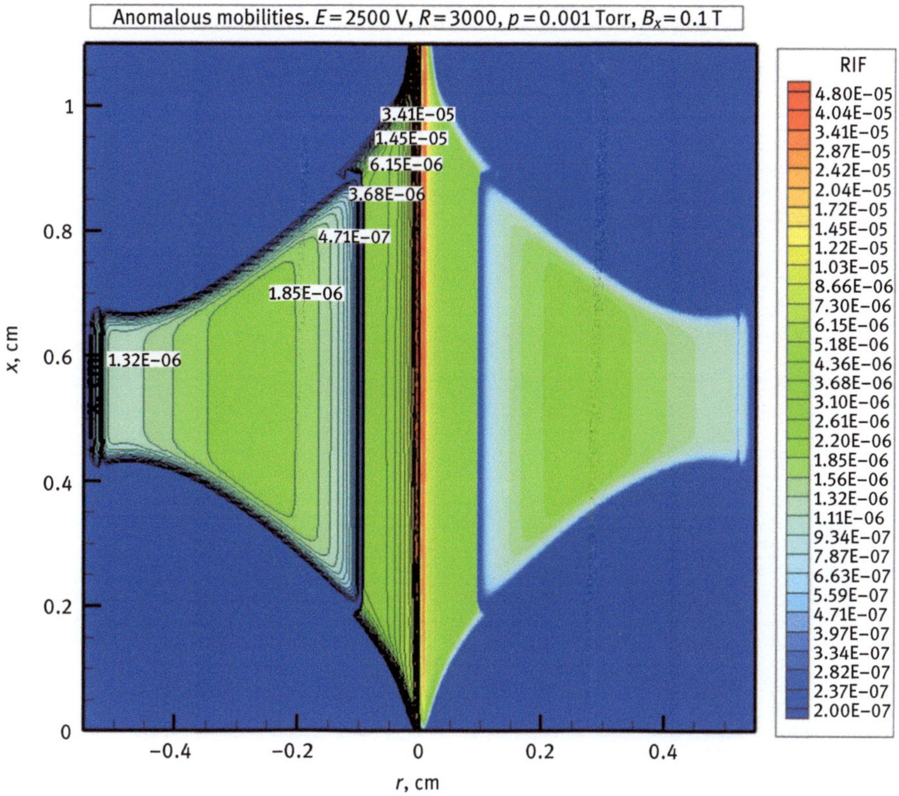

Figure 7.16: Rate of ionization by ionic collisions $\dot{\varpi}_{i,\varphi}$ predicted for radial direction of ions drift.

Boundary conditions for ions and electrons at $r < R_{ACi}$ (an outlet orifice of the chamber) should be formulated in view of electric field and concentration of charged particles outside of the discharge chamber. To simplify this problem, a special kind of boundary condition was used:

$$x = X, \quad r < R_{AC_i}: \qquad \frac{\partial n_i}{\partial x} = 0, \quad \mathbf{\Gamma}_e = \gamma^* \mathbf{\Gamma}_i, \quad \varphi = 0;$$

$$x = 0; \; x = X, \quad r > R_{AC_i}: \quad \frac{\partial n_i}{\partial x} = 0, \quad \mathbf{\Gamma}_e = \gamma \mathbf{\Gamma}_i, \quad \varphi = 0,$$

where γ is the coefficient of the second ion–electron emission; γ^* is the effective coefficient of the second ion–electron emission, which is taken of order $\gamma^* \sim (10^{-1} - 10^{-2})\gamma$.

For cylindrical surface of anode, the following boundary conditions were used:

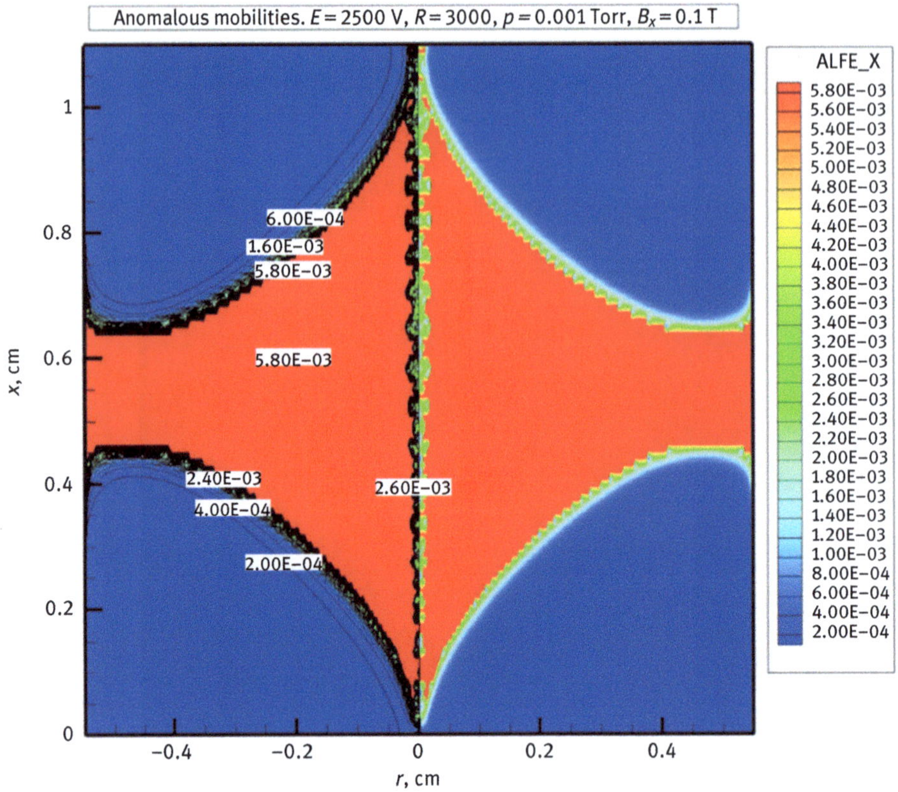

Figure 7.17: Coefficient of ionization α_e, calculated for axial flux of electrons (ALFE_X = α_e in 1/cm).

$$X_{A_1} < x < X_{A_2}, \quad r = R_C: \quad \frac{\partial n_e}{\partial r} = 0, \quad n_i = 0, \quad \varphi = V.$$

For cylindrical surface between anode and cathode:

$$X_{A_1} > x, \quad r = R_C: \quad \frac{\partial n_e}{\partial r} = \frac{\partial n_i}{\partial r} = \frac{\partial \varphi}{\partial r} = 0,$$

$$X_{A_2} < x, \quad r = R_C: \quad \frac{\partial n_e}{\partial r} = \frac{\partial n_i}{\partial r} = \frac{\partial \varphi}{\partial r} = 0.$$

A quasineutral plasma cloud of a spherical form in the center of the chamber is used as the initial condition.

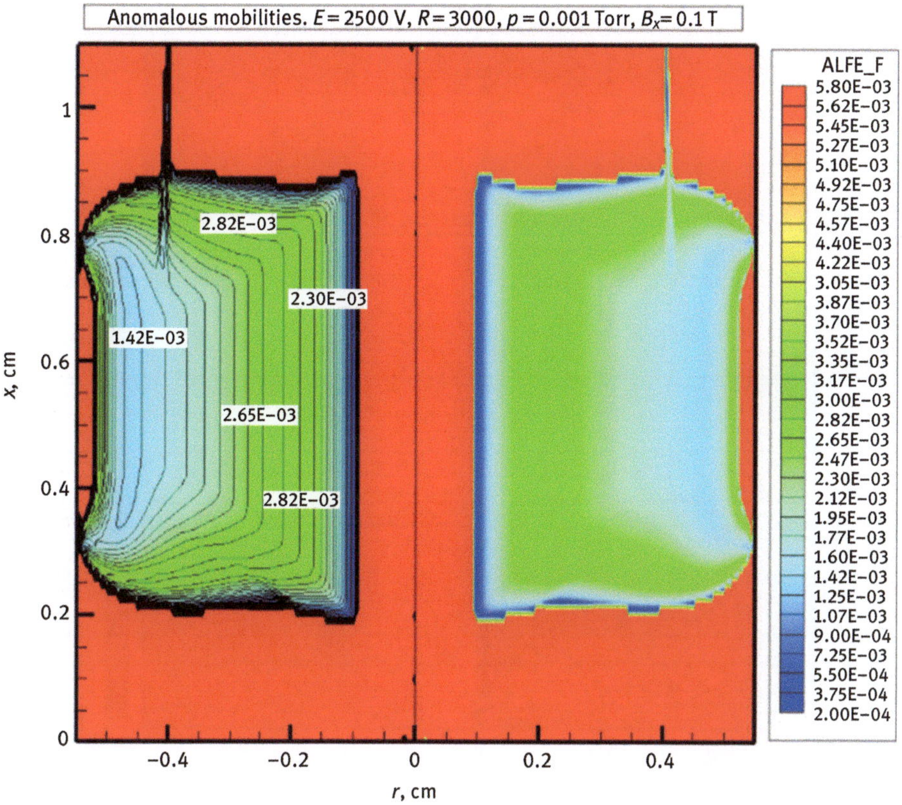

Figure 7.18: Coefficient of ionization α_e, calculated for azimuthal flux of electrons (ALFE_F = α_e in 1/cm).

7.2.2 Numerical simulation results

Calculations were performed for the following initial data: pressure in molecular hydrogen H_2 was $p = 1-5$ Torr; emf of power supply $E = 70-200$ V; $R_0 = 3$ kOhm; $\gamma = 0.33$; $B_x = 0.1$, $R_C = R_{AC_e} = 0.55$ cm, $R_{AC_i} = 0.4$ cm, $X_{A1} = 0.4$ cm, $X_{A2} = 0.7$ cm, $X_c = 1.1$ cm, $L = X_C = 1.1$ cm (this value was used as the spatial scale of the task).

Axial distributions of velocities of electrons and ions at different emf ($E = 70-200$ V), the radius of the outlet orifice ($R_{AC_i} = 0.2-0.4$ cm), and efficient coefficient of the second electronic emission ($\gamma^* = D_g = 0.1-0.01$) are shown in Figures 7.33 and 7.34.

Maximal velocities of electrons ($V_{e,x} \sim 5\times10^8$ cm/s) and ions ($V_{i,x} \sim 5\times10^8$ cm/s) are achieved for $E = 200$ V. Effects from other parameters are negligible.

Axial distributions of volume concentrations of ions and electrons for the same conditions are shown in Figure 7.35. The skewness in the concentration distributions

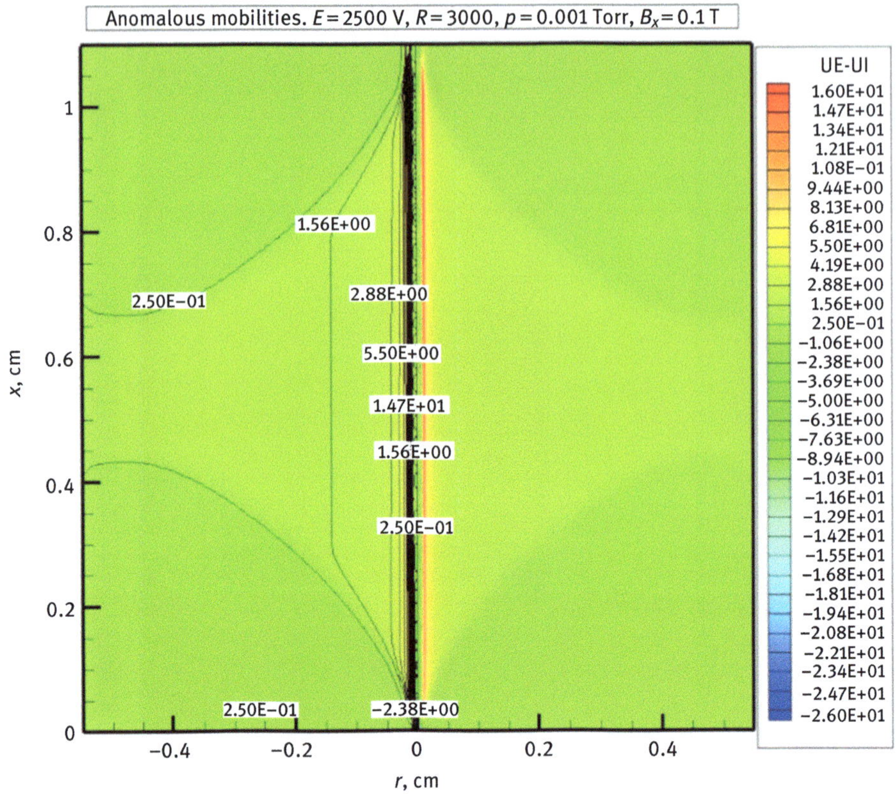

Figure 7.19: Difference between electronic and ionic concentrations (UE and UI are the concentration of electrons and ions in 10^{10} cm^3).

is observed because of the use of effective coefficients of the secondary ion–electron emission at the outlet orifice.

Concentrations of electrons and ions, as well as the current density, near surfaces of cathode and anticathode, are shown in Figures 7.36 and 7.37. These distributions are presented for minimal ($E = 70$ V) and maximal ($E = 200$ V) emf from the region investigated. Note that the decrease in effective coefficients γ^* gives a larger effect for the distribution of electrons.

Let us consider the electrodynamic structure of the Penning discharge at $E = 70$ V. In the conditions under consideration, this is the lower limit of E for this discharge existence.

Maximal electronic concentration is located near the anode (Figure 7.38; from the left). Concentrations of ions are achieved maximum near anode as well, where we can see the region of the quasineutral plasma. But another part of the gas discharge chamber is filled out by the ions (Figure 7.38, from the right).

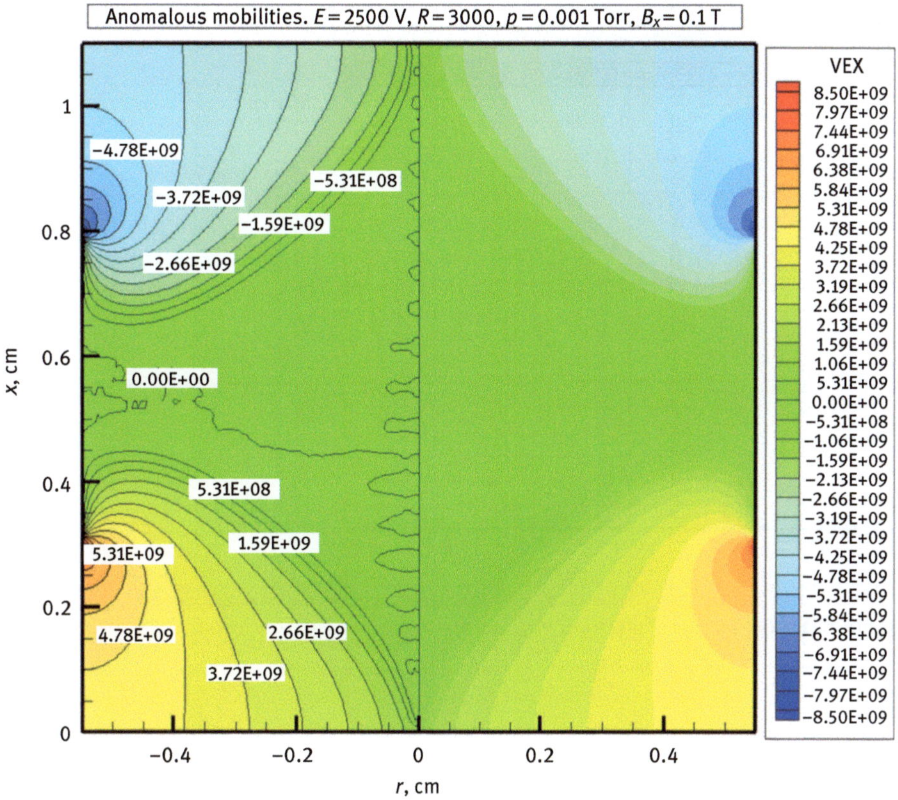

Figure 7.20: Axial projection of velocity of electrons (VEX – $V_{e,x}$ in cm/s).

So, there is a significant region of the volume charge in the chamber (see Figure 7.39, from the left). Maximal rate of ionization is observed near the anode (Figure 7.39, from the right), where the reduced electric field E/p is not so large, as in regions close to the cathode and anticathode (Figure 7.39, from the left).

Fields of axial and azimuthal velocities of electrons and ions are shown in Figures 7.40 and 7.41 (axial distributions of these velocities were presented also in Figures 7.33 and 7.34). One can see that there are two local regions near boundaries of anode, where maximal values of reduced electric field and axial and azimuthal velocities are observed. Note that the reduced electric field achieves in the chamber of a value of about 500 V/cm·V.

Analogous data for fields of electrodynamic characteristics were performed for $E = 200$ V (see Figures 7.42–7.46). The reduced electric field, in this case, achieves in the chamber a value of about 1,000 V/cm V. It means that the use of classical approximation for the first Townsend is not correct at larger emf.

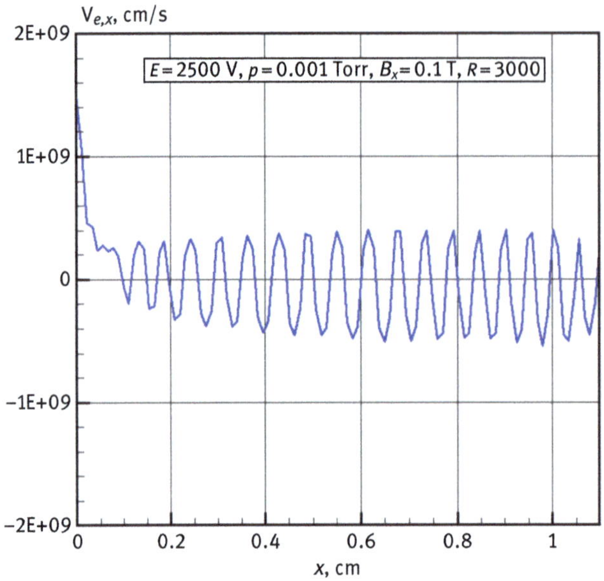

Figure7.21: Axial velocity of electrons $V_{e,x}$, cm/s.

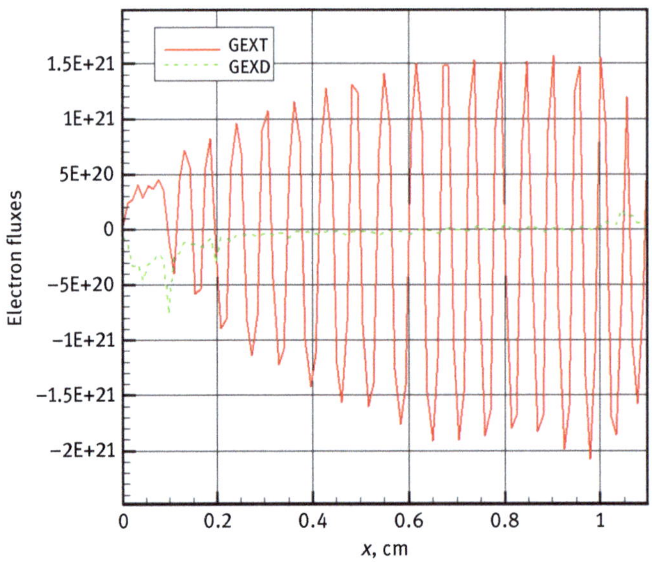

Figure 7.22: Axial flux of electrons in 1/(cm^{2}s).

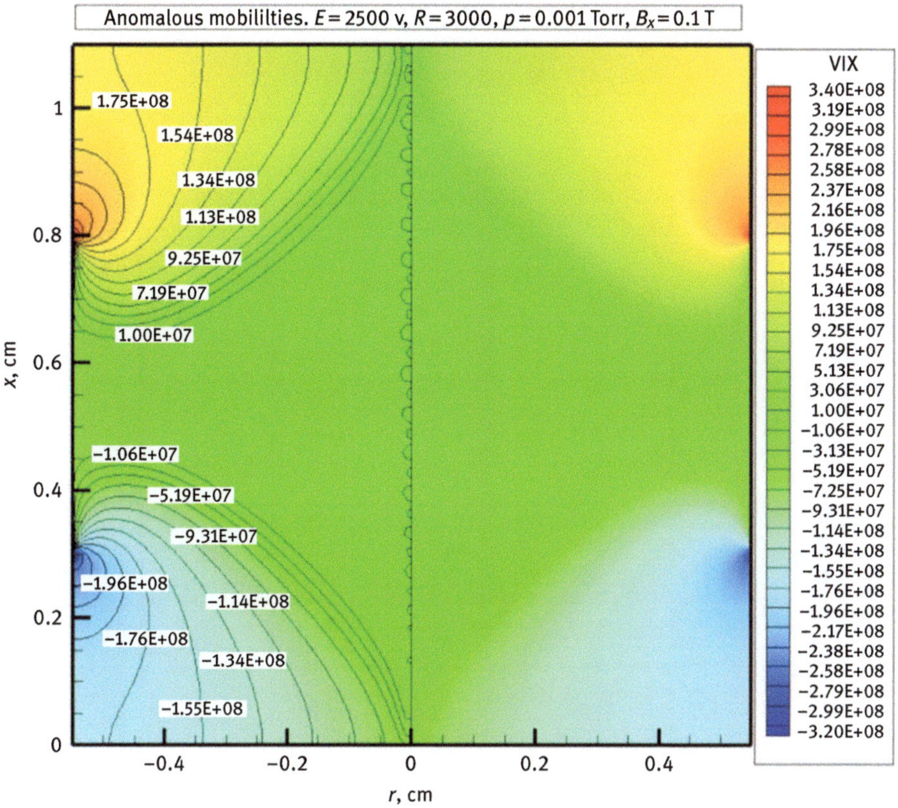

Figure 7.23: Axial projection of velocity of ions (VIX – $V_{i,x}$in cm/s).

Axial and azimuthal velocities of electrons and ions are shown in Figures 7.44 and 7.45. It is particularly remarkable that there is a region of extreme acceleration of electrons and ions, which is located near the outlet orifice. Probably it is connected with corresponding behavior of electron concentrations and with relatively large radial diffusion fluxes.

So, in this section, a drift-diffusion model of the Penning discharge at pressure 1 Torr has been tested in a wide region of its functioning with classical constitutive empirical coefficients. One of the significant elements of the model is the use of the effective coefficient of the secondary ion–electron emission for an outlet orifice of the chamber. This model allowed us to obtain information about the spatial structure of the Penning discharge in a gas-filled chamber at pressures of $p = 1 - 5$ Torr.

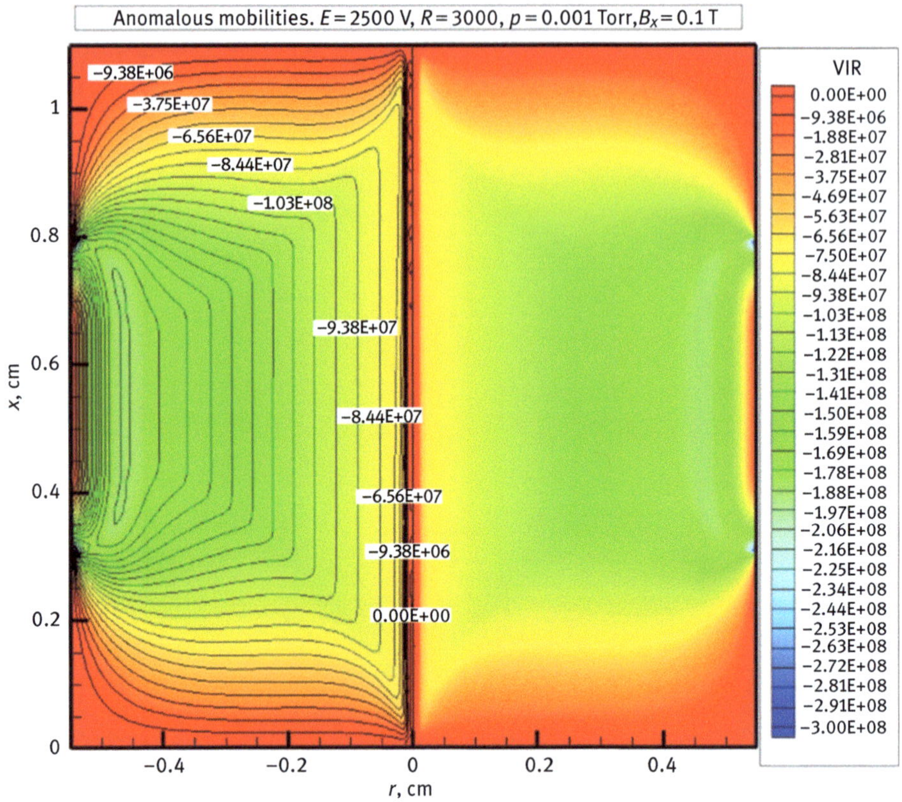

Figure 7.24: Radial projection of velocity of ions (VIR – $V_{i,r}$ in cm/s).

7.3 Numerical simulating the two-dimensional structure of the Penning discharge using the modified drift-diffusion model at intermediate pressures

In this section the spatial structure of the Penning discharge at pressures $p = 0.1 - 0.01\,\text{Torr}$ is investigated within the framework of the modified drift-diffusion model. This modification includes, as in previous section, three models of elementary physical processes, which take into account peculiarities of gas discharge processes at low pressures and large reduced electric fields. Presented results of numerical simulations are in reasonable agreement with the available experimental data.

A modified drift-diffusion MHD model for the calculation of drift velocities of ions and electrons under transition from low to high reduced electric fields was discussed in Section 7.1. The results of the calculation of drift velocities using this model are shown in Figure 7.47. It is seen that at $E/p \geq 100\,\text{V/cm} \cdot \text{Torr}$ there is

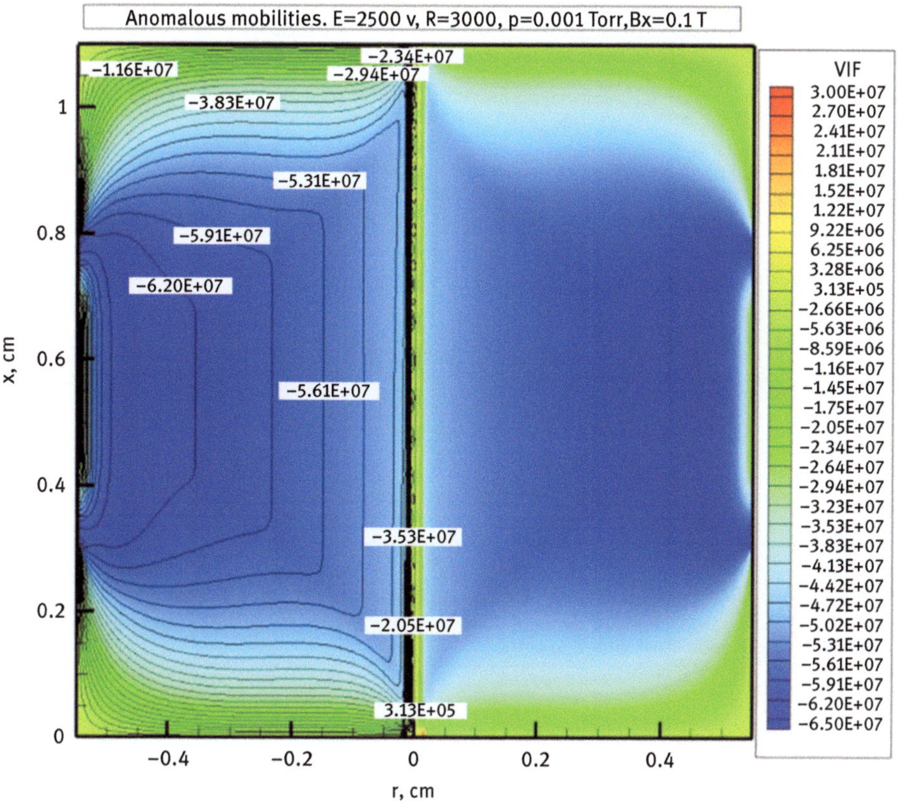

Figure 7.25: Azimuthal velocity of ions (VIF – $V_{i,\varphi}$ in cm/s).

observable slowing down of the drift velocity with an increase in E/p. However, it is obvious that when $E/p > 10^6\,\mathrm{V/cm \cdot Torr}$ this model does not work.

The second modification relates to the method for the calculation of ionizing processes at increasing E/p. It is well known that the Townsend formula gives acceptable results at $E/p < 1000\,\mathrm{V/cm \cdot Torr}$ (more precisely, for molecular hydrogen for $150 < E/p < 600\,\mathrm{V/cm \cdot Torr}$). The model of the ionization coefficient proposed in Section 7.1 is illustrated in Figure 7.48. It is assumed that at an energy of electrons less than the potential ionization of hydrogen, the Townsend formula is correct. At higher energies, the ionization coefficient $\alpha$is calculated as follows:

$$\alpha = N\sigma_i,$$

where σ_i is the cross section of ionizing collisions predicted by the Tompson formula

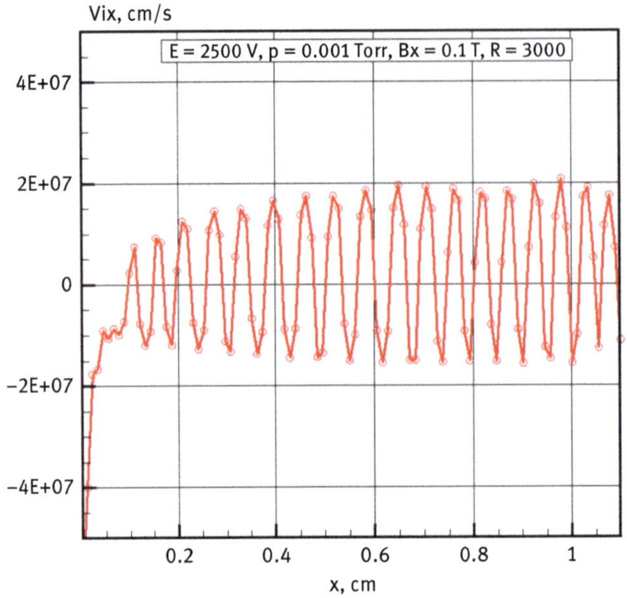

Figure 7.26: Axial velocity of ions $V_{i,x}$, cm/c.

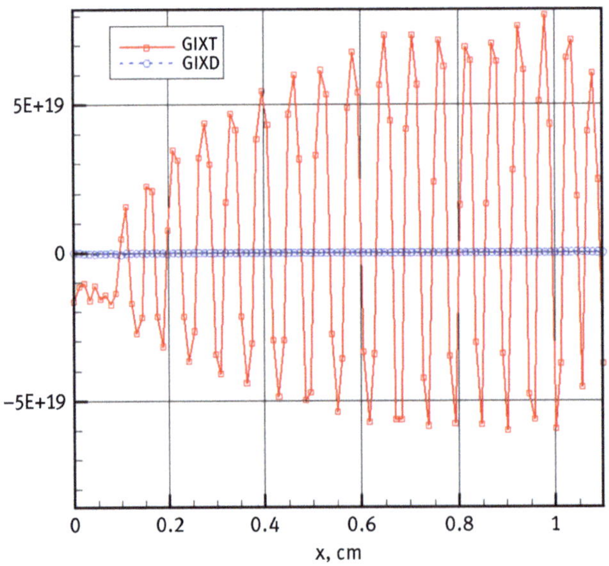

Figure 7.27: Axial flux of ions, in 1/(cm^{2}s).

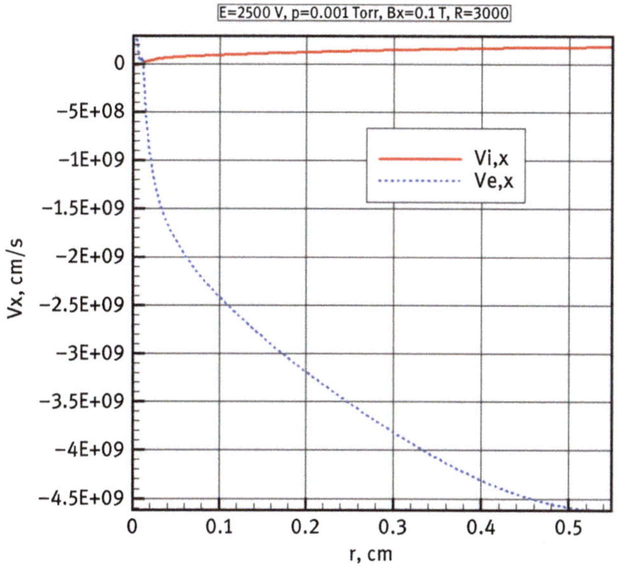

Figure 7.28: Radial distributions of axial velocities of electrons $V_{e,x}$ and ions $V_{i,x}$ in the plane of anticathode. Radius of outlet is $R_{AC_i} = 0.4$ cm (see Figure 7.1).

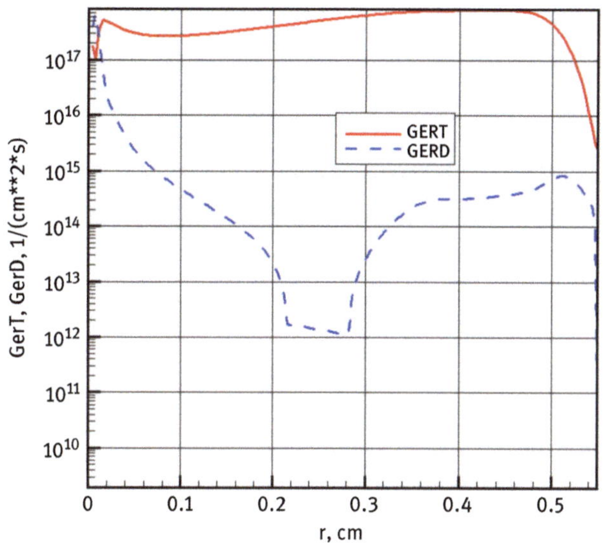

Figure 7.29: Radial distributions of axial flux of electrons due to drift (GerT) and diffusion (GerD) in cross section with axial coordinate $x = 0.55$ cm.

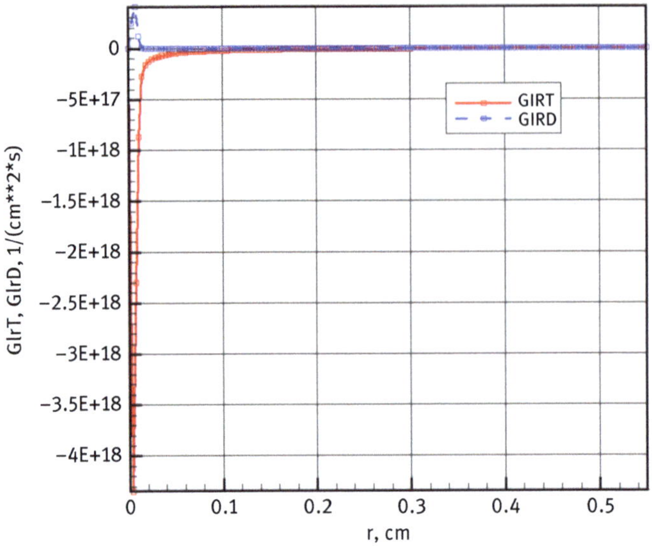

Figure 7.30: Radial distributions of axial ion fluxes due to drift (GlrT) and diffusion (GlrD) in cross section with axial coordinate $x = 0.55$ cm.

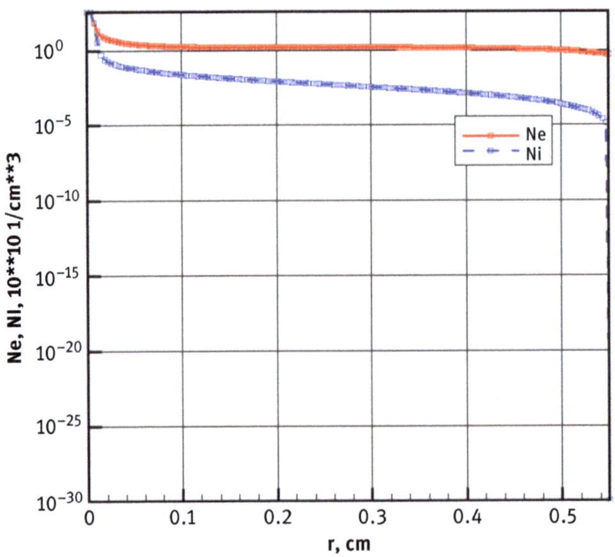

Figure 7.31: Radial distributions of concentration of electrons in cross section with axial coordinate $x = 0.55$ cm.

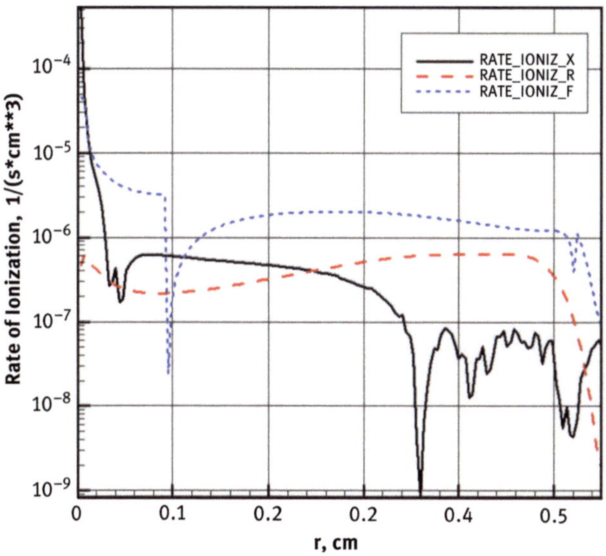

Figure 7.32: Radial distribution of total rate of ionization due to electronic collisions in cross section with axial coordinate $x = 0.55$ cm.

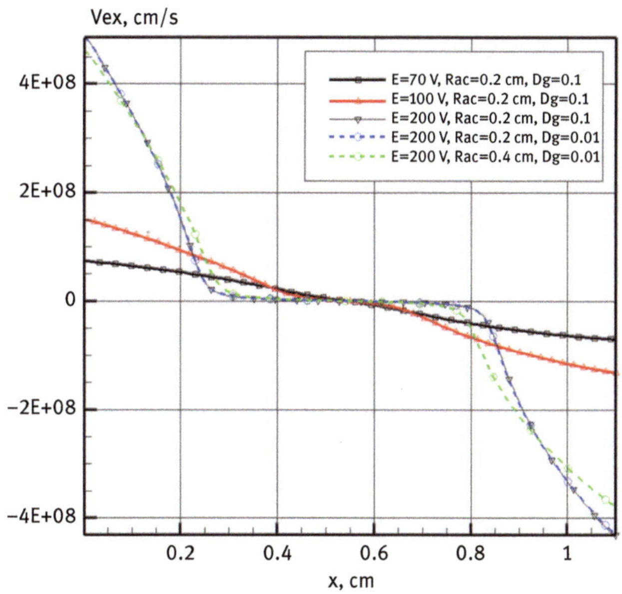

Figure 7.33: Axial distributions of velocity of electrons ($V_{e,x}$, cm/s) in the Penning discharge at $E = 70 - 200$ V, $R_0 = 3000$ Ohm, $p = 1$ Torr, $B_x = 0.1$ T, $R_{ACi} = 0.2$, and 0.4 cm.

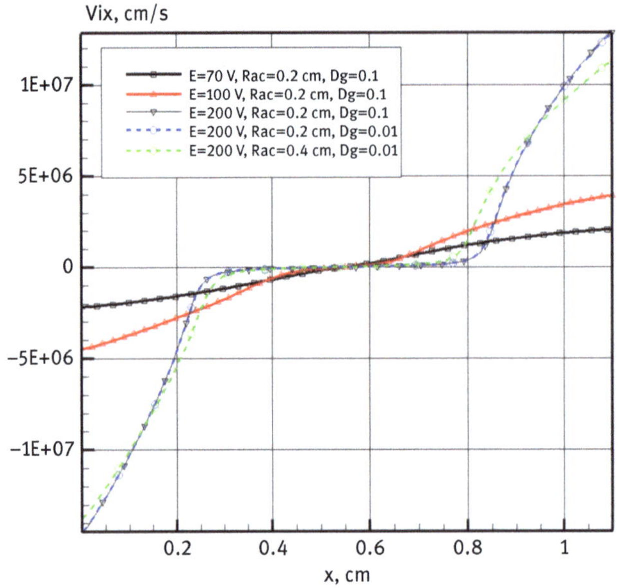

Figure 7.34: Axial distributions of velocity of electrons ($V_{i,x}$, cm/s) in the Penning discharge at $E = 70 - 200$ V, $R_0 = 3000$ Ohm, $p = 1$ Torr, $B_x = 0.1$ T, $R_{ACi} = 0.2$, and 0.4 cm.

$$\sigma_i = 4\pi a_0^2 \left(\frac{I_{H_2}}{\varepsilon}\right)^2 \cdot \frac{(\varepsilon - I_{H_2})}{I_{H_2}} \ \text{cm}^2.$$

Here, I_{H_2} is the potential of ionization of H_2, $4\pi a_0^2 = 3.52 \times 10^{-16}$ cm^2; a_0 is the Bohr radius and ε is the energy of electrons (as a rule, I_{H_2} and ε are measured in eV).

For cross-linking of numerical values of the first ionization coefficient, final formula is

$$\alpha = pA^* \exp\left[-\frac{B}{(E/p)}\right],$$

where $A^* = 2.751/\text{cm} \cdot \text{Torr}$; $B = 130\,\text{V}/\text{cm} \cdot \text{Torr}$.

According to Figure 7.48 the ionization coefficients first increase and then decrease as an increase in E/p.

Thus, in the Penning discharge used as the ions accelerator, there are two competitive mechanisms at applied voltage drop increasing, namely, a decrease of ionization efficiency and an increase of charged particle energies.

In this section, the modified drift-diffusion model is applied for the analysis of the Penning discharge at the pressure $p = 0.01 - 0.1$ Torr. At these pressures, the classical model cannot be applied for modelling the gas discharge chamber with a size of ~ 1 cm. On the other side, under these conditions, typically the reduced electric fields do not exceed the values of $E/p \sim 10^4\,\text{V}/\text{cm} \cdot \text{Torr}$.

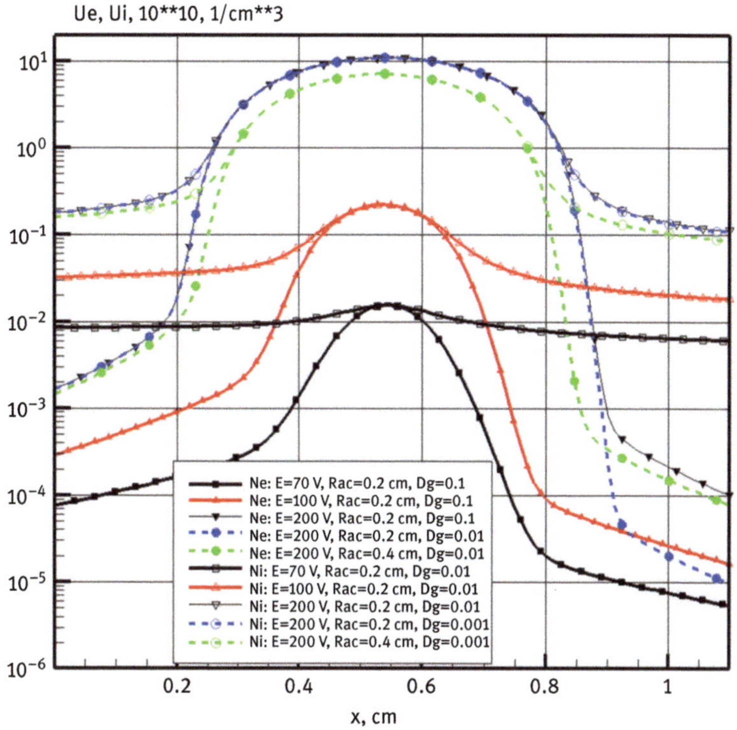

Figure 7.35: Axial distributions of concentration of electrons and ions (in 10^{10} cm^{-3}) in the Penning discharge at $E = 70 - 200$ V, $R_0 = 3000$ Ohm, $p = 1$ Torr, $B_x = 0.1$ T, $R_{ACi} = 0.2$, and 0.4 cm.

Note that the reduced electric field in a gas discharge chamber depends not only on applied voltage drop but also on the configuration of volume charge.

7.3.1 Governing equations

A cylindrical two-dimensional glow discharge in molecular hydrogen between flat electrodes with a presence of axial magnetic field is considered (Figure 7.49).

A modified drift-diffusion model (see Section 7.1) is used for the description of gas discharge processes. This theory is based on the continuity equations for electron and positive ion concentrations n_e and n_i together with the equations for the electrostatic field $\mathbf{E} = -\mathrm{grad}\varphi$. The governing equations of the drift-diffusion theory with axial magnetic field have the following form:

$$\frac{\partial n_e}{\partial t} + \frac{\partial \mathbf{\Gamma}_{e,x}}{\partial x} + \frac{1}{r}\frac{\partial r \mathbf{\Gamma}_{e,y}}{\partial r} = \dot{\omega}_i = \alpha(E)|\mathbf{\Gamma}_e| - \beta n_i n_e, \tag{7.29}$$

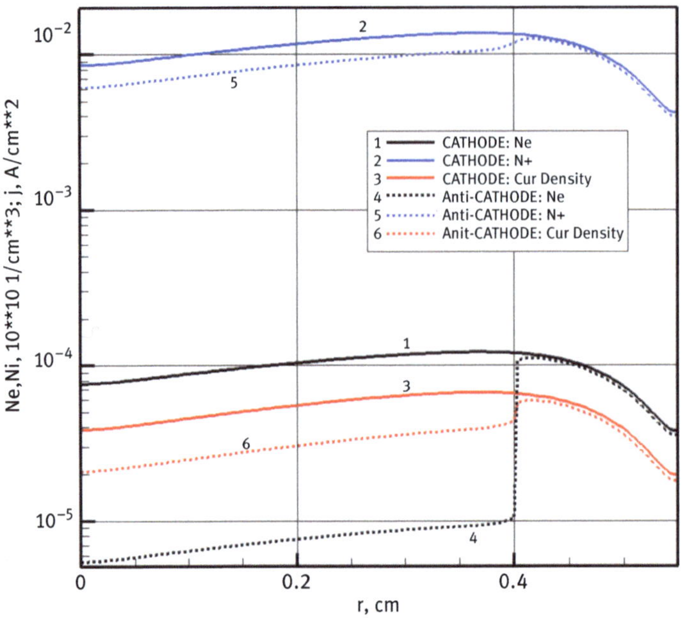

Figure 7.36: Distributions of concentrations of electrons and ions (in 10^{10} cm^{-3}), and current densities along cathode and anticathode in the Penning discharge at $E = 70$ V, $R_0 = 3000$ Ohm, $p = 1$ Torr, $B_x = 0.1$ T, $R_{ACi} = 0.4$ cm, $\gamma^* = 0.1\gamma$.

$$\frac{\partial n_i}{\partial t} + \frac{\partial \mathbf{\Gamma}_{i,x}}{\partial x} + \frac{1}{r}\frac{\partial r\mathbf{\Gamma}_{i,y}}{\partial r} = \dot{\omega}_i = \alpha(E)|\mathbf{\Gamma}_e| - \beta n_i n_e, \tag{7.30}$$

$$\frac{\partial^2 \varphi}{\partial x^2} + \frac{1}{r}\frac{\partial}{\partial r} r \frac{\partial \varphi}{\partial r} = 4\pi e(n_e - n_i)\,, \tag{7.31}$$

where

$$\mathbf{\Gamma}_e = n_e \mathbf{u}_e = -\hat{\mathbf{D}}_e \mathrm{grad} n_e - n_e \hat{\boldsymbol{\mu}}_e \mathbf{E}\,,\ \mathbf{\Gamma}_i = n_i \mathbf{u}_i = -\hat{\mathbf{D}}_i \mathrm{grad} n_i + n_i \hat{\mu}_i \mathbf{E}\,,$$

$$\hat{\mathbf{D}}_e = \begin{pmatrix} D_{e,x} & 0 \\ 0 & \frac{D_{e,r}}{1+b_e^2} \end{pmatrix},\ \hat{\mathbf{D}}_i = \begin{pmatrix} D_{i,x} & 0 \\ 0 & \frac{D_{i,r}}{1+b_i^2} \end{pmatrix},$$

$$\hat{\boldsymbol{\mu}}_e = \begin{pmatrix} \mu_{e,x} & 0 \\ 0 & \frac{\mu_{e,r}}{1+b_e^2} \end{pmatrix},\ \hat{\boldsymbol{\mu}}_i = \begin{pmatrix} \mu_{i,x} & 0 \\ 0 & \frac{\mu_{i,r}}{1+b_e^2} \end{pmatrix}, \tag{7.32}$$

$$\mathbf{j} = e(\mathbf{\Gamma}_i - \mathbf{\Gamma}_e),$$

$\alpha(E)$ and β are the ionization and recombination coefficients; $\mathbf{\Gamma}_e$ and $\mathbf{\Gamma}_i$ are the electron and ion flux densities; φ is the potential of electric field; $\mathbf{u}_e$ and $\mathbf{u}_i$ are the averaged velocities of electrons and ions; $\mu_e = \frac{e}{m_e v_e}$ and $\mu_i = \frac{e}{m_i v_{in}}$ are the electrons and

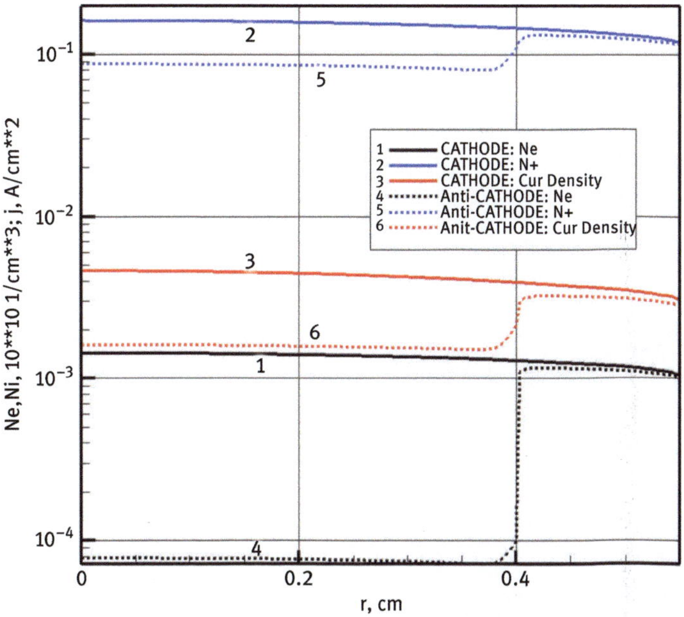

Figure 7.37: Distributions of concentrations of electrons and ions (in 10^{10} cm^{-3}), and current densities along cathode and anticathode in the Penning discharge at $E = 200$ V, $R_0 = 3000$ Ohm, $p = 1$ Torr, $B_x = 0.1$ T, $R_{ACi} = 0.4$ cm, $\gamma^* = 0.1\gamma$.

ions mobilities; $D_e = \frac{kT_e}{e}\mu_e$ and $D_i = \frac{kT_i}{e}\mu_i$ are the electrons and ions diffusion coefficients; ν_{en}, ν_{ei}, and ν_{in} are the frequencies of electron-neutral, electron-ion, and ion-neutral collisions; $\nu_e = \nu_{en} + \nu_{ei}$; T_e and T_i are the temperatures of electrons and ions.

Taking into account that the magnetic field has only x-component B_x (see Figure 7.49), one can write

$$\mathbf{\Gamma}_{e,x} = n_e u_{e,x} = \left(-D_e\frac{\partial n_e}{\partial x} - \mu_e n_e E_x\right),\ \mathbf{\Gamma}_{e,r} = n_e u_{e,r} = \frac{1}{1+b_e^2}\left(-D_e\frac{\partial n_e}{\partial r} - \mu_e n_e E_r\right),$$

$$\mathbf{\Gamma}_{e,\varphi} = n_e u_{e,\varphi} = -b_e \mathbf{\Gamma}_{e,r},$$

$$\mathbf{\Gamma}_{i,x} = n_i u_{i,x} = \left(-D_i\frac{\partial n_i}{\partial x} - \mu_i n_i E_x\right),\ \mathbf{\Gamma}_{i,r} = n_i u_{i,r} = \frac{1}{1+b_i^2}\left(-D_i\frac{\partial n_i}{\partial r} + \mu_i n_i E_r\right),$$

where $b_e = \frac{\mu_e}{c}B_x$ and $b_i = \frac{\mu_i}{c}B_x$ are the Hall parameters of electrons and ions.

The source summand in the right-hand side can be written as follows:

$$|\mathbf{\Gamma}_e| = \sqrt{\mathbf{\Gamma}_{e,x}^2 + \mathbf{\Gamma}_{e,r}^2 + \mathbf{\Gamma}_{e,\varphi}^2}. \tag{7.33}$$

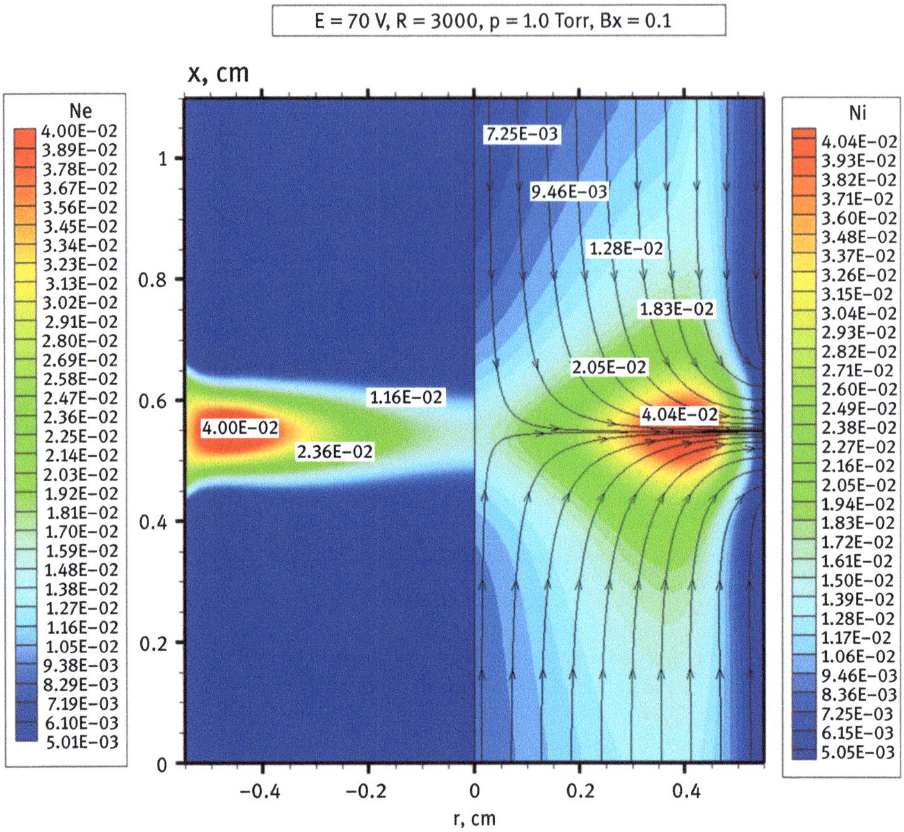

Figure 7.38: The concentration of electrons and ions (in 10^{10} cm^{-3}) in the Penning discharge at $E = 70$ V, $R_0 = 3000$ Ohm, $p = 1$ Torr, $B_x = 0.1$ T, $R_{ACi} = 0.4$ cm, $\gamma^* = 0.1\gamma$. Here and further, electric field lines are presented on the right-hand side of the figures.

Note that the azimuthal component of electronic flux $\mathbf{\Gamma}_{e,\varphi}$ can be significant due to the condition $b_e \sim 1$.

The introduced coefficients, which are taking into account a magnetic field, are written in the following form:

$$b_e = \frac{\mu_e B_x}{c} = \frac{\omega_e}{\nu_e}, \quad b_i = \frac{\mu_i B_x}{c} = \frac{\omega_i}{\nu_{in}},$$

where $\omega_e = \frac{eB_x}{m_e c}$ and $\omega_i = \frac{eB_x}{m_i c}$ are the Larmor frequencies of electrons and ions, m_e and m_i are the masses of electrons and ions.

The boundary conditions for charged particles and electrical potential are formulated as follows:

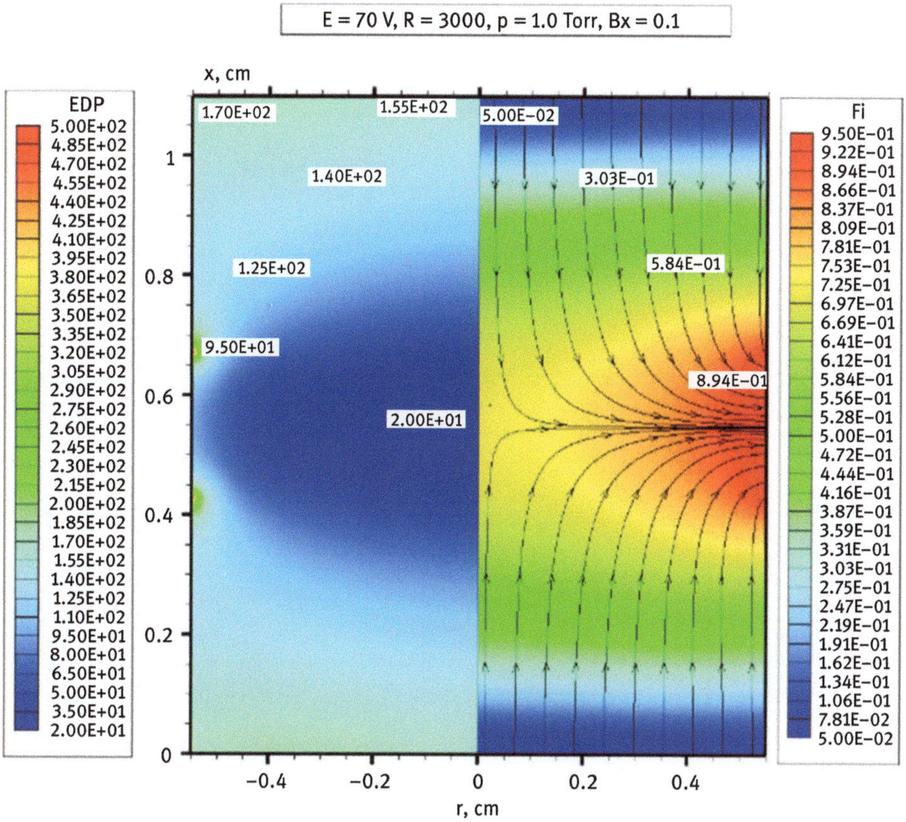

Figure 7.39: The reduced electric field E/p and electric potential in the Penning discharge at $E = 70$ V, $R_0 = 3000$ Ohm, $p = 1$ Torr, $B_x = 0.1$ T, $R_{ACi} = 0.4$ cm, $\gamma^* = 0.1\gamma$.

$$x = X,\quad r < R_{AC_1}: \qquad \frac{\partial n_i}{\partial x} = 0,\ \mathbf{\Gamma}_e = \gamma^* \mathbf{\Gamma}_i,\ \varphi = 0,$$

$$x = 0; x = X, r > R_{AC_1}: \quad \frac{\partial n_i}{\partial x} = 0, \mathbf{\Gamma}_e = \gamma \mathbf{\Gamma}_i, \quad \varphi = 0, \tag{7.34}$$

where γ is the coefficient of the secondly ion–electron emission; γ^* is the effective coefficient of the secondly ion–electron emission, which is taken of the order $\gamma^* \sim (10^{-1} - 10^{-2})\gamma$. This effective coefficient takes into account the meniscus properties of the anticathode.

For the cylindrical surface of anode, the following boundary conditions were used:

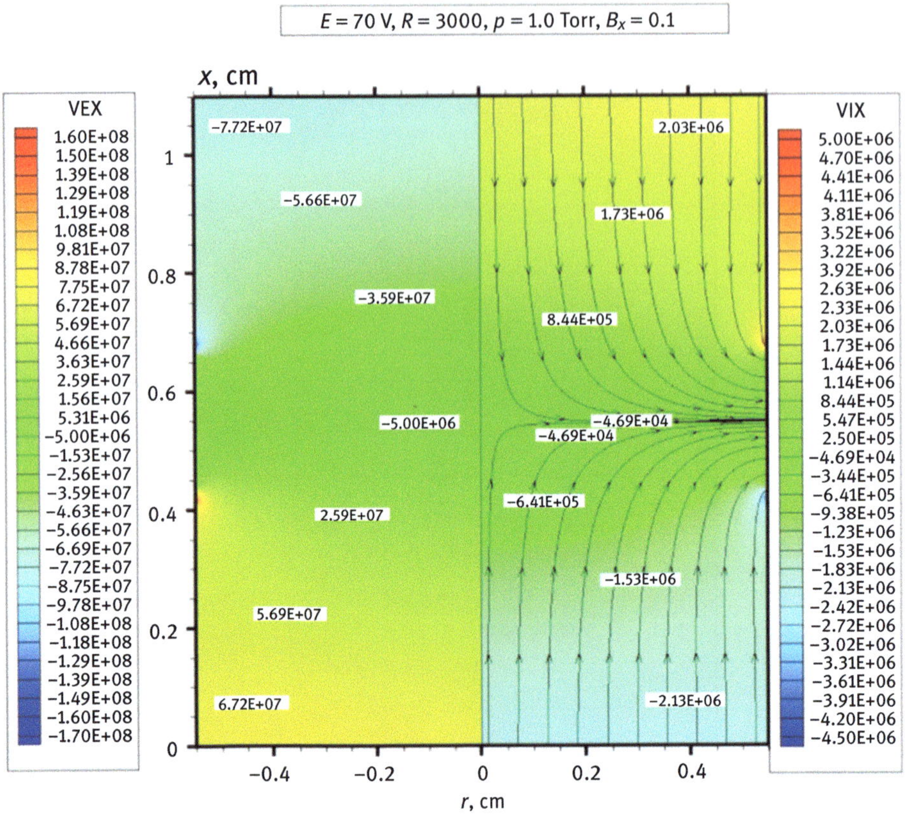

Figure 7.40: The averaged velocity of electrons and ions in the x-direction ($V_{e,x}$, $V_{i,x}$, cm/s) in the Penning discharge at $E = 70$ V, $R_0 = 3000$ Ohm, $p = 1$ Torr, $B_x = 0.1$ T, $R_{ACi} = 0.4$cm, $\gamma^* = 0.1\gamma$.

$$X_{A_1} < x < X_{A_2}, \quad r = R_C: \quad \frac{\partial n_e}{\partial r} = 0, \quad n_i = 0, \quad \varphi = V\,.$$

For cylindrical surface between anode and cathode

$$X_{A_1} > x, \quad r = R_C: \quad \frac{\partial n_e}{\partial r} = \frac{\partial n_i}{\partial r} = \frac{\partial \varphi}{\partial r} = 0\,,$$

$$X_{A_2} < x, \quad r = R_C: \quad \frac{\partial n_e}{\partial r} = \frac{\partial n_i}{\partial r} = \frac{\partial \varphi}{\partial r} = 0\,. \qquad (7.35)$$

A quasineutral plasma cloud of spherical shape in the center of the chamber is used as the initial condition.

The transport and thermophysics properties for H_2 were chosen as follows:

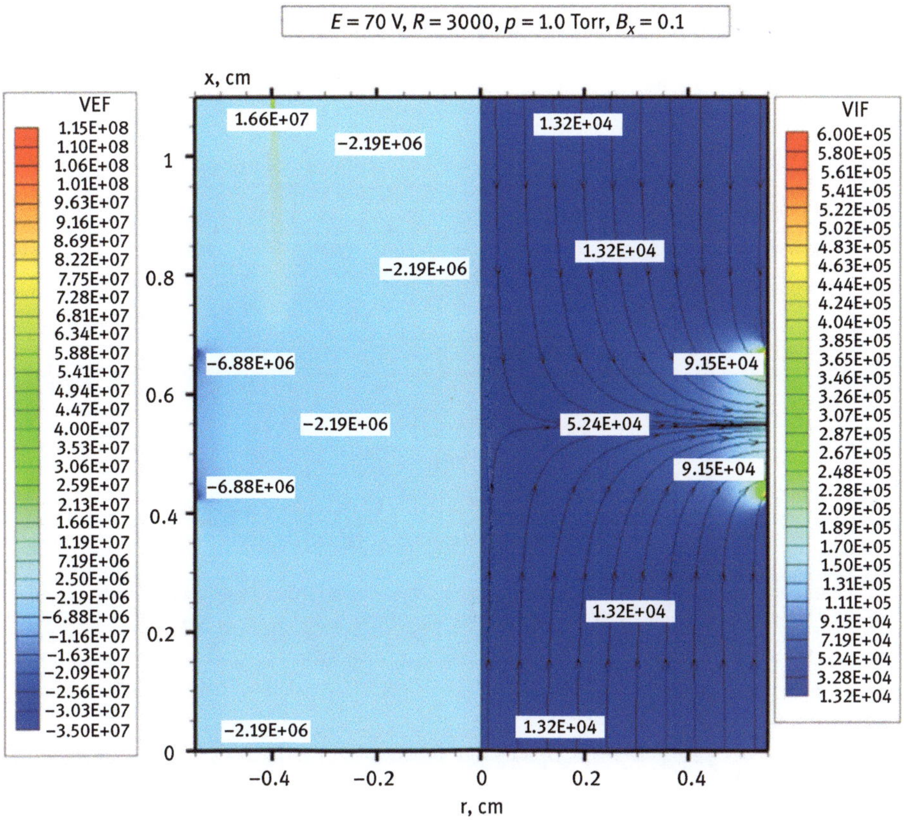

Figure 7.41: Averaged azimuthal velocity of electrons and ions ($V_{e,\varphi}$, $V_{i,\varphi}$, cm/s) in the Penning discharge at $E = 70$ V, $R_0 = 3000$ Ohm, $p = 1$ Torr, $B_x = 0.1$ T, $R_{ACi} = 0.4$cm, $\gamma^* = 0.1\gamma$.

$$\mu_e(p) = \frac{3.7 \times 10^5}{p} \frac{\text{cm}^2}{\text{V} \cdot \text{s}}, \quad \mu_i(p) = \frac{6.55 \times 10^3}{p} \frac{\text{cm}^2}{\text{V} \cdot \text{s}},$$

$$D_e = \mu_e(p) T_e, \quad D_i = \mu_i(p) T_i,$$

where p is the pressure in Torr.

The recombination coefficient β and electron temperature are taken as constants $\beta = 2 \times 10^{-7}$ cm^3/s and $T_e = 11610$ K.

The ionization coefficient for H_2 is determined as follows (the first Townsend formula):

$$\alpha(E) = p^* A^* \exp\left[-\frac{B}{(|\mathbf{E}|/p^*)}\right], \text{cm}^{-1},$$

where $$A^* = 2.75 \frac{1}{\text{cm} \cdot \text{Torr}}, \quad B = 130 \frac{\text{V}}{\text{cm} \cdot \text{Torr}}.$$

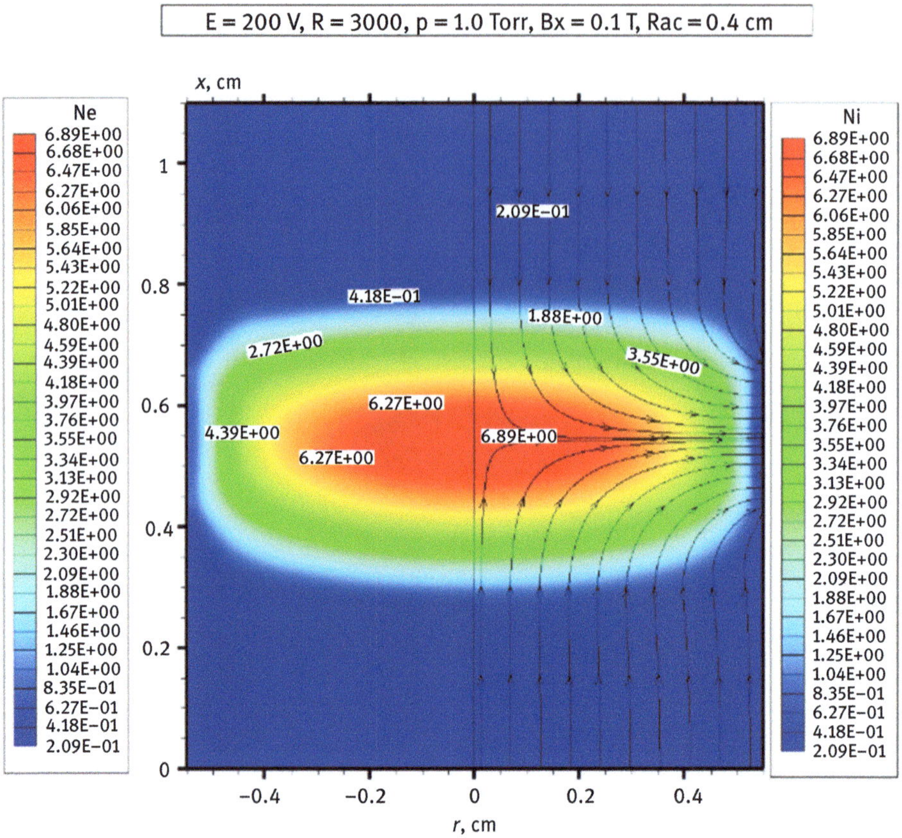

Figure 7.42: Concentration of electrons and ions (in 10^{10} cm^{-3}) in the Penning discharge at $E = 200$ V, $R_0 = 3000$ Ohm, $p = 1$ Torr, $B_x = 0.1$ T, $R_{ACi} = 0.4$cm, $\gamma^* = 0.1\gamma$.

Equations (7.29)–(7.31) are supplemented with the equation for the external electric circuit (see Figure 7.1), which is written for a stationary current as

$$\varepsilon = V + IR_0,$$

where V is the voltage on the electrodes; I is the total discharge current; ε is the emf in power supply; and R_0 is the external resistance.

In the case under consideration, we will use the modified drift-diffusion model, which includes two modifications of the classical model, namely, the modification of drift mobilities and the modification of velocity of ionization.

In Section 7.1 it was suggested to use the following relations for ion and electron mobilities:

$$\mu_{i,eff} = \min\left\{\frac{\mu_i p}{p}, \ \frac{9.2\times 10^4}{\sqrt{p_{\mathrm{Torr}} E_{\mathrm{V}}}}\right\}, \quad \mu_{e,eff} = \min\left\{\frac{\mu_e p}{p}, \ \frac{2.4\times 10^6}{\sqrt{p_{\mathrm{Torr}} E_{\mathrm{V}}}}\right\}, \tag{7.36}$$

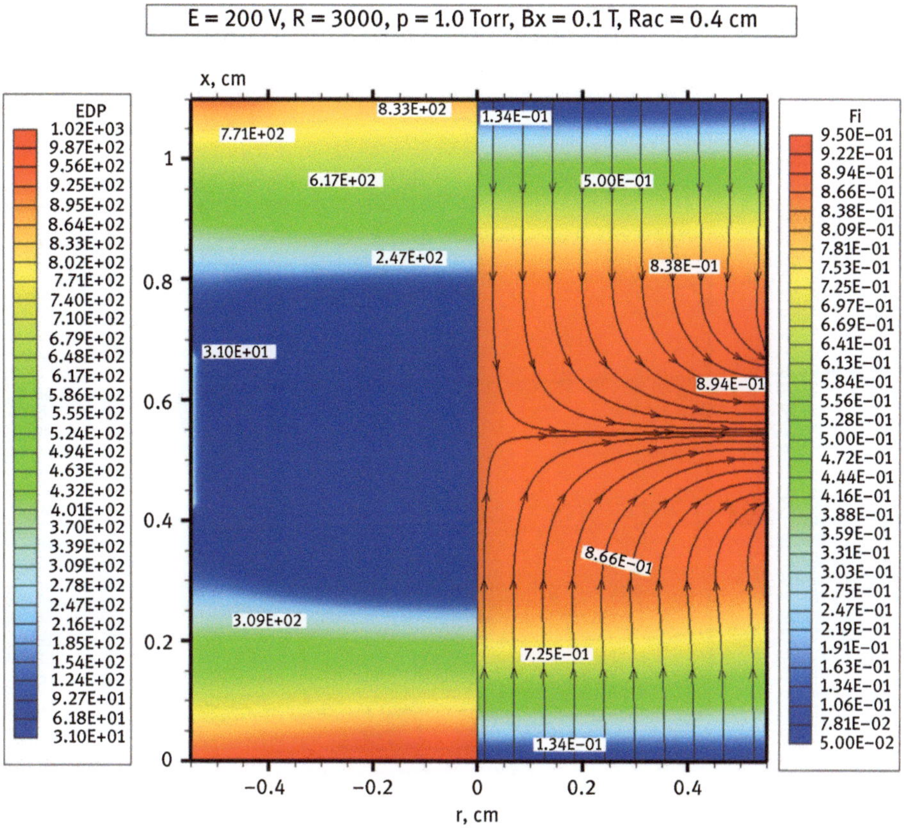

Figure 7.43: The reduced electric field E/p and electric potential in the Penning discharge at $E = 200$ V, $R_0 = 3000$ Ohm, $p = 1$ Torr, $B_x = 0.1$ T, $R_{ACi} = 0.4$cm, $\gamma^* = 0.1\gamma$.

and for corresponding diffusion coefficients

$$D_{i,eff} = \mu_{i,eff} T_i, \quad D_{e,eff} = \mu_{e,eff} T_e .$$

This gives a possibility to use classical definitions for drift velocities:

$$V_{dr,i} = \mu_{i,eff} \mathbf{E}, \quad V_{dr,e} = \mu_{e,eff} \mathbf{E} .$$

The ionization coefficient for H_2 is determined as follows:

$$\alpha_{low}(E) = pA^* \exp\left[-\frac{B}{(|\mathbf{E}|/p)}\right] \text{cm}^{-1}, \tag{7.37}$$

at low reduced electric fields and

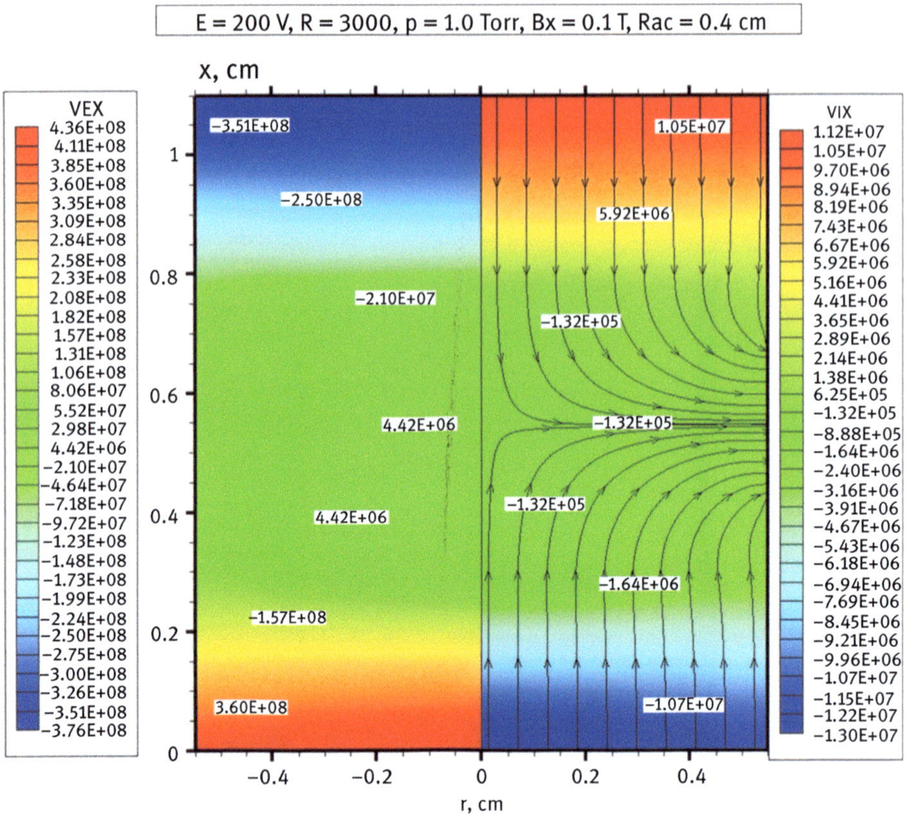

Figure 7.44: The averaged velocity of electrons and ions in the x-direction ($V_{e,x}$, $V_{i,x}$, cm/s) in the Penning discharge at $E = 200$ V, $R_0 = 3000$ Ohm, $p = 1$ Torr, $B_x = 0.1$ T, $R_{ACi} = 0.4$ cm, $\gamma^* = 0.1\gamma$.

$$\alpha_{high} = N_n \sigma_i\,, \ \text{cm}^{-1}, \tag{7.38}$$

for high reduced fields, where the ionization cross section σ_i could be calculated using the Tomson formula

$$\sigma_i = 4\pi a_0^2 \left(\frac{I_{H_2}}{\varepsilon}\right)^2 \cdot \frac{(\varepsilon - I_{H_2})}{I_{H_2}}\ , \text{cm}^2.$$

Instead of the reduced fields, we can use the dependence on the electron and ion energies, where kinetic energies are estimated as follows:

$$\varepsilon_{[\text{eV}]} = 2.84 \times 10^{-14} V^2_{\text{e[cm/s]}},\ \varepsilon_{[\text{eV}]} = 1.05 \times 10^{-12} V^2_{\text{i[cm/s]}}.$$

Now in the modified drift-diffusion model we can formulate constitutive relations for electrons:

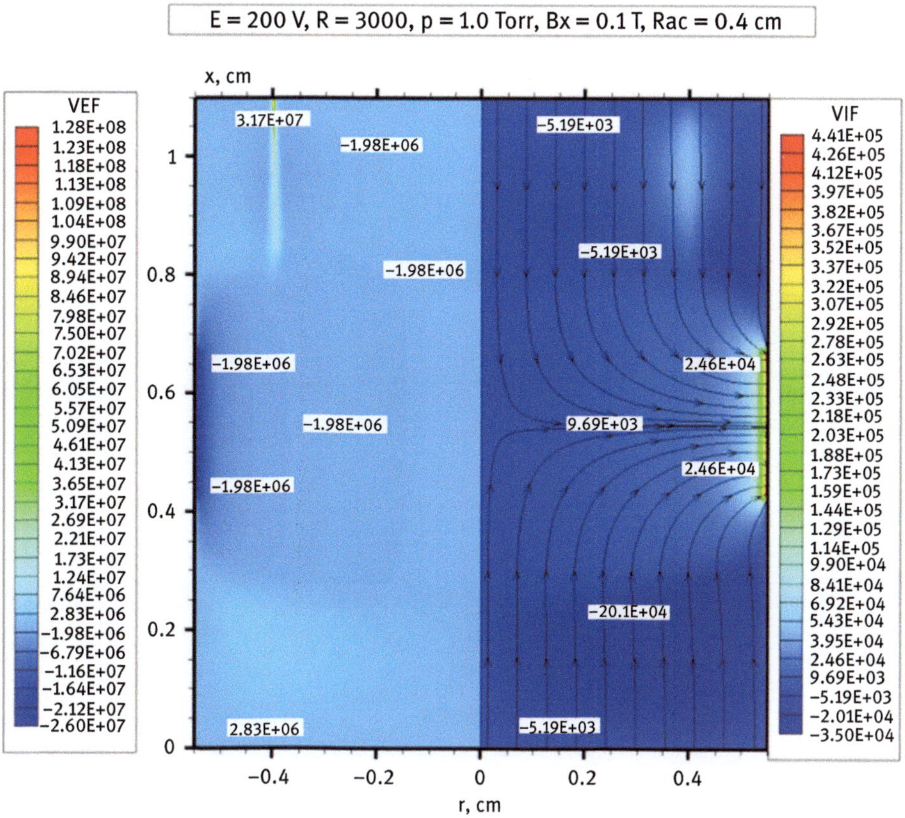

Figure 7.45: Averaged azimuthal velocity of electrons and ions ($V_{e,x}$, $V_{i,x}$, cm/s) in the Penning discharge at $E = 200\,\mathrm{V}$, $R_0 = 3000\,\mathrm{Ohm}$, $p = 1\,\mathrm{Torr}$, $B_x = 0.1\,\mathrm{T}$, $R_{ACi} = 0.4\,\mathrm{cm}$, $\gamma^* = 0.1\gamma$.

$$\mathbf{\Gamma}_{e,x} = -D_{e,eff}\frac{\partial n_e}{\partial x} - \mu_{e,eff} n_e E_x,\ \mathbf{\Gamma}_{e,r} = -\frac{D_{e,eff}}{1+b_{e,eff}^2}\frac{\partial n_e}{\partial r} - \frac{\mu_{e,eff}}{1+b_{e,eff}^2} n_e E_r,$$

$$\mathbf{\Gamma}_{e,\varphi} = -b_{e,eff}\mathbf{\Gamma}_{e,r},\ b_{e,eff} = \frac{\mu_{e,eff}B_x}{c}$$

and for ions:

$$\mathbf{\Gamma}_{i,x} = -D_{i,eff}\frac{\partial n_i}{\partial x} + \mu_{i,eff} n_i E_x\,,\ \mathbf{\Gamma}_{i,r} = -\frac{D_{i,eff}}{1+b_{i,eff}^2}\frac{\partial n_i}{\partial r} + \frac{\mu_{e,eff}}{1+b_{i,eff}^2} n_i E_r,$$

$$\mathbf{\Gamma}_{i,\varphi} = -b_{i,eff}\mathbf{\Gamma}_{i,r},\ b_{i,eff} = \frac{\mu_{i,eff}B_x}{c}\,.$$

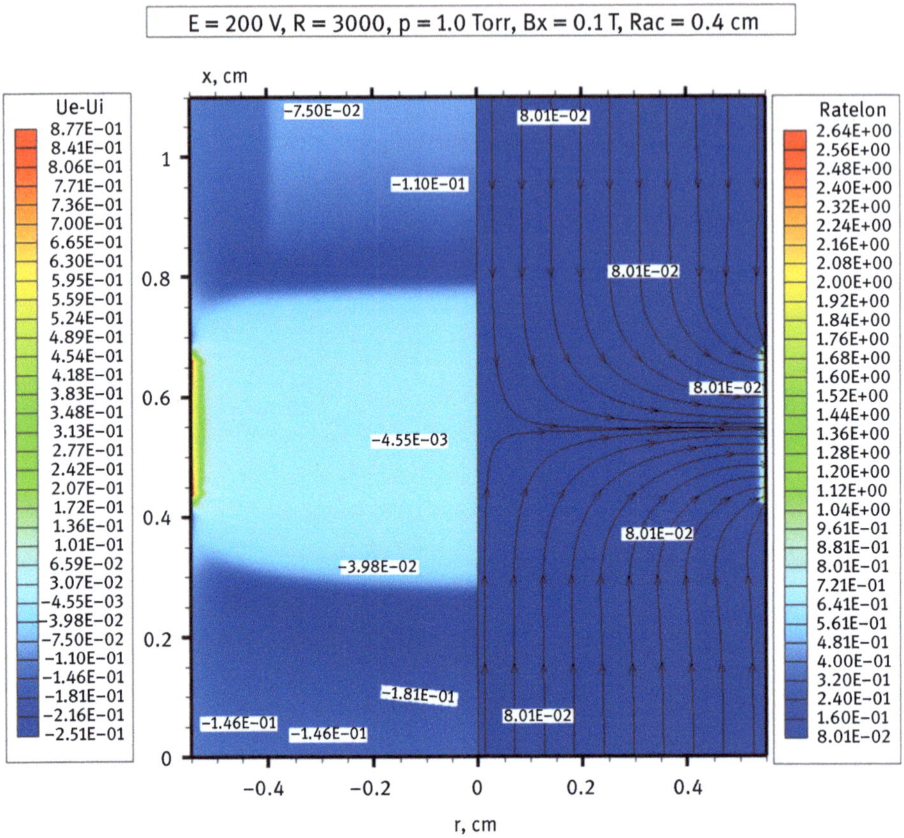

Figure 7.46: Difference of electrons and ions concentrations (in 10^{10} cm^{-3}) and rate of ionization (in 1.25×10^{18} $1/cm^3s$) in the Penning discharge at $E = 200$ V, $R_0 = 3000$ Ohm, $p = 1$ Torr, $B_x = 0.1$ T, $R_{ACi} = 0.4$ cm, $\gamma^* = 0.1\gamma$.

7.3.2 Results of numerical simulation

The calculations were performed for the following initial data: the pressure in molecular hydrogen H_2 was of $p = 0.1 - 0.01$ Torr; emf of power supply $E = 200$ V, $R_0 = 3000$ Ohm, $\gamma = 0.33$, $\gamma^* = 10^{-2}\gamma$, $B_x = 0.1$ T, $R_C = R_{AC_2} = 0.55$ cm, $R_{AC_1} = 0.2\,cm$, $X_{A_1} = 0.425$ cm, $X_{A_2} = 0.675$ cm, and $X_c = 1.1$cm (this value was used as a spatial scale of the task under consideration).

Figure 7.50 shows the electron and ion concentrations in discharge chamber at pressures $p = 0.1$ and 0.01 Torr. The main feature of these distributions is high concentrations of charged particles in the central region of the discharge chamber at high pressure and low concentrations when the pressure is low. Distributions of electron and ion concentrations at pressures $p = 0.1$ Torr (from the left) and $p = 0.01$ Torr

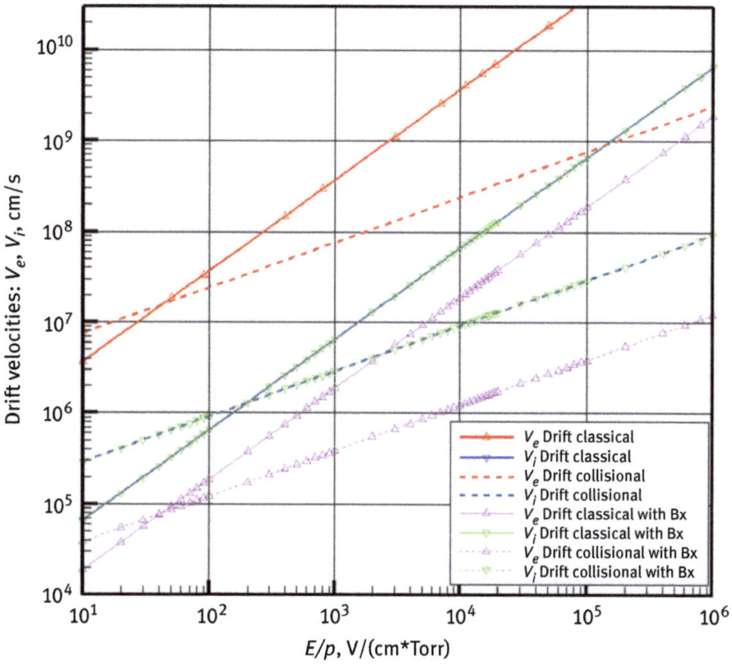

Figure 7.47: Drift velocities of electrons and ions versus reduce field.

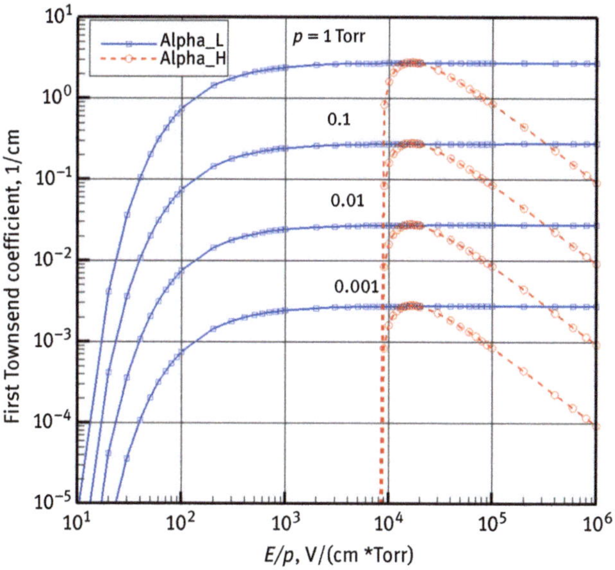

Figure 7.48: Ionization coefficient for modified DDM at pressure $p = 1.0 - 0.001$ Torr.

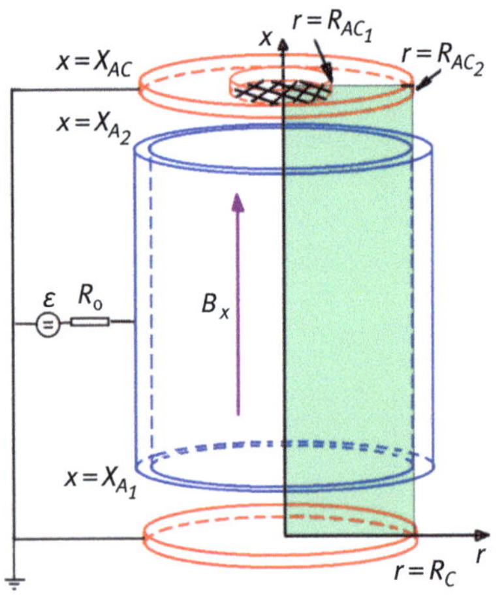

Figure 7.49: Schematic representation of a glow discharge in external magnetic field.

(from the right) are shown in this figure. At low pressure, the region of volume charge is shifted to internal cylindrical surface of the anode.

Decreasing the pressure leads to a significant change in the reduced electric field in the gas discharge chamber. From Figure 7.51 one can see that near anode the reduced field E/p sharply increases, whereas in the volume the reduced field is distributed quite smoothly. The distribution of the ionization coefficient (see Figure 7.52) corresponds to this distribution of the reduced field.

Differences between coefficients of ionization α_x and α_r correspond to the axial and radial fluxes of electrons points out to a "switching" from one dependence to another.

It is seen that at $p = 0.1$ Torr these differences are insignificant, whereas at $p = 0.01$ Torr the efficiency of ionization of neutral particles induced by electron collisions are different. The source term $\dot{\omega}_i$ in eqs. (7.29) and (7.30) is shown for two pressures in Figure 7.53. This term is decreased by two orders of the value when the pressure is decreased one order of the value. The volume of increased ionization is located near the anode. Thus, such a model of ionization in the Penning discharge gives low efficiency of ionization in the discharge volume. Nevertheless, presented results of numerical simulations are in reasonable qualitative agreement with available experimental data (Penning F.M., 1936; Hirsch E.N., 1964; Safronov B.G., et al., 1974).

Axial distributions of ion and electron velocities in the Penning discharge at pressures $p = 0.1$ Torr and $p = 0.01$ Torr are shown in Figure 7.54. The corresponding axial distributions of the reduced electric field and the energy of electrons, which were calculated using the axial velocity of electrons, are shown in Figure 7.55.

Thus, a spatial structure of the Penning discharge at pressures $p = 0.1 - 0.01$ Torr has been investigated within the framework of the modified drift-diffusion model. This modification includes three models of elementary physical processes, which

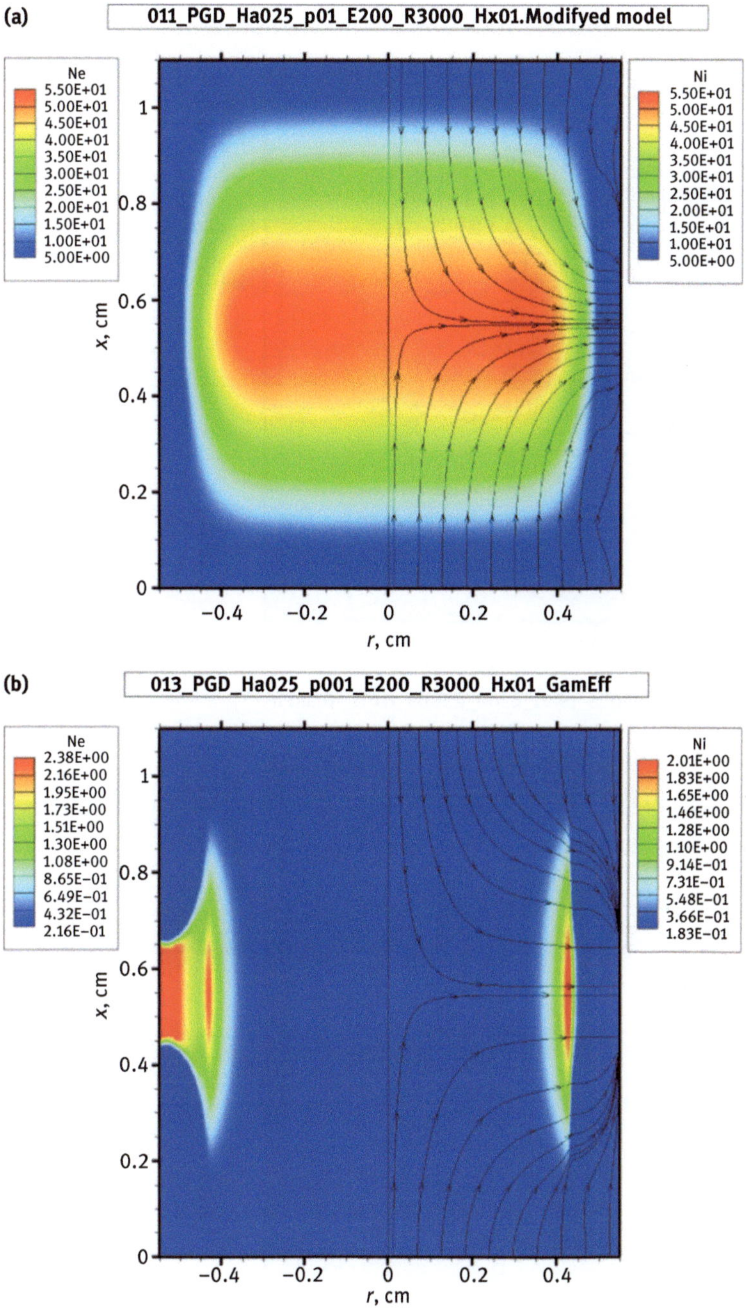

Figure 7.50: Concentration of electrons N_e and ions N_i (in 10^{10} cm^{-3}) at pressure (a) $p = 0.1$ Torr and (b) $p = 0.01$ Torr.

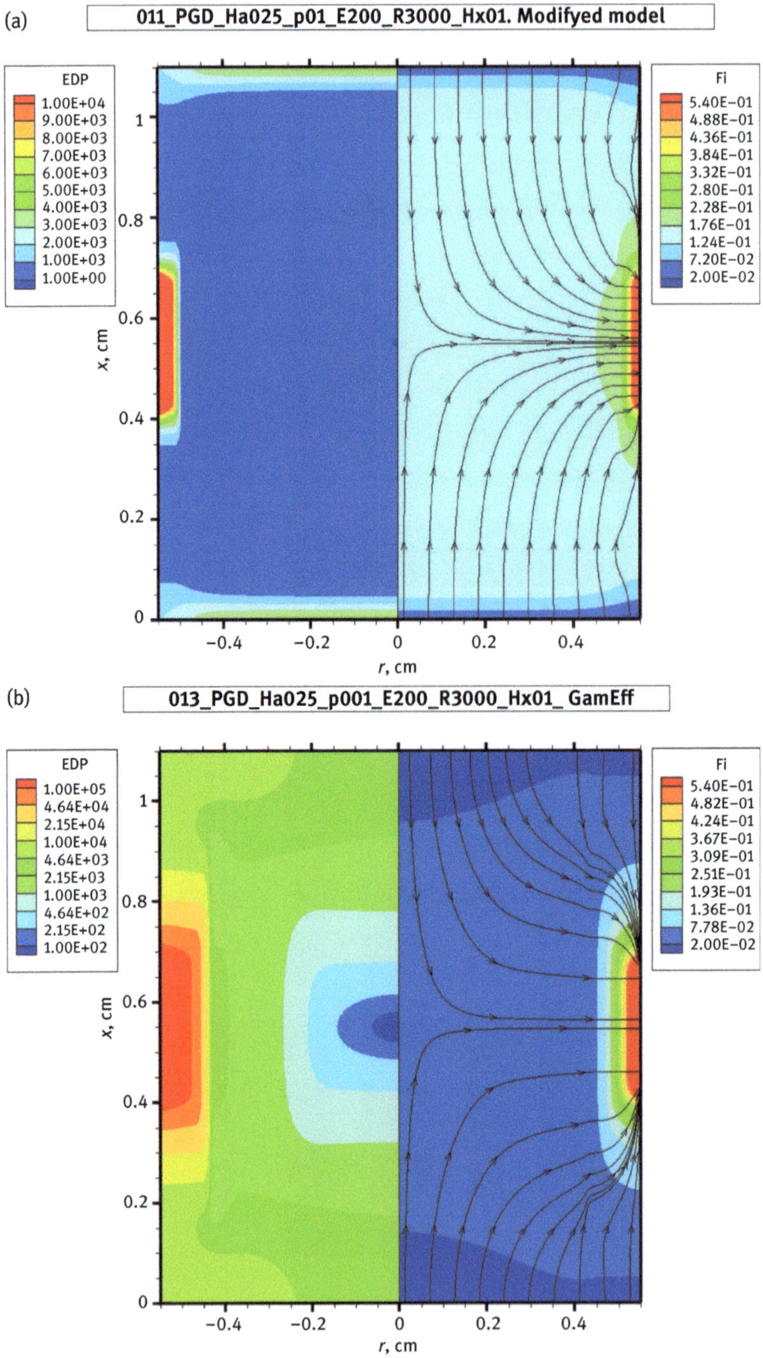

Figure 7.51: Reduced electric field (EDP = E/p in V/(cm*Torr)) and electric potential ($Fi = \varphi/\varepsilon$) at (a) $p = 0.1$ Torr and (b) $p = 0.01$ Torr.

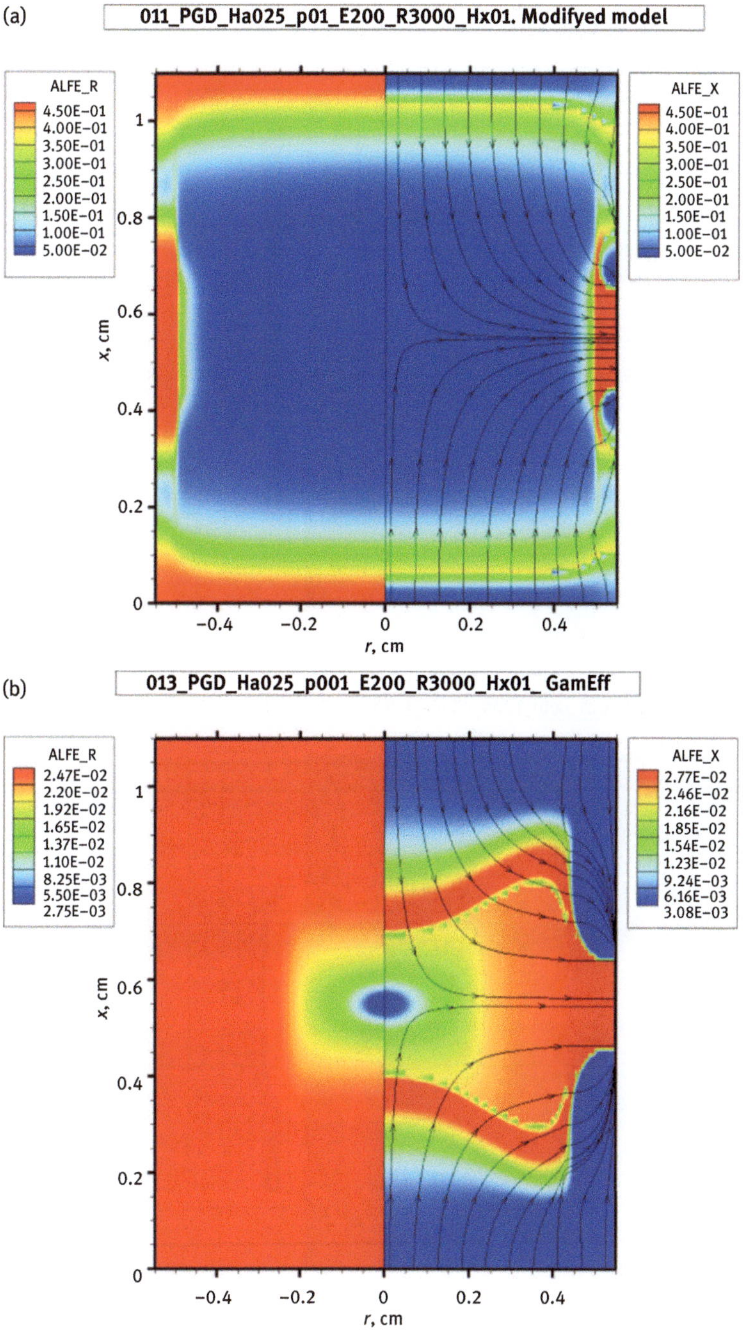

Figure 7.52: Coefficients of ionization (in 1/cm) at (a) $p = 0.1$ Torr and (b) at $p = 0.01$ Torr; ALFE_R = α_r is the coefficient of ionization calculated with E_r, ALFE_X = α_x is the coefficient of ionization calculated with E_x.

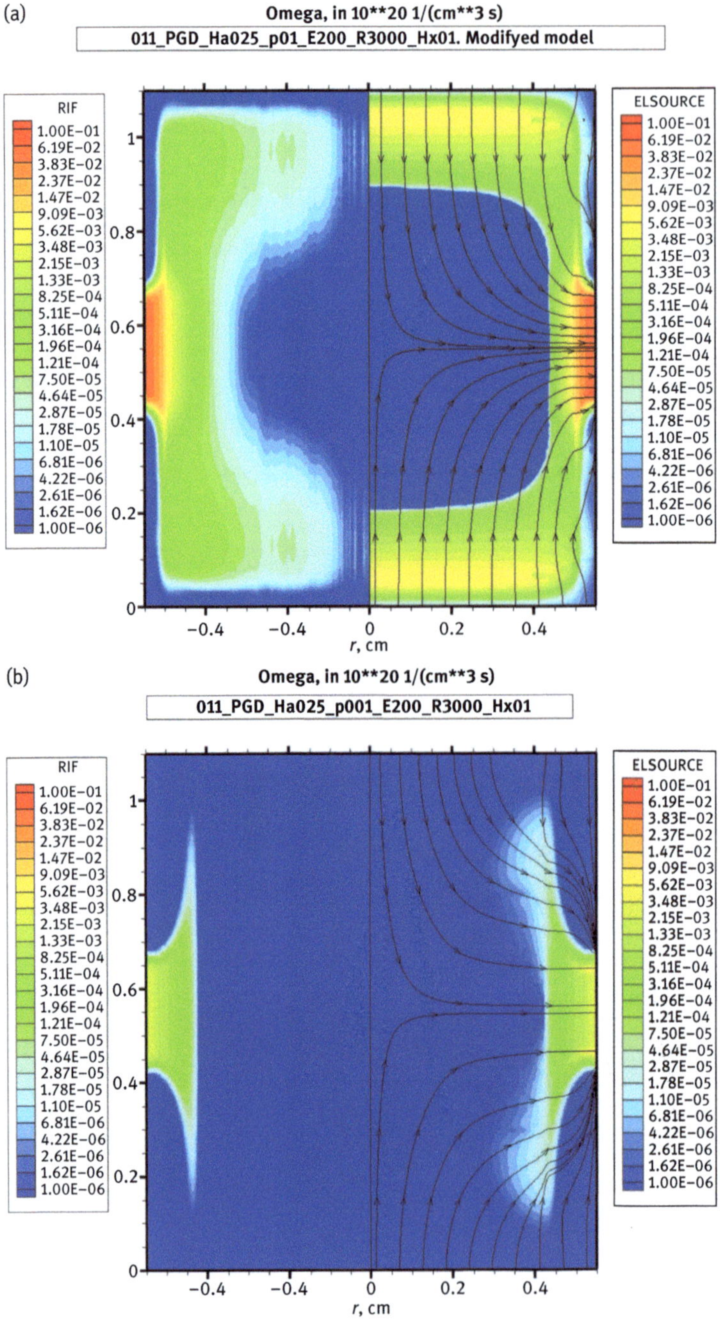

Figure 7.53: Source of ionization ELSOURCE = $\dot{\omega}_i$ (in $10^{20}/(\mathrm{cm}^3 \cdot \mathrm{s})$) at (a) $p = 0.1$ Torr and (b) $p = 0.01$ Torr. Left parts of the figures show source of ionization due to azimuthal movement of electrons.

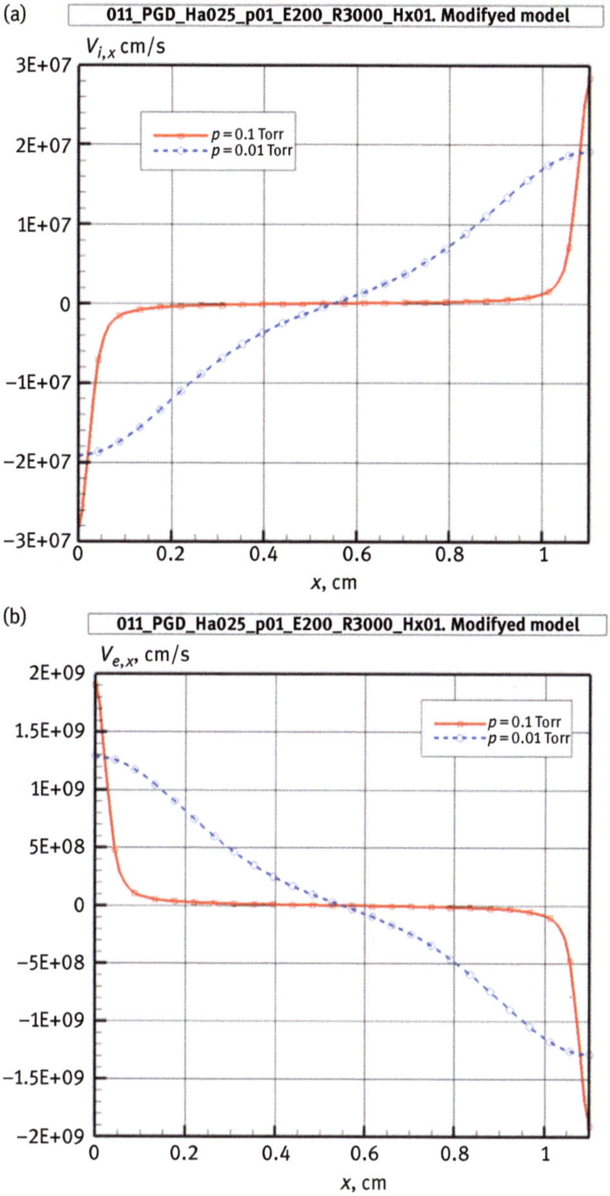

Figure 7.54: Axial distributions of (a) ion and (b) electron average velocities in *x*-direction at the pressure $p = 0.1$ Torr and 0.01 Torr.

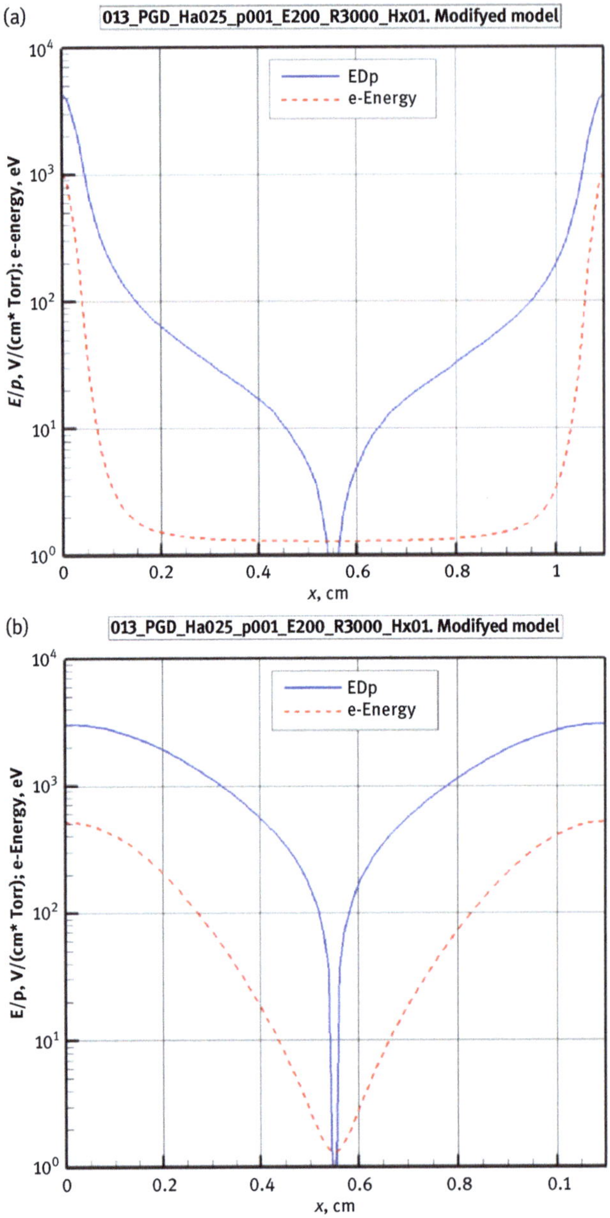

Figure 7.55: Axial distribution of reduced field (in V/(cm · Torr) and electron energy in eV at the pressure $p = 0.1$ Torr and 0.01 Torr.

take into account peculiarities of gas discharge processes at low pressures and large reduced electric fields. These are as follows:

1) the nonlinear model of the behavior of ion and electron drifts depending on the reduced electric field in a wide range of variation;
2) the model of ionization of neutral particles in collisions with electrons in a wide range of variation of the reduced electric field;
3) the anisotropic model of ionization by the axial, radial, and azimuthal fluxes of electrons.

The analysis of the presented predictions of spatial structure of the Penning discharge using the modified drift-diffusion model allows to suggest that the significant disadvantage of the model consists in the absence of the explicit source term connected with the low-energy electrons that were created in gas discharge volume after the first acts of ionizations by the high-energy electrons, being accelerated near cathode and anticathode. Presented modified drift-diffusion model should be improved in this aspect.

Part III: Ambipolar models of direct current discharges

8 Quasineutral model of gas discharge in external magnetic field and in gas flow

In the given part, there quasineutral model of partially ionized gas in the external magnetic field is considered. This model will be used for the creation of self-consistent computing models for numerical investigation of glow discharge parameters in hypersonic flows.

The constituent of the quasineutral plasma model is the model of the ambipolar diffusion of charged particles in a glow discharge, which will be described in Section 8.2 after defining some basic notations of plasma physics (Section 8.1).

8.1 The spatial scale of electric field shielding in plasma: the Debye radius

The characteristic spatial scale of an electric field shielding in plasma can be defined from the one-dimensional model of noncollisional plasma. The distribution functions of electrons f_e and ions f_i can be derived from the stationary kinetic equations of the following form (Chen F.F., 1984; Bittencourt J.A., 2004):

$$v_e \frac{\partial f_e}{\partial x} + \frac{e}{m_e} E \frac{\partial f_e}{\partial v_e} = 0, \tag{8.1}$$

$$v_i \frac{\partial f_i}{\partial x} - \frac{e}{m_i} E \frac{\partial f_i}{\partial v_i} = 0, \tag{8.2}$$

where x is the physical coordinate; v_e, v_i are the velocities of electrons and ions; E is the electric field strength which is defined with a potential φ from the equation of Poisson

$$-\frac{\mathrm{d}E}{\mathrm{d}x} = \frac{\mathrm{d}^2\varphi}{\mathrm{d}x^2} = \frac{e}{\varepsilon_0} \int (f_e - f_i)\mathrm{d}v, \tag{8.3}$$

where ε_0 is the dielectric permeability of the plasma, which is supposed constant.

For determination of f_e we will use the variable separation method

$$f_e = V(v_e)X(x). \tag{8.4}$$

Substituting (8.4) into (8.1) and changing the electric field strength on the electric potential, we obtain

$$\frac{1}{e(\mathrm{d}\varphi/\mathrm{d}x)} \frac{\mathrm{d}X}{\mathrm{d}x} = \frac{1}{m_e \mathrm{v}_e V} \frac{\mathrm{d}V}{\mathrm{d}v_e} = -\beta^2. \tag{8.5}$$

https://doi.org/10.1515/9783110648836-009

Integrating (8.5) separately by coordinates and velocity, one can find

$$X(x) = e^{-\beta^2 e\varphi}, \quad V(v) = Ae^{-\beta^2 (m_e v_e^2/2)} \tag{8.6}$$

or

$$f_e = Ae^{-\beta^2 [(m_e v_e^2/2) + e\varphi]}, \tag{8.7}$$

where the constant of integration A is defined from the condition of normalization of the distribution function

$$n_{e,0} = \int f_e \mathrm{d}v. \tag{8.8}$$

If $\varphi = 0$, then from (8.7) it follows that

$$f_e = Ae^{-\beta^2 (m_e v_e^2/2)}. \tag{8.9}$$

Let us admit that the motion of electrons is thermalized due to distant collisions, then the function f_e should look like the Maxwellian distribution function; hence, the separation constant β^2 can be assumed to be equal to

$$\beta_e^2 = (kT_e)^{-1}, \tag{8.10}$$

where T_e is the electron temperature.

Now, having used the normalization condition (8.8), we can define the constant of integration

$$A = n_{e,0} \left(\frac{m_e}{2\pi k T_e} \right)^{3/2}. \tag{8.11}$$

Then instead of (8.7) we have

$$f_e = n_{e,0} \left(\frac{m_e}{2\pi k T_e} \right)^{3/2} e^{-(m_e v_e^2/2kT_e) - (e\varphi/kT_e)}. \tag{8.12}$$

By analogy for ions

$$f_i = n_{i,0} \left(\frac{m_i}{2\pi k T_i} \right)^{3/2} e^{-(m_i v_i^2/2kT_i) - (e\varphi/kT_i)}. \tag{8.13}$$

Substituting (8.12) and (8.13) into (8.3), we obtain

$$\frac{d^2\varphi}{dx^2} = -\frac{e}{\varepsilon_0}\int\left[n_{i,0}\left(\frac{m_i}{2\pi kT_i}\right)^{3/2} e^{-(m_i v_i^2/2kT_i)-(e\varphi/kT_i)}\right.$$

$$\left. - n_{e,0}\left(\frac{m_e}{2\pi kT_e}\right)^{3/2} e^{-(m_e v_e^2/2kT_e)-(e\varphi/kT_e)}\right] dv =$$

$$= -\frac{e}{\varepsilon_0}\left(n_{i,0}e^{e\varphi/kT_i} - n_{e,0}e^{e\varphi/kT_e}\right).$$

Let us suppose $n_{i,0} \approx n_{e,0}$, $T_i \approx T_e = T$, then

$$\frac{d^2\varphi}{dx^2} = -\frac{e}{\varepsilon_0}n_{e,0}\left(e^{e\varphi/kT_i - e\varphi/kT_e}\right).$$

If to assume that the thermal energy of the charged particles motion surpasses the energy of electric field ($kT \gg e\varphi$) one can use an expansion of the two exponential functions into a series

$$\frac{d^2\varphi}{dx^2} \cong -\frac{e}{\varepsilon_0}n_{e,0}\left(1 + \frac{e\varphi}{kT} + \cdots - 1 + \frac{e\varphi}{kT} - \cdots\right) \approx -2\frac{e^2 n_{e,0}}{\varepsilon kT}\varphi. \tag{8.14}$$

The solution of the differential equation (8.14) at $x \to \infty$ is given as follows:

$$\varphi = Ae^{-\sqrt{2}(x/r_D)},$$

where $r_D = \sqrt{\varepsilon_0 kT/e^2 n_{e,0}}$ has a dimension of length, which refers to *the Debye shielding radius.*

This value defines the characteristic spatial scale, at which the Coulomb force action becomes small at distances $x > r_D$, which means that at distances $x > r_D$ it is possible to consider the plasma as the quasineutral one.

8.2 The ambipolar diffusion

The classical model of the quasineutral plasma of glow discharge is based on the following two equations (Brown S.C., 1966; Raizer Yu.P., 1991):

$$\frac{\partial n}{\partial t} + \operatorname{div}(\mathbf{V}n) = \operatorname{div}(D_a \operatorname{grad} n) + \dot{\omega}_e, \tag{8.15}$$

$$\operatorname{div}\mathbf{j} = 0, \tag{8.16}$$

where D_a is the coefficient of ambipolar diffusion; $\mathbf{V}$ is the velocity of gas flow; n is the volumetric concentration of the charged particles ($n_e \approx n_i = n$); $\mathbf{j}$ is the vector of current density; $\dot{\omega}_e = \alpha(E)|\Gamma_e| - \beta_e n^2$ is the velocity of charged particle formation, which includes the ionization (the first term) and the recombination.

Process of ambipolar diffusion arises in the inhomogeneous quasineutral plasma because of the large difference in masses between electrons and ions. Significantly lighter and consequently more mobile electrons quickly abandon any area of heterogeneity arising in plasma. The electric field of polarization, which arises at this moment between electrons and essentially more slowly moving ions, restrains the motion of electrons and accelerates the motion of ions. It was shown that the Debye radius can be considered as the characteristic spatial scale of such separation of the charged particles. The specified type of diffusion of particles refers to *ambipolar*.

Let us consider the coefficient of the ambipolar diffusion. For this purpose, we will consider fluxes of charged particles in the form used in the drift-diffusion model:

$$\Gamma_e = -\mu_e n_e \mathbf{E} - D_e \mathrm{grad} n_e, \tag{8.17}$$

$$\Gamma_i = +\mu_i n_i \mathbf{E} - D_e \mathrm{grad} n_i. \tag{8.18}$$

The Poisson equation will be used for the determination of the polarization electric field strength (for homogeneous external one):

$$\mathrm{div}\mathbf{E} = 4\pi e(n_i - n_e). \tag{8.19}$$

Let us estimate a typical spatial scale of quasineutrality in a positive column of glow discharge, having used the relation (see Section 8.1):

$$r_{\mathrm{D}} = 4.6\sqrt{\frac{T_e\,[\mathrm{K}]}{n_e\,[\mathrm{CM}^{-3}]}} = 525\sqrt{\frac{T_e\,[\mathrm{эB}]}{n_e\,[\mathrm{CM}^{-3}]}}.$$

If $T_e = 11610$ K, $n_e = 10^{11}$ cm^{-3}, then $r_{\mathrm{D}} \cong 1.7 \times 10^{-3}$ cm, which means the ambipolar approximation is true on scales of $L \gg r_{\mathrm{D}}$, for example, $L \sim 10^{-2}$cm.

Let us now consider a small heterogeneity of charges on the spatial scale L:

$$|n_i - n_e| \ll n_e \approx n_i = n. \tag{8.20}$$

Hence, the separation of charges would not increase up to a significant value, and the fluxes of electrons and ions should be approximately equal (further we shall consider such separation in the x-direction):

$$\Gamma_{e,x} \approx \Gamma_{i,x}, \tag{8.21}$$

but

$$\Gamma_{e,x} \approx -\mu_e E_x n - D_e \frac{\partial n}{\partial x}, \tag{8.22}$$

$$\Gamma_{i,x} \approx +\mu_i E_x n - D_i \frac{\partial n}{\partial x}. \tag{8.23}$$

To exclude the field of polarization E_x from (8.22) and (8.23), we will divide (8.22) on μ_e, and (8.23) on μ_i, and then sum up the outcomes

$$\frac{\Gamma_{e,x}}{\mu_e}+\frac{\Gamma_{i,x}}{\mu_i}=-\frac{D_e}{\mu_e}\frac{\partial n}{\partial x}-\frac{D_i}{\mu_i}\frac{\partial n}{\partial x},$$

or

$$\Gamma_x\frac{\mu_e+\mu_i}{\mu_e\mu_i}=-\left(\frac{D_e}{\mu_e}+\frac{D_i}{\mu_i}\right)\frac{\partial n}{\partial x},$$

or

$$\Gamma_x=-\frac{\mu_i D_e+\mu_e D_i}{\mu_e+\mu_i}\frac{\partial n}{\partial x}. \tag{8.24}$$

Thus, we have established that the flux of particles (electrons and ions) can be written in the form of diffusion flux

$$\Gamma_x=-D_a\frac{\partial n}{\partial x},\quad D_a=\frac{\mu_i D_e+\mu_e D_i}{\mu_e+\mu_i} \tag{8.25}$$

with the diffusivity D_a, which is called the coefficient of ambipolar diffusion.

Since $\mu_e \gg \mu_+$ and $D_e \gg D_+$, we can estimate the value of the ambipolar diffusion coefficient for typical conditions in a positive column of glow discharge ($T_i \sim 300\,\text{K}$, $T_e \sim 10,000\,\text{K}$, $n_e \sim n_i = n_n$).

Mobility of electrons is calculated under the formula

$$\mu_e=\frac{e}{m_e\nu_{e,n}}=\frac{1.76\times10^{15}}{\nu_{e,n}[c^{-1}]}\frac{\text{cm}^2}{\text{V}\cdot\text{s}} \tag{8.26}$$

where the typical value of collision frequency of electrons with neutral particles is

$$\nu_{e,n}=\frac{1}{\tau_{e,n}}=4.2\times10^9\,\frac{1}{\text{s}\cdot\text{Torr}}, \tag{8.27}$$

Therefore, at pressure of $p=1$ Torr in nitrogenous plasma with mobility of electrons

$$\mu_e(N_2)=4.5\times10^5\frac{1}{p[\text{Torr}]}=4.5\times10^5\frac{\text{cm}^2}{\text{V}\cdot\text{s}}, \tag{8.28}$$

the mobility of ions is equal to

$$\mu_i\approx1.45\times10^3\frac{1}{p[\text{Torr}]}\frac{\text{cm}^2}{\text{V}\cdot\text{s}}. \tag{8.29}$$

Thus

$$\mu_i = 1.45 \times 10^3 \, \frac{\text{cm}^2}{\text{V} \cdot \text{s}}.$$

From here the relation between the mobilities is as follows:

$$\frac{\mu_e}{\mu_i} = \frac{4.5 \times 10^5}{1.45 \times 10^3} \approx 300. \tag{8.30}$$

This enables to simplify the formula for the ambipolar diffusion coefficient

$$D_a = \frac{\mu_i \mu_e T_e + \mu_e \mu_i T_i}{\mu_e + \mu_i} \approx \mu_i T_e, \tag{8.31}$$

where T_e is expressed in eV.

When $T_e = 1\,\text{eV}$ it becomes

$$D_a \sim 2.28 \times 10^3 \, \frac{\text{cm}^2}{\text{s}}. \tag{8.32}$$

These estimations will be used later.

8.3 Ambipolar diffusion in an external magnetic field

Now we will consider ambipolar diffusion at the presence of an external magnetic field. Earlier it has been shown that in two-dimensional flat case, densities of particle fluxes are defined under formulas (see Chapter 5)

$$\Gamma_{e,x} = -\mu_e n E_{e,x} - \frac{D_e}{1 + b_e^2} \frac{\partial n}{\partial x}, \tag{8.33}$$

$$\Gamma_{i,x} = +\mu_i n E_{i,x} - \frac{D_i}{1 + b_i^2} \frac{\partial n}{\partial x}, \tag{8.34}$$

$$b_e = \omega_e \tau_{en}, \quad b_i = \omega_i \tau_{in}, \tag{8.35}$$

$$\omega_e = \frac{eB}{m_e c}, \quad \omega_i = \frac{eB}{m_i c}, \tag{8.36}$$

$$E_{e,x} = \frac{E_x - b_e E_y}{1 + b_e^2}, \quad E_{i,x} = \frac{E_x + b_i E_y}{1 + b_i^2}. \tag{8.37}$$

Let us estimate values of the Hall parameters b_e and b_i at the external magnetic field induction of

$$B = 0.5\,\text{T}.$$

Then

$$\omega_e = 17.5\times10^{10}B[T]\,c^{-1},\ \omega_e = 8.7\times10^{10}c^{-1}, \tag{8.38}$$

$$\omega_i = \omega_{N_2^+} = 3.3\times10^{6}B[T]\,c^{-1},\ \omega_i = 1.56\times10^{6}c^{-1}, \tag{8.39}$$

$$b_e = 17.5\times10^{10}B\frac{1}{4.2\times10^{9}} = 41.6B = 20, \tag{8.40}$$

$$b_i = 6\times10^{6}B\frac{1}{3.2\times10^{10}} = 1.875\times10^{-4}B\approx10^{-4}, \tag{8.41}$$

so $b_i \ll b_e$.

Let $E_x > b_e E_y$, then

$$E_{e,x} = \frac{E_x}{1+b_e^2},\ E_{i,x} = \frac{E_x}{1+b_i^2}, \tag{8.42}$$

then relations (8.33) and (8.34) can be rewritten as follows:

$$\Gamma_{e,x} = -\frac{\mu_e}{1+b_e^2}nE_x - \frac{D_e}{1+b_e^2}\frac{\partial n}{\partial x}, \tag{8.43}$$

$$\Gamma_{i,x} = +\frac{\mu_+}{1+b_i^2}nE_x - \frac{D_i}{1+b_i^2}\frac{\partial n}{\partial x}. \tag{8.44}$$

As before, for elimination of the polarization field from (8.43) and (8.44), we will divide eq. (8.43) by $\mu_e/(1+b_e^2)$, and (8.44) by $\mu_i/(1+b_i^2)$ and sum up their results

$$\Gamma_x\left(\frac{1+b_e^2}{\mu_e}+\frac{1+b_i^2}{\mu_i}\right) = -\left(\frac{D_e}{\mu_e}+\frac{D_i}{\mu_i}\right)\frac{\partial n}{\partial x},$$

or

$$\Gamma_x = -\frac{\mu_e\mu_i}{(1+b_e^2)\mu_i+(1+b_i^2)\mu_e}\frac{\mu_i D_e+\mu_e D_i}{\mu_e\mu_i}\frac{\partial n}{\partial x},$$

or

$$\Gamma_x \approx -\frac{\mu_i D_e+\mu_e D_i}{(1+b_e^2)\mu_i+\mu_e}\frac{\partial n}{\partial x} = -\mu_e^* D_a\frac{\partial n}{\partial x} = -\tilde{D}_a\frac{\partial n}{\partial x}, \tag{8.45}$$

where

$$\mu_e^* = \frac{\mu_e+\mu_i}{(1+b_e^2)\mu_i+\mu_e}; \tag{8.46}$$

D_a is the classical coefficient of ambipolar diffusion; $\tilde{D}_a$ is the effective coefficient of ambipolar diffusion in an external magnetic field. Note that the effective coefficient of ambipolar diffusion is used in the plasma physics for a long time (Chen F.F., 1984, Bittencourt J.A., 2004).

Let us estimate coefficient μ_e^* at first having made obvious simplification

$$\mu_e^* = \frac{\mu_e + \mu_i}{(1+b_e^2)\mu_i + \mu_e} \cong \frac{\mu_e}{(1+b_e^2)\mu_i + \mu_e}.$$

In the considered conditions (see (8.30)) $\mu_i \cong 0.3 \times 10^{-2}\mu_e$; therefore,

$$\mu_e^* = \frac{\mu_e}{(1+b_e^2) \cdot 0.3 \times 10^{-2}\mu_e + \mu_e} = \frac{1}{1+0.3\times 10^{-2}(1+b_e^2)}.$$

It means that

$$\begin{aligned} &\text{for} \quad B = 0.5\,\text{T}, \quad b_e = 20: \quad \mu_e^* = 0.45, \\ &\text{for} \quad B = 0.2\,\text{T}, \quad b_e = 8: \quad \mu_e^* = 0.84, \\ &\text{for} \quad B = 0.1\,\text{T}, \quad b_e = 4: \quad \mu_e^* = 0.952. \end{aligned}$$

Thus, it is possible to draw the following conclusions.

1. The fluxes of charged particles in the ambipolar diffusion approximation in an external magnetic field can be estimated under the formula

$$\Gamma = -\mu_e^* D_a \text{grad} n, \tag{8.47}$$

   where μ_e^* is the effective mobility of electrons (8.46).
2. At magnetic fields $B < 0.2\,\text{T}$, influence of an external magnetic field on ambipolar diffusion flux is small (certainly, for the considered conditions).

8.4 Two-dimensional model of ambipolar diffusion in an external magnetic field

To finish the description of the quasineutral model of glow discharge (8.15)–(8.16), we will consider the transformation of expression for a vector of the current density

$$\mathbf{j} = e(\Gamma_i - \Gamma_e) = e\left[\mathbf{e}_x(\Gamma_{i,x} - \Gamma_{e,x}) + \mathbf{e}_y\left(\Gamma_{i,y} - \Gamma_{e,y}\right)\right], \tag{8.48}$$

where $\mathbf{e}_x$, $\mathbf{e}_y$ are the unit vectors along axes x and y; the projections of particle flux densities should be taken from the following relations:

$$\Gamma_{e,x} = -\mu_e n_e E_{e,x} - \frac{D_e}{1+b_e^2}\frac{\partial n_e}{\partial x} + \frac{b_e}{1+b_e^2} D_e \frac{\partial n_e}{\partial y}, \tag{8.49}$$

$$\Gamma_{+,x} = +\mu_+ n_+ E_{+,x} - \frac{D_+}{1+b_+^2}\frac{\partial n_+}{\partial x} - \frac{b_+}{1+b_+^2} D_+ \frac{\partial n_+}{\partial y}, \tag{8.50}$$

$$\Gamma_{e,y} = -\mu_e n_e E_{e,y} - \frac{D_e}{1+b_e^2}\frac{\partial n_e}{\partial y} - \frac{b_e}{1+b_e^2} D_e \frac{\partial n_e}{\partial x}, \tag{8.51}$$

$$\Gamma_{+,y} = +\mu_+ n_+ E_{+,y} - \frac{D_+}{1+b_+^2}\frac{\partial n_+}{\partial y} + \frac{b_+}{1+b_+^2} D_+ \frac{\partial n_+}{\partial x}, \tag{8.52}$$

where effective components of the electric field are

$$E_{e,x} = \frac{E_x - b_e E_y}{1+b_e^2}, \qquad E_{e,y} = \frac{E_y - b_e E_x}{1+b_e^2}, \tag{8.53}$$

$$E_{i,x} = \frac{E_x + b_i E_y}{1+b_i^2}, \qquad E_{i,y} = \frac{E_y - b_i E_x}{1+b_i^2}. \tag{8.54}$$

First of all, we take into account that in the considered conditions $b_i \ll 1$; therefore, components of the effective electric field acting on ions are equal to

$$E_{i,x} = E_x, \; E_{i,y} = E_y, \tag{8.55}$$

$$\Gamma_{i,x} = +\mu_i n_i E_x - D_i \frac{\partial n_i}{\partial x}, \tag{8.56}$$

$$\Gamma_{i,y} = +\mu_i n_i E_y - D_i \frac{\partial n_i}{\partial y}. \tag{8.57}$$

The projection of the full density of the flux to coordinate axis x in view of $n_e \approx n_i \approx n$ is expressed in the form of

$$\Gamma_{i,x} - \Gamma_{e,x} = \mu_i n E_x - D_i \frac{\partial n}{\partial x} + \frac{\mu_e}{1+b_e^2} n E_x + \frac{D_e}{1+b_e^2}\frac{\partial n}{\partial x} - \frac{\mu_e}{1+b_e^2} n(b_e E_y) - \frac{D_e}{1+b_e^2} b_e \frac{\partial n}{\partial y}.$$

Let us introduce the effective mobility of electrons and the effective diffusivity of electrons

$$\tilde{\mu}_e = \frac{\mu_e}{1+b_e^2}, \; \tilde{D}_e = \frac{D_e}{1+b_e^2},$$

then

$$\Gamma_{i,x} - \Gamma_{e,x} = (\mu_i + \tilde{\mu}_e) n E_x + (\tilde{D}_e - D_i)\frac{\partial n}{\partial x} - \tilde{\mu}_e n(b_e E_y) - \tilde{D}_e b_e \frac{\partial n}{\partial y}. \tag{8.58}$$

By analogy, for the y-direction

$$\Gamma_{i,y} - \Gamma_{e,y} = (\mu_i + \tilde{\mu}_e) n E_y + (\tilde{D}_e - D_i)\frac{\partial n}{\partial y} - \tilde{\mu}_e n(b_e E_y) - \tilde{D}_e b_e \frac{\partial n}{\partial x}. \tag{8.59}$$

Let us now neglect the last two terms in (8.58) and (8.59) which means the vectors of particle fluxes are approximated in the form of

$$\Gamma_e = -\tilde{\mu}_e n\mathbf{E} - \tilde{D}_e \operatorname{grad} n, \tag{8.60}$$

$$\Gamma_i = +\mu_i n\mathbf{E} - D_i \operatorname{grad} n. \tag{8.61}$$

Then

$$\mathbf{j} = e(\Gamma_i - \Gamma_e) = e\left[\left(\mu_i + \tilde{\mu}_e\right)n\mathbf{E} + \left(\tilde{D}_e - D_i\right)\operatorname{grad} n\right]. \tag{8.62}$$

From here it is obvious that the vector of the electric field strength can be derived from the relation

$$\mathbf{E} = \frac{\mathbf{j}}{en\left(\mu_i + \tilde{\mu}_e\right)} - \frac{D_e - D_i}{n\left(\mu_i + \tilde{\mu}_e\right)} \operatorname{grad} n. \tag{8.63}$$

It allows to receive densities of fluxes in the quasineutral approximation. Substituting (8.63) into (8.60) and (8.61), we obtain

$$\Gamma_e = -\frac{\mathbf{j}}{e} \cdot \frac{\tilde{\mu}_e}{\tilde{\mu}_e + \mu_i} - \tilde{D}_a \nabla n, \tag{8.64}$$

$$\Gamma_i = +\frac{\mathbf{j}}{e} \cdot \frac{\mu_i}{\tilde{\mu}_e + \mu_i} - \tilde{D}_a \nabla n, \tag{8.65}$$

where

$$\tilde{D}_a = \frac{\tilde{\mu}_e D_i + \mu_i \tilde{D}_e}{\mu_i + \tilde{\mu}_e}. \tag{8.66}$$

Note that eqs. (8.64) and (8.65) can be rewritten using μ_e^*

$$\Gamma_e = -\frac{\mathbf{j}}{e} \mu_e^* - \mu_e^* D_a \nabla n, \tag{8.67}$$

$$\Gamma_i = +\frac{\mathbf{j}}{e} \cdot \frac{\mu_i}{\tilde{\mu}_e + \mu_i} - \mu_e^* D_a \nabla n. \tag{8.68}$$

As a result, the quasineutral model of glow discharge in an external magnetic field can be formulated in the following approximate form of

$$\frac{\partial n}{\partial t} + \operatorname{div}(\mathbf{V}n) = \operatorname{div}\left(\mu_e^* D_a \operatorname{grad} n\right) + \dot{\omega}_e, \tag{8.69}$$

$$\operatorname{div}\left[\left(\mu_i + \tilde{\mu}_e\right)n\mathbf{E} + \left(\tilde{D}_e - D_i\right)\operatorname{grad} n\right] = 0, \tag{8.70}$$

where

$$\tilde{\mu}_e = \frac{\mu_e}{1+b_e^2}; \tag{8.71}$$

$$\mu_e^* = \frac{\mu_e + \mu_i}{(1+b_e^2)\mu_i + \mu_e} \cong \frac{\mu_e}{(1+b_e^2)\mu_i + \mu_e}; \tag{8.72}$$

$$\tilde{D}_e = \frac{D_e}{1+b_e^2}. \tag{8.73}$$

Equation (8.70) can be used for the calculation of electric field E, and eq. (8.69) is used for evaluation of concentration of the charged particles.

The source term in (8.69) can be modified also in view of the magnetic field:

$$\dot{\omega}_e = \left(\frac{\alpha}{p^*}\right) p^* E \frac{\mu_e(p^*)}{1+b_e^2} - \beta n^2, \tag{8.74}$$

where the definition of frequency of ionization is used in the form of

$$v_i = \left(\frac{\alpha}{p^*}\right) v_{e,d} = \frac{\alpha}{p} \mu_e E, \tag{8.75}$$

where α/p^* is the coefficient of ionization; p^* is the effective pressure.

8.5 Illustrative results of numerical simulation

The presented theory of the quasineutral plasma of glow discharge in an external magnetic field was used for investigation of the surface glow discharge in the supersonic boundary layer with an external magnetic field. Schematic of the problem is shown in Figure 8.1.

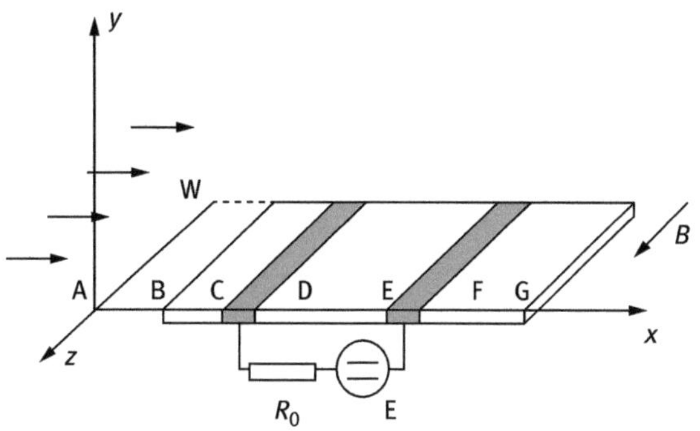

Figure 8.1: Aschematic of electromagnetic aerodynamic actuator.

Undisturbed input gas flow parameters in experiments are (Menart J., et al., 2003) statistical pressure $p_\infty = 78.4\,\text{Pa}(0.59\text{Torr})$, temperature $T_\infty = 43\,\text{K}$, and velocity $V_\infty = 675.5\,\text{m/s}$, which correspond to the Reynolds number $\text{Re} = 1.615 \times 10^5$ (related to a meter); therefore, the boundary layer was considered as laminar.

It is assumed that the surface glow discharge burns between two electrode sections arranged on a plane of a streamline plate. Such configuration of electrodes can be used as an electromagnetic aerodynamic actuator (Shang J.S., et al., 2005a). The temperature of the electrodes is $T_W = 600\,\text{K}$, and the dielectric surface is assumed to be heat-insulated. In the considered case, the discharge exists in a condition of fixed emf of a power supply of the electric circuit, $E = 1,200\,\text{V}$. Ohmic resistance of the external electric circuit is $R_0 = 12\,\text{kOhm}$.

Distribution of pressure along a streamlined surface with two heated up electrodes with the current flow of $I \sim 50\,\text{mA}$ is shown in Figure 8.2. The solid curve in this figure displays results of numerical integration of the Navier–Stokes equations and the equations of glow discharge electrodynamics. The dashed curve illustrates the distribution of pressure predicted by the asymptotic theory of the weak interaction (Hayes W., et al., 1959). From the presented data, it is shown that in the considered conditions, the pressure increases approximately to 1.7 times in the neighborhood

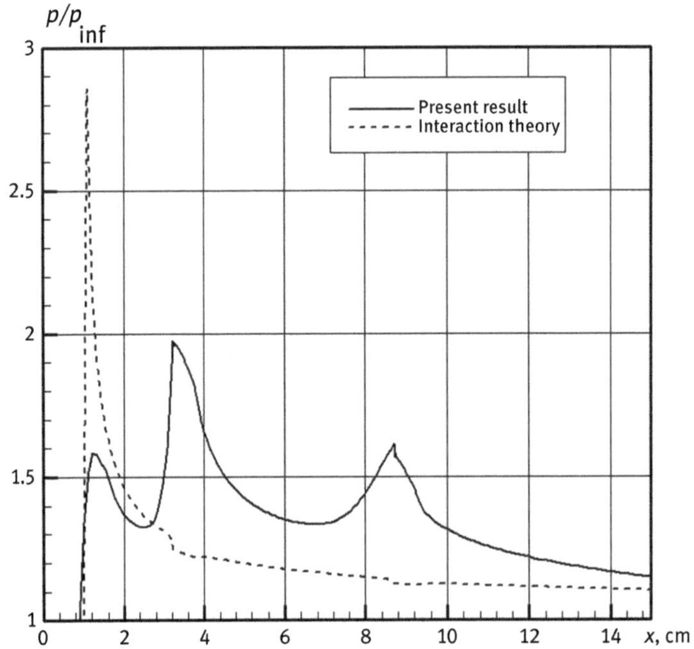

Figure 8.2: The pressure induced by surface glow discharge (continuous curve), and predicted by the theory of weak interaction (dashed curve).

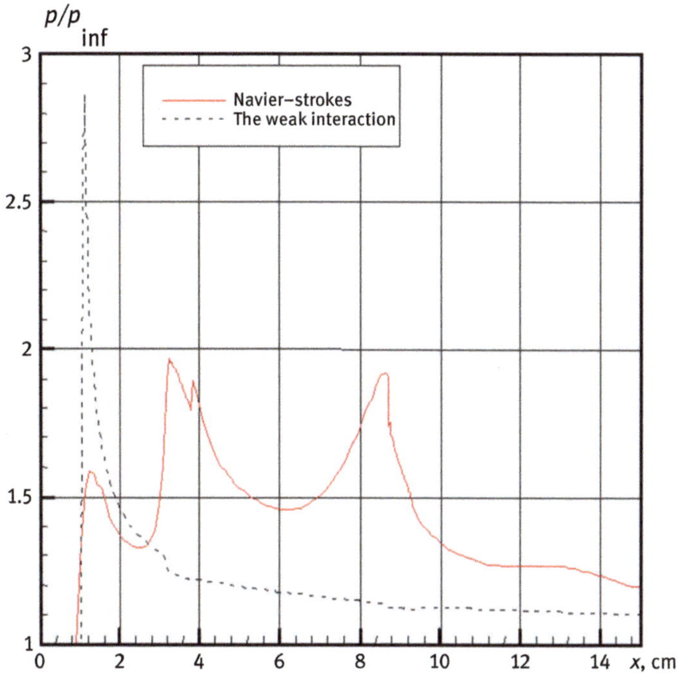

Figure 8.3: The pressure induced by surface glow discharge with magnetic field $B = -0.2\,\mathrm{T}$ (continuous curve), and predicted by the theory of weak interaction (dashed curve).

of the electrodes. In the cases considered further, the induction of the magnetic field varies within the limits of $0.2-0.5\,\mathrm{T}$. The pressure distribution above a streamline plate at $B = -0.2\,\mathrm{T}$ is shown in Figure 8.3.

The positive direction of the magnetic field corresponds to the negative z-direction, so that the induced volumetric force acts in the direction to the surface. Comparing Figures 8.2 and 8.3, it is possible to draw the conclusion that the external magnetic field appreciably influences on the distribution of the surface pressure, in particular above the anode ($x > 8\,\mathrm{cm}$).

It has been shown at the research of induced above surface pressure with the use of the elementary ambipolar diffusion model (Shang et al., 2005(a)) that with the growth of the magnetic field induction there is a monotonic increase of the induced pressure. However, later it was found that more correct calculation of the magnetic field within the limits of the ambipolar diffusion model leads to different results.

Figure 8.4 shows the distribution of pressure along the streamlined surface at $B = -0.5\,\mathrm{T}$. Comparing this result with the previous one, it is possible to draw a conclusion on a drop of the induced pressure at the growth of a magnetic field induction.

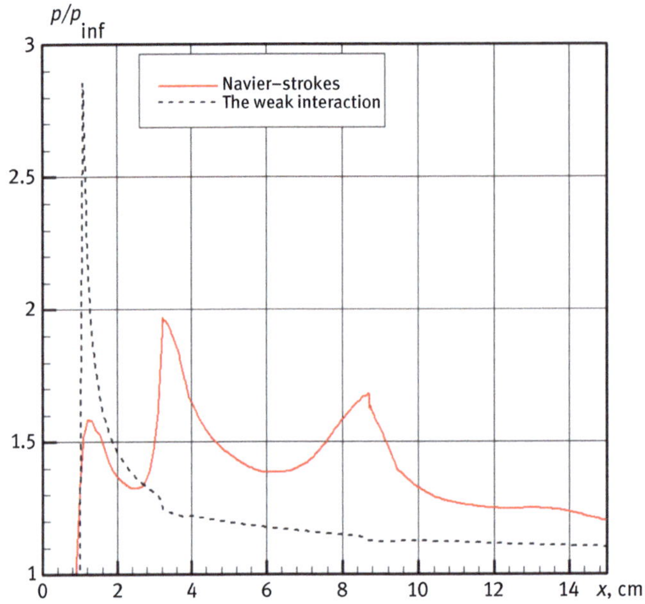

Figure 8.4: The pressure induced by surface glow discharge (continuous curve), and predicted by the theory of weak interaction (dashed curve), $B = -0.5\,\mathrm{T}$.

Thus, for deriving the greatest influence of a magnetic field on the distribution of pressure along a surface it is necessary to define optimum conditions, and the magnetic field should not be excessively great at all.

Note that the modified model of ambipolar diffusion at small inductions of a magnetic field (see Figure 8.3, $B = -0.2\,\mathrm{T}$) gives the numerical simulation results conterminous with the elementary theory of ambipolar diffusion, that is, numerical modeling proves the estimations made above to be true. It is obvious that the external magnetic field influences not only on the distribution of pressure but also on all other fields of gas dynamic and electrophysical functions. To illustrate this in Figures 8.5–8.7, comparisons are given for fields of electric potential, current densities, and volumetric concentration in two cases, which differ by magnetic inductions $B = -0.1\,\mathrm{T}$ and $B = -0.5\,\mathrm{T}$. From the presented results it is shown that an increase in the magnetic field leads to a drop in the concentration of the charged particles in the surface glow discharge.

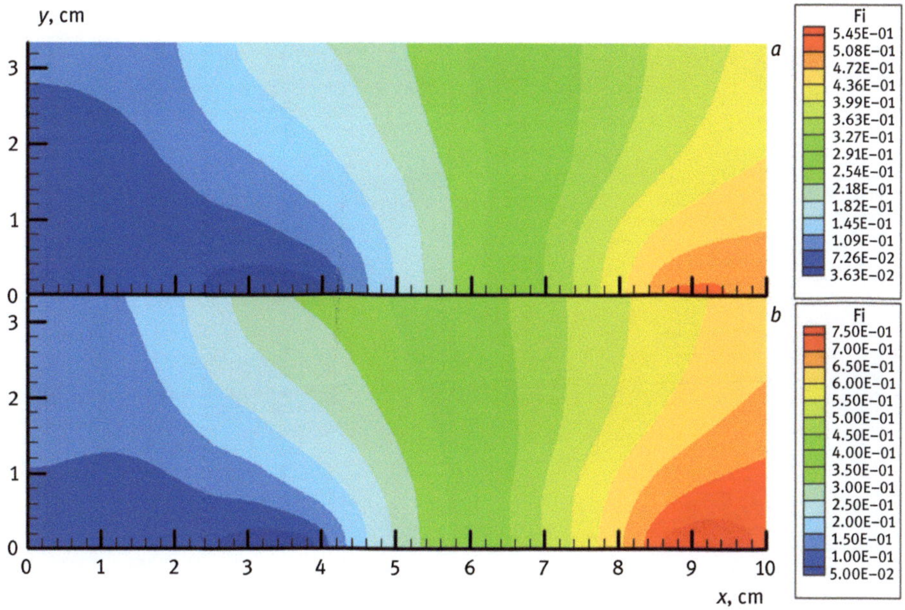

Figure 8.5: The electric potential in surface glow discharge with magnetic field: (a) $B = -0.1\,\mathrm{T}$ and (b) $B = -0.5\,\mathrm{T}$.

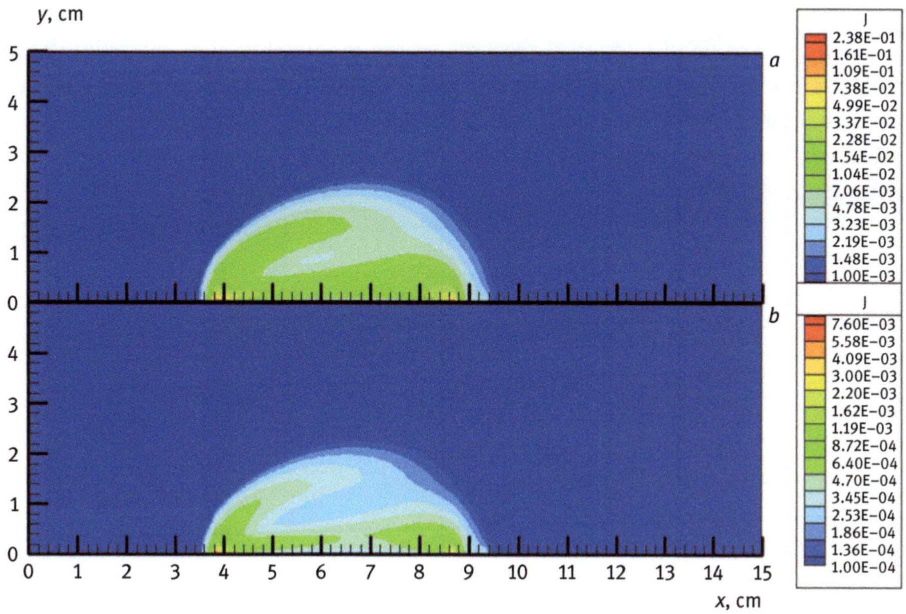

Figure 8.6: The module of a current density (A/cm^2) in surface glow discharge with magnetic field: (a) $B = -0.1\,\mathrm{T}$ and (b) $B = -0.5\,\mathrm{T}$.

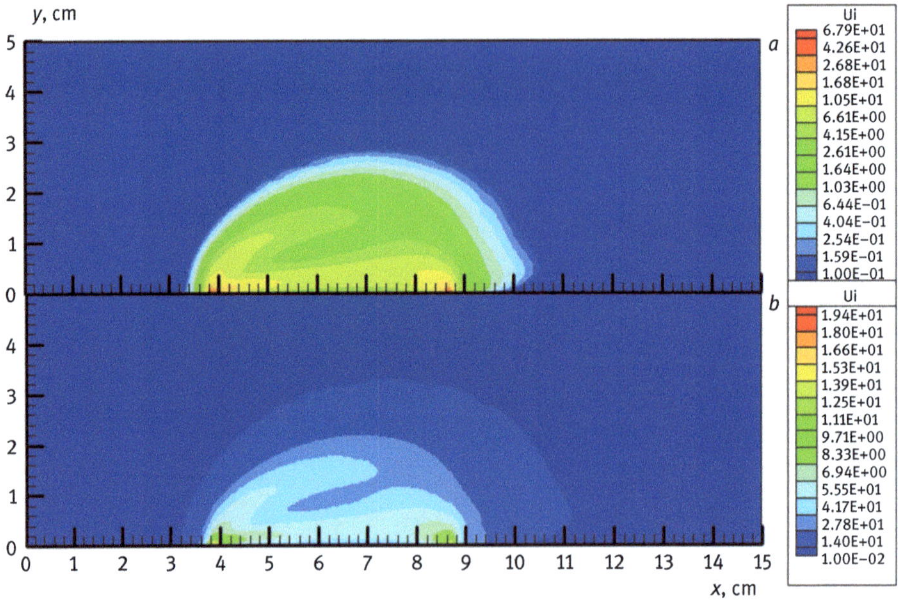

Figure 8.7: Volumetric concentration of particles (in $10^{10}\,cm^{-3}$) in surface glow discharge with magnetic field: (a) $B = -0.1\,T$ and (b) $B = -0.5\,T$.

Thus, the fulfilled numerical research of the modified model ambipolar diffusion has shown that the model can be used for searching optimum parameters of electromagnetic aerodynamic actuator and allows to predict qualitative tendencies of influence of a magnetic field on the glow discharge, which are known from experimental research.

9 Surface electromagnetic actuator in rarefied hypersonic flow

Hypersonic flow past the surface with sharp edge is investigated. This surface forms an obtuse angle; therefore, the shock wave generated by the leading edge interacts with the surface. The effect of influence on the surface direct current discharge and a transverse magnetic field on the gas dynamic characteristics is investigated. A numerical simulation is used to solve this problem. The calculation model includes the set of the Navier–Stokes and energy conservation equations, as well as equations of electrodynamics in the ambipolar approximation, and the Poisson equation.

Results of numerical modeling of the gas dynamics and electrodynamics of gas discharge with magnetic field show the change in the structure of the shock–wave interaction with surface far from location of the gas discharge. It is shown that the low-current glow discharge can be used as an electromagnetic actuator in hypersonic flows.

9.1 Introduction

In the late 1990s, a series of experimental studies were carried out to provide data on the mass and heat fluxes for aerodynamic configurations typical for hypersonic vehicles (Knight D., 2006). The results of these experiments were thoroughly documented, which allowed different groups to perform a validation of computational studies for author's computer codes developed by them. It is important that the results of the validation of these calculations were analyzed in detail by the group of highly qualified specialists in the field of computational aerodynamics (CFD), and the results of this analysis were published in the proceedings of a number of AIAA conferences. A review of these papers is presented in Knight D. (2006). Among the studied aerodynamic configurations, special attention was focused on

- blunt and sharp cones at Mach number $M = 8$;
- blunt cone with flare at $M = 8 - 10$;
- hollow sharp cylinder with flare at $M = 9 - 11$;
- interaction of shock waves at Mach ~ 10 (the fourth type of interaction in the Edney classification),
- reflection of stationary shock waves with forming the regular configuration or the Mach configuration, as well as the hysteresis phenomena.

A series of experimental and theoretical works was devoted to the interaction of shock waves with the turbulent boundary layers in various configurations of the flow. Each of these configurations is of great interest for the study of hypersonic

https://doi.org/10.1515/9783110648836-010

flows. In accordance to the purpose of this chapter, we draw attention to one of these tasks – hypersonic flow formed near surface which forms an obtuse angle with a sharp front edge. The detailed analysis of this problem was done in Knight D. (2006), MacLean M., et al., (2004), Holden M.S., et al. (2006), Gnoffo P. (2001), Wang W.-L., et al. (2002), and Kato H., et al. (2001), where the hollow cylinder flare flows were studied. Baseline data for one of the selected experiments (CUBRC Run # 14) are given in Table 9.1 (data are taken from Table 13.1 (Holden M.S., et al., 2006)).

Table 9.1: Freestream conditions for the test case Run#14.

Freestream conditions	Extended hollow cylinder flare
CUBRC run	14
Test gas	N_2
Mach number	10.3
V_∞, cm/s	2.304×10^5
p_∞, erg/cm^3	315.2
T_∞, K	120.4
ρ_∞, g/cm^3	8.810×10^7
T_w, K	295.2

Among other above-mentioned works, we highlight the work (Holden M., et al., 2006), where not only the results of calculations using two computer codes [WIND (*NPARC WIND CFD Code*) and DPLR (Wright M.J., et al., 1998)], but also the accomplished study of the properties obtained from numerical solutions on different meshes are presented. It was found that the quality of computational meshes and the quality of the calculation method applied are problems of particular importance. The good agreement between calculated and experimental data is obtained only on very detailed grids.

In this chapter by the example of the problem of streamline of the sharp edge and surface, forming an obtuse angle, the impact of the glow discharge placed on the surface of the structure of the flow is examined. Schematic picture of the problem is shown in Figure 9.1.

A general formulation of the problem of creating plasma actuators was discussed in Shang J.S., et al. (2005a,b,c). Earlier, the glow discharges localized on surface in the hypersonic flow was numerically studied in Surzhikov S.T. and Shang J.S. (2003, 2005). A detailed presentation of the numerical methods used to solve the problem of plasma actuators on the basis of the surface glow discharge of low power is given in previous chapters and in Shang J.S. (2016) and Shang J.S. and Surzhikov S.T. (2018).

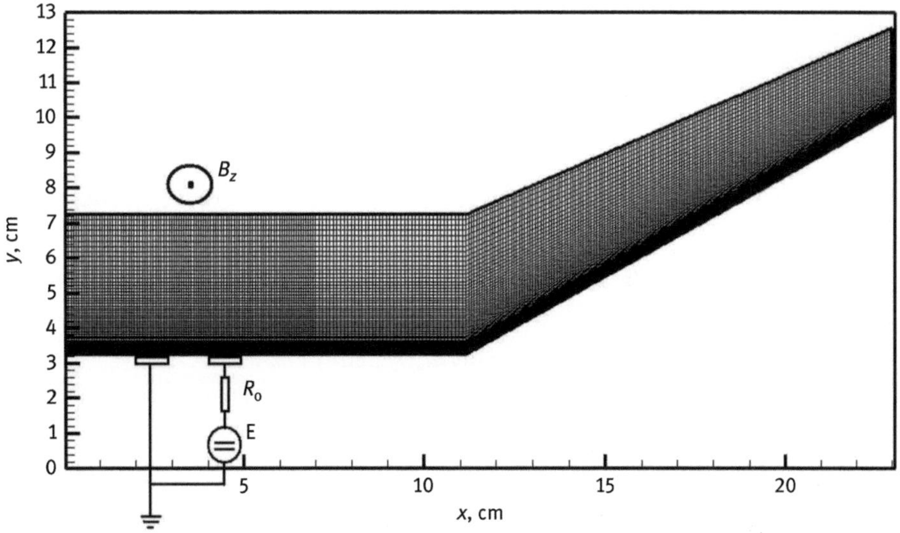

Figure 9.1: Schematic of the task.

In this chapter, an electromagnetic actuator formed by two electrodes made in the form of strips arranged transverse to the flow on the surface, as shown in Figure 9.1. The subject of a numerical study is to identify perturbations in the pressure and heat flux distributions along the surface of the obtuse angle. As an additional control parameter of the plasma actuator, a transverse magnetic field is considered, which is the polarization shown in Figure 9.1.

9.2 Gas dynamic model

We consider streamline of a sharp leading edge by a viscous, heat-conducting, and partially ionized gas. The surface forms an obtuse angle before which the glow discharge is lighted, and magnetic field is applied in the cross-flow direction. The glow discharge is organized between two rectangular sections of the electrode at the surface, as shown in Figure 9.1. An external electric circuit is taken into account. It consists of a power source and an ohmic resistance. In this chapter, it has been shown that the order in arrangement of electrodes in the flow is important. In this chapter, it is assumed that the cathode is located upstream (ground electrode in Figure 9.1).

The set of governing equations of the dynamics of a viscous, heat-conducting, and partially ionized gas takes into account the current flowing through that as well as the transversal magnetic field. It looks as follows:

$$\frac{\partial \rho}{\partial t} + \operatorname{div}(\rho \mathbf{V}) = 0, \tag{9.1}$$

$$\frac{\partial \rho u}{\partial t} + \operatorname{div}(\rho u \mathbf{V}) = -\frac{\partial p}{\partial x} - \frac{2}{3}\frac{\partial}{\partial x}(\mu \operatorname{div}\mathbf{V}) + \frac{\partial}{\partial y}\left[\mu\left(\frac{\partial u}{\partial y} + \frac{\partial v}{\partial x}\right)\right] + 2\frac{\partial}{\partial x}\left(\mu \frac{\partial u}{\partial x}\right) + f_{M,x}, \tag{9.2}$$

$$\frac{\partial \rho v}{\partial t} + \operatorname{div}(\rho v \mathbf{V}) = -\frac{\partial p}{\partial y} - \frac{2}{3}\frac{\partial}{\partial y}(\mu \operatorname{div}\mathbf{V}) + \frac{\partial}{\partial x}\left[\mu\left(\frac{\partial u}{\partial y} + \frac{\partial v}{\partial x}\right)\right] + 2\frac{\partial}{\partial y}\left(\mu \frac{\partial v}{\partial y}\right) + f_{M,y}, \tag{9.3}$$

$$\rho c_p \frac{\partial T}{\partial t} + \rho c_p \mathbf{V} \operatorname{grad} T = \frac{\partial p}{\partial t} + \operatorname{div}(\lambda \operatorname{grad} T) + \mathbf{V} \operatorname{grad}(p) + Q_\mu + Q_J, \tag{9.4}$$

$$\frac{\partial n}{\partial t} + \operatorname{div}(\mathbf{V} n) = \operatorname{div}\left(\mu_e^* D_a \operatorname{grad} n\right) + \dot{\omega}_e, \tag{9.5}$$

$$\operatorname{div}\left[\left(\mu_i + \tilde{\mu}_e\right) n \mathbf{E} + \left(\tilde{D}_e - D_i\right) \operatorname{grad} n\right] = 0, \tag{9.6}$$

where x and y are the Cartesian coordinates connected with unit vectors $\mathbf{i}$ and $\mathbf{j}$; $\mathbf{V} = (u, v)$ is the velocity vector and its projections on x- and y-axes; ρ and p are the density and pressure; μ is the dynamic viscosity coefficient; c_p is the heat capacity at constant pressure; T is the temperature of the gas; λ is the heat conductivity coefficient; $f_{M,x}$, $f_{M,y}$ are the components of the ponderomotive force; Q_J is the heat release power due to the electric current; Q_μ is the dissipative function

$$Q_\mu = \mu\left[2\left(\frac{\partial u}{\partial x}\right)^2 + 2\left(\frac{\partial v}{\partial y}\right)^2 + \left(\frac{\partial v}{\partial x} + \frac{\partial u}{\partial y}\right)^2 + \frac{2}{3}\left(\frac{\partial u}{\partial x} + \frac{\partial v}{\partial y}\right)^2\right], \tag{9.7}$$

$\dot{\omega}_e = \alpha(E)|\mathbf{\Gamma}_e| - \beta_e n^2$ is the source term related to the ionization (the first summand) and to the recombination (the second summand); $\alpha(E)$ is the coefficient of ionization (the first Townsend coefficient); $\alpha(E)$ is the recombination coefficient; $|\mathbf{\Gamma}_e|$ is the module of the vector of electron flux; E is the module of the electric field strength; D_a is the ambipolar diffusivity; $\mathbf{j} = e(\mathbf{\Gamma}_i - \mathbf{\Gamma}_e)$ is the vector of current density; $\mathbf{\Gamma}_i$ and $\mathbf{\Gamma}_e$ are the vectors of the ion and electron flux densities.

An approximation of the quasineutral plasma is used; therefore, $n = n_i = n_e$, where n_i, n_e are the concentrations of ions and electrons;

$$\tilde{\mu}_e = \frac{\mu_e}{1 + b_e^2}, \quad \mu_e^* = \frac{\mu_e + \mu_i}{\left(1 + b_e^2\right)\mu_i + \mu_e} \cong \frac{\mu_e}{\left(1 + b_e^2\right)\mu_i + \mu_e}, \quad \tilde{D}_e = \frac{D_e}{1 + b_e^2},$$

$b_e = \frac{\mu_e B_z}{c}$ is the Hall parameter, B_z is the magnetic field induction, and c is the speed of light.

When one takes into account the magnetic field, the term on the right-hand side in eq. (9.5) should be modified as follows:

$$\dot{\omega}_e = \alpha(E)\tilde{\mu}_e E - \beta n^2.$$

Equation (9.6) is the Maxwell equation for conservation of electric current under steady-state conditions. Equation (9.5) expresses the continuity of the charged components of the partially ionized gas. To obtain this equation, the following assumptions were used:

- Velocities of the charged and neutral particles in partially ionized gas are equal, $\mathbf{V} = \mathbf{V}_e = \mathbf{V}_i$ ($\mathbf{V}$ is the mean velocity of neutral particles, $\mathbf{V}_e$, $\mathbf{V}_i$ are the mean velocities of electrons and ions);
- The concentrations of ions and electrons are equal ($n = n_e = n_i$);
- Charged particles diffuse with a factor of ambipolar diffusivity D_a.

Taking into account that the electric field strength is connected to the electric potential φ, $\mathbf{E} = -\mathrm{grad}\varphi$, instead of eq. (9.6) it is more preferable to use the following approximate equation for potential:

$$\frac{\partial}{\partial x}\left[(\mu_i + \tilde{\mu}_e) n \frac{\partial \varphi}{\partial x}\right] + \frac{\partial}{\partial y}\left[(\mu_i + \tilde{\mu}_e) n \frac{\partial \varphi}{\partial y}\right] - \\ - \frac{\partial}{\partial x}\left[(\tilde{D}_e - D_i)\frac{\partial n}{\partial x}\right] - \frac{\partial}{\partial y}\left[(\tilde{D}_e - D_i)\frac{\partial n}{\partial y}\right] = 0. \tag{9.8}$$

Equations (9.1)–(9.6) are recorded in the Cartesian coordinate system. To calculate the structure of gas flow the new variables in the curved geometry are introduced $\xi = \xi(x, y)$, $\eta = \eta(x, y)$. These variables are connected with Cartesian coordinates by the one-to-one transformations. Now, the coordinate ξ is directed along the surface, and the η-coordinate is directed normally to the surface.

The boundary conditions for systems (9.1)–(9.6) are presented in the following form (see Figure 9.1):

$$\xi = 0: \quad u = V_\infty, \ v = 0, \ T = T_\infty, \ p = p_\infty, \ \rho = \rho_\infty, \ n = n_\infty, \ \frac{\partial \varphi}{\partial \xi} = 0;$$

$$\xi = L: \quad \frac{\partial u}{\partial \xi} = \frac{\partial v}{\partial \xi} = \frac{\partial T}{\partial \xi} = \frac{\partial n}{\partial \xi} = \frac{\partial \rho}{\partial \xi} = \frac{\partial \varphi}{\partial \xi} = 0;$$

$$\eta = 0, \quad \xi > \xi_{Edge}, \ u = v = 0, \quad \frac{\partial p}{\partial \eta} = 0;$$

$$\eta = 0, \ \xi_{C_1} \le \xi \le \xi_{C_2} \ \text{(the cathode section)}: \frac{\partial n}{\partial \eta} = 0, \ \varphi = 0;$$

$$\eta = 0, \ \xi_{A_1} \le \xi \le \xi_{A_2} \ \text{(the anode section)}: \frac{\partial n}{\partial \eta} = 0, \ \varphi = V_c;$$

$$\eta = 0; \ \xi_{\mathrm{Edge}} < \xi < \xi_{C_1}, \ \ \xi_{C_2} < \xi \langle \xi_{A_1}, \ \ \xi \rangle \xi_{A_2} \ \text{(the dielectric surface)}:$$

$$n = 10^{-5} n_0, \ \frac{\partial \varphi}{\partial \eta} = 0;$$

$$y = H, \quad x > 0: \qquad u = V_\infty, \ v = 0, \ T = T_\infty, \ p = p_\infty, \ \rho = \rho_\infty, \ n = n_\infty, \ \frac{\partial \varphi}{\partial \eta} = 0 .$$

Here ξ_{Edge} is the longitudinal coordinate of the leading sharp edge; ξ_{C_1}, ξ_{C_2}, ξ_{A_1}, ξ_{A_2} are the longitudinal coordinates of the cathode and anode boundaries; n_0 is the typical concentration of charged particles in glow discharge ($n_0 \sim 10^{10}\text{см}^{-3}$); V_c is the voltage drop between cathode and anode. Surface temperature $T_w = 300\,\text{K}$ is assumed constant for all calculations.

The electrotechnical equation for external electric circuit is used in the following form:

$$IR_0 + V_c = E, \tag{9.9}$$

from which the voltage drop V_c can be found, V_c is the ohmic resistance, I is the total current in electric circuit, which is determined by integral over the surfaces of electrodes:

$$I = \int_0^L (\mathbf{jn})_c \mathrm{d}x = \int_0^L (\mathbf{jn})_a \mathrm{d}x ,$$

where $\mathbf{n}$ is the unit vector normal to cathode (c) and anode (a) surfaces; $\mathbf{j}$ is the vector of current density. All dimensions in the task are related to the lengths of the electrodes in z-direction.

Gas dynamic profiles without electric discharge as well as plasma cloud with charged particles concentration of above cathode sections were used as initial conditions for the calculations. The time-asymptotic method was used for integration of governing equations; therefore, the precise type of initial conditions was not very significant. Nevertheless, due to nonlinear character of the problem under consideration, the initial conditions should be quite accurate.

In accordance with Holden M.S., et al. (2006) the hypersonic flow of molecular nitrogen is considered. The constitutive relationships for N_2 are based on the molecular-kinetic theory (Hirschfelder J.O., et al., 1964):

$$\mu_m = 2.67 \cdot 10^{-5} \sqrt{M_A T} \frac{1}{\sigma^2 \Omega^{(2,2)*}}, \ \text{g}/(\text{cm} \cdot \text{s})$$

is the dynamic viscosity,

$$\lambda = 8.334 \times 10^{-4} \sqrt{\frac{T}{M_A}} \frac{\left(0.115 + 0.354 \frac{c_p}{R_0} M_A\right)}{\sigma^2 \Omega^{(2,2)*}}, \text{W}/(\text{cm} \cdot \text{K})$$

is the heat conductivity,

$$c_p = 8.317 \times 10^7 \frac{7}{2} \frac{1}{M_A}, \quad \rho = \frac{p}{T} \frac{M_A}{R_0}, \quad M_A = 28 \ \text{g/mol},$$

$$R_0 = 8.314 \times 10^7 \text{ erg}/(\text{mol} \cdot \text{K}),\ \sigma = 3.68\,\mathring{A},\ \Omega^{(2,2)*} = 1.157\left(T^*\right)^{-0.1472},$$

$$T^* = \frac{T}{(\varepsilon/k)},\ (\varepsilon/k) = 71.4\,.$$

Coefficient of ionization was calculated as follows:

$$\nu_i = \left(\frac{\alpha}{p^*}\right) p^* E \mu_e(p^*),\ \left(\frac{\alpha}{p^*}\right) = A \exp\left[-\frac{B}{(E/p^*)}\right],\ (\text{cm} \cdot \text{Torr})^{-1},$$

where

$$p^* = p\frac{293}{T},\ D_e = \mu_e(p^*)T_e,\ D_+ = \mu_+(p^*)T_+,\ \text{cm}^2/\text{s},$$

$$\mu_e(p^*) = 4.2 \times 10^5 \frac{1}{p^*},\ \mu_+(p^*) = 1{,}450\frac{1}{p^*},\ \text{cm}^2/(\text{s} \cdot \text{V}),$$

$$\beta = 2 \times 10^{-7}\ \text{cm}^3/\text{c},\ T_e = 11{,}610\ \text{K},\ A = 12\,(\text{cm} \cdot \text{Torr})^{-1},\ B = 342\,\text{V}/(\text{cm} \times \text{Torr}),$$

μ_+ is the drift of ions.

Heat release due to the electric current (the Joule heat) is calculated by using the calculated current density and electric field strength

$$Q_J = \eta(\mathbf{j} \cdot \mathbf{E}) = 1.6 \times 10^{-19} \eta\left[nE^2(\mu_+ + \tilde{\mu}_e) + (\tilde{D}_e - D_+)\mathbf{E}\,\text{grad}\,n\right], \tag{9.10}$$

where η is the coefficient of efficiency of gas heating. Setting this ratio in the range $\eta \sim 0.1 - 0.3$ allows to partially take into account the shortcomings of the model used in the description of the kinetics of physical and chemical processes, as well as the processes of vibrational relaxation.

An additional force acted on gas due to the presence of magnetic field has two components:

$$\mathbf{F}_M = f_{M,x}\mathbf{i} + f_{M,y}\mathbf{j} = c^{-1}[\mathbf{jB}]\,. \tag{9.11}$$

These two projections of the magnetic field are taken into account in eqs. (9.2) and (9.3). The set of eqs. (9.1)–(9.6) describing a motion of a viscous, heat-conducting, and partially ionized gas along the surface with glow discharge at the presence of external magnetic field was solved by the type-asymptotic method up to achieving a numerical convergence with a relative error of 10^{-5}, which is calculated over the entire field of flow for functions $u,\ v,\ p,\ T,\ n,\ \varphi$. At each step, the following equations were integrated numerically:

- the continuity equation and the Navier–Stokes equations;
- the energy conservation equation;
- the set of electrodynamic equations.

The AUSM - advective upstream splitting method was used for integrating the gas dynamic equations. The energy conservation equation was integrated with the use of the implicit finite-difference method of the second-order accuracy in time and in space. The same numerical approach was used for integration of the electrodynamic equations. This part of the numerical simulation procedure was the most time-consuming due to exponential character of the source of ionization.

Second internal iterative process (at each time step) consisted in achieving the mutual convergence of functions obtained at solving the Fourier–Kirchhoff equations (the energy conservation equation) and the set of electrodynamic equations. The need in this iterative process was caused by the fact that the gas temperature is very strongly influenced by its ionization rate, which in turn produces strong perturbations on the right-hand side of eq. (9.5).

9.3 Results of numerical simulation

Two configurations of mutual location of electrodes were investigated. Coordinates of electrodes for these configurations are given in Table 9.2. The choice of these two configurations follows from the obvious fact that creation of an artificial disturbance to the leading edge leads to significant consequences for the entire flow field.

Table 9.2: Coordinates of electrodes.

	Configuration #1	Configuration #2
x_{C_1}, cm	2	3
x_{C_2}, cm	3	4
x_{A_1}, cm	4	5
x_{A_2}, cm	5	6

Figure 9.2 shows the distribution of the gas dynamic parameters of the streamlined surface without an electrical discharge. It is assumed that the edge of the plate is located at a distance of 1 cm from the computational domain boundaries.

Distributions of Mach number (Figure 9.2a) and the longitudinal velocity (Figure 9.2f) show that near the sharp bend of the streamlined surface (cm), low speeds are observed. Calculations on detailed computational grids with high accuracy methods show that there is a fairly extensive area of the vortex motion of the gas. In the present case, the length of this zone does not exceed of about 1 cm.

From Figure 9.2f one can see that two areas of the flow spread are in the positive direction. The first area is from the leading edge, and the second area is from the

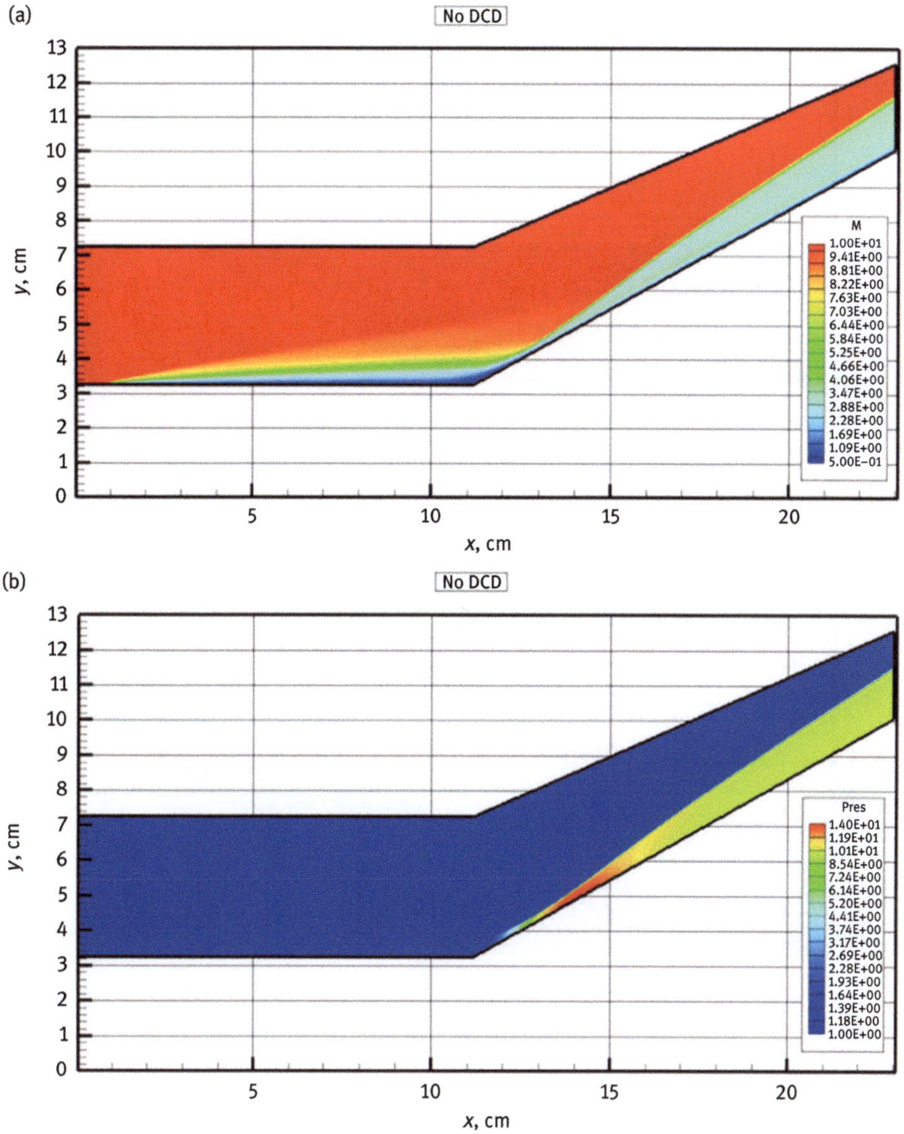

Figure 9.2: Gas dynamic parameters without gas discharge: (a) Mach numbers; (b) pressure (p/p_∞); (c) density (ρ/ρ_∞); (d) temperature (in K); (e) longitudinal velocity (u/V_∞); and (*f*) transversal velocity (v/V_∞).

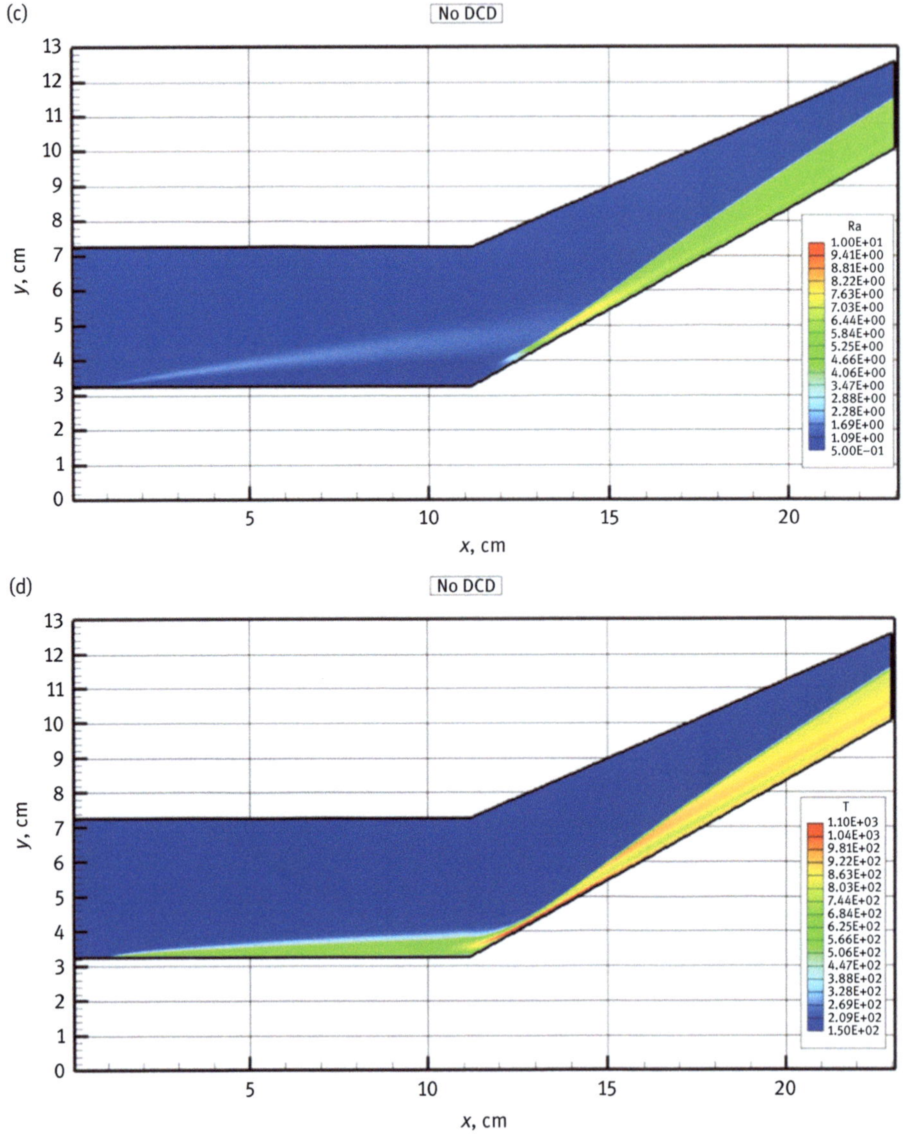

Figure 9.2 (continued)

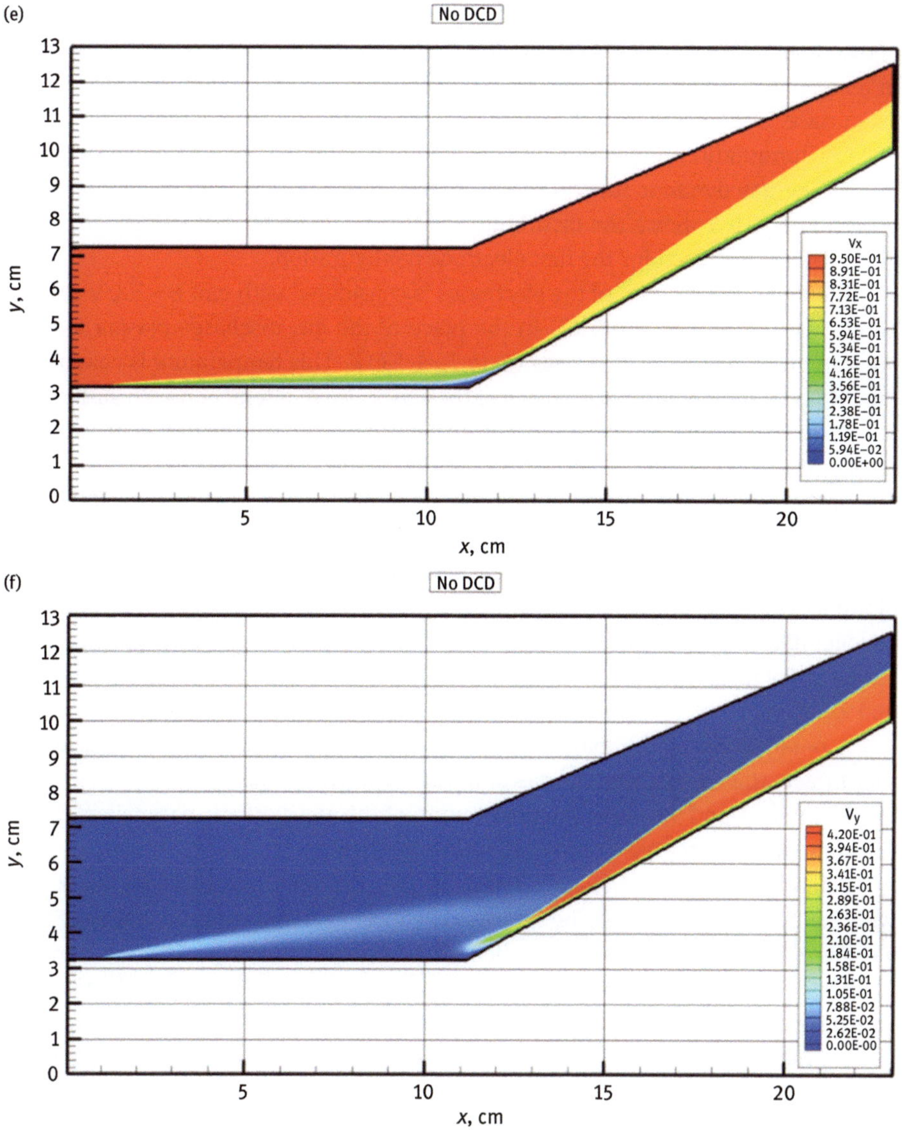

Figure 9.2 (continued)

sharp turn of the surface. The most high pressure is observed on the surface in place of the fall of the bow shock. In the vicinity of this place, there is a noticeable increase in temperature (Figure 9.2d) and, as a consequence, an increase in the heat flux to the surface. Here there is also a natural increase in density (Figure 9.2c).

The formulation of present task includes an examination of the possibility to modify the gas dynamic functions by the use of the surface glow discharge of low power. Figure 9.3 shows the distribution of the gas dynamic parameters when the glow discharge is used for the first electrode configuration.

The first configuration of the electrodes is considered with emf $E = 200$ V and the ohmic resistance of $R_0 = 300$ kΩ. In the place of the gas discharge plasma location one can observe heating of the gas up to $T \sim 1,100$ K. This temperature is commensurable with the temperature near place of the shock–wave interaction with surface.

Perturbations in pressure and density distributions are insignificant, whereas changes in the longitudinal and transversal velocities are visible.

In the case of the second configuration of electrodes with emf $E = 200$ V and $R_0 = 300$ kΩ, the similar perturbation in distributions of gas dynamic functions close to the new arrangement of the electrodes are observed. In this case, the results of numerical simulation are shown in Figure 9.4.

Significantly more noticeable perturbations of gas dynamic parameters are observed in the magnetic field. Corresponding numerical results are shown in Figures 9.5 and 9.6.

A polarization of the magnetic field produces a force component $f_{M,y}$ toward the surface, that is, pushing the gas to the streamlined surface.

Figures 9.5 and 9.6 show the distribution of the gas dynamic parameters for the first and second electrode configurations, respectively.

Before analyzing the characteristics of gas dynamic perturbations associated with the interaction between the surface glow discharge and external magnetic field, we consider the appropriate distribution of the electrodynamic parameters shown in Figures 9.7 and 9.8. Figures 9.7a and 9.8a show the distribution of electric potential, which clearly shows the location of the anode and cathode (the largest potential).

Figures 9.7b and 9.8b show the axial component of the ponderomotive force. Near the cathode (or rather, near its right border) it is positive, and at the left border of the anode it is negative, which leads to a marked braking of gas flow. The normal component of the ponderomotive force is directed to the surface. The highest value of the heat release power occurs to be between the cathode and the anode (Figures 9.7d and 9.8d). In the same area, the greatest concentration of charged particles (Figures 9.7e and 9.8e) is observed.

Perturbations of the gas dynamic parameters shown in Figures 9.5 and 9.6 are conditioned by the presence of glow discharge with magnetic field whose general characteristics are shown in Figures 9.7 and 9.8.

The localized domain of deceleration of the gas flow between the cathode and the anode is shown in Figures 9.5a, e and 9.6a, e. In this domain, the normal

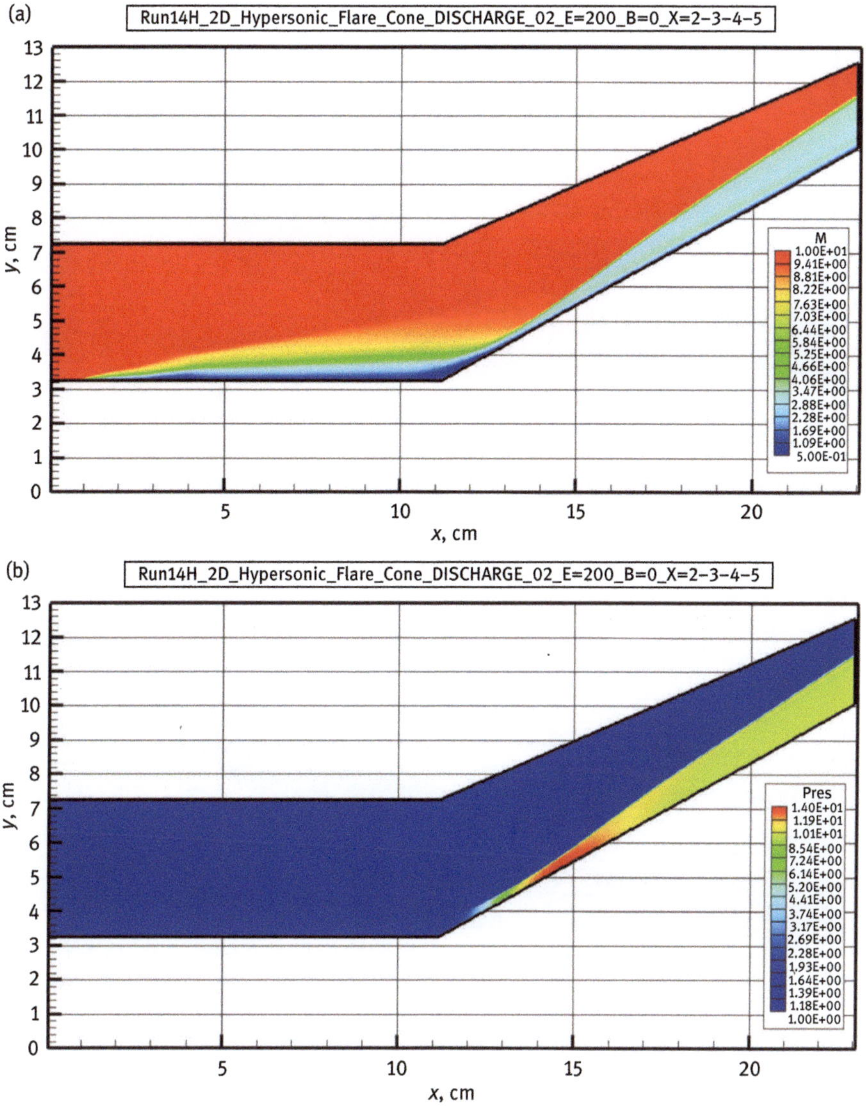

Figure 9.3: No magnetic field. Gas dynamic parameters with gas discharge of the first configuration: (a) Mach numbers; (b) pressure (p/p_∞); (c) density (ρ/ρ_∞); (d) temperature (in K); (e) longitudinal velocity (u/V_∞); (f) transversal velocity (v/V_∞).

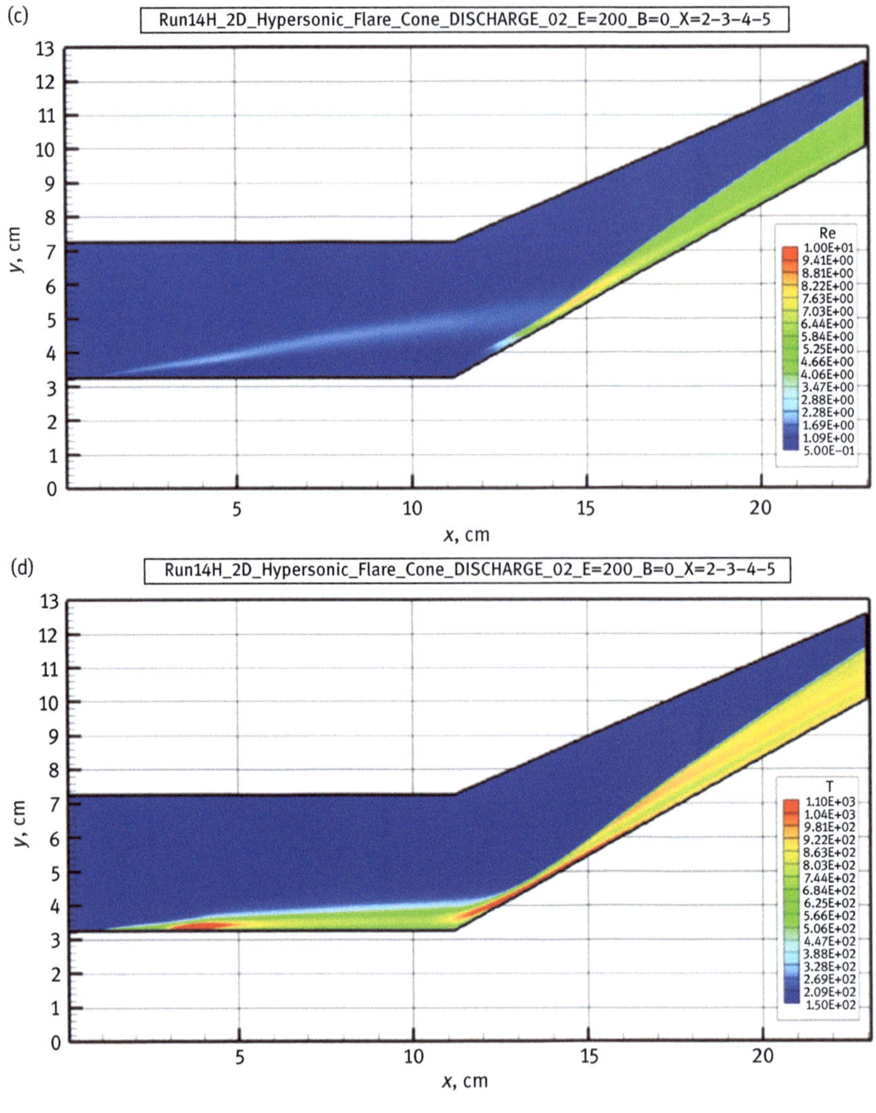

Figure 9.3 (continued)

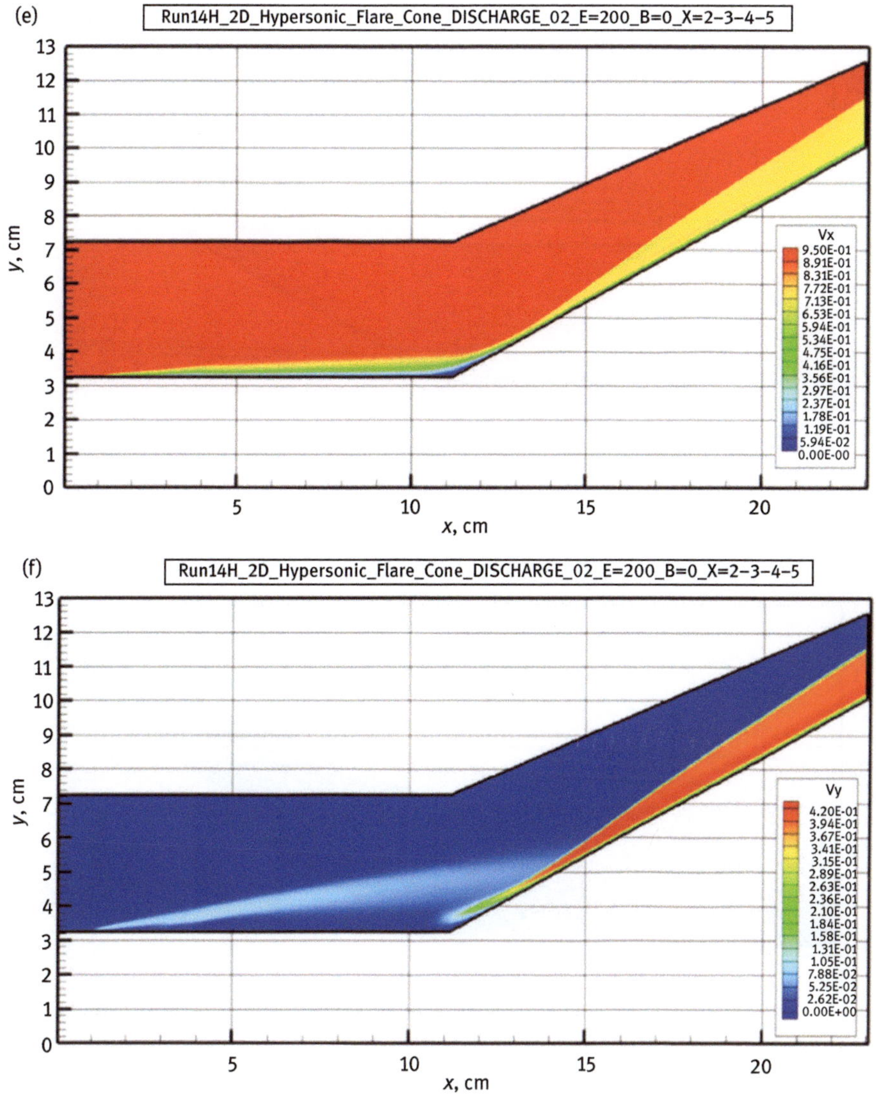

Figure 9.3 (continued)

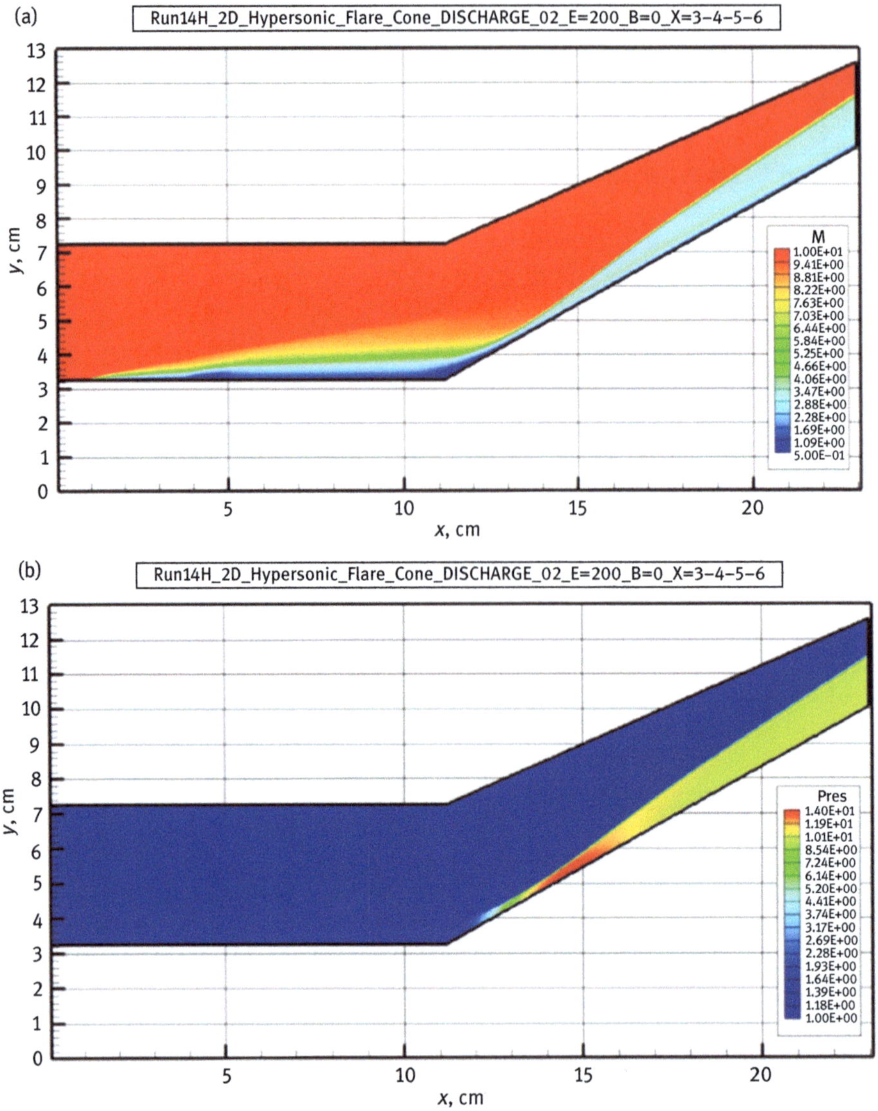

Figure 9.4: No magnetic field. Gas dynamic parameters with gas discharge of the second configuration: (a) Mach numbers; (b) pressure (p/p_∞); (c) density (ρ/ρ_∞); (d) temperature (in K); (e) longitudinal velocity (u/V_∞); (f) transversal velocity (v/V_∞).

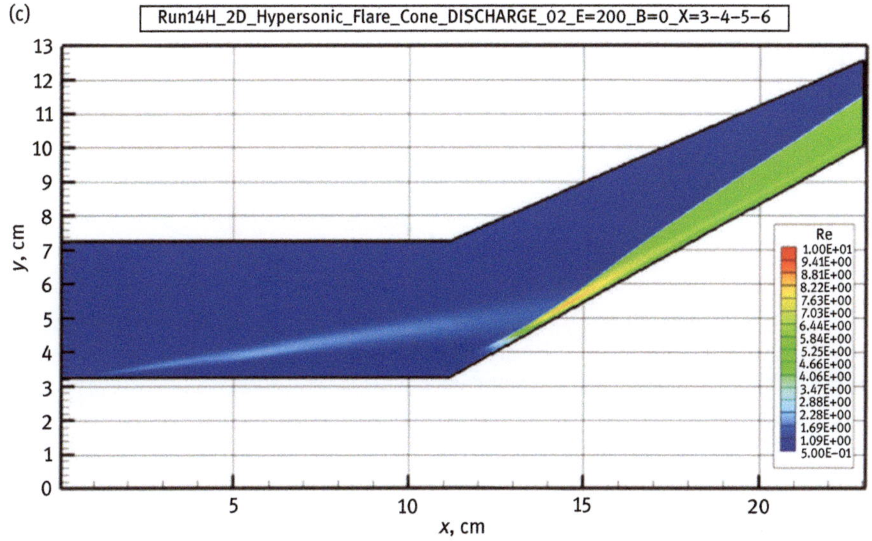

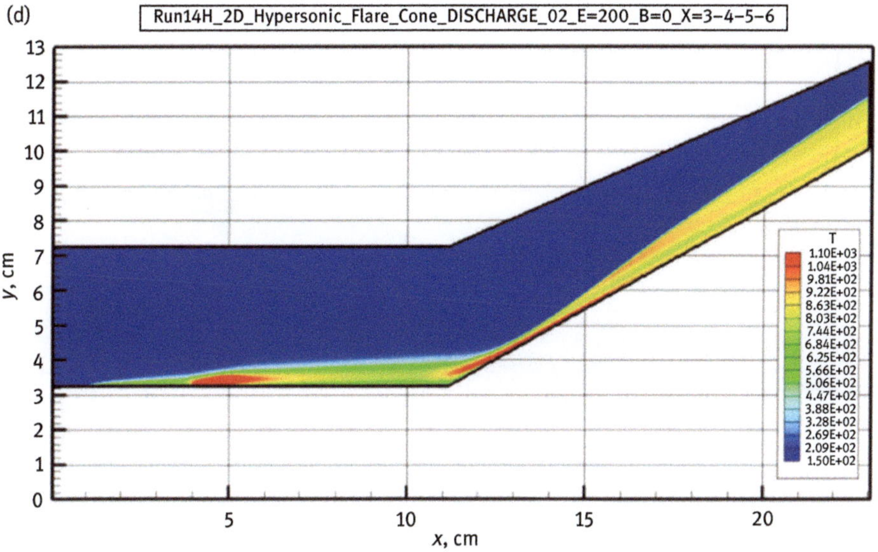

Figure 9.4 (continued)

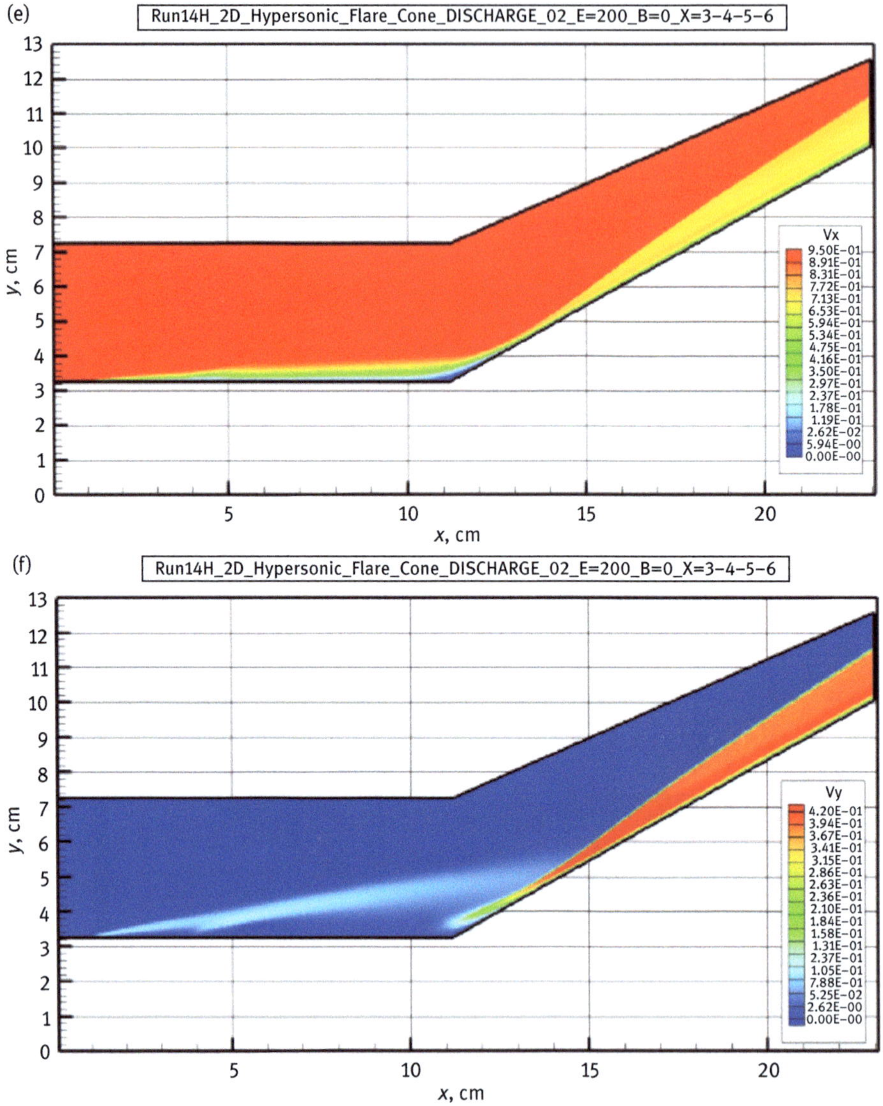

Figure 9.4 (continued)

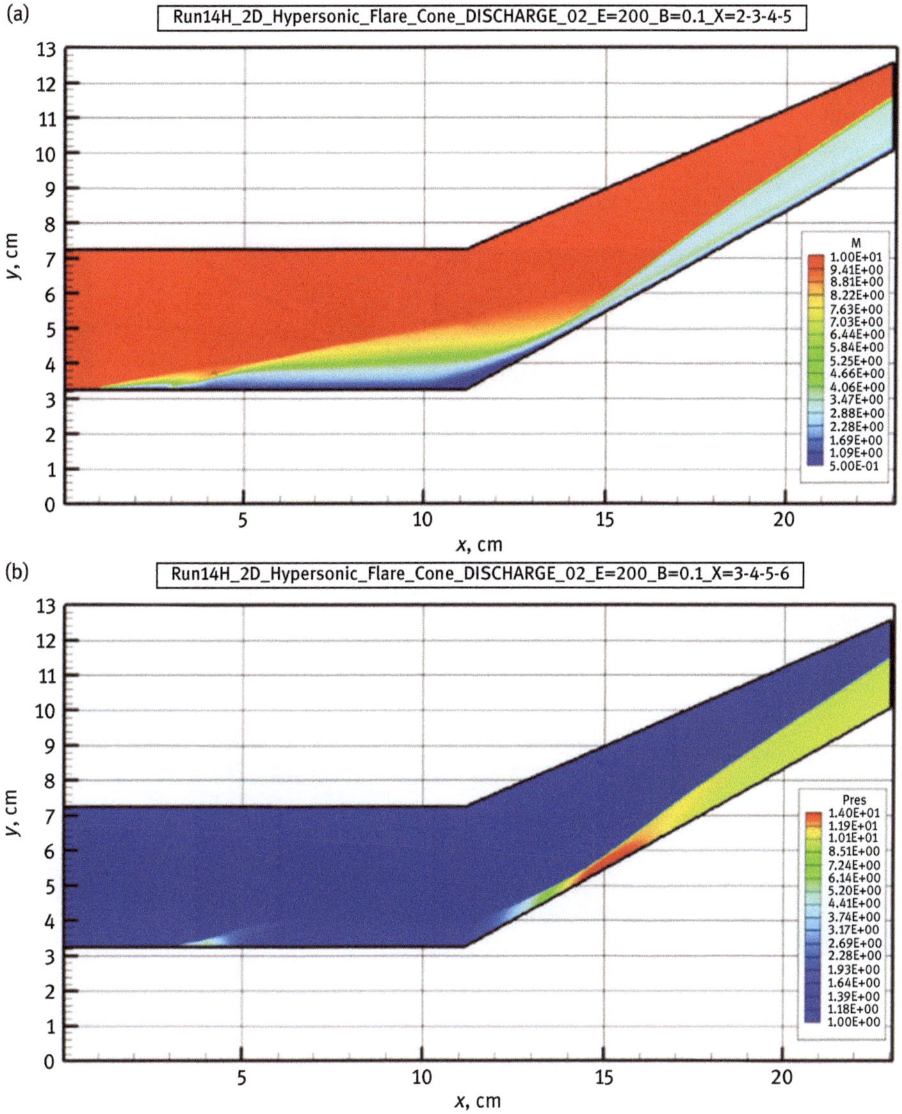

Figure 9.5: Gas dynamic parameters with gas discharge of the first configuration and with magnetic field of $B_z = 0.1\,\text{T}$: (a) Mach numbers; (b) pressure (p/p_∞); (c) density (ρ/ρ_∞); (d) temperature (in K); (e) longitudinal velocity (u/V_∞), (f) transversal velocity (v/V_∞).

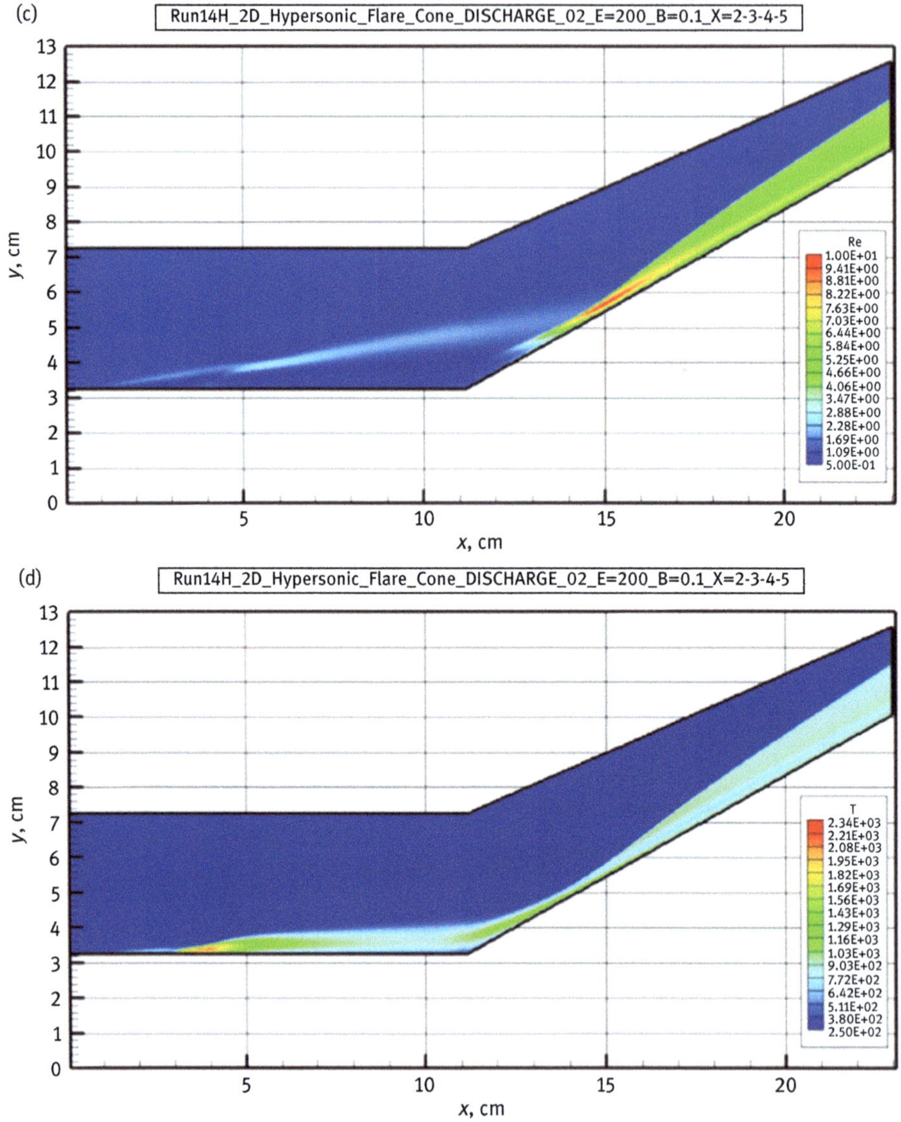

Figure 9.5 (continued)

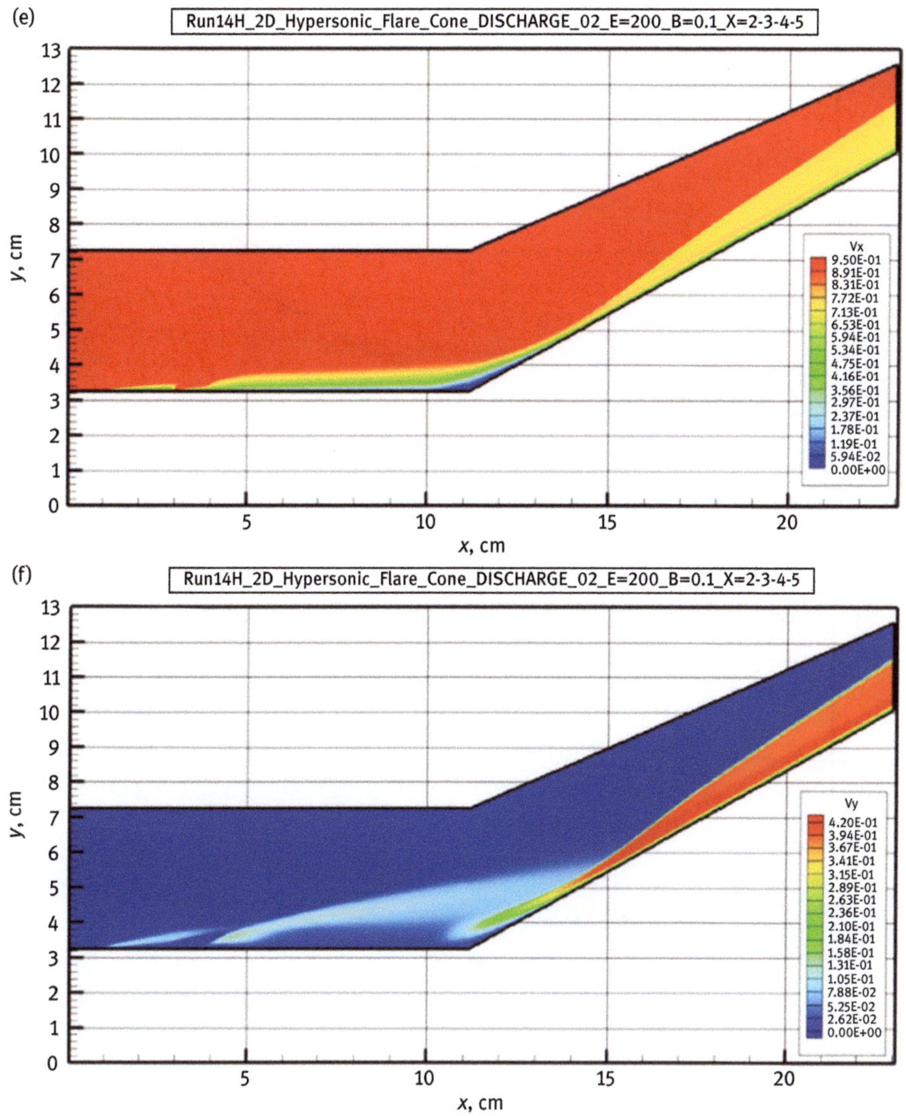

Figure 9.5 (continued)

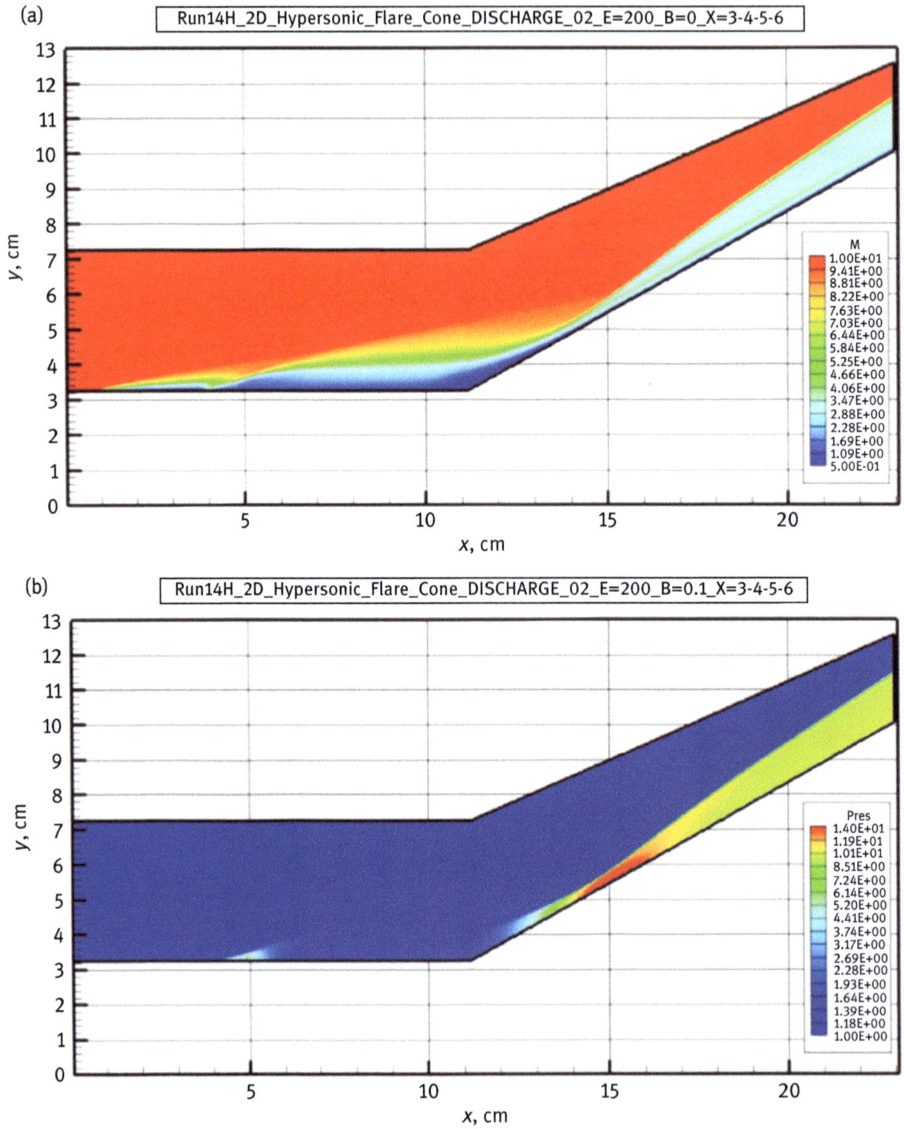

Figure 9.6: Gas dynamic parameters with gas discharge of the second configuration and with magnetic field of $B_z = 0.1\,T$: (a) Mach numbers; (b) pressure (p/p_∞); (c) density (ρ/ρ_∞); (d) temperature (in K); (e)longitudinal velocity (u/V_∞); (f) transversal velocity (v/V_∞).

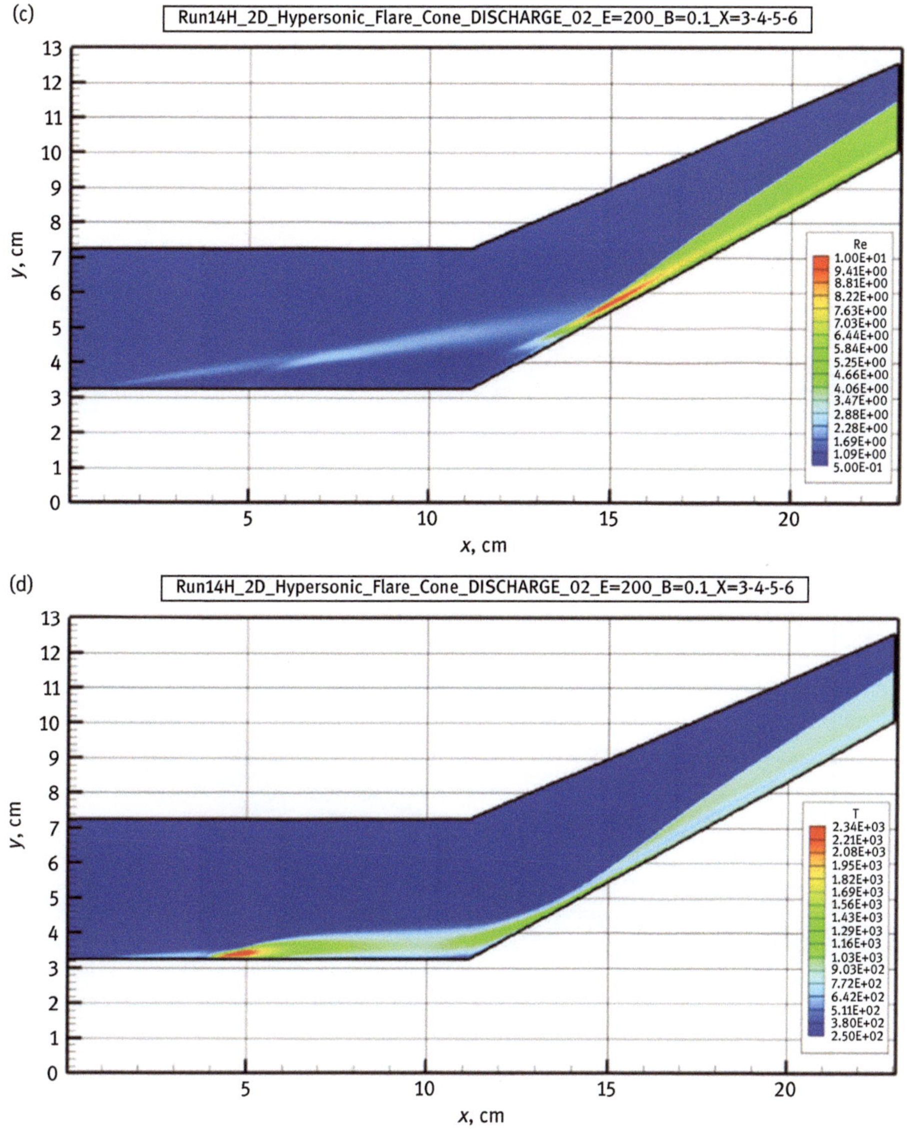

Figure 9.6 (continued)

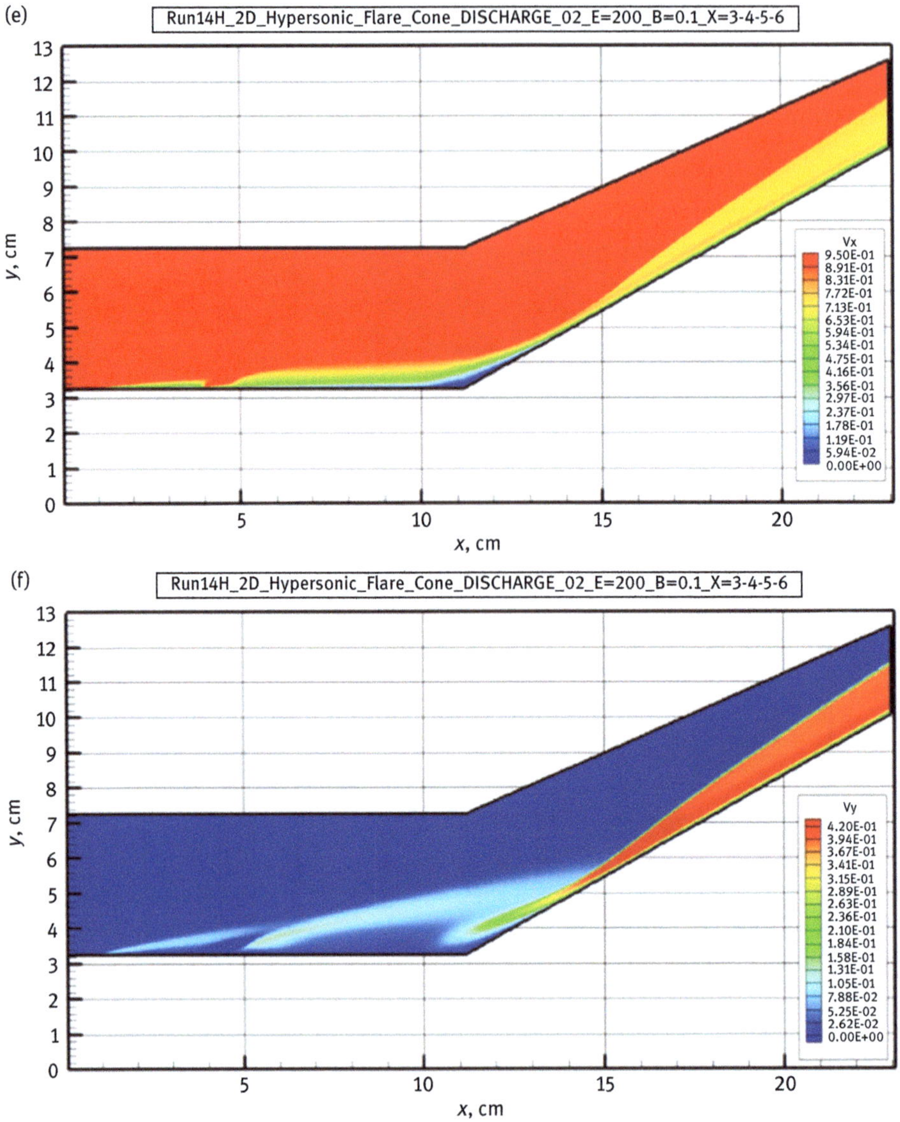

Figure 9.6 (continued)

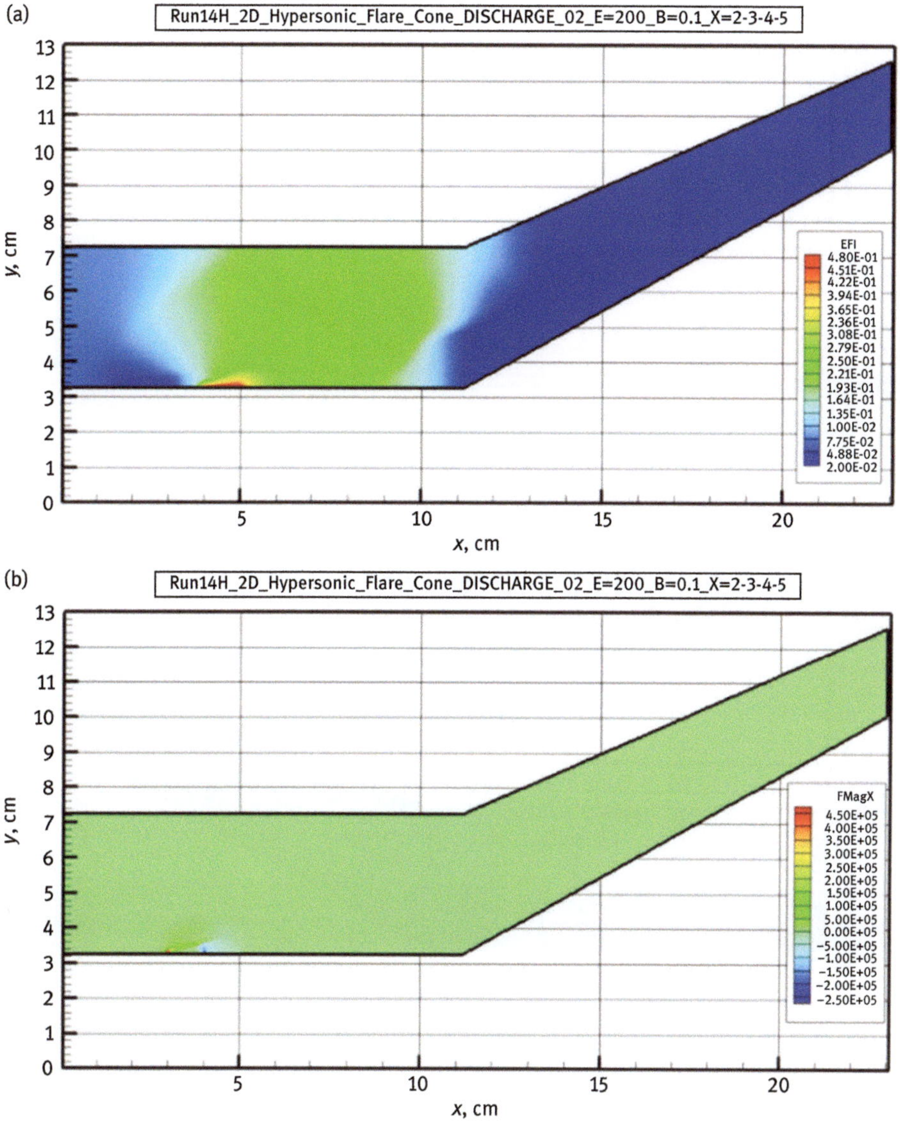

Figure 9.7: Electrodynamic parameters with gas discharge of the first configuration and with magnetic field of $B_z = 0.1\,\text{T}$: (a) electric potential (φ/ε); (b) x-component of the magnetic force (in dyn); (c) y-component of the magnetic force (in dyn); (d) current density (mA/cm^2); (e) concentration of electron in 10^{10} cm^{-3}.

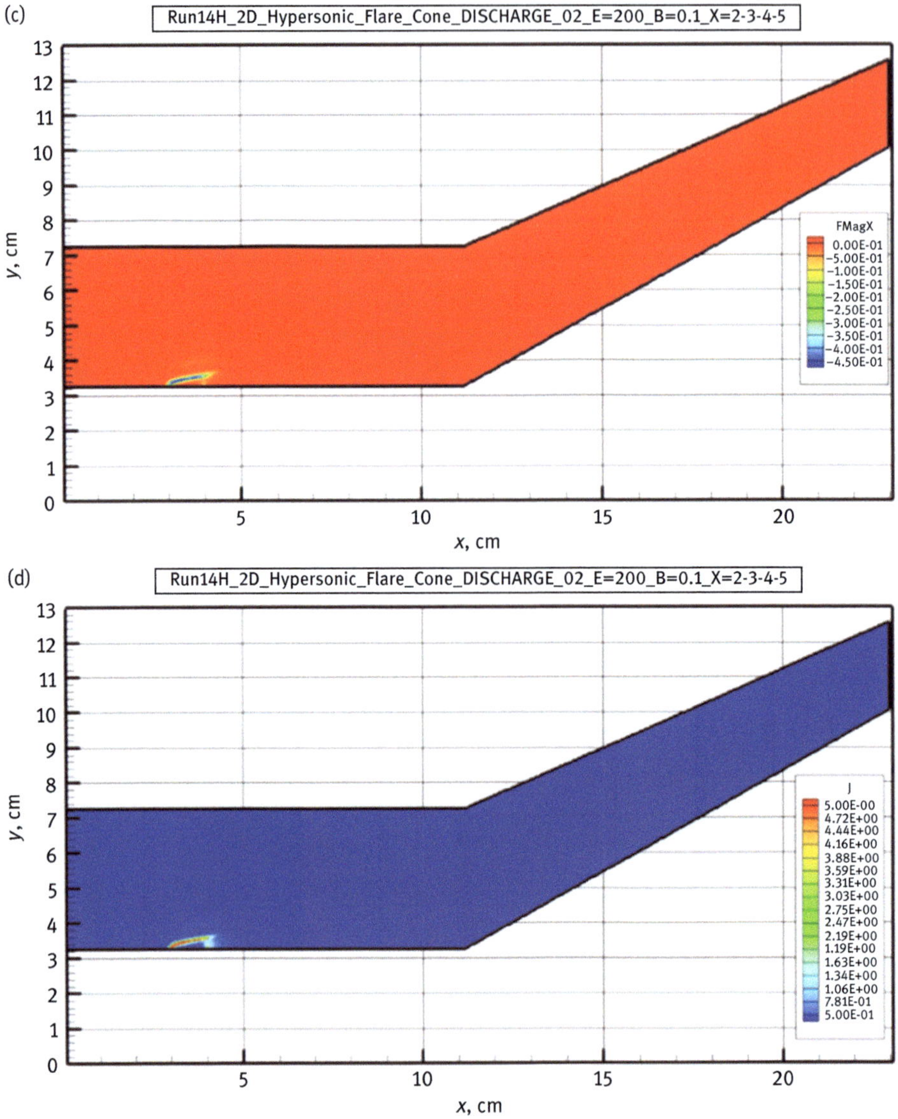

Figure 9.7 (continued)

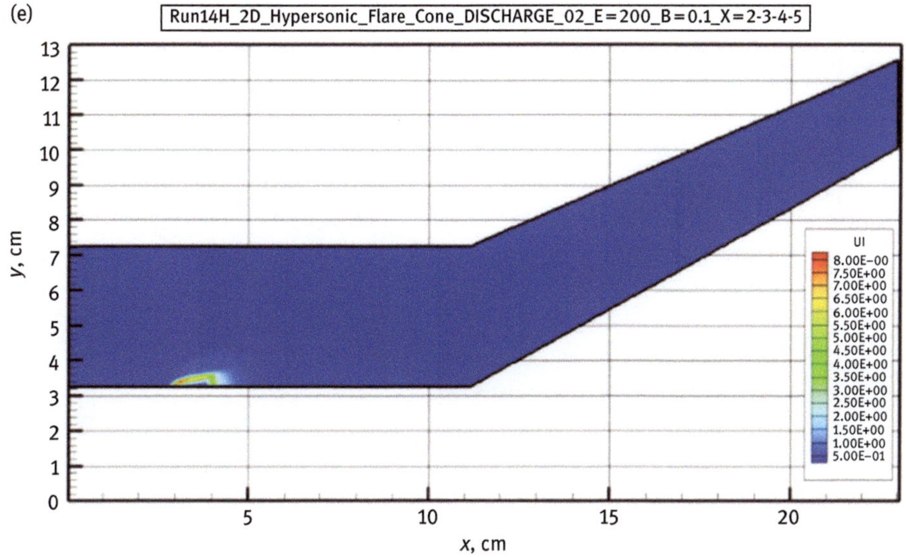

Figure 9.7 (continued)

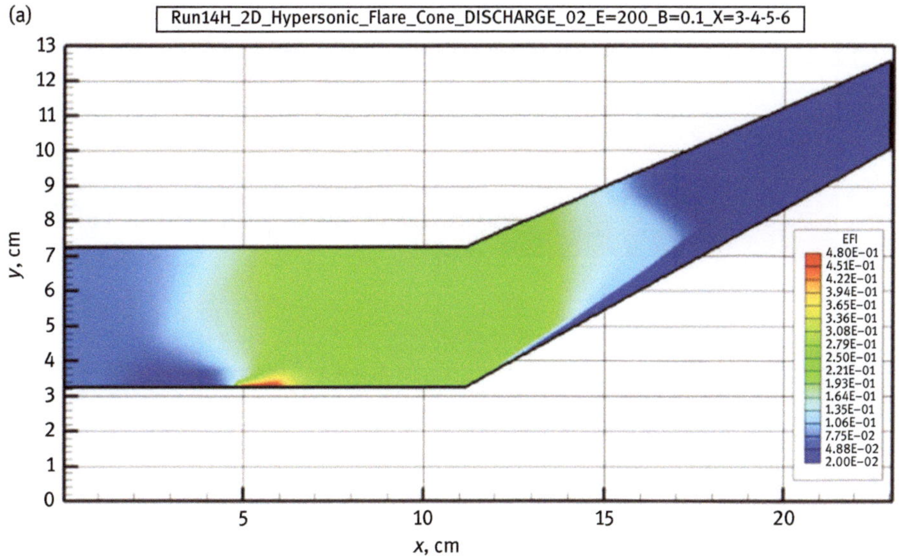

Figure 9.8: Electrodynamic parameters with gas discharge for the second configuration and with magnetic field of $B_z = 0.1\,\mathrm{T}$: (a) electric potential (φ/ε); (b) *x*-component of the magnetic force (in dyn); (c) *y*-component of the magnetic force (in dyn); (d) current density (mA/cm^2); (e) concentration of electron in 10^{10} cm^{-3}.

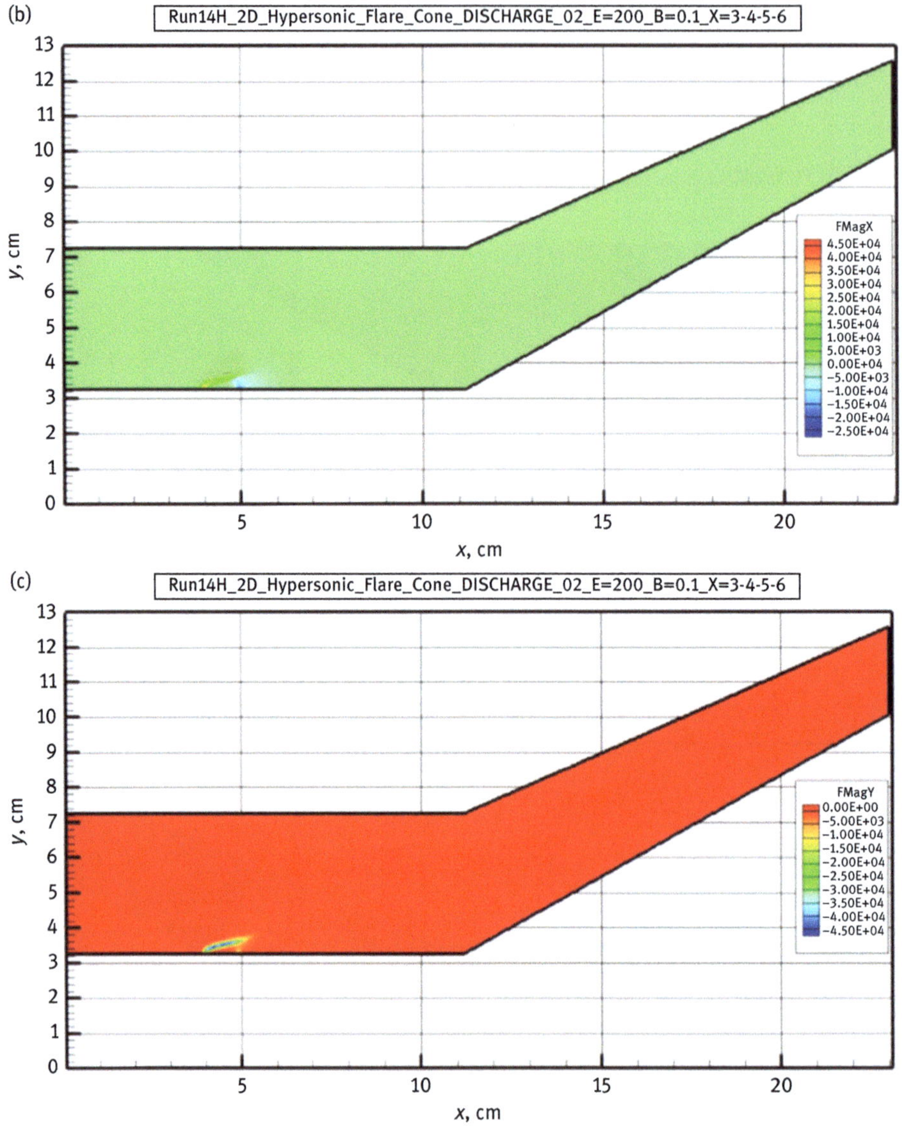

Figure 9.8 (continued)

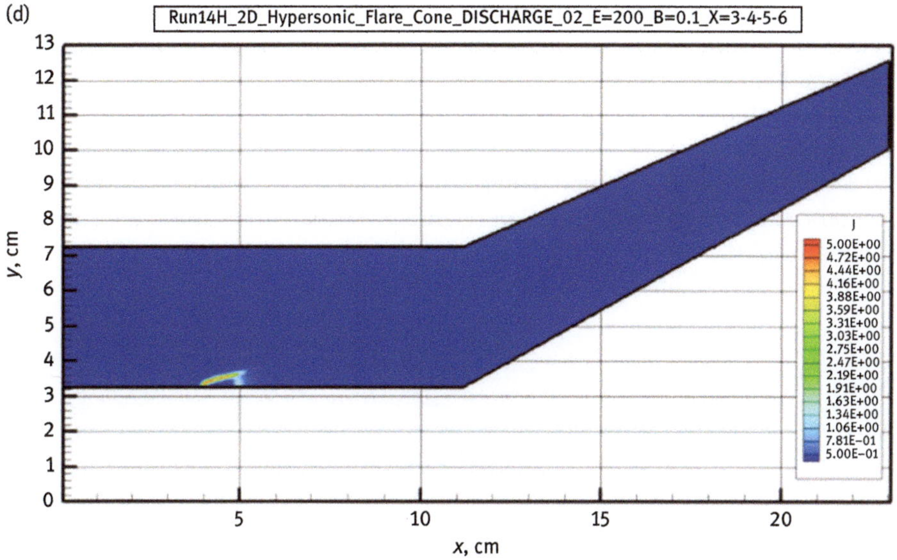

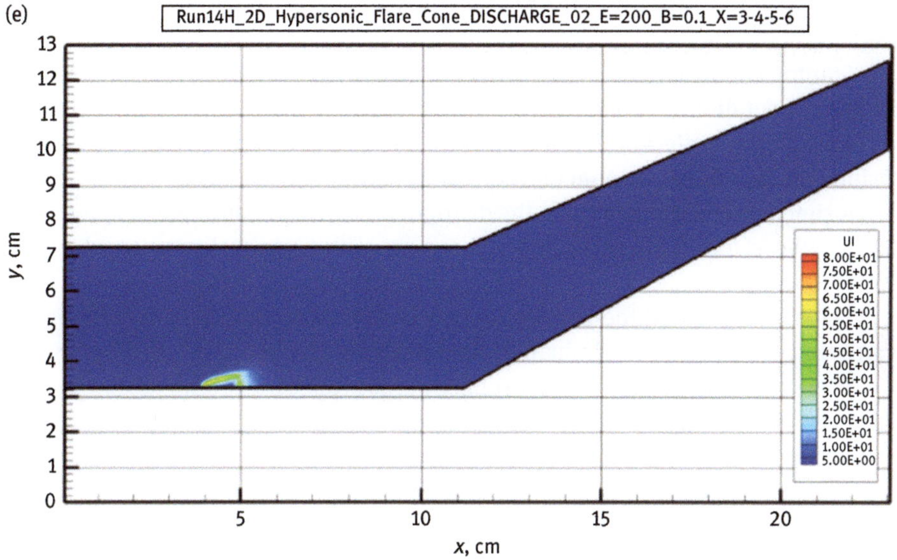

Figure 9.8 (continued)

velocity component dramatically increases from the surface (Figures 9.5f and 9.6f), as well as the local domain of high pressure and density are seen (Figures 9.5b and 9.6b, and Figures 9.5c and 9.6c). It is noteworthy that in this area the gas is heated to a temperature of ~2,300 K, which is higher than that in the discharge without magnetic field.

Distributions of the Stanton numbers, of the pressure coefficient, as well as the friction coefficient of surface tension (Figure 9.11) along the surface are shown in Figures 9.9–9.11.

The Stanton number was calculated by the formula

$$St = \frac{q_{w,c}}{\frac{1}{2}\rho_\infty V_\infty^3},$$

where $q_{w,c}$ is the convective heat flux.

The coefficients of the pressure and friction are calculated as follows:

$$c_p = \frac{p - p_\infty}{\frac{1}{2}\rho_\infty V_\infty^2}, \quad c_f = \frac{\tau_w}{\frac{1}{2}\rho_\infty V_\infty^2} = \frac{\left(\mu \frac{\partial u}{\partial y}\right)_w}{\frac{1}{2}\rho_\infty V_\infty^2}.$$

The Stanton coefficient very strongly increases in the domain between cathode and anode without discharge, especially with the magnetic field.

An important effect of the electric discharge and magnetic field is a shift of the heating domain by the amount of ~3 cm along the inclined surface.

A similar effect is observed in the distribution of the pressure coefficient c_p with the difference that it increases at displacement of the maximal pressure along the surface.

The distribution of the surface friction coefficient c_f points to increasing the vortex in gas discharge with magnetic field (Figure 9.11a, b).

Therefore, the problem of hypersonic flow past the surface with sharp edge has been considered. This surface forms an obtuse angle; thus, the shock wave generated by the leading edge interacts with the surface.

The effect of influence of the surface direct current discharge and a transverse magnetic field on the gas dynamic characteristics has been investigated. To solve this problem, a numerical simulation model has been used. This model is based on the set of the Navier–Stokes and energy conservation equations, as well as on equations of electrodynamics in the ambipolar approximation, and the Poisson equation.

The results of numerical modeling of the gas dynamics and electrodynamics of gas discharge with magnetic field point out to a change in the structure of the shock–wave interaction with surface far from the location of the gas discharge.

Hence, by the calculations it is shown that the low-current glow discharge can be used as an electromagnetic actuator in hypersonic flows.

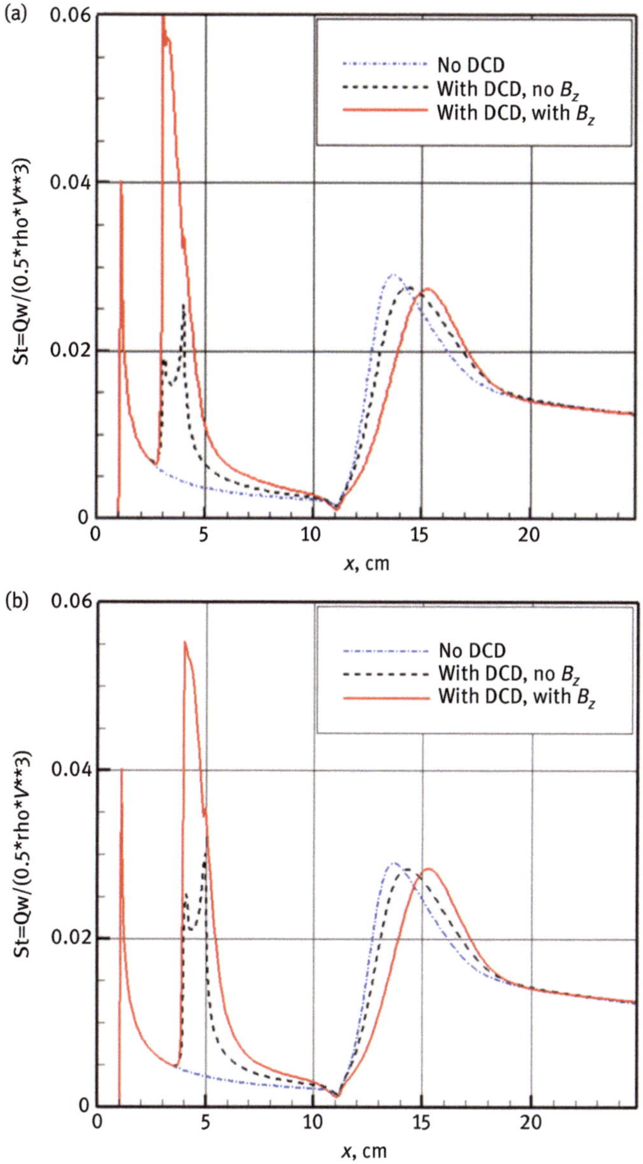

Figure 9.9: Distribution of the St number along the surface.

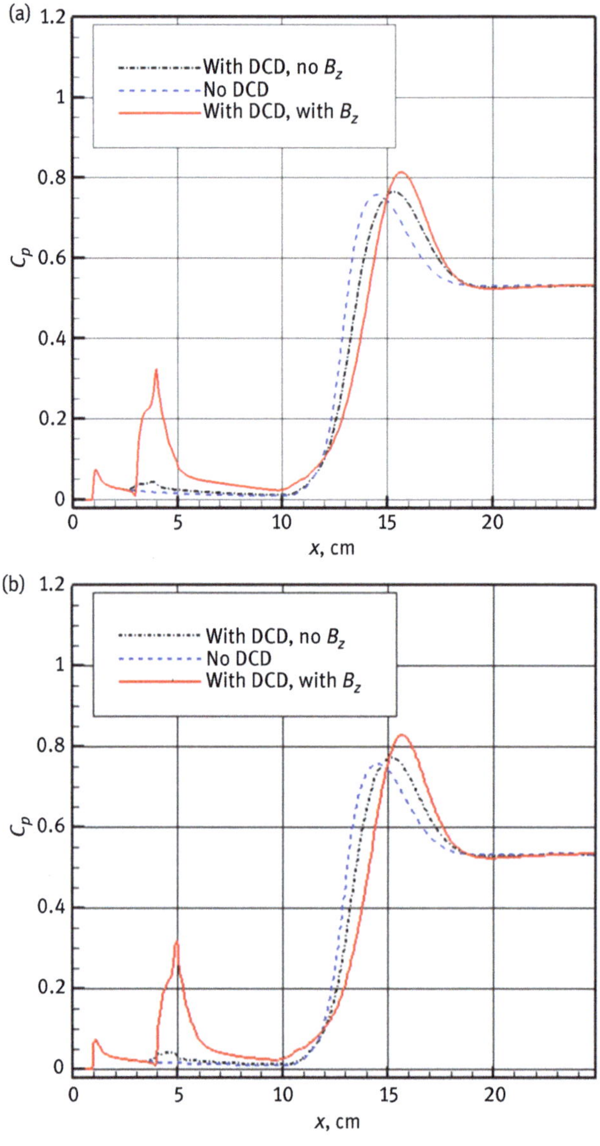

Figure 9.10: Distribution of the C_p-coefficient along the surface.

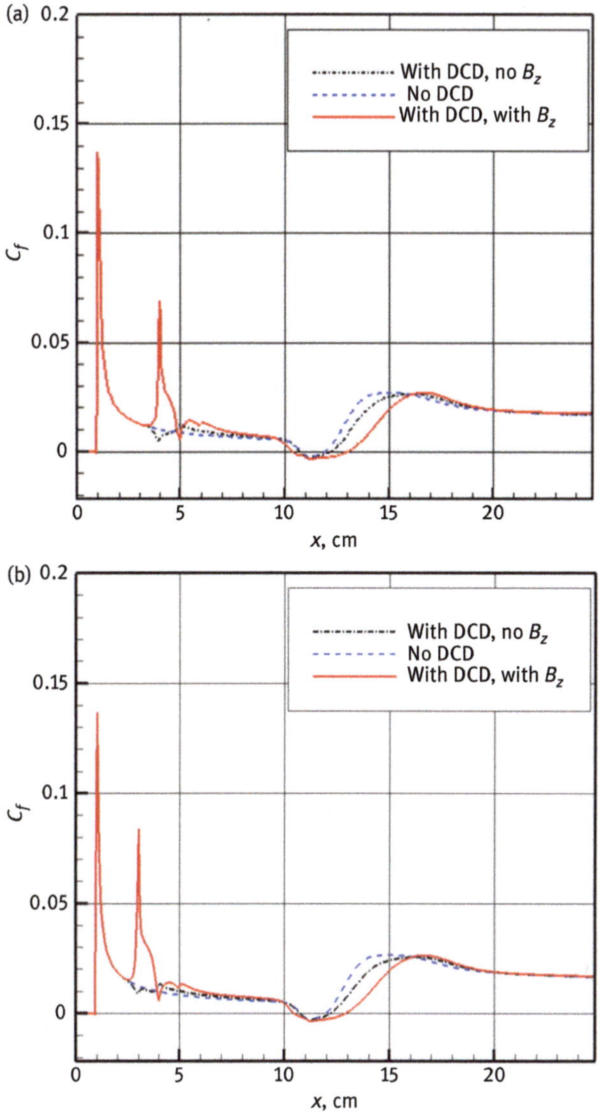

Figure 9.11: Distribution of C_f-coefficient along the surface.

References

Almeida P.G.C., Benilov M.S., and Faria M.J. Three-dimensional modeling of self-organization in DC glow microdischarges. IEEE Trans. Plasma Sci. 2011. V.39. No. 11. pp. 2190–2191.

Anderson D.A., Tannehill J.C., and Pletcher R.H. Computational fluid mechanics and heat transfer. Hemisphere Publishing Corporation. New York. 1997.

Anderson J.D. Hypersonic and high temperature gas dynamics. McGraw-Hill, Inc. 1989. 690 p.

Aslanbekov R.R., and Kolobov V.I. Two-dimensional simulation of the transition from Townsend to glow discharge and subnormal oscillations. J. Phys. D.: Appl. Phys. 2003. V.36. pp. 2986–2994.

Baranov I.Ya. Low-current moderate-pressure RF-discharge with secondary electron photoemission. Plasma Physics Reports. 2002. V.28. No. 1. pp. 71–77.

Birdsall C.K., and Langdon A.B. Plasma physics, via computer simulation. McGraw-Hill, Inc. New York. 1985. pp. 464.

Bittencourt J.A. Fundamentals of plasma physics. Pergamon Press. 2004.

Boeuf J.P. Numerical model of RF glow discharges. Phys. Rev. A. 1987.V.36. No. 6. pp. 2782–2792.

Boeuf J.P., and Marode E. A Monte Carlo analysis of an electron swarm in a nonuniform field: the cathode region of a glow discharge in helium. J. Phys. D.: Appl. Phys. 1982. V.15. No. 11. pp. 2169–2170.

Boni A.A., and Su F.Y. Propagation of laser supported deflagration waves. Phys. Fluids. 1974. V.17. pp. 340–342.

Boulos M.I. Flow and temperature fields in the fire-ball of an inductively coupled plasma. IEEE Transaction on Plasma Science. 1976. V.PS-4. pp. 28–39.

Boulos M.I., Gagne R., and Barnes R.M. Effect of swirl and confinement on the flow and temperature fields in an inductively coupled r.f. plasma. Canad. J. Chem. Eng. 1980. V.58. No. 3. pp. 367–375.

Boyd I.D., Keidar M., and McKeon W. Modeling of a pulsed plasma thruster from plasma generation to plume far field. Journal of Spacecraft and Rockets. 2000. V.37. No. 3. pp. 399–407.

Bracco F.V., and O'Rourke P.J. A review of initial comparisons of computed and measured two-dimensional unsteady flame fields. Progr. Energy and Combust. Sci. 1981. V.7. No. 2. pp. 103–124.

Brio M., and Wu C.C. An upwind differencing scheme for the equations of ideal magnetohydrodynamics. J. Comput. Phys. 1988. V.75. pp. 400–422.

Brown S.C. Introduction to electrical discharges in gases. Wiley. N.Y. 1966.

Capitelli M., Ferreira C.M., Gordiets B.F., and Osipov A.I. Plasma kinetics in atmospheric gases. Springer. 2000. pp. 300.

Capitelli, M., Colonna, G., Gorse, C., Minelli, P., Pagano, D., and Giordano, D. Thermodynamic and transport properties of two temperature H_2plasmas. AIAA. .2001. V.11. pp. 2001–3018.

Cebeci T. Analysis of turbulence flows. Elsevier. 2004. pp. 376.

Chapman S., and Cowling T.G. The mathematical theory of non-uniform gases. Cambridge University Press. 1964. pp. 134–150.

Chen F.F. Introduction to plasma physics. Plenum Press. New York. 1984.

Chetverushkin B.N. Mathematical modeling of problems in the gas dynamics of a radiating gas. Nauka. 1985

Chew G.F., Goldberger M.L., and Low F.E. The Boltzmann equation and the one-fluid hydromagnetic equations in the absence of particle collisions. Proc. Roy. Soc. 1956. A 236. pp. 112.

Chorin A.J. A numerical method for solving incompressible viscous flow problems. J. Comput. Phys. 1967. V.2. No. 1. pp. 12–36.

https://doi.org/10.1515/9783110648836-011

Chorin A.J. Numerical solution of the Navier–Stokes equations. Math. Comput. 1968. V.22. No. 4. pp. 745–762.
Clarke J.F., and McChesney M. The dynamics of real gases. Butterworth Inc. Washington D.C. 1964. pp. 66–86.
Cloutman L.D., Dukowicz J.K., and Ramshaw J.D. Numerical simulation of reactive flow in internal engine. Lect. Notes Phys. 1981. V.141. pp. 119–124.
Conrad R., Raizer Yu.P., and Surzhikov S.T. Continuous optical discharge stabilised by gas flow in weakly focused laser beam. AIAA Journal. 1996. V.34. No. 8. pp. 1584–1588.
Courant R., and Friedrichs K.O. Supersonic flow and shock waves. Intersciences. New York. 1948.
Delcroix J.L. Introduction to the theory of ionized gases. Interscience Publishers. Inc. New York. 1960.
Devoto R.S. Transport properties of ionized monatomic gases. Physics of Fluids. 1966. V.9. No. 6. pp. 1230–1240.
Devoto R.S. Transport coefficients of ionized argon. Phys. Fluids. 1973. V.16. No. 5. pp. 616–623.
Dresvin S.V. (ed.) Physics and technology of low-temperature plasma. Iowa State University. 1977. pp. 471.
Eddy T.L., and Cho K.Y. A. Multitemperature model for plasmas in chemical nonequilibrium. Heat Transfer in Thermal Plasma Processing, American Society of Mechanical Engineers. 1991. HTD-V. V.161. pp. 195–210.
Edwards J.R., and Liou M.S. Low-diffusion flux-splitting methods for flows at all speeds. AIAA J. 1998. V.36. No. 9. pp. 1610–1617.
Engel A.v., and Steenbeck M. Elektrische Gasentladungen. Springer. Berlin. 1932.
Farlow S.J. Partial differential equations for scientists and engineers. John Wiley & Sons, Inc. 1982.
Ferziger J.H., and Kaper H.G. Mathematical theory of transport processes in gases. North-Holland Publishing Company. Amsterdam – London. 1972.
Fiala A., Pitchford L.C., and Boeuf J.P. Two-dimensional, hybrid model of low-pressure glow discharge. Phys. Rev. 1994. V. E. 49. No. 6. pp. 5607–5623.
Filipskii M.V., and Surzhikov S.T. Calculation of radiation fluxes to the surface of a space vehicle with the aid of discreet ordinates method. Journal of Engineering Physics and Thermophysics. 2007. V.80. No. 2. pp. 284–291.
Fletcher C. Computational Techniques for Fluid Dynamics. Springer-Verlag Berlin Heidelberg. 1998. 401 p.
Fowler M.C., and Smith D.C. Ignition and maintenance of subsonic plasma waves in atmospheric pressure air by CW CO_2 laser radiation and their effect on laser beam propagation. J. Appl. Phys. 1975. V.46. pp. 138–150.
Gaitonde D.V., and Poggie J. An implicit technique for 3-D turbulent MGD with the generalized Ohm's law. AIAA. 2001. V.2001–2736. pp. 17.
Gaitonde D.V., and Poggie J. Preliminary analysis of 2-D scramjet flowpath with MGD control. AIAA. 2002. V.2002-2134. pp. 18.
Garscadden A. Conduction of electricity in gases. In book: Atomic. Molecular and Optical Physics. APC. 1986.
Gatsonis N.A., and Yin X. Hybrid (particle/fluid) modeling of pulsed plasma thruster plumes. J. of Propulsion and Power. 2001. V.17. No. 4. pp. 945–958.
Gatsonis N. A., Eckman R., Yin X., Pencil E., and Myers R. Experimental investigations and modeling of pulsed plasma thruster plumes. J. of Spacecraft and Rockets. 2001. V.38. No. 3. pp. 454–464.
Gentry A.A., Martin R.E., and Daly B.J. An Eulerian differencing method for unsteady compressible flow problems. J. of Comput. Phys. 1966. V.1. pp. 87–118.
Giordano D., Capitelli M., Colonna G., and Gorse C. Tables of internal partition functions and thermodynamic properties of high-temperature air species from 50 to 100 000 K. ESA STR-237. 1994.

Gladush G.G., and Samokhin A.A. Stability of a high-frequency glow discharge in the normal combustion regime. Journal of Applied Mechanics and Technical Physics. 1986. V.27. No. 6. pp. 803–811.
Gladush G.G., and Samokhin A.A. A numerical investigation of current lacing on the electrodes in a glow discharge. Journal of Applied Mechanics and Technical Physics. 1981. V.22. No. 5. pp. 608–615.
Gnoffo P.A. CFD validation studies for hypersonic flow prediction. AIAA. 2001. 2001-1025. pp. 13.
Graves D.B., and Jensen K.E. A continuum model of DC and RF discharges. IEEE Transactions on Plasma Science. 1986. V.PS-14. No. 2. pp. 78–91.
Habashi W.G., Dompierre J., Bourgault Y., et al. Certifiable computational fluid dynamics through mesh optimization. AIAA J. 1998. V.36. No. 5. pp. 703–711.
Harlow F.H., and Amsden A.A. A numerical fluid dynamics calculation method for all flow speeds. J. Comput. Phys. 1971. V.8. No. 2. pp. 197–213.
Harlow F.H., and Welch J.E. Numerical calculation of time-dependent viscous incompressible flow of fluid with free surface. J. Phys. Fluids. 1965. V.8. No. 12. pp. 2182–2189.
Hayes W., and Probstein R. Hypersonic flow theory. Academic Press. New York. 1959. pp.333–374.
Hirsch E.N. On the mechanism of the penning discharge. Brit. J. Appl. Phys. 1964. V.15. pp. 1535–1544.
Hirsch C. Numerical computation of internal and external flows. V.1. 2. John Wiley & Sons. Chichester, New York. 1990.
Hirschfelder J.O., Curtis C.F., and Bird R.B. Molecular theory of gases and liquids. Wiley. New York. 1964. pp. 1219.
Hirt C.W., Amsden A.A., and Cook J.L. An arbitrary Lagrangian-Eulerian computing method for all flow speeds. J. Comput. Phys. 1974. V.14. No. 3. pp. 227–253.
Hockney R.W., and Eastwood J.W. Computer simulation using particles. McGraw-Hill Int. Book Company. 1981.
Hoffmann K.A., and Chiang S.T. Computational fluid dynamics. V.1. 2. Publication of Engineering Educational System. Wichita, Kansas State University. 2000. pp. 3.
Hoh F.C. Low-temperature plasma diffusion in a magnetic field. Rev. Modern Physics. 1962. V.34. No. 2. pp. 267–286.
Holden M.S., Wadhams T.P., Harvey J.K., and Candler G.V. Comparison between measurements in regions of laminar shock wave boundary layer interaction in hypersonic flows with Navier-Stokes and DSMC simulations RTO-TR-AVT-007-V3. 2006. pp. 4.1–4.56.
Hughes W.F., and Young F.J. The electromagnetodynamics of fluids. Robert E. Krieger Publishing Company. Malabar, Florida. 1989.
Huxley L.G.H., and Crompton R.W. The diffusion and drift of electrons in gases. John Wiley & Sons. New York. 1974. pp. 669.
Itakawa Y. Momentum transfer cross sections for electron collisions with atoms and molecules. Atomic Data and Nuclear Data Tables. 1974. V.14. No. 1. pp. 1–10.
Itakawa Y. Momentum transfer cross sections for electron collisions with atoms and molecules. Revision and Supplement. Atomic Data and Nuclear Data Tables. 1978. V.21. No. 1. pp. 69–75.
Jackson J.P., and Nielsen P.E. Role of radiative transport in the propagation of laser supported combustion waves. AIAA Journal. 1974. V.12. No. 11. pp. 1498–1501.
Jones L.W., and Keefer D.S. NASA's laser-propulsion project. Astronautics & Aeronautics. 1982. V.20. No. 9. pp. 66–73.
Jones W.P., and Launder B.E. The computation of low Reynolds number phenomena with a two-equation model of turbulence. Int. J. Heat Mass Transfer. 1973. V.16. pp. 1119.

Kato H., and Tannehill J.C. Computation of hypersonic laminar separated flows using an iterated PNS algorithm. AIAA. 2001. 2001–1028. pp. 16.

Keefer D.R., Henriksen B.B., and Braerman W.F. Experimental study of a stationary laser-sustained air plasma. J. Appl. Phys. 1975. V.46. pp. 1080–1083.

Klosterman E.L., and Byron S.R. Measurement of subsonic laser absorption wave propagation characteristics at 10.6 mm. J. Appl. Phys. 1974. V.45. No. 11. pp. 4751–4759.

Knight D. Test cases for CFD validation of hypersonic flight. RTO-TR-AVT-007-V3. 2006

Kolobov V.I., and Tsendin L.D. Analytic model of the cathode region of a short glow discharge in light gases. Phys Rev. 1992. A. No. 46. pp. 7837–7843.

Koonin S.E. Computational physics. Addison Wesley Publishing Company Inc. New York, Amsterdam. 1986.

Kortshaden U., and Tsendin L.D. (Eds.) Electron kinetics and applications of glow discharges. NATO ASI series B.: Physics. V.367. Kluwer Academic Publishers. New York, Boston. 2002. pp. 556.

Kossyi I.A., Kostinsky A.Yu., Matveyev A.A., and Silakov V.P. Kinetic scheme of the non-equilibrium discharge in nitrogen-oxygen mixtures. Plasma Sources Sci. Technol. 1992. V.1. pp. 207–220.

Kotov M.A., Kozlov P.V., Ruleva L.B., Solodovnikov S.I., Surzhikov S.T., and Tovstonog V.A. The spectral characteristic investigations of normal glow discharge. J. of Physics: Conference Series. 2017. V.815. pp. 012026.

Kozlov P.V., Kotov M.A., Ruleva L.B., Solodovnikov S.I., and Surzhikov S.T. Normal glow discharge in nitrogen and inert gases between flat electrodes with distance of 10 mm. J. of Physics: Conf. Series. 2019. V.1250. pp. 012015.

Kozlov P.V., Kotov M.A., Ruleva L.B., Solodovnikov S.I., and Surzhikov S.T. Normal glow discharge investigation in inert gases and nitrogen between flat electrodes with 20 mm of discharge gap. J. of Phys.: Conf. Series. 2018. V.1009. pp. 012021.

Krall N.A., and Trivelpiece A.W. Principles of plasma physics. McGraw-Hill Book Company. 1973.

Kulikovskii A. G., Pogorelov N.V., and Semenov A. Yu. Mathematical aspects of numerical solution of hyperbolic systems. Chapman & Hall/CRC monographs and surveys in pure and applied mathematics. V.118. Chapman & Hall. CRC, Boca Raton. FL. 2001. pp. 540.

Kumar H., and Roy S. Multidimensional hydrodynamic plasma-wall model for collisional plasma discharges with and without magnetic fields effect. Physics of Plasmas. 2005. V.12. pp. 093508.

Landau L.D., Lifshitz E.M. Electrodynamics of Continuous Media (Volume 8 of A Course of Theoretical Physics). Pergamon Press. 1960.

Levitskii S.M. Investigation of the ignition potential of a high-frequency discharge in gas in the transitional frequency and pressure range. Journal of Technical Physics. 1957. V.27. No.5. pp. 970–977.

Losev S.A. Gasdynamic laser. Springer-Verlag Berlin Heidelberg. 1981. 300 p.

Macheret S.O. Physics of magnetically accelerated nonequilibrium surface discharge in high-speed flow. AIAA paper. 2006. 2006-1005.

Macheret S.O., Shneider M.N., and Miles R.B. Magnetohydrodynamic and electrodynamic control of hypersonic flows of weakly ionized plasmas. AIAA J. 2004. V.42. No. 7. pp. 1378–1387.

Macheret S.O., Shneider M.N., and Miles R.B. Modeling of air plasma generation by repetitive high voltage nanosecond pulses. IEEE Transactions on Plasma Sciences. 2002. V.20. pp. 1301–1314.

MacLean M., and Holden M. Validation and comparison of WIND and DPLR results for hypersonic laminar problems. AIAA. 2004. 2004-0529. pp. 17.

McDaniel E.W. Collision phenomena in ionized gases. John Wiley & Sons Inc. New York. 1964. pp. 797.

McDaniel E.W., and Mason E.A. The mobility and diffusion of ions in gases. Willey and Sons. New York, London, Sydney, Toronto. 1973. pp. 381.
Menart J., Shang J.S., Kimmel R., and Hayes J.R. Effect of surface plasma discharges on boundary layers at Mach 5. AIAA. 2003. 2003-4165. pp. 15.
Metrogul A.E., and Krier H. Two-temperature modeling of laser sustained hydrogen plasmas. J. of Thermophysics and Heat Transfer. 1994. V.8. No. 4. pp. 781–790.
Mihalas D., and Mihalas B.W. Foundations of radiation hydrodynamics. Oxford Univ. Press. New York. 1984.
Mirabo L., Raizer Yu.P., and Surzhikov S.T. Laser combustion waves in Laval nozzles. High Temperature. 1995. V.33. No. 1. pp. 11–20.
Mitcher M., and Kruger Ch.H., Jr. Partially ionized gases. A Wiley-Interscience publication. 1992.
Morozov A.I. The conceptual development of stationary plasma thrusters. Plasma Physics Reports. 2003. V.29. No. 3. pp. 235–250.
Mostaghimi J., Proulx P., and Boulos M.I. An analysis of the computer modeling of the flow and temperature fields in an inductively coupled plasma. Numer. Heat Transfer. 1985. V.8. No. 2. pp. 187–201.
Nagpal R., and Garscadden A. Electron and heavy particle energy redistribute ions in glow discharges in gas mixtures. Contrib. Plasma Phys. 1995. V.35. No. 4–5. pp. 301–330.
NPARC WIND CFD Code, Version 5. http://www.grc.nasa.gov/www/winddocs
Oran E.S., and Boris J.P. Numerical simulation of reactive flow. Elsevier. New York, Amsterdam. 1987. pp. 660.
Ozisik M.N. Radiative transfer and interaction with conduction and convection. Willey-Interscience Publication. John Willey & Sons. New York. 1973.
Penning F.M. Die glimmentladungbeiniedrigendruckzwischenkoaxialenzylinder in einemaxialenmagnetfeld. 1937. V.l.4. No. 2. pp. 71–72.
Penning F.M. The glow discharge at low pressure between coaxial cylinders in an axial magnetic field. Physica III. 1936. V.9. pp. 873–894.
Petrusev A.S., Surzhikov S.T., and Shang J.S. A two-dimensional model of glow discharge in view of vibrational excitation of molecular nitrogen. High Temperature. 2006a. V.44. No. 6. pp. 814–822.
Petrusev A.S., Surzhikov S.T., and Shang J.S. Chemical processes in air glow discharge for aerospace applications. AIAA. 2006b. 06-1460. pp. 11.
Petrusev A.S., Surzhikov S.T., and Shang J.S. Some peculiarities of direct current discharges for aerospace applications. AIAA. 2005. 2005-5305. pp. 11.
Petukhov I.V. Numerical calculation of two-dimensional flows in the boundary layer. In the book. "Numerical methods of solving differential and integral equations". Moscow.: USSR Academy of Sciences. 1964, S. 304–324 [in Russian].
Peyret R. (Ed.) Handbook of computational fluid mechanics. Elsevier Science & Technology books. 1996. pp. 467.
Phillips C.E.S. The action of magnetized electrodes upon electrical discharge phenomena in rarefied gases. V.197. Philosophical Trans. Royal Soc. A. London. 1901. pp. 135–150.
Poggie J. Numerical Simulation of DC and RF glow discharges. AIAA. 2007. 2007-0632. pp. 10.
Poggie J. Numerical exploration of flow control with glow discharges. AIAA 2004-2658. 2004. 13 p.
Pomraning G.C. The equations of radiation hydrodynamics. Pergamon Press. Oxford, New York. 1973. pp. 287.
Potter D. Computational physics. John Wiley & Sons. London, New York. 1973. pp. 393.
Powell K.G. An approximate Riemann solver for magnetohydrodynamics. unpublished. 1994.
Raizer Yu. P., and Surzhikov S.T. High–altitude explosions and their magnetohydrodynamic description. AIAA J. 1995. V.33. No. 3. pp. 479–485.

Raizer Yu. P., Surzhikov S.T. Magnetohydrodynamic description of collisionless plasma expansion in upper atmosphere. AIAA J. 1995. V.33. No. 3. pp. 486–490.
Raizer Yu.P. Gas discharge physics. Springer-Verlag. 1991. pp. 449.
Raizer Yu.P., and Surzhikov S.T. More on the nature of the normal current density at the cathode of a glow discharge. Sov. Tech. Phys. Lett. 1987. V.13. No. 3. pp. 5.
Raizer Yu.P., and Surzhikov S.T. Two-dimensional structure of a normal glow discharge and the role of diffusion in the formation of cathode and anodic current spots. High Temperature. 1988. V.25. No. 3. pp. 428–435.
Raizer Yu.P., Shneider M.N., and Yatsenko N.A. Radio-frequency capacitive discharges: physics. experimental technique. Applications. MIPT Publishers. 1995. pp. 310.
Resler E.L., and Sears W.R. The prospects for magneto-aerodynamics. J. Aero Sci. 1958. V.25. pp. 45–235.
Roache P.J. Computational fluid dynamics. Hermosa Publishers. Albuquerque. 1980.
Roe P.L. Approximate Riemann solvers, parameter vectors and difference schemes. J. of Comput. Phys. 1981. V.43. pp. 357–372.
Safronov B.G., and Konovalov V.G. Investigation of a Penning discharge at low pressure. NASA Technical Translation NASA TT F-15. V.523. 1974. pp. 14.
Shang J.S. Computational electromagnetic-aerodynamics. Wiley& Sons, Inc. Hoboken, New Jersey. 2016. pp. 424.
Shang J.S., Huang P.G., Yan H., and Surzhikov S.T. Computational electrodynamic simulation of direct current discharge. Journal of Applied Physics D. 2009. V.105. pp. 023303.
Shang J.S., Kimmel R., Menart J., and Surzhikov S.T. Hypersonic flow control using surface plasma actuator. J. of Propulsion and Power. 2008. V.24. No. 5. pp. 923–934.
Shang J.S., Surzhikov S.T., Kimmel R., Gaitonde D., Menart J., and Hayes J. Mechanisms of plasma actuators for hypersonic flow control. Progress in Aerospace Sciences. 2005. V.41. pp. 642–668.
Shang S., Chang C., and Surzhikov S. Simulating hypersonic magneto-fluid dynamic compression in rectangular inlet. AIAA J. 2007. V.45. No. 11. pp. 2710–2720.
Shang J.S. Recent research in magneto-aerodynamics. Progress in Aerospace Sciences. 2001. V.37. pp. 1–20.
Shang J.S., Menart J., Kimmel R., and Hayes J. Hypersonic inlet with plasma induced compression. AIAA. 2006. 06-0764. pp. 11.
Shang J.S., and Surzhikov S.T. Magneto-aerodynamic actuator for hypersonic flow control. AIAA J. 2005. V.43. No. 8. pp. 1633–1643.
Shang J.S., and Surzhikov S.T. Magneto-fluid-dynamics interaction for hypersonic flow control. AIAA. 2004. 04-0508. pp. 11.
Shang J.S., and Surzhikov S.T. Plasma dynamics for aerospace engineering. Cambridge University Press. Cambridge. 2018. pp. 387.
Shang J.S., Surzhikov S.T., Kimmel R., Gaitonde D., Menart J., and Hayes J. Plasma actuator for hypersonic flow control. AIAA. 2005. 05-0563. pp. 14.
Shang S., Chang C., Surzhikov S. Simulating hypersonic magneto-fluid dynamic compression in rectangular inlet. AIAA J. 2007. V.45. No. 11. pp. 2710 –2720.
Shin J., Naraganswamy V., Raja L.L., and Clemens N.T. Characterization of a direct-current glow discharge plasma actuator in low-pressure flow. AIAA J. 2007. V.45. No. 7. pp. 1596–1605.
Shkarofsky I.P., Johnston T.W., and Bachynski M.P. The particle kinetics of plasmas. Addison-Wesley Publishing Company. 1966.
Siegel R., and Howell J.R. Thermal radiation heat transfer. McGraw-Hill Book Company. New York. 1972.
Smirnov A., Raitses Y., and Fisch N.J. Experimental and theoretical studies of cylindrical Hall thrusters. Physics of Plasmas. 2007. V.14. pp. 057106.

Smirnov A.S. Near-electrode layers in capacitive RF discharge. Journal of Technical Physics. 1984. V.53. No. 1. pp. 61–65.

Spitzer L., Jr. Physics of fully ionized gases. Interscience Publishers, Inc. New York. 1956. pp. 105.

Stratton J.A. Electromagnetic Theory. McGraw-Hill Book Company, Inc. 1941.

Su F.Y., and Boni A.A. Nonlinear model of laser supported deflagration waves. Phys. Fluids. 1976. V.19. pp. 960–966.

Sun M.T., Wu S.T., and Dryer M. On the time-dependent numerical boundary conditions of magnetohydrodynamics flows. J. Comput. Phys. 1995. V.116. pp. 330–342.

Surzhikov S .T., Petrusev A.S., and Shang J.S. Some peculiarities of direct current discharges for aerospace applications. AIAA Paper. 2005a. 05-5305. pp. 11.

Surzhikov S.T. Application of the modified drift-diffusion theory to study of the two-dimensional structure of the Penning discharge. AIAA. 2015. 2015-1832. pp. 27.

Surzhikov S.T. Computational physics of electric discharges in gas flows. Walter de Gruyter. Berlin/Boston. 2013. pp. 428.

Surzhikov S.T. Computing system for mathematical simulation of selective radiation transfer. AIAA. 2000a. 00-2369. pp. 15.

Surzhikov S.T. Drift-diffusion model of normal glow discharge in an axial magnetic field. Doklady Physics. 2016. V.61. No. 12. pp. 596–600.

Surzhikov S.T. Expansion of multi-charged plasma clouds into ionospheric plasma with magnetic field. J. of Spacecraft and Rockets. 1999. V.36. No. 6. pp. 907–911.

Surzhikov S.T. Glow discharge in external magnetic field in hypersonic flow of rarefied gas. High Temperature. 2009a. V.47. No. 4. pp. 459–471.

Surzhikov S.T. Interaction of a glow discharge with a rarefied hypersonic flow in a curvilinear channel. Russian Journal of Physical Chemistry B. 2009c. V.3. No. 3. pp. 383–390.

Surzhikov S.T. Laser-plasma generator with artificial turbulence of the input gas stream. J. of Thermophysics and Heat Transfer. 2001. V.15. No. 2. pp. 239–242.

Surzhikov S.T. Mathematical models of subsonic Laval nozzles of laser-plasma accelerators. High Temperature. 1995. V.33. No. 3. pp. 435–448.

Surzhikov S.T. Numerical analysis of laser-supported combustion waves Sov. J. Quantum Electronics. 2000b. V.30. No. 5. pp. 416–420.

Surzhikov S.T. Numerical study of acceleration of the hydrogen ions in the penning discharge at pressures around 1 Torr. AIAA. 2017a. 2017-0155. pp. 13.

Surzhikov S.T. Radiation-convection heat transfer in the process of slow motion of a radiating gas in a region of local energy release. Fluid Dynamics. 1997. V.32. No. 3. pp. 431–440.

Surzhikov S.T. Rarefied hypersonic flow in a plane channel with a surface glow discharge. Fluid Dynamics. 2009b. V.44. No.2. pp. 302–313.

Surzhikov S.T. Subsonic radiation gas dynamics in a laser plasma generator channel. Fluid Dynamics. 2005. V.40. No. 3. pp. 446–461.

Surzhikov S.T. The bifurcation of subsonic gas flow past a localized volume of low-temperature plasma. High Temperature. 2002. V.40. No. 4. pp. 546–556.

Surzhikov S.T. The two-dimensional structure of the penning discharge in a cylindrical chamber with axial magnetic field at pressure of about 1 Torr. Technical Physics Letters. 2017b. V.43. No. 2. pp. 169–172.

Surzhikov S.T. Thermo-convective instability in slow laser-induced combustion waves. Doklady Physics. 2008. V.53. No. 6. pp. 305–309.

Surzhikov S.T., and Gatsonis N.A. Plasma flow through a localized heat release region. AIAA. 1999. 99-3738. pp. 11.

Surzhikov S.T., and Gatsonis N.A. Splitting method for unsteady 3D magnetogasdynamic flows applied to pulsed plasma thruster plumes. AIAA. 2001. 01-2743. pp. 11.

Surzhikov S.T., Capitelli M., Colonna G., and Gorse C. Radiative gas dynamic model of hydrogen laser-supported plasma generator. J. of Thermophysics and Heat Transfer. 2008. No. 1. pp. 62–70.

Surzhikov S.T., and Krier H. Unsteady dynamic variables method for heterogeneous solid propellant burning. AIAA J. 2001. V.39. No. 12. pp. 2343–2350.

Surzhikov S.T., and Shang J.S. Normal glow discharge in axial magnetic field. Plasma Sources Sciences and Technology. 2014. V.23. pp. 054017.

Surzhikov S.T., and Shang J.S. Numerical simulation of subsonic gas flows with glow discharge and magnetic field. AIAA. 2003a. 03-3759. pp. 11.

Surzhikov S.T., and Shang J.S. Supersonic internal flows with gas discharge and external magnetic field. AIAA. 2003b. 03-3625. pp. 11.

Surzhikov S.T., and Shang J.S. Viscous interaction on a flat plate with a surface discharge in magnetic field. High Temperature. 2005b. V.43. No. 1. pp. 19–30.

Surzhikov S.T., and Shang J.S. Glow discharge in magnetic field with heating of neutral gas. AIAA. 2003c. 03-3654. pp. 11.

Surzhikov S.T., and Shang J.S Multi-fluid model of weakly ionized electro-negative gas. 2004a. AIAA. 04-2659. pp. 11.

Surzhikov S.T., and Shang J.S. Physics of the surface direct current discharge in magnetic field. AIAA. 2004b. 04-0176. pp. 11.

Surzhikov S.T., and Shang J.S. Two-component plasma model for two-dimensional glow discharge in magnetic field. J. of Comput. Phys. 2004c. V.199. pp. 437–464.

Thompson K.W. Time-dependent boundary conditions for hyperbolic systems. J. of Comput. Phys. 1987. V.68. pp. 1–24.

Thompson K.W. Time-dependent boundary conditions for hyperbolic systems, II. J. of Comput. Phys. 1990. V.87. pp. 439–461.

Vanajakshi J.C., Thompson K.W., and Black D.C. Boundary value problems in magnetohydrodynamics (and fluid dynamics). I. Radiation boundary condition. J. of Comput. Phys. 1989. V.84. pp. 343–350.

Wang W.-L., and Boyd I.D. Continuum breakdown in hypersonic viscous flows. AIAA. 2002. 2002–0651. pp. 11.

Ward A.L. Calculations of cathode-fall characteristics. J. of Applied Physics. 1962. V.33. No. 9. pp. 2789–2794.

Ward A.L. Effect of space charge in cold-cathode gas discharges. Physical Review. 1958. V.112. No. 6. pp. 1852–1857.

Wilberts A.T.M., Kroesen G.M.W., Timmermans T.J., and Schram D.C. The continuum emission of an arc plasma. JQSRT. 1991. V.45. No. 1. pp. 1–10.

Wilcox D.C. Turbulence modeling for CFD. DCW Industries., Inc. 1993. pp. 460.

Wright M.J., Candler G.V., and Bose D. Data parallel line relaxation method for the Navier-Stokes equations. AIAA J. 1998. V.36. No. 9. pp. 1603–1609.

Yatsenko N.A. Link high DC plasma potential mode of combustion of high-frequency capacitive discharge at medium pressure. J. Technical Physics. 1981. V.51. No. 6. pp. 1195–1204.

Zel'dovvich Ya.B., and Raizer Yu.P. Physics of shock waves and high-temperature hydrodynamics phenomena. Dover Publications Inc. Mineola, NY. 2002.

Appendix

Fundamental constants

Speed of light	$c = 2.998 \times 10^{10}$ cm/s
Stefan–Boltzmann's constant	$\widetilde{\sigma} = 5.67 \times 10^{-8} \frac{\text{W}}{\text{m}^2\text{K}^4} = 5.67 \times 10^{-5} \frac{\text{erg}}{\text{s} \cdot \text{cm}^2\text{K}^4}$
Universal gas constant	$R_0 = 8.3145 \frac{\text{J}}{\text{mol} \cdot \text{K}} = 8.3145 \times 10^7 \frac{\text{erg}}{\text{mol} \cdot \text{K}}$
Electron charge	$e = 4.802 \times 10^{-10}\ \text{g}^{1/2}\ \text{cm}^{3/2}/\text{s}$ (CGSE electrostatic system of units)
Boltzmann's constant	$k = 1.38 \times 10^{-23} \frac{\text{J}}{\text{K}} = 1.38 \times 10^{-16} \frac{\text{erg}}{\text{K}}$
Electron mass	$m_\text{e} = 9.109 \times 10^{-28}$ g
Proton mass	$m_\text{p} = 1.673 \times 10^{-24}$ g
Atomic mass unit	$m_0 = 1.66 \times 10^{-24}$ g
Bohr's radius	$a_0 = 0.529 \times 10^{-8}$ cm
Avogadro number	$N_\text{A} = 6.022 \times 10^{23} \frac{1}{\text{mol}}$
Loschmidt's number (the number of molecules in 1 cm^3 at $T = 273.15$ K, $p = 1$ atm = 101, 325 Pa)	$N_0 = 2.687 \times 10^{19}\ \text{cm}^{-3}$
Planck's constant	$h = 6.625 \times 10^{-27}$ erg × s

Ratios between units of electricity and magnetism

Charge

$$1\,\text{C} = 3 \times 10^9 \text{ SGSE units} = 9 \times 10^{11}\ \text{V} \times \text{cm} = 6.25 \times 10^{18} \text{ electron charges}$$

Current

$$1\,\text{A} = 1\,\frac{\text{C}}{\text{s}} = 3 \times 10^9 \text{ SGSE units} = 9 \times 10^{11}\,\frac{\text{V} \cdot \text{cm}}{\text{s}} = 6.25 \times 10^{18}\,\frac{\text{electron charges}}{\text{s}}$$

Voltage

$$1\,\text{V} = \frac{1}{300} \text{ SGSE units}$$

$$E\,[\text{V/cm}] = 300E\,[\text{SGSE units}]$$

Resistance

$$1 \text{ Ohm} = \frac{1}{9 \times 10^{11}} \text{ SGSE units} = \frac{1}{30}\,\text{s}$$

https://doi.org/10.1515/9783110648836-012

Conductivity	$\sigma\left[\frac{1}{\text{Ohm}\cdot\text{cm}}\right]=\frac{1}{9\times10^{11}}\sigma\left[\text{s}^{-1}\right]$
Capacitance	$1\,\text{Farad}=9\times10^{11}\,\text{SGSE units}$
Strength of magnetic field	$1\,\text{Oersted}=1\,\text{SGSE units}$
Inductance	$1\,\text{Henry}=10^{9}\,\text{SGSE units(cm)}$
Absolute magnetic permeability of vacuum	$\mu_0=4\pi\times10^{-7}\frac{\text{Henry}}{\text{m}}=12.566\times10^{-7}\frac{\text{Henry}}{\text{m}}$
Absolute dielectric permeability of vacuum	$\varepsilon_0=\frac{1}{\mu_0c^2}=8.854\times10^{-12}\frac{\text{Farad}}{\text{m}}$
	$\frac{E}{p}\left[\frac{\text{V}}{\text{cm}\cdot\text{Torr}}\right]=3.3\times10^{16}\frac{E}{N}\left[\text{V}\cdot\text{cm}^2\right]=0.33\frac{E}{N}\,[\text{Td}]$
	$1\,\text{Td}=10^{-17}\,\text{V}\cdot\text{cm}^2$

The temperature $E/k=11,610\text{k}$ corresponds to energy $E=1$ eV.
The energy $\varepsilon=1\,\text{eV}=1.602\cdot10^{-2}$ erg corresponds to temperature $T=\varepsilon/K=11,600$ K,
frequency $f=\varepsilon/h=2.418\cdot10^{-14}\,\text{s}^{-1}$,
wavelength $\lambda=hc/\varepsilon=1.24\cdot10^{-4}\,\text{cm}=12.4\,\text{Å}$, and
wave number $\omega=\varepsilon/hc=8,067\,\text{cm}^{-1}$.

Index

https://doi.org/10.1515/9783110648836-013

Also of interest

Statistical Physics
Michael V. Sadovskii, 2019
ISBN 978-3-11-064510-1, e-ISBN (PDF) 978-3-11-064848-5

Quantum Field Theory
Michael V. Sadovskii, 2019
ISBN 978-3-11-064515-6, e-ISBN (PDF) 978-3-11-064852-2

Quantum Systems, Channels, Information: A Mathematical Introduction
Alexander S. Holevo, 2019
ISBN 978-3-11-064224-7, e-ISBN 978-3-11-064249-0

Coherent Quantum Physics: A Reinterpretation of the Tradition
Arnold Neumaier, 2019
ISBN 978-3-11-066729-5, e-ISBN: 978-3-11-066738-7

Bei Fragen zur Produktsicherheit wenden Sie sich bitte an:
If you have any questions regarding product safety,
please contact:

Walter de Gruyter GmbH
Genthiner Straße 13
10785 Berlin
productsafety@degruyterbrill.com